Neutron Applications in Earth, Energy and Environmental Sciences

Neutron Scattering Applications and Techniques

Series Editors:

Ian S. Anderson
Neutron Sciences Directorate
Oak Ridge National Laboratory
Building 8600, MS 6477
Oak Ridge, TN 37831
USA
andersonian@ornl.gov

Alan J. Hurd
Lujan Neutron Scattering Center at LANSCE
Los Alamos National Laboratory
PO Box 1663, MS H805
Los Alamos, NM 87545
USA
ajhurd@lanl.gov

Robert L. McGreevy
ISIS
Science & Technology Facilities Council
Rutherford Appleton Laboratory
Harwell Science & Innovation Campus
Chilton, Didcot OX11 0 QX
UK
robert.mcgreevy@stfc.ac.uk

Neutron Applications in Earth, Energy, and Environmental Sciences
Liyuan Liang, Romano Rinaldi, and Helmut Schober, eds.
ISBN 978-0-387-09415-1, 2009

Liyuan Liang · Romano Rinaldi ·
Helmut Schober
Editors

Neutron Applications
in Earth, Energy
and Environmental Sciences

Springer

Editors

Liyuan Liang
Oak Ridge National Laboratory
Oak Ridge, TN
USA
liangl@ornl.gov

Romano Rinaldi
Università di Perugia
Italy
rrinaldi@unipg.it

Helmut Schober
Institut Laue-Langevin
Grenoble
France
schober@ill.fr

100566+360

ISBN 978-0-387-09415-1   e-ISBN 978-0-387-09416-8
DOI 10.1007/978-0-387-09416-8

Library of Congress Control Number: 2008932522

Printed on acid-free paper

springer.com

Preface

The use of neutrons to investigate the fundamental properties of materials began with the search for new ways of exploiting the knowledge and instrumentation acquired during the all-out scientific efforts of the Manhattan Project in the 1940s. Neutron applications were explored in the 1950s by Bertram N. Brockhouse at the Chalk River Laboratory in Ontario and by Clifford G. Shull at the Graphite Reactor at Oak Ridge National Laboratory (ORNL) in Tennessee. When the Royal Swedish Academy of Sciences awarded them the 1994 Nobel Prize in Physics, for pioneering contributions to the development of neutron scattering techniques, the citation noted that they had between them helped to answer the questions of where atoms "are" and what atoms "do." The saying "neutrons see where atoms are and what they do" has become the motto of neutron scattering science.

Brockhouse's and Shull's pioneering applications were essentially limited to studies of the physical properties of matter, and in particular to phase transitions, magnetic structures, phonons, and especially the hydrogen bond. In the last 20 years or so, the use of neutrons has expanded tremendously following the development of new technology for the production of cold, thermal, and epithermal neutrons. This has resulted in orders-of-magnitude improvements to brilliance, energy resolution, and detector efficiencies compared with the original sources and measuring devices. Many of the scientific applications that now employ neutrons were quite unanticipated. Neutrons, with wavelengths on the order of angstroms, are capable of probing molecular structures and motions and increasingly find applications in a wide array of scientific fields, including biochemistry, biology, biotechnology, cultural heritage materials, earth and environmental sciences, engineering, material sciences, mineralogy, molecular chemistry, and solid state and soft matter physics.

This volume surveys the diversity of present day applications of neutron methods in the fields of Earth, Energy, and Environmental Sciences. Neutron applications in Earth sciences, including mineralogy, petrology, geochemistry, volcanology, structural geology, and sedimentology, are presented first for structural studies. The second set of applications is in the area of energy, particularly the sources of energy, upon which our modern civilization depends. In view of the inevitable exhaustion of natural fossil fuels, energy alternatives are needed for sustainable advancement. The last aspect of the Earth system dealt with, the environment, of course, becomes prominent in any of the studies addressed under the other two denominations, for energy use is linked to impacts to the natural environment of the planet Earth.

The book intends to provide the novice with an inspiring introduction to the use of neutrons in material science and technology and to stimulate the expert to consider these non-conventional techniques as readily available problem-solving tools in the fields of application considered. The International Year of Planet Earth, as declared by the United Nations Educational, Scientific and Cultural Organization and the International Union of Geological Sciences for 2008, testifies to the increased awareness of the strong ties linking these three fields.

The international scientific community has been engaged in a major effort to produce the next generation of large-scale neutron sources and instruments, along with the newest synchrotron-based X-ray facilities, in response to an ever growing demand for knowledge of the fundamental properties of materials (natural or man-made, organic or inorganic) and their scientific and technological implications.

The coming on-line of a new generation of spallation neutron sources, for example, at Oak Ridge National Laboratory (ORNL) (Fig. 1) in the United States and at Tsukuba, Japan (Fig. 2), represent the most tangible aspects of this effort. Additionally, the construction of ISIS-TS2, the second target station at the spallation neutron source (ISIS) of the Rutherford Laboratory in the United Kingdom, is well on its way to completion (Fig. 3); and construction is expected to begin soon for the Chinese source (CSNS) in Dong-guan, People's Republic of China. At the same time, efforts have been completed to renew the reactor-based neutron sources for the ORNL High-Flux Isotope Reactor and the research reactor at Grenoble, France (Fig. 4). The latter, supported by the Institut Laue-Langevin Millennium Programme (www.ill.fr), provided an order-of-magnitude gain at the experimental stations. The European Spallation Source has been planned for over 12 years, and construction seems imminent. This project is to build and operate the world's most intense

Fig. 1 Arial view of the pulsed Spallation Neutron Source at Oak Ridge National Laboratory, USA, completed and in operation in 2008. *Photo courtesy of ORNL*

Fig. 2 The new Japanese pulsed Spallation Neutron Source (JSNS) close to completion as a Materials and Life Science Facility at J-PARC, Tsukuba, Japan. *Photo courtesy of JAERI*

Fig. 3 Aerial view of ISIS and TS2 (the *box-shaped* building in foreground *right*) at the Rutherford Appleton Laboratory, UK. Part of the new synchrotron ring Diamond is visible on the *upper right side. Photo courtesy of RAL*

neutron source. It is currently one of the largest research and development infrastructures slated to be built in Europe during the next 10 years, with an estimated construction cost of 1.0–1.5 G€.

The editors—in the hope that these present day neutron applications in studies of the Earth, energy, and the environment help stimulate novel uses of neutrons to unveil otherwise unobtainable information—are grateful to all the contributors whose insights, diligence, and timeliness made this volume possible. Our thanks also go to the ORNL and Springer editorial staff: D. Counce, V. J. Ewing, C. Horak, E. Tham, and L. Danahy, whose assistance is critical to this volume. We are indebted to the following reviewers and the series editors (Ian Anderson, Alan Hurd, and Robert McGreevy) whose suggestions helped us maintain balance and accuracy: Alberto Albinati, Muhammad Arif, Miguel Ángel Castro Arroyo, Craig Brown, Michele Catti, Jack Carpenter, Gabrial Cuello, Wulf Depmeier, A. N. Fitch,

Fig. 4 Aerial view of the Institut Laue-Langevin reactor building, Grenoble, France. Part of the European Synchrotron Radiation Facility ring is visible to the *right* of the reactor dome. *Photo Courtesy of ILL*

Jean Louis Fourquet, Yoshiyuki Inaguma, A. Johs, Walter Kob, David Lennon, E. Lehmann, Li Liu, Chun Loong, Geoff Lloyd, Yuri Melnichenko, Laurent Michot, D. A. Neumann, Michael Prager, Werner Press, Alessandro Pavese, Keith Ross, Paul Schofield, Holger Stunitz, Jose Teixeira , Wolfgang Treimer, Costas Tsouris, Sven Vogel, Peter Votonbel, Thomas Voigtmann, Hanna Wacklin, and Michele Zucali.

Special thanks go to our families: Robert, Mark, and Katherine; Jenny, Victoria and Alexandra; Anita, Alexander, Rafaela and Carmen for their patience, support, and understanding during the many months of our involvement in this endeavour.

Oak Ridge, TN, USA   Liyuan Liang
Perugia, Italy   Romano Rinaldi
Grenoble, France   Helmut Schober

Contents

Contributors

Michael Agamalian Neutron Scattering Science Division, Oak Ridge National Laboratory, MS-6475, Oak Ridge, TN 37831-6475, USA
magamalian@ornl.gov

Peter W. Albers AQura GmbH, AQ-EM, Rodenbacher Chaussee 4, D-63457 Hanau, Germany
peter.albers@aqura.de

Jan Albus Geologisches Institut, Nussallee 8, D-53115 Bonn, Germany
Jan.Albus@freenet.de

Ian Anderson Neutron Scattering Sciences Division, Oak Ridge National Laboratory, MS-6475, Oak Ridge, TN 37831-6475, USA
andersonian@ornl.gov

John F. Ankner Neutron Scattering Science Division, Oak Ridge National Laboratory, MS 6475, 1 Bethel Valley Road, Oak Ridge, TN 37831-6475, USA
anknerjf@ornl.gov

Isabelle Bihannic Laboratoire Environnement et Minéralurgie, UMR 7569 CNRS-INPL-ENSG, 15, Av. du Charmois, BP 40, F-54501 Vandoeuvre Lès Nancy Cedex, France
isabelle.bihannic@ensg.inpl-nancy.fr

Michele Catti Dipartimento di Scienza dei Materiali, Università di Milano Bicocca, Via Cozzi 53, I-20125 Milano, Italy
catti@mater.unimib.it

Milva Celli CNR-Istituto Sistemi Complessi, Via della Madonna del Piano, 10, I-50019 Sesto Fiorentino, Italy
milva.celli@isc.cnr.it

Daniele Colognesi CNR-Istituto Sistemi Complessi, Via della Madonna del Piano, 10, I-50019 Sesto Fiorentino, Italy
daniele.colognesi@isc.cnr.it

Samrath Lal Chaplot Solid State Physics Division, Bhabha Atomic Research
Centre, Mumbai 400085, India
chaplot@magnum.barc.gov.in

Laurent Charlet Environmental Geochemistry Group, University of Grenoble-I,
LGIT, B.P. 53, F-38041 Grenoble Cedex, France
laurent.charlet@obs.ujf-grenoble.fr

Narayani Choudhury Solid State Physics Division, Bhabha Atomic Research
Centre, Mumbai 400085, India
dynamics@barc.gov.in; chaplot@barc.gov.com

David R. Cole Chemical Sciences Division, Oak Ridge National Laboratory,
MS-6110, Oak Ridge, TN 37831-6110, USA
coledr@ornl.gov

Stephen J. Covey-Crump School of Earth, Atmospheric and Environmental
Sciences, University of Manchester, Oxford Road, Manchester, M13 9PL, UK
S.Covey-Crump@manchester.ac.uk

Gabriel J. Cuello Diffraction Group, Institut Laue Langevin, 6, Rue Jules
Horowitz, B.P.156, F-38042 Grenoble Cedex 9, France
cuello@ill.fr

Alfred Delville Centre de Recherche sur la Matière Divisée, CNRS-Orléans
University, 1b rue de la Férollerie, F-45071 Orléans Cedex 2, France
delville@cnrs-orleans.fr

Bruno Demé Institut Laue-Langevin, 6 rue Jules Horowitz, F-38042 Grenoble
Cedex 9, France
deme@ill.fr

**Jan F. Derks IES Integrated Exploration Systems, Ritterstr. 23, 52072 Aachen,
Germany**
j.derks@ies.de

Alejandro Fernández-Martínez Institut Laue-Langevin, B.P. 156, F-38042
Grenoble, France
fernande@ill.fr

Nikolaus Froitzheim Geologisches Institut, Nussallee 8, D-53115 Bonn, Germany
niko.froitzheim@uni-bonn.de

Hermann Gies Institut für Geologie, Mineralogie and Geophysik, Ruhr-
Universität Bochum, Universitätsstr. 150, D-44780 Bochum, Germany
Hermann.Gies@ruhr-uni-bochum.de

Baohua Gu Environmental Sciences Division, Oak Ridge National Laboratory,
MS-6036, Oak Ridge, TN 37831-6036, USA
gub1@ornl.gov

Andrew Harrison Institut Laue Langevin (ILL), 6, Rue Jules Horowitz, B.P.156, F-38042 Grenoble Cedex 9, France
Harrison@ill.fr

Richard J. Harrison Department of Earth Sciences, University of Cambridge, Downing Street, Cambridge, CB2 3EQ, UK
rjh40@esc.cam.ac.uk

Jürgen Horbach Institut für Materialphysik im Weltraum, Deutsches Zentrum für Luft- und Raumfahrt (DLR), D-51170 Köln, Germany
juergen.horbach@dlr.de

Ekkehard Jansen Steinmann-Institut der Universität Bonn, Abteilung Endogene Prozesse, Poppelsdorfer Schloß, D-53115 Bonn, Germany
e.jansen@fz-juelich.de

Alexander Johs Environmental Sciences Division, Oak Ridge National Laboratory, MS-6038, Oak Ridge, TN 37831-6038, USA
johsa@ornl.gov

Florian Kargl Institute of Mathematics and Physics, Aberystwyth University, Aberystwyth SY23 3BZ, UK
ffk@aber.ac.uk

Alexander I. Kolesnikov Intense Pulsed Neutron Source Division, Argonne National Laboratory, Argonne IL 60439, USA
akolesnikov@anl.gov

Michael M. Koza Institut Laue Langevin, 6 rue Jules Horowitz, B.P. 156, F-38042 Grenoble Cedex 9, France
koza@ill.fr

Walter Kurz Institute of Applied Geosciences, Graz University of Technology, Rechbauerstr. 12, A-8010, Graz, Austria
walter.kurz@tugraz.at

Eberhard H. Lehmann Spallation Neutron Source Division, Paul Scherrer Institut, CH-5232 Villigen, Switzerland
eberhard.lehmann@psi.ch

Andrey A. Levchenko Peter A. Rock Thermochemistry Laboratory and NEAT ORU, University of California, Davis, CA 95616, USA
alevchenko@ucdavid.edu

Liyuan Liang Environmental Sciences Division, Oak Ridge National Laboratory, MS-6038, Oak Ridge, TN 37831-6038, USA
liangl@ornl.gov

Eugene Mamontov Neutron Scattering Science Division, Oak Ridge National Laboratory, MS-6475, Oak Ridge, TN 37831-6475, USA
mamontove@ornl.gov

Andreas Meyer Institut für Materialphysik im Weltraum, Deutsches Zentrum für Luft- und Raumfahrt (DLR), 51170 Köln, Germany
andreas.meyer@dlr.de

Laurent J. Michot Laboratoire Environnement et Minéralurgie, UMR 7569 CNRS-INPL-ENSG, 15, Av. du Charmois, BP 40, F-54501 Vandoeuvre Lès Nancy Cedex, France
Laurent.michot@ensg.inpl-nancy.fr

Stewart F. Parker ISIS Facility, Rutherford Appleton Laboratory, Chilton, Didcot, Oxfordshire, OX11 0QX, UK
s.f.parker@rl.ac.uk

Marie Plazanet Institut Laue-Langevin, 6 rue Jules Horowitz, 38042 Grenoble Cedex 9, France
plazanet@ill.fr

Jan Pleuger Geologisches Institut der ETH Zürich, CH-8092 Zürich, Switzerland
Jan.pleuger@erdw.ethz.ch

Roger Pynn Indiana University Cyclotron Facility, 2401 Milo B. Sampson Ln, Bloomington, IN 47408-1398, USA
rpynn@indiana.edu

Simon A. T. Redfern Earth Sciences, University of Cambridge, Cambridge, CB2 3EQ, UK
satr@esc.cam.ac.uk

Romano Rinaldi Dipartimento di Scienze della Terra, Universita' di Perugia, I-06100 Perugia, Italy
rrinaldi@unipg.it

Gabriela Román-Ross Environmental Geochemistry Group, LGIT, Université Joseph Fourier, F-38401, France
gabriela.Roman-Ross@obs.ujf-grenoble.fr

Nancy L. Ross Department of Geosciences, Crystallography Laboratory, Virginia Tech, Blacksburg, VA 24061, USA
nross@vt.edu

Gernot Rother Chemical Sciences Division, Oak Ridge National Laboratory, MS-6110, Oak Ridge, TN 37831-6110, USA
rotherg@ornl.gov

Helmut Schober Time-of-Flight and High Resolution Group, Institut Laue-Langevin, 6, rue Jules Horowitz BP 156, F-38042 Grenoble Cedex 9, France
schober@ill.fr

Paul F. Schofield Department of Mineralogy, Natural History Museum, Cromwell Road, London SW7 5BD, UK
pfs@nhm.ac.uk

Neal T. Skipper Department of Physics and Astronomy, University College London, Gower Street, London WC1E 6BT, UK
n.skipper@ucl.ac.uk

Oleg Sobolev Laboratoire de Géophysique Interne et Tectonophysique 1381, rue de la Piscine, F-38041 Grenoble, France
sobolev@ill.eu

Elinor C. Spencer Department of Geosciences, Crystallography Laboratory, Virginia Tech, Blacksburg, VA 24061, USA
espence@vt.edu

Sarah S. Staniland School of Biological Sciences, University of Edinburgh, The King's Buildings, Edinburgh, EH9 3JR, UK
s.staniland@ed.ac.uk

Roberto Triolo The University of Palermo, I-90128 Palermo, Italy
triolo@unipa.it

Frédéric Villiéras Laboratoire Environnement et Minéralurgie, UMR 7569 CNRS-INPL-ENSG, 15, Av. du Charmois, BP 40, F-54501 Vandoeuvre Lès Nancy Cedex, France
frederic.villieras@ensg.inpl-nancy.fr

Lukas Vlcek Department of Chemical Engineering, Vanderbilt University, Nashville, TN 37235, USA
lukas.vlcek@Vanderbilt.Edu

Jens M. Walter Geowissenschaftliches Zentrum der Universität Göttingen, Goldschmidtstraße 3, D-37077 Göttingen, Germany
jwalter@gwdg.de

Wei Wang Environmental Sciences Division, Oak Ridge National Laboratory, MS-6036, Oak Ridge, TN 37831-6036, USA
wnagw@ornl.gov

Bruce Ward School of Chemistry and EaStChem, The University of Edinburgh, The King's Buildings, Edinburgh, EH9 3JJ, UK
b.ward@ed.ac.uk

David J. Wesolowski Chemical Sciences Division, Oak Ridge National Laboratory, MS-6110, Oak Ridge, TN 37831-6110, USA
wesolowskid@ornl.gov

Marco Zoppi Consiglio Nazionale delle Ricerche (CNR)-Istituto Sistemi Complessi, Via della Madonna del Piano, 10, I-50019 Sesto Fiorentino, Italy
marco.zoppi@isc.cnr.it

Chapter 1
Neutron Applications in Earth, Energy, and Environmental Sciences

Romano Rinaldi, Liyuan Liang, and Helmut Schober

Abstract Neutron-based studies permit the determination of the structural details and the dynamics of atomic arrangements in materials from simple measurements of scattering and absorption processes. Neutrons are scattered by atomic nuclei, are sensitive to the atomic magnetic moment, and have scattering and absorption cross-sections independent of atomic number and mass. They therefore have a complementary role to X-rays, scattered by the electrons in atoms. A prominent aspect of this lies in the sensitivity of neutrons to light elements, in particular hydrogen, a ubiquitous component of organic and inorganic matter and a key component of Earth, energy, and environment-related materials. Furthermore, thanks to the low absorption of neutrons by most substances, neutron scattering allows good quality data to be obtained over a wide range of non-ambient environments. This permits studies of transformations and fundamental properties of materials in situ, while they are still subject to the physical–chemical conditions in the diverse environments in which they normally exist and function, from the Earth's surface to its deep interior, and to laboratory conditions of one's choice. The limitations traditionally connected with modest neutron flux, and consequent need for relatively large samples, are being overcome by current advances in neutron sources and instrumentation. As a result, the potential of neutron-based methods in the examination of materials in Earth, energy, and environmental studies has gained momentum and opened up diverse new possibilities in these fields of scientific and technological research.

1.1 Introduction

The intrinsic properties of neutrons, as discussed in Chapter 2, make them a versatile probe for the study of materials encountered in Earth, energy, and environmental sciences. Applications of neutrons in the Earth Sciences are relatively recent, but the prospect is promising because neutrons provide a unique, non-destructive method of obtaining information ranging from the angstrom-scale of atomic structures and

R. Rinaldi (✉)
Dipartimento di Scienze della Terra, Universita' di Perugia, I-06100 Perugia, Italy
e-mail: rrinaldi@unipg.it

L. Liang et al. (eds.), *Neutron Applications in Earth, Energy and Environmental Sciences*, Neutron Scattering Applications and Techniques,
DOI 10.1007/978-0-387-09416-8_1, © Springer Science+Business Media, LLC 2009

1

related motions to the micron-scale of material strain, stress, and texture, as well as to the meso-scale of porous media and defects in materials and functional components. A full understanding of basic phenomena such as crustal subduction, earthquakes, and volcanic eruptions depends on the physical, chemical, and rheological properties of the materials involved (crustal and mantle rocks, magmas and fluids). These in turn depend on the structure and properties of the constituent minerals and the associated hydrous components, which can be determined using neutron techniques as described in this volume.

Similarly, materials involved in the technological developments required in the search for energy alternatives, and the fundamental processes involved in the sequestration and transformation of toxic wastes to minimize environmental impact, can also be profitably investigated by methods based on neutron sources. The modern world is primarily dependent on energy from fossil fuels and mineral resources originating from geological processes and materials within sediments and crustal rocks. The environment of living organisms is in turn threatened by the wastes generated as well as by the rapid consumption of these energy reserves. This volume intends to provide a representative number of examples of neutron applications bridging these many aspects of research related with the system Earth.

1.2 Neutron-Matter Interactions: Intrinsic Properties, Advantages, Disadvantages, and Complementarity

Of the many properties of neutrons (Chapter 2) as probes for the study of materials, the following are particularly relevant here:

- Neutrons have no electric charge.
- They interact with the nuclei rather than with the charge distribution of atoms in matter.
- They have wavelengths in the range of interatomic distances. They have magnetic moment and interact with the magnetic moment of atoms in matter.
- Their mass ($\sim$1 a.m.u.) is similar to that of atomic nuclei (1–240 a.m.u.), hence they have energy and momentum similar to those of atoms in solid and fluid materials.

From these properties, a number of features emerge when considering the use of neutrons in scattering and absorption experiments [1].

- Neutron scattering results from interactions with atomic nuclei, i.e., over scattering lengths (distances) of the order of 10^{-15} m (1 fm). By comparison, X-ray interactions (with electrons) occur over distances of 10^{-10} m (1 Å), i.e., five orders of magnitude larger. Furthermore, scattering and absorption cross-sections of neutrons do not dependent in a systematic way upon atomic number and mass, allowing discrimination between neighbors and isotopes in the periodic table (Chapter 2).

- Although scattering amplitude decays greatly with the scattering vector Q (the inverse of the scattering length) for X-rays, there are insignificant variations of scattering amplitude in the same range of Q (Bragg angle) for neutrons [2]. Consequently, powder diffraction with neutrons can resolve very fine structural and textural details of complex atomic structures, having access to a large number of reflections all the way to very short d spacings (very high Bragg angles).
- The weak interaction of neutrons with matter results in very low attenuation (see Fig. 2.1 in Chapter 2) offering a unique advantage for non-destructive, in situ work and bulk analyses (with high grain statistics on large objects) of undisturbed samples. For polycrystalline materials, no crushing is required to obtain powder patterns.
- Resonance absorption of neutrons yields several useful applications, from the measurement of temperature through the Doppler broadening of resonance lines [See Chapter 4, and Ref. 3] to the detection of trace element components of materials by neutron resonance capture analysis [4].
- Direct imaging techniques, based on selective absorption of neutrons by different atomic species and materials, also profit from the high penetration and offer a range of novel applications for bulk materials and apparatus, in energy, environmental, and plant sciences, etc., although still with less than optimum resolution. Lehmann et al. (Chapter 11) review such applications and the state of the art.

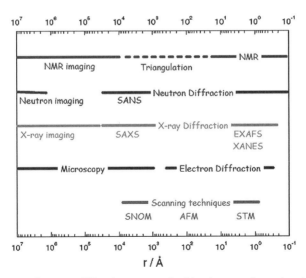

Fig. 1.1 The range of neutron diffraction compared with other experimental techniques. While there is considerable overlap in terms of the length scale, the information obtained is often complementary due to different element specificity. Furthermore, magnetic structure determination is nearly entirely a domain of neutron scattering. Techniques in the lower half of the diagram are typically only applied to very small samples [12]

Descriptions of the formalisms for neutron scattering (coherent and incoherent, elastic, magnetic and inelastic), and the principles of their application in neutron instruments are found in [5] and in Chapters 2, 3, and 4 of this volume.

1.2.1 Advantages

A neutron probe offers the following advantages over more traditional methods of material analysis:

- The nuclear and non-linear dependence of neutron scattering for different atomic species readily permits discrimination between iso- or quasi-iso-electronic species. Neutrons have no fundamental limitations in probing light elements and in distinguishing between isotopes.
- Neutron powder diffraction produces information beyond the range of X-ray diffraction and complementary to X-ray diffraction. The high penetration capability of neutrons allows in situ analysis of samples by environmental equipment, such as high-temperature furnaces, low-temperature cryostats, high-pressure reaction cells, and differential loading frames, etc. Diffraction information is obtained from the bulk of the sample rather than from the surface only. This is also relevant in the study of displacive phase transitions where surface behavior can be different from the bulk. The few very strong absorbers (e.g., B, Cd, Gd) can effectively be used for screening devices.
- A large neutron beam allows good sampling of the object under investigation, i.e., the data collected can be considered to be representative of the whole object.
- Systematic effects, such as absorption and preferred orientation, play an insignificant role in neutron diffraction analysis.
- Neutrons are produced via moderation in a thermal bath. They thus have by definition energies similar to the energies of excitations in solids and liquids at those moderator temperatures. This makes neutrons unique tools for investigations of energies in the range of nano-eV to several eV, corresponding to time scales ranging from attoseconds to microseconds. This unique capacity combined with all the other advantages already outlined above makes neutrons in many cases the only probe capable of accessing dynamic processes, with a space resolution at the atomic scale.

Other advantages are offered by time-of-flight (TOF) neutron diffraction at pulsed spallation sources (see Chapter 3). In the TOF mode, from a combination of De Broglie's relation ($\lambda = h/mv$, see Chapter 2) with Bragg's law ($\lambda = 2d_{hkl} \sin \theta$), one obtains the following TOF relation:

$$t(\mu \sec) = 252.78 \, L(m)2d_{hkl}(\text{Å}) \sin \theta,$$

where L represents the flight path (in meters). The d spacing of the diffracting crystal plane (hkl) is thus resolved in time. The method is essentially wavelength dispersive (time), with position-sensitive detectors having a time resolution of the order of 1 μs and a spatial resolution of the order of 1 mm. The *advantages* are therefore:

- A complete diffraction pattern is collected at any given scattering angle and over a large range of angles from forward- to back-scattering directions. Measurement is made with a stationary experimental setup, i.e., neither sample nor detector movements are required.
- Complete objects of variable shapes and sizes (even fairly large ones) can be investigated without prior preparation, by simple "immersion" in the neutron beam. Some rotation may only be needed if the detectors do not cover a variety of azimuth angles.
- TOF diffractometers provide optimum resolution in backscattering mode where the line width of diffraction peaks is largely independent of sample thickness; allowing line-width analyses to be performed for stress/strain measurements [6].
- Several banks of position-sensitive detectors used in TOF diffractometers cover a large *d*-spacing range, permitting accurate determination of scale factors, i.e., phase fractions in Rietveld analysis [7, 8] A multitude of detector elements provides remarkably robust TOF neutron diffraction Rietveld refinements for the analysis of fine structural details.
- Preferred orientation or texture effects are easily recognized, and the textural and microstructural information may even be an important part of the characterization of the object [9].

Furthermore, the high peak intensity of a modern pulsed spallation source opens the way to time-resolved kinetic measurements and pulse-probe techniques allowing materials to be investigated in non-equilibrium, or transient, conditions.

1.2.2 Disadvantages

Most disadvantages stem from the intrinsic relative limitations of neutron flux available at conventional reactor sources, or at first-generation spallation sources. The brilliance of modern neutron sources is approximately 14 orders of magnitude lower than that of a third-generation synchrotron source. High-flux research reactors such as the Institut Laue-Langevin (ILL, Grenoble, France) and the High-Flux Isotope Reactor (HFIR, Oak Ridge, TN) are currently operating close to the practical limit of brilliance for research reactor sources. Further increments can only be expected from newer pulsed spallation sources [10]. The relatively low flux available at present at operating neutron beam lines obligates the use of relatively large samples (see Chapter 5 for examples of inelastic neutron scattering applications to single crystals) and very efficient detectors. This is being partly overcome in the most recent generation of pulsed neutron sources, such as the Spallation Neutron Source at Oak Ridge, USA [11], where high-brilliance neutron beams, large detector arrays, and TOF techniques have been implemented to great advantage.

1.2.3 Complementarity with Other Material Analysis Techniques

Neutron methods complement many other techniques to probe the intrinsic properties of materials and compounds. Figures 1.1 and 1.2 show the ranges of elastic and inelastic neutron scattering applications, compared with a number of other techniques. There are clearly large areas of overlap as well as definite areas of unique applications. Furthermore, when considering specific applications, such as light elements or bulk magnetic scattering, neutrons have a unique role. Neutrons couple directly to mass and spin, and photons couple preferentially to and through charge distribution; these techniques are therefore complementary at a fundamental level.

One area of apparent competition might be in the high-energy, high-momentum transfer region (top right corner of Fig. 1.2). Previously, this was only within reach of inelastic neutron scattering, but developments of (triple-axis) X-ray inelastic scattering have now produced a large region of overlap. In specific cases, for example, the measurement of phonon dispersion curves at high energies, neutrons will always "struggle" because of their lower velocity. Consequently, in the overlap region, photons would be the probe of choice. However, for comparable measurements in liquids and glasses, the different element specificity of the two techniques makes them highly complementary for all but the simplest (elemental) systems. Developments of photon and neutron sources will further enhance this complementarity.

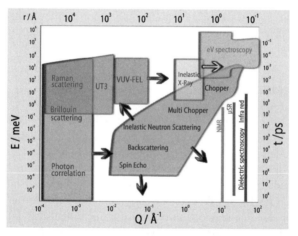

Fig. 1.2 Inelastic neutron scattering provides information that covers a large area in space and time. Other techniques mainly cover different areas. With latest generation planned neutron sources in Europe, the ranges will extend in the directions indicated, filling in the "missing areas". The diagram covers typical atomic length and time scales. Techniques that do not directly provide distance information are indicated only as bars along the time axis. The time scale only refers to equilibrium phenomena. Non-equilibrium effects, such as those studied in "pump-probe" experiments at very short time intervals (fs) will always remain the domain of photon-based techniques [12]

1.3 Recent Neutron Applications in Earth, Energy, and Environmental Research

Neutron scattering and spectroscopy applications in the study of minerals, rocks, and environmental and energy-related materials stem from some unique merits of neutrons, experimental techniques, and present-day neutron sources [13]. These advantages become evident with the examples of applications reported in this volume. The intent is to provide a guided introduction to the potential of these techniques in a variety of these related fields.

1.3.1 High Penetration Power

The low attenuation of neutron beams by many metals and materials has made extreme sample environments (high temperature, high pressure, reaction cells, differential loading frames, etc.) easier to work with for neutron scattering. The extreme conditions are often needed to simulate Earth surface and interior conditions (Fig. 1.3), and these conditions can be reproduced by environmental equipment that has little interference with the neutron beam, as compared with other experimental probes. High-temperature furnaces with Vanadium or Zirconium heating elements are virtually transparent to thermal neutrons. High-pressure cells (also in combination with high temperature) allow access to large portions of the reciprocal space through low-absorbing gaskets. Sample cells fitted with pressure- and

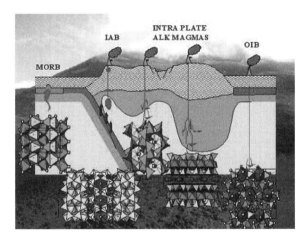

Fig. 1.3 Mineral crystal structures carry a record of their genesis, especially when studied under the conditions at which they formed deep in the Earth. Structural models of common rock-forming minerals shown from the left are olivine, amphibole, pyroxene, mica, and garnet. Various geodynamic environments are represented. MORB = Mid-Ocean Ridge Basalt; IAB = Island Arc Basalt; OIB = Ocean Island Basalt. The hydrous component of minerals and rocks has a great impact on all geodynamic processes (*Picture courtesy of R. Oberti*)

temperature-measuring devices can deliver accurate measurement conditions. A good example of this is given by pressure cells with anvils made of BN lined with Cd foil (which absorb the beam quite effectively) and gaskets made of TiZr alloy, which transmit neutrons directly through to the sample [14].

Other equipment for in situ experiments include reaction cells, differential loading frames, cryostats, etc., making it possible to study minerals, and other Earth, energy, and environment materials in their natural or non-ambient conditions and in a *time-resolved* mode. The transformations and reactions occurring in the diverse environments of formation can thus be explored and monitored directly and, importantly, with access to details all the way down to atomic-scale resolution.

On the other hand, the weak interaction results in a low flux of the scattered neutrons; hence, weak signals from small samples. This can be a limiting factor, especially in high-pressure studies where such limitations demand an increase in sample size, which in turn increases the size of the apparatus toward practical and mechanical limits. The most suitable objects for neutron applications are therefore either bulky or gram-scale samples. The in situ and time-resolved studies described in this monograph encompass a large number of possible applications in a variety of fields.

Studies on natural minerals and their magnetism at high pressure and temperature are reviewed by Redfern and Harrison (Chapter 4). Some fundamental features, such as cation order–disorder and phase transitions in rock-forming minerals, are important in modeling the geological phenomena. The pressure range of current apparatus is still inadequate to match the first transition zone within the Mantle. However, there are no real limits to either temperature or reciprocal space exploration in high-temperature furnaces. The emphasis is therefore on simultaneous temperature and pressure cells, which offer otherwise unobtainable sets of conditions compared with other presently available techniques.

Due to the special instrumental requirements imposed by high-pressure experiments, special beam lines and devices have been or are planned at the new neutron spallation sources in the USA [SNAP, SNS] and Japan [15] to match pressures and temperatures typical of crust to upper mantle environments. Pressures up to tens of GPa and temperatures in excess of 2000°K on $\sim$1-mm^3 samples are envisaged on a routine basis. Many opportunities exist with regard to both technical improvements and scientific advancements in this field.

Clays, their water content, and the transformations they undergo upon hydration and dehydration are reviewed by Bihannic et al. (Chapter 18). The interlayer water dynamics at the Ängstrom scale are linked to the scale of landslides and slope instability, and the arrangement of water in swelling clay is important to sorption of trace pollutants by clay liners. The chemical and physical behavior of contaminant sorption materials requires an approach typical of a hierarchical structure (Triolo and Agamalian, Chapter 20), with properties ranging from those of solid nano-porous media to dispersed colloidal gels. Uptake of radioactive waste and other pollutants by clay-like and nano-porous compounds is reviewed by Cuello et al. (Chapter 17). In situ studies are expected to provide a better understanding of the sorption properties of clays and other minerals for radionuclide species and other organic pollutants.

1.3.2 Scattering Power for ^1H and ^2H

Neutrons, unlike X-rays, are efficiently and strongly scattered by hydrogen ^1H and deuterium ^2H atoms. The hydrous component is of paramount importance in Earth compounds and related processes. All biological materials as well as many minerals have a structural hydrous component in the form of hydroxyls or water molecules. Several energy-related materials, from zeolites to gas hydrates, whether natural or man-made, also fall into this category. The water carried by minerals in subduction processes [16], and existing at depth in the lower Mantle, could be equivalent to that of several oceans [17–19]. Water in minerals (even nominally anhydrous minerals), rocks, magmas, and in all Earth materials generally strongly influences the behavior and properties of these materials at scales from atomic to continental levels. The character of volcanic eruptions, ranging from highly destructive (Plinian) to relatively benign (Hawaiian), depends on the water content and overall chemistry of the magma involved.

Isomorphic substitution of deuterium (^2H) for hydrogen (^1H) may have to be adopted to limit the high background resulting from the strong incoherent scattering of hydrogen; a problem still felt in neutron powder diffraction at conventional sources, but more tolerable in single-crystal work and at present-day neutron sources.

The study of the hydrous component in minerals and Earth materials by neutron scattering methods, reviewed by Gies (Chapter 7) and addressed in several other chapters, ranges from nominally anhydrous mantle minerals to highly hydrated natural and industrial zeolites and clay minerals. Neutron scattering of these latter materials provides accurate determinations of water molecules in the structure (distributed in channels, cages, and interlayer spaces) and their dynamics in response to changes in physical and chemical environments.

Hydrogen has a fundamental role in many environmental issues and materials, from surface reactions in rocks and minerals (Ross et al., Chapter 8) to the compounds related to the "Hydrogen Economy" (Celli et al., Chapter 14), and the immense gas hydrate deposits within continental shelf sediments (Koza and Schober, Chapter 12). The stability of gas hydrates must be precisely determined by in situ structure studies under differing physical and chemical conditions to derive accurate predictive models of their stability in the geological environment, and to devise methods for tapping these very extensive energy resources.

The proton dynamics of highly dispersed materials is addressed by Albers and Parker (Chapter 13). Applications include determining the proton-related surface chemistry of nanomaterials used as reinforcing fillers of tires where improved safety and reduced fuel consumption are at stake. Surface science studies consider adsorption of hydrogen on nanodispersed precious metal particles as fuel cell catalysts. The occupancy of catalytically relevant sites provides essential information for tailoring better catalysts. Surface deactivation of industrial catalysts by coke deposition, chemical transformation of deposits, and other processes results in significant operational loss at large-scale plants, which can be minimized by understanding the mechanism of catalyst deactivation.

Hydrogen is also ubiquitous in biological materials, and its presence and distribution, either as an atomic species or in bulk, can be accurately investigated with neutrons. Johs et al. (Chapter 16) review neutron reflectivity methods for probing interfacial interactions between proteins and minerals, which is significant in elucidating the structural changes involved in the microbiologic reduction of toxic metals in the environment. Cuello et al. (Chapter 17) review the use of neutrons in water pollutant speciation of both heavy metals and organic contaminants. Notable examples are the hydration structure around Hg as a solute, or around lanthanides adsorbed on clay minerals. In conjunction with X-ray methods, the uptake of As by common sedimentary minerals has been characterized. Discussion of these problems in the context of water strategies from governmental authorities provides direct links with practical applications.

Sensitivity to hydrogen can also be used to good advantage for following water distribution and dynamics in many environments. This can include studies of porosity in rocks, concretes, and other media, as well as elucidating structures of planetary ices and hydrocarbon molecular structures. Triolo and Agamalian (Chapter 20) address the former using a combination of SANS and USANS techniques. Such studies may include phase transitions, such as the freeze-thaw cycles and other dynamics in relation to sorbent and adsorbate properties. Other applications may regard the experimental determination of the fractal dimension of volume pore systems in hydrocarbon or water-containing rocks.

Cole et al. (Chapter 19) show how the behavior of fluids (polar and non-polar) in the confined geometries of pores and fractures differs from the bulk. Many parameters, such as pore size, connectivity, and hydrophilic or hydrophobic character, must be taken into account, and both structural modifications and dynamic behavior can be addressed by SANS and USANS techniques. Problems addressed range from the design of micro- or meso-porous media for industrial applications and the enhancement of oil recovery to CO_2 sequestration in spent oil reservoirs.

In addition to hydrogen, other light elements are also amenable to accurate structural determination by neutrons. Catti (Chapter 15) gives examples of lithium compounds in energy applications. The geochemical properties of such light elements merit investigation in natural minerals and geological materials, an opportunity that will certainly draw increased attention.

1.3.3 Iso-Electronic Species and High Q

The contrast in neutron-scattering cross-sections between isotopes (which is either none or very little in X-ray diffraction) offers many opportunities for structural studies in mineralogical and environmental research. Additionally, since the scattering cross-section for neutrons does not change with scattering vector (i.e., does not fall off as the inverse of the atomic radius), this permits the collection of diffraction data up to large scattering vectors, thus providing a significant increase in the amount of information available in a diffraction pattern and to decouple the information on thermal motion and site occupancies.

Many common ions in minerals have equal or similar numbers of electrons. For instance, O^{2-}, Na^+, Mg^{2+}, Al^{3+}, and Si^{4+} present in most rock-forming silicate minerals possess, in these (mostly ionic) compounds, an effective maximum scattering power for X-rays corresponding to the same number of 10 electrons. The same applies to Cl^-, K^+, Ca^{2+}, Ti^{4+}, all with 18 electrons. Other elements, such as Fe and Mn, have several oxidation states, and are difficult to discriminate by X-ray scattering. All of these are readily resolved by neutron diffraction with direct determination of their site occupancies and order–disorder distributions in the crystal structures. The "high-Q" advantage also provides structural parameters (i.e., occupancy and thermal motion) less affected by correlations, and offers high precision and accuracy for the determination of bonding geometries in crystal structures. Recent studies of mantle minerals have revealed fundamental properties that underpin major geodynamic events. Redfern and Harrison (Chapter 4) give examples of these studies. Much more work in conjunction with other techniques (X-ray diffraction, IR, NMR, etc.) is needed in this field to provide fundamental, complementary information.

1.3.4 Fundamental Structural Properties

Inelastic Neutron Scattering (INS), which is not subject to tight selection rules on mode symmetries and wave vectors, can be used to determine phonon-dispersion curves and phonon densities of states to reveal the fundamental structural properties of minerals and phase transformations under pressures and temperatures of the Earth's interior. Applications of INS to in situ studies offer a unique opportunity for solving fine structural details (atomic and proton dynamics, soft modes, etc.) and the modeling and interpretation of fundamental thermodynamic parameters. Choudhury and Chaplot (Chapter 5) survey these studies and future needs with examples from a large variety of rock-forming minerals. These data are invaluable for geochemical modeling of Earth interior conditions that are beyond the reach of other experimental techniques.

The properties of silicate melts are of fundamental importance for a thorough understanding of some geological processes and to investigate the nature of glass, a material known for thousands of years and still deserving close scientific and technological attention. Meyer et al. (Chapter 6) review the physical and chemical properties of glass and silicate melts by using quasi-elastic neutron scattering and molecular dynamics simulations. The interplay of intermediate range order, atomic dynamics, and properties of mass transport in alkali silicate melts give insights into the mechanisms relevant to geology, petrology, and volcanology, as well as to glass technology.

Nanostructures and time-resolved surface reactions in minerals and their dynamics are reviewed by Ross et al. (Chapter 8). A combination of inelastic and quasi-elastic neutron scattering with molecular dynamics techniques provides insights into the thermodynamic properties of adsorbed water on mineral surfaces and nanoparticles. This complementary approach potentially offers a complete description of the energy, structure, and dynamics of the hydration layers present on the surface of mineral nanoparticles that are an intrinsic part of the particles themselves.

1.3.5 Texture and Residual Stress Analysis

In quantitative texture analyses (and also for stress and strain measurements), the high penetration of neutrons and the availability of wide beams allow the investigation of large specimens, which produce global volume textures with high grain statistics even on coarse-grained materials. Position-sensitive detectors and time-of-flight techniques provide reflection-rich diffraction patterns of polymineralic rocks containing low-symmetry mineral constituents. Residual stress and strain analysis of geological materials requires the high accuracy and sensitivity that neutrons can offer, since natural effects on rocks are orders of magnitude smaller than in technological materials.

Stress and strain analysis and the mechanical behavior of geological materials studied in situ by the use of engineering apparatus (Covey-Crump and Schofield, Chapter 9) reveal the fundamental parameters of rock behavior motivated by the search for predictive seismological and tectonic models. The concomitant mechanics and chemistry of carbonate rocks and subsurface hydrates could tie in with global water and carbon cycles. Neutron diffraction applied to the study of the orientation distribution function (texture) of crystallites within the bulk of multi-mineral rock samples (Pleuger et al., Chapter 10) show the sequence of complex tectonic events in geologic time. The use of large samples, afforded by neutrons, provides high grain statistics even on coarse-grained material and data from the undisturbed interior of a three-dimensional sample rather than from a polished surface.

1.3.6 Magnetism

Natural ferri/o-magnets are common compounds in Solar System planets. Geologists have long utilized the magnetic signatures of rocks and minerals to reconstruct stratigraphic sequences, and much remains to be discovered regarding the magnetic properties of many minerals. Neutrons are very sensitive to magnetic moment, making them effective probes to determine magnetic structure, collective magnetic excitation, and crystal field energy levels in many magnetic elements. Neutron diffraction of magnetic minerals and materials was recently reviewed by Harrison [20]. Recent applications to minerals, and a future outlook including cation and magnetic ordering in ilmenite-hematite solid solution and exolution mechanisms, are reviewed by Redfern and Harrison (Chapter 4). They investigate these properties and report that slowly cooled rocks containing finely exsolved members of the hematite-ilmenite series have strong and extremely stable magnetic remanence, suggesting an explanation for some magnetic anomalies in the deep crust and on planetary bodies that no longer retain a magnetic field, such as Mars. Magnetic diffraction with polarized neutrons on single crystals reveals two generations of lamellae with different magnetic properties within the same crystal. Future developments are expected by magnetic diffraction studies at pressures and temperatures of the Earth interior. Only neutrons with a wide coverage of reciprocal space, even in the very taxing conditions of such bulky environmental apparatus, can provide sufficiently accurate data to extract this information.

Many technological applications are based on the magnetic properties of materials, and many recent discoveries have implications in the fields of Earth, energy, and environmental sciences. Staniland et al. (Chapter 21) provide an outlook of the possibilities offered by neutron scattering for understanding the magnetic properties of bacteria in natural environments.

1.3.7 Direct Imaging

Although not strictly a scattering technique and still of somewhat limited resolution, imaging by radiography and tomography with neutrons offers the advantage of a very high penetrating power. The trade-off in resolution, when compared with X-rays, can be compensated for by the precision of measurements made in large-scale experiments. Applications in the Earth sciences include studies of structural and rheological properties of molten systems reproducing natural magmas, directly investigated inside custom-built apparatus by neutron scattering and imaging methods [21, 22], and the study of rock permeability, moisture transport, and porosity determinations (Lehmann, Chapter 11). Digital neutron imaging is fast replacing the traditional film technique, and energy selection by TOF techniques opens up new perspectives. Advances are expected in many fields of application, from the direct observation of functional properties in fuel cells to the adhesive properties of cold welding, to time-resolved studies and energy-selective imaging with Bragg edge enhanced features, to stroboscopic imaging by coupling the process with the pulse frequency of the source. Earth science, energy, and environmental studies can greatly benefit from these developments

In actual fact, the greatly expanding application of neutron imaging techniques, encompassing a large variety of fields, many of which are outside the scope of this book, merits a special treatment that will be covered by the next monograph in the Series.

1.4 Concluding Remarks

Studies utilizing neutron scattering techniques have many diverse applications in Earth, energy, and environmental sciences. They derive from a number of advantages offered by neutrons over other experimental techniques. Among these the sensitivity of neutrons to hydrogen plays a prominent role. By complementing and extending other research techniques, including X-ray analysis, these applications reveal many promising areas of enquiry. The aim of this monograph is to provide examples of recent progress, evaluate current developments, and consider future advances in this expanding field of scientific research.

References

1. R. Rinaldi, Neutron scattering in Mineral Sciences, Eur. J. Mineral. **14**, 195 (2002)
2. C. G. Schull, E. O. Wollan, X-ray, electron, and neutron diffraction, Science. **108**, 69 (1948)
3. S. A. T. Redfern, Neutron powder diffraction of minerals at high pressures and temperatures: some recent technical developments and scientific applications, Eur. J. Mineral. **14**, 251 (2002)
4. M. Blaauw, H. Postma, P. Mutti, Quantitative neutron capture resonance analysis verified with instrumental neutron activation analysis, Nucl. Instrum. Methods Phys. Res. A **505**, 508–511 (2003)
5. M. T. Dove, An introduction to the use of neutron scattering methods in mineral sciences, Eur. J. Mineral. **14**, 203 (2002)
6. W. Schaefer, Neutron diffraction allied to geological texture and stress analysis, Eur. J. Mineral. **14**, 263 (2002)
7. H. M. Rietveld, A profile refinement method for nuclear and magnetic structures, J. Appl. Cryst. **2**, 65 (1969)
8. A. C. Larson, R. B. Von Dreele, GSAS: general structure analysis system LAUR 86-748 Report (Los Alamos National Laboratories, USA) (1986), http//www.ccp14.ac.uk
9. R. B. Von Dreele, Quantitative texture analysis by Rietveld refinement, J. Appl. Cryst. **30**, 517 (1997)
10. F. Mezei, New perspectives from new generations of neutron sources, C. R. Physique **8**, 909–920 (2007)
11. K. N. Clausen, J. Mesot, The route forward for Europe: the European spallation source (ESS)!, Neutron News **18/4**, 2 (2007)
12. F. Boue, R. Cywinski, A. Furrer, H. Glattli, S. Kilcoyne, R. L. McGreevy, D. McMorrow, D. Myles, H. Ott, M. Rübhausen, G. Weill. Neutron scattering and complementary experimental techniques. The ESS Project, Volume II, Chapter 5. (2002) ISBN 3-89336-302-5. http://neutron.neutron-eu.net/n_documentation
13. R. Rinaldi, H. Schober, Neutrons at the Frontiers of Earth Sciences and Environment, Neutron News. **17/2**, 3 (2005)
14. S. Klotz, Th. Straessle, G. Rousse, G. Hamel, V. Pomjakushin, Angle-dispersive neutron diffraction under high pressure to 10 GPa, Appl. Phys. Let. **86**, 031917 1–3 (2005)
15. H. Kagi, T. Nagai, K. Komatsu, W. Utsumi, M. Arai, High-pressure beamline at the new neutron source, J-PARC, in Japan, EGU-05, Vienna A-06979, 230 (2005), http://www.cosis.net/abstracts/EGU05/06979/EGU05-A-06979.pdf and http://neutron.neutron-eu.net/n_nmi3/n_networking_activities/n_nese/855
16. T. Kawamoto, Hydrous phases and water transport in the subducting slab, Rev Mineral. Geochem. **62(1)**, 273 (2006)
17. J. R. Smyth, A crystallographic model for hydrous wadsleyite: an ocean in the Earth's Interior?, Am. Mineral. **79**, 1021 (1994)
18. J. R. Smyth, Hydrogen in high pressure silicate and oxide mineral structures. Rev. Mineral. Geochem. **62(1)**, 85 (2006)
19. E. Ohtani, Water in the mantle, Elements, **1**, 25 (2005)
20. R. J. Harrison, Neutron diffraction of magnetic materials, Rev. Mineral. Geochem. **63**, 113 (2006)
21. B. Winkler, K. Knorr, A. Kahle, P. Vontobel, E. Lehmann, B. Hennion, G. Bayon, Neutron imaging and neutron tomography as non-destructive tools to study bulk rock samples, Eur. J. Mineral. **14**, 349 (2002)
22. B. Winkler, Applications of neutron radiography and neutron tomography, Rev. Mineral. Geochem. **63**, 459 (2006)

Chapter 2
Neutron Scattering—A Non-destructive Microscope for Seeing Inside Matter

Roger Pynn

How can we determine the relative positions and motions of atoms in a bulk sample of solid or liquid? Somehow we need to be able to see inside the sample with a suitable magnifying glass. It turns out that neutrons provide us with this capability. They have no charge, and their electric dipole moment is either zero or too small to measure. For these reasons, neutrons can penetrate matter far better than charged particles. Furthermore, neutrons interact with atoms via nuclear rather than electrical forces, and nuclear forces are very short range—on the order of a few femtometers (i.e., a few times 10^{-15} m). Thus, as far as the neutron is concerned, solid matter is not very dense because the size of a scattering center (i.e., a nucleus) is typically 100,000 times smaller than the distance between centers. As a consequence, neutrons can travel large distances through most materials without being scattered or absorbed. The attenuation, or decrease in intensity, of a beam of neutrons by aluminum, for example, is about 1% per millimeter compared with 99% per millimeter or more for X-rays. Figure 2.1 illustrates just how easily neutrons penetrate various materials compared with X-rays or electrons.

Like so many other things in life, the neutron's penetrating power is a double-edged sword. On the plus side, the neutron can penetrate deep within a sample even if it first has to pass through a container (such as would be required for a liquid or powder sample, for example, or if the sample had to be maintained at low temperature or high pressure). The corollary is that neutrons are only weakly scattered once they do penetrate. To make matters worse, available neutron beams have inherently low intensities. X-ray instruments at synchrotron sources can provide fluxes of 10^{18} photons per second per square millimeter compared with 10^4 neutrons per second per square millimeter in the same energy bandwidth even at the most powerful continuous neutron sources.

The combination of weak interactions and low fluxes make neutron scattering a signal-limited technique, which is practiced only because it provides information about the structure of materials that cannot be obtained in simpler, less expen-

R. Pynn (✉)
Indiana University Cyclotron Facility, 2401 Milo B. Sampson Ln,
Bloomington, IN 47408-1398, USA
e-mail: rpynn@indiana.edu

L. Liang et al. (eds.), *Neutron Applications in Earth, Energy and Environmental Sciences*, Neutron Scattering Applications and Techniques,
DOI 10.1007/978-0-387-09416-8_2, © Springer Science+Business Media, LLC 2009

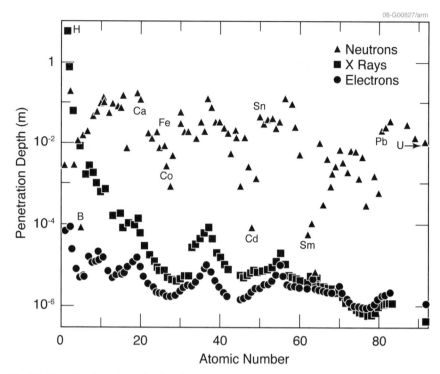

Fig. 2.1 The plot shows how deeply a beam of electrons, X-rays, or thermal neutrons penetrates a particular element in its solid or liquid form before the beam's intensity has been reduced by a factor $1/e$, which is about 37% of its original intensity. The neutron data are for neutrons having a wavelength of 0.14 nm

sive ways. This feature also means that no generic instrument can be designed to examine all aspects of neutron scattering. Instead a veritable zoo of instruments has arisen, each of which trades off the ability to resolve some details of the scattering in order to maintain sufficient scattered neutron intensity for a meaningful measurement.

In spite of its unique advantages, neutron scattering is only one of a battery of techniques for probing the structures of materials. All of the techniques, such as X-ray scattering, electron microscopy, and nuclear magnetic resonance (NMR), are needed if scientists are to understand the full range of structural properties of matter. In many cases, the different methods used to probe material structure give complementary information because the nature of the interactions between the probe and the sample are different.

To understand the neutron scattering technique, we first examine the *scattering by a single nucleus* and then *add up scattering from all of the nuclei* within the solid or liquid we are interested in. This allows us to describe phenomena like *neutron diffraction*, used to determine the atomic arrangement in a material, and *inelastic neutron scattering*, which measures the vibrations of atoms. Minor modifications of the theory allows us to describe **Small-Angle Neutron Scattering (SANS)** that is used to study larger structures such as polymers and colloids, as well as **surface reflection** of neutrons (often called **reflectometry**) in which layered materials and interfaces are probed. The consequences of the neutron's magnetic moment can also be explored. It leads to **magnetic scattering** of neutrons as well as to the possibility of **polarized neutron beams** that can provide enhanced information about vector magnetization in materials.

2.1 Scattering by a Fixed Nucleus

The scattering of neutrons by nuclei is a quantum mechanical process. Formally, the process has to be described in terms of the wavefunctions of the neutron and the nucleus. Fortunately, the results of this calculation can be understood without going into all of the details involved. It is useful, though, to be able to switch to and fro between thinking about the **wavefunction of a neutron**—the squared modulus of which tells us the probability of finding a neutron at a particular point in space—and a particle picture of the neutron. This wave-particle duality is common in describing subatomic particles, and we will use it frequently, sometimes referring to neutrons as particles and sometimes as waves.

The neutrons used for scattering experiments usually have energies similar to those of atoms in a gas such as air. Not surprisingly, the velocities with which they move are similar to the velocities of gas molecules—a few kilometers per second. Quantum mechanics tells us that the wavelength of the neutron wave is inversely proportional to the speed of the neutron. For neutrons used in scattering experiment, the wavelength, λ, is usually between 0.1 and 1 nm. Often, we work in terms of the neutron wavevector $\vec{k}$, which is a vector of magnitude $2\pi/\lambda$ that points along the neutron's trajectory. The magnitude of the wavevector, $\vec{k}$, is related to the neutron velocity, v, by the equation

$$|\vec{k}| = 2\pi m v/h, \tag{2.1}$$

where h is Planck's constant and m is the mass of the neutron.

The scattering of a neutron by a free nucleus can be described in terms of a **cross section**, σ, measured in barns (1 barn $= 10^{-28}$ m^2), that is equivalent to the effective area presented by the nucleus to the passing neutron. If the neutron hits this area, it is scattered isotropically, that is, with equal probability in any direction. The scattering is isotropic because the range of the nuclear interaction between the neutron and the nucleus is tiny compared with the wavelength of the neutron, so the nucleus essential looks like a **point scatterer**.

Suppose that at an instant in time we represent neutrons incident on a fixed nucleus by a wavefunction $e^{i\vec{k}.\vec{r}}$, in other words, a plane wave of unit amplitude.

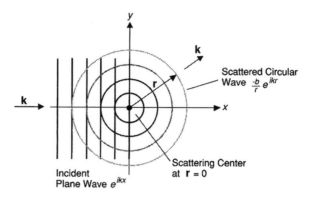

Fig. 2.2 A neutron beam incident on a single scattering center and traveling in the x direction can be represented by a plane wave e^{ikx} with unit amplitude. Because the neutron sees the scattering center (a nucleus) as a point, the scattering will be isotropic. As a result, the scattered neutron beam spreads out in spherical wavefronts (here drawn as circles) of amplitude b/r. The $1/r$ part of this amplitude factor, when squared to get neutron intensity, accounts for the $1/r^2$ decrease in intensity with distance that occurs as the scattered wavefront grows in size. Because we have taken the scattering center to be rigidly fixed, the scattering is elastic—that is, there is no change in the neutron's energy—so the incident and scattered wavevectors both have magnitude k

Note that the squared modulus of this wave function is unity for all positions $\vec{r}$ so the neutron has the same probability of being found anywhere but has a definite momentum $mv = hk/2\pi$. For a wave traveling in the x direction (i.e., for $\vec{k}$ parallel to the x axis), the nodes of the wavefunction are straight wavefronts, as shown in Fig. 2.2.

The amplitude of the neutron wave scattered by the nucleus depends on the strength of the interaction between the neutron and the nucleus. Because the scattered wave is isotropic, its wavefunction can be written as $(-b/r)e^{ikr}$ if the scattering nucleus is at the origin of our coordinate system. The spherical wavefronts of the scattered neutrons are represented by the circles spreading out from the nucleus in Fig. 2.2. The factor $1/r$ in the wavefunction of the scattered neutron takes care of the inverse square law that applies to all wave motions: the intensity of the neutron beam, given by the square of the wavefunction, decreases as the inverse square of the distance from the source, in this case the scattering nucleus. The constant b, referred to as the **scattering length** of the nucleus, measures the strength of the interaction between the neutron and the nucleus. The minus sign means that b is a positive number for repulsive interaction between neutron and nucleus.

Because we have specified that the nucleus is fixed and because the energy of the neutron is too small to change the internal state of the nucleus, the scattering occurs without any change in the neutron's energy and is said to be **elastic**. Because the neutron's energy is unchanged by the collision, the same wavevector k appears in the incident and scattered wavefunctions. It turns out that the cross section, σ, is given by $\sigma = 4\pi b^2$—as if the scattering length were half the radius of the nucleus as seen by the neutron. In the majority of cases of interest to neutron scatterers, b is a real energy-independent quantity that has to be determined by experiment or, these days, simply looked up in a table called the ***barn book***[1]. The scattering length is not correlated with atomic number in any obvious systematic way and varies even from isotope to isotope of the same element. For example, hydrogen and deuterium, both of which interact weakly with X-rays because they only contain one electron, have neutron scattering lengths that are relatively large and quite different from one another. The differences in scattering length from one isotope to another can be used in various ***isotope-labeling*** techniques to increase the amount of information obtained from some neutron scattering experiments.

2.2 Scattering of Neutrons by Matter

To work out how neutrons are scattered by matter, we need to ***add up the scattering from each individual nucleus***. The result of this lengthy quantum mechanical calculation is quite simple to explain and understand, even if the details are abstruse.

When neutrons are scattered by matter, the process can alter both the momentum and the energy of the neutrons and the matter. The scattering is not necessarily elastic as it is for a single fixed nucleus because the atoms in matter are free to move to some extent. They can therefore recoil during a collision with a neutron or, if they are moving when a neutron arrives, they can impart energy to the neutron. As is usual in a collision, the total energy and momentum are conserved: the energy, E, lost by the neutron in a collision is gained by the scattering sample. From Equation (2.1) it is easy to see that the amount of momentum given up by the neutron during its collision, the ***momentum transfer***, is $h\vec{Q}/2\pi = h(\vec{k} - \vec{k}')/2\pi$, where $\vec{k}$ is the wavevector of the incident neutrons and $\vec{k}'$ is that of the scattered neutrons. The quantity $\vec{Q} = \vec{k} - \vec{k}'$ is known as the ***scattering vector***, and the vector relation between $\vec{Q}, \vec{k}$, and $\vec{k}'$ can be displayed pictorially in the so-called ***scattering triangle*** (Fig. 2.3). The angle 2θ shown in the scattering triangles is the angle through which the neutron is deflected during the scattering process. It is referred to as the ***scattering angle***. For elastic scattering, for which $|\vec{k}| = |\vec{k}'|$, a little trigonometry applied to the triangles shows that $Q = 4\pi \sin\theta/\lambda$.

In all neutron scattering experiments, scientists measure the intensity of neutrons scattered by matter (per incident neutron) as a function of the variables Q and E. The scattered intensity, denoted $I(\vec{Q}, E)$, is often referred to colloquially as the ***neutron-scattering law*** for the sample material. In 1954, Van Hove [1] showed that the scattering law can be written in terms of time-dependent correlations between the positions of pairs of atoms in the sample. Van Hove's result implies that $I(\vec{Q}, E)$ is proportional to the Fourier transform of a function that gives the probability of finding two atoms a certain distance apart. It is the simplicity of the result that

---

[1] Thermal neutron cross-sections have been tabulated in a variety of places. The most accessible source these days is likely the NIST web site www.ncnr.nist.gov, which lists values tabulated in *Neutron News*, **3**(3) (1992), pp. 29–37. An alternative source is the *Neutron Data Booklet* (A. J. Dianoux and G. H. Lander Eds) published by the Institut Laue-Langevin, Grenoble, France. This can be ordered at http://www-llb.cea.fr/menl/neutronlist/msg00063.html.

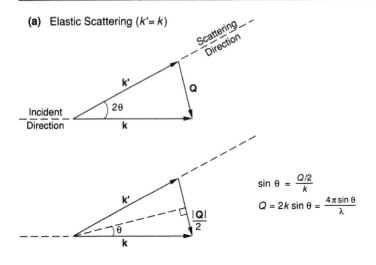

(a) Elastic Scattering ($k' = k$)

$$\sin \theta = \frac{Q/2}{k}$$

$$Q = 2k \sin \theta = \frac{4\pi \sin \theta}{\lambda}$$

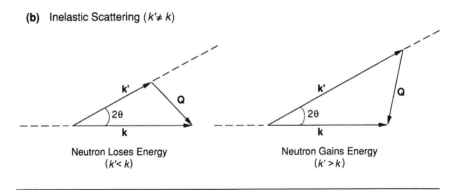

(b) Inelastic Scattering ($k' \neq k$)

Neutron Loses Energy
($k' < k$)

Neutron Gains Energy
($k' > k$)

Fig. 2.3 Scattering triangles are depicted here for both (**a**) an elastic scattering event in which the neutron is deflected but does not gain or lose energy (so that $k' = k$) and (**b**) inelastic scattering in which the neutron either loses energy ($k' < k$) or gains energy ($k' > k$) during the interaction with the sample. In both elastic and inelastic scattering events, the neutron is scattered through the angle 2θ, and the scattering vector is given by the vector relationship $\vec{Q} = \vec{k} - \vec{k}'$. For elastic scattering, simple trigonometry shows (lower triangle in (**a**)) that $Q = 4\pi \sin \theta / \lambda$

is responsible for the power of neutron scattering. If nature had been unkind and included correlations between triplets and quadruplets in the expression for the scattering law, neutron scattering would never have been used to probe the structure of materials.

Van Hove's work makes use of an observation made by Fermi that the actual interaction between a neutron and a nucleus may be replaced by an effective potential that is much weaker than the actual interaction. This so-called pseudo-potential causes the same scattering as the actual potential but is weak enough to be used in a perturbation treatment of scattering originally derived by Max Born. The Born approximation [2] says that the probability of a neutron wave of wavevector $\vec{k}$ being scattered by a potential $V(\vec{r})$ to become an outgoing wave of wavevector $\vec{k}'$ is proportional to

$$\left| \int e^{i\vec{k}.\vec{r}} V(\vec{r}) e^{-i\vec{k}'.\vec{r}} \mathrm{d}\vec{r} \right|^2 = \left| \int e^{i\vec{k}.\vec{Q}.\vec{r}} V(\vec{r}) \mathrm{d}\vec{r} \right|^2, \tag{2.2}$$

where the integration is over the volume of the sample. The potential $V(r)$ to be used in this equation is the Fermi pseudo-potential which, for an assembly of nuclei situated at positions $\vec{r}_j$, is given by

$$V(\vec{r}) = \frac{2\pi \hbar^2}{m} \sum_j b_j \delta(\vec{r} - \vec{r}_j), \tag{2.3}$$

where $\delta(\vec{r})$ is a Dirac delta function, which takes the value unity at position $\vec{r}$ and is zero everywhere else. The b_j that appear in Equation (2.3) are the scattering lengths that we encountered earlier. Van Hove was able to show that the scattering law $I(\vec{Q}, E)$ could be written as

$$I(\vec{Q}, E) = \frac{1}{h} \frac{k'}{k} \sum_{i,j} b_i b_j \int_{-\infty}^{\infty} \left\langle e^{-i\vec{Q}.\vec{r}_i(0)} e^{i\vec{Q}.\vec{r}_j(t)} \right\rangle e^{-i(E/\hbar)t} \mathrm{d}t. \tag{2.4}$$

In this equation, the nucleus labeled i is at position $\vec{r}_i$ at time zero, while the nucleus labeled j is at position $\vec{r}_j$ at time t. Equation (2.4) is a double sum over all of the positions of the nuclei in the sample, and the angular brackets $\langle \ldots \rangle$ indicate that we need to do a thermodynamic average over all possible configurations that the sample could take. The steps in Van Hove's calculation are given in detail in several texts, for example, ***The Theory of Thermal Neutron Scattering*** by G. L. Squires [3].

The position vectors $\vec{r}_i(0)$ and $\vec{r}_j(t)$ are quantum mechanical position operators so, while Equation (2.4) looks simple, it is not as simple to evaluate in practice as one might imagine. To get a feeling for what the equation is telling us, we can try ignoring this subtlety and treat the vectors as classical quantities. In this case, the double sum in Equation (2.4) becomes

$$\sum_{i,j} b_i b_j \left\langle e^{-\vec{Q}.\{\vec{r}_i(0) - \vec{r}_j(t)\}} \right\rangle = \sum_{i,j} b_i b_j \int_{\text{sample}} \delta[\vec{r} - \vec{r}_i(0) + \vec{r}_j(t)] e^{-i\vec{Q}.\vec{r}} \mathrm{d}\vec{r}. \tag{2.5}$$

Let us suppose further that all of the nuclei in our sample have the same scattering length so that $b_i = b_j = b$. Then the right-hand side of Equation (2.5) becomes

$$Nb^2 \int_{\text{sample}} G(\vec{r}, t) e^{-i\vec{Q}.\vec{r}}, \tag{2.6}$$

where

$$G(\vec{r}, t) = \frac{1}{N} \sum_{i,j} \delta(\vec{r} - [\vec{r}_i(0) - \vec{r}_j(t)]), \tag{2.7}$$

and N is the number of nuclei in the sample. Evidently, the function $G(\vec{r}, t)$ is zero unless the separation between nucleus i at time zero and nucleus j at time t is equal to the vector $\vec{r}$. Thus, the function tells us the probability that, within our sample, there will be a nucleus at the origin of our coordinate system at time zero *as well as* a nucleus at position $\vec{r}$ at time t. For this reason, $G(\vec{r}, t)$ is called the **time-dependent pair correlation function** because it describes how the correlation between the positions of nuclei evolves with time.

In terms of $G(\vec{r}, t)$, Van Hove's scattering law can be written as

$$I(\vec{Q}, E) = \frac{Nb^2}{h} \frac{k'}{k} \int_{-\infty}^{\infty} dt \int_{\text{sample}} G(\vec{r}, t) e^{-i\vec{Q}\cdot\vec{r}} e^{-i(E/\hbar)t} d\vec{r}, \tag{2.8}$$

which allows us to see that the scattering law is proportional to the space and time Fourier transform of the time-dependent correlation function. This general result provides a unified description of all neutron scattering experiments. By inverting Equation (2.8)—easier said than done in many cases—we can obtain from neutron scattering information about both the equilibrium structure of matter and the way in which this structure evolves with time.

Even for a sample made up of a single isotope, all of the scattering lengths that appear in Equation (2.4) will not be equal. This occurs because the interaction between the neutron and the nucleus depends on the nuclear spin, and most isotopes have several nuclear spin states. Generally, however, there is no correlation between the spin of a nucleus and its position in a sample of matter. For this reason, the scattering lengths that appear in Equation (2.4) can be averaged without interfering with the thermodynamic average denoted by the angular brackets. Two averages come into play: the average value of b (denoted $\langle b \rangle$) and the average value of b^2 (denoted $\langle b^2 \rangle$). In terms of these quantities, the sum in Equation (2.4) can be averaged over the nuclear spin states to read

$$\sum_{i,j} \langle b_i b_j \rangle A_{ij} = \sum_{i,j} \langle b \rangle^2 A_{ij} + \sum_{i} (\langle b^2 \rangle - \langle b \rangle^2) A_{ii}, \tag{2.9}$$

where A_{ij} is shorthand for the integral in Equation (2.4). The first term is a sum over all pairs of nuclei in the sample. While i and j are occasionally the same nucleus, in general they are not because there are so many nuclei in the sample.

The first term in Equation (2.9) is **coherent scattering** in which neutron waves scattered from **different** nuclei interfere with each other. This type of scattering depends on the distances between atoms (through the integral represented by A_{ij})

and on the scattering vector $\vec{Q}$, and it thus gives information about the structure of a material. ***Elastic coherent*** scattering tells us about the **equilibrium structure**, whereas ***inelastic coherent scattering*** (with $E \neq 0$) provides information about the collective motions of the atoms, such as those that produce ***phonons*** or vibrational waves in a crystalline lattice. In the second type of scattering, ***incoherent scattering***, represented by the second term in Equation (2.9), there is no interference between the waves scattered by different nuclei. Rather the intensities scattered from each nucleus just add up independently. Once again, one can distinguish between elastic and inelastic scattering. ***Incoherent elastic scattering*** is the same in all directions, so it usually appears as unwanted background in neutron scattering experiments. ***Incoherent inelastic scattering***, on the other hand, results from the interaction of a neutron with the same atom at different positions and different times, thus providing information about atomic diffusion.

The values of the coherent and incoherent scattering lengths given by $b_{coh} = \langle b \rangle$ and $b_{inc} = \sqrt{\langle b^2 \rangle - \langle b \rangle^2}$ for different elements and isotopes do not vary in any systematic way across the periodic table. For example, hydrogen has a large incoherent scattering length (25.18 fm) and a small coherent scattering length (-3.74 fm). Deuterium, on the other hand, has a small incoherent scattering length (3.99 fm) and a relatively large coherent scattering length (6.67 fm).

The simplest type of coherent scattering to understand is ***diffraction***. As the incident neutron wave arrives at each atom, the atomic site becomes the center of a scattered spherical wave that has a definite phase relative to all other scattered waves. As the waves spread out from a regular array of sites in a crystal, the individual disturbances will reinforce each other only in particular directions. These directions are closely related to the symmetry and spacing of the scattering sites. Consequently, one may use the directions in which constructive interference occurs and its intensity to deduce both the symmetry and the lattice constant of a crystal.

Even though diffraction is predominantly an elastic scattering process (i.e., one for which $E = 0$), most neutron diffractometers integrate over the energies of scattered neutrons. Accordingly, in the Van Hove formalism, rather than setting $E = 0$ in Equation (2.4), we integrate the equation over E to obtain the diffracted intensity. The integral over E gives us another Dirac delta function, this time $\delta(t)$, which tells us that the pair correlation function $G(\vec{r}, t)$ has to be evaluated at $t = 0$. For a sample containing a single isotope, the result is

$$I(\vec{Q}) = b_{coh}^2 \sum_{i,j} \left\langle e^{i\vec{Q}.(\vec{r}_i - \vec{r}_j)} \right\rangle, \tag{2.10}$$

where the positions of the nuclei labeled i and j are now to be evaluated at the same instant. If the atoms in the crystal were stationary, the thermodynamic averaging brackets in Equation (2.10) could be removed because $\vec{r}_i$ and $\vec{r}_j$ would be constants. In reality, the atoms in a crystal oscillate about their equilibrium positions and spend only a fraction of the time at these positions. When this is taken into account, the thermodynamic averaging introduces another factor, so that

$$I(\vec{Q}) = b_{\text{coh}}^2 \sum_{i,j} e^{i\vec{Q}.(\vec{r}_i - \vec{r}_j)} e^{-Q^2 \langle u^2 \rangle / 2} = \left| b_{\text{coh}} \sum_i e^{i\vec{Q}.\vec{r}_i} \right|^2 e^{-Q^2 \langle u^2 \rangle / 2} \equiv S(\vec{Q}), \quad (2.11)$$

where $e^{-Q^2 \langle u^2 \rangle / 2}$ is called the **Debye–Waller factor** and $\langle u^2 \rangle$ is the average of the squared displacement of nuclei from their equilibrium lattice sites. Equation (2.11) gives the intensity that would be measured in a neutron scattering experiment with a real crystal. This quantity is often denoted by $S(\vec{Q})$ and called the **structure factor**.

Because there are so many atoms in a crystal, and each contributes a different phase factor to Equation (2.11), it is perhaps surprising that $S(\vec{Q})$ is non-zero for any value of $\vec{Q}$. We can determine the values of Q that give non-zero values by consulting Fig. 2.4. Suppose $\vec{Q}$ is perpendicular to a plane of atoms such as Scattering Plane 1 in this figure. If the value of $\vec{Q}$ is an integral multiple of $2\pi/d$, where d is the distance between parallel neighboring planes of atoms (Scattering Planes 1 and 2 in the figure), then $\vec{Q}.(\vec{r}_i - \vec{r}_j)$ is a multiple of 2π and $S(\vec{Q})$ is non-zero because each exponential term in the sum in Equation (2.11) is unity. Thus, for diffraction to occur, $\vec{Q}$ must be perpendicular to a set of atomic planes. Using the relationship between Q, θ, and λ shown in Fig. 2.3, the condition for diffraction is easily rewritten as **Bragg's Law** (discovered in 1912 by William and Lawrence Bragg, father and son [4]).

$$n\lambda = 2d \sin \theta. \tag{2.12}$$

Bragg's law can also be understood in terms of the path-length difference between the waves scattered from neighboring planes of atoms (Fig. 2.4), and it is often taught that way in undergraduate physics classes, even though the planes of atoms do not really reflect neutrons like mirrors. For constructive interference to occur between the waves scattered from adjacent planes, the argument goes, the path-length difference must be a multiple of the wavelength, λ. Applying this condition to Fig. 2.4 immediately yields Equation (2.12).

Diffraction, or **Bragg scattering** as it is sometimes called, may occur for any set of planes that we can imagine in a crystal, provided the neutron wavelength, λ, and the angle θ satisfy Equation (2.12). To obtain diffraction from a set of planes, the crystal must be rotated to the correct orientation so that $\vec{Q}$ is perpendicular to the scattering planes—much as a mirror is adjusted to reflect the sun at someone's face. The signal that is observed in a neutron detector as the crystal is rotated in this way is called a **Bragg peak** because the signal appears only when the crystal is in the correct orientation. According to Equation (2.11), the intensity of the scattered neutrons is proportional to the square of the density of atoms in the atomic planes responsible for the scattering. Thus an observation of Bragg peaks allows us to deduce both the spacing of the planes and the density of atoms in the planes. To measure Bragg peaks corresponding to many different atomic planes, we have to pick the neutron wavelength and scattering angle to satisfy Bragg's Law and then rotate the crystal until the Bragg diffracted beam falls on the detector.

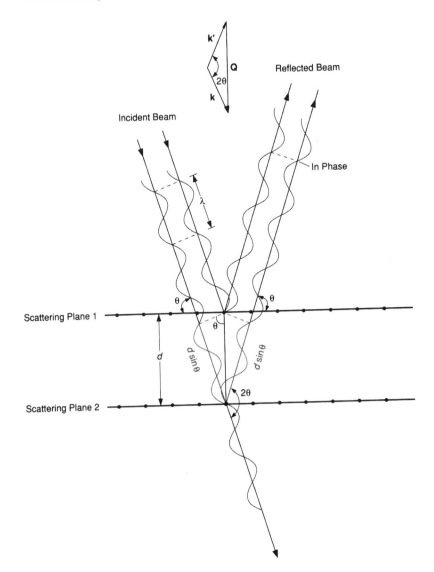

Fig. 2.4 Constructive interference occurs when the waves scattered from adjacent scattering planes of atoms remain in phase. This happens when the difference in distance traveled by waves scattered from adjacent planes is an integral multiple of wavelength. The figure shows the extra distance traveled by the wave scattered from Scattering Plane 2 is $2\,d\,\sin\theta$. When that distance is set equal to $n\lambda$ (where n is an integer), the result is Bragg's Law $n\lambda = 2d\sin\theta$. Primary scattering occurs when $n = 1$, but higher-order Bragg peaks are also observed for other values of n

To this point we have been discussing a simple type of crystal with one type of atom. However, the materials of interest to scientists almost invariably contain many different types of atoms. It is often not trivial to deduce the atomic positions even from a measurement of the intensities of many Bragg peaks because some of the exponential factors that contribute to $S(\vec{Q})$ are now complex and the phases of these quantities cannot be obtained directly from the measurement of Bragg scattering. $S(\vec{Q})$, given by Equation (2.11), is equal to the square of the modulus of a complex quantity, so the phase of this quantity simply cannot be obtained by measuring scattered intensities. This difficulty is referred to as the ***phase problem*** in diffraction. In practice, crystallographers generally have to resort to modeling the structure of crystals, shifting atoms around until they find an arrangement that accurately predicts the measured Bragg intensities when plugged into Equation (2.11). There are now many computer packages available to help them. One well known package is that by Larson and Von Dreele [5].

In diffraction experiments with single crystals, the sample must be correctly oriented to obtain Bragg scattering. On the other hand, polycrystalline powders, which consist of many randomly oriented single-crystal grains, will diffract neutrons whatever the relative orientation of the sample and the incident neutron beam. There will always be grains in the powder that are correctly oriented to diffract. However, many diffracting planes may "reflect" under the same or very similar angle, hence the need for high resolution (see Chapter 1 and Chapter 4). This observation is the basis of a widely used technique known as ***neutron powder diffraction***.

In a ***powder diffractometer at a reactor neutron source***, a beam of neutrons with a single wavelength is selected by a device called a ***monochromator*** and directed toward a powder sample. The monochromator is usually an assembly of single crystals—often made of either pyrolytic graphite, silicon, or copper—each correctly oriented to diffract a mono-energetic beam of neutrons toward the scattering sample. The neutrons scattered from a powder sample are counted by suitable neutron detectors and recorded as a function of the angles through which they were scattered by the sample. Each Bragg peak in a typical diffraction pattern (Fig. 2.5) corresponds to diffraction from atomic planes with different interplanar spacings, d.

In a ***powder diffractometer at a pulsed neutron source***, the sample is irradiated by a pulsed beam of neutrons with a wide spectrum of energies. Scattered neutrons are recorded in banks of detectors located at different scattering angles, and the time at which each scattered neutron arrives at the detector is also recorded. At a particular scattering angle, the result is a diffraction pattern very similar to that measured at a reactor, but now the independent variable is the neutron's ***time of flight*** rather than the scattering angle. The time of flight is easily deduced from the arrival time of the neutron because we know when the pulse of neutrons left the source. Because the neutron's time of flight is inversely proportional to its velocity, it is linearly related to the neutron wavelength.

At both reactors and pulsed sources, powder diffraction patterns can be plotted as a function of the interplanar separations, d, by using Bragg's law to derive the value of d from the natural variables—scattering angle at a reactor and neutron wavelength at a pulsed source. Using these patterns, the atomic structure of a polycrystalline sample may be deduced by using Equation (2.11) and shifting the atoms

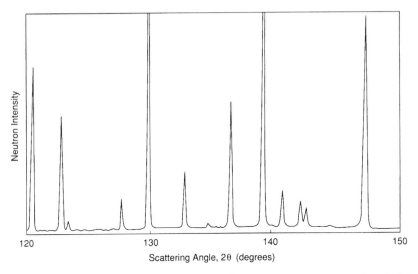

Fig. 2.5 A typical powder diffraction pattern obtained at a reactor neutron source gives the intensity, or number of neutrons, as a function of the scattering angle 2θ. Each peak represents neutrons that have been scattered from a particular set of atomic planes in the crystalline lattice

around until a good fit to the experimental pattern is obtained. In practice, one needs to account carefully for the shape of the Bragg peaks in carrying out this refinement. These shapes result from the fact that the diffractometers really use a small range of wavelengths or scattering angles rather than a single value as we have assumed—an effect called ***instrumental resolution***. The simultaneous refinement of the atomic positions to obtain a powder diffraction pattern that is the same as the measured pattern is often referred to as ***Rietveld analysis***, after the inventor of the technique [6].

2.3 Probing Larger Structures

We have seen already in our discussion of diffraction that elastic scattering at a scattering vector $Q = (4\pi/\lambda)\sin\theta$ results from periodic modulations of the neutron scattering lengths with period $d = 2\pi/Q$. Combining these two expressions gives us Bragg's Law (Equation (2.12)). In the case of a crystal, d was interpreted as the distance between the planes of atoms. To measure structures that are larger than typical interatomic distances, we need to arrange for Q to be small, either by increasing the neutron wavelength, λ, or by decreasing the scattering angle. Because we do not know how to produce copious fluxes of very-long-wavelength neutrons, we always need to use small scattering angles to examine larger structures such as polymers, colloids, or viruses. For this reason, the technique is known as ***small-angle neutron scattering***, or SANS.

For SANS, the scattering wavevector, Q, is small, so the phase factors in Equation (2.11) do not vary greatly from one nucleus to its neighbor. For this reason, the sum in Equation (2.11) can be replaced by an integral and the structure factor for SANS can be written as

$$S(\vec{Q}) = \left| \int_{\text{sample}} \rho(\vec{r}) e^{i\vec{Q}\cdot\vec{r}} \mathrm{d}\vec{r} \right|^2 , \qquad (2.13)$$

where $\rho(\vec{r})$ is a spatially varying quantity called the **scattering length density** calculated by summing the coherent scattering lengths of all atoms over a small volume (such as that occupied by a single molecule) and dividing by that volume. In many cases, samples measured by SANS can be thought of as particles with a constant scattering length density, ρ_p, which are dispersed in a uniform medium with scattering length density ρ_m. Examples include pores in rock, colloidal dispersions, biological macromolecules in water, and many more. The integral in Equation (2.13) can, in this case, be separated into a uniform integral over the whole sample and terms that depend on the difference, $(\rho_p - \rho_m)$, a quantity often called the **contrast factor**. If all of the particles are identical and their positions are uncorrelated, Equation (2.13) becomes

$$S(\vec{Q}) = N_p(\rho_p - \rho_m)^2 \left| \int_{\text{particle}} e^{i\vec{Q}\cdot\vec{r}} \mathrm{d}\vec{r} \right|^2 , \qquad (2.14)$$

where the integral is now over the volume of one of the particles and N_p is the number of particles in the sample. The integral of the phase factor $e^{i\vec{Q}\cdot\vec{r}}$ over a particle is called the **form factor** for that particle. For many simple particle shapes, the form factor can be evaluated analytically [7], whereas for complex biomolecules, for example, it has to be computed numerically. Usually, the intensity one sees plotted for SANS is normalized to the sample volume so that $S(Q)$ is reported in units of cm^{-1} (as compared with Equation (2.14), which has units of length squared).

Equation (2.14) allows us to understand an important technique used in SANS called **contrast matching**. The total scattering is proportional to the square of the scattering contrast between the particle and the matrix in which it is embedded. If we embed the particle in a medium whose scattering length density is equal to that of the particle, the latter will not scatter—it will be invisible.

Suppose that the particles we are interested in are spherical eggs rather than uniform spheres: they have a core (the yolk) and a shell (the white) whose scattering lengths are different. If such particles are immersed in a medium whose scattering length density is equal to that of the egg white, then a neutron scattering experiment will only "see" the yolk. The form factor in Equation (2.14) will be evaluated by integrating over this central region only. On the other hand, if the dispersing medium has the scattering length density of the yolk, only the egg white will be visible: the form factor will be that of a thick hollow shell. The scattering patterns will be different in the two cases, and from two experiments, we will discover the structure of both the coating and the core of the particle.

Variation of the scattering length density of the matrix is often achieved by choosing a matrix that contains hydrogen (such as water). By replacing different fractions of the hydrogen with deuterium atoms, a large range of scattering length densities can be achieved for the matrix. Both DNA and typical proteins can be contrast matched by water containing different fractions of deuterium. *Isotopic labeling* of this type can also be applied to parts of molecules in order to highlight them for neutron scattering experiments.

2.4 Inelastic Scattering

In reality, atoms are not frozen in fixed positions inside a crystal. Thermal energy causes them to oscillate about their lattice sites and to move around inside a small volume with the lattice site as its center. Since an atom can contribute to the constructive interference of Bragg scattering *only* when it is located exactly at its official position at a lattice site, this scattering becomes weaker the more the atoms vibrate and the less time they spend at their official positions. The factor by which the Bragg peaks are attenuated because of the atomic motion is called the ***Debye–Waller factor***. It was introduced above in connection with Equation (2.11).

Although such weakening of the scattering signal is the only effect of the thermal motions of atoms on elastic Bragg scattering, it is not the only way to use neutrons to study atomic motions. In fact, one of the great advantages of neutrons as a probe of condensed matter is that they can be used to measure details of atomic and molecular motions by measuring *inelastic scattering* in which the neutron exchanges energy with the atoms in a material.

Atoms in a crystal lattice behave a bit like coupled pendulums—if we displace one pendulum, the springs that couple the pendulums together will cause the neighboring pendulums to move as well and a wave-like motion propagates down the line of pendulums. The frequency of the motion depends on the mass of the pendulums and the strength of the spring constants coupling them. Similar waves pass through a lattice of atoms connected by the binding forces that are responsible for the cohesion of matter. Of course, in this case, the atomic vibrations occur in three dimensions and the whole effect becomes a little hard to visualize. Nevertheless, it is possible to prove that any atomic motion in the crystal can be described by a superposition of waves of different frequencies and wavelengths traveling in different directions. These lattice waves are known as ***phonons***. Their energies are quantized so that each phonon has an energy $h\nu$, where ν is the frequency of atomic motion associated with the phonon. Just as in the pendulum analogy, the frequency of the phonon depends on the wavelength of the distortion, the masses of the atoms, and the stiffness of the "springs", or binding forces, that connect them.

When a neutron is scattered by a crystalline solid, it can absorb or emit an amount of energy equal to a quantum of phonon energy, $h\nu$. This gives rise to *inelastic coherent scattering* in which the neutron energy before and after the scattering differ by an amount, E—the same quantity that appeared in Equation (2.8)—equal to the phonon energy. In most solids ν is a few terahertz (THz), corresponding to phonon energies of a few meV (1 Thz corresponds to an energy of 4.18 meV). Because the thermal neutrons used for neutron scattering also have energies in the meV range, scattering by a phonon causes an appreciable fractional change in the neutron energy, allowing accurate measurement of phonon frequencies.

For inelastic scattering, the neutron has different velocities, and thus different wavevectors, before and after it interacts with the sample. To determine the phonon energy and the scattering vector, $\vec{Q}$, we need to determine the neutron wavevector before and after the scattering event. Different methods are used to accomplish this at reactor and pulsed neutron sources. At reactors the workhorse instruments for this type of measurement are called ***three-axis spectrometers***. They work by using assemblies of single crystals both to set the incident neutron wavevector and to analyze the wavevector of the scattered neutrons. Each measurement made on a three-axis spectrometer—usually taking several minutes—records the scattered neutron intensity at a single wavevector transfer and a single energy transfer. To determine the energy of a phonon, a sequence of measurements is usually made at a constant value of the wavevector transfer. The peak that is obtained during this ***constant-Q*** scan yields the energy of the phonon. Constant-Q scans were invented by B. N. Brockhouse [8], one of the co-winners with C. Shull in 1994 of the Nobel Prize for their work on neutron scattering. At pulsed spallation sources, there is no real equivalent of the three-axis spectrometer and constant-Q scans cannot be performed using hardware. Rather, the time-of-flight technique is used with a large array of detectors and various cuts through the data in $(\vec{Q}, E)$ space are performed after the experiment using appropriate computer software.

2.5 Magnetic Scattering

So far we have discussed only the interaction between neutrons and atomic nuclei. But there is another interaction between neutrons and matter—one that results from the fact that the neutron has a magnetic moment. Just as two bar magnets either attract or repel one another, the neutron experiences a force of magnetic origin whenever it moves in a magnetic field, such as that produced by unpaired electrons in matter.

Ferromagnetic materials, such as iron, are magnetic because the moments of their unpaired electrons tend to align spontaneously. For many purposes, such materials behave as if small magnetic moments were located at each atomic site with all the moments pointing in the same direction. These moments give rise to ***magnetic Bragg scattering*** of neutrons in the same manner as the nuclear interactions. Because the nuclear and magnetic interactions experienced by the neutron are of similar magnitude, the corresponding Bragg reflections are also of comparable intensity.

One difference between the two types of scattering, however, is that the magnetic interaction, unlike the nuclear interaction is not isotropic. The magnetic interaction is dipolar, just like that between two bar magnets. And just like two bar magnets, the strength of the interaction between the neutron and a nucleus depends on the relative orientations of their magnetic moments and the line joining their centers. For neutrons, the dipolar nature of the magnetic interaction means that only the component of the sample's magnetization, which is perpendicular to the scattering vector, $\bar{Q}$, is effective in scattering neutrons. Neutron scattering is therefore sensitive to the spatial distribution of both the direction and the magnitude of magnetization inside a material.

The anisotropic nature of the magnetic interaction can be used to separate nuclear and magnetic Bragg peaks in ferromagnets, for which both types of Bragg peaks occur at the same values of $\bar{Q}$. If the electronic moments can be aligned by an applied magnetic field, magnetic Bragg peaks for which $\bar{Q}$ is parallel to the induced magnetization vanish, leaving only the nuclear component. On the other hand, an equivalent Bragg peak for which $\bar{Q}$ is perpendicular to the magnetization will manifest both nuclear and magnetic contributions.

2.6 Polarized Neutrons

Usually, a neutron beam contains neutrons with magnetic moments pointing in all directions. If we measured the number of neutrons with moments parallel and antiparallel to a given direction—say an applied magnetic field—we would find equal populations. However, various special techniques can generate a polarized neutron beam, that is, one where the majority of neutron moments are in the same direction. The polarization of such a beam, once it is created, can be maintained by applying a small, slowly varying magnetic field (a milli-Tesla is enough) parallel to the neutron moments. Such a field is called a *guide field*.

There are several ways to polarize neutron beams: Bragg diffraction from suitable magnetized crystals [9], reflection from magnetized mirrors made of CoFe or from supermirrors [10], and transmission through polarized ^{3}He [11], for example. Each of these methods selects neutrons whose magnetic moments are either aligned parallel or antiparallel to an applied magnetic field and removes other neutrons from the beam in some way. If the neutron moments are parallel to the applied field, they are said to be *"up"*: if the moments are antiparallel they are said to be *"down"*. An "up" polarizer will not transmit "down" neutrons, and a "down" polarizer blocks "up" neutrons. Thus, by placing "up" polarizers before and after a sample, the neutron scattering law can be measured for those scattering processes in which the direction of the neutron moments are not changed by the scattering process. To measure the other combinations—such as "up" neutron being flipped to "down" neutrons—requires either a combination of "up" and "down" polarizers or a *flipper*, a device that can change a neutron spin from "up" to "down" or vice versa. Various types of flippers have been developed over the years, but the most common these days are direct-current coils (often called Mezei flippers) and adiabatic rf flippers [12].

If flippers are inserted on either side of a sample, we can measure all of the neutron-spin-dependent scattering laws—up to down, up to up, and so forth—simply by turning the appropriate flipper on or off. This technique, known as ***polarization analysis***, is useful because some scattering processes flip the neutron's magnetic moment, whereas others do not. Scattering from a sample that is magnetized provides a good example. It turns out that magnetic scattering will flip the neutron's moment if the magnetization responsible for the scattering is perpendicular to the magnetic field used to maintain the neutron's polarization at the sample position. If the scattering magnetization is parallel to the guide field, no flipping of the neutron spin occurs. Thus, polarization analysis can be used to determine the direction of the magnetic moments in a sample that is responsible for particular contributions to the neutron scattering pattern.

Incoherent scattering (cf. Equation (2.9)) that arises from the random distribution of nuclear spin states in materials provides another example of the use of polarization analysis. Most isotopes have several nuclear spin states, and the scattering cross section for neutrons varies with the spin states of both the nucleus and the neutron. The random distribution of nuclear spins in a sample gives rise to incoherent scattering of neutrons as we have described above. It turns out that two-thirds of the neutrons scattered by this incoherent process have their spins flipped, whereas the moments of the remaining third are unaffected. This result is independent of the isotope that is responsible for the scattering and of the direction of the magnetic guide field at the sample. Although incoherent scattering can also arise if the sample contains a mixture of isotopes of a particular element, neither this second type of incoherent scattering nor coherent scattering flip the neutron's spin. Polarization analysis thus becomes a useful tool for sorting out these different types of scattering, allowing nuclear coherent scattering to be distinguished from magnetic scattering and nuclear spin-incoherent scattering.

The polarization of neutron beams can also be used to improve the resolution of neutron spectrometers using a technique known as ***neutron spin echo*** [13]. If the magnetic moment of a neutron is somehow turned so that it is perpendicular to an applied magnetic field, the neutron moment starts to precess around the field direction, much like the second hand on a stopwatch rotates about the spindle in the center of the watch. This motion of the neutron moment is called ***Larmor precession***. The sense of the precession—clockwise or anticlockwise—depends on the sense of the magnetic field, and the rate of precession is proportional to the magnitude of the applied magnetic field. If a neutron is sent through two identical regions in which the magnetic fields are equal but oppositely directed, it will precess in one direction in the first field and in the opposite direction in the second field. Because the fields are equal in strength, the number of forward rotations will be equal to the number of backward rotations. This is the phenomenon of spin echo, first discovered by Hahn for nuclear spins [14]. If the neutron's speed changes between these field regions—because it is inelastically scattered, for example—the number of forward and backward rotations will not be equal. In fact, the difference will be proportional to the change in the neutron's speed. By measuring the final polarization of the neutron beam, we can obtain a sensitive measure of the change of neutron speed and hence the change in its energy, E, caused by a scattering event. Spin echo spectrometers based on this method provide the best energy resolution obtainable with neutrons—they can measure neutron energy changes less than a nano-electron-volt.

2.7 Reflectometry

So far we have described only experiments in which the structure of bulk matter is probed. One may ask whether neutrons can provide any information about the structure at or close to the surfaces of materials. At first sight, one might expect the answer to be a resounding "No!" After all, one of the advantages of neutrons is that they can penetrate deeply into matter without being affected by the surface. Furthermore, because neutrons interact only weakly with matter, large samples are generally required. Because there are far fewer atoms close to the surface of a sample than in its interior, it seems unreasonable to expect neutron scattering to be sensitive to surface structure.

In spite of these objections, it turns out that neutrons are sensitive to surface structure when they impinge on surfaces at sufficiently low angles. In fact, for smooth, flat surfaces, reflection of neutrons occurs for almost all materials at angles of incidence, α_i (defined for neutrons as the angle between the incident beam and the surface), less than a **critical angle**, denoted γ_c. Surface reflection of this type is **specular**; that is, the angle of incidence, α_i, is equal to the angle of reflection, α_r, just as it is for a plane mirror reflecting light. In this case, the wavevector transfer, $\bar{Q}$, is perpendicular to the material surface, usually called the z direction, and is given by (refer to the scattering triangle Fig. 2.3)

$$Q_z = k_{iz} - k_{rz} = \frac{2\pi}{\lambda}(\sin\alpha_i + \sin\alpha_r) = \frac{4\pi\sin\alpha_i}{\lambda}, \qquad (2.15)$$

where k_{iz} is the z component of the incident neutron wavevector and k_{rz} is the z component of the reflected neutron wavevector. There is no change of the neutron wavelength on reflection, so the process is elastic and the modulus of the neutron wavevector is not changed. Just like light incident on the surface of a pond, some of the neutron energy is reflected and some is transmitted into the reflecting medium with a wavevector given by $n \cdot k$ where n is the **refractive index** of the material and k the incident wavevector. The kinetic energy of the neutron before it strikes the surface is just $\hbar^2 k^2/2m$ (m is the neutron mass), while the kinetic energy inside the medium is $\hbar^2 n^2 k^2/2m$. The difference is just the neutron scattering potential of the medium given by Equation (2.3), averaged over the sample volume. Conservation of energy immediately tells us that the refractive index is given by

$$n = 1 - \rho\lambda^2/2\pi, \qquad (2.16)$$

where ρ is the scattering length density we introduced earlier in connection with Eqn (2.13). Because the surface cannot change the component of the neutron's velocity parallel to the surface, we can also use the conservation of energy to write down the z-component of the neutron's wavevector inside the medium, k_{tz}:

$$k_{tz}^2 = k_{iz}^2 - 4\pi\rho. \tag{2.17}$$

Critical external reflection of neutrons occurs when the transmitted wavevector reaches zero, allowing us to express the critical angle as $\gamma_c = \sin^{-1}(\lambda\sqrt{\rho/\pi})$. For a strongly reflecting material, such as nickel, the critical angle measured in degrees is roughly equal to the wavelength in nanometers—well under a degree for thermal neutrons. As the angle of incidence increases above the critical angle, fewer of the incident neutrons are reflected by the surface. In fact, reflectivity, $R(Q_z)$—the fraction of neutrons reflected by the surface—obeys the same law, discovered by Fresnel, that applies to the reflection of light: reflectivity decreases as the fourth power of the angle of incidence at sufficiently large grazing angles.

However, Fresnel's law applies to reflection of radiation from smooth, flat surfaces of a homogeneous material. If there is a variation of the scattering length density close to the surface—for example, if there are layers of different types of material—the neutron reflectivity measured as a function of the angle of incidence shows a more complicated behavior. Approximately, we may write [15]

$$R(Q_z) = \frac{16\pi^2}{Q_z^4} \left| \int \frac{d\rho(z)}{dz} e^{iQ_z z} dz \right|^2, \tag{2.18}$$

where the pre-factor is the Fresnel reflectivity mentioned above. Equation (2.18) tells us that the reflectivity depends on the gradient of the average scattering length density perpendicular to the surface (i.e., in the z direction). If we have a layered material, this gradient will have spikes at the interfaces between layers and one may use a measurement of the reflectivity to deduce the thickness, sequence, and scattering length densities of layers close to the surface. This method is usually referred to as *neutron reflectometry*. Often, the reflectivity is analyzed using a more complicated formula than Equation (2.18), one in which the scattering length density, averaged over dimensions parallel to the surface, is split into thin layers and each density is refined until the measurement can be fitted. This method is called the *Parratt formalism* [16] and has been programmed into a number of widely available software packages. A certain amount of caution is needed in computing $\rho(z)$, however, Equation (2.18) has the same phase problem as we discussed for diffraction, and this means that $\rho(z)$ cannot be uniquely obtained from a single reflectivity curve. Indeed, remarkably different versions of $\rho(z)$ can yield very similar values of $R(Q_z)$.

In some ways, surface reflection is very different from conventional neutron scattering. After all, Van Hove's assumption in deriving Equation (2.4) was that neutron scattering was weak enough to be described by a perturbation theory—the Born Approximation—yet critical reflection from a surface is anything but a weak process because all of the incident neutrons are reflected at angles of incidence below γ_c. At large enough angles of incidence, the scattering is weak enough for the Born Approximation to apply, but between the so-called critical edge and these large angles, a different theory called the Distorted Wave Born Approximation (DWBA) is needed. It turns out that the effect of surface roughness on neutron reflection, for example, is different from the result one gets from Equation (2.18), and only the DWBA can account for the observed experimental effects.

Reflectometry has been applied to soft matter [17]—where deuterium contrast enhancement can often be used to highlight particular layers—as well as to man-made multilayer structures of metals and alloys. In the latter case, *polarized neutron reflectometry* provides a powerful tool for determining the depth profile of *vector magnetization* close to the surface of a sample [18].

2.8 Conclusion

The conceptual framework for neutron scattering experiments was established over 50 years ago by Van Hove, who showed that the technique measures a time-dependent pair correlation function between scattering centers—either nuclei or magnetic moments. Because thermal neutrons have wavelengths similar to inter-atomic distances in matter and energies similar to those of excitations in matter, neutrons turn out to be a useful probe of both static structure and dynamics. The weak interaction between neutrons and matter is both a blessing and a curse. On the positive side, it is what allows us to describe scattering experiments so completely by the Born approximation and therefore to interpret the data we obtain in an unambiguous manner. It also permits us to contain interesting samples in relatively massive containers so that we can do neutron scattering experiments at a variety of temperatures and pressures. On the negative side of the balance sheet, the weak interaction of neutrons means that the scattering signals we measure are small—of the more than 10^{18} neutrons produced per second in the best research reactors, less than 1 in 10^{10} are used in a typical experiment and very few of these ever provide useful information! The weak interaction also means that we need to start out with a copious supply of neutrons if we are ever to have a chance of recording any that are scattered, and this, in turn, means that we are obliged to centralize the production of neutrons at a few specialized laboratories, restricting access to a technique that has been extremely useful in scientific fields from condensed matter physics to chemistry and from biology to engineering. But, in spite of all these penalties of a signal-limited technique, neutron scattering continues to occupy an important place among the panoply of tools available to study materials structure because it can often provide information that cannot be obtained in any other way. Were it not so, the technique would long have followed the dodo.

Acknowledgments Much of the material included here was first published in an edition of *Los Alamos Science* in 1990. I am grateful to Necia Cooper and her editorial staff for their patience during the preparation of the original manuscript and for their insistence that I avoid jargon and concentrate on underlying principles. David Delano drew the original cartoons for *Los Alamos Science* and some of his work is included in the on-line edition of this chapter. David invented the characters shown in his cartoons and did a wonderful job of translating science into pictures.

References

1. L. Van Hove, Phys. Rev. **95**, 249 (1954)
2. Messiah, *Quantum Mechanics*, Dover Publications, New York (1999)
3. G. L. Squires, *Introduction to Thermal Neutron Scattering*, Dover Publications, New York (1996)
4. W. L. Bragg, *The Diffraction of Short Electromagnetic Waves by a Crystal*, Proc Camb. Phil. Soc. **17**, 43 (1914)
5. C. Larson and R. B. Von Dreele, *General Structure Analysis System (GSAS)*, Los Alamos National Laboratory Report LAUR 86-748 (2000).]. See also http://www.ncnr.nist.gov/xtal/software/gsas.html.
6. H. M. Rietveld, Acta Crystallogr. **22**, 151 (1967)
7. Guinier, and G. Fournet, *Small-Angle Scattering of X-Rays*, John Wiley and Sons, New York, 1955. A more accessible source may be http://www.ncnr.nist.gov/resources/
8. N. Brockhouse, Bull. Amer. Phys. Soc. **5**, 462 (1960)
9. Freund, R. Pynn, W. G. Stirling, and C. M. E. Zeyen, Physica **120B**, 86 (1983)
10. F. Mezei, P. A. Dagleish, Commun. Phys. **2**, 41 (1977)
11. K. H. Andersen, R. Chung, V. Guillard, H. Humblot, D. Jullien, E. Lelièvre-Berna, A. Petoukhov, and F. Tasset, Physica **B 356**, 103 (2005)
12. S. Anderson, J. Cook, G. P. Felcher, T. Gentile, G. L. Greene, F. Klose, T. Koetzle, E. Lelievre-Berna, A. Parizzi, R. Pynn, and J. K. Zhao, J. Neutron Res. **13**, 193 (2005)
13. F. Mezei, Z. Phys. **255**, 146 (1972)
14. E. L. Hahn, Phys. Rev., **80**, 580 (1950)
15. J. Als-Nielsen and D. McMorrow, *Elements of Modern X-Ray Physics*, Wiley, New York (2000)
16. L. G. Parratt, Phys. Rev. **95**, 359 (1954)
17. J. S. Higgins and H. C. Benoit, *Polymers and Neutron Scattering*, Clarendon Press, Oxford (1994)
18. G. F. Felcher, Physica B **192**, 137 (1993)

Chapter 3
Neutron Scattering Instrumentation

Helmut Schober

Abstract This chapter gives a short introduction into neutron scattering instrumentation to allow the non-specialist reader to acquire the basics of the method necessary to understand the technical aspects in the topical articles. The idea is not to go into details but to elaborate on the principles as general as possible. We start with a short discussion of neutron production at large-scale facilities. We then present the main characteristics of neutron beams and show how these can be tailored to the specific requirements of the experiment using neutron optical devices and time-of-flight discrimination. This will allow us in the final section to present a non-exhaustive selection of instrument classes. Emphasis will be given to the design aspects responsible for resolution and dynamic range, as these define the field of scientific application of the spectrometers.

3.1 Introduction

Free neutrons are elementary particles discovered by James Chadwick [1, 2] in 1932. The quantum mechanical state of a free neutron is determined by the momentum $|\vec{p}>$ ($\vec{p} = \hbar\vec{k}, k = h/\lambda$) and spin state $|s>$, where, for all practical purposes, it will be sufficient to work with non-relativistic energies ($E = \hbar^2 k^2/2m$). The mass of the neutron[1] is given as 1.008 amu, which leads us to the relation between energy and wavelength of a neutron:

$$E[\text{meV}] \equiv 2.0725k^2 \left[\text{Å}^{-2}\right] \equiv 81.8204\lambda^{-2} \left[\text{Å}^2\right]. \tag{3.1}$$

---

[1]We use the constants as given by the National Institute of Standards and Technology (http://physics.nist.gov/cuu/Constants/): $h = 6.62606896 \times 10^{-34}$ Js and $m = 1.674927211 \times 10^{-27}$ kg. As $e = 1.602176487 \times 10^{-19}$ C, one Joule is equivalent to $6.24150965 \times 10^{18}$ eV. A wealth of interesting information on neutron and on neutron instrumentation can be found in the Neutron Data Booklet [3].

H. Schober (✉)
Institut Laue-Langevin, F-38042 Grenoble, France
e-mail:Schober@ill.fr

L. Liang et al. (eds.), *Neutron Applications in Earth, Energy and Environmental Sciences*, Neutron Scattering Applications and Techniques, DOI 10.1007/978-0-387-09416-8_3, © Springer Science+Business Media, LLC 2009

The spin of the neutron is $1/2$ with an associated magnetic moment μ_n of -1.9132 nuclear magnetons. A more detailed discussion of the spin properties is given in the section on spin - echo spectroscopy.

Neutrons interact with the nuclei of the sample via the strong interaction and with the electrons via their magnetic dipole moment. In both cases the scattering is weak for thermal and cold neutrons. Neutrons thus constitute a simple, non-destructive probe of matter. As neutrons penetrate deeply into matter, it is rather straightforward to use even complex and bulky sample environments. This fortunately does not prevent neutrons from being highly sensitive, which is capable of investigating minority components of a sample down to a few ppm. Neutrons are equally very well suited to the study of films and interfaces of atomic thickness.

To profit fully from these extraordinary analytic capacities, tentatively summarized in Table 3.1, requires optimized instrumentation. This is the more important as even at the strongest sources neutron scattering remains a flux-limited technique. It is the aim of this chapter to outline the principles of neutron production and neutron instrumentation. It is one of our main objectives to demonstrate how by profiting from modern technology it is possible to considerably improve the performance of both neutron sources and instruments.

The philosophy underlying this chapter is to discuss instruments in terms of functional building blocks. From this functional point of view, a monochromator or a time-of-flight chopper cascade are quite similar as both constitute filters that slice wavelength bands from the beam. Speaking in more general terms, instrument devices reshape the phase space elements that describe what kind of neutrons are present with what probability at any time t and at any place $\vec{r}$ along the beam trajectory. Optical elements, for example, reshape phase space elements by concentrating the beam intensity in space or by rendering the beam more parallel. It is the transformation of phase space elements that fully characterizes the performance of a single device. When assembling building blocks to create an instrument, it is important to realize that the phase space element shaped by one component constitutes the raw material for the subsequent transformation stage. The whole instrument can, therefore, be viewed as a production line with the individual instrument components shaping and consecutively transferring the beam. In this approach the sample is just

Table 3.1 Present performance parameters of neutron scattering instrumentation. The values are to be taken as indicative. In an actual experiment the specificities of the sample and the exact nature of the scientific question are decisive

In space	In time
Sample volume for structure studies: 10^{-3} mm^3	Shortest oscillation periods: 10^{-15} s
Mass for dynamic studies: 5 mg (protonated), 50 mg (other)	Longest observable relaxation time: 1 μs
Maximum accuracy in distances: 10^{-3} nm	Maximum resolution in real time: 0.1 ms
Largest observable object size: 0.1 mm	Stability in time: years
Chemical sensitivity 10 ppm	
Magnetic sensitivity $0.01 \mu_B$ / atom	
Accuracy on direction of magnetic moment: $0.5°$	

one of the many components that make up the instrument.[2] Its beam shaping and beam transfer capacities are described by the double differential scattering cross section or scattering law. Determining this scattering law with high precision is the whole purpose of the experiment. It is evident that there has to be a perfect match of device characteristics to achieve this goal in an efficient manner. As we hopefully will be able to demonstrate, optimizing and matching phase space elements are the whole art of instrument design.

3.2 Neutron Sources

While neutrons are omnipresent in the universe, they are very rare in the free state. Most of them are actually bound in nuclei or neutron stars. We, therefore, have to free the neutrons from the nuclei via technological processes like fission or spallation. Once produced, free neutrons do not live forever, but have a limited lifetime of about 886 seconds.

3.2.1 Method of Production

A good neutron source produces a high density of neutrons of appropriate wavelength and time structure. It is essential that the neutrons can be extracted from the source and transported efficiently to the spectrometers. This requires a two-stage process. First, neutrons bound in nuclei are liberated via nuclear excitation and subsequent nuclear decay processes. Due to the energies involved in these nuclear reactions (several MeV), this produces high-energy, that is, very short wavelength neutrons that are not well suited for the purpose of investigating condensed matter on the nanometer length scale (a few meV). The necessary slowing down of the neutrons is achieved in a second stage via *moderation*, that is, scattering of the neutrons by the moderator atoms or molecules.

There are many nuclear reactions that count neutrons among their final products. The yield is, however, in most cases insufficient for neutron scattering applications. Today, there are only two processes in use that extract a sufficient number of neutrons from nuclei: fission and spallation. Recently, laser-induced fusion was proposed as an alternative way of neutron production [4]. As controlled and sustained fusion involves a high degree of technical complexity, this concept would, however, reach maturity and applicability only after decades of development.

In the case of *fission*, slow neutrons are absorbed by metastable ^{235}U. The excited nucleus decays in a cascade of fission products. On the average 2.5 neutrons are produced by the fission of one ^{235}U nucleus. These neutrons possess very elevated

---

[2] Bragg scattering from a sample provides the necessary information about its atomic structure. If the same Bragg reflection is used for monochromating the beam, then the sample becomes a technical instrumental device.

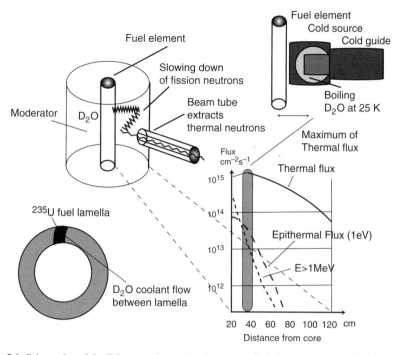

Fig. 3.1 Schematics of the ILL research reactor. A compact fuel element is surrounded by a D_2O moderator. Due to the collision with the D_2O molecules, the highly energetic neutrons from the ^{235}U fission are slowed down until they are in thermal equilibrium with the moderator. These processes produce a spectrum of thermal neutrons with a maximum in flux of 1.5×10^{15} neutrons per cm^2 per second, peaking at about 40 cm from the core center. Thermal beam tubes with noses placed close to the flux maximum extract these neutrons from the reactor core. One of these tubes feeds a suite of thermal neutron guides. In order to shift the neutron spectrum even further down in energy, two cavities with boiling D_2O are placed as close as technically possible to the thermal flux maximum. Cold neutron guides inserted into the beam tubes extract the spectrum from the cold source. A γ-heated graphite block (not shown) at 2400 K transposes the spectrum for certain beam tubes toward the higher energy end

energies of about 1 MeV and are unsuited for inducing further fission processes in ^{235}U. With the help of moderators, the fast neutrons are slowed down to meV energies (see Fig. 3.1). Light atoms are preferable as they take up an appreciable amount of the neutron energy at each collision. For H_2O we need about 18 collisions, while for the heavier D_2O 25 collisions are necessary.[3] These slow neutrons sustain the

---

[3] It is evident that the physics involved in moderation changes drastically along the cascade from MeV to meV. At high energies the collisions are inelastic, that is, they involve excitations of the nuclei that scatter the neutron. At low energies the equilibrium is attained via the exchange of low-frequency excitations of the moderator with the neutron. The character and spectrum of these excitations influence the moderation. In the case of H_2O and D_2O, the collisions involve, for example, predominantly rotational motions around the center of mass. This is the reason why it is the mass of H or D that counts for the moderation efficiency and not that of the molecule.

chain reaction in a nuclear reactor. By allowing some of the moderated neutrons to escape from the core region, free neutrons for scientific use are obtained. A high neutron density is achieved by using very compact core designs. A good moderator slows the neutrons down quickly. It creates a thermal flux that peaks at a short distance from the core, but far enough to allow for efficient beam extraction. The neutrons should have a long thermal diffusion length, that is, a long lifetime in the moderator. This precludes H_2O as a moderator material for very high-flux reactor neutron sources like the Institut Laue Langevin (ILL). Due to the absorption cross section of hydrogen, the thermal diffusion length of a neutron in H_2O is only about 3 cm while a mean free path of about 6 cm would be required for full moderation. In the case of D_2O, the thermal diffusion length is 1 m while full moderation is achieved within 10 cm.[4]

In the case of *spallation*, high-energy protons produced by an accelerator hit metallic targets like uranium, tungsten, lead, or mercury [5]. The thus excited nuclei "boil off" particles. Among these we encounter up to 20 high-energy neutrons.[5] As in the case of fission, the high-energy spallation neutrons have to be moderated to be useful for scattering applications. However, due to the different dimensions of the target, the size and shape of the moderators differ markedly from those used in reactors. The spallation process is about an order of magnitude more efficient than fission in producing neutrons; that is, a spallation source working with a 5 MW particle beam will produce about the same number of neutrons as a fission reactor of 50 MW thermal power. In terms of the overall energy balance, this is not a real advantage as the production of a 5 MW particle beam needs at least an order of magnitude higher electrical power.

3.2.2 Characteristics of the Source

In a standard reactor, neutrons are produced at a constant rate. The flux of neutrons thus has no explicit time structure. We are dealing with a *continuous* or *steadystate* neutron source. Typical examples of continuous neutron sources are the ILL reactor in Grenoble (see Fig. 3.1), France, the High-Flux Isotope Reactor (HFIR) at Oak Ridge National Laboratory (ORNL, Oak Ridge Tennessee), or the new Munich

---

[4] If the dimensions of the moderator are small, as in the case of a cold source, then H_2 may be a useful and easier - to - handle alternative to D_2 [6], despite the fact that D_2 would offer the optimum cold spectrum.

[5] The nuclear reactions following the impact of the proton beam depend on the proton energy. Below an excitation energy of 250 MeV, the boiling-off of neutrons is dominant. Most of them have energies in the 2 MeV range. This boiling-off is the main neutron production channel even for 1 GeV proton beams as normally only part of the proton energy is deposited in the target nuclei. There is, however, also an appreciable amount of faster neutrons. Their spectrum reaches up to the incident proton energy. These very fast neutrons require extremely heavy shielding around the target. The neutrons and the remaining excited nucleus engage in a cascade of secondary decay processes. Per useful neutron about 30 MeV of energy have to be evacuated in the case of spallation, compared to about 200 MeV in the case of fission.

reactor FRM-II in Germany. The unperturbed neutron thermal flux of the ILL is about 1.5×10^{15} neutrons s^{-1} cm^{-2}. Quasi-continuous neutron beams can equally be obtained via spallation. This is the case of the neutron source SINQ at the Paul Scherrer Institut (PSI) in Villigen, Switzerland.

In many cases it can be advantageous to work with pulsed neutron beams, as this allows the neutron energies to be determined by simply measuring the time of flight (see Section 3.3.6). At a continuous neutron source, this is achieved by the mechanical chopping devices incorporated into the spectrometer design. The production of pulsed neutron beams directly at the source is more or less straight forward in the case of spallation. It is sufficient to bunch the protons in the accelerator and storage ring. Depending on the time structure of the proton beam and the characteristics of the moderator, neutron pulses as short as a few microseconds can be produced at adapted rates of 50 or 60 Hz. This principle is used at *pulsed spallation sources* like ISIS in the UK, the SNS in the USA (see Fig. 3.2), or J-SNS in Japan.

It is important to keep the *incoherent nature* of all neutron sources in mind. Both production and moderation are accomplished via completely random events. Neutron sources thus have to be compared to light bulbs in optics. The lack of coherence, if translated into quantum mechanics, implies that there is no correlation of the phases of the neutron wave fields emanating from different regions of the source. Only the incoherent nature of the source makes it possible to describe the neutron beam by a probability distribution $p(\vec{k})$ in the wave-vector. For a coherent source like a LASER, the phase relation between different waves making up the wave field has to be taken into consideration.

As the moderation of the neutrons is done via collisions in a thermal bath (for example, D_2O, H_2, or D_2 molecules in the liquid state), the spectrum of the fully moderated neutrons can be considered a classical gas in thermal equilibrium and thus follows statistically a Maxwell–Boltzmann distribution governed by the temperature T of the bath.[6] The probability of finding a neutron in the state $|\vec{k} >$ is thus given by

$$p(\vec{k}) = \frac{1}{k_T^3 \sqrt{\pi^3}} e^{-\frac{k^2}{k_T^2}}, \tag{3.2}$$

with the mean neutron momentum and energy defined as

$$\hbar k_T = \sqrt{2mk_b T}, \qquad E_T = \frac{1}{2m}\hbar^2 k_T^2 = k_b T \tag{3.3}$$

and the Bolzmann constant

---

[6] In practice, the spectrum will differ from the ideal Maxwell–Boltzmann distribution. This is due to leakage of fast neutrons or incomplete moderation processes. These corrections become more important in the case of pulsed sources. The detailed shape of the source matters for the flux transmitted to the guides [7].

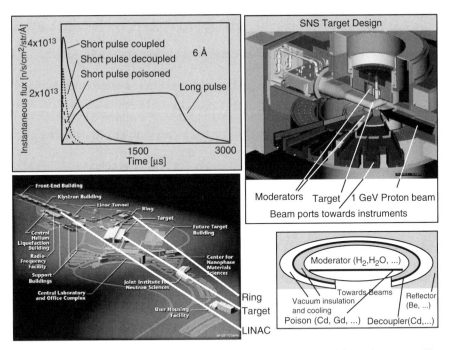

Fig. 3.2 Schematics of the SNS spallation source. H^- ions are produced by an ion source. The linac accelerates the H^- beam from 2.5 MeV to 1 GeV. The linac is a superposition of normal conducting and superconducting radiofrequency cavities that accelerate the beam and a magnetic lattice which provides focusing and steering. The H^- beam is stripped off its electrons, and the proton beam is compressed in a ring in time by about a factor 1000. After an accumulation of approximately 1200 turns, these protons are kicked out in a bunch, producing a microsecond pulse that is delivered to the liquid mercury target with a 60 Hz frequency. The high-energy neutrons coming out of the target are slowed down by passing them through cells filled with water (to produce room-temperature neutrons) or through containers of liquid hydrogen at a temperature of 20 K (to produce cold neutrons). These moderators are located above and below the target. The pulse shape depends strongly on the design of the moderator. Fully coupled moderators ensure that the neutrons get into full thermal equilibrium with the scattering medium. As this takes time, the corresponding pulses are long but the integrated intensity is high. Decoupled moderators give sharper pulses with less intensity. In the case of a poisoned decoupled moderator, neutrons that fall below a certain energy are confined to a layer at the moderator surface. This sharpens the pulse further. The basic principle of decoupling and poisoning a moderator is shown schematically in the lower right corner. The different pulse shapes as calculated for the ESS project [8, 9] are compared in the diagram. SNS is a short pulse spallation source. A long pulse source, as proposed for the next European source ESS, does not need an accumulator ring. Apart from the pulse shape, the repetition rate of the pulses is a crucial design parameter. The ratio of pulse length τ over pulse separation or repetition period T is called the duty cycle of the source. In the case of the SNS coupled moderator, it is about 2%. In principle, one would like to have very intense pulses at large intervals, that is, a low duty cycle. The lower limit on the repetition rate is set by the technological means required to produce more and more neutrons within a single pulse in order to maintain a maximum time averaged flux. These increase the cost and complexity of the accelerator and target. For example, at 8 Hz and 5 MW one would need 600 kJ/pulse, versus 23 kJ/pulse at SNS (60 Hz at 1.4 MW)

$$k_b = 0.08617 \, \text{meV/K} \quad \text{or} \quad k_b^{-1} = 11.60 \, \text{K/meV}. \quad (3.4)$$

The normalization is such that the integral over the phase space density

$$dN = p(\vec{k}) d^3 r d^3 k = \frac{N}{k_T^3 \sqrt{\pi}^3} e^{-\frac{k^2}{k_T^2}} d^3 r d^3 k \quad (3.5)$$

gives the total neutron density, which is assumed [10] to vary only slowly with $\vec{r}$. From the phase space density, we can calculate the energy distribution with

$$\Phi(E)dE = \Phi_{\text{thermal}} \frac{2}{\sqrt{\pi}} \frac{\sqrt{E}}{(k_b T)^{3/2}} \exp\left(-\frac{E}{k_b T}\right) dE, \quad (3.6)$$

and

$$\Phi_{\text{thermal}} = \frac{2N}{\sqrt{\pi}} v_T = \frac{2N}{m\sqrt{\pi}} \hbar k_T, \quad (3.7)$$

denoting the *thermal flux* of the source. Both at reactors and spallation sources, different moderators at different temperatures are used to give optimized flux distributions for the various scientific applications. At the ILL the thermal spectrum is moderated to lower energies by using boiling D_2 sources [7] at 25 K (see Fig. 3.1). The up-moderation is achieved via a graphite block heated to 2400 K via the γ-radiation.

At a short pulse source, the extremely narrow pulse of high-energy neutrons spreads in time as it is cooling down. This process is highly non-linear and, therefore, influences the pulse shape (see Fig. 3.2). In the beginning the neutrons loose a high percentage of their respective energies, while toward the end collisions get less and less efficient. A moderator is termed fully *coupled* if neutrons of practically any energy can pass freely between the reflector surrounding the target and the moderator. If the shape of the moderator is optimized, then this procedure allows thermalizing all neutrons that are hitting the moderator. Depending on wavelength full moderation leads to pulses between 100 and 350 μs. If shorter pulses are required, then the moderator has to be *decoupled* from the reflector. This is done by surrounding the moderator on all sides that are exposed to the reflector by a shield of absorbing material. All neutrons below a certain threshold velocity, which is determined by the absorption edge of the absorber, are denied entrance to the moderator. This decoupling cuts off the tail of intensity that would otherwise arise from those late arriving neutrons which were already pre-moderated in the reflector. The moderation volume can be further reduced by placing an absorbing plate directly inside the moderator (see Fig. 3.2). The volume opposite to the beam ports can then no longer contribute to the final stages of the moderation process. Reduced moderation volume translates into shorter residence times for the neutrons and thus shorter pulses. It is obvious that this so-called *moderator poisoning* has a high price in intensity. Due to this unavoidable trade-off of pulse length and intensity, the design

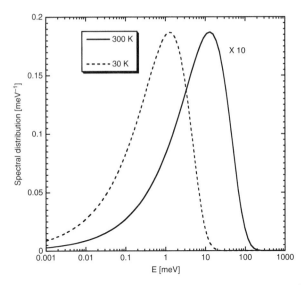

Fig. 3.3 Maxwell–Boltzmann energy distribution for moderator temperatures of 300 and 30 K, respectively. The y-axis has to be scaled by the integrated flux $\Phi_{thermal}$. For better comparison, the 300 K curve has been multiplied by a factor 10

of the moderator around the target is of paramount importance for the performance of a pulsed source.

The colder the moderator, the lower the mean energy of the neutrons and the narrower the energy distribution (see Fig. 3.3). A colder spectrum thus has a higher flux for a given energy interval ΔE in the region of the spectrum's maximum.

As moderated neutrons are in thermal equilibrium with a bath at temperature T, their mean energy as given by Eq. (3.3) will automatically match the mean kinetic energy of any material at that temperature. In quantum mechanical terms, the neutron energy is comparable to the energy of excitations getting thermally populated at that temperature. This is a very important fact for neutron spectroscopy. It means that the energy exchange with the sample during scattering is often[7] of the same order of magnitude as the neutron energy itself. The relative energy resolution $\Delta E / E$ can thus be rather relaxed. The above argument holds for any fully moderated particle. The peculiarity of the neutron resides in the fact that it has a concomitant wavelength (from about 0.5 to 20 Å) which probes equally the right length scales, that is, inter-atomic distances and crystal lattice spacings. This is due to the mass of the neutron. A generally accepted classification scheme for neutron energies is given as follows: *epithermal* above 500 meV, *hot* from 100 to 500 meV, *thermal* from 10 to 100 meV, and *cold* below 10 meV. Apart from the moderated spectrum described

---

[7] There are naturally exceptions to this. Low-energy excitations are populated at temperatures very much lower than the lowest moderation temperature. Their investigation requires a high relative energy resolution $\Delta E / E$ using, for example, a backscattering spectrometer (see Section 3.4.2).

by Eq. (3.6), beams contain varying amounts of fast and epithermal neutrons that escape the moderator before they become fully thermalized. As these neutrons contribute to the background, special care is taken to eliminate them before the beam arrives at the instrument.

3.3 Neutron Beams

Once the neutrons are produced and moderated to appropriate temperatures, they have to be transported with the right characteristics to the sample. Before going into the description of the technical possibilities for beam transport and beam shaping, we would like to introduce a statistical description of the beams. This will allow us to give precise meaning to the concepts of nominal beam parameters, monochromaticity and divergence. The correct mathematical description of the beam is equally important for correctly relating the measured intensities to their theoretical interpretation in terms of the scattering function. A description of the beam with the help of distribution functions has the additional merit that it follows the logic of modern simulation packages. This description is based on the incoherent nature of the neutron beam (see above) and thus cannot describe interference phenomena, as observed when splitting a beam into coherent parts with the help of perfect crystals [11].

3.3.1 Statistical Description

The neutrons extracted from the moderator through a beam port can be considered an expanding gas. As the neutrons do no longer interact with the bath, they are out of thermal equilibrium. For such a beam of neutrons to be fully defined requires that we know at any point in space and time the probability $p(\vec{k}; (\vec{r}, t))$ of detecting a neutron with direction $\hat{k} = \vec{k}/|\vec{k}|$ and energy $E(k)$; that is, we know the phase space density dN of the beam at any point along its trajectory. In steady - state operation, there is no explicit time dependence and thus the parameter t can be ignored. This is not at all the case for pulsed beams.

The probability distribution has to be normalized such that the integral

$$N(t) = \int_K \int_R p(\vec{k}; (\vec{r}, t)) d^3 k d^3 r \tag{3.8}$$

gives the number of neutrons detectable at a given time in the real space element $R = \Delta x \Delta y \Delta z$, possessing wave-vectors falling into the k-space element $K = \Delta k_x \Delta k_y \Delta k_z$. Therefore, $p(\vec{k}; (\vec{r}, t))$ incorporates the neutron production (see Eq. 3.2) and all beam shaping that has occurred upstream from $\vec{r}$.

The *particle current* of a beam passing through an area S is determined by its density and velocity distribution. It is given in the most general case by the surface integral

$$I(t) = \int \left[\int_S p(\vec{k}; (\vec{r}, t))(\vec{v}_k \cdot d\vec{a}) \right] d^3k = \frac{\hbar}{m} \int \left[\int_S p(\vec{k}; (\vec{r}, t))(\vec{k} \cdot d\vec{a}) \right] d^3k, \quad (3.9)$$

where $d\vec{a}$ is the infinitesimal surface vector, which is locally perpendicular to S.

In the case of a rectangular area A perpendicular to $\hat{z}$, Eq. (3.9) reduces to

$$I(t) = \frac{\hbar}{m} \int_A dx dy \int p(\vec{k}; (x, y, t))k_z d^3k, \qquad (3.10)$$

which in case of a homogeneous distribution of wave-vectors over the beam cross section A further simplifies to

$$I(t) = A \frac{\hbar}{m} \int p(\vec{k}; t)k_z d^3k = A \cdot n \cdot \bar{v}_z. \qquad (3.11)$$

In the last equation we have introduced the mean velocity of the beam along $\hat{z}$ and the particle density n. If we divide the current by the area A, we get the *flux* $\Phi(t)$ of the beam

$$\Phi(t) = \frac{\hbar}{m} \int p(\vec{k}; t)k_z d^3k = n \cdot \bar{v}_z, \qquad (3.12)$$

that is, the number of particles passing along $\hat{z}$ per unit area and second. A beam is the better defined the narrower the distribution function. A *monochromatic beam* contains, for example, only neutrons within a well-defined energy or wavelength band. In general, the monochromaticity is not perfect, and the energies are distributed around a nominal energy[8] defined as

$$E_0(\vec{r}, t) = \frac{1}{N} \int_K \int_R p(\vec{k}; (\vec{r}, t)) \frac{\hbar^2 k^2}{2m} d^3k d^3r, \qquad (3.13)$$

with the standard deviation

$$\Delta E(\vec{r}, t) = \sqrt{\frac{1}{N} \int_K \int_R p(\vec{k}; (\vec{r}, t)) \left(\frac{\hbar^2 k^2}{2m} - E_0 \right)^2 d^3k d^3r}. \qquad (3.14)$$

ΔE is the absolute and

$$\frac{\Delta E}{E_0} = 2 \frac{\Delta \lambda}{\lambda}, \qquad (3.15)$$

---

[8] Throughout this chapter we will denote the nominal values by the subscript zero and the standard deviations or full widths half maximum (FWHM) by the prefix Δ.

the relative energy spread of the beam. Both evolve in general along the beam as a function of position and time. In the same way we have defined the mean energy, it is possible, to define the nominal wave-vector direction $\hat{k}_0(\vec{r}, t)$ of the beam. The angular deviations $(\Delta\hat{k}_x(\vec{r}, t), \Delta\hat{k}_y(\vec{r}, t), \Delta\hat{k}_z(\vec{r}, t))$ from the mean direction of the beam are called the *divergence* of the beam.

3.3.2 Scattering of the Beam

The task of neutron scattering consists of measuring the changes in the probability distribution $p(\vec{k})$ due to the interaction with a sample. In simple terms, we want to determine the angular deviation of a neutron beam concomitantly with its change in energy and spin. As neutrons cannot be labeled,[9] this requires the knowledge of the direction, energy, and spin of the neutron beam both before and after the scattering event.

$$\frac{\mathrm{d}^2\sigma}{\mathrm{d}\Omega\mathrm{d}E_f} = \frac{k_f}{k_i} S(\vec{Q}, \hbar\omega), \tag{3.16}$$

that is, the fraction of neutrons scattered per second into a solid angle element $\mathrm{d}\Omega$ with energies comprised between E_f and $E_f + \mathrm{d}E_f$.[10] The scattering cross section has thus the dimension of an area. In Eq. (3.16) we have related the double differential scattering cross section to the *scattering function* $S(\vec{Q}, \hbar\omega)$ with

$$\vec{Q} = \vec{k}_i - \vec{k}_f, \tag{3.17}$$

$$Q^2 = k_i^2 + k_f^2 - 2k_i k_f \cos 2\theta, \tag{3.18}$$

and

$$\hbar\omega = E_i - E_f = \frac{\hbar^2}{2m} \left(k_i^2 - k_f^2 \right). \tag{3.19}$$

The triple of vectors $(\vec{Q}, \vec{k}_i, \vec{k}_f)$ forms a triangle in the scattering plane, which is called the *scattering triangle* (see Fig. 3.5). The intensity in the detector per final energy interval can be written as [12]

---

[9] If the interaction with the sample produces a deterministic change in one of the variables, e.g., if it does not change the spin state of the neutron, then we can use this variable as a label to encode one of the others, e.g., the energy. This principle is at the origin of the spin - echo technique (see Section 3.4.8).

[10] This is equal to the number of neutrons scattered per second into a solid angle element $\mathrm{d}\Omega$ with energies comprised between E_f and $E_f + \mathrm{d}E_f$ and normalized to the incoming flux.

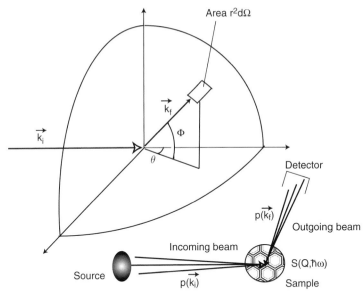

Fig. 3.4 Scattering of a beam by the sample. In the classical approach, the scattering is described by the double differential cross section giving the probability that a normalized flux of neutrons of incident wave-vector $\vec{k}_i$ is scattered into a solid angle element $d\Omega$ perpendicular to the direction $\hat{k}_f$ with energies comprised between $E_f = \hbar^2 k_f^2/2m$ and $E_f + dE_f$, that is, between $(\hbar^2/2m)k_f^2$ and $(\hbar^2/2m)(k_f^2 + 2k_f \, dk_f)$. If we choose the coordinate system such that $\hat{k}_f = \hat{k}_z^f$, then $d\Omega = dk_x^f \, dk_y^f / k_f^2$. In practice the incoming beam of neutrons has a distribution $p(\vec{k}_i; t)$ in energy and direction, depending on how the neutrons were moderated and then shaped by the beam transport. In the same way the neutrons detected will have a distribution $p(\vec{k}_f; t)$ arising from the beam shaping during the transport from the sample to the detector and from the detector efficiency. The beam characteristics described by $p(\vec{k}_i; t)$ and $p(\vec{k}_f; t)$ are responsible for both the statistics and the resolution of the measurements

$$I(E_f)dE_f \propto \int \int p(\vec{k}_i) S(\vec{Q}, \hbar\omega) p(\vec{k}_f) d^3 k_i d^3 k_f. \tag{3.20}$$

For simplicity the distributions $p(\vec{k}_i)$ and $p(\vec{k}_f)$ have been assumed to be homogeneous over the sample volume and not to depend on time, making the space integration trivial.

Equation. (3.20) has a simple interpretation if we think in terms of transport of neutrons to and from the sample. The distribution $p(\vec{k}_i)$ describes the probability of neutrons with wave-vector $\vec{k}_i$ to arrive at the sample after having been produced at the source and shaped by the various optical elements (guides, choppers, monochromators). It includes thus, for example, the source brilliance. The distribution $p(\vec{k}_f)$ describes the probability of neutrons leaving the sample being detected in the detector element corresponding to the solid angle $d\Omega$ and energy interval between E_f and $E_f + dE_f$. Possible obstacles are filtering devices like slits and analyzers or

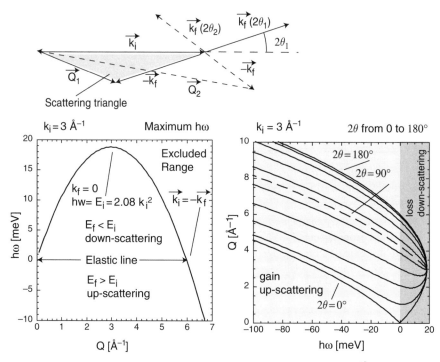

Fig. 3.5 Dynamic range of a scattering experiment. The wave-vector transfer $\vec{Q}$ is given by the vector addition $\vec{k}_i - \vec{k}_f$, while the energy transfer is determined by $\hbar\omega = E_i - E_f$. Both quantities are not independent. For a given k_i there are always combinations $(\vec{Q}, \hbar\omega)$ for which the scattering triangle does not close, that is, for which the *kinematic conditions* are not fulfilled. The example presented here shows on the left - hand side the maximum energy transfer that can be obtained as a function of $Q = |\vec{Q}|$ with a k_i of $3\,\text{Å}^{-1}$. On the right - hand side, we plot the energy transfer associated with a given Q under specific scattering angles 2θ (angles increase from $0°$ to $180°$ in $20°$ steps). These are the $(Q, \hbar\omega)$ trajectories that one would measure with a direct geometry time-of-flight spectrometer (see Section 3.4.3). Everything outside the 0 and $180°$ lines is inaccessible to the experiment. These lines, thus, limit the field of view of the measurement in reciprocal space

inefficient detectors. $S(\vec{Q}, \hbar\omega)$ describes the transmutation of the beam via the sample. It gives the flux normalized probability of an incident neutron with wave-vector $\vec{k}_i$ being scattered into an outgoing neutron with wave-vector $\vec{k}_f$. $S(\vec{Q}, \hbar\omega)$ is completely independent from how we set up our spectrometer. It is a function entirely determined by the sample and is in this sense more fundamental than the scattering cross section.[11] The determination of $S(\vec{Q}, \hbar\omega)$ is the goal of any experiment. The description of its properties is the main task of the theory of neutron scattering as, e.g., described by Squires [13], Lovesey [14], or Price and Sköld [15].

---

[11] The scattering function as defined here does still contain the interaction potential of the neutron with matter. In other words, it contains the scattering length b_l of the nuclei and has, therefore, the unit of area per energy.

Equation (3.20) can be cast in the form [16].

$$I(E_f)\mathrm{d}E_f \propto \int \int R(\vec{Q} - \vec{Q}_0, \hbar\omega - \hbar\omega_0)S(\vec{Q}, \hbar\omega)\mathrm{d}^3 Q\mathrm{d}\hbar\omega. \qquad (3.21)$$

with

$$R(\vec{Q}, \hbar\omega) = \int \int p(\vec{k}_i)p(\vec{k}_f)\delta\left(\vec{Q} - (\vec{k}_i - \vec{k}_f)\right)\delta\left(\hbar\omega - \frac{\hbar^2}{2m}(k_i^2 - k_f^2)\right)\mathrm{d}^3 k_i\mathrm{d}^3 k_f. \qquad (3.22)$$

The *resolution function* $R(\vec{Q}, \hbar\omega)$ gives the probability of finding combinations of neutrons $\vec{k}_i$ and $\vec{k}_f$ in the beam, such that $\vec{k}_i - \vec{k}_f = \vec{Q}$ and $E_i - E_f = \hbar\omega$. The variables $\vec{Q}_0$ and $\hbar\omega_0$ denote the most probable $\vec{Q}$ and $\hbar\omega$ values. They are the centers of gravity of the resolution function about which the measurement is conducted. In order to retrieve the physically relevant quantity $S(\vec{Q}_0, \hbar\omega_0)$, the measurement has to be de-convoluted with the resolution function. In particular, changes in the overall number of neutrons present in the incoming and detected beams and described mathematically by the resolution volumes

$$V_i = \int p(\vec{k}_i)\mathrm{d}^3 k_i \qquad \text{and} \qquad V_f = \int p(\vec{k}_f)\mathrm{d}^3 k_f \qquad (3.23)$$

have to be corrected for.

The scattering of a neutron by matter can to the first order be described as the superposition of spherical waves

$$\psi_{\text{out}}(\vec{r}) \propto \sum_l \frac{b_l}{|\vec{r} - \vec{r}_l|}e^{ik|\vec{r} - \vec{r}_l|}\psi_{\text{in}}(\vec{r}_l) \qquad (3.24)$$

$$\propto \int p(\vec{k}_i) \sum_l \frac{b_l}{|\vec{r} - \vec{r}_l|}e^{ik|\vec{r} - \vec{r}_l|}e^{i\vec{k}_i\vec{r}_l}\mathrm{d}^3 k_i, \qquad (3.25)$$

with $\vec{r}_l$ denoting the position of the scatterer l and b_l the scattering length. For simplicity, we assumed static scatterers leading to elastic scattering $k = k_f = k_i$. The source of the scattered spherical waves is the incoming neutron wave ψ_{in}, which in our description of the beam is given as a distribution in $\vec{k}_i$. Thus atoms along the wavefront, that is, atoms finding themselves at positions of constant $\vec{k}_i \cdot \vec{r}_l$, scatter in phase.[12] Due to the distribution in direction and wavelength of $\vec{k}_i$ (and $\vec{k}_f$) the interference of the scattered waves is smeared out if the distance between

---

[12] If we observe the scattered wave ψ_{out} at the detector, we must include in Eq. (3.25) the description of the scattered beam as a function of $\vec{k}_f$. This includes as outlined above the probability of a neutron $\vec{k}_f$ reaching the detector and leads us to an expression equivalent to Eq. (3.21).

the scatterers becomes too large. In this picture, the resolution function describes the extension in space and time of the so-called *coherence volume*. Scattering from atoms placed within the coherence volume has to be added coherently, that is, by respecting the phase [17]. Scattering originating from atoms belonging to different coherence volumes can be simply added as intensity. This implies that we have no access to correlations from a given atom to the atoms found outside the coherence volume. The coherence volume is thus the instrument's *field of view* in direct space.

We will see in the following sections that the resolution $\Delta Q_i, i = x, y, z$ and $\hbar \Delta \omega$ and thus the associated coherence volumes increase strongly with longer wavelengths. This is not surprising as the beam definition is always done relative to the nominal neutrons; that is, relative resolutions like $\Delta Q / Q$ and $\Delta \omega / \omega$ are similar over a large range of instrument variables. Longer wavelengths, however, imply a reduced range of accessible $(\vec{Q}, \hbar \omega)$ points. For example, in the case of elastic scattering, $|\vec{k}_i| = |\vec{k}_f|$, the maximum Q-transfer that can be obtained is $2\vec{k}_i$ for $\vec{k}_f = -\vec{k}_i$. Equally neutrons cannot lose more than their total energy so $\hbar \omega < E_i$. By deducing $|\vec{k}_i|$, we reduce the *dynamic range* of the scattering experiment; that is, we reduce the *field of view* of our experiment in reciprocal space (see Fig. 3.5). Large fields of view in direct and reciprocal space are thus conflicting requirements.

3.3.3 Beam Shaping

As $S(\vec{Q}, \hbar \omega)$ is the relevant quantity to be extracted from the data, it is obvious from Eq. (3.20) that the better defined the beam, that is, the narrower the distributions $p(\vec{k}_i)$ and $p(\vec{k}_f)$, the more accurate the measurement, that is, the higher the resolution and the larger the coherence volume. This requires intense beams of highly collimated (sharp distribution in $\vec{k}/|\vec{k}|$) and highly monochromatic beams (sharp distribution in E).[13] The whole art of neutron scattering instrumentation consists of optimizing the neutron beams for the scientific task at hand. Ideally, we want an instrument that offers high luminosity at high resolution over a broad range of energies and momentum transfers, that is, with a large dynamical range.

---

[13] One may ask why we do not use energy - sensitive detection. This would avoid tailoring the outgoing beam and thus increase the overall luminosity of the experimental set-up. Energy - sensitive detection requires a correlation between the detector output signal and the neutron energy. A typical energy - sensitive detector is a gas chamber for charged particles. The principle works because the energies of the particles are orders of magnitude larger than the ionization energies required to leave a trace of their trajectory in the gas. For a similar technique to work with thermal neutrons, we would require a system that possesses trackable excitations restricted to the sub-meV range as any meV excitation would completely ruin the energy sensitivity. Such systems are difficult to conceive. An alternative to tracking the trajectory via local inelastic excitation of a medium is a system based on elastic scattering. Incoming neutrons are scattered into different angular regions depending on their energies. The correlation between angle and energy is based on the Bragg-law (see Section 3.3.5) and the scattering can be produced by oriented powders surrounded by position - sensitive detectors. Some pioneering work is currently carried out in this area. The problem with such devices lies with the fact that they are quite cumbersome and difficult to realize for large-area detectors.

The economic way of doing this would consist of producing only those free neutrons that suit the purpose, that is, those that have the right energy and the right direction. This is, however, not possible with current neutron sources, which, as we have seen, produce a gas of free neutrons with little to no directionality and a rather broad distribution in energy. Despite this fact, instrument design cannot be dissociated from the neutron source. In particular, the moderation process of the neutrons and their time structure in the case of pulsed sources are of prime importance for the performance of an instrument suite [18].

Dealing with a neutral particle, the only experimental possibility of manipulating a neutron in vacuum is by acting onto its magnetic moment via strong magnetic fields. This technically demanding method of creating strong continuous or alternating magnetic fields is presently only employed for polarized neutron instruments. In the case of normal, that is, unpolarized beams, the neutron is manipulated via the absorption or nuclear and/or magnetic scattering it experiences in matter. Absorption is a brute force method that allows a beam to be tailored in time and space by simply eliminating undesirable neutrons through screens. A more efficient shaping of the beam can be achieved by exploiting the specular reflection of neutrons at an interface or the Bragg reflection of neutrons from crystal lattices. In particular, this allows spacial or angular beam compression (one always at the expense of the other, as the phase space density is a conserved quantity) and energetic filtering. If the beam is pulsed, efficient way of filtering neutrons is provided by discriminating them according to their velocities by measuring the difference in time of flight.[14]

3.3.4 Mirrors and Guides

Specular reflection of neutrons has been observed as early as 1944 by Fermi and Zinn [19] and can be described via the *index of refraction n* of the neutrons.[15] The index of refraction of a medium is defined as the ratio of the k-vector modulus in the medium with respect to its value k_0 in vacuum and reflects the fact that the speed of the particle and thus its kinetic energy has to adapt to a variation in the potential energy as a consequence of overall energy conservation. For a neutron, the interaction of the neutron with the nuclei can be described by the *Fermi pseudo-potential*

$$V(\vec{r}) = \frac{2\pi\hbar^2}{m} \sum_l b_l \delta(\vec{r} - \vec{R}_l), \qquad l = 1, \dots, N, \qquad (3.26)$$

with $\vec{R}_l$ denoting the site of the nuclei l. If the medium can be considered homogeneous, then the *scattering density function* reduces to

---

[14] A more detailed discussion of the different filters can, for example, be found in [20].
[15] For a historical account of the early experiments with neutrons, see [21].

$$\sum_l b_l \delta(\vec{r} - \vec{R}_l) = \sum_i N_i b_i, \tag{3.27}$$

with N_i denoting the number density of scatterers of type i. The *index of refraction* n can be directly related to the density of scattering amplitude $\sum_i N_i b_i$ via

$$n = \frac{k}{k_0} = \sqrt{1 - \frac{4\pi}{k_0^2} \sum_i N_i b_i} = \sqrt{1 - \frac{\lambda^2}{\pi} \sum_i N_i b_i}. \tag{3.28}$$

The quantity $\sum_i N_i b_i$ is equally called the *scattering length density* (SLD) of the material. Values of b are typically in the range of 10^{-15} m and $N \approx 10^{29}$ per m^3. For $\lambda = 1$ Å we can estimate the order of magnitude of the expression $\lambda^2 \sum_i N_i b_i$ as $\approx 10^{-6}$. Using $\sqrt{1-x} \approx 1 - \frac{1}{2}x$ for small x leads to

$$n = \frac{k}{k_0} = 1 - \frac{\lambda^2}{2\pi} \sum_i N_i b_i. \tag{3.29}$$

The index of refraction of thermal and even cold neutrons is, therefore, extremely close to one if compared to that of ordinary light. This reflects the weak interaction of neutrons with matter. As in the case of light, we get total reflection if

$$\cos \theta_c = n. \tag{3.30}$$

As $|n - 1|$ is small for neutrons we may write

$$n^2 \approx 1 - \frac{\lambda^2}{\pi} \sum_i N_i b_i \tag{3.31}$$

to obtain

$$1 - (\sin \theta_c)^2 = 1 - \frac{\lambda^2}{2\pi} \sum_i N_i b_i, \tag{3.32}$$

and thus

$$\sin \theta_c = \sqrt{\frac{\sum_i N_i b_i}{\pi}} \lambda. \tag{3.33}$$

θ_c is called the *critical angle* of the material. Total external reflection (contrary to light $n < 1$) inside a channel is used to transport neutrons over large distances.

Nickel, due to its large coherent scattering cross section and thus rather large critical angle of $\theta_c = 0.1°/\text{Å}$, is a material of choice for these *neutron guide* applications. The transport is the more efficient the longer the wavelength, since as due to the increased critical angle at longer wavelength larger divergences can be tolerated. The critical angle can be greatly enhanced by using multi-layered materials, such as those made up of Ni and Ti. Using modern technology, these so-called *supermirrors* have elevated reflectivity of above 80% up to angles several times larger than that of natural Ni. Supermirror guides are usually classified according to their m value, with m defined via the relation $\theta_c = m \cdot \theta_c^{\text{Ni}}$. An $m = 2$ guide thus transports a divergence twice as high as natural Ni.[16]

Specular reflection does not only allow transporting neutrons, but also is an equally good means of *focusing* beams in space at the expense of angular divergence. This is achieved by bending the mirrors. The form of the neutron channel is decisive for its performance. Due to material imperfections, reflection is never 100% even below the critical angle. The number of reflections in the guide is to be minimized. This can, e.g., be achieved by using the so-called *ballistic guides* [8, 22] (see Fig. 3.6). These guides open up considerably in cross section over the main part of the length, reducing the divergence of the beam at the expense of spacial dilution. This reduces the number of reflections. Shortly before the sample the neutrons are refocussed.

An important task of neutron guides is to reduce the amount of unwanted fast neutrons at the instrument. They are, therefore, often curved in such a way that there is no direct view of the moderator; that is, any neutron arriving at the instrument has been forced to reflect at least once.

It is not trivial to predict the transmission properties of a neutron guide exactly. The distribution $p(\vec{k})$ at the exit of the guide does not only depend on the reflectivity of the mirrors but also equally on its filling at the entrance. It, therefore, cannot be generally assumed that the divergence transported is only a function of the critical angle–as we will do for convenience in some of the following sections. $p(\vec{k})$ may not even be homogeneous over the guide cross section. This is necessarily the case for a curved guide where the curvature introduces a symmetry break between the inner and the outer side of the guide.

Any straight neutron guide can be considered a *collimator* as it limits the divergence of the beam at the outlet to the critical angle $\pm\theta_c$. This holds provided that every neutron with a larger divergence cannot pass the guide without reflection. This is the case if the ratio of the width a to the length l of the guide satisfies the relation $a/l \leq \tan\theta_c$. If the coating is not reflecting at all, that is, if the walls consist of neutron absorbing materials like Cd or Gd, then the collimation is achieved purely by geometric means and we speak of a Soller collimator. To avoid excessively long collimation distances to achieve a compact design, multichannel collimator

---

[16] The appropriate choice of the multi-layer material allows constructing a guide that in a magnetic field allows to reflect only one spin component of the material. It thus can be used as a polarizing device.

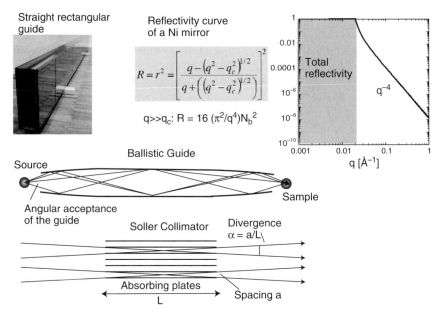

Fig. 3.6 Neutron guides. The reflection occurs on coated glass substrates that are assembled to form closed channels. The reflectivity is one below the critical momentum transfer q_c and drops off rapidly with larger q (that is, with larger reflection angles). Natural Ni is the reference coating material. Supermirror guides are guides featuring a multi-layer coating that pushes the critical angle to higher values. Every reflection is in practice connected with losses. The geometry of the guide is, therefore, of prime importance for its performance. Ballistic guides are very efficient transport media over large distances [8, 22]. They feature an anti-trumpet section at the beginning of the guide to reduce the beam divergence. This leads to fewer reflections that in addition take place at small angles in the long straight section. A trumpet before the sample focuses the beam to the original cross section, increasing the divergence to its original value

constructions are used, which allow a small a/l ratio to be kept over large beam cross sections. The individual channels are delimited by neutron absorbers coated onto thin spacers. The disadvantage of a Soller collimator with respect to a guide is the fact that the transmission of a neutron depends on the point of entry into the collimator giving a triangular transmission[17] as a function of divergence angle θ. This reduces the overall transmitted flux in comparison to a guide that ideally has a transmission equal to one for $|\theta| \leq \theta_c$.

3.3.5 Crystal Monochromators

Neutrons are particle waves and as such are diffracted by gratings. If the grating is composed by the lattice planes of a crystal, then the diffraction can be described as

---

[17] In practice, it turns out that due to imperfections the transmission can be well described by a Gaussian distribution.

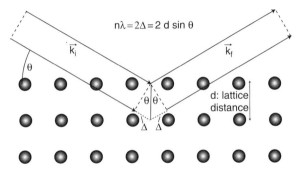

Fig. 3.7 Bragg scattering. In the case of a continuous beam, energy selection is achieved by exploiting the fixed relation between the wavelength of neutrons reflected from the crystal planes and the direction of reflection. The reflection is said to be wavelength dispersive. Under a given Bragg angle θ, a particular wavelength $\lambda(\theta)$ is observed. The d-spacing has to be chosen with care, as it determines for a given wavelength the Bragg angle and thus, according to Eq. (3.37), the resolution of the filter.

a reflection governed by the *Bragg equation*

$$n\lambda = 2 \cdot d \cdot \sin\theta, \tag{3.34}$$

with the *scattering angle* 2θ denoting the angle spanned by the incoming $\vec{k}_i$ and scattered $\vec{k}_f$ neutron wave vectors. d is the distance between the lattice planes and n denotes the order of the Bragg reflection. The Bragg equation states the familiar fact that positive interference from planes of scattering is obtained when the optical path difference is a multiple n of the wavelength (see Fig. 3.7). The crystal placed at a given angle with respect to the beam thus acts as an energy or wavelength filter. The Bragg reflection of neutrons by a single crystal has been observed as early as 1936 [23] and used by Zinn in 1946 as a monochromator [24].

As we will see in the sections on instrumentation, it is very helpful to reformulate the Bragg equation in reciprocal space

$$\vec{\tau} = \vec{k}_f - \vec{k}_i, \qquad |\vec{k}_i| = |\vec{k}_f|, \tag{3.35}$$

with $\vec{\tau}_{hkl} = h\vec{a}_1^* + k\vec{a}_2^* + l\vec{a}_3^*$ denoting the lattice vector in reciprocal space, that is, the vector that is perpendicular to the planes and has a length $|\vec{\tau}_{hkl}| = 2\pi/d_{hkl}$. In other words, Bragg scattering occurs when $\vec{k}_i$, $\vec{k}_f$, and τ form an isosceles triangle with τ as the base (see Fig. 3.8). The manifold of $\vec{k}$-vectors that can be reflected is a two-dimensional surface in $\vec{k}$-space defined by the requirement that the projection of $\vec{k}$ onto $\vec{\tau}$ equals the constant $\tau/2$.

The band that is cut out from the incoming spectrum has a width defined by the beam divergences $\Delta\theta$ and uncertainties in the lattice spacing Δd:

$$\Delta\lambda = 2 \cdot d \cdot \cos\theta\Delta\theta + 2 \cdot \sin\theta\Delta d = (\lambda\cot\theta)\Delta\theta + \frac{\lambda}{d}\Delta d. \tag{3.36}$$

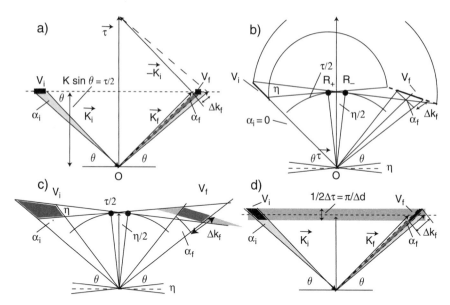

Fig. 3.8 Description of Bragg scattering in reciprocal space. The Bragg equation is satisfied, if both the incoming and the reflected beam have a projection of $k \sin \theta = \tau/2$ onto the lattice vector $\vec{\tau}$. This gives a one-to-one match of $\vec{k}_i$ onto $\vec{k}_f$. The incoming beam has necessarily a finite divergence α_i leading to a divergent reflected beam. This is indicated by the *lightly shaded areas* in figure (**a**). The finite divergence entails a spread in the length of the outgoing wavevectors Δk_f, that is, a finite width in energy of the produced beam. This width can be reduced by decreasing the divergence of the outgoing beam with the help of a collimator (*dark shaded area*). The incoming phase space element V_i will translate into a phase space element V_f after the reflection, defined by the interesction of the dark shaded area with the $\tau/2$-line. It becomes immediately evident that for given divergences the monochromaticity is increasing with the Bragg angle θ and thus depends strongly on the lattice spacing d. If the crystal is not perfect, then there will be a distribution of the orientation of the lattice planes called mosaic described by η. Graphically, this corresponds to tilting the diagram of figure (**a**) by $\pm \eta/2$ around the origin O. As shown in figure (**b**) the image of a tightly collimated incoming phase space element V_i can be constructed by mirroring the shortest and longest $\vec{k}_i$ with respect to the lines $O - R_-$ and $O - R_+$, respectively. It is important to note that the obtained phase space element V_f is inclined with respect to $\vec{k}_f$. In figure (**c**) we show the general situation of an incoming beam of divergence α_i being reflected by a mosaic crystal into an outgoing beam of divergence α_f. The area of V_f reflects the intensity of the outgoing beam, provided the phase space element V_i is fully filled. The experimental set-up is optimized if V_f is maximum for a required $p(\vec{k}_f)$, that is, for a required $\vec{k}_f$ and Δk_f. Figure (**d**) shows the situation in phase space for a crystal with varying d-spacing. This can be achieved via a gradient in the chemical composition of the crystal or by straining the crystal mechanically or thermally. The volume of V_f is given by the cross section of the *shaded area* arising from the divergence with the one arising from the distribution in τ

The width $\Delta\lambda$ or in terms of relative resolution,

$$\frac{\Delta\lambda}{\lambda} = \cot\theta\,\Delta\theta + \frac{\Delta d}{d}, \tag{3.37}$$

gives the width of the distribution function $p(k)$ defining the beam after the crystal monochromator in terms of the length of the $\vec{k}$-vector.[18] For a beam divergence of $1°$ and scattering under $2\theta = 90°$, we get a wavelength resolution of about 1% from a perfect crystal. The resolution can be improved by better defining the angular distribution of the beam. This is done with the help of collimators, that is, arrays of absorbing neutron channels (see above), which allow defining independently the divergence $\Delta\vec{k}_i$ and $\Delta\vec{k}_f$ of the incoming and scattered beam. This naturally comes at the cost of a strongly reduced beam intensity (see Fig. 3.8). For a given divergence, $\Delta\theta$. The best resolution, that is, the narrowest wavelength band, is obtained at $2\theta = 180°$, that is, in backscattering geometry. We will discuss this in more detail in Section 3.4.2.

As the Bragg scattering is dispersive, it introduces a strong correlation between wavelength and angle into the beam. In reciprocal space, this means that the phase space volume is *inclined* with respect to the outgoing wave-vector $\vec{k}_f$. In practical terms, this break in rotational symmetry about the beam direction implies that left and right are no longer equivalent. This has important consequences for setting up the instruments.

For most practical applications the angular acceptance of an ideal crystal is too small to make efficient use of a divergent large-size beam. To overcome the problem, the crystals are processed in order to introduce either a variation in the lattice spacing d or a distribution of crystal orientations created by defects in the single crystal. In reciprocal space (see Fig. 3.8), this implies introducing a spread either in the length or direction of $\vec{\tau}$. The Bragg equation (Eq. (3.35)) is, therefore, satisfied for rings of finite thickness in $\vec{k}$-space. Crystals composed of crystallites are called *mosaic crystals*. The angular distribution of crystallites can be determined experimentally by turning (rocking) the crystal through the Bragg angle in a monochromatic beam with sufficiently high angular resolution. The probability of finding a crystallite with an orientation $\theta + \Delta\theta$ is given by the thus measured intensity as a function of angle (rocking curve). It normally is well described by a Gaussian function

$$W(\Delta\theta) = \frac{1}{\eta\sqrt{2\pi}} e^{-\frac{\Delta\theta^2}{2\eta^2}}. \tag{3.38}$$

η is called the *mosaicity* of the crystal and for most monochromators varies between 10 and 40 min of arc. Mosaicity is responsible for the fact that the distribution in wavelength after reflection from the crystal has a Gaussian form. This is very

---

[18] The shape of $p(\lambda)$ will naturally reflect the shape of the distributions $p(\theta)$ and $p(d)$. For simplicity we may assume both to be described by Gaussian functions.

important for the resolution of spectrometers based on filtering wavelength with crystals (see Section 3.4.1). The mosaicity does not, however, necessarily have to be isotropic. In many cases it is even desired to be low in the vertical direction where it would only lead to beam loss.

The above formulae apply to flat crystals. Modern crystal monochromators are large in size and composed of a large number of single plates that are oriented such as to direct the beam onto small sample areas (see Fig. 3.9).[19] This is particularly important for thermal or hot neutrons, where converging neutron guides are inefficient as focusing devices due to the low critical angle even for super-mirrors. These *vertically* and *horizontally focusing* monochromators accept by definition a large divergence. Despite this fact, the energy resolution can be preserved provided the source size (eventually limited by slits) is comparable to the focus spot (sample size) and the right geometry is chosen (*monochromatic focusing*) for the distances from source to monochromator and from monochromator to sample [25].

It is not trivial to predict the intensity of a beam reflected from a crystal. The crystallographic quantity playing the role of a linear reflection coefficient is given by

$$I_{hkl} = \frac{\lambda^3}{N_c^2 \sin^2 \theta} |F_{hkl}|^2, \qquad (3.39)$$

where the symbols N_c and F_{hkl} are the number of unit cells per unit volume of the crystal and the structure factor for the reflecting plane (*hkl*), respectively. The structure factor measures the strength of the interference arising from the scattering by various lattice planes and is given by

$$F_{hkl} = \sum_i b_i e^{i\vec{\tau}_{hkl} \cdot \vec{R}_i} = \sum_i b_i e^{2\pi i (hx_i + ky_i + lz_i)} \quad , i = 1, \ldots, r, \qquad (3.40)$$

with $\vec{R}_i = x_i \vec{a}_1 + y_i \vec{a}_2 + z_i \vec{a}_3$ giving the positions of the r atoms in the primitive cell of the crystal. For a crystal of thickness l, the scattered fraction of the neutron beam is in first approximation given by $I_{hkl} \cdot l$; that is, it increases linearly with the crystal thickness.

In reality, this is not the case due to the problem of *extinction*. In a perfect crystal already the first few thousand lattice planes exhaust the original incoming spectrum. Planes lying deeper in the crystal see a strongly modified incoming beam consisting of the original plus multiply reflected beams that overlay coherently. This coherent superposition is not taken into account by the simple *kinematic theory* used to obtain Eq. (3.39) and leads to *dynamic diffraction theory* [11, 28]. Increasing the depth

---

[19] A good overview of focusing Bragg optics is given in [26].

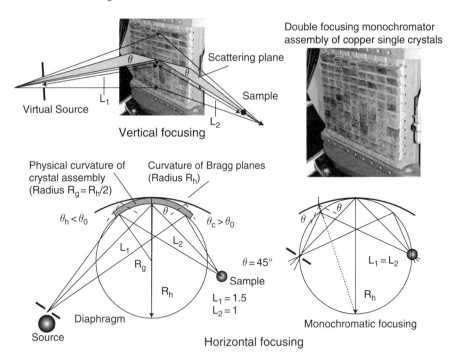

Fig. 3.9 Crystal monochromators. Large assemblies of crystals are used to focus the beam onto the sample. The example shown here is double focusing; that is, it compresses the beam both vertically and horizontally. The vertical curvature $\rho_v = 1/R_v$ satisfies the lens equation $2\sin\theta/R_v = 1/L_1 + 1/L_2$. The Bragg angle stays constant over the height of the monochromator. Vertical focusing is such always monochromatic. It produces an image of the vertical source of height $h_{\mathrm{im}} = h_s(L_2/L_1)$, with a concomitant increase of the vertical divergence of the beam. The source height h_s is often virtual and is defined by diaphragms. For a guide the distance of the virtual source is defined by the divergence of the guide. For zero divergence, that is, for a fully parallel beam, $L_1 = \infty$. The size of the image is in practice always limited by the size of the crystallites used to build the monochromator. The horizontal focusing is more complex. The curvature $\rho_h = 1/R_h$ for the planes is determined via the relation $2/(R_h \sin\theta) = 1/L_1 + 1/L_2$. The Bragg angle generally varies over the monochromator. In the configuration shown, neutrons passing to the left of the center line have longer wavelengths; that is, they are hotter than the nominal ones ($\theta_h < \theta$), while to the right they are colder ($\theta_c > \theta$). Only in the case where $L_1 = L_2 = 1/R_h \sin\theta$ do we obtain a Bragg angle that is constant over the monochromator. This is called monochromatic focusing. The two focal points are said to lie on the *Rowland circle*. For every wavelength λ, there is exactly one possible set of positions for source and sample. As both the vertical curvature R_v and the horizontal curvature R_h depend on the Bragg angle θ, they have to be adjusted mechanically if the device is to be used for varying wavelengths. The phase space element impinging onto the sample can be further shaped by using a physical curvature of the crystal alignment with a radius different from the radius of the Bragg plane inclination. A particularly favorable case is $R_g = R_h/2$ [27]. Please note that for clarity of the graphical presentation, the dimensions of the monochromator are largely exaggerated. In practice, the distance from the monochromator to the sample is on the order of 1–2 m while the monochromators have typical dimensions of $20 \times 20 \, \mathrm{cm}^2$. Exceptions are analyzers for backscattering instruments (see Section 3.4.2), which at distances of 1–2 m are up to 2 m high. In backscattering $\theta = 90°$, and thus the situation becomes completely symmetric with respect to horizontal and vertical curvature ($L_1 = L_2 = L$ and $1/R_h = 1/R_v = 2/L$). In that case the Bragg angle is constant in all directions; that is, the focussing is necessarily monochromatic

of the crystal thus does not simply produce to a further increase in the scattered beam but may, on the contrary, redirect the neutrons into transmission. This effect is called primary extinction. Primary extinction is responsible for the finite wavelength acceptance [29] of a perfect crystal ($\Delta d = 0$) for an ideally collimated beam $\Delta\theta = 0$, which is called the Darwin width.[20] In the case of silicon, the *Darwin width* amounts to a few seconds of arc. In the same way as the angular acceptance is finite for an ideal monochromatic beam, the monochromaticity of a backscattered beam is finite.

It is evident that crystal monochromators work best for neutrons with wavelengths in the vicinity of the lattice distances, allowing for reasonably large Bragg angles. In particular for long wavelength neutrons ($\lambda > 6$ Å), large d-spacings are required in order to maintain a reasonable resolution. Such crystals of good reflectivities are not available naturally. We, therefore, have to use artificially layered materials or the time-of-flight technique. A similar problem is encountered for very short wavelengths where reflectivities for small d-spacings decrease rapidly, among other reasons, because of the thermal vibrations of the atoms. Here a cooling of the monochromator can be an efficient means of increasing the diffracted intensity.

3.3.6 Time-of-Flight Filters

If the dimensions of the slits that the neutron beam encounters along its path are large compared to the neutron wavelength, then the beam propagation may be treated *ballistically*, that is, as the ensemble of trajectories of non-interacting classical particles. Apart from the specificities related to the finite mass of the neutron, this treatment is equivalent to the ray optics of electromagnetic radiation. A classical particle of velocity v spends a well-defined time t traveling a distance L. By measuring the time-of-flight t, we can determine v and thus the energy of the particle. In a beam with no possibility of labeling the individual particles, this technique works only if all neutrons have the same starting time, that is, if the beam is pulsed. Pulsing can be produced at the source, which is the most economic way, as right from the start unwanted neutrons are not produced. In the case of continuous sources, the pulsing can be achieved via turning devices that open the beam for a limited amount of time Δt, eliminating all the neutrons produced outside this time window. According to whether the pulsing is done (i) via collimators turning about an axis perpendicular to the beam or by (ii) absorbing disks with transparent slits rotating about an axis parallel to the beam, we speak of *Fermi* [30] or *disk choppers*, respectively (see Fig. 3.10).

We will now derive some useful relations for neutron time-of-flight spectroscopy. The energy of a free neutron is related to its classical speed $\vec{v}$ via

---

[20] This effect cannot be described by the Bragg equation, as it was derived on the assumption of an infinitely large homogeneously illuminated scattering volume.

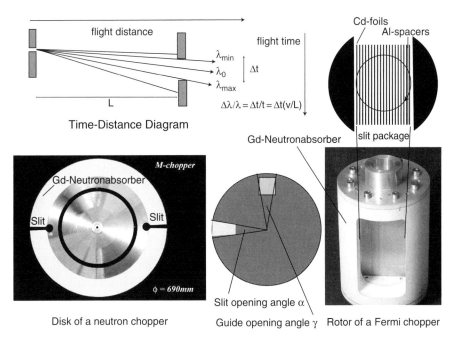

Fig. 3.10 Filtering wavelength bands from a neutron beam via the time-of-flight method. A pulse of neutrons with a distribution in velocity $v = \frac{h}{m\lambda}$ separates in time. Only those neutrons arriving within the time interval Δt at the chopper are transmitted. The relative wavelength resolution $\Delta\lambda/\lambda$ is proportional to the opening time Δt divided by the flight time t. The pulsing is achieved either by the source or by mechanical devices, the so-called neutron choppers. Disk choppers turn with their axis of rotation parallel to the beam. Open slits alternate with absorbing sections. For a disk with an opening angle α corresponding to the angular opening γ of the guide, the transmission as a function of time is triangular. For $\alpha > \gamma$ we get a trapezoid. The FWHM is given for single choppers to good approximation by $\tau = (\alpha + \gamma)/2\omega$, with ω denoting the angular speed of the chopper. If we work with a so-called counter-rotating chopper, that is, with two chopper disks right next to each other and turning in opposite direction, then the opening time is halved ($\tau = \alpha/2\omega$). Fermi-choppers have the axis of rotation perpendicular to the beam. The chopping is achieved via a rotating collimator made of absorbing sheets alternating with transparent spacers. The opening time is defined by the divergence α of the slip pack. Its FWHM is given as $\tau = \alpha/\omega$

$$E = \frac{1}{2}mv^2 = \frac{\hbar^2 k^2}{2m}, \qquad k = \frac{2\pi}{\lambda}. \qquad (3.41)$$

$\vec{k}$ is the wavevector of the matter wave associated with the free neutron of well-defined linear momentum $\vec{p} = \hbar\vec{k}$. This means (see Eq. (3.1)) that neutrons of a few Å wavelength possess an energy of a few meV, which corresponds to typical excitations in solids. This perfect match of microscopic length and energy scales with the wavelength and energy of the neutron is one of the many reasons why neutrons are an ideal tool for the investigation of microscopic processes. From the instrument design point of view, the velocities are more important than the energies as they determine the opening times required for filtering the energies.

$$v = \frac{\hbar k}{m} = \frac{h}{m\lambda} = \frac{6.6261 \cdot 10^{-34}\text{Js}}{1.6749 \cdot 10^{-27}\text{kg}} = \frac{3956\frac{\text{m}}{\text{s}}}{\lambda[\text{Å}]}. \qquad (3.42)$$

Neutrons of $\lambda = 4\,\text{Å}$ thus have a speed of about $1\,\text{km s}^{-1}$. The maximum in the energy distribution of a thermal moderator at room temperature ($T = 293\,\text{K}$) corresponds to neutrons of speed $2200\,\text{m/s}$. The fundamental equation governing time-of-flight spectroscopy relates the travel time t over a distance L to the neutron wavelength λ:

$$t = \alpha L\lambda\,, \quad \alpha = m_\text{n}/h = 252.77\mu\text{s}/(\text{Åm}). \qquad (3.43)$$

A pulsed beam of width τ that has travelled down a distance L will have at any time t a relative wavelength spread of

$$\frac{\Delta\lambda}{\lambda} = \frac{\tau}{t}. \qquad (3.44)$$

If we take $L = 10\,\text{m}$ and $v_0 = 1000$ m/s, we can achieve a 1% wavelength resolution for a pulse width of $\tau = 100\,\mu\text{s}$. To produce such a pulse for typical guide cross sections, we need a device turning at $6000\,\text{rpm}$, which is a reasonable frequency for cyclic mechanical motion.[21] The shape of the distribution $p(\lambda)$ after the filter depends on the transmission T of the opening mechanism as a function of time. For mechanical devices like disk choppers, the transmission T corresponds at any time t to the folding of the slit cross section with the guide cross section (see Fig 3.10). For rectangular slits in front of rectangular guides $T(t)$ is a trapezoid, which becomes a triangle if the slit has the same width as the guide. Single disks open and close at the edge of the guide. Counter-rotating disks open and close the passage for the beam in the center of the guide. $T(t)$ has for mechanical devices well-defined cut-offs. Statistical contributions like fluctuating chopper phases add small Gaussian contributions. This clean resolution without wings is one of the main advantages of a time-of-flight filter.

In contrast to wavelength dispersive filters, time-of-flight filters do not introduce a correlation between angle and wavelength. In particular, they do not break left–right symmetry. This can be very advantageous when working with large detector arrays (see Fig. 3.16). They, however, introduce by definition a correlation between wavelength and time. As time is taken with respect to the opening of the choppers, this leads to a correlation of wavelength and position over the transmitted beam cross section.

The energy filtering performance depends strongly on the wavelength. From Eqs. (3.41 and 3.42) we deduce that the energy uncertainty ΔE connected with a time uncertainty Δt goes with the third inverse power of the wavelength:

---

[21] The task of pulsing the beam would be considerably more difficult, if we were dealing with lighter particles (e.g., electrons) of the same kinetic energy and spread out over similar guide cross sections.

$$\Delta E = \frac{h^3}{m^2 \lambda^3} \frac{\Delta t}{L}. \tag{3.45}$$

From an energy resolution point of view, it is thus recommended to work with the longest wavelength possible.

For a pulses source and a particular moderator, the pulse width is given. In that case we have to invert the above argument. To obtain a specific wavelength resolution from a given pulse width τ, we need a certain flight time t. This is no problem for a single pulse provided we can transport the neutrons efficiently over large distances, such as with the help of ballistic guides. For a sequence of pulses, we are confronted with the *frame overlap* problem. If we start out with a very broad spectrum of wavelengths, then the fast neutrons of a given pulse will sooner or later take over the slower ones of previous pulses. This frame overlap will pollute the characteristics of a time-of-flight filter. The problem can be overcome by limiting the acceptable wavelength band, by adding supplementary choppers to the time-of-flight filter. If this is not possible, for example, because the full wavelength band is needed for the experiment, then the only solution is increasing the time between the pulses or in reducing the original pulse width. The ration τ/t thus can be regarded as the intrinsic wavelength resolution capability of the filter [18]. In the case of a pulsed source, it is identical to the *duty-cycle* of the source, that is, to the percentage of time the source is on. A long pulse source with a low repetition rate thus has a similar intrinsic wavelength resolution as a short pulse source with a higher repletion rate but equal duty cycle.

To obtain a monochromatic pulsed beam, it is sufficient to place a second so-called *monochromating chopper* at a distance L_{pm} downstream from the *pulse creating chopper*. This chopper cuts a certain wavelength band from the incoming spectrum, producing a monochromatic pulsed beam. If the neutron source itself is pulsed with an adapted pulse length, the source can in principle take over the role of the pulsing chopper. However, in practice it turns out that a pulse-shaping chopper is still needed in most cases to define a perfect pulse shape without wings.

3.3.7 Velocity Selectors

We have seen that time-of-flight filters allow neutrons to be sorted in time according to their energy. Monochromatic pulsed beams are obtained by selecting a slice of the sorted energy spectrum. In many cases the correlation of energy with time is not required; that is, we are completely satisfied with a continuous but coarsely monochromatized beam. This can be achieved via *velocity selectors* [31–33]. In a velocity selector the neutrons are sorted into helical channels cut into a cylinder that rotates with a speed ω (see Fig. 3.11). Neutrons traveling parallel to the axis of rotation of the cylinder will keep a constant distance from the channel walls provided they have a velocity given by

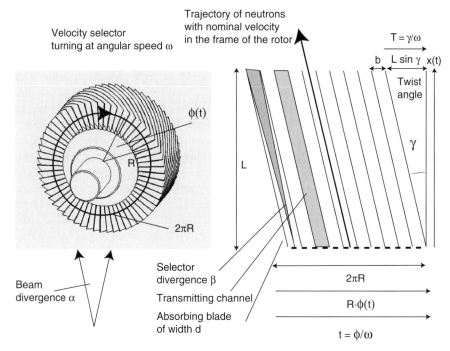

Fig. 3.11 Velocity selectors. The diagram to the right is obtained by rolling the selector at a given radius R onto a flat surface. This distance–distance diagram can be mapped onto a time–distance diagram $x(t)$ for the neutron. The time t corresponds to the angle Φ covered during the period t by the selector. Neutrons traveling parallel to the axis of rotation of the cylinder will keep a constant distance from the channel walls, provided they travel the distance L during the time T, which is necessary to change the orientation of the selector by the angle γ. This leads to Eq. (3.46). The finite width d of the blades leads to transmission losses. Neutrons traveling parallel to the axis of rotation but with different velocity can pass the detector, provided their trajectories fall into the divergence of the selector defined by the angle $\beta \approx b \cos \gamma / L$. If the neutrons impinge onto the selector with a divergence α, this leads to further broadening of the wavelength band. Both transmission and resolution depend on the distance from the center of the rotor; that is, they change over the height of the beam (Drawing of velocity selector courtesy of Lukas Födinger.)

$$v = \frac{L}{\gamma}\omega, \tag{3.46}$$

that is, a wavelength

$$\lambda[\text{Å}] = \frac{h}{mv} = \frac{h\gamma}{mL\omega} = 3956\frac{\gamma}{\omega[\text{s}^{-1}]L[\text{m}]}, \tag{3.47}$$

with γ denoting the pitch or twist angle of the helix (the phase change between the two faces of the rotor of length L). Thus a 30-cm long velocity selector running at 10,000 rpm and featuring a twist angle of 45° would have maximum transmission for neutrons of about 10 Å. A velocity selector can be described as a cascade of an infinite number of disk choppers [34]. As neutrons are fully confined to one channel, there is no cross-talk between the channels as in the case of a finite chopper cascade. The selector can thus work quasi-continuously; that is, the channels can cover the whole volume of the cylinder, one lying next to the other along the circumference. A neutron missing one channel will immediately find the following one so that there is no loss in flux apart from that unavoidable due to the finite thickness d of the channel walls. As can be deduced from the diagram of Fig. 3.11, a given channel is open for a time $\tau = b/(2\pi \cdot R \cdot \omega)$. As the flight path L for a parallel beam is given by the length of the selector, neutrons will path the selector if they are faster than $v_0 - L/\tau$ or slower than $v_0 + L/\tau$. The shape of the distribution $p(\lambda)$ after the selector is triangular and

$$\Delta\lambda = \lambda \frac{b}{L \sin\gamma} = \frac{\beta}{\tan\gamma}, \tag{3.48}$$

gives the FWHM of this triangle. The triangular shape can be understood by considering that neutrons with the nominal speed will be transmitted over the full width of the channels, while the extreme cases (the neutrons at the very fast or very slow end) are transmitted only if they enter the channels on the left and exit on the right or vice versa. In between, the transmission decreases linearly on both sides of the maximum. The argument is similar to the one used for disk choppers. In the case of a divergent beam, the divergence α has to be added quadratically to β in Eq. (3.48), decreasing the wavelength resolution of the chopper further.

$$\frac{\Delta\lambda}{\lambda} = \frac{\sqrt{\beta^2 + \alpha^2}}{\tan\gamma}. \tag{3.49}$$

Velocity selectors are used mainly in applications that require a rather relaxed resolution. To reach a 10% wavelength resolution with a selector of the above mentioned characteristics, we require an opening of about $4.5°$. This selector thus has to feature at least 80 channels, that is, 80 blades separating the different compartments. For a 1% resolution we would require 800 blades, which is not only technically impossible but also equally would lead to a selector with hardly any transmission.

3.4 Instruments

We have reviewed in the last section the most important devices for shaping and filtering non-polarized neutron beams. These devices are the necessary and sufficient elements to design the most common non-polarized neutron spectrometers. We will illustrate this fact by presenting a number of instrument classes that allow to determine the scattering function $S(\vec{Q}, \hbar\omega)$ under various conditions. In the last part of this section we will deal with the spin-echo technique. It has the peculiarity of determining the time Fourier transform $I(\vec{Q}, t)$ of the scattering function by encoding of the velocity change of the neutrons in the beam polarization. Only at this point

we will be obliged to add polarization manipulation devices to our instrumentation
tool box.

3.4.1 Three-Axis Spectrometers

The three-axis spectrometer (see Fig. 3.12) can be considered the mother of all
crystal spectrometers [25]. It allows both the momentum $\vec{Q} = \vec{k}_i - \vec{k}_f$ and the energy
transfer $\hbar\omega = E_i - E_f$ of the neutrons to be determined. This is achieved by shaping
monochromatic beams of neutrons before and after the scattering event with the help
of Bragg scattering from crystals (see Section 3.35), called *monochromator* and

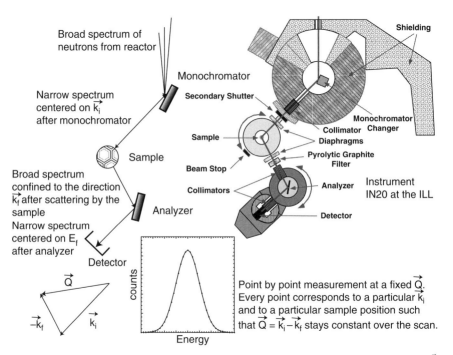

Fig. 3.12 Schematics of a three-axis spectrometer. During the point-by-point measurement, $\vec{k}_i$
(Bragg angle of the monochromator) and $\vec{k}_f$ (angular position of the analyzer as well as Bragg-
angle of the analyzer) are permanently readjusted such as to follow a precise trajectory in $\vec{Q} = \vec{k}_i - \vec{k}_f$ and $\hbar\omega = E_i - E_f$. The sample orientation adapts to the changing spectrometer configurations
in such a way that the $(Q, \hbar\omega)$ setting of the spectrometer is mapped onto a precise $(Q, \hbar\omega)$ point of
the sample's reciprocal space. The resolution of the spectrometer is determined by the divergence
of the respective beams as well as by the mosaic of the monochromator and analyzer crystals. Due
to the breaking of right-left symmetry by Bragg scattering, configurations with alternating sense
of scattering at monochromator, sample and analyzer so-called *W-configuration* (shown here is
a right–left–right arrangement) offer better resolutions than so-called *U-configurations*. In many
cases it is sufficient to keep a rather tight energy resolution. The relaxed Q-resolution can be
attained by curving the crystals and working with diaphragms but no collimators. This leads to
appreciable gains in intensity particularly important for small samples

analyzer, respectively. Typical crystals are graphite, copper or silicon. The primary and secondary beams are characterized by well-defined wavevector distributions $p(\vec{k}_i)$ and $p(\vec{k}_f)$.

By using focusing geometries, monochromators and analyzers not only filter energy bands but equally concentrate the beam spatially onto rather small sample volumes. This naturally comes at the price of increased beam divergence and thus lower resolution in $\vec{Q}$. The analyzer turns around the sample position to allow for varying momentum transfer $\vec{Q} = \vec{k}_i - \vec{k}_f$. In the classical form only one analyzer crystal is used, while multi-analyzer spectrometers allow to monitor simultaneously a whole array of $\vec{k}_f$-values [35]. The sample is normally of single crystalline form and its orientation in the classical mode of operation is permanently readjusted such that the momentum transfer of the neutrons coincides with a particular point in the $(\vec{Q}, \hbar\omega)$ space of the sample. In this way the scattering function $S(\vec{Q}, \hbar\omega)$—once the data corrected for resolution—can be determined in a point-by-point mapping.

It can be shown [16, 36] that for mosaic crystals the resolution (see Eq. 3.22) of a three-axis spectrometer is given by

$$R(\vec{Q}_0 + \Delta\vec{Q}, \hbar(\omega_0 + \Delta\omega)) = R(\vec{Q}_0, \hbar\omega_0)\exp\left\{-\frac{1}{2}\sum_{k=1}^{4}\sum_{l=1}^{4}M_{kl}X_kX_l\right\}, \quad (3.50)$$

with $X_1 = \Delta Q_x$, $X_2 = \Delta Q_y$, $X_3 = \Delta Q_z$, and $X_4 = \hbar\Delta\omega$. The (4×4)-matrix (M_{kl}) is a complex function of the beam divergences and mosaicities of the crystals. It fully describes the resolution properties of a particular instrument set-up. The nice feature about Eq. (3.50) is that despite the complexity of (M_{kl}) any straight cut of the resolution function through $(\vec{Q}_0, \hbar\omega_0)$ leads to a Gaussian function.[22] The surfaces in $(\vec{Q}, \omega)$ space, for which the resolution function, that is, for which

$$\sum_{k=1}^{4}\sum_{l=1}^{4}M_{kl}X_kX_l \quad (3.51)$$

is constant are *ellipsoids* in four-dimensional space. When graphically illustrating the ellipsoids, one normally uses the line that corresponds to a drop in intensity of 50%. In the case of scans through dispersive single crystal excitations like phonons, the orientation of the ellipsoid with respect to the dispersion relation is important. If at a given $(\vec{Q}, \hbar\omega)$ point the long axis of the ellipsoid is aligned parallel to the dispersion $\hbar\omega(\vec{q})$, the resolution in energy is better (focusing configuration) than in the case of a perpendicular orientation (anti-focusing configuration).[23]

Using a large variety of crystals for both monochromator and analyzer on cold, thermal and hot sources energy transfers from 0 up to 200 meV can be covered

---

[22] This relies on the fact that the transmission of both collimators and crystals is a very good approximation Gaussian (see Sections 3.3.5 and 3.3.4, as well as [36]).

[23] We have used the notation $\vec{q} = \vec{Q} - \vec{\tau}_{hkl}$ with $\vec{\tau}_{hkl}$ indicating the closest reciprocal lattice vector.

with a relative resolution of a few percent. The choice of the instrument configuration is motivated by the requirement to close a scattering triangle for a particular $(\vec{Q}_0, \hbar\omega_0)$ point, while retaining sufficient resolution in both $\vec{Q}$ and $\hbar\omega$ to determine the structure of $S(\vec{Q}, \hbar\omega)$ in the vicinity of $(\vec{Q}_0, \hbar\omega_0)$. Naturally, all the effort is in vain unless the dynamic range and resolution are matched by sufficient intensity and signal-to-noise allowing for satisfactory data statistics within reasonable measuring time.

One problem of three-axis spectrometers is higher-order Bragg scattering. Apart from the nominal energy E_i, the monochromator equally transmits $E_n = n^2 E_i, n = 2, 3, \ldots$, and the same holds for the analyzer. This can lead to *spurious scattering*. Elastic scattering may, for example, spuriously appear in the inelastic channels when the relation $n_1^2 E_i = n_2^2 E_f, n_i, n_f = 1, 2, 3, \ldots$ is fulfilled. The probability of such events is particularly high when $E_i \gg E_f$. The effect can be reduced by using a velocity selector or adapted energy filters. An often used low-pass filter is Be-powder [37]. It deflects all neutrons with wavelength shorter than 3.96 Å, that is, with energies higher than 5.2 meV into Debye-Scherrer rings (see Section 3.4.4). Longer-wavelength neutrons do not meet the Bragg condition even in backscattering and are thus transmitted unless they are scattered inelastically by crystal vibrations. To reduce the inelastic attenuation of the beam, Be-filters are cooled to liquid nitrogen temperatures.

3.4.2 Backscattering Spectrometers

We have seen (Eq. (3.37)) that a crystal monochromator has its highest resolution in backscattering, that is, for Bragg angles of $90°$. A backscattering spectrometer is a three-axis spectrometer where both the monochromator and the analyzer work in backscattering geometry [38]. We can deduce from Fig. 3.13 the lengthening of the wave-vector when going out of backscattering by an angle ϵ for the value of highest divergence $\Delta\theta/2$ as

$$\Delta k = k_0 \left(\frac{1}{\cos\left(\epsilon + \frac{\theta}{2}\right)} - 1 + \frac{\Delta\tau}{\tau} \right), \tag{3.52}$$

which for ideal backscattering and reasonably small divergences leads to

$$\frac{\Delta k}{k_0} = \frac{1}{8}\Delta\theta^2. \tag{3.53}$$

$\Delta\tau = 2\pi/\Delta d$ gives the spread in lattice constant due to strain and other crystal imperfections. As we have already mentioned in Section 3.3, even perfect crystals do have a finite band width in backscattering. Therefore, the energy resolution cannot be better than

$$\frac{\Delta E}{E} = 2 \left[\frac{1}{8} \Delta \theta^2 + \frac{16\pi N_c |F_{hkl}|}{|\vec{G}_{hkl}|^2} \right]. \tag{3.54}$$

For reflection from perfect Si(111) planes ($\lambda = 6.27\text{Å} \approx 2\pi \text{Å}$, $E = 2.08\,\text{meV}$) and a beam with the divergence arising at this λ from a Ni-coated guide ($\Delta\theta = 1.25° = 0.02\,\text{rad}$), we get $\Delta E = (0.24 + 0.08)\,\mu\text{eV}$. The contribution from the beam divergence is thus three times more important than that due to primary extinction. To retain a reasonable neutron flux it is not recommended to decrease the divergence in order to fully match the two terms. To the contrary, in practice slightly imperfect crystals are accepted leading to an overall resolution (monochromator plus analyzer) of a bit less than 1 μeV.[24]

As both the monochromator and the analyzer are set in backscattering, the energy transfer is fixed. If both the analyzer and detector use the same reflection, then the energy transfer is zero and only neutrons falling into the resolution window (about 1 μeV for Si111 reflections) will be counted. This allows determining the strictly, elastic scattering of the sample. If registered as a function of temperature, the drop of the elastic intensity signals the onset of motion faster than a few nanoseconds, e.g., because the sample starts melting. This kind of instrument operation is called *fixed* (energy) *window scan*.

The whole idea of backscattering spectroscopy is based on the fact that the wavelength does to first order not depend on the scattering angle. It is, thus excluded to scan the energy in the classical three-axis mode by turning the monochromator. This is best demonstrated by a simple calculation. If we want to achieve a dynamic range of about 10 times the resolution, then we have to change the incoming energy by $\pm 10\,\mu$eV. Using the Bragg equation this leads to a Bragg angle of $\theta = 86°$ for Si111. At this angle we would, however, already have a linear contribution to the resolution (see Eq. (3.37)) of $\Delta E = 5\,\mu$eV, i.e., 50% of the dynamic range, which is not acceptable. As a variation of the angle is excluded we are left with the possibility of acting upon the lattice spacing d. It can be changed either thermally or by moving the crystal parallel to the neutron trajectory (Doppler broadening). From basic considerations of energy and momentum conservation, the *Doppler broadening* induced by a crystal with speed v_D is given as

$$\Delta E = 2E \frac{v_D}{v} + O\left(\left(\frac{v_D}{v} \right)^2 \right). \tag{3.55}$$

If we take $\lambda = 6.27\,\text{Å}$ and a Doppler velocity of 5 m/s, then (see Eq. (3.42)) $v_D/v \approx 1.6 \cdot 10^{-2}$, that is, we achieve an energy transfer of about $\pm 30\mu$eV (provided that the deflector is capable of transmitting such a 3% wavelength band). Thermal expansion in crystalline materials is of the order of $\frac{\Delta d}{d} \approx 10^{-5}$ per K. By heating

---

[24] On IN16 at the ILL this 1 μeV resolution is accompanied by a Gaussian profile of $p(\hbar\omega)$. This is not a trivial fact as the Darwin distribution arising from primary extinction is not Gaussian.

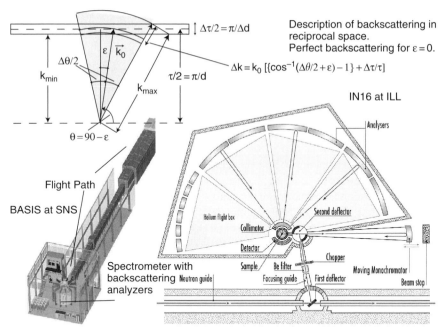

Description of backscattering in reciprocal space.
Perfect backscattering for $\varepsilon = 0$.

$\Delta k = k_0 \left[\{\cos^{-1}(\Delta\theta/2 + \varepsilon) - 1\} + \Delta\tau/\tau\right]$

Fig. 3.13 Backscattering spectrometers. IN16 is the classical type of a backscattering spectrometer. The neutrons from the guide are deflected twice before they hit the monochromator (at a guide end position the first deflector would not be necessary). A narrow wavelength band $\Delta\lambda_i$ comes back from the monochromator to the sample. The second graphite (002) deflector is mounted as two reflecting quarter circle segments on a chopper. This insures that the monochromatic neutrons returning from the Doppler-driven monochromator can pass to the sample by the remaining two open segments of the chopper. The sample is placed directly behind this chopper. The background arising during the opening of the graphite deflector is suppressed by an additional chopper. Higher-order reflections from the graphite are cut off by a cooled beryllium filter. From the sample the neutrons are scattered into the analyzers (The detectors placed close to the sample do not collect data at that moment.). The analyzers select a narrow wavelength band $\Delta\lambda_f$ that is transmitted back through the sample into the detector. As the flux is low, large spherical analyzers are used leading to a rather poor Q-resolution. The incoming energy is varied by moving the monochomator parallel to the neutron trajectory. The neutrons experience positive and negative Doppler shifts proportional to the monochromator speed (Eq. (3.55)). In the case of a pulsed source instrument like BASIS at SNS backscattering is used only for the energy filtering after the sample. The incoming energy is selected using the time-of-flight technique. As an energy resolution close to that of the backscattering analyzers is required, it is necessary to use a short pulse in conjunction with a long flight path. At SNS this is achieved at 84 m using the poisoned moderator $\Delta t \approx 50 - \mu$s at 6 Å. The diagram shows backscattering in the space of the neutron wave-vectors. τ is perpendicular to the surface and of length $2\pi/d$. Primary extinction or crystal imperfections lead to a relative wavelength uncertainty ($\frac{\Delta\tau}{\tau} = 1.86 \times 10^{-5}$ for perfect Si 111), which can be considered a radial mosaic

a CaF_2 ($\lambda = 6.307\,\text{Å}$) crystal (as used on the backscattering instrument IN13 at the ILL) from 77 to 700 K, the energy can be varied by $\frac{\Delta E}{E} = 2\frac{\Delta d}{d} \approx 3.3\%$. The heating of the monochromator is a slow process. The result of the scan is thus only available after hours of data accumulation, while in the case of the Doppler drive the spectrum is collected simultaneously.

3.4.3 Time-of-Flight Spectrometers

The instruments discussed so far filter the energy of the neutrons using Bragg optics. While this works very well in the thermal range, the method has limitations for very hot (loss in reflectivity of the crystal) and very cold neutrons (insufficient d-spacing) as we have seen in Section 3.3.5. In addition, an array of many crystals is geometrically cumbersome if a large solid angle Ω of scattered neutrons has to be analyzed. The time-of-flight filter discussed in Section 3.3.6 has the advantage that it works for any wavelength on the condition that suitable mechanical devices for closing and opening the beam are available. Time-of-flight spectrometers can be grouped into two main classes: (i) *direct* and (ii) *indirect geometry spectrometers*. Direct geometry spectrometers work with a pulsed monochromatic beam at the sample. The monochromatization can be achieved with a crystal followed by a Fermi-chopper taking care of the pulsing. In that case the primary spectrometer resembles very much that of a three-axis instrument. The energy after the scattering is analyzed by measuring the time it takes the neutron to travel the known distance L_{sd} from the sample to the detector. In this way we obtain a spectrum of intensity $I(t)$ with $E_f(t) = \frac{m}{2}\frac{L_{sd}^2}{t^2}$ for every detector pixel (see Fig. 3.14). As the neutrons arriving from the monochromator at the Fermi-chopper possess a very strong correlation of wavelength and angle, the speed of rotation of the chopper has a strong influence on the resolution [39]. If chosen correctly, then the slower neutrons pass the Fermi-chopper first catching up with the faster ones at the detector. This so-called *time-focusing* naturally depends on the energy transfer the neutrons experience at the sample. Faster chopper rotation leads to a fulfillment of the focusing condition at higher energies in up-scattering (neutron energy gain).

When working with very cold neutrons, both the crystal monochromator and the Fermi-chopper run into performance problems. It is then necessary to replace the primary spectrometer by a cascade of disk choppers. The chopper closest to the source creates pulses.[25] A second chopper is placed as close to the sample as possible. It performs the energy filtering using the fact that neutrons will disperse along the path from the pulsing to the monochromating chopper. The intensity at the detector of such a spectrometer is proportional to the product $\tau_p \cdot \tau_m$ of the opening times of the pulsing and monochromating choppers [40, 41].

---

[25] In the case of a pulsed source, this function can in principle be assured by the source itself. In most cases it is, however still necessary to shape the pulse.

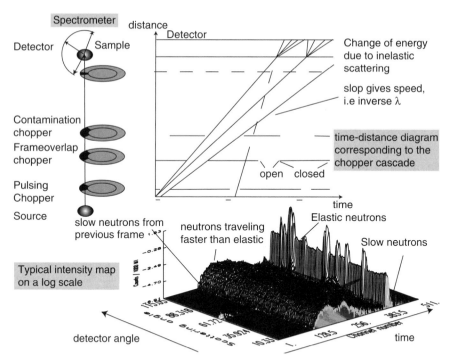

Fig. 3.14 Schematic lay-out of a generic chopper spectrometer. The pulsed source—or in the case of a continuous source the pulsing chopper—produces bursts of polychromatic neutrons. These neutrons disperse as they travel to the sample. Within the resolution of the instrument a particular neutron velocity is selected by the monochromating chopper. Additional choppers are necessary to avoid the contamination of the beam by neutrons originating from other source pulses. If the frequency of the pulses is sufficiently low, like in the case of a long-pulse spallation source, several wavelength packages can be selected from one source pulse. This scheme of *repetition rate multiplication* [6, 42] allows using the full time between source pulses for efficient data collection in time-of–flight spectroscopy. A good overview of the beam trajectories can be obtained via a time–distance diagram. The data acquisition covers the accessible $(\vec{Q}, \hbar\omega)$-range simultaneously. Every detector pixel described by angles (θ, Φ) is associated with a spectrum $I(t)$. Neutrons having gained energy at the sample (up-scattering) arrive prior to those having been scattered elastically, while the slower ones follow later. Neutrons slowed down too strongly by the scattering from the sample will be detected simultaneously with the fast ones from the successive pulse. This frame overlap can be reduced at the expense of intensity by increasing the time between pulses using a suppressor or frame-overlap chopper

The energy resolution of a time-of-flight spectrometer is determined by the accuracy of measuring the flight time and flight path. The width in time Δt of a pulsed beam when arriving at the detector can be derived directly from the time–distance diagram using purely geometric arguments. Converting the time uncertainty into an apparent energy width ΔE on the basis of Eq. (3.45) we obtain

$$\Delta E[meV] = 0.6472 \frac{\left[\Delta t_p^2 + \Delta t_m^2 + \Delta t_l^2\right]^{1/2}}{\lambda_f^3 L_{pm} \cdot L_{sd}} \tag{3.56}$$

with

$$\Delta t_p = \tau_p \left(L_{ms} + L_{sd}\frac{\lambda_f^3}{\lambda_i^3}\right) \tag{3.57}$$

$$\Delta t_m = \tau_m \left(L_{pm} + L_{ms} + L_{sd}\frac{\lambda_f^3}{\lambda_i^3}\right) \tag{3.58}$$

$$\Delta t_l = L_{pm} \cdot \lambda \cdot \alpha \cdot \Delta L. \tag{3.59}$$

L_{pm}, L_{ms}, and L_{sd} denote the distance from the pulsing to the monochromating chopper, from the monochromating chopper to the sample, and from the sample to the detector, respectively. The equation simplifies considerably in the case of elastic scattering, that is, for $\lambda_i = \lambda_f$. The first contribution Δt_p describes the spreading of the initial pulse τ_p at the detector in the case of a hypothetical infinitely sharp opening of the monochromating chopper. The second contribution Δt_m arises from the spread that an infinitely sharp pulse from the pulsing chopper (or source) experiences at the detector due to the finite opening time of the monochromating chopper τ_m. The flight path uncertainties ΔL include finite sample and detector size. They translate into a spread of arrival times, which is not related to differences in velocity.

Using Eq. (3.44) we get a time uncertainty of about 25 μs for a typical value of $\Delta L = 20$ mm and $\lambda = 5$ Å. This sets the scale for the chopper openings, which should produce time uncertainties in the same range to achieve balanced resolution contributions to Eq. (3.56).

Indirect geometry time-of-flight spectrometers use crystal analyzers to determine the final energy. The incoming beam is energy dispersive, that is, neutrons arrive with energies $E_i(t) = \frac{m}{2}L_{ps}^2/t^2$ at the sample with L_{ps} denoting the distance to the pulse creating device. The intensity measured in the detector reflects the time dependence of the incoming spectrum. Indirect geometry spectrometers are particularly useful when large energy transfers are to be investigated with high resolution like in the case of vibrational spectroscopy. A typical example is TOSCA at ISIS. The resolution of the secondary spectrometer being high and constant the overall resolution is determined mainly by the length of the primary flight path from the pulsing chopper to the sample. Time-of-flight spectrometers give access to energy transfers up to several hundred meV. The best resolution of about 10 μeV is obtained with disk chopper spectrometers at long wavelengths. Time-of-flight spectrometers thus connect in energy to the dynamic range of backscattering spectrometers, however, with a reduced Q-range. At higher energies the resolution varies usually between 2% (high resolution) and 5% (high intensity).

A particular type of indirect time-of-flight spectrometer is that using the crystal analyzers in backscattering geometry (see Section 3.4.2). The time uncertainty $\Delta t/t$ has to match the resolution $\Delta\lambda/\lambda \approx 0.25 \cdot 10^{-3}$ of backscattering. This implies long flight times and short pulses. A typical example is the spectrometer BASIS at SNS (see Fig. 3.13) . It combines a flight path of 84 m with a short 50 μs pulse from the poisoned moderator leading at 6 Å to a theoretical $\Delta t/t \approx 0.5 \cdot 10^{-3}$. The advantage of this time-of-flight backscattering spectrometer with respect to the classical three-axis type is the large dynamical range due to the fact that the incoming energy varies simply as a function of time without the need for a Doppler drive or heated monochromator. In practice it has, however, difficulties in attaining the ultimate backscattering energy resolution.

3.4.4 Fixed-Wavelength Diffractometers

If we use a three-axis spectrometer without the analyzer, then we will integrate the signal over all final energies, that is, we determine—after corrections for resolution and integration over the solid angle Ω spanned by the detector area—the diffraction pattern [43]

$$I(2\theta) = \int_\Omega \frac{d\sigma}{d\Omega} d\Omega = \int_\Omega d\Omega \int \frac{d^2\sigma}{d\Omega dE_f} dE_f. \qquad (3.60)$$

The structural information of the sample, that is, the correlation of the atomic positions at any time t is given by the energy integrated scattering function

$$S(\vec{Q}) = \hbar \int S(\vec{Q}, \hbar\omega) d\omega. \qquad (3.61)$$

Both expressions could be directly converted into each other if the integration over energy done by the spectrometer was (i) complete and (ii) not mixing inelastic intensity corresponding to various $\vec{Q}$ values in the same detector pixel. In general, the corrections due to these deficiencies in the energy integration can be handled and thus the structural information can be fully retrieved from the diffraction pattern [44].

If the sample is a single crystal, the scattering will consist of Bragg peaks. The experimental task consist in measuring the Bragg intensities (see Eq. (3.39)) with sufficient accuracy to extract the structure factors $|F_{hkl}|$ corresponding to the (hkl) lattice planes. In a single counter mode this is achieved by scanning the intensity as a function of $(\omega, 2\omega)$, with ω denoting the rotation of the crystal in the equatorial plane of the spectrometer and 2ω the concomitant detector rotation. The orientation of the crystal is assured by Eulerian cradles that allow attaining arbitrary

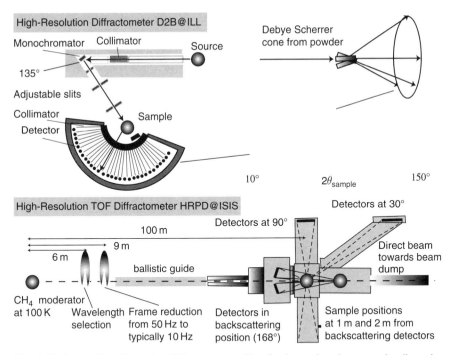

Fig. 3.15 Schematics of powder diffractometers. The fixed wavelength or angular-dispersive diffractometer uses the Bragg reflection from a crystal to define a monochromatic incoming beam. The better defined the divergence before and after the monochromator and the larger the scattering angle $2\theta_{mono}$ (135° in the case of D2B@ILL), the higher the wavelength resolution $\Delta\lambda/\lambda$ (see Eq. (3.64)). The overall Q-resolution depends on the scattering angle $2\theta_{sample}$ at the sample and on the collimation between sample and detector. It varies strongly with $2\theta_{sample}$ (see Eq. 3.64). The scattered intensity is recorded over the full angular range using a large position-sensitive detector (PSD). As for three-axis spectrometers, configurations with alternating sense of scattering at monochromator and sample offer better $\vec{Q}$-resolution. Powder samples lead due to the random orientation of the crystallites to scattering concentrated in cones around the incoming wavevector. Each cone properly integrated over the PSD gives a powder line in the diffraction pattern $I(2\theta_{sample})$. In the case of a time-of-flight or wavelength-dispersive spectrometer, the incoming wavelength varies as a function of time. One, therefore, obtains diffraction patterns as a function of the d-spacing in any pixel of the large area PSDs. This is particularly useful when heavy sample environment like pressure cells limit the available scattering angles. The longer the flight path the higher the wavelength resolution (see Eq. (3.70)). Highest overall resolution is obtained in backscattering at the sample (Eq. (3.37)). Efficient neutron transport is achieved by ballistic neutron guides.

(hkl)-points. The efficiency of the measurement can be enhanced by using two-dimensional position-sensitive detectors (PSD). The $\vec{Q}$-resolution of a single crystal spectrometer is sufficient, if it allows the clear separation of the Bragg peaks of interest. The density of Bragg peaks increases with the size of the unit cell. For larger unit cells it is preferable to work with longer wavelengths. The resolution in real space is determined by the shortest d_{hkl}-spacings that can be measured. It thus improves with shortening the wave-length. Good Q and d resolution are, therefore, conflicting requirements.

In an ideal powder sample, small crystallites are oriented randomly (see Fig. 3.15). The scattering from a particular set of lattice planes corresponds to the scattering obtained by turning a single crystal oriented in Bragg scattering geometry about the direction of the incoming beam, that is, instead of Bragg-peaks we obtain *Debye-Scherrer cones*, whose non-uniform intensity thereby indicates preferred orientation of diffracting grains. Intensity from the cones can be determined simultaneously using large-area detector arrays.

The deviation from ideal randomization is called *texture* (that is, preferred orientation) and a particularly interesting subject of investigation for earth sciences (see Fig. 3.16). It is the goal of the texture investigation to determine the distribution of crystal orientations

$$f(\mathbf{G}(\psi_1, \phi, \psi_2)), \qquad \mathbf{G} = [G_{i,j}], \qquad (3.62)$$

that is, the probability of finding a crystal in an orientation described by the rotation matrix $\mathbf{G}$ with respect to a fixed reference frame. (ψ_1, ϕ, ψ_2) denote the three Euler angles. Neutron diffraction probes the orientation of planes. For a given (hkl) a texture measurement is thus insensitive to the part of $\mathbf{G}$ that maps the plain (hkl) onto itself. The full three-dimensional orientational distribution of the crystal $f(\mathbf{G}(\psi_1, \phi, \psi_2))$ has to be constructed mathematically from the two-dimensional projections,

$$P_{hkl}(\alpha, \beta) = \frac{1}{2\pi} \int_{\psi} f(\mathbf{G}) d\psi, \qquad (3.63)$$

of a set of lattice plains $[hkl]$. ψ denotes the angle of rotation about an axis perpendicular to the planes. α and β denote the tilting and rocking of the plane (see Fig. 3.16). While a single crystal is fully oriented when two planes are identified [26], the inversion for arbitrary P_{hkl} has, generally, no unique solution. The measured reduced distributions are normally presented as projections onto a flat surface giving so-called *pole figures* [45].

The resolution of a fixed wavelength diffractometer depends according to Eq. (3.37) on the collimation of the beams and the scattering angles at the monochromator and the sample [46]. The full width at half maximum of the powder peaks can be well described by the equation

$$\Delta^2_{\mathrm{FWHM}} = U \tan^2 \theta_{\mathrm{sample}} + V \tan \theta_{\mathrm{sample}} + W, \qquad (3.64)$$

with the prefactors U, W and V, being complex functions of the beam divergences and of the monochromator angle θ_{mono}. The minimum of that function occurs for

$$\tan \theta_{\mathrm{sample}} = -\frac{V}{2U} \approx \tan \theta_{\mathrm{mono}}, \qquad (3.65)$$

---

[26] In that case $P_{hkl}(\alpha, \beta) = \delta(\alpha - \alpha_0)\delta(\beta - \beta_0)$.

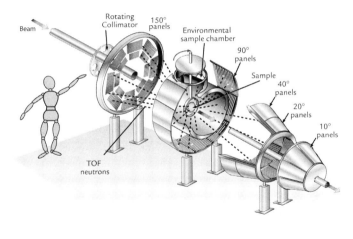

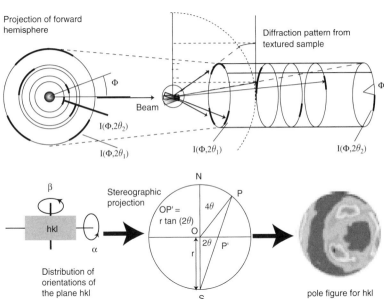

Fig. 3.16 Texture. Schematic view of the time-of-flight diffractometer HIPPO at Los Alamos Neutron Science Center (LANSCE). The detector tubes are arranged in panels on rings of constant diffraction angle. The intensity can be gathered with equal quality for all Φ simultaneously. The schematic drawing shows the scattering arising from a distribution of crystal orientations. When the crystallites in the sample are not randomly oriented, the diffraction pattern deviates from isotropic Debye–Scherrer rings. The anisotropy can, for example, be created by shear or uniaxial pressure during the material's formation. To retrieve the orientational information we have to determine the intensity I as a function of the rotation angle Φ about the incoming beam. For graphical display $I(\Phi)$ can be projected onto a sphere and from the sphere onto the sphere's equatorial plane. The mechanism used here is the Lambert projection, which has the nice property of conserving areas. By using different sample orientations it is possible to determine the distribution of a given lattice plain (h, k, l). This distribution can be displayed by projecting it from a sphere onto a plain. This produces so-called pole figures. If the stereographic projection is used, then angles are conserved

that is, when the scattering angle at the sample is close to the monochromator angle. It is possible with high-resolution diffractometers like D2B at the ILL (see Fig. 3.15) to attain resolutions of $0.1°$ equivalent to $\Delta d/d \approx 5 \cdot 10^{-4}$. This is in the order of the limit set by the line broadening due to the *primary extinction* in the crystallites. Opening up the beam divergence and going to larger d-spacings for the monochromator (that is, lower take-off angles for a given wavelength) increases flux at the expense of resolution. This can be important in the case of kinetic studies.

Using adapted collimation systems it is possible to illuminate only small, so-called *gauge volumes*, of the sample.[27] This allows to scan large samples for variations in scattering patterns. The variation of the lattice spacing, for example, monitors the strain in a material. In the same way one can follow the spacial variations of crystallographic phases or texture. The method equally works for small angle scattering (see Section (3.4.6)) and can be used together with radiography and tomography to *image* samples.

3.4.5 Time-of-Flight Diffractometers

An indirect time-of-flight spectrometer can be turned into a diffractometer in the same way as a three-axis spectrometer, that is, by taking out the analyzers [47] (see Figure 3.15). This allows determining the energy integrated differential cross section $d\sigma/d\Omega$ by recording the intensity as a function of the arrival time in the detector. It is instructive to calculate the observed spectrum for simple Bragg scattering. To this end we design a simple time-of-flight diffractometer consisting of an initial flight path L_{ps} from the pulse creation device to the sample plus a single detector at a scattering angle 2θ at a distance L_{sd}. Neutrons will be deflected into the detector, if the wavelength (see Eq. (3.43))

$$\lambda = \frac{h}{mL}t \tag{3.66}$$

satisfies the Bragg equation, that is, if

$$d = \frac{h}{2mL\sin\theta}t_{hkl}, \tag{3.67}$$

with $L = L_{ps} + L_{sd}$ denoting the total flight path. If we turn the crystal in the beam or if we use a powder sample, then we expect for every lattice plan (hkl) a peak in the detector at time t_{hkl}, which is linked to the respective lattice spacing d_{hkl} via the simple relation

---

[27] The shape of the gauge volume depends on the scattering angle. It is a rectangular solid with a square base in the case of $2\theta = 90°$.

$$d_{hkl}[\text{Å}] = \frac{1.978 \cdot 10^{-3}}{\sin\theta} \frac{1}{L[\text{m}]} t_{hkl}[\mu\text{s}]. \qquad (3.68)$$

Inverting the above equation leads to the spectrum observed under a particular detector angle as a function of time:

$$t_{hkl}[\mu\text{s}] = 5.0556 \cdot 10^2 \cdot L[\text{m}] \cdot d_{hkl}[\text{Å}] \cdot \sin\theta. \qquad (3.69)$$

So in the backscattering mode ($\theta = 90°$) and with a flight path of 100 m, a 1 Å difference in lattice spacing will lead to a peak separation of about 50 ms. This assumes that the necessary wavelengths are present in the spectrum. According to Eq. (3.67) this implies a minimum bandwidth of 0.5 Å. During the 50 ms measuring time it is clear that no neutrons from another source pulse can be allowed to arrive in the detector. The 0.5 Å wavelength band thus translates into a maximum source frequency of 20 Hz.

Equation (3.67) depends on the three variables t, L, and θ. If we assume that they vary independently, then the relative resolution is given as

$$\frac{\Delta d}{d} = \sqrt{\left(\frac{\Delta t}{t}\right)^2 + \left(\frac{\Delta L}{L}\right)^2 + \Delta\theta^2 \cot^2\theta}. \qquad (3.70)$$

In a time-of-flight diffraction pattern, contributions varying with wavelength have to be corrected for. The most prominent is the effect on the intensity of a Bragg reflection, which goes as

$$I_{hkl} \propto I(\lambda)|F_{hkl}|^2 \lambda^4 \cot\theta \Delta\theta. \qquad (3.71)$$

Absorption is normally a linear function of λ. Extinction and, in particular, inelastic corrections are more complicated.

Like in the case of the fixed wavelength diffractometers, data on time-of-flight diffractometers are collected over large solid angles using position-sensitive detectors. The resolution is best in backscattering geometry. Lateral detector banks $2\theta = 90°$ are particularly useful in the case of heavy sample environment, like pressure cells.

3.4.6 SANS Instruments

There is a continuously increasing interest in the study of objects from the nanometer to micrometer range. This requires a particular instrumentation capable of measuring $S(\vec{Q})$ with good precision for small $\vec{Q}$ values. The need for small Q may be easily seen in the case of Bragg scattering. Using the relation

$$Q = \frac{4\pi}{\lambda}\sin\theta, \qquad (3.72)$$

we may rewrite the Bragg equation (Eq. 3.34) in the form

$$Q = \frac{2\pi}{d}. \tag{3.73}$$

This demonstrates that smaller Q vectors probe larger lattice plane distances d. This statement holds more generally as the scattering function $S(Q, \hbar\omega)$ is determined by scattering density correlations on the scale $2\pi/d$. In order to investigate objects extending from about 1 nm to 100 nm, we thus need rather small Q-values in the range of $10^{-3} to 10^{-1}$ Å^{-1}. For a given wavelength λ this can only be achieved by going to very small scattering angles, hence the name of *Small Angle Neutron Scattering* or SANS.[28] The classical small-angle instrument uses a *pin-hole geometry*, that is, it defines the beam divergence through apertures placed at a well-chosen distance from each other.[29]

The $\vec{Q}$-resolution of a small-angle instrument is given by the uncertainties in wavelength λ and scattering angle 2θ. The wavelength distribution of the incoming beam is determined either by the velocity selector (see Section 3.3.7) or in the case of a pulsed source by the pulse width Δt and flight time t of the beam [48]. The uncertainty in the scattering angle is due to the finite character of the beam divergence, sample size, and detector resolution. Using the notation introduced in Fig 3.17 we get

$$\frac{\Delta\theta}{\theta} = \frac{\sqrt{\arctan^2\left(\frac{d_1+d_2}{2(L_1+L_2)}\right) + \arctan^2\left(\frac{d_2+\Delta D}{4L_2}\right)^2}}{\arctan\frac{D}{2L_2}}. \tag{3.74}$$

For small angles and a primary flight path L_1 not too different from the secondary flight path L_2, this can be simplified to

$$\frac{\Delta\theta}{\theta} \approx \frac{(d_1 + d_2)^2 + (d_2 + \Delta D)^2}{4D}. \tag{3.75}$$

Typically, apertures, sample size, and detector pixels are in the centimeter range. This leads to uncertainties in $\Delta\theta/\theta$ that vary with $1/D$ and range from close to 50% near a 5×5 cm^2 beam stop to 5% at the periphery of a 1×1 m^2 detector. It may seem surprising that $\Delta\theta/\theta$ does not, to a first approximation, depend on L. The length L

---

[28] If there were means of moderating the neutron spectrum efficiently down to even lower temperatures, that is, of achieving high neutron flux at wavelengths ranging from 10 to 1000 Å, large objects could in principle be investigated at wider angles. In practice one would, however, reach limits due to the high absorption and finally weak penetration (see Section 3.4.7) of long-wavelength neutrons.

[29] We will assume in the following circular apertures. In practice one often uses the guide opening with a rectangular aperture in front of the sample. This introduces an asymmetry in the divergence contribution to the resolution.

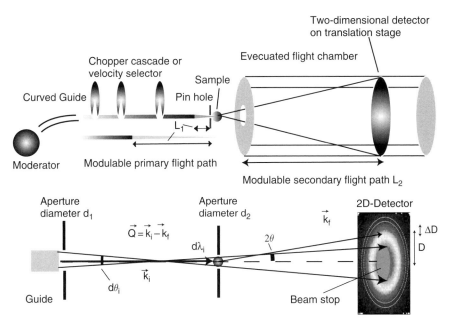

Fig. 3.17 Schematics of a small-angle scattering instrument. The wavelength selection is either performed by a velocity selector or by a chopper cascade, depending on whether one prefers to work in continuous or time-of-flight mode. In the latter case $\vec{Q}(\theta;t)$ is a function of time, while in the former case it is constant. The incoming beam is defined by a set of apertures d_1 and d_2 placed at a collimation distance L_1 typically between 10 and 30 m. The size of the apertures defines the divergence that can be transmitted through the system without scattering at the sample. This direct beam creates a dark area on the detector covered by a beam stop. If the instrument is well designed, the intensity due to parasitic scattering should drop off by several orders of magnitude within a few millimeters from the beam stop. The beam is transported up to the first aperture by a guide system. The guide system itself is highly adjustable allowing an easy variation of L_1. It assures an optimal filling of the beam defined by the apertures. If for example the maximum divergence $\Delta\theta_i < 0.5°$, then for $\lambda > 5\,\text{Å}$ a normal Ni coating of the guide is sufficient. The second aperture is placed close to the sample. The detector is placed in an evacuated flight tube. It has a typical size between $50 \times 50\,\text{cm}^2$ and $1 \times 1\,\text{m}$. It is mounted on a translation stage that allows varying the distance L_2 to adapt it to L_1. Some SANS machines offer off-center positioning of the beam to cover higher Q-values

has, however, a very strong influence on the absolute Q-resolution. This is due to the fact that L defines—for a given wavelength λ—at what distance D the signal corresponding to a particular Q will be observed in the detector.

$\Delta\lambda/\lambda$ is primarily determined by the velocity selector or time-of-flight system and thus basically independent of $\Delta\theta/\theta$. We thus can add both contributions quadratically to get:

$$\frac{\Delta Q}{Q} = \sqrt{\left(\frac{\Delta\lambda}{\lambda}\right)^2 + \left(\frac{\Delta\theta}{\theta}\right)^2}. \tag{3.76}$$

Given the relaxed resolution in angle it is possible to work with rather large wavelength bands in the range from 10% to 15%. The instrument is usually arranged so that $\Delta\lambda/\lambda = \Delta\theta/\theta$ about half way out from the beam centre to the edge of the detector.

The rather coarse wavelength resolution required for SANS measurements is a problem for pulsed sources. As $\Delta\lambda/\lambda = \Delta t/t \approx \tau_p h/mL\lambda$ and as the flight path length L is fixed by the angle required to reach Q, the wavelength resolution is generally too good even for long pulse lengths τ_p. This implies an unavoidable loss of intensity. Time-of-flight SANS instruments at continuous sources have the advantage that they can use arbitrarily broad pulses created by choppers.

The relative resolution of the SANS diffractometer lying in the 10% range it is essential to tune the Q-value correctly to the size of the object under investigation. For 1000 Å objects, 15 m sample-detector distance and 5 Å neutrons ($10^{-3} < Q < 10^{-1}$) may be a good choice, while for 10,000 Å objects, 30 m and 20 Å ($10^{-4} < Q < 10^{-2}$) are certainly recommended, despite the fact that the flux from the cold source will be drastically reduced in this range (see Section 3.2.2).

3.4.7 Reflectometers

All instruments discussed so far study samples in the bulk. We have seen in Section 3.3.4 that the neutron has an index of refraction. This property can be converted into a powerful tool for studying surfaces and interfaces (see Fig. 3.18). Specular reflection gives information on the profile of the scattering-length density $\sum_i N_i b_i$ perpendicular to the surface. The information is contained in the drop of reflectivity for reflection angles larger than the critical angle. As the drop is drastic (see Fig. 3.6), the instrument performance relies on high sensitivity. Good reflectometers can determine reflectivity changes from 1 to 10^{-8}. As can be seen in Fig. 3.18 the interference fringes in the reflectivity arising from layers of different scattering density are to a good approximation separated by $2\pi/d$, with d denoting the thickness of the layers. As the Q-range is equally set by $2\pi/d$, the relative resolution requirement $\Delta Q/Q$ is similar to the SANS case, that is, it is in the order of a few percent%. Therefore, higher resolution and lower Q are needed to determine the scattering length density profile on a longer length scale, while thin films can be studied with rather modest resolution at larger Q. In the way particle size dispersion smears out the SANS pattern, interface roughness smears out the specular reflectivity curve.

When neutrons are scattered from laterally homogeneous stratified media, only specular reflection is observed. Sample inhomogeneities, such as interfacial roughness or voids, give rise to off-specular scattering. The easiest way to describe this scattering theoretically is based on the distorted-wave Born approximation (DWBA) [49, 50], which uses the neutron wavefunctions that describe reflection from a smooth surface as the basis functions for perturbation theory.[30]

---

[30] For a nice example see [51].

If the beam impinges on the surface below the critical angle, then the corresponding evanescent wave field will penetrate the surface layer. Within this layer it can be scattered laterally (for the geometry see Fig. 3.18) leading in the case of large objects to a small-angle signal (*Grazing Incidence Small Angle Scattering* or GISANS). If the layer is laterally ordered, the evanescent wave may be diffracted (*Grazing Incidence Diffraction*).

As the angles of reflection are always small (see Eq. (3.33)) the resolution of a reflectometer is given to a good approximation by the expression Eq. (3.76) used for small-angle scattering. Reflectometers need a beam with very small divergence (in the order of 1–10 min of arc) in the plane of reflection. This is achieved with the help of slits. The Q value can be varied either by using a monochromatic beam

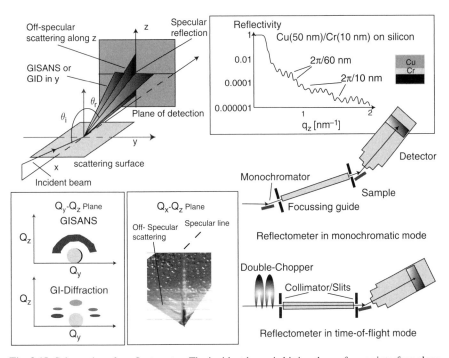

Fig. 3.18 Schematics of a reflectometer. The incident beam is hitting the surface or interface close to the critical angle. Under specular reflection $\theta_i = \theta_r$ we can follow the scattering length density as a function of depth below the surface. The reflectivity drops off quickly as soon as we are above the critical q_z. With neutron reflectometry it is typically possible to cover 6–8 orders of magnitude in reflectivity. The shown reflectivity profile of a layered thin film gives an idea of the signal. The distance between the fringes is directly related to the film thickness. High resolution is required for thick films. Off-specular scattering probes correlations in the x-direction. If the beam is impinging below the critical angle, then the corresponding wave-field explores a surface layer with a thickness corresponding to the penetration length. Within this layer it can be scattered laterally. If there is order like in a crystal, we will get Bragg-scattering called grazing incidence diffraction. In the case of larger randomly arranged objects we end up with grazing incidence small-angle scattering (GISANS). Reflectometers like the one shown here (D17 at the ILL) work either in a monochromatic or in a time-of-flight mode

and changing the angle of incidence or by using an energy-dispersive beam and the time-of-flight method. Reflectometers offer in this way a range from about 0.005 to $1.5\,\text{Å}^{-1}$. They thus can probe length scales perpendicular to the surface from $4\,\text{Å}$ to about $1000\,\text{Å}$.

In the case of grazing incidence the volume probed is defined by the penetration of the *evanescent wave*

$$\psi_e \propto e^{-\frac{1}{2}\sqrt{q_c^2 - q^2}\cdot z}, \qquad q < q_c, \tag{3.77}$$

which can be taken as

$$d_{\text{penetration}} = \frac{2}{\sqrt{q_c^2 - q^2}}. \tag{3.78}$$

For example, on a silicon surface this penetration is about $100\,\text{Å}$ at $q = 0$ and diverges at $q = q_c$. The divergence has the surprising consequence that the q_c is not altered when a thin film—let us say $100\,\text{Å}$ of Ni on Si—is put on a substrate.

For off-specular scattering Q_x is the relevant parameter. As $Q_x = Q_z \cdot \sin \epsilon$, with ϵ denoting the deviation of the reflected beam from the specular line the accessible length scales are considerably longer ranging from about 0.1 to $100\,\mu\text{m}$.

On a continuous source the monochromatic option allows to select the wavelength corresponding to the maximum in the spectrum of the source, while in the time-of-flight mode a large part of the spectrum has to be utilized. The time-of-flight option has the advantage that the full Q-range can be covered simultaneously, which is very important for kinetic studies. It equally allows for an easy variation of the wavelength resolution. In addition, the reflection angles are invariant. In the monochromatic mode the reflection angle has to be varied, implying a readjustment of the sample position along the $\hat{z}$-direction. Most neutron guides are higher than wide. It is, therefore, easier to build a reflectometer like a classical diffractometer with a horizontal scattering plane focusing the beam in the vertical direction, where the divergence has little influence on the resolution, onto the sample (in the case of GISANS or GI-Diffraction this may no longer be the case and a tighter collimation has to be introduced). This implies a vertical sample position, which is not suited for studying liquid surfaces. Reflectometers for horizontal sample positions use dedicated guides that create a beam with a large horizontal divergence at the sample position.

3.4.8 Spin-Echo Spectrometers

All the spectrometer concepts that we have discussed so far incorporate mechanisms for adjusting the energy resolution. As with the standard devices the phase space volumes can be reshaped but not compressed (Liouville theorem), this implies trimming the wavelength bands of both the incoming and the scattered beam. Even if this is done in the most economic way, it leads to a loss in intensity that goes at least with the inverse square of the energy resolution. One of the main keys to successful high-resolution neutron spectroscopy in the Liouville limited domain is the decoupling of energy and Q-resolution. This permits, as we have demonstrated for the backscattering spectrometer (see Section 3.4.2), sub-μeV resolution with acceptable count rates for experiments that tolerate a rather poor Q-resolution.

In soft matter or liquids we are very often interested in studying very slow processes. An exponential relaxation of the form

$$I(t) = I_0 \exp(-t/\tau) \tag{3.79}$$

leads to a Lorentzian form of the corresponding scattering function

$$S(\hbar\omega) = I_0 \int_0^\infty e^{-i\omega t} e^{-\frac{t}{\tau}} \, dt = \frac{I_0 \tau}{1 + (\omega\tau)^2} = \frac{I_0 \tau}{1 + (E\tau/\hbar)^2}. \tag{3.80}$$

The half-width half-maximum of this function is given by

$$\Delta E_{\text{HWHM}} = \frac{\tau}{\hbar}. \tag{3.81}$$

This leads to the relation,

$$\tau[\text{ns}] = 0.66/\Delta E_{\text{HWHM}}[\mu\text{eV}], \tag{3.82}$$

between the measurable relaxation time τ of the system and the required energy resolution ΔE of the spectrometer. Therefore, in backscattering we are sensitive to time scales of the order of 1 ns.

To study even longer times requires the use of spin-echo spectrometers. This technique makes a very efficient use of neutrons. It encodes the change of the velocity distribution function $\Delta p(v)$ occurring at the sample in the polarization $\vec{P}$ of the beam. The encoding of the change $\Delta p(v)$ is to a first approximation independent of the velocity distribution function $p(v)$ itself. This method thus allows for the use of a rather broad wavelength band with correspondingly high beam intensities.

We will in the following develop the main ideas of classical spin-echo spectroscopy.[31] Deducing the formula relating the measured signal to its interpretation as the time Fourier transform of the scattering function will require a sufficiently detailed discussion of polarized beam manipulation. This explains the length and the more mathematical character of this section. We will, however, try to stay close to the experimental observables.

---

[31] Spin-echo spectroscopy was invented by Feri Mezei in 1972 [52]. We adopt in our presentation an approach similar to that used in the introduction to spin-echo spectroscopy by Mezei [53, 54], Cywinski [55], and Lechner–Longeville [41].

We start by giving a short overview of the basic formalism of *beam polarization*. The neutron is a fermion that carries a spin $s = 1/2$. This translates into the fact that the quantum mechanical state of the neutron evolves within a two-dimensional space.[32] The link of the quantum states to the three components of the spin vector $\vec{S}$, which is a quantum mechanical observable in real space, is provided through the Pauli spin operators

$$\sigma_x = \begin{bmatrix} 0 & 1 \\ 1 & 0 \end{bmatrix}, \; \sigma_y = \begin{bmatrix} 0 & -i \\ i & 0 \end{bmatrix}, \; \sigma_z = \begin{bmatrix} 1 & 0 \\ 0 & -1 \end{bmatrix}, \tag{3.83}$$

with

$$\vec{S} = \frac{\hbar}{2}\vec{\sigma}. \tag{3.84}$$

Any quantum state of the neutron can be written as a coherent superposition

$$|\chi> = a_+|+> + a_-|->, \quad |a_+|^2 + |a_-|^2 = 1 \tag{3.85}$$

of two orthogonal basis vectors $|+>$ and $|->$, which we take for matters of practicality as the eigenvectors of the Pauli matrix σ_z with eigenvalues of $+1$ and -1. The coefficients $a_\pm$ are to be identified with the wave functions $\psi_\pm(\vec{r}, t)$ of the particle. They contain, for example, the energetic Zeeman splitting of the two spin components in an applied magnetic field (see below). To render notation more fluent

---

[32] This is a direct consequence of special relativity. The Dirac equation

$$\mathbf{H}\psi = (c \cdot \vec{\alpha} \cdot \mathbf{p} + \beta mc^2)\psi \tag{3.86}$$

is relativistically invariant if, and only if,

$$\mathbf{H}^2 \psi = c^2(\mathbf{p} \cdot \mathbf{p} + c^2 m^2)\psi. \tag{3.87}$$

$\mathbf{p} = -i\hbar\vec{\nabla}$ is the momentum operator. This implies that the coefficients α_i and β satisfy the anti-commutation relations

$$\alpha_i \alpha_j + \alpha_j \alpha_i = 2\delta_{ij}\mathbf{1}, \tag{3.88}$$
$$\alpha_i \beta + \beta \alpha_i = 0, \tag{3.89}$$
$$\beta^2 = \mathbf{1}, \tag{3.90}$$

with $\mathbf{1}$ denoting the unit matrix and $i, j = x, y, z$. It can be shown that at least four dimensions are required to represent such an algebra. A possible choice for the (4×4) matrices is given by

$$\alpha_i = \begin{bmatrix} 0 & \sigma_i \\ \sigma_i & 0 \end{bmatrix}, \; \beta = \begin{bmatrix} 1 & 0 \\ 0 & -1 \end{bmatrix}, \tag{3.91}$$

with σ_i denoting the (2×2) Pauli matrices. In vacuum the only symmetry breaking direction is give by the momentum $\vec{p}$ of the neutron. Without loss of generality the $\hat{z}$-direction can be choosen to coincide with $\hat{p}$. The four components of the Dirac wave function then correspond, respectively, to particle and anti-particle of positive (spin aligned parallel to $\hat{p}$) and negative (spin aligned anti-parallel to $\hat{p}$) helicity.

the space-time dependence of the wave function will not be explicitly mentioned in those cases where we are dealing with quantum mechanical problems consisting in manipulating the spin variables of free neutrons.

The neutron spin manifests itself experimentally via its *coupling to magnetic fields*. The interaction potential of the neutron with matter[33] comprises, therefore, a magnetic contribution

$$V_{\mathrm{m}} = -\vec{\mu}_n \cdot \vec{B}(\vec{r}) \tag{3.92}$$

given by the scalar product of the magnetic field $\vec{B}$ with the magnetic moment operator of the neutron

$$\vec{\mu}_n = (\gamma_n \mu_N)\vec{\sigma} = (\gamma_n \mu_N)\frac{2}{\hbar}\vec{\mathbf{S}} = \gamma_L\,\vec{\mathbf{S}}. \tag{3.93}$$

$\gamma_n = -1.91304275(45)$ denotes the neutron's magnetic moment in units of the nuclear magneton μ_N and $\gamma_L = -1.832 \times 10^8$ rad·s^{-1}·T^{-1} stands for the gyromagnetic ratio of the neutron.

Inside magnetic materials the magnetic field felt by the neutron includes, in addition to the applied external field, the magnetization $\vec{M}$ of the sample[34]

$$\vec{B} = \mu_0(\vec{H}_{\mathrm{applied}} + \vec{M}). \tag{3.94}$$

We will now discuss how the interaction with magnetic fields allows the manipulation of a polarized neutron beam [56]. We start by giving a precise meaning to the term polarization. When leaving the moderator, the neutrons can be considered a paramagnetic non-interacting gas of fermions. Polarized neutron beams are characterized by a finite magnetization given by the expression

$$< \vec{\mathbf{M}}_n(\vec{r}, t) > = \gamma_n \mu_N < \vec{\sigma}(\vec{r}, t) >$$
$$= \gamma_n \mu_N \int \mathrm{d}^3 k \sum_{\chi} p(\vec{k}, \chi) < \vec{k}, \chi | \vec{\sigma}(\vec{r}, t) | \vec{k}, \chi > \tag{3.95}$$

with $p(\vec{k}, \chi)$ describing the distribution of the beam in terms of k-vector and spin state χ at any point and time. The expectation value

$$\vec{P}(\vec{r}, t) = (1/V) \int \mathrm{d}^3 k \sum_{\chi} p(\vec{k}, \chi) < \vec{k}, \chi | \vec{\sigma}(\vec{r}, t) | \vec{k}, \chi > \tag{3.96}$$

---

[33] This contribution adds to the nuclear interaction described by the Fermi pseudo-potential introduced in Eq. 3.26

[34] As the magnetization of the neutron gas is very small, we are not obliged to make a distinction in between the magnetic field $\vec{H}$ and the magnetic flux density $\vec{B}$ outside magnetic materials.

is called the beam polarization, with $V = \int d^3k \sum_\chi p(\vec{k}, \chi)$ denoting the phase space volume.

If we restrict ourselves to one particular quantum state $|\chi>$, then we can relate the orientation of $\vec{P}_\chi$ to the coefficients of the quantum state introduced in Eq. (3.85) by using the explicit form of the Pauli spin-matrices of Eq. (3.83):

$$a_+ = e^{-i\frac{\phi}{2}} \cos\frac{\theta}{2}, \tag{3.97}$$

$$a_- = e^{i\frac{\phi}{2}} \sin\frac{\theta}{2}, \tag{3.98}$$

with the spherical coordinates $0 \le \theta \le \pi$ and $0 \le \phi < 2\pi$ defining the orientation of $\vec{P}_\chi$ with respect to the magnetic field (see Fig. 3.19). For example, a neutron polarized along the direction $\hat{x}$ perpendicular to the magnetic field is described by the quantum state

$$|+>_x = \frac{1}{\sqrt{2}}(|+>_z + |->_z), \tag{3.99}$$

as can be easily verified by calculating the expectation value of $\vec{P}$ according to Eq. (3.96).

A polarized beam can be produced from the paramagnetic neutron gas exploiting magnetic interaction. In neutron spin-echo spectroscopy, the standard technique of polarizing a beam consists in reflection from a magnetically birefringence medium. The index of refraction, which we had introduced in the section on neutron mirrors and guides (see Eq. (3.28)), reads in the magnetic case as

$$n_\pm = 1 - \frac{\lambda^2}{2\pi} \sum_i N_i b_i \pm \frac{m\lambda^2}{h^2} \vec{\mu}_n \cdot \vec{B}, \tag{3.100}$$

where the indices $\pm$ refer to the two quantum eigenstates of the neutron in the applied field. This difference in n allows polarizing the beam in reflection (only one spin state is reflected) or transmission (only one spin state is transmitted) (see Fig 3.19). Other possibilities of beam polarization are spin discriminating Bragg reflection (Bragg scattering is active only for one spin component), spin-dependent absorption in a polarized ^{3}He-filter (only one spin state is absorbed) or separation of spin components in a strong magnetic field gradient like in a Stern-Gerlach type set-up or in strong hexapolar magnets. The later method is, for example, used for ultra cold neutrons or helium.

A polarized neutron beam can be manipulated using magnetic fields. The evolution of the magnetization of the neutron beam if subjected to a magnetic field $\vec{B}$ is described by the equation

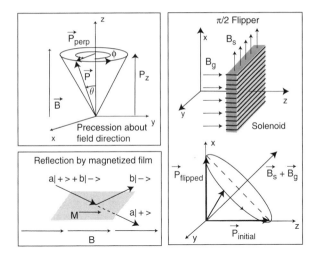

Fig. 3.19 Polarized neutron beams. The neutron interacts with magnetic fields via its magnetic moment. A polarized neutron beam is characterized by a finite magnetization corresponding to an alignment of the magnetic moments. Polarized beams can be created, for example, via reflection from a magnetized multilayer (super-mirror). As the reflectivity depends on the alignment of the neutron's magnetic moment with respect to the magnetization of the film, it is possible to design devices that allow selecting one of the two spin components of the neutron either in transmission or in reflection. The main problem of such polarizers is the limited angular acceptance. Polarized neutron beams are transported via guide fields. The component of the beam polarization $\vec{P}_\perp$ that is perpendicular to the field direction precesses around the field with the Larmor frequency, while the component parallel to the field is stationary. For a slow change of the guide field, the parallel component of the polarization will adiabatically follow the field. The orientation of the polarization with respect to the field, therefore, is preserved. If the field changes abruptly, the polarization $\vec{P}$ has no time to adapt and will start to precess about the new direction. This fact is exploited to reorient the polarization of the beam with respect to the field. In the case of a $\pi/2$-flipper a solenoid creates a field $\vec{B}_s$ that together with the guide field $\vec{B}_g$ leads to a total field $\vec{B}_t = \vec{B}_s + \vec{B}_g$ at $45°$ with respect to the initial beam polarization (along $\vec{B}_g$). The neutron beam spends just enough time in the field region of the solenoid to perform a $180°$ rotation about $\vec{B}_t$. When leaving the solenoid non-adiabatically, the polarization is now perpendicular to the guide field, that is, it has performed a $90°$ or $\pi/2$ turn. In the spin-echo technique this flip corresponds to $t = 0$ as it starts the precession motion within the spin-echo coils (see Fig. 3.20). A π-flipper (not shown) needs a horizontal field close to zero and a vertical field strong enough to turn the polarization by $180°$ during the flight-time of the neutron within the flipper coil

$$\frac{d < \vec{\mathbf{M}} >}{dt} = -\gamma_L (< \vec{M} > \times \vec{B}), \qquad (3.101)$$

which reads in terms of the polarization

$$\frac{d\vec{P}}{dt} = -\gamma_L (\vec{P} \times \vec{B}). \qquad (3.102)$$

For $\vec{B}$ directed along $\hat{z}$ we find the solutions

$$P_x(t) = \cos(\omega_L t)P_x(0) - \sin(\omega_L t)P_y(0),$$
$$P_y(t) = \sin(\omega_L t)P_x(0) - \cos(\omega_L t)P_y(0),$$
$$P_z(t) = P_z(0). \tag{3.103}$$

The components of $\vec{P}$ perpendicular to the field thus precess about $\hat{B}$ with the *Larmor frequency* $\omega_L = -\gamma_L B$ (see Fig 3.19).

Once the neutrons have left the moderator they are a priori in a collision-less regime. Despite this fact, a *guide field* is required to maintain the magnetization of the neutron beam as even small magnetic perturbations would quickly lead to depolarization.[35] The coordinate system is chosen such that the direction of the guide field coincides with the $\hat{z}$-axis.

If the direction of the guide field is changed slowly, then the neutrons will follow this change *adiabatically* meaning that the polarization of the beam will always precess about the local magnetic field direction with the projection of the polarization upon $\vec{B}$ remaining unchanged. If, to the contrary, the change of field is abrupt, then the neutron cannot adapt to it and will carry its original polarization into the new field region, where it will start precessing about the new field direction according to Eq. (3.103). *Non-adiabatic* field change over a well-defined distance L is thus a perfect tool to reorient the polarization with respect to the guide field.[36] Devices doing this are called neutron *spin flippers* (see Fig 3.19). They are a vital component of any spin-echo spectrometer. The most common flippers change the polarization by 90° or 180° with respect to the guide field. They are called $\pi/2$ and π flippers, respectively.

The ratio of the Larmor frequency over the angular frequency of the field change as seen in the neutron's frame of reference

$$C = \omega_L \left/ \frac{d\theta_B}{dt} \right. = -\gamma_L B \left/ \frac{d\theta_B}{dz} \right. v \tag{3.104}$$

determines whether a field change may be considered adiabatic or not. An adiabadicity parameter of $C = 10$ implies that the polarization of the beam precesses 10 times before the guide field $\vec{B}$ has completed a full turn. For a beam centered on

---

[35] There is a whole area of neutron scattering devoted to creating stable polarized beams in field-free regions generated either via super-conducting (cryo-pad) or $\mu - metal$ shielding (μ-pad). We will not further discuss this technique here, but refer to the literature [57].

[36] The rotation angle of the polarization is a function of the time the neutrons spend within the field. It thus depends on the neutron wavelength. Such devices have, therefore, diminishing performance for broad wavelength bands.

a nominal wavelength of 5 Å such a C can be achieved provided the field gradient along $\hat{z}$ does not exceed 0.13 T (or 1.3 kGauss) per cm.[37]

We now will consider in more detail the evolution of the polarization of a beam with a finite component perpendicular to the field ($\vec{P}_\perp \perp \vec{B}$). According to Eq. (3.102) and (3.103) the polarization $\vec{P}_\perp$ will accumulate a precession angle proportional to the field integral

$$\phi = \gamma_{\rm L} \int_0^L \frac{1}{v(l)} \vec{B} \cdot d\vec{l}. \tag{3.105}$$

In general, the trajectory can be assumed rectilinear and the velocity constant. In addition, we may restrict ourselves without loss of generality to the case where $\vec{P}_\perp = \vec{P}$. This allows describing the precession of the polarization vector by the number N_p of turns it performs about the field axis. N_p is given by the simple expression

$$N_p = \frac{\phi}{2\pi} = 7361 \cdot B[\text{Tesla}] \cdot L[\text{m}] \cdot \lambda[\text{Å}], \tag{3.106}$$

where L denotes the length of the trajectory in the field region. For a field of 0.1 T and 5 Å neutrons, this amounts to 3680 turns per meter. A neutron with a different velocity will acquire a different phase angle. Thus a distribution in the neutron velocity will lead quickly to a complete smearing of the polarization because

$$\vec{P} = \int d^3k \, p(\vec{k}) \, \vec{P}(\phi(\vec{k})) = 0, \tag{3.107}$$

if the orientation of the polarization vectors given by the precession angle

$$\phi(\vec{k}) = \frac{\gamma_{\rm L} m}{\hbar k} \int_0^L \vec{B} \cdot d\vec{l} \tag{3.108}$$

is distributed randomly over 2π. This would mean that we have missed our goal of increasing the intensity by accepting a broad wavelength band for the measurement. Fortunately, a loss of beam polarization does not mean a loss of quantum coherence of single neutrons. To the contrary, as the neutrons do not encounter statistical perturbations along their trajectories, the quantum state of any single neutron is at

---

[37] To get an idea about field strength: The earth magnetic field has a strength of about 0.5 Gauss, which is equivalent to 0.05 mT. A strong permanent neodymium magnet ($\text{Nd}_2\text{Fe}_{14}\text{B}$) used in hard-disk drives can reach values above 1 T, and strong superconducting laboratory magnets achieve nearly 20 T in continuous operation. The highest magnetic fields ($B > 10^8$) Tesla that are observed in the universe stem from neutron stars. The question of whether in these stars the magnetic field could move only with the charged particles, leaving the neutrons behind, is a matter of current scientific debate.

any time fully correlated with its state at $t = 0$, that is, at the point of entry into the precession region.[38] It is thus in principle possible to restore the polarization of the beam even for a distribution of wavelengths. The practical realization of such a polarization restoration is called *spin echo*. According to Eq. (3.102) the polarization $\vec{P}$, its variation as a function of time $d\vec{P}/dt$ and the field $\vec{B}$ form a right-handed orthogonal system (remember that γ_L is negative). The precession angle in the laboratory frame thus changes sign upon reversal of either the field or the polarization direction. Therefore, if we can identify a point L' along the trajectory such that for all neutrons

$$\int_0^{L'} \vec{B} \cdot d\vec{l} = \int_{L'}^{L} \vec{B} \cdot d\vec{l}, \tag{3.109}$$

then by flipping $\vec{P}$ by $180°$ at L' we will obtain a full echo of the polarization at L

$$\vec{P}(L) = \vec{P}(0). \tag{3.110}$$

Practically, the echo can be achieved with a beam traveling down two magnetic coils of identical characteristics with a π-flipper in between (see Fig. 3.20).[39]

The spin-echo set-up becomes a neutron spectrometer[40] by (i) placing a sample[41] after the π-flipper and (ii) analyzing the component of the final polarization $\vec{P}_f$ along its initial value $\vec{P}_i$:

$$P = \vec{P}_f \cdot \vec{P}_i = \vec{P}(L) \cdot \vec{P}s(0). \tag{3.111}$$

This so-called *linear polarization analysis* is achieved by flipping the polarization after the second spin-echo coil by $\pi/2$ and passing the beam through a spin analyzer. The intensity I_{Det} measured in the detector is related to P via the expression

$$I_{Det} = \frac{I_0}{2}(1+ <P>), \tag{3.112}$$

---

[38] For a single neutron $|\vec{P}|$ is preserved. The depolarization of the beam arises from the fact that for every neutron with polarization vector $\vec{P}$ we find another one in the beam with $\vec{P}' = -\vec{P}$.

[39] In principle, one could equally work with two coils with opposite field directions. In practice, one prefers often the version with identical fields and a π-flipper. In that case, the two coils can be put into series assuring identical currents. In addition, opposite field directions would make it more difficult to create homogeneous field components along the trajectory.

[40] In resonance spin-echo spectroscopy, the two precession solenoids are substituted by two pairs of radiofrequency coils. As this does not change the basic principle, we refer the reader to the literature for further insight into this very interesting technique [58].

[41] We assume for the moment that the sample does not alter the polarization of the beam. This is the case for nuclear coherent and isotope incoherent scattering.

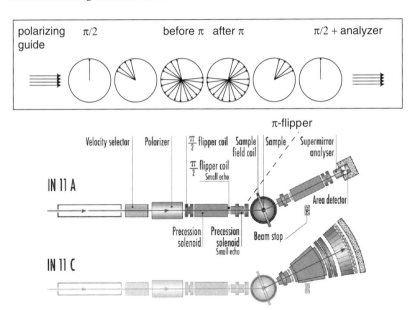

Fig. 3.20 The archetypical spin-echo spectrometer IN11. A guide delivers neutrons from the cold source. The nominal wavelength can be chosen between 3.8 and 12 Å. A velocity selector filers out a 15–22% wide wavelength band. A FeAg suppermirror polarizer creates a beam with 99% polarization parallel to the beam direction. At the entrance of the first precession solenoid, the spin is flipped by $\pi/2$ so that now $\vec{P} \perp \vec{B}$. This flip corresponds to the start of the precession in the first spectrometer arm. Just before the beam reaches the sample, the polarization is reversed by a π turn and the neutrons start precessing in the opposite sense in the second arm of the spectrometer. The polarization is thus restored for elastic scattering provided the field integrals in the two arms are equal. The polarization is flipped back along the beam direction at the end of the second precession solenoid. The projection of the polarization onto the field axis after the $\pi/2$ flip corresponds to the component of the polarization before the $\pi/2$ flip parallel to the initial polarization (see Eq. (3.111)). It is analyzed with the help of a CoTi supermirror. The maximum field integral that can be achieved in this set-up over 2 m is 0.27 T·m. This corresponds according to Eq. (3.126) to a spin-echo time of about 3 ns at 4 Å. When studying relaxations it is important to cover several orders of magnitude in time. For a fixed wavelength this is achieved by varying the field integral. On IN11 very short times, that is, very small field integrals, are produced with a second set-up consisting of shorter solenoids. The reason resides in the fact that it becomes increasingly difficult to have stable spin-echo conditions at very low fields in large coils due to stray fields from the flippers and ultimately the earth magnetic field. Combining the two set-ups it is possible to cover three orders of magnitude in spin-echo time at one wavelength. Field homogeneity is important for the performance of a spin-echo spectrometer. This limits the accessible solid scattering angle that can be covered simultaneously. The IN11C spectrometer offers a $30° \times 1.5°$ solid angle by using a fan-shaped secondary spectrometer. For comparison, the solid angle sustained by IN11A is only $0.9° \times 0.9°$. The larger opening has a certain price in resolution. The maximum spin-echo time measurable with IN11C is only 0.5 ns at 4 Å

where the brackets indicate averaging over all neutrons. I_0 is the mean intensity of the detected beam and as such—within the approximation discussed in Section 3.4.4—proportional to the static structure factor for the scattering vector under consideration:

$$I_0 \propto \int S(\vec{Q}, \hbar\omega)\mathrm{d}\omega \equiv S(\vec{Q}). \tag{3.113}$$

Under the condition that (i) the beam is monochromatic and that (ii) the scattering is elastic we obtain according to Eq. (3.105) a sinusoidal response if we scan the current and thus the field integral in the second arm of the spectrometer.[42]

$$P = \cos\phi = \cos\left(\gamma_L(B_1L_1 - B_2L_2)/v\right) = \cos(\epsilon\lambda) \tag{3.114}$$

with

$$\epsilon = \gamma_L \frac{m(B_1L_1 - B_2L_2)}{h}. \tag{3.115}$$

For a distribution of velocities $p(v)$, the result is a simple superposition of the monochromatic response as long as we stay with elastic scattering

$$P = <\cos\phi> = \int_0^\infty p(v)\cos\left(\gamma_L(B_1L_1 - B_2L_2)/v\right)\mathrm{d}v, \tag{3.116}$$

$$= \int_0^\infty p(\lambda)\cos(\epsilon\lambda)\,\mathrm{d}\lambda. \tag{3.117}$$

This expression, which defines the shape of the spin-echo curve, is simply the Fourier-transform[43] of the wavelength distribution of the beam in terms of the field integral difference expressed via the variable ϵ. For a 10–15% wavelength spread we can expect to observe in the order of 10 oscillations (see Fig. 3.21).

We will now consider the case of inelastic scattering. The velocity of the neutrons changes from v_1 in the first arm to v_2 in the second arm with

$$\hbar\omega = \frac{m}{2}\left(v_2^2 - v_1^2\right). \tag{3.118}$$

For a single scattering channel $v_1 \rightarrow v_2$, the resulting polarization is given by

$$P = \cos\left(\gamma_L\left[\frac{B_1L_1}{v_1} - \frac{B_2L_2}{v_2}\right]\right). \tag{3.119}$$

---

[42] In practice, this is done by a small extra coil in one of the spectrometer arms.

[43] Strictly speaking, we get the cosine-transform of the wavelength distribution. For symmetric functions the cosine transform is, however, identical to the Fourier transform.

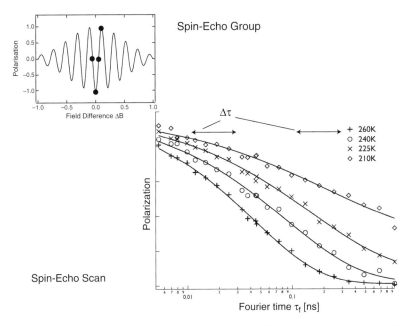

Fig. 3.21 Spin-echo experiment. The aim of a spin-echo experiment is to measure the loss of polarization as a function of the Fourier time τ_f. The fundamental scan parameter is the field integral in the two main precession coils (symmetric scan). For every setting of the spectrometer the various correction fields are tuned such as to produce a maximum spin echo for an elastic scatterer according to Eq. (3.109). The spin echo is determined by applying a small field difference ΔB to one of the spectrometer arms (asymmetric scan). This leads to an oscillating spin-echo response called the spin-echo group. The envelop of this group is the cosine transform of the wavelength distribution of the beam according to Eq. (3.117). In practice, one refrains from measuring the full spin-echo group as this would be too time consuming. Four points as indicated in the figure are sufficient to determine the amplitude and phase of the group with good precision. The spin-echo group delivers the polarization for a particular spectrometer setting. The presence of inelastic processes leads to a drop of the polarization. Relaxation processes become active polarization suppressors as their decay time falls into the window of the spin-echo time. In general, it is sufficient to consider the amplitude of the spin-echo signal. In the example given (the glass-former cis-Decalin measured on the spin-echo spectrometer IN11C at the ILL), the evolution of the polarization amplitude as a function of Fourier time and sample temperature shows that the relaxation time of the liquid becomes longer as the temperature decreases. The spin-echo technique integrates the relaxation function over rather broad time windows Δt (graphics courtesy of P. Fouquet and S. Eibl)

If we are dealing with a distribution of incoming neutrons and a multitude of scattering channels, then

$$P = \int_0^\infty \int_0^\infty p(v_1)T(v_1, v_2) \cos\left(\gamma_L\left[\frac{B_1L_1}{v_1} - \frac{B_2L_2}{v_2}\right]\right) dv_1 dv_2, \quad (3.120)$$

$$\equiv \int_0^\infty \int_0^\infty p(\lambda_1)T(\lambda_1, \lambda_2) \cos(\epsilon_1\lambda_1 - \epsilon_2\lambda_2) \, d\lambda_1 d\lambda_2, \quad (3.121)$$

with the transfer function $T(v_1, v_2)$ expressing the normalized probability that a neutron of velocity v_1 is scattered into a neutron with velocity v_2. If properly transformed into energy space, then $T(v_1, v_2)$ is nothing else but the scattering function $S(\hbar\omega)$ correctly integrated over the accessible Q-range. Although accurate, the above expression, which could be considered the master equation of spin-echo spectroscopy, does not lend itself easily to a physical interpretation. This holds even if we restrict ourselves to the spin-echo point for elastic scattering ($\epsilon_1 = \epsilon_2 = \gamma_L m B L / h$), in which case the above equation simplifies to

$$P = \int_0^\infty p(\lambda_1)d\lambda_1 \int_0^\infty T(\lambda_1, \lambda_2) \cos\left(\epsilon_1(\lambda_1 - \lambda_2)\right) d\lambda_2. \tag{3.122}$$

The situation changes if we restrict ourselves to small energy transfers. In that case we may approximate v_2 by $v_1 + \hbar\omega/mv_1$. This leads for a specific scattering channel to

$$P = \cos\left(\gamma_L \left[\frac{B_1 L_1}{v_1} - \frac{B_2 L_2}{v_1 + \hbar\omega/mv_1}\right]\right), \tag{3.123}$$

which to leading order in $\hbar\omega$ results in

$$P = \cos\left(\gamma_L \left[\frac{B_1 L_1 - B_2 L_2}{v_1} + \frac{B_2 L_2}{mv_1^3} \hbar\omega\right]\right). \tag{3.124}$$

Under the spin-echo condition for purely elastic scattering[44], the first term in the above equation is zero and we are left with

$$P = \cos\left(\gamma_L \frac{B_2 L_2}{mv_1^3} \hbar\omega\right) = \cos\left(\gamma_L \frac{m^2 B_2 L_2}{2\pi h^2} \lambda_0^3 \omega\right) = \cos(\tau_f \cdot \omega), \tag{3.125}$$

where we have introduced the Fourier time

$$\tau_f = \gamma_L \frac{m^2 B_2 L_2}{2\pi h^2} \lambda_0^3,$$
$$= 0.1863 \,[\text{ns}] \,(B_2 \cdot L_2) \,[\text{Tesla} \cdot \text{m}] \, \lambda_0^3 \,[\text{Å}^3]. \tag{3.126}$$

Summing over all scattering channels on the basis of Eq. (3.121) we obtain

$$P(\vec{Q}, \tau_f) = \;<\cos(\omega\tau_f)> \;= \frac{\int S(\vec{Q}, \hbar\omega)\cos(\omega\tau_f)d\omega}{\int S(\vec{Q}, \hbar\omega)d\omega} \equiv \tilde{I}(\vec{Q}, \tau_f). \tag{3.127}$$

[44] We may choose in principle any fixed value for the energy transfer for the spin-echo condition [59]. This is, for example, useful when measuring the width of an excitation with high precision using a combination of spin-echo and three-axis techniques. This is realized at the TRISP instrument at the FRM-II in Munich [60].

Here we have used the fact that for small energy transfers the momentum transfer is well approximated by its elastic value $\vec{Q} \approx k_i(\hat{k}_i - \hat{k}_f)$. The measured polarization $P(\vec{Q}, \tau_f)$ thus corresponds to the Fourier transform of the scattering function[45] normalized to the static structure factor $S(\vec{Q})$. This statement is true under the assumption that the energy integration covers all active fluctuation processes. We have discussed this problem already in Section 3.4.4. The measured polarization thus can be identified with the normalized *intermediate scattering function* $\tilde{I}(\vec{Q}, \tau_f)$. The Fourier time τ_f depends according to Eq. (3.126) linearly on the field integral. The wavelength enters to the third power. To attain large times, that is, slow relaxations, it is thus preferable to work with long wavelengths provided that the scattering triangle (see Section 3.2) for the $\vec{Q}$ to be investigated can be closed. The static structure factor is normally determined by performing a measurement at low temperatures where the inelastic processes are frozen out. If we take as a concrete example the Lorentzian scattering function given by Eq. (3.80), then the intermediate scattering function derived from the polarizability using Eq. (3.127) leads us back to the exponential relaxation of Eq. (3.79).

At this point we have not yet included the averaging over the wavelength distribution of the incoming beam. We have, therefore, not really gained with respect to the spectrometers working in $(\vec{Q}, \hbar\omega)$-space. Incorporating a wavelength spread in the incoming beam leads to the following expression for the polarizability:

$$P(\vec{Q}, \tau_f) = \int p(\lambda)\, \tilde{I}(\vec{Q}, \tau_f(\lambda))\, d\lambda. \qquad (3.128)$$

This integral corresponds effectively to a smearing of the Fourier time τ_f. For a Gaussian distribution of wavelengths centered on the nominal wavelength λ_0,

$$p(\lambda) = \frac{1}{\sqrt{2\pi a^2}} \exp\left(-\frac{(\lambda - \lambda_0)^2}{2a^2}\right), \qquad (3.129)$$

we can calculate the mean Fourier time and mean standard deviation as

$$\tau_0 = \int p(\lambda)\, \tau_f(\lambda)\, d\lambda \approx 1.02\, \tau_f(\lambda_0) \qquad (3.130)$$

$$\Delta\tau = \sqrt{\int p(\lambda)\, \tau_f^2(\lambda)\, d\lambda - \tau_0^2} \approx 0.3\, \tau_0, \qquad (3.131)$$

where the numerical values [41] correspond to a typical 15% wavelength band. The mean Fourier time is thus practically not affected by the time smearing. The relative Fourier time uncertainty $\Delta\tau/\tau_0$, however, is rather large and constant for all Fourier times. This is, however, generally not a problem as relaxation processes take place

---

[45] As outlined before this is true for symmetric functions. As we are, however, interested in small energy transfers, $S(\vec{Q}, \hbar\omega)$ is symmetric for all but the lowest temperatures.

on exponential scale and thus do not vary rapidly with time. The important feature of spin echo is not the time resolution but the dynamic range in particular toward very long times. Using strong solenoids in conjunction with short wavelengths, it is currently possible to attain Fourier times close to 0.5 μs. According to Eq. (3.82) this translates into an energy of 1 neV. It is completely unthinkable to obtain such a resolution for a classical scattering experiment in $(\vec{Q}, \hbar\omega)$-space by tailoring the distribution functions of the beam. We would be obliged to sacrifice for identical detector solid angle four to five orders of magnitude in flux.

So far we have treated the spin-echo instrument as ideal. In the real world we have to live with imperfections. Among these we find, for example, variations of the field integral as a function of the neutron trajectory. These variations are the more important the more divergent the beam. They can be corrected for by using special Fresnel coils [61]. Despite all these efforts fields and trajectories will never be ideal. It is thus important to measure the resolution as a function of Fourier time. The best candidate is the sample itself at it has exactly the same geometry and thus will produce the right beam dimensions and divergences after the scattering. To freeze out the slow motions the resolution measurement has to be done at low temperatures. Under these elastic conditions any drop in polarization $P_{\text{elastic}}(Q, \tau_f)$ must be attributed to instrumental imperfections. The physical signal must be taken with respect to this baseline. The corrected polarization arising from the sample $P_S(Q, \tau_f)$ is thus given by

$$P_S(Q, \tau_f) = \frac{P_{\text{measured}}(Q, \tau_f)}{P_{\text{elastic}}(Q, \tau_f)}. \tag{3.132}$$

Contrary to spectrometers operating in energy space, the process of resolution correction is thus not a complex de-convolution but a simple division. In most cases it is the resolution and not the technically achievable field integrals that limits the maximum Fourier times. We usually consider that this limit has been achieved when the polarization $P_{\text{elastic}}(Q, \tau_f)$ has dropped to a value of $1/e$.

When deriving the spin-echo relations we have tacitly assumed that the spin of the neutrons is not affected by the scattering process. This is strictly true only for coherent and isotope incoherent nuclear scattering. In the case of spin incoherent scattering, two-third of the events lead to a spin flip irrespective of the precise scattering geometry. These neutrons will not restore their original polarization traveling down the secondary spectrometer. Therefore, in a general experiment with the π-flipper active the polarization measured at any Fourier time will correspond to

$$P_{\text{measured}}(Q, \tau_f) = P_{\text{coh}} \frac{I_{\text{coh}}}{I_{\text{coh}} + I_{\text{inc}}}(Q, \tau_f) - \frac{1}{3} P_{\text{inc}} \frac{I_{\text{inc}}}{I_{\text{coh}} + I_{\text{inc}}}(Q, \tau_f), \tag{3.133}$$

where $\vec{P}_{\text{coh}}$ and $\vec{P}_{\text{inc}}$ denote the values for the polarization that one would obtain from hypothetical samples of purely coherent or incoherent character, respectively. The normalization of this function is no longer trivial as the coherent (collective) and incoherent (single particle) dynamic response of the sample will not necessarily take place on the same time scale. Therefore, the ratios of scattering intensities $I_{\mathrm{coh}}/I_{\mathrm{coh}} + I_{\mathrm{inc}}$ and $I_{\mathrm{inc}}/I_{\mathrm{coh}} + I_{\mathrm{inc}}$ are functions of $\vec{Q}$ and τ_f. The situation becomes even more complex in the case of magnetic scattering. For a paramagnetic sample the magnetic scattering can take over the role of a π-flipper of proper orientation, that is, it turns the polarization by $180°$ about a well-defined axis. In consequence, by doing a measurement with and without a π-flipper, it is possible to separate the dynamic paramagnetic from the coherent and isotope incoherent nuclear response. The investigation of ferro-magnetic samples is possible under certain conditions, but the discussion of these techniques is beyond the scope of this introduction.

The requirement of very precise field integrals makes it difficult to achieve very large solid angles with the conventional spin-echo technique. This is a real drawback, for example, compared to backscattering spectrometers with very large analyzers. The problem can be alleviated for medium-resolution spin-echo spectrometers by using a spherically symmetric design. This principle was successfully implemented with the SPAN [62, 63] spectrometer at the Hahn Meitner Institut in Berlin, Germany. A similar project called WASP [64] is part of the instrument upgrade program at the Institut Laue Langevin in Grenoble, France.

If the beam is pulsed, then it is a priori possible to combine time-of-flight with the spin-echo technique [65]. For slow processes the neutrons arriving in the detector as a function of time can be characterized according to their wavelength. The measured polarization as a function of time thus pertains to a wavelength band, the width of which can be controlled via the time structure of the beam. The technical challenge of time-of-flight spin echo resides in the fact that the fields in the solenoids as well as in the flippers and other correction elements have to follow the evolution of the wavelength as a function of time. This has, however, been demonstrated technically feasible [65]. The advantages of such a technique are obvious in the case of a pulsed source. Time-of-flight spin-echo spectroscopy can, however, equally be of great interest at continuous sources, in particular when the signal to be measured is spread over a large dynamic range.

3.5 Concluding Remark

It cannot be the aim of this introduction to neutron instrumentation to be comprehensive. There are many other important aspects, such as polarization analysis [56] or Laue diffraction [66], that have not even been touched upon. Imaging techniques will be dealt with in another book in this series. In all cases there is a large amount of literature that deals specifically with these concepts.

Acknowledgments I would like to thank Ken Andersen, Bob Cubitt, Charles Dewhurst, Björn Fåk, Henry Fischer, Peter Fouquet, Bernhard Frick, Arno Hiess, and Jacques Ollivier for a critical reading of the manuscript and Hannu Mutka for clarifying questions involving 2π. I would like to present my apologies to Anita, Carmen, Rafaëla, and Alexander for the evenings and weekends spent in front of the computer instead of having had fun with them.

References

1. J. Chadwick, Nature **129**, 312 (1932).
2. J. Chadwick, Proc. Roy. Soc. A **136**, 692 (1932).
3. *Neutron Data Booklet*, A-J. Dianoux and G. Lander (Eds.), Institut Laue-Langevin, Grenoble, France (2002).
4. A. Taylor et al., Science **315**, 1092 (2007).
5. J.M. Carpenter and W.B. Yelon, *Neutron Sources*, in *Neutron Scattering*, K. Sköld and D.L. Price (Eds.) *Methods of Experimental Physics* **23** A, Academic Press, New York (1986).
6. P.A. Kopetka, J.M. Rowe, and R.E. Williams, *Cold Neutrons at NIST*, Nuclear Engineering and Technology **38**, 427 (2006).
7. P. Ageron, Nuclear Instr. and Meth. A **284**, 197 (1989).
8. F. Mezei, J. Neutron Res. **6**, 3 (1997).
9. The ESS Project, Volume I-IV, Jülich, Germany (2002).
10. H. Maier-Leibnitz, Nukleonik **2**, 5 (1966).
11. H. Rauch and S. Werner, *Neutron Interferometry*, Oxford University Press, Oxford (2000).
12. B. Dorner, J. Neutron Res **13**, 267 (2005).
13. G.L. Squires, *Introduction to the Theory of Thermal Neutron Scattering*, Cambridge University Press, Cambridge (1978).
14. S.W. Lovesey, *Theory of Neutron Scattering from Condensed Matter*, Vol. 1: *Nuclear Scattering*, Vol. 2: *Polarization Effects and Magnetic Scattering*, Oxford University Press, Oxford (1984).
15. D.L. Price and K. Sköld, in *Neutron Scattering*, K. Sköld and D.L. Price (Eds.) *Methods of Experimental Physics* **23** A, Academic Press, New York (1986).
16. B. Dorner, Acta Cryst. A **28**, 319 (1972).
17. J. Felber, R. Gähler, R. Golub, and K. Prechtl, Physica B **252**, 34 (1998)
18. H. Schober, E. Farhi, , F. Mezei, P. Allenspach, K. Andersen, P.M. Bentley, P. Christiansen, B. Cubitt, R.K. Heenan, J. Kulda, P. Langan, K. Lefmann, K. Lieutenant, M. Monkenbusch, P. Willendrup, J. Saroun, P. Tindemans, and G. Zsigmond, Nuclear Instr. Methods A **589**, 34 (2008).
19. E. Fermi and W.H. Zinn, Phys. Rev. **70**, 103 (1946).
20. L. Dobrzynski, K. Blinowski, and M. Cooper, *Neutrons and Solid State Physics*, Ellis Horwood, New York, London, Toronto, Sydney, Tokyo, Singapore (1994).
21. A.I. Frank, Sov. Phys. Usp. **25**, 280 (1982).
22. C. Schanzer, P. Böni, U. Filges, and T. Hils, Nuclear Instr. Methods A **529**, 63 (2004).
23. D.P. Mitchell and P.N. Powers, Phys. Rev. **50**, 486 (1936).
24. W.H. Zinn, Phys. Rev. **71**, 752 (1947).
25. R. Currat, *Three-Axis Inelastic Neutron Scattering*, in *Neutron and X-ray Spectroscopy*, F. Hippert, E. Geissler, J.L. Hodeau, E. Lelièvre-Berna, and J.R. Regnard (Eds.), Springer, Grenoble (2006).
26. A. Magerl and V. Wagner (Eds.) *Focusing Bragg Optics*, Nuclear Instr. Methods A, **338**, 1-152 (1994).
27. RE. Lechner, R.v. Wallach, H.A. Graf, F.-J. Kasper, and L. Mokrani, Nuclear Instr. Meth A, **338**, 65 (1994).
28. V.F. Sears, *Neutron Optics*, Oxford University Press, New York (1989).
29. C.G. Darwin, Phil. Mag. **27**, 315, 675 (1914).
30. E. Fermi, J. Marshall, and L. Marshall, Phys. Rev. **72**, 193 (1947).
31. N. Holt, Rev. Sci. Instrum. **28**, 1 (1957).
32. H. Friedrich, V. Wagner, and P. Wille, Physica B, **156**, 547 (1989).
33. L. Dohmen, J. Thielen, and B. Ahlefeld, J. Neutron Res **13**, 275 (2005).
34. C.D. Clark, E.W. Mitchell, D.W. Palmer, and I.H. Wilson, J. Sci. Instrum. **43**, 1 (1966).
35. M. Kempa, B. Janousova, J. Sarouna, P. Flores, M. Böhm, F. Demmel, and J. Kulda, Physica B, **385-386** 1080 (2006).

36. M.J. Cooper and R. Nathans, Acta Cryst. **23**, 357 (1967).
37. A.K. Freund, Nuclear Instr. Methods **213**, 495 (1983).
38. B. Frick, *Neutron Backscattering Spectroscopy*, in *Neutron and X-ray Spectroscopy*, F. Hippert, E. Geissler, J.L. Hodeau, E. Lelièvre-Berna, and J.R. Regnard (Eds.), Springer, Grenoble (2006).
39. H. Mutka, Nuclear Instr. Methods A, **338**, 145 (1994).
40. R.E. Lechner, Proc. Workshop on Neutron Scattering Instrumentation for SNQ, Maria Laach, Sept. 3–5, 1984; R. Scherm and H. Stiller (Eds.), Jül-1954, p. 202 (1984).
41. R.E. Lechner and S. Longeville, *Quasielastic Neutron Scattering in Biology, Part I: Methods* in Neutron Scattering in Biology, Techniques and Applications, J. Fitter, T. Gutberlet, and J. Katsaras, (Eds.), Springer Verlag, Berlin (2006), pp. 309–354.
42. K. Lefmann, H. Schober, and F. Mezei, Meas. Sci. Technol. **19**, 034025 (2008).
43. G.E. Bacon, *Neutron Diffraction* 3rd edn., Clarendon, Oxford (1975).
44. H.E. Fischer, A.C. Barnes, and P.S. Salmon, Rep. Prog. Phys. **69**, 233 (2006).
45. W. Schäfer, Eur. J. Mineral. **14**, 263 (2002).
46. A.W. Hewat, Nuclear Instr. Methods **127**, 361 (1975).
47. C.G. Windsor, *Pulsed Neutron Scattering*, Taylor and Francis, London (1981).
48. K. Lieutenant, T. Gutberlet, A. Wiedenmann, and F. Mezei, Nuclear Instr. Methods A **553**, 592 (2005).
49. S.K. Sinha, E.B. Sirota, S. Garoff, and H.B. Stanley, Phys. Rev. B **38**, 2297 (1988).
50. J. Daillant and A. Gibaud *X-ray and Neutron Reflectivity: Principles and Applications*, Springer Verlag , Berlin (1999).
51. V. Lauter-Pasyuk, H.J. Lauter, B.P. Toperverg, L. Romashev, and V. Ustinov, Phys. Rev. Lett. **89** 167203 (2002).
52. F. Mezei, Z. Physik **255**, 146 (1972).
53. F. Mezei, *The Principles of Neutron Spin Echo* in *Neutron Spin Echo*, Lecture Notes in Physics, Vol. 128, Ed. F. Mezei, Springer Verlag, Berlin (1980).
54. F. Mezei, *Fundamentals of Neutron Spin Echo Spectroscopy* in *Neutron Spin Echo Spectroscopyy, Basics, Trends and Applications*, F. Mezei, C. Pappas, and T. Gutberlet (Eds.), Springer Verlag, Berlin, Heidelberg, New York (2003).
55. R. Cywinski, *Neutron Spin Echo Spectroscopy*, in *Neutron and X-ray Spectroscopy*, F. Hippert, E. Geissler, J.L. Hodeau, E. Lelièvre-Berna, and J.R. Regnard (Eds.), Springer, Berlin (2006).
56. W.G. Williams, *Polarized Neutrons*, Clarendon, Oxford (1988).
57. E. Lelièvre-Berna, P.J. Brown, F. Tasset, K. Kakurai, M. Takeda, L.-P. Regnault, Physica B **397**, 120 (2007).
58. M. Bleuel, F. Demmel, R. Gähler, R. Golub, K. Habicht, T. Keller, S. Klimko, I. Köper, S. Longeville, and S. Prodkudaylo, *Future Developments in Resonance Spin Echo* in *Neutron Spin Echo Spectroscopyy, Basics, Trends and Applications*, F. Mezei, C. Pappas, and T. Gutberlet (Eds.), Springer Verlag, Berlin, Heidelberg, New York (2003).
59. K. Habicht, R. Golub, R. Gähler, and T. Keller, *Space-Time View of Neutron Spin-echo, Correlation Functions and Phonon Focusing* in *Neutron Spin Echo Spectroscopyy, Basics, Trends and Applications*, F. Mezei, C. Pappas, and T. Gutberlet (Eds.), Springer Verlag, Berlin, Heidelberg, New York (2003).
60. T. Keller, K. Habicht, H. Klann, M. Ohl, H. Schneider, and B. Keimer, Appl. Phys. A **74** [Suppl.], S332 (2002).
61. T. Keller, R. Golub, F. Mezei, and R. Gähler, Physica B **241**, 101 (1997) and references therein.
62. C. Pappas, G. Kali, T. Krist, P. Böni, and F. Mezei, Physica B **283**, 365 (2000)
63. C. Pappas, A. Triolo, F. Mezei, R. Kischnik, and G. Kali, *Wide Angle Neutron Spin Echo and Time-of-Flight Spectrometer* in *Neutron Spin Echo Spectroscopyy, Basics, Trends and Applications*, F. Mezei, C. Pappas, and T. Gutberlet (Eds.), Springer Verlag, Berlin, Heidelberg, New York (2003).

64. R. Hölzel, P.M. Bentley, and P. Fouquet, Rev. Sci. Instrum. **77**, 105107 (2006).
65. B. Farago, *Time-of-Flight Neutron Spin Echo: Present Status* in *Neutron Spin Echo Spectroscopyy, Basics, Trends and Applications*, F. Mezei, C. Pappas, and T. Gutberlet (Eds.), Springer Verlag, Berlin, Heidelberg, New York (2003).
66. C.C. Wilson and D.A. Myles, *Single Crystal Neutron Diffraction and Protein Crystallography* in Neutron Scattering in Biology, Techniques and Applications, J. Fitter, T. Gutberlet, and J. Katsaras, (Eds.), Springer Verlag, Berlin (2006), pp. 21–42.

Part I
Applications: Earth Sciences

Chapter 4
Structural and Magnetic Phase Transitions in Minerals: In Situ Studies by Neutron Scattering

Simon A. T. Redfern and Richard J. Harrison

Abstract The application of neutron scattering to the study of structural and magnetic phase transitions in minerals is discussed. The advantages of neutrons for structural characterization of phase transitions are enumerated and compared with the data that might be obtained from X-ray methods. Elements that are difficult to distinguish by X-ray diffraction can show huge contrasts in neutron diffraction experiments; this contrast has been exploited in studies of site occupancies and intra-mineral partitioning of elements difficult to distinguish by other methods, such as Mg–Al and Fe–Mn pairs. Selected examples of the use of these methods in recent studies are outlined. These include the study of cation order–disorder phase transitions in minerals, ranging in complexity from rather simple silicate structures such as olivine and spinel (where ordering may occur between two sites) to more complex double-chain silicates (where partitioning studies by neutron diffraction have identified the trends over as many as four different crystallographic sites). The ability to build complex sample environments around the minerals under study has been beneficial in cases where extreme high temperatures (as great as 2000 K) are of interest, or where buffering of oxidation states is required. The magnetic moment of the neutron provides a unique tool for the study of the magnetic structures of oxide minerals, and the identification of magnetic ordering schemes in minerals such as magnetite were some of the first examples of the application of this method to magnetic minerals. The principles of magnetic scattering of neutrons are briefly outlined; and applications of magnetic studies by powder diffraction using both unpolarized and polarized neutrons are discussed, including recent studies of nanoscale hematite exsolution in ilmenite by polarized neutron scattering. Finally, the extension of the entire family of such studies in mineralogy to conditions pertinent to deep planetary interiors is considered as an example of the sophisticated high-pressure and high-temperature sample environment apparatus that has been developed for mineralogical studies.

S.A.T. Redfern (✉)
Earth Sciences, University of Cambridge, Cambridge, CB2 3EQ, UK
e-mail: satr@esc.cam.ac.uk

L. Liang et al. (eds.), *Neutron Applications in Earth, Energy and Environmental Sciences*, Neutron Scattering Applications and Techniques,
DOI 10.1007/978-0-387-09416-8_4, © Springer Science+Business Media, LLC 2009

4.1 Introduction

One of the major applications of neutron scattering in the earth and environmental sciences (in particular, mineral sciences) has been in the study of phase transitions. Neutron scattering methods provide unique insights into the origins and mechanisms of these processes and have enabled mineralogists to develop new models of phase transformation behavior. Furthermore, neutron scattering is a powerful tool for studying magnetic materials. The time–temperature–pressure dependence of processes such as phase transformations, exsolution, cation ordering, transitions in magnetic structure, and disordering in minerals has considerable potential geophysical, geochemical, and petrological importance.

Neutron scattering has proved invaluable in determining atomic occupancies over sites. The power of neutrons to discern anion occupancies of relatively light elements, including (most important) hydrogen, places neutron diffraction in a unique position. It is an essential tool in the arsenal of the mineralogist, and the use of such methods is likely to increase further for determining order–disorder phenomena and partial occupancies over sites. Take atomic ordering as an example. Substitutional order–disorder transformations and similar structural phase transitions are typically some of the most efficient ways a mineral can adapt to changing temperature, pressure, or chemical composition. Disorder of distinct species across different crystallographic sites at high temperature provides significant entropic stabilization of mineral phases relative to low-temperature ordered structures. Positional or orientational disordering can have similar drastic effects. For example, the calcite–aragonite phase boundary shows a significant curvature at high temperature, which is due to the disorder of CO_3 groups within the calcite structure, associated with an orientation order–disorder phase transition [1]. This leads to an increased stability of calcite with respect to aragonite over that predicted by a simple Clausius–Clapeyron extrapolation of the low pressure–temperature thermochemical data. Understanding of the structural characteristics of the phase transition in calcite developed in tandem with studies of the analogous transition in nitratine, $NaNO_3$; but it was not until Dove and Powell [2] carried out high-temperature neutron diffraction experiments on powdered calcite that there was direct experimental evidence linking the thermodynamic and structural nature of the transition in $CaCO_3$. The key to the success of their study was the fact that neutrons penetrate the entire volume of samples held at extreme conditions (in this case, very close to the melting temperature and under a confining CO_2 pressure). The latest neutron study of this phenomenon [3] shows that the rotational disorder of the CO_3 groups is analogous to Lindemann melting (Fig. 4.1). Similarly, neutron diffraction has been used to investigate molecular orientational disorder in ammoniated silicates to great effect [4].

Order–disorder in minerals may occur over a variety of length scales and via a number of mechanisms. Substitutional disorder is typically observed, usually of cations over shared sites (such as Al/Si order–disorder in aluminosilicates), but certain molecular groups may display orientational order–disorder behavior, for example, carbonate oxy-anions or ammonium ions in relevant phases. Most commonly, only the long-range order is considered, because this is what is most obviously

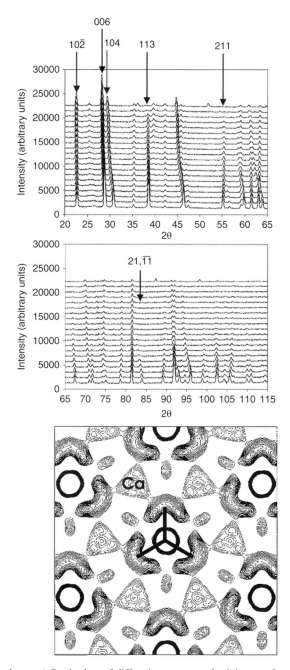

Fig. 4.1 (*Top* and *center*) Stack plots of diffraction patterns of calcite as a function of temperature, collected at the DUALSPEC diffractometer, Chalk River. (*Bottom*) Observed Fourier maps at height $z = 0$ of the hexagonal unit cell of calcite at 1189 K. The *open circles* show the carbon positions and the *solid line* shows the location of the ordered bond to oxygen. Oxygen distributions are banana-shaped around this vector, demonstrating the spread in orientations of the CO_3 group at this temperature [3]

observed by structural diffraction methods, either through the direct measurement of scattering amplitudes at crystallographic sites or bond-lengths in the solid or less directly through the measurement of coupled strains that may arise through the elastic interplay between the degree of order and the shape and size of the unit cell. Ordering over short-length scales can also be detected through neutron methods, including inelastic neutron scattering. Recently, computational methods have also been employed successfully to elucidate and illuminate experimental observations of ordering and to begin to separate and compare short- and long-range ordering effects [5–8].

Furthermore, we recall that neutrons have a magnetic moment and are scattered by the magnetic moments of atoms in a sample (Chapter 2). The cross section for magnetic scattering is sensitive to the relative orientation of the neutron magnetic moment, the atomic magnetic moment, and the scattering vector. This sensitivity allows magnetic structures to be determined from the intensities of magnetic diffraction peaks, in much the same way that crystal structures are determined from the intensities of nuclear diffraction peaks [9]. Small-angle neutron scattering and polarized-neutron reflectometry can yield magnetic information across a range of length scales spanning structure to nanostructure and microstructure. It can be applied to the study of magnetic nanoparticles, spin glasses, and magnetic multilayers [10–13]. Inelastic magnetic scattering can be used to probe a range of magnetic excitations, providing quantitative information about the magnetic exchange forces between neighboring spins, particularly valuable in understanding the magnetic properties of signature minerals in planetary interiors [14–22].

The time–temperature dependence of cation order–disorder and structural and magnetic transformations in minerals have considerable petrological importance. Measurements of time-, temperature-, and pressure-dependent phenomena by neutron diffraction methods generally employ in situ techniques, using high-temperature (T) furnaces, pressure (P) cells, combined P and T apparatus, and, potentially, sample environments that incorporate gas or vapor controls. Time-dependent studies usually involve the measurement of the response of a system to a perturbation in the external conditions, and they demand rapid data collection and high radiation fluxes (or at least counts into the detector system), which is often a limitation for neutron studies. Problems of dead time and data capture can become important for particularly rapid processes, and there are possibilities for measurements at pulsed sources that adopt a stroboscopic approach, although such a methodology has not yet been exploited for mineralogical studies (see [23] for an example of rapid transformation process studies in a ferroelectric system).

This chapter highlights the advantages of using neutron powder diffraction for the study of typical structural and magnetic transformations in minerals at high temperatures and pressures. It reviews recent progress, focusing particularly on the results obtained on non-convergent ordering in olivines, convergent order–disorder in Fe–Ti oxides and non-convergent ordering in spinels, and potential for the same approach to study the inter-site partitioning of cations in the amphibole structure. In particular, we find that in situ studies made possible by the use of neutron methods have proved very valuable in determining the processes responsible for these sorts of phase transitions at the temperatures and pressures at which they occur. The chapter finishes by suggesting further possible future routes to study ordering in mineral systems at high temperatures and high pressures using a novel apparatus designed to allow in situ high-T/P neutron diffraction.

4.2 Advantages of Neutrons for Mineralogical Studies

It is worthwhile to begin a discussion of the use of neutrons for studying transformation processes in minerals by first reviewing the roles and natures of diffraction, both of X-rays (a tool for mineralogists approaching a century now) and of neutrons (the theme of this volume), from periodic and aperiodic solids. These methods can be used to provide information about the atomic-scale structure of materials. Each has been used substantially in developing our understanding of the nature of materials' response to changes in variables such as temperature, pressure, and composition, important in interpreting phase stabilities and phase transitions in minerals. It is clear that the two techniques have different and somewhat complementary advantages. The differences are apparent from first considerations (Chapters 1 and 2). In contrast to X-rays, neutrons have significant mass (1.675×10^{-27} kg). But neutrons in thermal equilibrium adhere to a Maxwellian distribution of energies such that around room temperature there is a peak flux at around 25 meV, an energy that may be translated, by considering the kinetic energy of the neutron ($E = h^2/[2m\lambda^2]$), to a wavelength of $\lambda = 1.8$ Å. Furthermore, magnetic scattering results from the dipole–dipole interaction between the magnetic moment of the neutron and the magnetic moment of an atom. A neutron has a spin of 1/2 and generates a magnetic moment of $\gamma = -1.913\,\mu_N$, where μ_N is the nuclear magneton ($1\,\mu_N = 5.05 \times 10^{-27}$ Am^2). This moment is around a factor of 1000 smaller than that of an electron. Nonetheless, Bragg scattering of neutrons from crystals may be considered in much the same way as X-rays.

In X-ray diffraction, the X-rays are scattered by electrons surrounding atoms. The atomic scattering factor, f, is determined by summing the contributions from all electrons, taking into account the path difference of the scattered waves (Fig. 4.2); and the electron density is obviously spread out over the entire volume of the atom. Figure 4.2 shows that the phase difference between waves scattered from different parts of the atom increases with scattering angle or scattering vector $\mathbf{Q}$ (Fig. 4.3), where the modulus $Q = 4\pi \cdot \sin\theta/\lambda$ (when we apply this relation to Bragg scattering from lattice planes of spacing d [$\lambda = 2d \sin\theta$]), we clearly have $Q = 2\pi/d$, and this difference influences the $\mathbf{Q}$-dependence of f, $f(\mathbf{Q})$.

When we consider scattering of X-rays by atoms, we can think of scattering from the continuous distribution of electrons around the atom. If we denote the electron density as $\rho_{el}(\mathbf{r})$, the X-ray atomic scattering factor is given as

$$f(\mathbf{Q}) = \int \rho_{el}(\mathbf{r}) \exp(i\mathbf{Q} \cdot \mathbf{r}) d\mathbf{r}$$

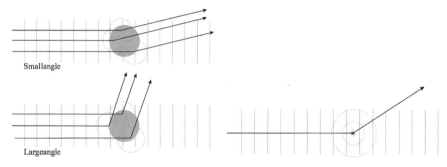

Fig. 4.2 The phase difference of waves scattered from a large scattering object increases with scattering angle (*left*). For scattering of radiation from small scattering objects (e.g., the scattering of neutrons from the nucleus), the path differences will be minimal (*right*) and the scattering strength will not suffer destructive interference at large angles

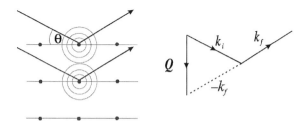

Fig. 4.3 The relationship between Bragg angle, θ, and scattering vector, Q, for diffraction from a lattice. Q may be defined as the difference between the wave vectors of the incoming and scattered rays, k_i and k_s

In the limit $\mathbf{Q} - 0$, where the X-rays are scattered without deflection, we have

$$f(\mathbf{Q} = 0) = \int \rho_{el}(\mathbf{r})d\mathbf{r} = Z$$

where Z is the total number of electrons in the atom or ion. For all atoms and ions of interest, $\rho_{el}(\mathbf{r})$ can be calculated using quantum mechanics in order to obtain the scattering factor. In practice there will not be an analytical function for $f(\mathbf{Q})$, and so for practical use the numeric values of $f(\mathbf{Q})$ are fitted to an appropriate functional form.

This fall-off, or form factor, results in attenuation of the diffracted intensity at high scattering angles, θ, for constant wavelength (λ) diffraction from crystalline solids, giving rise to the characteristic weakening of signal at high scattering vectors $\mathbf{Q}$ (Fig. 4.4). At zero scattering angle, all the electrons are in-phase and the atomic scattering factor is equal to the number of electrons. Therefore, the heavier the element, the higher the X-ray atomic scattering factor; this relationship leads to domination of X-ray scattering from a mineral by the heavier elements. It is, therefore, quite a challenge to obtain information about lighter atoms, such as hydrogen and other light elements, using X-ray diffraction. Furthermore, in considering ionic

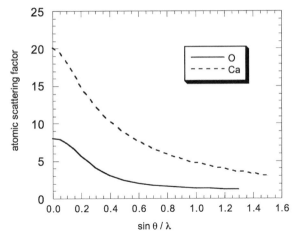

Fig. 4.4 The variation of X-ray atomic scattering factor with angle for oxygen and calcium atoms

compounds, as is relevant in much of mineralogy, many constituent atomic species are effectively isoelectronic and share the same scattering power (Chapter 1). In addition, there are inherent limitations in X-ray scattering because of the form factor: as scattering angle increases, the X-ray atomic scattering factor decreases (see Fig. 4.4) because of increasing destructive interference between X-rays scattered from electrons in different parts of the atom.

In application of these ideas to the X-ray scattering function, the width of the scattering function is given by the inverse of the atomic/ionic radius. Transferring these ideas to neutron scattering, where the neutrons are scattered by the nuclei, which are typically 10^{-5} times the size of the atom, the scattering function is so wide in Q that it can be treated as a constant for all values of Q of practical interest. This means that information in diffraction space out to high scattering vectors (small d-spacings) may be obtained without the problem of form factor attenuation.

The constancy of the neutron scattering length to high scattering vectors is particularly useful in powder diffraction studies where the increase in the number of observable reflections that can be obtained at good signal-to-noise levels means that temperature factors and occupancies may be refined with greater confidence. In addition, for neutrons, the scattering power (equivalent quantity to atomic scattering factor) varies irregularly across the periodic table (shown in Fig. 4.5 for natural abundance isotopes). It turns out that this is very important when it comes to considering how neutrons may be employed to study cation ordering processes. From Fig. 4.5 it can be seen that it is therefore easier to obtain information about lighter atoms, and even different isotopes of the same atom, using neutron diffraction.

Neutrons are also scattered by the magnetization density of an atom. Since the magnetic moment originates from the electrons, interference between neutrons scattered from different parts of the electron cloud causes the amplitude of magnetic scattering to decrease with increasing Q. This decrease is described by the magnetic

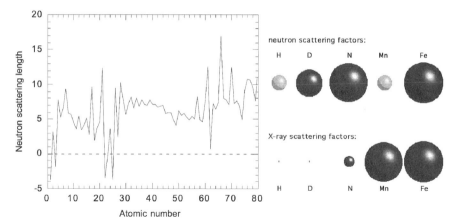

Fig. 4.5 (*Left*) The variation of mean coherent neutron scattering length (in fm, 10^{-15} m) with atomic number for natural abundance isotopes. (*Right*) These data are represented graphically for a few examples and compared with scattering factors for X-rays. We see that iron and manganese have sharply contrasting neutron scattering lengths (negative for manganese, positive for iron), but very similar X-ray atomic scattering factors

form factor $f(Q)$. Unlike the nuclear contribution, which remains constant as a function of Q, the magnetic contribution to a neutron diffraction pattern is restricted to small Q. The magnetic form factors vary with the valence state of the ion. One should be aware that the default form factors used by Rietveld refinement programs such as GSAS are often for the neutral atom. Ionic values of the form factor coefficients may have to be entered manually. In general, the periodicity of the magnetic structure will be different from that of the nuclear structure. In this case, the scattering vector takes the form

$$Q = 2\pi(H + k),$$

where k is referred to as the propagation vector. When the nuclear and magnetic unit cells are the same (i.e., $k = 0$), the nuclear and magnetic Bragg peaks coincide. The total intensity for unpolarized neutrons is then simply the sum of nuclear and magnetic contributions

It is worth reiterating here that the nuclear scattering length for neutrons may involve a spin-dependent component. While for many nuclei the nuclear spin is zero (and thus spin-dependent components of the nucleus–neutron interactions are absent), even when there is a spin the effects are relatively weak. Isotope effects are usually ignored. For many important elements, only one isotope occurs in significant quantities. For example, the natural abundances of the isotopes of oxygen—^{15}O, ^{16}O, and ^{17}O—are 1%, 98%, and 1%, respectively; thus we may assume (valid to a very good approximation) that the scattering lengths of all oxygen atoms in a structure are constant. There are, however, a few notable exceptions to this assumption. One such is nickel, for which there are two major isotopes, ^{58}Ni (relative abundance

68.3%) and ^{60}Ni (relative abundance 26.1%), with respective scattering lengths of 14.4 and 2.8 fm, with neither of these nuclei having a non-zero spin. Potential problems are associated with particular isotopes. For example, lithium is naturally 92.5% ^{7}Li, which absorbs neutrons weakly, but the 7.5% abundant ^{6}Li has an absorption cross section that is more than 20,000 times greater and has a strong influence on the overall absorption of lithium.

In certain cases the isotopic variability of neutron scattering cannot, therefore, be wholly ignored, and in some cases there is cause for using specific isotopes of elements of interest when synthetic samples are to be studied. Perhaps, the most significant of these in the application of neutrons to mineralogy is the behavior of hydrogen (^{1}H), in which the spin-dependence of the nucleus–neutron interaction (or proton–neutron interaction) is very important; this is even more significant because ^{1}H is more than 99.9% abundant in natural hydrogen. The proton and neutron both have spin $\frac{1}{2}$ and can be aligned in four ways. The scattering function for atoms like hydrogen depends upon averages of the scattering lengths for the parallel and anti-parallel configurations of spins that can occur. Deviations from these averages lead to a high incoherent background in diffraction experiments involving hydrous phases (Fig. 4.6). This can be avoided by substituting deuterium for hydrogen ^{1}H, perhaps the most common example of the use of isotopic substitution employed in neutron diffraction studies. Such steps are strongly recommended whenever possible in powder diffraction to limit the background due to incoherent scattering.

In transferring these concepts to geosciences, we can first observe that the application of neutron scattering to the study of earth materials is rather less mature than

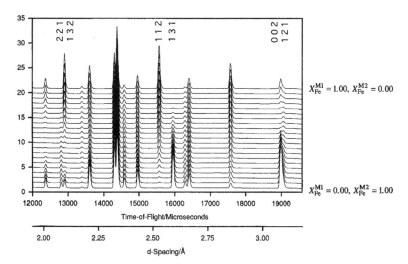

Fig. 4.6 Simulated neutron powder diffraction patterns of (Fe, Mn)$_2$SiO$_4$ olivines (kneblites) as a function of M-site occupancy over the two sites of the olivine structure. Note the large changes in intensity of certain reflections due to the large contrast in coherent scattering length for Mn and Fe

its use in other branches of the physical sciences; however, this volume is testament to the increasing recognition now being given to the role neutron scattering may play in solving problems of mineral behavior. Neutrons provide a unique probe for the study of minerals, since their wavelengths (sub-Ångstrom to tens of Ångstroms) are of the same order of magnitude as the inter-atomic spacings in minerals and correspond to energies similar to many electronic and atomic processes. The fact that the scattering power of neutrons depends not upon the number of electrons surrounding the atom but rather upon the nuclear cross section makes neutrons particularly suitable for studying order–disorder and mixing processes in minerals, where chemically similar atoms (which may also therefore tend to have similar X-ray scattering powers) substitute on crystallographic sites. For example, the nuclear scattering contrast for magnesium and aluminum is almost five times greater than the contrast in X-ray atomic scattering factors. For manganese and iron, the difference is even more marked: the neutron scattering contrast is more than 36 times greater than the X-ray scattering contrast, since manganese has a negative neutron scattering length while that of iron is large and positive, whereas their X-ray atomic scattering factors differ in magnitude by less than 4%.

Thus elements that are difficult to distinguish by X-ray diffraction can show huge contrasts in neutron diffraction experiments. For example, one anticipates significant changes in the intensities of reflections within a powder diffraction pattern of a mineral in which manganese and iron may interchange between sites (Fig. 4.6). Of particular additional interest to earth scientists is the possibility of detecting hydrogen (or more correctly, deuterium) in crystal structures using neutron diffraction, a task that is well nigh impossible by X-ray diffraction.

Aside from strong scattering contrasts between chemically similar pairs of substituting atoms, the characteristics of neutron scattering provide further advantages that may be exploited in studies of high-temperature order–disorder. As the fall-off in scattering power with scattering vector Q is negligible, data may be obtained out to high-scattering vectors (corresponding to small d-spacings), and complex structures may confidently and routinely be refined from powder data using Rietveld methods with high precision and accuracy. Combining these facets of neutron powder diffraction (using fixed geometry time-of-flight methods) with the fact that stable sample environments may be constructed around the sample without the worry of overly attenuating the incident and diffracted beams, the conclusion is apparent that neutron powder diffraction is a powerful tool for observing structural changes at extremes of temperature. This much has been demonstrated by recent studies of the temperature dependence of inter-site partitioning of metal cations in olivines and spinels [24–29] and the work on more complex hydrous silicates that is presented below. That said, the interaction of neutrons with matter is very weak. This has some positive consequences as well as the inherent disadvantage that neutron scattering is generally weaker than for X-rays. In particular, we find that neutrons probe the bulk of samples (penetrating 1 cm or more through the sample), they do not damage the sample, they will cause only a small perturbation on the system, and systems respond linearly so that neutron scattering theory is very quantitative.

4.3 Examples

4.3.1 Non-Convergent Cation Order–Disorder in Olivines and Spinels

As early as 1983, Nord and co-workers had employed neutron diffraction for the study of Ni–Fe cation distributions in the olivine-related phosphate sarcopside [30, 31]. The temperature dependence of non-convergent cation exchange between the M1 and M2 octahedral sites of silicate olivines (Fig. 4.7) has also been the subject of a number of neutron diffraction studies, from the single-crystal studies of members of the forsterite–fayalite solid solution [32–35] to powder diffraction studies of the same system [36]. The strong contrast between manganese (negative scattering length) and other cations has led to interest in the Fe–Mn, Mg–Mn, and Mg–Ni systems as model compounds [24–27,37], but there have also been studies of Fe–Zn olivines [38] and most recently the Co–Mg system [39]. Millard et al. [40] used neutron diffraction to determine cation distributions in a number of germanate olivines.

The high-temperature behavior of Fe–Mg order–disorder appears to be complicated by crystal field effects and M-site vibrational modes, which influence the site preference of Fe^{2+} at high temperature for either M1 or M2 [35, 36]; but the cation exchange of the Fe–Mn, Mg–Mn, and Mg–Ni olivines is dominantly controlled by

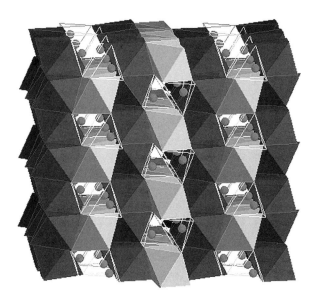

Fig. 4.7 The structure of olivine can be thought of as edge-sharing slabs of M1 (*dark*) and M2 (*light*) octahedra, shown here aligned vertically with a completely ordered arrangement of cation occupation. Upon heating, exchange between these sites may occur to give rise to disordered configurations. At no point do the two sites become symmetrically equivalent, and hence the disordering is non-convergent

size effects: the larger M2 site accommodating the larger of the two cations in each pair (manganese or nickel, in these cases).

In all of the recent in situ experiments, the use of time-of-flight neutron powder diffraction allowed the measurement of states of order at temperatures in excess of 1000°C under controlled oxygen fugacities (especially important given the variable oxidation states that many of the transition metal cations of interest can adopt). In the majority of the powder diffraction studies mentioned above, diffraction patterns were collected on the POLARIS time-of-flight powder diffractometer at the ISIS spallation source [41]. The diffraction patterns of the Fe–Mn and Ni–Mn olivines [24–27] were collected in four 30-min time bins over 2 h at each isothermal temperature step on heating, and over a single 30-min period on cooling. Diffraction patterns of the Mg–Mn olivine sample were collected over 1-h time intervals at each isothermal step on heating. Thus time–temperature pathways were investigated. Data were collected to rather large scattering vectors, corresponding to refineable information d-spacings at 0.5 Å or less (Fig. 4.8). Structural data were then obtained by Rietveld refinement of the whole patterns, giving errors in the site occupancies of around 0.5% or less. The low errors in refined occupancies result principally from the fact that the contrast between manganese (with a negative scattering length) and the other cations is very strong for neutrons (cf. Fig. 4.6).

All experiments showed the same underlying behavior of the degree of order as a function of temperature. This can be modeled according to Landau [42] expansion for the excess Gibbs free energy of ordering, of the type

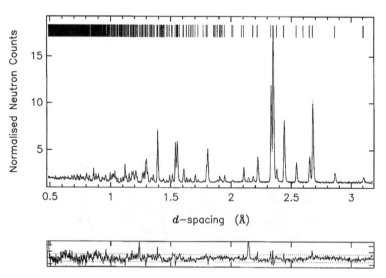

Fig. 4.8 Time-of-flight neutron powder diffraction pattern of $(Fe_{0.3}Mn_{0.7})_2SiO_4$. Vertical bars represent the positions of reflections. The difference between the fit and the experiment is shown in the lower part of the figure (amplified: the peak at 2.1 Å is due to scattering from the vanadium sample can). Data down to 0.5 Å yield useable information and aid the refinement of site occupancies across the structure [27]

$$\Delta G = -h\eta + \frac{a}{2}(T - T_C)\eta^2 + \frac{b}{4}\eta^4,$$

where h, a, b, and T_c are material-dependent parameters and an order parameter, η, describes the degree of cationic order–disorder over the two sites. This expression, chosen to describe the free energy change due to ordering, is formally equivalent to the reciprocal solution model at lowest order, although the manner in which free energy is partitioned between entropy and enthalpy differs between the two approaches [43, 44]. The Landau formulation essentially treats entropy as vibrational rather than configurational. Kroll et al. [44] have shown that the addition of a configurational entropy term models the entropy at high η more accurately, in particular for the non-convergent ordering behavior of magnesium and iron on M-sites in pyroxene, a topic that has been the subject of recent study by Gatta et al. [45].

In each case studied by neutrons (Fig. 4.9), the order parameter remains constant at the start of the heating experiment and then increases to a maximum before following a steady decline with T to the highest temperatures. This general behavior reflects both the kinetics and the thermodynamics of the systems under study: at

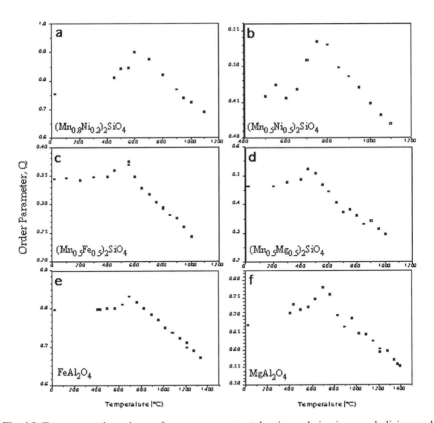

Fig. 4.9 Temperature dependence of non-convergent metal cation-ordering in several olivines and spinels, all measured by Rietveld refinement of neutron powder diffraction data

low temperatures, the samples are not in equilibrium and the results of the refinements reflect the kinetics of order–disorder; at high temperatures, the states of order are equilibrium states, reflecting the thermodynamic drive toward high-temperature disorder. The initial increase in order results from the starting value being lower than equilibrium; as soon as the temperature is high enough for thermally activated exchange to commence (on the time scale of the experiments), the occupancies of each site begin to converge toward the equilibrium order–disorder line. Using Ginzburg–Landau theory, which relates the driving force for ordering to the rate of change of order, one can obtain a kinetic description of the expected t-T-η pathway that relates to the thermodynamic description of the non-convergent disordering process:

$$\frac{d\eta}{dt} = -\frac{\gamma \exp(-\Delta G^*/RT)}{2RT}\frac{\partial G}{\partial \eta}$$

Since the low-temperature data, which lie below the thermodynamic disordering curve, give information about the kinetics of order–disorder, and the high-temperature data on the equilibrium ordering curve provide the thermodynamic description of the process, the entire ordering process may be derived from a single neutron diffraction experiment [26]. In other words, the cation occupancies provided by these sorts of time–temperature in situ measurements provide data on the free energy surface for the order–disorder process; and the minima of this surface denote the equilibrium ordering behavior, while its gradient is indicative of the kinetic behavior.

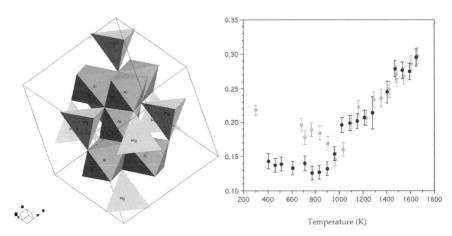

Fig. 4.10 (*Left*) Al–Mg order–disorder in spinel is non-convergent, involving exchange between the tetrahedral and the octahedral sites occupied by magnesium and aluminum, respectively, in fully ordered normal spinel. (*Right*) Data collected by in situ neutron powder diffraction of $MgAl_2O_4$ spinel, demonstrating the increased disorder (occupancy of aluminum into tetrahedral sites) on heating (*dark data points*). The cooling path (*lighter data points*) in this experiment resulted in a more ordered spinel at the end of the experiment, since cooling in the diffractometer was slower than in the original synthesis quench [28]

Similar behavior has been observed in spinels (Fig. 4.10), on which a large number of studies have been carried out exploiting the scattering characteristics of neutrons. Neutrons have been valuable for determining Mg–Al distributions, since atoms in this pair have very similar X-ray atomic scattering factors but quite different neutron scattering lengths. Neutron scattering studies of this phase date back to the mid-1970s, with the investigation by Rouse et al. [46]. Peterson et al. [47] were the first to carry out a time-of-flight powder diffraction study of the nature of disordering in this phase, and their investigations have been extended and built upon in subsequent studies at higher T as well as high P [28, 48, 49].

These studies of cation ordering in olivines and spinels have shown that, in most cases, the degree of cation order measured at room temperature is an indication of the cooling rate of a sample, rather than the temperature from which it has cooled. Calculated η-T cooling pathways for an Fe–Mn olivine are shown in Fig. 4.11, in which it is shown that variations in cooling rate over 13 decades might be ascertained from the degree of order locked in to room temperature.

The influence of processes of intra-mineral partitioning (order–disorder) on inter-mineral partitioning in olivine solid solutions has been pointed out by Bish [50]. Data obtained by neutron powder diffraction on the composition dependence of Ni–Mg ordering in olivines [51] illustrate his argument. In Fig. 4.12, the results are presented in terms of a distribution coefficient, K_D, for disorder, which relates to the exchange reaction $Ni^{(M2)}Mg^{(M1)}SiO_4\ Mg^{(M2)}Ni^{(M1)}SiO_4$, such that $K_D = [Mg^{(M2)}Ni^{(M1)}]/[Ni^{(M2)}Mg^{(M1)}]$. We see that the magnesium-rich sample shows a higher degree of order, with nickel ordering onto M1.

If the partitioning of nickel between, say, a melt and olivine is considered, then equations can be written, which depend upon the state of order of nickel in olivine.

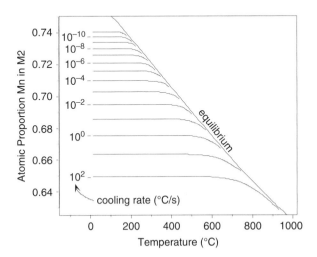

Fig. 4.11 Calculated cooling paths over 13 decades of cooling rate showing the dependence of the room-temperature site occupancy of FeMnSiO$_4$ on cooling rate. The room-temperature site occupancy given by neutron diffraction is a direct measure of the cooling rate of the sample

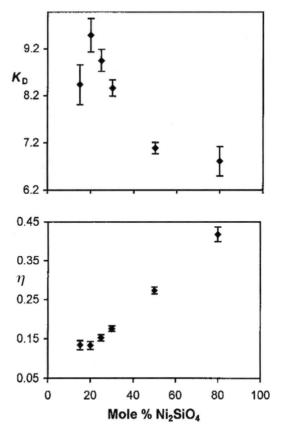

Fig. 4.12 Trends for K_D and degree of order η vs. bulk composition at room temperature in $(Ni,Mg)_2SiO_4$ olivines, measured by neutron powder diffraction [51]

For a disordered sample,

$$2NiO_{liquid} + Mg_2SiO_{4\,olivine} \Leftrightarrow 2MgO_{liquid} + Ni_2SiO_{4\,olivine};$$

but for an olivine in which all the nickel resides in M1, this must be rewritten as

$$NiO_{liquid} + Mg^{(M1)}Mg^{(M2)}SiO_{4\,olivine} \Leftrightarrow MgO_{liquid} + Ni^{(M1)}Mg^{(M2)}SiO_{4\,olivine}.$$

Thus for a disordered olivine, the activity is simply equal to the molar proportion of the $NiSi_{0.5}O_2$ component, $X[NiSi_{0.5}O_2]$; but for an (intermediately) ordered olivine, the system deviates from Raoult's law. Neutron powder diffraction has been used to measure such changes in K_D with composition directly from determinations of the site scattering (and hence cation occupancies) at each site (Fig. 4.12). It is clear that the extent of ordering must be quantified in order to generate accurate models of mineral behavior. The in situ studies that have been performed in recent years have allowed the temperature dependence of this ordering to be determined accurately to high T. In these cases, in situ study has been essential, since high-temperature disordered states are generally non-quenchable as a result of the fast kinetics of cation exchange in olivines and spinels and the unavoidable re-equilibration of samples on quenching from annealing conditions. Thus neutron diffraction techniques are invaluable for directly determining the long-range ordering characteristics of these important rock-forming minerals.

4.3.2 Cation Ordering in Crystal-Chemically Complex Minerals: Pushing the Limits of Powder Data

Consider two studies of cation ordering that show the strengths of neutron diffraction in tackling low-contrast element pairs in crystal-chemically complex minerals. In these studies, the ordering involves Mg/Al on octahedral sites and Al/Si on tetrahedral sites. The neutron (scattering-length) contrasts for Mg/Al and Al/Si are not as great as for (say) Mn/Fe or Mn/Mg exploited in olivines (above), but are rather better than is the case for X-rays (scattering factors). The possibility arises, therefore, of being able to determine site occupancies directly from site scattering; this is not possible by X-ray diffraction, where recourse must be made to mean bond length arguments. While these studies have not resorted to in situ methodologies, they mark the types of complex mineralogical systems that one would ultimately wish to investigate at high pressures and temperatures. At present, they mark a goal for this type of study.

4.3.2.1 Mg/Al Ordering in Dioctahedral Micas

Micas (hydrous sheet silicates) with a high phengite component, $K[MgAl][Si_4]O_{10}(OH)_2$, are characteristic of high pressures of formation (>1 GPa). These micas are "dioctahedral," having two out of three sites in the octahedral sheet occupied by divalent or trivalent cations, commonly Mg^{2+}, Fe^{2+}, and Al^{3+}. Various polytypes exist, which arise from different ways of stacking the 2:1 (tetrahedral:octahedral sheet ratio) slabs. A segment of the trigonal $P3_12$ phengite structure ($3T$ polytype) is shown in Fig. 4.13. There are two non-equivalent octahedral sites (M2, M3) and two non-equivalent tetrahedral sites (T1, T2). The site topology of the octahedral layer consists of six-membered rings of alternating M2 and M3 sites surrounding a central vacant octahedral site. Similarly, the tetrahedral layers comprise six-membered rings of alternating T1 and T2 tetrahedra. The distribution of cations over the octahedral sites may be a function of pressure and/or temperature and may also be correlated with polytype.

The long-range ordering of octahedral cations in these micas is not well understood. This uncertainty is largely a consequence of the very similar X-ray scattering factors of magnesium and aluminum. Single-crystal X-ray structure refinements of phengitic micas imply a contradiction between octahedral-site occupancies (Mg, Al) refined from site-scattering values (electrons per site) and those

Fig. 4.13 The dioctahedral structure of phengite, viewed parallel to the triad axis. M2 and M3 octahedra are marked and lie between two tetrahedral sheets of T1 (*light*) and T2 (*dark*) tetrahedra. The lower sheet is omitted for clarity

indicated by <M–O> bond lengths. For example, in the single-crystal X-ray study by Amisano-Canesi et al. [52] of a typical high-pressure phengitic mica of composition $K_{0.9}[Mg_{0.58}Al_{1.43}][Si_{3.57}Al_{0.43}]O_{10}(OH)_2$, site-scattering values require all magnesium to be at M2 (with some aluminum) and M3 to be fully occupied by aluminum, whereas the <M2–O> average bond length is considerably shorter than <M3–O>. The site-scattering values were not well constrained because of the very similar X-ray atomic scattering factors of magnesium and aluminum. The tetrahedral-site occupancies were also not well defined because of the low contrast between aluminum and silicon for X-rays and the variable correlations between <T–O> and Al/Si on sites. The scattering contrasts between aluminum and magnesium and between aluminum and silicon are better for neutrons than X-rays. Hence, in principle, neutron diffraction could resolve the site occupancy problem of phengites. With this in mind, Pavese et al. [53] used neutron powder diffraction to investigate cation ordering in the same phengite as studied by Amisano-Canesi et al. [52]. The powder route was used because crystals of phengite large enough for single-crystal neutron diffraction are unavailable. For micas, neutron powder diffraction has the added advantage of reducing preferred orientation, because a large volume of powder (3–4 cm^3) can be loosely packed into the sample holder. Using a starting model for Rietveld refinement in which (a) the tetrahedral bond lengths were constrained to be those found in the single-crystal X-ray study of Amisano-Canesi et al. [52], and (b) refined site occupancies had to be consistent with the total octahedral composition as given in the chemical formula, it proved possible to refine the occupancies of M2 and M3 sites [53]. It did not prove possible to refine occupancies for the T1 and T2 sites (unstable refinements). The neutron experiment clearly showed that, in contrast to all X-ray refinements of phengites, magnesium orders at M3 (not M2) and M2 are fully occupied by aluminum. The refined occupancies they obtained at 293 K are M2(Al) = 1.06(5) and M3(Al) = 0.36(5). The <M2–O> and <M3–O> bond lengths of 1.964(5) Å and 1.979(5) Å are qualitatively consistent with these occupancies. At 893 K the occupancies are M2(Al) = 0.95(6) and M3(Al) = 0.47(6); so, within error, there was little or no Mg/Al disordering on the time scale of the experiment. This was the first experiment that indicated neutron powder diffraction is capable of providing valuable information on octahedral site occupancies in such complex silicates, information that was unobtainable at that point by X-ray diffraction.

Subsequent high-temperature studies [54] on phengite (Fig. 4.14) have focused on the behavior of the hydroxyl group and have been able to correlate infrared observations with structural changes at the hydroxyl site at extreme temperatures using constant wavelength angle dispersive neutron powder diffraction at Institut Laue-Langevin, D2B. Here, a large-angle dispersive bank of detectors is arranged on an air bed and sweeps through a small range of angles to collect an entire diffraction pattern relatively rapidly, in contrast to the fixed arrangement of detectors and wide distribution of neutron energies at the ISIS spallation source. Recent improvements ("super" D2B) have increased the number, resolution, and height of the detectors so that the detector bank operates as a quasi-2D system, resulting in an order of magnitude increase in counts at the detector, essential for the investigation of subtle effects in complex structures.

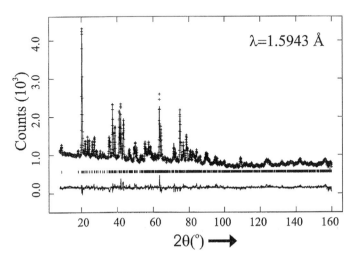

Fig. 4.14 Constant wavelength neutron powder diffraction data for phengite (at 100°C) collected at D2B, ILL [54]. Data were collected at 100°C intervals up to 1000°C. Note the high background that results from the incoherent scattering off hydrogen in this natural (undeuterated) sample

4.3.2.2 Mg/Al and Al/Si Ordering in Amphiboles

Amphiboles are a major group of hydrous minerals, occurring in a wide range of geological environments and Earth history. Their chemistry and cation ordering behavior is often diagnostic of the conditions of crystallization (pressure, temperature, cooling rates). Therefore, they have enormous potential as geological "indicators," provided that their cation ordering behavior can be quantified. However, these minerals are very complex structurally and chemically. A very successful approach to understanding the major crystal–chemical principles controlling cation ordering in amphiboles and their pressure–temperature stability has been to use synthetic analogues, in which the complex chemistry of natural amphiboles is modeled using key condensed phases in systems of components such as Na_2O-CaO-MgO-Al_2O_3-SiO_2-H_2O. Unfortunately, synthetic amphibole crystals are almost always too small for single-crystal X-ray structure refinement. Experience has shown that Rietveld refinement of X-ray powder diffraction data does not give reliable site occupancies unless very contrasting element pairs are involved (e.g., Mg/Co, Mg/Si, Ga/Si). These compositions are often of little direct relevance to natural systems. Subtle problems such as Mg/Al and Al/Si ordering are certainly beyond the capabilities of X-ray powder diffraction because occupancies from site scattering are likely to be unresolved, and bond lengths are not sufficiently well constrained to derive meaningful occupancies from established bond length vs. site occupancy relationships.

Al/Si and Mg/Al ordering in the geologically important high-temperature amphibole pargasite, $NaCa_2[Mg_4Al]$ $[Si_6Al_2]O_{22}(OH)_2$, has been studied by neutron powder diffraction. There are four T1 and four T2 tetrahedral sites and two M1, two M2, and one M3 octahedral sites per formula unit. Cation ordering involves the distribution of $4Mg + 1Al$ over the five octahedral sites and $2Al + 6Si$ over the eight tetrahedral sites. Pargasite has space group $C2/m$ and 15 atoms in its asymmetric unit. Therefore, it presents a considerable challenge to Rietveld refinement if the state of Mg/Al and Al/Si order is to be determined. It is one of the most structurally complex minerals yet studied by neutron powder diffraction. To give an idea of the experimental challenge, it is worth noting that pargasite was synthesized hydrothermally at 0.1 GPa, 1200 K, but that the sample size for neutron diffraction was 2.4 g, which is large from a synthesis viewpoint but modest for neutron diffraction. The neutron experiment was done at 295 K and lasted 16 h. The experimental, simulated, and difference patterns are shown in Fig. 4.15.

Notice that the powder diffraction pattern of pargasite (Fig. 4.15) shows many well-resolved peaks below 1 Å. However, it did not prove possible to refine site occupancies of tetrahedra and octahedra. Evidently, the scattering contrast for Al/Si and Mg/Al is not high enough to enable this to be done for this complex structure, compared with simple structures such as $MgAl_2O_4$ spinel for which site occupancies can be found from site-scattering values [28]. However, the bond lengths are very well constrained (±0.002 Å to ±0.003 Å) and allow occupancies for tetrahedral and octahedral sites to be deduced using the well-defined correlations between site occupancy and mean T1–O and M–O bond lengths [56, 57]. The standard errors on bond lengths obtained from the neutron powder diffraction data approach those of single-crystal refinements and inter-nuclear distances can be determined very

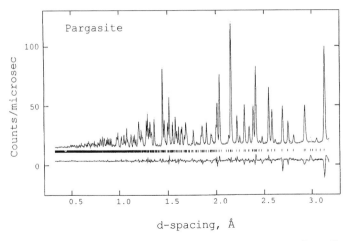

Fig. 4.15 Time-of-flight powder diffraction pattern of the amphibole pargasite, collected at the POLARIS diffractometer, ISIS. Note the high-quality data below 1 Å d-spacing [55]

precisely, even from powder data. It was found that Al orders at M2 and M3 (and not M1) in both natural and synthetic pargasites, indicating that they share a common crystal chemistry. Nonetheless, the fact that the distribution of Al is different in natural pargasites and synthetics may point to fundamentally different growth mechanisms. Hence, neutron powder diffraction has provided important new information about the relationship between cation distributions in natural amphiboles and their synthetic analogs.

4.3.3 Magnetic Structure and Transitions in Minerals

Two studies of magnetic ordering processes are considered, which show the strengths of neutron scattering for this class of mineral behavior.

4.3.3.1 Magnetite (Fe_3O_4)

Magnetite at room temperature has the cubic inverse spinel structure, with an Fe^{3+} ion occupying one tetrahedral site and Fe^{2+} and Fe^{3+} ions occupying two octahedral sites per formula unit. Néel [58] proposed a ferrimagnetic structure for magnetite, with ferromagnetic ordering of moments within tetrahedral and octahedral sub-lattices and anti-ferromagnetic ordering of the tetrahedral sub-lattice with respect to the octahedral sub-lattice. Since the tetrahedral and octahedral sites are not symmetry-related to each other, this ordering scheme yields a magnetic unit cell that is equal to the nuclear unit cell ($k = 0$). Figure 4.16a shows a comparison of the powder X-ray and neutron diffraction patterns of magnetite at room temperature [59]. Neutron peaks occur in the same positions as the X-ray peaks as a result of the coincidence of magnetic and nuclear scattering contributions for $k = 0$ (Fig. 4.17a).

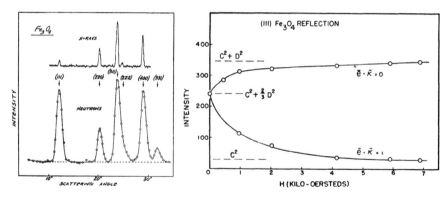

Fig. 4.16 (**a**) Comparison between X-ray and neutron diffraction patterns of magnetite at room temperature [59]. The large difference in the (111) intensities is caused by the magnetic contribution to the neutron diffraction pattern. (**b**) Plot of the (111) neutron intensity as a function of magnetic field applied either perpendicular (*upper curve*) or parallel (*lower curve*) to the scattering vector. C^2 and D^2 refer to the nuclear and magnetic contributions to the peak, respectively. The magnetic contribution is reduced to zero when the sample is saturated parallel to the scattering vector

The magnetic contribution to the neutron diffraction pattern is evident from the enhanced intensity of the (111) peak. For collinear magnetic structures, the intensity of the magnetic Bragg peak depends on α, the angle between the magnetic structure factor, $M(Q)$, and the scattering vector. Here, $M(Q)$ is given by:

$$M(Q) = p \sum_{j=1}^{n_m} f_j(Q)\mu_j e^{iQ.r_j},$$

where $M(Q)$ is a vector and the sum involves only the n_m magnetic atoms in the unit cell, each with moment μ_j in Bohr magnetons. The intensity of the Bragg reflection is determined by the magnetic interaction vector, $M_\infty(Q)$, defined as the component of $M(Q)$ perpendicular to Q.

Structure factor calculations indicate that the magnetic contribution to the intensity of the (111) peak in magnetite is a factor of 30 higher than the nuclear contribution. Figure 4.16b illustrates the effect of an applied magnetic field on the intensity of the (111) peak. In zero field, the intensity is given by $C^2 + 2/3 D^2$, where C and D refer to the nuclear and magnetic structure factors, respectively, and the factor of 2/3 is the average value of $\sin^2(\alpha)$ for a cubic material (a consequence of powder averaging). The intensity of the peak reduces to C^2 as the sample is gradually saturated in a direction parallel to the scattering vector. This is a direct consequence of the fact that M_∞ is zero when the magnetic moments lie parallel to the scattering vector Q. Conversely, the intensity rises to $C^2 + D^2$ when the sample is saturated in a direction perpendicular to Q. Such measurements can be used to study the mechanisms of magnetization reversal in multi-domain and single-domain magnetite powders.

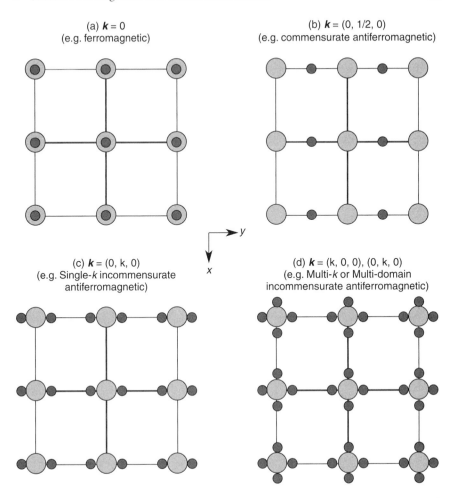

Fig. 4.17 Schematic illustration of the relationship between nuclear (*light grey*) and magnetic (*dark grey*) diffraction peaks for a range of different magnetic structures. (**a**) For $k = 0$ structures, the nuclear and magnetic peaks coincide. (**b**) If the magnetic cell is a supercell of the nuclear structure, pure magnetic reflections appear as superlattice peaks between the nuclear reflections. (**c**) Single-k incommensurate structures are characterized by the presence of satellite peaks at positions $+k$ and $-k$ from each nuclear reflection. (**d**) Multi-k structures and single-k structures with multiple domains yield identical diffraction patterns consisting of multiple sets of satellite peaks

Other studies of magnetite using neutron diffraction include investigations of the crystal and magnetic structure below the Verwey transition [60–63], determination of spin-wave dispersion curves [14, 15, 64, 65], studies of magnetite thin films using polarized neutron reflectometry [66], studies of the effect of zinc-doping on the Verwey transition [67], and small-angle neutron scattering of magnetite from magnetotactic bacteria [68] and Chapter 21.

4.3.3.2 Fe_2O_3–Cr_2O_3

Fe_2O_3 and Cr_2O_3 are iso-structural, and there is complete solid solution between the two end members. Despite the similar size and identical charge of Fe^{3+} and Cr^{3+}, the solid solution displays non-ideal behavior as a result of the presence of competing magnetic interactions. In Fe_2O_3, all spins lie parallel to the (006) layers. The (006) intralayer super exchange interactions are positive and the interlayer interactions are negative, leading to the magnetic structure shown in Fig. 4.18f [69].

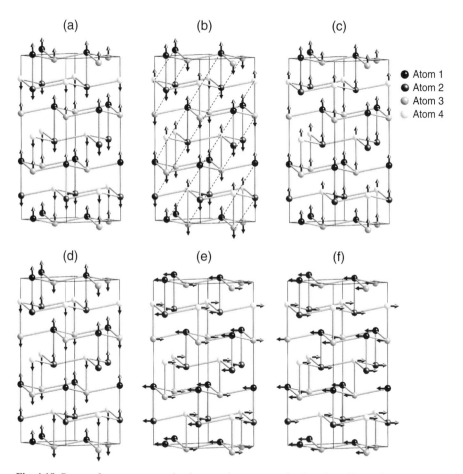

Fig. 4.18 Range of symmetry-permitted magnetic structures for the mineral hematite with $k = 0$. (**a**) Structure corresponding to IR Γ_1. This is the observed magnetic structure of hematite below the Morin transition. (**b**) Structure corresponding to IR Γ_2. This is the observed magnetic structure of Cr_2O_3. *Dashed lines* indicate ferromagnetically coupled layers. (**c**) Structure corresponding to IR Γ_3. (**d**) Structure corresponding to IR Γ_4. (**e**) One possible structure based on a linear combination of basis vectors ν_{1-4} of IR Γ_5. Symmetry dictates antiferromagnetic coupling of moments within the (006) layers. (**f**) One possible structure based on a linear combination of basis vectors ν_{1-4} of IR Γ_6. Symmetry dictates ferromagnetic coupling of moments within the (006) layers. The structure shown is the observed magnetic structure of hematite above the Morin transition

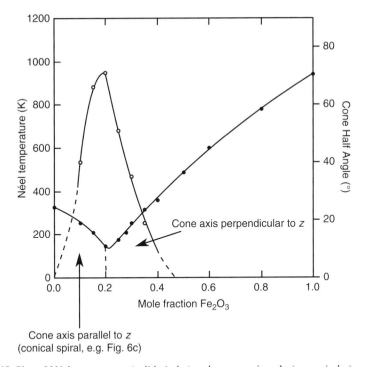

Fig. 4.19 Plot of Néel temperature (*solid circles*) and cone semi-angle (*open circles*) as a function of composition in the solid solution Cr_2O_3–Fe_2O_3 [70]. The magnetic structures of the end members are incompatible with each other, leading to non-ideal mixing and the occurrence of incommensurate helical magnetic structures in the solid solution

In Cr_2O_3, all spins lie perpendicular to the (006) layers, and the signs of the dominant intra- and interlayer super exchange interactions are interchanged, leading to the magnetic structure shown in Fig. 4.18b. The different exchange interactions are caused by the different electronic configurations of Cr^{3+} and Fe^{3+} (Cr^{3+} has a less-than-half-filled 3d shell, whereas Fe^{3+} has a more-than-half-filled shell). The incompatibility of the Fe_2O_3 and Cr_2O_3 magnetic structures can be seen from the fact that the Néel temperature of each phase decreases with increasing substitution of the other phase, reaching a minimum at a composition of 20% Fe_2O_3–80% Cr_2O_3 (Fig. 4.19).

Powder neutron diffraction patterns measured at 77 K are shown in Fig. 4.20 for a range of compositions spanning the Néel temperature minimum [70]. There are three prominent Bragg reflections in this 2θ range: (003), (101), and (012), The (003) and (101) peaks are purely magnetic in origin, whereas the (012) peak contains both magnetic and nuclear contributions. In order to remove the nuclear contribution to the (012) peak, the corresponding room-temperature (paramagnetic) diffraction pattern has been subtracted from each low-temperature diffraction pattern. The different magnetic structures of end members Fe_2O_3 and Cr_2O_3 can be readily distinguished by their different magnetic structure factors: for Fe_2O_3 the

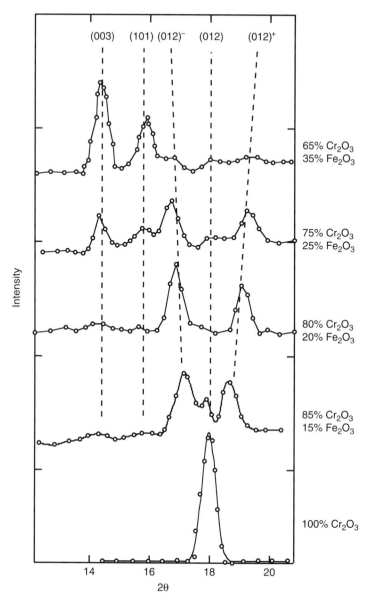

Fig. 4.20 Neutron diffraction patterns as a function of composition in the solid solution Cr_2O_3–Fe_2O_3 [70]. The diffraction pattern of pure Cr_2O_3 consists of a single (012) peak, which contains both nuclear and magnetic contributions. The nuclear contribution has been removed by subtracting the room-temperature paramagnetic diffraction pattern. The diffraction pattern of pure Fe_2O_3 (not shown) consists of (003) and (101) magnetic peaks only. The diffraction patterns of intermediate compositions contain a fundamental contribution from either the Cr_2O_3-like magnetic structure (for $x < 20\%$ Fe_2O_3) or the Fe_2O_3-like magnetic structure (for $x > 20\%$ Fe_2O_3) plus two satellite peaks $(012)^+$ and $(012)^-$. The relative intensity of the fundamental and satellite components can be used to determine the cone semi-angle of the corresponding conical spiral and conical cycloid structures

(003) and (101) peaks are strong and the (012) is absent; for Cr_2O_3 the (003) and (101) peaks are absent and the (012) peak is strong. For intermediate compositions, satellite peaks appear on either side of the (012) nuclear peak position, indicating the presence of an incommensurate component to the magnetic structure. For a composition 85% Cr_2O_3–15% Fe_2O_3, the satellites surround a small (012) magnetic peak.

Cox et al. [70] demonstrated that the magnetic structure in this case is the sum of two components: a fundamental component equivalent to that of the Cr_2O_3 end member and an incommensurate component with k parallel to z. This yields a conical spiral structure in which the cone axis is parallel to the propagation vector (Fig. 4.21a). The spacing of the satellite peaks indicates a spiral periodicity of 80 Å. The cone semi-angle can be determined from the relative intensities of the satellite and fundamental peaks. As the fundamental peak intensity tends to zero, the cone semi-angle tends towards 90° and the structure becomes a screw spiral. The variation in the cone semi-angle with composition is shown in Fig. 4.19 (open circles). The rapid decrease in the intensity of the (012) peak with increasing composition corresponds to a rapid increase in the cone semi-angle, which reaches a maximum for a composition of 80% Cr_2O_3–20% Fe_2O_3. For a composition 75% Cr_2O_3–25% Fe_2O_3, the fundamental component switches from (012) to (003) and (101), indicat-

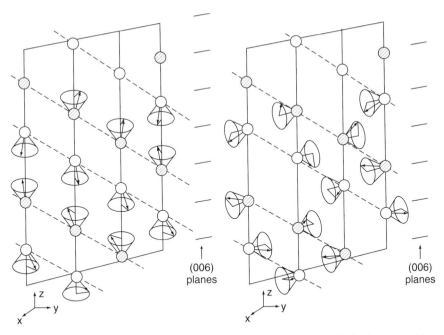

Fig. 4.21 (**a**) Conical spiral structure adopted in the solid solution Cr_2O_3–Fe_2O_3 for compositions <20% Fe_2O_3. The cone axes are oriented parallel and antiparallel to the z axis (equivalent to the orientation of moments in the endmember Cr_2O_3 magnetic structure). *Dashed lines* indicate the planes of ferromagnetically coupled spins. (**b**) Conical cycloid structure adopted in the solid solution Cr_2O_3–Fe_2O_3 for compositions >20% Fe_2O_3. The cone axes are oriented perpendicular to the z axis (equivalent to the orientation of moments in the end member Fe_2O_3 magnetic structure)

ing a switch in magnetic structure from a Cr_2O_3-like conical spiral to an Fe_2O_3-like conical cycloid, with the cone axis now perpendicular to the propagation vector (Fig. 4.21b). The intensity of the (003) and (101) peaks increases with increasing Fe_2O_3 content. This corresponds to a decrease in the cone semi-angle and a gradual transition to end member Fe_2O_3 structure.

4.3.4 Polarized Neutron Diffraction

The main disadvantage of using unpolarized neutrons to study magnetic materials is that it is often difficult to separate out the nuclear and magnetic contributions to the diffraction signal. If measurements are made above and below the Néel temperature, then it may be possible to remove the nuclear contribution by subtracting the diffraction signal of the paramagnetic phase. This may cause problems, however, because of the different lattice parameters and Debye–Waller factors for measurements made at different temperatures. A more reliable approach is to use polarized neutrons. By determining the change in the polarization state of the scattered beam relative to the incident beam (polarization analysis), the magnetic and nuclear contributions can be fully separated at any temperature. Polarization analysis can yield information about the magnetic interaction vector that cannot be obtained using unpolarized neutrons. For some complex magnetic structures, polarization analysis provides the only method to determine the magnetic structure unambiguously [71].

Polarization analysis requires not only a polarized beam but also the ability to detect only those neutrons with a given spin. This requires the addition of both a polarizing monochromator and a polarizing analyzer. In order to preserve the polarization of the neutrons as they travel through the instrument, it is necessary to apply a guide magnetic field along the entire length of the beam path. This can be achieved, for example, using a series of closely spaced permanent magnets. The guide field at the sample position is often maintained using Helmholtz coils. Any inhomogeneity in the guide field can lead to depolarization of the beam, and care must be taken that any magnetic field applied to the sample is in the same direction as the guide field. The measured intensity is divided into two channels: the non-spin-flip (nsf) channel, containing those neutrons whose spin is preserved during scattering, and the spin-flip (sf) channel, containing those neutrons whose spin is flipped during scattering. The ratio of the nsf to sf intensity is referred to as the flipping ratio. Whether or not a neutron spin is flipped depends on the relative orientation of $M_\perp$ and P_i, the polarization vector of the incident beam. If $M_\perp$ is parallel to P_i, precession of the neutron spin around $M_\perp$ preserves the z component (nsf scattering). If $M_\perp$ is perpendicular to P_i, precession of the neutron spin around $M_\perp$ reverses the z component (sf scattering). In order to determine the sf states, it is necessary to be able to flip the polarization of the incident and scattered beams relative to the polarization of the monochromator and analyzer. This can be achieved using radio-frequency spin flippers, which use the precession of the neutron spin around a radio-frequency magnetic field to reverse the spin of any incoming neutron. One mineralogical example of this approach is given below.

4.3.4.1 Nanoscale Hematite Exsolution in Ilmenite

The magnetic properties of the ilmenite–hematite solid solution are profoundly influenced by nanoscale microstructures associated with subsolvus exsolution and cation ordering. Slowly cooled rocks containing finely exsolved members of the hematite–ilmenite series have strong and extremely stable magnetic remanence, suggesting an explanation for some magnetic anomalies in the deep crust and on planetary bodies that no longer retain a magnetic field, such as Mars [72–77]. This remanence has been attributed to a stable ferrimagnetic substructure originating from the coherent interface between nanoscale ilmenite and hematite exsolution lamellae (the so-called lamellar magnetism hypothesis [78–80]). The characteristic nanoscale exsolution microstructure of a natural sample with bulk composition 84% $FeTiO_3$–16% Fe_2O_3 is shown in Fig. 4.22. Nanoscale precipitates of hematite appear as thin white lines surrounded by dark "strain shadows" in Fig. 4.22.

A single crystal extracted from the same sample was investigated using longitudinal polarization analysis on the TASP triple-axis spectrometer at the Paul Scherrer Institute in Zürich, Switzerland. Scans across the (003) peak were recorded at 280 K and in fields of 0.05 and 2 T (Fig. 4.23). Although the resolution of the spectrometer is relatively poor, the use of polarization analysis enables the diffraction signals from ilmenite and hematite to be separated. Ilmenite has the same rhombohedral crystal structure as hematite, except that the iron and titanium cations are ordered onto

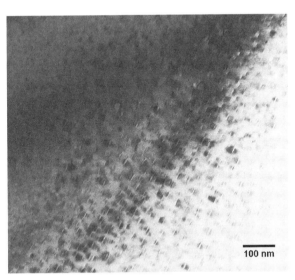

Fig. 4.22 Bright-field transmission electron micrograph of a natural intergrowth of nanometer-scale hematite precipitates (*bright lines* surrounded by dark shadows) within a matrix of ilmenite. A different crystal from the same sample was used for the neutron diffraction work

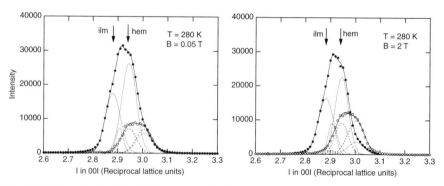

Fig. 4.23 Non-spin-flip (*closed circles*) and spin-flip (*open circles*) intensity across the (003) peak in a natural intergrowth of ilmenite and hematite. Measurements were made at 280 K and in applied magnetic fields of (**a**) 0.05 T and (**b**) 2 T. *Dotted and dashed curves* show the four fitted Gaussian contributions to the non spin-flip and spin-flip intensities, respectively. Black arrows indicate the expected positions of the dominant ilmenite and hematite reflections, based on lattice parameters determined via X-ray diffraction

alternating (006) layers [8, 81, 82]. This cation ordering lowers the symmetry of the structure from R-3c to R-3. Consequently, there is a strong nuclear contribution to the (003) peak from ilmenite, whereas the nuclear contribution is systematically absent from hematite because of the presence of the c-glide plane. Ilmenite is paramagnetic at 280 K and contributes only to the nsf intensity. Hematite is magnetic at 280 K and, in the chosen experimental geometry, can contribute to both the nsf and the sf intensities. The ratio of nsf to sf scattering varies with the orientation of spins within the basal plane.

Figure 4.23a was recorded in the guide field of 0.05 T. The expected positions of the ilmenite and hematite contributions—based on the lattice parameters determined by X-ray diffraction—are indicated by the arrows. There is a broad peak in the nsf intensity that encompasses both the ilmenite and the hematite positions. The peak in the sf intensity, however, is centered on the hematite position. As expected, the sf contributions to the two ilmenite peaks are within error of being zero, consistent with the nuclear origin of the (003) ilmenite peak. The two hematite peaks have very similar sf scattering components but very different nsf scattering components. Analysis of these observations led to the conclusion that the two hematite peaks correspond to different generations of exsolution lamellae. This sample contains a mixture of micron-scale first-generation hematite lamellae and nanoscale second-generation hematite lamellae. The micron-scale lamellae themselves contain nanoscale exsolution precipitates of ilmenite. Peak 2 behaves in a manner consistent with the nanoscale hematite lamellae, whose magnetization is dominated by a spin-canted moment with a smaller lamellar contribution. Such small lamellae are close to the superparamagnetic limit and are more likely to be demagnetized initially. The magnetic moments of the nanoscale hematite precipitates will be easily reoriented in a magnetic field, leading to an increase in the sf scattering and a decrease in the nsf scattering, as observed. Peak 1 behaves in a manner consistent with single-domain hematite containing nanoscale ilmenite exsolution lamellae, whose magnetization is dominated by the defect moment associated with spin imbalance at the ilmenite lamellar interfaces. Such regions are less influenced by the application of a magnetic field, since the orientation of magnetization is controlled by the physical placement of the ilmenite within the hematite host [80]. If this interpretation is correct, the neutron data support the hypothesis that the natural remanent magnetization in this sample is predominantly carried by the lamellar moment associated with the fine scale ilmenite precipitates within the larger hematite exsolution lamellae [83].

4.4 Extension to High Pressure and Temperature

Neutron scattering has proved invaluable in determining atomic occupancies over sites. Aside from the examples cited earlier, the power of neutrons to discern anion occupancies, including (most important) oxygen, places neutron diffraction in a unique position. These methods are essential tools in the arsenal of the mineralogist, and their use is likely to increase for determining order–disorder phenomena and partial occupancies over sites.

Although the importance of neutron powder diffraction for the study of orien-tational order–disorder transitions has been noted and much time devoted to the discussion of substitutional (alloy) order–disorder of cations in minerals, it should be noted that neutron diffraction has also been employed in the study of positional disorder of both cations and anions in earth materials. Notable examples include the studies of Li-ion motion and disorder in β-eucryptite [84, 85]. While lithium, a light element, is difficult to detect using X-ray methods, it shows strong scattering by neutrons, although isotopically enriched samples are used to avoid strong incoherent scattering. Other areas of study of disorder in minerals that neutron powder diffrac-tion has addressed include numerous investigations of molecular and proton disorder in hydrous minerals, including a number of such studies of double-layer hydroxides as well as of interlayer water in clay minerals (Chapter 18). Such investigations have not usually involved the pursuit of phase transition phenomena; therefore, they are outside the scope of this chapter.

The extension of high-temperature studies, such as those outlined above, has begun with the development of in situ high-P/T methodologies adapted for use at neutron sources. Time-of-flight diffraction in $90°$ scattering geometry is particularly well suited for this purpose, and it has been exploited for the study of cation order–disorder in spinel and in ilmenite as a function of both P and T simultaneously (Fig. 4.24 [49, 82]). These studies have revealed that the thermodynamic controls on order–disorder transformations in minerals are not limited to the effects of the volumes of ordering/disordering on enthalpies, but that cation–cation (non-bonded) interactions, which drive ordering, change substantially with pressure and modify the order–disorder behavior significantly in the two examples investigated to date.

Such high-P/T experiments are particularly taxing and push at the experimental limits of neutron powder diffraction (Chapter 3). The samples studied in assem-

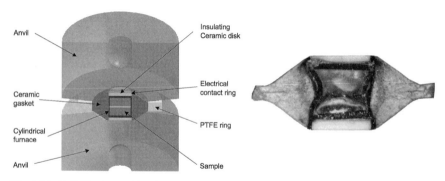

Fig. 4.24 (*Left*) The internal heating system employed in the Paris-Edinburgh loading frame high-P/T apparatus [86]. Neutrons pass through the sample parallel to the vertical axis of the figure, scattering at 90° horizontally. (*Right*) Recovered sample assembly containing spinel heated to 1500°C and pressurized to 3 GPa, used for in situ study by Méducin et al. [49]

blies such as that shown in Fig. 4.24 are typically very small (of the order of 4 mm in diameter and approximately the same dimension in length) and are surrounded by additional materials associated with the gasketing, anvils, and loading frame. Problems of small sample volumes are, therefore, compounded with problems of incident beam attenuation, incoherent scattering, and diffracted beam contamination in the case of poorly collimated set-ups. Such problems can be overcome by careful experiments, but the quality of data collected is necessarily compromised compared with the best ambient pressure datasets. Nevertheless, this approach can ensure an accurate control of the P/T sample environment and allows data collection from relatively large portions of reciprocal space.

The extension of these studies to pressures and temperatures approaching the transition zone and lower mantle of the earth will require an order of magnitude improvement in neutron beam flux. Such an improvement is, fortunately, on the horizon with the next generation of neutron sources currently under construction around the globe. Clearly, the prospects for increased use of extreme sample environments in the application of neutron diffraction to systems at real Earth interior conditions are likely to be substantial in the coming decade with the advent of new spallation sources, including the high-pressure beam line Spallation Neutrons and Pressure Diffractometer at the Spallation Neutron Source in the United States and the extreme conditions beam line Extreme Sample Environments Spectrometer (**EXESS**) planned for target station 2 at ISIS.

Acknowledgments The authors gratefully and freely acknowledge the fruitful "neutron" collaborations with many colleagues both in Cambridge and elsewhere (notably, but not exclusively, Martin Dove and Michael Carpenter), who have illuminated our studies of mineral behavior. The polarized neutron study of ilmenite–hematite was performed in collaboration with Luise Theil Kuhn, Kim Lefmann, and Bertrand Roessli, with a sample kindly provided by Suzanne McEnroe and Peter Robinson. Much of the work described in this chapter was funded through the UK Council for the Central Laboratory of the Research Councils (CCLRC) at ISIS. In addition, Richard Harrison was funded via a UK Natural Environment Research Council (NERC) Advanced Fellowship

(NE/B501339/1 "Mineral magnetism at the nanometer scale"), and Simon Redfern acknowledges support from NERC via grants GR3/11741 and NE/B506086/1.

References

1. Redfern SAT, Salje E, Navrotsky A (1989) High-temperature enthalpy at the orientational order-disorder transition in calcite: implications for the calcite/aragonite phase equilibrium. Contrib. Mineral. Petrol. 101:479–484
2. Dove MT, Powell BM (1989) Neutron diffraction study of the tricritical orientational order-disorder phase transition in calcite at 1260 K. Phys. Chem. Miner. 16:503–507
3. Dove MT, Sawinson IP, Powell BM, Tennant DC (2005) Neutron powder diffraction study of the orientational order-disorder phase transition in calcite, $CaCO_3$. Phys. Chem. Miner. 32:493–503
4. Mookherjee M, Redfern SAT, Swainson I, Harlov DE (2004) Low-temperature behaviour of ammonium ion in buddingtonite [$N(D,H)_4AlSi_3O_8$] from neutron powder diffraction. Phys. Chem. Miner. 31:643–649
5. Meyers ER, Heine V, Dove M (1998) Thermodynamics of Al/Al avoidance in the ordering of Al/Si tetrahedral framework structures. Phys. Chem. Miner. 25:457–464
6. Warren MC, Dove MT, Redfern SAT (2000a) Ab initio simulation of cation ordering in oxides: application to spinel. J. Phys.: Condens. Matter 12:L43–L48
7. Warren MC, Dove MT, Redfern SAT (2000b) Disordering of $MgAl_2O_4$ spinel from first principles. Mineral. Mag. 64:311–317
8. Harrison RJ, Becker U, Redfern SAT (2000b) Thermodynamics of the $R\bar{3}$ to $R\bar{3}c$ transition in the ilmenite-hematite solid solution. Am. Mineral. 85:1694–1705
9. Rodríguez-Carvajal J (1993) Recent advances in magnetic structure determination by neutron powder diffraction. Physica B 192:55–69
10. Arai M, Ishikawa Y, Saito N, Takei H (1985a) A new oxide spin glass system of $(1-x)FeTiO_3-xFe_2O_3$. II. Neutron scattering studies of a cluster type spin glass of $90FeTiO_3-10Fe_2O_3$. J. Phys. Soc. Jpn. 54:781–794
11. Arai M, Ishkawa Y, Takei H (1985b) A new oxide spin glass system of $(1-x)FeTiO_3-xFe_2O_3$. IV. Neutron scattering studies on a reentrant spin glass of 79 $FeTiO_3-21$ Fe_2O_3 single crystal. J. Phys. Soc. Jpn. 54:2279–2286
12. Arai M, Ishikawa Y (1985) A new oxide spin glass system of $(1-x)FeTiO_3-xFe_2O_3$. III. Neutron scattering studies of magnetization processes in a cluster type spin glass of $90FeTiO_3-10Fe_2O_3$. J. Phys. Soc. Jpn. 54:795–802
13. Ott F, Cousin F, Menelle A (2004) Surfaces and interfaces characterization by neutron reflectometry. J. Alloys Compd. 382:29–38
14. Brockhouse BN (1957) Scattering of neutrons by spin waves in magnetite. Phys. Rev. 106:859–864
15. Alperin HA, Steinsvoll O, Nathans R, Shirane G (1967) Magnon scattering of polarised neutrons by the diffraction method: measurements on magnetite. Phys. Rev. 154:508–514
16. Samuelsen EJ (1969) Spin waves in antiferromagnets with corrundum structure. Physica 43:353–374
17. Samuelsen EJ, Shirane G (1970) Inelastic neutron scattering investigation of spin waves and magnetic interactions in α-Fe_2O_3. Phys. Status Solidi 42:241–256
18. Hansen MF, Bodker F, Morup S, Lefmann K, Clausen KN, Lindgård P-A (1997) Dynamics of magnetic nanoparticles studied by neutron scattering. Phys. Rev. Lett. 79:4910–4913
19. Hansen MF, Bodker F, Morup S, Lefmann K, Clausen KN, Lindgård P-A (2000) Magnetic dynamics of fine particles studied by inelastic neutron scattering. J. Magn. Magn. Mater. 221:10–25

20. Lefmann K, Bødker F, Hansen MF, Vázquez H, Christensen NB, Lindgård P-A, Clausen KN, Mørup S (1999) Magnetic dynamics of small α-Fe_2O_3 and NiO particles studied by neutron scattering. Eur. Phys J. D 9:491–494

21. Klausen SN, Lefmann K, Lindgård P-A, Clausen KN, Hansen MF, Bødker F, Mørup S, Telling M (2003) An inelastic neutron scattering study of hematite nanoparticles. J. Magn. Magn. Mater. 266:68–78

22. Klausen SN, Lefmann K, Lindgård P-A, Theil Kuhn L, Bahl CRH, Frandsen C, Mørup S, Roessli B, Cavadini N, Niedermayer C (2004) Magnetic anisotropy and quantized spin waves in hematite nanoparticles. Phys. Rev. B 70:214411

23. Eckold G, Hagen M, Steigenberger U (1998) Kinetics of phase transitions in modulated ferroelectrics: time-resolved neutron diffraction from Rb_2ZnCl_4. Phase Transitions 67:219–244

24. Henderson CMB, Knight KS, Redfern SAT, Wood BJ (1996) High-temperature study of cation exchange in olivine by neutron powder diffraction. Science 271:1713–1715

25. Redfern SAT, Henderson CMB, Wood BJ, Harrison RJ, Knight KS (1996) Determination of olivine cooling rates from metal-cation ordering. Nature 381:407–409

26. Redfern SAT, Henderson CMB, Knight KS, Wood BJ (1997) High-temperature order-disorder in $(Fe_{0.5}Mn_{0.5})_2SiO_4$ and $(Mg_{0.5}Mn_{0.5})_2SiO_4$ olivines: an in situ neutron diffraction study. Eur. J. Mineral. 9:287–300

27. Redfern SAT, Knight KS, Henderson CMB, Wood BJ (1998) Fe-Mn cation ordering in fayalite-tephroite $(Fe_xMn_{1-x})_2SiO_4$ olivines: a neutron diffraction study. Mineral. Mag. 62:607–615

28. Redfern SAT, Harrison RJ, O'Neill HStC, Wood DRR (1999) Thermodynamics and kinetics of cation ordering in $MgAl_2O_4$ spinel up to 1600°C from in situ neutron diffraction. Am. Mineral. 84:299–310

29. Harrison RJ, Redfern SAT, O'Neill HStC (1998) The temperature dependence of the cation distribution in synthetic hercynite $(FeAl_2O_4)$ from in-situ neutron structure refinements. Am. Mineral. 83:1092–1099

30. Nord AG (1983) Neutron-diffraction studies of the olivine-related solid-solution $(Ni_{0.75}N_{0.25})_3(PO_4)_2$. Neues Jahrb. Mineral. Monatsh. 9:422–432

31. Ericsson T, Nord AG (1984) Strong cation ordering in olivine-related (Ni,Fe)-sarcopsides—a combined Mössbauer, X-ray and neutron diffraction study. Am. Mineral. 69:889–895

32. Untersteller E, Hellner E, Heger G, Sasaki S, Hosoya S (1986) Determination of cation distribution in synthetic (Mg,Fe)-olivine using neutron diffraction. Zeit. Kristallogr. 174:198–198

33. Artioli G, Rinaldi R, Wilson CC, Zanazzi PF (1995) High temperature Fe-Mg cation partitioning in olivine: in-situ single-crystal neutron diffraction study. Am. Mineral. 80:197–200

34. Rinaldi R, Wilson CC (1996) Crystal dynamics by neutron time-of-flight Laue diffraction in olivine up to 1573 K using single frame methods. Sol. State Commun. 97:395–400

35. Rinaldi R, Artioli G, Wilson CC, McIntyre G (2000) Octahedral cation ordering in olivine at high temperature. I: in situ neutron single crystal diffraction studies on natural mantle olivines (Fa12 and Fa10). Phys. Chem. Miner. 27:623–629

36. Redfern SAT, Artioli G, Rinaldi R, Henderson CMB, Knight KS, Wood BJ (2000) Octahedral cation ordering in olivine at high temperature. II: an in situ neutron powder diffraction study on synthetic $MgFeSiO_4$ (Fa50). Phys. Chem. Miner. 27:630–637

37. Ballet O, Fuess H, Fritzsche T (1987) Magnetic structure and cation distribution in $(Fe,Mn)_2SiO_4$ olivine by neutron diffraction. Phys. Chem. Miner. 15:54–58

38. Krause MK, Sonntag R, Kleint CA, Ronsch E, Stusser N (1995) Magnetism and cation distribution in iron zinc silicates. Physica B 213:230–232

39. Rinaldi R, Gatta GD, Artioli G, Knight KS, Geiger CA (2005) Crystal chemistry, cation ordering and thermoelastic behaviour of $CoMgSiO_4$ olivine at high temperature as studied by in situ neutron powder diffraction. Phys. Chem. Miner. 32:655–664

40. Millard RL, Peterson RC, Swainson IP (2000) Synthetic $MgGa_2O_4$-Mg_2GeO_4 spinel solid solution and beta-$Mg_3Ga_2GeO_8$: chemistry, crystal structures, cation ordering, and comparison to Mg_2GeO_4 olivine. Phys. Chem. Miner. 27:179–193

41. Hull S, Smith RI, David WIF, Hannon AC, Mayers J, Cywinski R (1992) The POLARIS powder diffractometer at ISIS. Physica B 180:1000–1002
42. Landau LD (1937) On the theory of phase transitions, Part I. Sov. Phys. JETP 7:19
43. Carpenter MA, Salje EKH (1994) Thermodynamics of non-convergent cation ordering in minerals: II Spinels and orthopyroxene solid solution. Am. Mineral. 79:770–776
44. Kroll H, Schlenz H, Phillips MW (1994) Thermodynamic modelling of non-convergent ordering in orthopyroxenes: a comparison of classical and Landau approaches. Phys. Chem. Miner. 21:555–560.
45. Gatta GD, Rinaldi R, Knight KS, Molin G, Artioli G (2007) High temperature structural and thermoelastic behaviour of mantle orthopyroxene: an in situ neutron powder diffraction study. Phys. Chem. Miner. 34 185–200
46. Rouse KD, Thomas MW, Willis BTM (1976) Space group of spinel structure—neutron diffraction study of $MgAl_2O_4$. J. Phys. C: Solid State Phys. 9:L231–L233
47. Peterson RC, Lager GA, Hitterman RL (1991) A time-of-flight neutron powder diffraction study of $MgAl_2O_4$ at temperatures up to 1273 K. Am. Mineral. 76:1455–1458
48. Pavese A, Artioli G, Hull S (1999) In situ powder neutron diffraction of cation partitioning vs. pressure in $Mg_{0.94}Al_{2.04}O_4$ synthetic spinel. Am. Mineral. 84:905–912
49. Méducin F, Redfern SAT, Le Godec Y, Stone HJ, Tucker MG, Dove MT, Marshall WG (2004) Study of cation order-disorder in $MgAl_2O_4$ spinel by in situ neutron diffraction up to 1600 K and 2.6 GPa. Am. Mineral. 89:981–986
50. Bish DL (1981) Cation ordering in synthetic and natural Ni-Mg olivine. Am. Mineral. 66:770–776
51. Henderson CMB, Redfern SAT, Smith RI, Knight KS, Charnock JM (2001) Composition and temperature dependence of cation ordering in Ni–Mg olivine solid solutions: a time-of-flight neutron powder diffraction and EXAFS study. Am. Mineral. 86:1170–1187
52. Amisano-Canesi A, Chiari G, Ferraris G, Ivaldi G, Soboleva SV (1994) Muscovite- and phengite-3T: crystal structure and conditions of formation. Eur. J. Mineral. 6: 489–496
53. Pavese A, Ferraris G, Prencipe M, Ibberson R (1997) Cation site ordering in phengite 3T from the Dora-Maira massif (western Alps): a variable-temperature neutron powder diffraction study. Eur. J. Mineral. 9:1183–1190
54. Mookherjee M, Redfern SAT, Zhang M (2001) Thermal response of structure and hydroxylation of phengite 2M1: an in situ neutron diffraction and FTIR study. Eur. J. Mineral. 13:545–555
55. Welch MD, Knight KS (1999) A neutron powder diffraction study of cation ordering in high-temperature synthetic amphiboles. Eur. J. Mineral. 11:321–331
56. Oberti R, Ungaretti L, Canillo E, Hawthorne FC, Memmi I (1995b) Temperature-dependent Al order-disorder in the tetrahedral double chain of C2/m amphiboles. Eur. J. Mineral. 7:1049–1063
57. Hawthorne FC (1983) The crystal chemistry of the amphiboles. Can. Mineral. 21:173–480
58. Néel L (1948) Propriétés magnetiques des ferrites; ferrimagnétisme et antiferromagnétisme. Ann. Phys. 3:137–198
59. Shull CG, Wollan EO, Koehler WC (1951a) Neutron scattering and polarization by ferromagnetic materials. Phys. Rev. 84:912–921
60. Wright JP, Bell AMT, Attfield JP (2000) Variable temperature powder neutron diffraction study of the Verwey transition in magnetite Fe_3O_4. Solid State Sciences 2:747–753
61. Wright JP, Attfield JP, Radaelli PG. (2001) Long range charge ordering in magnetite below the Verwey transition. Phys. Rev. Lett. 87:266401
62. Wright JP, Attfield JP, Radaelli PG (2002) Charge ordered structure of magnetite Fe_3O_4 below the Verwey transition. Phys. Rev. B 66:214422
63. Yang JB, Zhou XD, Yelon WB, James WJ, Cai Q, Gopalakrishnan KV, Malik SK, Sun XC, Nikles DE (2004) Magnetic and structural studies of the Verwey transition in $Fe_{3-\delta}O_4$ nanoparticles. J. Appl. Phys. 95:7540–7542

64. McQueeney RJ, Yethiraj M, Montfrooij W, Gardner JS, Metcalf P, Honig JM (2005) Influence of the Verwey transition on the spin-wave dispersion of magnetite. J. Appl. Phys. 97: 10A902
65. Watanabe H, Brockhouse BN (1962) Observation of optical and acoustical magnons in magnetite. Phys. Lett. 1:189–190
66. Morrall P, Schedin F, Langridge S, Bland J, Thomas MF, Thornton G (2003) Magnetic moment in an ultrathin magnetite film. J. Appl. Phys. 93:7960–7962
67. Kozlowski A, Kakol Z, Zalecki R, Knight K, Sabol J, Honig JM (1999) J. Phys.: Condens. Matter 11:2749–2758
68. Krueger S, Olson GJ, Rhyne JJ, Blakemore RP, Gorby YA, Blakemore N (1990) Small-angle neutron scattering from bacterial magnetite. J. Appl. Phys. 67:4475–4477
69. Shull CG, Strauser WA, Wollan EO (1951b) Neutron diffraction by paramagnetic and antiferromagnetic substances. Phys. Rev. 83:333–345
70. Cox DE, Takei WJ, Shirane G (1963) A magnetic and neutron diffraction study of the Cr_2O_3–Fe_2O_3 system. J. Phys. Chem. Solids 24:405–423
71. Brown PJ (1995) *In*: International Tables for Crystallography, Vol. C. Wilson AJC (ed), Kluwer Academic Publishers, Dordrecht, p. 391
72. McEnroe SA, Harrison RJ, Robinson P, Golla U, Jercinovic MJ (2001) The effect of fine-scale microstructures in titanohematite on the acquisition and stability of NRM in granulite facies metamorphic rocks from Southwest Sweden: implications for crustal magnetism. J. Geophys. Res. 106:30523–30546
73. McEnroe SA, Harrison RJ, Robinson P, Langenhorst F (2002) Nanoscale hematite-ilmenite lamellae in massive ilmenite rock: an example of "Lamellar Magnetism" with implications for planetary magnetic anomalies. Geophys. J. Int. 151:890–912
74. McEnroe SA, Langenhorst F, Robinson P, Bromiley G, and Shaw C (2004a) What's magnetic in the lower crust? Earth Planet. Sci. Lett. 226:175–192
75. McEnroe SA, Brown LL, Robinson P (2004b) Earth analog for Martian magnetic anomalies: remanence properties of hemo-ilmenite norites in the Bjerkreim-Sokndal Intrusion, Rogaland, Norway, J. Appl. Geophys. 56:195–212
76. McEnroe SA, Skilbrei JR, Robinson P, Heidelbach F, Langenhorst F, Brown LL (2004c) Magnetic anomalies, layered intrusions and Mars. Geophys. Res. Lett. 31:L19601
77. Kasama T, McEnroe SA, Ozaki N, Kogure T, Putnis A (2004) Effects of nanoscale exsolution in hematite-ilmenite on the acquisition of stable natural remanent magnetization. Earth Planet. Sci. Lett. 224:461–475
78. Harrison RJ, Becker U (2001) Magnetic ordering in solid solutions. In C. Geiger (ed) "Solid solutions in silicate and oxide systems", EMU Notes in Mineralogy vol. 3 (Eötvös University Press, Budapest, Hungary), Chapter 13, 349–383
79. Robinson P, Harrison RJ, McEnroe SA, Hargraves RB (2002) Lamellar magnetism in the haematite-ilmenite series as an explanation for strong remanent magnetization. Nature 418:517–520
80. Robinson P, Harrison RJ, McEnroe SA, Hargraves R (2004) Nature and origin of lamellar magnetism in the hematite-ilmenite series. Am. Mineral. 89:725–747
81. Harrison RJ, Redfern SAT, Smith RI (2000a) In situ study of the $R\bar{3}$ to $R\bar{3}c$ transition in the ilmenite-hematite solid solution using time-of-flight neutron powder diffraction. Am. Mineral. 85:194–205
82. Harrison RJ, Stone HJ, Redfern SAT (2006) Pressure dependence of Fe-Ti order in the ilmenite-hematite solid solution: implications for the origin of lower crustal magnetization. Phys. Earth Planet. Int. 154:266–275
83. Robinson P, Heidelbach F, Hirt AM, McEnroe SA, Brown LL (2006) Crystallographic–magnetic correlations in single-crystal haemo-ilmenite: new evidence for lamellar magnetism. Geophys. J. Int., 165: 17–31.
84. Xu HW, Heaney PJ, Yates DM, Von Dreele RB, Bourke MA (1999) Structural mechanisms underlying near-zero thermal expansion in β-eucryptite: a combined synchrotron X-ray and neutron Rietveld analysis. J. Materials Res. 14:3138–3151

85. Sartbaeva A, Redfern SAT, Lee WT (2004) A neutron diffraction and Rietveld analysis of cooperative Li motion in β-eucryptite. J. Phys. Cond. Matter 16:5267–5278
86. LeGodec Y, Dove MT, Francis DJ, Kohn SC, Marshall WG, Pawley AR, Price GD, Redfern SAT, Rhodes N, Ross NL, Schofield PF, Schooneveld E, Syfosse G, Tucker MG, Welch MD (2001) Neutron diffraction at simultaneous high temperatures and pressures., with measurement of temperature by neutron radiography. Mineral. Mag. 65:737–748

Chapter 5
Inelastic Neutron Scattering and Lattice Dynamics: Perspectives and Challenges in Mineral Physics

Narayani Choudhury and Samrath Lal Chaplot

Abstract An understanding of the fundamental physics of the Earth's interior requires information about the phase transitions and thermodynamic properties of key mantle-forming mineral phases. Inelastic neutron scattering (INS) is an indispensable tool for determining key lattice dynamics properties like the phonon dispersion relation (PDR) and density of states, which govern a wide range of material behaviors including structural phase transitions, thermodynamic properties, elasticity, and melting. In this chapter we review recent reported studies involving INS and lattice dynamics calculations of geophysically important minerals. We also review recent applications of INS involving experimental and theoretical ab initio and atomistic studies of the phonon spectra and thermodynamic properties of minerals and of other novel phenomena like high-pressure phonon softening, structural phase transitions, pressure-induced amorphization, magnetic excitations, melting, etc. We discuss the current understanding of the dynamical behavior, thermodynamic properties, and phase transitions of key mantle components like the olivine and pyroxene end members forsterite and enstatite, the mineral zircon, important silica polymorphs, and magnesium oxide; recent results on water and ice; other complex silicates; hydrogen storage materials; etc. Inelastic neutron scattering and complementary techniques like inelastic X-ray scattering have been used to explore the high-pressure PDR and density of states of iron, diamond, and magnesium oxide; to study magnetic excitations; to estimate the magnetic contributions to thermodynamic properties; etc. The theoretical calculations enable fruitful microscopic interpretations of complex experimental data and provide an atomic-level understanding of vibrational and thermodynamic properties.

5.1 Introduction

An understanding of the fundamental physics of the Earth's interior requires information about the phase transitions and thermodynamic properties of key constituent

N. Choudhury (✉)
Solid State Physics Division, Bhabha Atomic Research Centre, Mumbai 400085, India
e-mail: dynamics@barc.gov.in; chaplot@barc.gov.in

L. Liang et al. (eds.), *Neutron Applications in Earth, Energy and Environmental Sciences*, Neutron Scattering Applications and Techniques,
DOI 10.1007/978-0-387-09416-8_5, © Springer Science+Business Media, LLC 2009

mineral phases [1–6]. Information about the Earth's interior is only inferred from seismic observations. Compositional modeling based on accurate data about the structure and thermodynamic properties of minerals is essential to interpreting the complex seismic data. Of particular interest is an understanding of mineral behavior in terms of its microscopic structure and dynamics [7–60]. These characteristic properties are conveniently studied using neutron scattering–based techniques. The current generation of diffractometers and spectrometers at modern neutron sources permit accurate determination of the structural and dynamical behavior of minerals and has contributed to the understanding of high pressure–high temperature crystal structures, lattice vibrations, anomalous thermodynamic behavior, structural phase transitions, etc.

In this chapter, we review (1) the current developments and prospects for novel inelastic neutron experiments in mineral and earth sciences [14–21], which include experiments at reactors and spallation sources at high temperatures and pressures, and (2) the application of theoretical methodologies for the planning, analysis, and interpretation of complex inelastic neutron data [11–18], including atomistic lattice dynamics [11–15], computer simulations [12, 13], advanced ab initio electronic structure methods [61–70], phonon spectra and thermodynamic properties [14–18, 33], etc. The integration of neutron scattering and other vibrational spectroscopic experiments with these theoretical simulation methods has been extremely successful in providing microscopic insights [14–152] relevant to high-pressure mineral and materials sciences, seismology, geochemistry, petrology, etc., and provides a new means for exploring planetary interiors.

State-of-the-art experimental methods permit re-creation of the extreme high pressure–high temperature conditions prevalent in the Earth's interior [6]. Experimental data are, however, often very limited, particularly for the high-pressure phases, which can be synthesized in very small amounts only. The phase transitions of minerals give rise to important seismic discontinuities in the Earth's interior [1–5] that characterize the basic structure of the Earth. A major geophysical goal is therefore to develop reliable models for predicting the thermodynamic properties and phase transitions of constituent minerals. A key requirement for the prediction of thermodynamic properties is an accurate description of the phonon density of states (PDOS) [14–18, 33]. Inelastic neutron scattering (INS) measurements using single-crystal and powder samples provide complete measurements of vibrational properties, including the phonon dispersion relation (PDR) and PDOS. Studies of these phonon properties are of immense interest as they provide valuable quantitative information concerning elasticity, piezoelectric and dielectric behavior, and thermodynamic properties and govern the dynamics of phase instabilities. Neutron scattering experiments have played a key role in validating theoretical models, which in turn provide microscopic insights to a variety of novel phenomena like structural phase transitions and melting of minerals at the extreme conditions they experience in the Earth's interior; pressure-induced changes in atomic coordination, bonding, and elastic properties; seismic wave discontinuities; competing interactions [144]; etc. Lattice dynamics calculations in turn have played a key role in the planning, analysis, and interpretation of INS experiments in these structurally complex solids.

Quantum mechanical first principles calculations have been very effective in elucidating microscopic details about the vibrational spectra, elastic behavior, and thermodynamic properties of several minerals like α-quartz [25, 65–67], forsterite (Mg_2SiO_4) [51], perovskite ($MgSiO_3$) [52], alumina (Al_2O_3) [59], iron [60], magnetite (Fe_3O_4) [84], etc. Because of the structural complexity of several mantle minerals (e.g., pyroxenes, garnets), involving a large number of atoms in the unit cell, an atomistic approach involving interatomic potentials is still very useful for studying high pressure–high temperature properties [14–18]. Even for simple minerals like silica, which form prototype systems, development of interatomic potentials has helped understand complex phenomena like the role of phonon instability in causing pressure-induced amorphization (PIA) at high pressures, anomalous high pressure–high temperature phase transitions, melting, etc. [25, 47–49]. The success of atomistic models in predicting thermodynamic properties depends crucially on their ability to explain a variety of microscopic and macroscopic dynamical properties. These include crystal structures, elastic constants, equations of state, phonon frequencies, dispersion relation, density of states, and thermodynamic properties like specific heat and thermal expansion. In this context, INS and complementary techniques such as Raman and infrared (IR) spectroscopy provide some of the most valuable tools in assessing models of mineral properties. Inelastic neutron scattering is unique as it provides information about the complete phonon spectra arising from the phonons of all wavelengths, whereas optical techniques directly probe only the optically active phonons, limited to the long wavelengths (in one-phonon measurements).

Several recent works on inelastic X-ray scattering [84–89] using synchrotrons and inelastic nuclear resonant scattering from select isotopes like iron-57 have been used effectively to understand the high pressure–high temperature PDR, density of states, and thermodynamic properties of magnesium oxide (MgO), iron, magnetite (Fe_3O_4), diamond, etc. Neutron scattering techniques, however, offer certain unique advantages over other probes, like X-rays, as they can measure subtle changes in structure and dynamics, especially involving low-Z elements like hydrogen and oxygen, which are important constituents of most minerals [14–18], ceramics [19], ferroelectrics [90, 146] hydrogen storage materials [20, 21], etc. Neutron-based methods therefore continue to be the technique of choice for most reported studies [14–40] and are well described in several important review papers [14–21, 90].

The inelastic scattering of a monochromatic beam of neutrons is described by the scattering cross section, which consists of an incoherent and a coherent part. Coherent INS measurements record peaks in scattered neutron groups whenever the energy and momentum conservation laws are simultaneously satisfied, while incoherent scattering requires only energy conservation. For nonmagnetic materials, the intensity in neutron scattering is determined by the nuclear scattering cross sections. Unlike in X-ray scattering, nuclear scattering cross sections are not related in a simple fashion to the atomic number and in general are quite different for different isotopes. Coherent scattering essentially has contributions from spatially and/or temporally correlated atoms, while the incoherent cross sections are due to single-particle excitations. The incoherent cross sections of predominantly abundant mineral constituents like O, Si, Mg, Fe, Ca, Al, etc., are small. Several minerals have molecular water. The incoherent cross section of hydrogen ($\sigma_{inc}^{H} \sim 80$ barns) is very large. In powder samples of minerals with ≥ 1 wt% hydrogen, neutron scattering is predominantly due to hydrogen atoms, and samples might need deuteration to reveal the true vibrational spectra involving motions of all the atoms. The extremely large incoherent neutron scattering cross section of hydrogen has been effectively used to study the dynamics of water molecules in several minerals like beryl, gypsum, cordeirite, bassanite, layer silicates, and zeolites [91, 92]. Although we briefly discuss some incoherent INS results for clathrates, this chapter deals with phonon modes obtained mainly from coherent INS techniques. We are not discussing quasielastic neutron scattering studies of the dynamics of water molecules in minerals but these aspects are described in Chapter 8 of this book. For hydrogen-bearing samples, quasielastic neutron scattering arise from single-particle stochastic or diffusive dynamics of the hydrogen atom due to its extremely large incoherent cross section. Studies of variations in the width of the quasielastic peak with momentum transfer allow one to make inferences about the nature of diffusion (rotational or translational). Quasielastic neutron scattering studies of several minerals aimed at understanding the dynamics of water molecules have been reported [91, 92].

Inelastic neutron scattering experiments require much larger size samples, typically single crystals on the order of $1\,cm^3$ and powder samples of about $10\,cm^3$ and upward, than those used in optical spectroscopies. The experimental determination of the PDOS is complementary to the determination of the PDR. The latter is not always possible because a suitable single crystal may not be available. Moreover, depending on the complexity of the structure, it is often difficult to determine all the phonon branches in the PDR separately; usually a complete PDR is obtained only for wave vectors along the main symmetry directions of the Brillouin zone, where the phonon symmetry is also high. Therefore, to obtain a complete picture of the dynamics, it is useful to determine the PDOS. The interpretation of both the PDR and the PDOS data may be carried out on the basis of reliable lattice dynamics calculations.

5.2 Techniques

5.2.1 Inelastic Neutron Scattering

Two kinds of neutron sources suitable for INS experiments are available: the thermal neutron reactors that provide a continuous beam of neutrons and spallation neutron sources providing intense pulses of neutrons. Measurements of the PDR and PDOS can in principle be carried out using both reactors as well as spallation sources. However, thermal neutrons from a nuclear reactor are best suited for the measurements of the acoustic and low-frequency optic modes (spanning 0–80 meV spectral range) in a single crystal. The energies of these low-frequency modes are particularly important for the estimation of the low-temperature specific heat. On the other hand, the high energies of neutrons from a spallation source enable measurements over the entire spectral range (typically, 0–150 meV for the silicate mantle minerals) and are best exploited for the measurements of the PDOS.

5.2.1.1 Inelastic Neutron Scattering from Single Crystals: Phonon Dispersion Relation

The basics of neutron scattering are covered in earlier chapters of this book [Chapters 1, 2 and 3]. The measurement of PDR from reactors providing a continuous beam of neutrons is usually carried out using a triple-axis spectrometer. Usually, these experiments are carried out keeping either the momentum transfer $\mathbf{Q}$ (Figs. 5.1 and 5.2) or the energy transfer $E = \hbar\omega$, constant [14–17]. Such experiments on single crystals aim to probe the scattering from one phonon at a time. In crystalline solids, the momentum transfer $\mathbf{Q}$ is related to the wave vector of the excitation by a reciprocal lattice vector, $\mathbf{Q} = \mathbf{G} \pm \mathbf{q}$, where $\mathbf{G}$ is a reciprocal lattice vector and $\mathbf{q}$ is the reduced wave vector in the Brillouin zone.

The neutron scattering structure factor due to a one-phonon inelastic process is given by

$$S(\mathbf{Q}, \omega) = A \sum_{qj} \frac{\hbar}{2\omega(\mathbf{qj})} \left\{ n(\omega) + \frac{1}{2} \pm \frac{1}{2} \right\} \left| F_j^{(1)}(\mathbf{Q}) \right|^2 \delta(\mathbf{Q} - \mathbf{G} + \mathbf{q})\delta(\omega \mp \omega(\mathbf{qj})),$$

(5.1)

where

$$F_j(\mathbf{Q}) = \sum_k b_k^{coh} \frac{\mathbf{Q}.\boldsymbol{\xi}(\mathbf{qj}, k)}{\sqrt{m_k}} \exp\{-W_k(\mathbf{Q})\} \exp(i\mathbf{G}.\mathbf{X}(k)),$$

(5.2)

where b_k and m_k are the neutron scattering lengths and masses, respectively, of the atoms k; $X(k)$ is the coordinate; $\exp[-W_k(\mathbf{Q})]$ is the Debye–Waller factor; and A is a proportionality constant. The variables $\hbar\mathbf{Q}$ and $\hbar\omega$ are, respectively, the momentum and energy transfer on scattering of the neutron, while $n(\omega)$ is the phonon population factor given by

$$n(\omega) = \frac{1}{\exp(\hbar\omega/k_B T) - 1}.$$

(5.3)

The upper and lower minus (−) and plus (+) signs correspond to energy loss and energy gain of the neutrons, respectively. The two delta functions in Eq. (5.1) stand for the conservation of momentum and energy. Due to these conservation laws, it is possible to choose specific regions in $\mathbf{Q}$-ω space (with constant $\mathbf{Q}$ or ω) where one can scan for peaks in $S(\mathbf{Q},\omega)$. From a large number of such measurements on a single crystal, one can identify several points of the PDR $\omega_j(\mathbf{q})$.

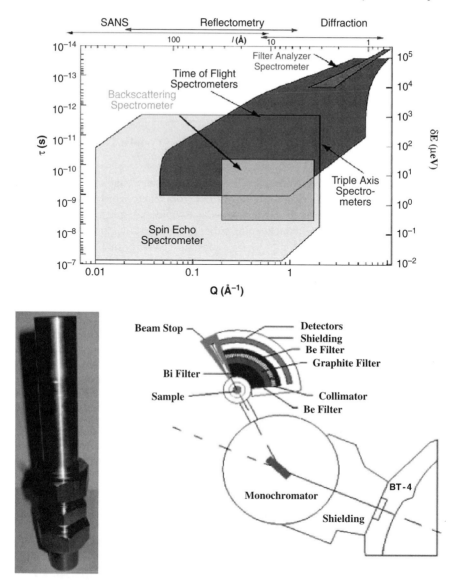

Fig. 5.1 (*Top*) **Q**-ω phase space regions covered by typical neutron inelastic scattering instruments (D.A. Neumann [20]). The length and timescale of the dynamical motions are respectively related to the wave vector and energy transferred in the scattering process through Fourier transform. (*Bottom*) Neutron scattering pressure cell and filter-analyzer neutron spectrometer layout at the National Institute of Standards and Technology (USA) (Struzhkin et al. [21])

The phase term $\exp[i\mathbf{G}.\mathbf{X}(k)]$ or $\exp[i\mathbf{Q}.\mathbf{X}(k)]$ in Eq. (5.2) depends on the phase terms contained in the eigenvector $\boldsymbol{\xi}$, which in turn depends on the phase terms in the dynamical matrix. For a consistent description of the lattice dynamics, see reference [14].

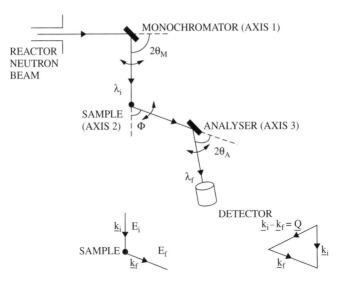

Fig. 5.2 A schematic diagram of the triple-axis spectrometer (TAS) at the Dhruva reactor, Trombay, India (Chaplot et al. [14]). In a typical experiment, a monochromatic beam of neutrons is incident on a sample and the energy spectrum of the scattered beam is measured in different directions. The different wave vectors are chosen by a suitable choice of the Bragg angles of the monochromator and analyzer crystals, the scattering angle between the incident and the scattered beams, and the sample orientation. At the Dhruva reactor, a Cu(111) monochromator is used to monochromatize thermal neutrons emerging from a tangential beam. By changing the angle of the Bragg reflection of the monochromator, neutron beams of continuously variable incident wavelength are obtained. These monochromatic neutrons fall on the sample contained in a sample holder and can be observed over a scattering angle range of 10–90°, and the inelastically scattered neutrons are energy analyzed using a pyrolitic graphite (002) analyzer. The intensity of the scattered neutrons as a function of energy is measured by a BF_3 detector. The Dhruva TAS instrument has been used for studies of phonon dispersion relation and density of states of several minerals (Table 5.1)

The phonon cross sections depend strongly on **Q** and the eigenvector $\boldsymbol{\xi}$ of the excitation. For simple structures, the eigenvectors are determined entirely from the symmetry of the space group; the structure factors are then entirely determined from the crystal structure. For more complex structures [like those of the minerals olivine or pyroxene (Fig. 5.3)], the space group symmetry only classifies the phonons into a number of irreducible representations. The eigenvectors could be any linear combinations of the symmetry vectors associated with the irreducible representation, and their understanding involves extensive lattice dynamics calculations. The lattice dynamical model can then be used to compute the structure factors, and the regions in reciprocal space where the neutron scattering cross sections are large may be identified. The calculated structure factors have been used as guides for our measurements of PDR from single crystals of the minerals forsterite [23], fayalite [40], zircon [24, 37], and the aluminum silicate minerals sillimanite and andalusite [34, 43].

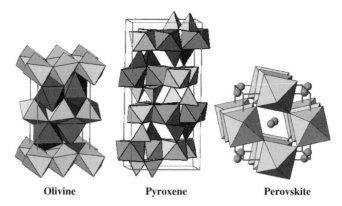

| Olivine | Pyroxene | Perovskite |

Fig. 5.3 Polyhedral representation of the crystal structures of selected mantle minerals drawn using the code Xtaldraw [140]. Olivine and pyroxene are principal components of the Earth's upper mantle, while perovskite ($MgSiO_3$) is the predominant constituent of the Earth's lower mantle. The olivine structure has distinct M1 and M2 octahedral sites with isolated silicate tetrahedra. The pyroxene structure has distinct M1 and M2 octahedral sites with two types of corner-shared SiO_4 tetrahedra. In the $MgSiO_3$ structure, the silicate atoms are in octahedral coordination, while the magnesium atoms occupy distorted eight-coordinated structures. The Mg atoms are shown as *spheres*

The neutron energy is usually measured either by orienting a crystal at the proper Bragg diffraction angle or by using the neutron time-of-flight (TOF) method. In the TOF method [Chapter 1], the transit time of neutrons over a flight path, from which the neutron velocity can be determined, is electronically measured. While the triple-axis spectrometers in reactors allow for scans with constant $\mathbf{Q}$ or ω, such scans are not efficiently carried out at a pulsed source where one obtains the complete TOF energy spectrum of the neutrons scattered from the sample along a given direction. In the latter case, the trajectory of the scan in $(\mathbf{Q},\omega)$ space is not along constant $\mathbf{Q}$ or constant ω, but determined by the geometrical and other instrumental settings. In the PRISMA type spectrometer at ISIS, a number of $\mathbf{Q}$-ω scans can be simultaneously obtained using several independent analyzers and detectors, thus partly compensating for the inability to carry out constant $\mathbf{Q}$ scans. While in general, these $\mathbf{Q}$-ω scans have severe limitations as observations at arbitrary $\mathbf{Q}$ and E (e.g., a pure longitudinal scan) is not possible, excellent data may be obtained in favorable cases. Very good data have been obtained for zircon, $ZrSiO_4$, using this spectrometer [37].

5.2.1.2 Inelastic Neutron Scattering from Powder Samples: The Density of States

Whereas in the case of scattering from single crystals, the scattering vector has a definite orientation with respect to the reciprocal space of the single crystal, in powder samples the reciprocal space axes belonging to the different grains of the powder have different orientations, and in fact, ideally, all possible orientations are possible with equal probability [14–16]. Thus, averaging over the various grains is equivalent to averaging over all orientations of the scattering vector. Further averaging over the magnitude of $\mathbf{Q}$, or in the limit of large $\mathbf{Q}$, where the correlations between atomic motions become small and $S(\mathbf{Q},\omega)$ becomes independent of $\mathbf{Q}$ (except for some smoothly varying $\mathbf{Q}$-dependent factors), one gets the density of states $S(\omega)$, the density of excitations at the frequency ω, although weighted with the neutron scattering cross sections. The averaging over $\mathbf{Q}$ should be carried out over a suitable chosen range of $\mathbf{Q}$ because the relative heights of the various peaks in $S(\mathbf{Q},\omega)$ would be normally $\mathbf{Q}$-dependent.

The coherent INS data from a powder sample are usually analyzed in the incoherent approximation. In this approximation, the correlation between the motions of atoms is neglected and one treats the scattering as incoherent with the scattering amplitude b_k^{coh}. This is valid only for large $\mathbf{Q}$. In this approximation, the scattering expression will not contain the phase terms and the directional averaging over $\mathbf{Q}$ of the term $|\mathbf{Q}.\boldsymbol{\xi}|^2$ gives $\mathbf{Q}^2\boldsymbol{\xi}^2/3$.

The one-phonon scattering from a powder sample is given by [14]

$$S_{coh}^{(1)}(\mathbf{Q},\omega) = \sum_k \exp(-2W_k(\mathbf{Q})) \frac{(b_k^{coh})^2}{m_k} \sum_{Gqj} \frac{1}{3} Q^2 |\boldsymbol{\xi}(qj,k)|^2$$

$$\frac{\hbar}{2\omega(\mathbf{q},j)} \left\{ n(\omega) + \frac{1}{2} \pm \frac{1}{2} \right\} \delta(\mathbf{Q} - \mathbf{G} + \mathbf{q})\delta(\omega \mp \omega(\mathbf{qj})). \quad (5.4)$$

Incoherent inelastic scattering also contributes to the scattering from a powder sample in almost the same way as coherent scattering in the incoherent approximation. Thus, the so-called neutron-weighted density of states involves weighting by the total scattering cross section of the constituent atoms. The measured scattering function in the incoherent approximation is therefore given by [14–16, 93]

$$S_{inc}^{(1)}(Q,E) = \sum_k b_k^2 e^{-2W_k(Q)} \frac{Q^2}{2m_k} \frac{g_k(E/\hbar)}{E} \left(n(\omega) + \frac{1}{2} \pm \frac{1}{2} \right), \quad (5.5)$$

where the partial density of states $g_k(E/\hbar)$ is given by

$$g_k(\omega) = \int \sum_j |\boldsymbol{\xi}(\mathbf{qj},k)|^2 \delta(\omega - \omega_j(\mathbf{q}))d\mathbf{q}. \quad (5.6)$$

The scattering thus provides the density of one-phonon states weighted by the scattering lengths and the population factor. We note that the neutrons are sensitive to the partial density of states due to various species of atoms. The observed neutron-weighted PDOS is the sum of the partial components of the density of states weighted by their scattering length squares and is given by [14–16, 93]

$$g^n(\omega) = A \sum_k \left\{ \frac{4\pi b_k^2}{m_k} \right\} g_k(\omega). \quad (5.7)$$

The weighting factors $4\pi b^2/m$ for the various atoms in the units of barns/amu are Mg: 0.150, Al: 0.055, Fe: 0.201, Ca: 0.075, Si: 0.077, and O: 0.265. By comparing the experimental phonon spectra with the calculated neutron-weighted density of states obtained from a lattice dynamical model, the dynamical contribution to frequency distribution from various atomic and molecular species can be understood.

Any inelastic neutron spectrum contains a contribution from multiphonon scattering that has to be estimated and subtracted from the experimentally observed spectrum G(E) to obtain the one-phonon density of states. Usually, multiphonon scattering contributes a continuous spectrum and effectively increases the background. The neutron scattering measures the scattering function in terms of $\mathbf{Q}$ and E, which in the conventional harmonic phonon expansion can be written as

$$S(Q, E) = S^{(0)} + S^{(1)} + S^{(n)}, \tag{5.8}$$

where $S^{(0)}$, $S^{(1)}$, and $S^{(n)}$ represent elastic, one-phonon, and multiphonon scattering, respectively. The multiphonon contribution to the total scattering is usually estimated in the incoherent approximation using Sjolander's formalism [14, 24], in which the total contribution is treated as a sum of the partial components of the density of states from various species of atoms. The coherent scattering due to multiphonon excitations involves the energy and wave vector conservation rules similar to those for one-phonon excitations, namely

$$E_i - E_f = \pm\hbar\omega(\mathbf{q}1, j1) \pm \hbar\omega(\mathbf{q}2, j2) \pm \ldots, \tag{5.9}$$

and

$$\mathbf{k}_i - \mathbf{k}_f = \mathbf{G}\text{-}\mathbf{q}1 - \mathbf{q}2 - \ldots, \tag{5.10}$$

where the plus or minus ($\pm$) corresponds to either creation or annihilation of any of the phonons. Note that for any of the parameters (E_i, E_f, k_i, k_f) it would usually be possible to obtain several combinations of $\mathbf{q}$ and ω satisfying the above conservation rules. It therefore turns out that the multiphonon contribution to the neutron inelastic spectrum does not give rise to very sharp peaks but only a rather continuous spectrum unlike the one-phonon contribution. Further, the interatomic correlations are expected to be small; the multiphonon contributions are therefore analyzed in the incoherent approximation. In this approximation,

$$S^{(m)}(Q, E) = \sum_k \frac{A_k b_k^2}{m_k} S_k^m(Q, E), \tag{5.11}$$

where the total multiphonon scattering cross section is a weighted sum of the multiphonon contribution from each atomic species. Contribution from the incoherent multiphonon scattering is taken into account by using the total scattering lengths squares of the atoms. The computation of $S_k^m(\mathbf{Q}, \omega)$ is carried out using Sjolander's formalism. $S_k^m(\mathbf{Q}, \omega)$ is given by

$$S_k^m(Q, \omega) = \exp(-2W_k) \sum_{n=2}^{\infty} G_n(\omega) \frac{(2W_k)^n}{n!}, \qquad (5.12)$$

with $G_0(\omega) = \delta(\omega)$ and $G_1(\omega) = g_k(\omega)$, where $g_k(\omega)$ is the partial density of states and

$$G_n(\omega) = \int_{-\infty}^{\infty} g_k(\omega - \omega')G_{n-1}(\omega')d\omega', \qquad (5.13)$$

gives the higher-order terms. The multiphonon contribution is usually estimated on the basis of a lattice dynamical model.

Inelastic neutron scattering studies of the PDOS usually require about $10\,cm^3$ of powder sample. In addition to the triple-axis spectrometer, other spectrometers are often conveniently used for PDOS experiments, such as those based on the measurement of the neutron TOF or a neutron filter corresponding to a certain filter neutron energy as the neutron-energy analyzer. These latter techniques are particularly convenient at pulsed neutron sources, whereas all the above techniques are used at reactors. Different spectra are obtained for several different Q, which may differ due to the coherent scattering, and then averaged over Q to obtain the neutron-weighted PDOS. Various corrections to the data are needed due to the background scattering, geometrical factors, efficiency of the monochromator and analyzer, second-order reflections from the monochromator and analyzer crystals, etc. The measured spectra also deviate from the true PDOS because different atoms contribute differently in proportion to their scattering cross sections divided by their masses. Lattice dynamical models are very useful in providing a microscopic understanding of the complex data and in estimating multiphonon components.

5.2.2 Lattice Dynamics Calculations

Lattice dynamics calculations [9–15] enable the calculation of one-phonon structure factors in order to select the regions in reciprocal space most appropriate for the detection of particular phonons. Furthermore, they are very important for assignment of the various inelastic signals to specific phonon branches. The computed eigenvectors provide detailed information about the pattern of atomic displacements characterizing the various phonon modes. Lattice dynamics calculations require information about the interatomic forces that can be obtained either by using a quantum-mechanical *ab initio* formulation or by using semi-empirical interatomic potentials. The mantle minerals have extremely complex structures (Fig. 5.3), and a multiscale approach using a combination of first principles and atomistic calculations is very useful to study the physics over various lengths and timescales.

Density functional theory [61–66] permits calculation of the total energy of solids without any parameterization to experimental data. The phonon spectra and elastic, piezoelectric, and dielectric tensors are related to the second derivatives of the total energy with respect to variables like atomic displacements, macroscopic strain, and electric field. All these material properties can be efficiently computed using recent advances in density functional perturbation theory (DFPT) [62–66]. Excellent review papers [14–15, 62] describe the theoretical techniques in great detail, including the calculation of the PDR, density of states, and associated thermodynamic properties using both atomistic and first principles-based approaches. Ab initio calculations have also been very successful in high-pressure crystal structure predictions [25, 54–56, 143], using metadynamics approaches [54, 55] and other novel evolutionary techniques [56]. Group theoretical symmetry analysis [11, 14, 15] helps classify the phonon modes belonging to various representations, which enables direct comparisons with observed single crystal Raman, IR, and inelastic neutron data. Because of the selection rules, only phonon modes belonging to certain group theoretical representations are active in typical single-crystal Raman, IR, and INS measurements. These selection rules are governed by the symmetry of the system and the scattering geometry used. The theoretical scheme for the derivation of the symmetry vectors is based on irreducible multiplier representations and described in detail in reference [11]. This involves construction of symmetry-adapted vectors, which are used for block diagonalizing the dynamical matrix and for assignment of the phonon modes belonging to various representations.

5.3 Results and Discussions

Magnesium silicate in various polymorphic forms constitutes a major component of the Earth's mantle. The upper mantle, which extends to a depth of 440 km, contains olivine, pyroxene, and garnet phases, whereas the lower mantle, below a 660-km depth, is largely made of the perovskite phase [1–5]. Apparently, the most important difference between the upper and lower mantle silicates is that the silicon coordination increases from four to six and the magnesium coordination increases from six to eight. This kind of information is indirectly inferred from seismic observations and compositional modeling of the Earth's interior based on accurate information about the structure and thermodynamic properties of the constituent phases [1–5]. Accurate modeling of mantle minerals is therefore of utmost importance and simultaneously is also a major challenge in condensed matter physics. In Table 5.1, we summarize selected research efforts involving inelastic neutron experiments and theoretical calculations in understanding the phonon spectra and associated thermodynamic properties of key minerals of geophysical interest. In this chapter, using typical examples, we review and illustrate some results of current interest.

5.3.1 Olivine Minerals: Forsterite (Mg_2SiO_4) and Fayalite (Fe_2SiO_4)

The olivine minerals form principal constituents of the Earth's upper mantle. The structures of the olivine minerals are fairly complex, comprising edge-shared MgO_6/FeO_6 octahedra and isolated silicate tetrahedra (Fig. 5.3) with 28 atoms/unit cell

Table 5.1 Selected research efforts on key minerals of geophysical interest[a]

Materials	INS experiments $[\omega_j(\mathbf{q})]$[b]	INS experiments $[g(\omega)]$[b]
Olivine [22, 23, 40, 41]		
Forsterite (Mg_2SiO_4) [22, 23]	BNL	ANL
Fayalite (Fe_2SiO_4) [40, 41]	BNL	ANL
Pyroxene/Enstatite/Perovskite (MgSiO$_3$) [15, 26, 27]		
Orthoenstatite[c] [26]		ANL
Protoenstatite[c] [15, 101]		
Low-clinoenstatite [15, 27]		
High-clinoenstatite [15, 27]		
$MgSiO_3$ Perovskite [15, 30–32]		
Aluminum silicate (Al$_2$SiO$_5$) [29, 43]		
Sillimanite [29, 34]	ORNL	RAL
Andalusite [43, 102]	Dhruva, PSI	
Kyanite [29]		RAL
Garnets [14, 16, 36, 38]		
Almandine [16, 36]		ANL
Pyrope [16, 38, 103, 104]	RAL	RAL
Grossular [16, 38, 105]		RAL
Spessartine [38]		
Yttrium aluminum garnet [46],		
Yttrium iron garnet [46]		
Zircon structured compounds		
Zircon ($ZrSiO_4$) [16, 24, 37]	Dhruva, RAL, LLB	ANL
$LuPO_4$, $YbPO_4$, etc. [106, 111]	LLB	ANL
Perovskites		
$PbTiO_3$, $BaTiO_3$, and $SrTiO_3$ [146]		ANL
Oxides/carbonates		
α-quartz (SiO_2) [25, 47–49, 99]	ILL	
Quartz structured $FePO_4$, $GaPO_4$, $AlPO_4$ [16]		ANL
Sapphire	ILL	
α-Al_2O_3 [107]		
α-Cr_2O_3 [108]	ILL	
Calcite ($CaCO_3$) [109–110]]	NRU	
Rhodochrosite ($MnCO_3$) [43, 44]	Dhruva	Dhruva
Sodium carbonate (Na_2CO_3) [43]		Dhruva
Zirconia (ZrO_2) [42], Yttria (Y_2O_3) [42]		Dhruva
Li_2O [147, 148], UO_2 [148, 149, 152]		

[a] Selected list of minerals for which inelastic neutron scattering measurements of phonon dispersion relation, $\omega_j(\mathbf{q})$, and/or density of states, $g(\omega)$, have been measured and the facilities that performed the measurements. In several of these studies [14–47], lattice dynamics calculations have been used for the planning, analysis, and interpretation of experimental neutron data and for the prediction of various microscopic and macroscopic thermodynamic properties.

[b] BNL—Brookhaven National Laboratory, USA; ANL—Argonne National Laboratory, USA; ORNL—Oak Ridge National Laboratory, USA; RAL—Rutherford Appleton Laboratory, UK; Dhruva—Dhruva reactor, Bhabha Atomic Research Centre, India; PSI—Paul Scherrer Institut, Switzerland; LLB—Laboratoire Léon Brillouin, Saclay, France; ILL—Institut Laue-Langevin, France; NRU—NRU reactor, Chalk River, Canada.

[c] Polarized single-crystal Raman spectroscopic measurements [26, 106] were also carried out at the University of Washington, USA

(Figs. 5.3–5.7). Using large single crystals of forsterite and fayalite, the PDR for
a number of acoustic and optic phonon branches were measured [23, 40] using
the triple-axis spectrometer at the Brookhaven National Laboratory (Table 5.1). We
developed interatomic potentials that were fruitfully used to study the elastic, vibrational, and thermodynamic properties of forsterite [23, 33]. The calculated neutron
cross sections were used as guides for coherent INS experiments [14–17]. Recently,
Li et al. [51] have reported first principles calculations of the PDR, density of states,
and thermodynamic properties of forsterite. Their first principles results [51] are in
good agreement with our inelastic-neutron single crystal and powder data (Figs. 5.4
and 5.8) [23] and rigid-ion model calculations of macroscopic variables like specific heat, thermal expansion, etc. [33]. The computed first-principles-derived PDOS
(Fig. 5.4) reveal a band gap around 90–110 meV in agreement with the observed
PDOS data obtained from INS experiments (Fig. 5.8). Our studies [26] indicate that
the separation between the external and the internal silicate group vibrations gives
rise to the band gap in forsterite.

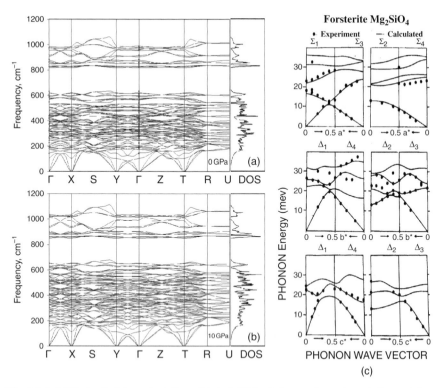

Fig. 5.4 (a,b) First-principles calculations (Li et al. [51]) of phonon dispersion relation and density of states of forsterite at $P = 0$ and $P = 10$ GPa. **(c)** The observed inelastic neutron data
[23] of dispersion relation (*symbols*) of low-frequency phonons of forsterite compared with model
calculations (*full lines*)

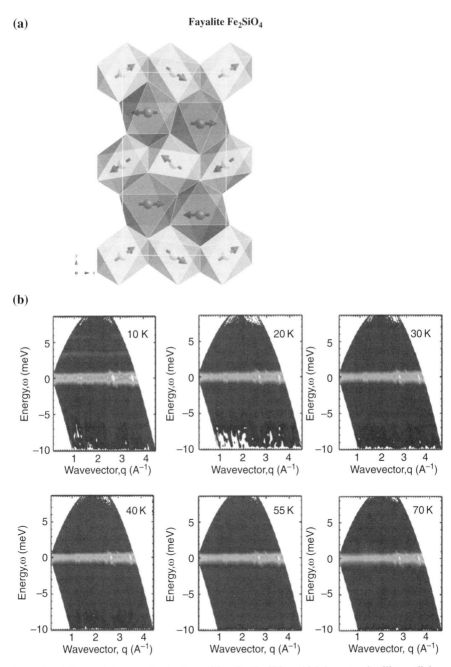

(a)

Fayalite Fe$_2$SiO$_4$

(b)

Fig. 5.5 (**a**) The canted magnetic structure of fayalite Fe$_2$SiO$_4$, which has an orthosilicate olivine structure [83]. Figure adapted from Aronson et al. [83]. The divalent iron ions occupy two different sites. The M1 sites have approximate local tetragonal symmetry and M2 sites have approximate local trigonal symmetry [83, 136–139]. Contour plots of the intensity of scattered neutrons as functions of the wave vector **q** and neutron energy transfer ω for fayalite [83] at temperatures above and below the $T_{\mathrm{N}} = 65$ K Neel temperature are shown in (**b**)

(a)   (b)

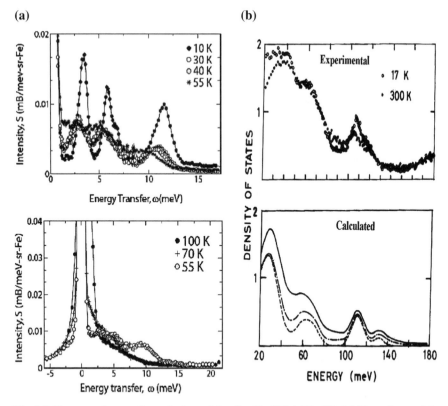

Fig. 5.6 Phonon and magnetic excitations of fayalite (Fe_2SiO_4) [41, 83]. (a) Dependence of the scattering on temperature for a fixed wave vector of 2 ± 0.25 Å (Aronson et al. [83]). (b) The observed and computed phonon density of states obtained from INS measurements and theoretical calculations [41]. The experimental neutron spectra of fayalite at T = 300 K and 17 K, inclusive of multiphonon contributions [41] is compared with the calculated phonon spectra having one-phonon and multiphonon contributions at 300 K (*full line*) and at 17 K (*dash-dot line*). The dashed line gives the calculated one phonon contribution to the phonon spectra at T = 300 K

The end member fayalite [17] also has a magnetic contribution to the INS data at temperatures below 65 K due to a paramagnetic to antiferromagnetic transition. Aronson et al. [83] have recently used low-temperature INS measurements (Figs. 5.5 and 5.6) to study the magnetic excitations in the antiferromagnetic and paramagnetic phases of polycrystalline fayalite (Fe_2SiO_4). Sharp, nondispersing excitations were found in the ordered state, at 3.3, 5.4, 5.9, and 11.4 meV and were interpreted as arising from the spin-orbit manifold of the Fe^{2+} ions. They have calculated the contribution of the heat capacity arising from these magnetic excitations and found that it compares favorably with the magnetic heat capacity deduced experimentally. Their analysis indicates that the M1 and M2 sites behave distinctly [83]. The M1 site behaves quasi-locally and appears in the heat capacity as a Schottky anomaly, which explains the shoulder in the heat capacity curve near 20 K, while the M2 site contributes predominantly to the critical lambda anomaly observed in the specific heat (Fig. 5.7). These results are complementary to our earlier INS experimental and theoretical studies giving the PDOS and phonon contributions to the heat capacity of fayalite (Fig. 5.7).

(a)

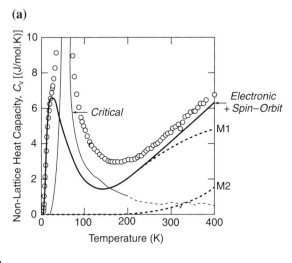

(b)

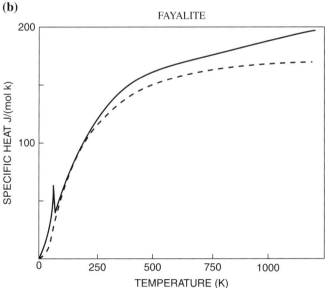

Fig. 5.7 (**a**) The non-lattice heat capacity of fayalite (*circles*) compared with the computed contributions from the M1 and M2 sites (*dashed*) and their sum (*bold solid line*) considering magnetic and electronic contributions (Aronson et al. [83]). The thin solid line [83] is the critical contribution and is shown dashed when its value falls below nominal uncertainties ($\sim$ 1 J/(mol K)) in the phonon contribution. (**b**) The phonon contributions of specific heat (*dashed line*) compared with the observed [135] specific heat (*full line*) of fayalite [41]. INS measurements of the phonon dispersion relation of fayalite have also been reported [40]

5.3.2 The MgSiO₃ Polymorphs Enstatite and Perovskite

Enstatite is the magnesium end member of pyroxene ($Mg_{(1-x)}$ $Fe_x)_2Si_2O_6$, which is an important rock-forming silicate and, with olivine, an abundant constituent of the upper mantle. Enstatite has several polymorphs: orthoenstatite; protoenstatite; and the low and high clinoenstatites, whose structures are closely related. The basic building blocks of the enstatite polymorphs comprise single SiO_4 chains and double MgO_6 octahedral bands (Fig. 5.3). Structurally, the polymorphs are distinguished by the different stacking sequences of the orientations of the MgO_6 octahedra with respect to the silicate chains. Inelastic neutron scattering measurements [26] of the PDOS in polycrystalline orthoenstatite ($Mg_2Si_2O_6$) have been carried out. The data are interpreted based on lattice dynamical models using interatomic potentials validated by X-ray diffraction, Raman, IR, and inelastic neutron data [26]. The computed thermodynamic properties (i.e., specific heat, equation of state, and thermal expansion at high pressure and high temperature) of the $MgSiO_3$ polymorphs enstatite and perovskite (Fig. 5.8) are in excellent agreement with reported experimental data [31]. It is worth noting that the model of interatomic potential, once tested satisfactorily, can be used to provide microscopic insights on a wide variety of complex phenomena. Due to their complex structures, first principles calculations of phonon properties have not been reported for the enstatite polymorphs.

The computed phase diagram of enstatite obtained by comparison of the Gibbs free energies of various phases [27] is found to be in qualitative agreement with experiments (Fig. 5.9). While there is some ambiguity from experiments as to whether orthoenstatite or $P2_1/c$-clinoenstatite (the low-pressure phase) is the stable ambient phase, lattice dynamics calculations predict essentially similar ambient state-free energies for the two phases [27]. Detailed lattice dynamics calculations indicate that the PDOS (as a function of pressure) of the various enstatite polymorphs [ortho, proto, $P2_1/c$-clinoenstatite, and $C2/c$-clinoenstatite (the high-pressure phase)] have gross similarities [27] and are in overall agreement with the measured PDOS of orthoenstatite. In the low-frequency region, the PDOS of the proto phase is shifted to lower frequencies, resulting in a higher vibrational entropy of the proto phase as compared to the ortho phase. The lowering of the low-frequency density of states in the proto phase is consistent with the observed softening of the elastic constant C_{55} in the proto phase as compared to orthoenstatite.

The phase transitions of minerals give rise to important seismic discontinuities (Fig. 5.9) in the Earth's mantle [1, 2, 30], which have been used to make inferences about the structure of the Earth. Molecular dynamics (MD) studies using these potentials have been very useful in understanding the high pressure–high temperature phase transitions of enstatite [31] and resultant seismic discontinuities therefrom (Fig. 5.9). With increasing pressure, enstatite (with silicon in tetrahedral coordination) transforms [31] first to a new, novel five-coordinated silicon phase and then to the lower-mantle perovskite phase involving six-coordinated silicon atoms. Although the occurrence of crystalline five coordinated silicate phases is somewhat rare, the new intermediate phase obtained through the simulations is found to be crystalline. The calculated seismic velocities and densities [31] across

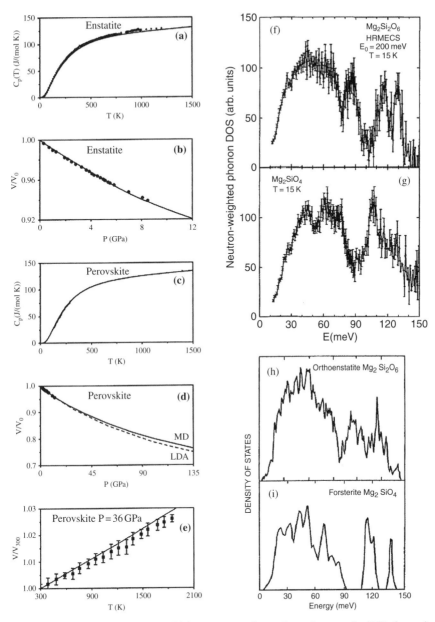

Fig. 5.8 The calculated high pressure–high temperature thermodynamic properties [30] of enstatite and perovskite (MgSiO$_3$). (**a–e**) The specific heat $C_P(T)$, equation-of-state, and thermal expansion are compared with reported experiments [121–135]. The inelastic neutron scattering data of phonon density of states (**f,g**) [26] are compared with the calculated (**h,i**) phonon density of states for orthoenstatite and forsterite

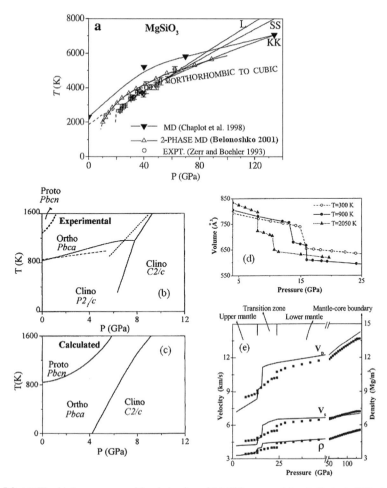

Fig. 5.9 (**a**) The high-pressure melting behavior of MgSiO$_3$ perovskite studied via MD simulations [31, 32, 151], compared with experiments [129]. (**b** and **c**) The computed phase diagram of enstatite [27] obtained by comparing the Gibb's free energies is in qualitative agreement with experimental [123, 133–135] observations. (**d** and **e**) Using transferable interatomic potentials [30], we simulated the transverse V_s and longitudinal V_P wave velocities (*full lines*) (**e**), which are in agreement with Preliminary Reference Earth Model (*symbols*) estimates [2]. These reveal that with increasing pressure, enstatite (having four-coordinated silicate groups) undergoes phase transitions to the lower-mantle perovskite phase (having six-coordinated silicon) via a novel intermediate five-coordinated silicate phase [30] (**d** and **e**)

the phase transitions (Fig. 5.9) for a pure MgSiO$_3$ mantle were overall in good agreement with Preliminary Reference Earth Model estimates based on seismic and other data [2]. These MD studies suggest that the major discontinuities between the upper mantle, transition zone, and lower mantle could arise partially as a result of changes in silicon coordination. The simulations have enabled a microscopic visualization of the key mechanisms of these transitions and provide useful insights about the variations in the vibrational properties due to these changes in silicate coordination.

Another important application of these potentials was understanding the high-pressure melting [31, 32] in $MgSiO_3$ perovskite (Fig. 5.9(a)). The MD simulations of $MgSiO_3$ perovskite revealed an orthorhombic to cubic transition accompanied by a sharp increase in diffusion of the oxygen atoms. The phase transition and melting temperature were found to depend heavily on the level of defects in the solid [31, 32]. At pressures like those found in the Earth's lower mantle, the transition is found to occur at temperatures substantially higher than the mantle temperatures. More recently, first principles techniques have been successfully applied to understand the complex melting of perovskite [94].

Recently, a first-principles-based approach [96] helped discover a new high-pressure post-perovskite phase. The post-perovskite orthorhombic *Cmcm* phase at high pressure was subsequently observed experimentally [95]. This post-perovskite phase is stable above 125 GPa at 2,500 K and exhibits a positive Clapeyron slope [96] such that the transformation pressure increases with temperature. Post-perovskite is considered to be a primary suspect for inducing seismic discontinuities [96, 97] corresponding to a depth of about 2,700 km, where the highly variable D'' (D double-prime) seismic discontinuity occurs. Post-perovskite holds great promise for mapping experimentally determined information regarding the temperatures and pressures of its transformation into direct information regarding temperature variations in the D'' layer [96–98]. Various workers have reported studies of the vibrational spectra and thermodynamic properties of post-perovskite [98].

5.3.3 High-Temperature Inelastic Neutron Scattering Measurements, Soft Mode-Driven Phase Transitions, and Pressure-Induced Amorphization of Silica

Silica is an important mineral occurring in the Earth's crust. Silica in its various crystalline and amorphous forms finds several industrial applications including being a raw material for glasses, ceramics, silicon, etc. The silica polymorph α-quartz exhibits several interesting properties including PIA, high pressure–high temperature phase transitions, anomalous elastic properties, negative Poisson ratios, and soft-mode behavior [25, 47–49]. Quartz oscillators and optical waveguides are used extensively in long-distance telecommunications and industry.

Inelastic neutron scattering measurements of the complete PDR of α-quartz along various high symmetry directions (Fig. 5.10) have been reported [99]. DFPT calculations of the PDR (Fig. 5.10) of α-Quartz [66] are in very good agreement with the INS data [99]. Ab initio DFPT calculations [67] indicate that the PDOS of α-quartz with tetrahedrally coordinated silicate is significantly different from that of stishovite, involving octahedrally coordinated silicon (Fig. 5.10). These differences in the PDOS give rise to important differences in their heat capacities (Fig. 5.11) [119, 120].

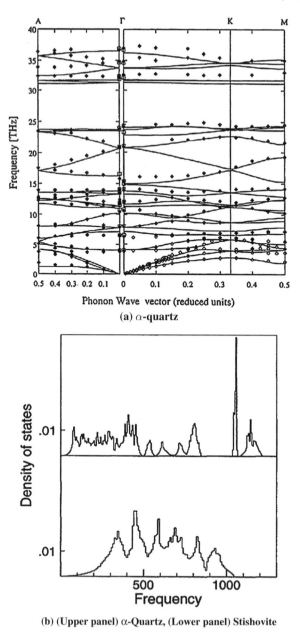

(a) α-quartz

(b) (Upper panel) α-Quartz, (Lower panel) Stishovite

Fig. 5.10 (a) The phonon dispersion relation of α-quartz (*symbols* give measured INS) data of Strauch and Dorner [99]) compared with the first-principles calculations (*lines*) of Gonze et al. [66]. (b) Ab initio density functional perturbation theory calculations (Lee and Gonze [67]) of the phonon density of states of α-quartz (*upper panel*) and stishovite (*lower panel*). The horizontal axis in (b) is Frequency in cm^{-1}

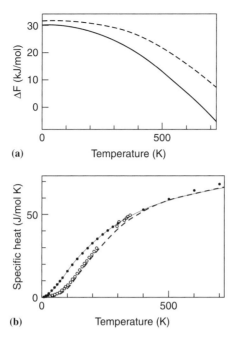

Fig. 5.11 (**a**) The phonon contributions to the Helmholtz free energies (Lee and Gonze [67]) of α-quartz (*full line*) and stishovite (*dashed line*). (**b**) The computed [67] and observed specific heat of α-quartz (*full line* calculated [67], *filled symbols* experiments [119]) and stishovite (*dashed line*, calculations [67]; *open symbols*, experiments [120])

Several high-temperature INS studies aimed at understanding the α–β phase transition of quartz (Fig. 5.12) and the PDR of the high-temperature β-phase and incommensurate phases of silica have been reported [71, 72]. High-temperature INS measurements [71, 72] reveal that α-quartz undergoes a classic displacive transition to a β-quartz phase at 573°C. Measurements of the relative intensity of the critical scattering about various reciprocal points establish that the general pattern of displacements associated with the α–β transition is determined by soft phonon mode (Fig. 5.12). Hyper-Raman studies indicate that the square of the phonon frequency (Fig. 5.12) and the integrated intensity of the soft phonon obey Landau–Cochran's soft mode theory [73].

Quartz amorphizes at pressures of around 18–35 GPa [25, 47–49]. Although PIA occurs in a variety of solids like α-quartz, coesite, ice, etc., its origin is not clearly understood. Molecular dynamics simulations [47] using interatomic potentials fitted to first principles total energy surfaces reveal PIA at around 22 GPa, in agreement with experiments. Around the amorphization pressure, α-quartz displays soft phonon modes [48] with a K-point phonon mode becoming unstable above 21 GPa. First principles DFPT calculations [25] of the PDR and elastic properties of α-quartz at high pressures (Figs. 5.13 and 5.14) confirm the K-point phonon instability at high pressure. The calculated crystal structure, equation of state, and long-wavelength phonon frequencies (Fig. 5.13) were found to be in good agreement with

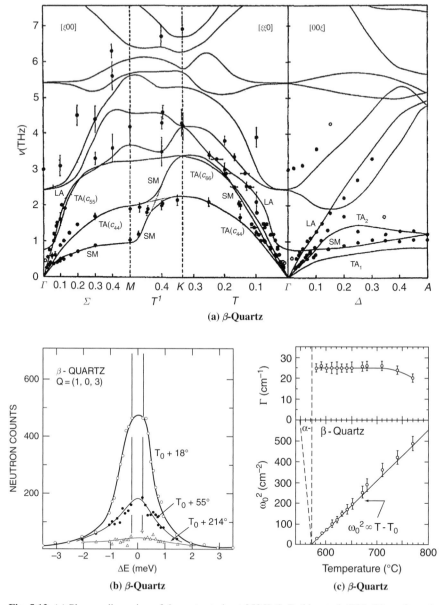

Fig. 5.12 (a) Phonon dispersion of β-quartz at about 950 K (J. Bethke et al. [72]). The soft mode (SM) is indicated. *Symbols* correspond to measured INS data and lines are the result of model calculations [72]. (b) Energy analysis of scattering about the (1,0,3) reflection in β-quartz at several temperatures above T_0 (Axe and Shirane [71]). The strong central peak is superimposed upon weaker critical inelastic scattering. (c) Variation of the soft-mode phonon frequency and width with temperature as obtained using hyper-Raman studies (Tezuka et al. [73])

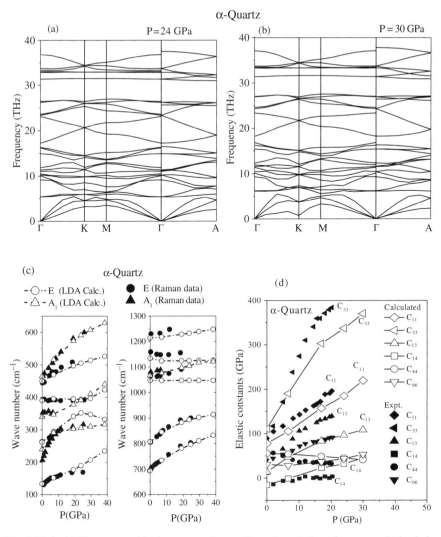

Fig. 5.13 (**a–b**) The computed high-pressure phonon dispersion relation of α-quartz obtained via first-principles calculations [25]. Although the computed zone center phonon frequencies [25] shown in (**c**) harden with pressure, in agreement with reported Raman and infrared data [112], around 30 GPa, a K-point zone boundary phonon mode becomes unstable, in qualitative agreement with earlier theoretical predictions [48]. The first principles [25] derived calculated elastic constants (**d**) are in good agreement with ambient-pressure Brillouin data [113], but deviate from experimental [113] values at high pressures. These deviations between first principles theory and experiments at high pressure are due to the extreme sensitivity of the obtained elastic constants to hydrostatic conditions [25]

the reported high-pressure experimental data and the available first principles LDA calculations. The calculations reveal that the zone boundary (1/3, 1/3, 0) K-point phonon mode becomes unstable for pressures above 32 GPa (Fig. 5.14).

Around the same pressure, studies of the Born stability criteria reveal that the structure is mechanically unstable [25]. The phonon and elastic softening (Fig. 5.14) are related to the high-pressure phase transitions and amorphization of quartz and these studies suggest that the mean transition pressure is lowered under nonhydrostatic conditions. Application of uniaxial pressure results in a post-quartz crystalline monoclinic $C2$ structural transition [25]. This structure, intermediate between quartz and stishovite, has two-thirds of the silicon atoms in octahedral coordination while the remaining silicon atoms remain tetrahedrally coordinated. This monoclinic $C2$ polymorph of silica, which is found to be metastable under ambient conditions, is possibly one of the several competing dense forms of silica containing octahedrally

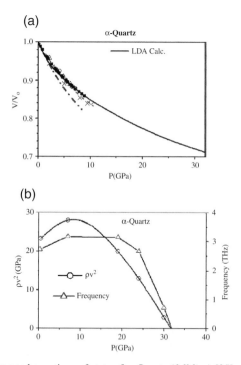

Fig. 5.14 (**a**) The computed equations of state of α-Quartz (*full line*) [25] are in good agreement with experimental (*symbols*) data [114–117] and reported calculations (*dashed* and *dash-dot lines*) [118]. (**b**) The calculated K-point soft-mode phonon frequency and the corresponding elastic constant as a function of pressure [25]. There is simultaneous elastic and phonon softening at high pressure in α-quartz around 32 GPa when the Born stability criteria gets violated [25], and the K-point phonon mode becomes unstable. A new post-quartz $C2$ structure [25], which is found to be structurally and dynamically stable at 32 GPa, is obtained. The calculated X-ray diffraction spectra of the post-quartz $C2$ phase [25] compares favorably with an unidentified high-pressure dense silica phase observed previously [141, 142]

coordinated silicon. The calculated X-ray intensities of this phase compare satisfactorily with a hitherto unidentified high-pressure phase of silica [25].

5.3.4 Zircon (ZrSiO₄)

Zircon is an important mineral found in igneous rocks and sediments. Because it is a natural host for the radioactive elements uranium and thorium in the Earth's crust, it is a potential candidate for nuclear waste storage. The high pressure–high temperature stability of zircon is therefore of considerable interest. A phase transition from the zircon to scheelite structure has been observed in static high-pressure and shock experiments.

Inelastic neutron scattering measurements of the PDR of zircon [24] were carried out at various reactors (Dhruva [37], Laboratoire Léon Brillouin [24]) and at the ISIS pulsed neutron source [37]. The observed INS data from reactors and spallation sources (Fig. 5.15) are in good agreement. The high symmetry of this mineral (*bct* (body centered tetragonal), space group *I4₁/amd*, 12 atoms in the primitive cell), relative to other minerals, was well exploited for selectively measuring phonons of specific group theoretical representations. The PDR data on zircon [24] may be seen as a rare example of extensive measurements of the PDR up to 85 meV. First principles DFPT calculations of the PDR were completed and found to be in agree-

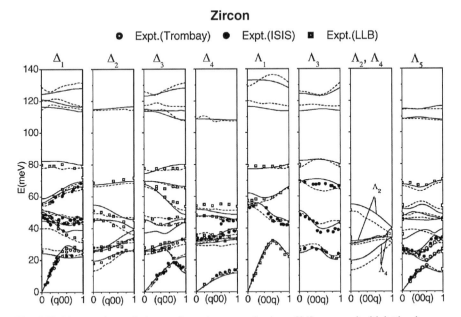

Fig. 5.15 The experimental phonon dispersion curves in zircon [24] compared with lattice dynamics calculations (*solid lines* for shell model, *dashed lines* for ab initio calculations). The *open squares, solid circles,* and *open circles* give the phonon peaks identified in the experiments at Laboratoire Léon Brillouin (France), ISIS (UK), and Dhruva (India), respectively [37]

ment with the measurements [24]. The experimental data of the phonon frequencies and the crystal structure have been used to refine an interatomic potential with a lattice dynamics shell model [24]. The shell model calculations produce a very good description of the available data on the PDOS measured on a polycrystalline sample. The model is used further to calculate the structure, dynamics, equation of state, specific heat, thermal expansion, and Gibb's free energy as a function of pressure in the zircon and scheelite phases. The free energy calculation reproduces the stability of the scheelite phases above 10 GPa [24] as compared to the zircon phase, which is in qualitative agreement with experimental observations.

5.3.5 High-Pressure Inelastic Neutron Scattering Studies

5.3.5.1 High-Pressure Inelastic Neutron Scattering of Body-Centered Cubic Iron

Iron is a major component of the Earth's core. Extensive efforts have been directed toward understanding the phase diagram of iron and the transition mechanisms inherent in it [74]. Recent progress in high-pressure techniques allows the measurement of phonon dispersion curves to $\sim$12 GPa by INS on triple-axis spectrometers. Klotz and Braden [74] have reported INS measurements of the phonon dispersion of body-centered cubic (bcc) iron under high pressure (up to 10 GPa), which sheds light on the bcc-hexagonal close-packed (hcp) transition in iron (Fig. 5.16). The complete determination of the phonon dispersion of any bcc metal under high pressure is of considerable importance to the understanding of iron under extreme pressure and temperature. The pressure dependence of phonon spectra was found to be surprisingly uniform. Contrary to the behavior found in other bcc elements, there was a lack of significant pre-transitional behavior close to the martensitic bcc–hcp transition, which could be related to the Burgers mechanism. Klotz et al. [74] have interpreted this finding as a confirmation of predictions from spin-polarized total energy calculations that explained the transition by the effect of pressure on the magnetism of iron. High-pressure frequencies were used to develop a lattice dynamics model from which thermodynamic quantities have been determined. The experiments provide highly detailed and accurate reference data for ab initio calculations that give inconsistent results on iron at core conditions. Furthermore, they provide excellent constraints and benchmarks for the more recently developed X-ray techniques that are able to derive and estimate elastic and dynamical properties of iron to much higher pressures.

Lubbers et al. [85] have studied the lattice dynamics of the hcp phase of iron with nuclear inelastic absorption of synchrotron radiation at pressures from 20 to 42 GPa. Subsequently, the PDOS of iron up to 153 GPa was also determined [86]. A variety of thermodynamic parameters were derived from the measured density of phonon states for hcp iron such as Debye temperatures, Gruneisen parameters, mean sound velocities, and lattice contributions to entropy and specific heat.

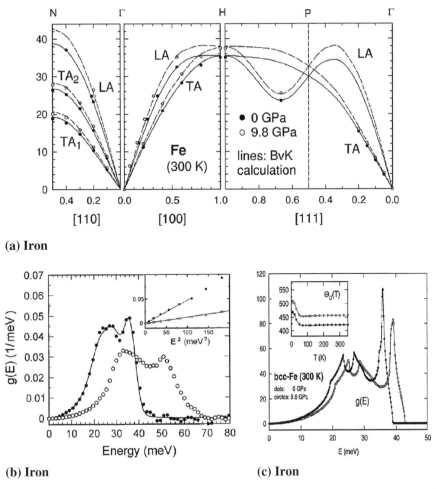

(a) Iron

(b) Iron   **(c) Iron**

Fig. 5.16 High-pressure inelastic neutron scattering (INS) studies of iron. (**a**) Phonon dispersion of body-centered cubic (*bcc*) iron to 10 GPa obtained via high-pressure INS (Klotz and Braden [74]). The INS data [74] of the PDR of *bcc* iron at 0 GPa (*full circles*) and 9.8 GPa (*open circles*) are compared with fits using theoretical model (*full line*). [Vertical axis: energy in meV.] (**b**) Nuclear resonant scattering (Lubbers et al. [85]) of the PDOS of iron at ambient pressure (*solid circles*) and at 42 GPa (*open circles*). The inset in (**b**) shows a plot of the low-energy part of $g(E)$ versus E^2 up to 15 meV [74] used to derive the sound velocity. (**c**) The phonon density of states $g(E)$ at 300 K and Debye temperature θ_D of *bcc* iron (**c**) at 0 and 9.8 GPa (Klotz and Braden [74])

5.3.5.2 High-Pressure Inelastic Neutron Scattering Studies: Applications to Negative Thermal Expansion Materials

High-pressure INS measurements have contributed to the understanding of negative thermal expansion (NTE) in a wide variety of materials including zirconium tungstate (ZrW_2O_8) [16, 39] and ice [76]. Isotropic NTE [16] has been reported in cubic ZrW_2O_8 over a wide range of temperatures (0–1,050 K). Mittal et al.

[16, 39] reported direct experimental determination of the Grüneisen parameters of phonon modes as a function of energy, averaged over the whole Brillouin zone by measuring the powder INS spectra at ambient pressure and at 1.7 kbar. They observed a pronounced softening of the phonon spectrum at 1.7 kbar compared to that at ambient pressure by about 0.1–0.2 meV for phonons of energy below 8 meV. This unusual phonon softening on compression, corresponding to large negative Grüneisen parameters, was able to account for the observed large NTE.

Water and ice are important constituents of the Earth. Strassle et al. [76] have reported INS single-crystal measurements of the phonon dispersion of ice, Ih, under hydrostatic pressure up to 0.5 GPa, at 140 K. These reveal a pronounced softening of various low-energy modes (Fig. 5.17). In particular, the transverse acoustic phonon branch in the [ζ00] direction, having polarization in the hexagonal plane, reveals pronounced softening. With the help of a lattice dynamics model [76], they demonstrate that these anomalous features in the phonon dispersion are at the origin of the NTE coefficient in ice below 60 K. Moreover, extrapolation to higher pressures shows that the mode frequencies responsible for the NTE approach zero at 2–5 GPa [76], which explains the known PIA in ice. These results [76] provided the first clear experimental evidence that PIA in ice was due to a lattice instability (i.e., mechanical melting supported by earlier theoretical calculations on α-quartz [48, 49]). Recent neutron scattering data [58] reveal the central role of phonon softening, which leads to a negative melting line, solid-state amorphization, and negative thermal expansion of ice. These studies [58] reveal that PIA is due to mechanical melting at low temperatures, while at higher temperatures amorphization is governed by thermal melting (violations of Born's and Lindemann's criteria, respectively). This confirms earlier conjectures of a crossover between two distinct amorphization mechanism and provides a natural explanation for the strong annealing observed in high-density amorphous ice [58].

Quasi-one-dimensional water encapsulated inside single-walled carbon nanotubes was studied by neutron diffraction and INS by Kolesnikov et al. [79]. Their results [79] reveal an anomalously soft dynamics characterized by pliable hydrogen bonds, anharmonic intermolecular potentials, and large-amplitude motions in nanotube ice (Fig. 5.18). The soft dynamics of nanotube water/ice arises mainly from the drastic change in hydrogen-bond connectivity of the central water chain. Anomalously enhanced thermal motions along the chain direction, interpreted by a low-barrier, flattened, highly anharmonic potential well, explain the large mean square hydrogen displacements and fluid-like behavior of nanotube water at temperatures far below the nominal freezing point.

5.3.5.3 High-Pressure Inelastic Neutron Scattering Studies of Iron Oxide, Germanium, Zinc, Gallium Antimonide, and Lead Telluride

High-pressure INS studies of several other simple systems such as germanium, iron oxide, zinc, gallium antimonide, lead telluride, etc., have also been reported [75]. The pressure-induced frequency shifts of the acoustic branches have been studied

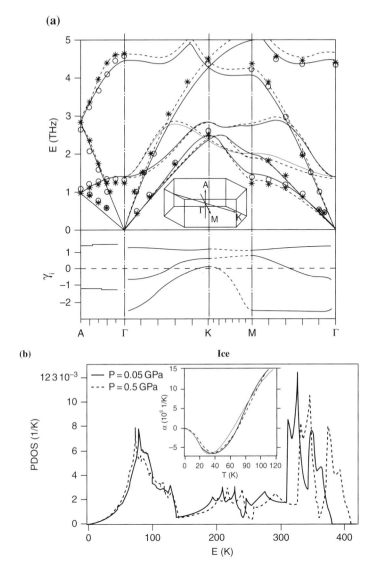

Fig. 5.17 (**a**) PDR of ice (Strassle et al. [76]) at $T = 140\,\text{K}$ at $0.05\,\text{GPa}$ (*circles*) and $0.5\,\text{GPa}$ (*stars*). The *solid* and *dashed lines* are fits using the LD model. (**a**) Bottom: Mode Gruneisen parameters γ_i of the transverse and longitudinal acoustic branches [76]. (**b**) PDOS ($T = 140\,\text{K}$) at 0.05 and $0.5\,\text{GPa}$ (Strassle et al. [76]). Inset: Linear thermal expansion coefficient of D_2O ice [76], Ih, reconstructed (*solid line*) from the measured phonon dispersion in comparison with literature data (*dashed lines*) [132, 145]

in considerable detail. In several of these solids pronounced "mode softening" is found under pressure. The mode Grüneisen parameters and elastic constants have been determined and are found to be in good agreement with predictions from first-principle calculations.

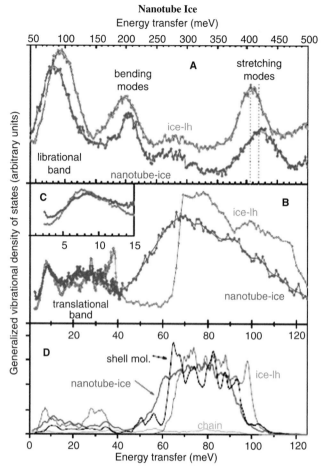

Fig. 5.18 Inelastic neutron scattering measurements of the generalized vibrational density of states (Kolesnikov et al., [79]) of nanotube ice and Ih at 9 K for (**A**) intramolecular bending (hydrogen–oxygen–hydrogen) and stretching (oxygen–hydrogen) modes and (**B**) intermolecular vibrations. A part of the observed density around 120 meV is due to multiphonon neutron scattering. The excess density below 5 meV in nanotube ice can be seen more clearly in the inset (**C**). The $G(E)$ obtained from molecular dynamics simulations at 50 K for nanotube ice [79], chain molecules in nanotube ice, and shell molecules shown separately (Kolesnikov et al. [79])

5.3.6 Incoherent Inelastic Neutron Scattering and Hydrogen Storage in Molecular Clathrates and Alanates

Clathrates are believed to occur in large quantities on some outer planets, moons, and trans-Neptunian objects, binding gas at fairly high temperatures [21, Chapter 12]. Clathrates have also been discovered in large quantity on Earth [21, Chapter 12]. Application of hydrogen as an environmentally clean and efficient fuel has evoked considerable interest in the use of hydrates and other simple molecular solids for hydrogen storage applications. Clathrate hydrates (or alternatively gas clathrates) are a class of solids in which gas molecules occupy "cages" made up of hydrogen-bonded water molecules. These cages are unstable when empty, collapsing into conventional ice crystal structure, but they are stabilized by the inclusion of appropriately sized molecules within them.

Clathrates serve as prototypical models for a class of crystalline framework materials (Fig. 5.19) with glass-like thermal conductivity [100, Chapter 12]. The

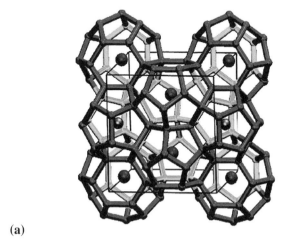

(a)

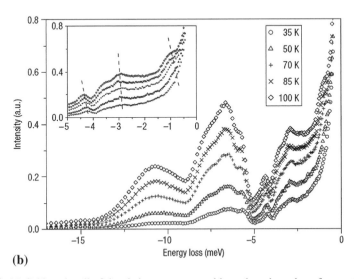

(b)

Fig. 5.19 (**a**) Cubic unit cell of the clathrate structure with a selected number of cages (Struzkin et al. [21]). (**b**) Incoherent inelastic neutron scattering (INS) spectra of krypton clathrate hydrate (Tse et al. [100]). The inset shows the low-frequency "rattling" vibrations due to the guests' motions. The lines are guides to the eye showing the change of the vibrational energies with temperature

anomalous thermal properties of clathrates are believed to arise because of the scattering of thermal phonons of the framework by "rattling" motions of the "guests" in the clathrate cages. Neutron diffraction studies have provided valuable quantitative information on the enclathration of deuterium under different temperature and

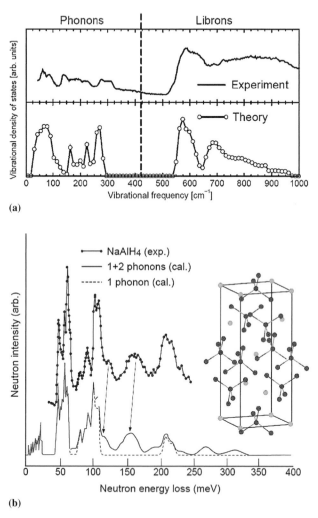

(a)

(b)

Fig. 5.20 (a) Comparison of INS data measured for hydrogen clathrate with the calculations for H_2O clathrate type *sII* cages (Struzkin et al. [21]). Calculations reproduce the librational modes of H_2O above $500\,cm^{-1}$ ($\sim70\,meV$) and the cage deformational modes (phonons) below $300\,cm^{-1}$ ($\sim40\,meV$) [21]. (b) Figure adapted from Ref. [20]. Measured (*top*) and calculated (*bottom*) neutron vibrational spectra of sodium alanate ($NaAlH_4$) (Neumann [20]; Iniguez et al. [81]). The calculated 1 and $1+2$ phonon contributions are shown as the *dashed* and *solid line*, respectively. The structure of $NaAlH_4$ determined from neutron diffraction data [20] is shown in the inset

pressure conditions [21]. Using incoherent INS in combination with site-specific krypton-83 nuclear resonant inelastic scattering spectroscopy and MD simulations, Tse et al. [100] provide unambiguous evidence and characterization of the effects on these guest–host interactions in a structure-II krypton clathrate hydrate. Incoherent INS data imply that there is a very strong coupling between the guest and the ice-like lattice vibrations (Fig. 5.19), as seen from the distinct features in the data resulting from the effect of the guest vibrations on the cage dynamics. The resonant scattering of phonons led to unprecedented large anharmonic motions of the guest atoms, which contribute to the anomalous thermal transport in this system. The explanation of the unusual MD has wide implications for the understanding of the thermal properties of disordered solids and structural glasses [100].

INS scattering studies of the PDOS of hydrogen clathrate using the filter-analyzer neutron spectrometer (FANS) (Fig. 5.1) at the National Institute of Standards and Technology (USA) have been reported (Fig. 5.20) and compared with lattice dynamics calculations [21]. Calculations for the empty cages show band gaps between the librational modes of H_2O above 70 meV and the cage deformational modes below 40 meV. The low-frequency part of the phonon band consists of vibrational modes where the clathrate cages move as a whole and get distorted at the same time. The calculations are in good overall agreement with experiments; the distinct lower- and higher-frequency vibrations correspond to the phonon and libron bands, respectively [21].

The discovery that titanium enhances reversible hydrogen sorption in sodium alanates has opened up new prospects for lightweight hydrogen storage. Inelastic neutron scattering measurements [81] of the vibrational spectrum of sodium alanate ($NaAlH_4$) reveal several sharp phonon bands up to 250 meV (Fig. 5.20). First principles DFPT calculations of the one-phonon spectra fail to explain several features in the observed data. Inclusion of two-phonon scattering, however, yields a spectrum that is in very good agreement with the experimental data. The results from DFPT are therefore very important as they help identify the multiphonon contributions and enable microscopic interpretations of the observed data. The low-energy modes are whole-body translations of the alanate (AlH_4) tetrahedra, while the strong features around 50 meV arise from the torsional modes of the tetrahedra. The peaks above 200 meV arise from various aluminum-hydrogen stretching modes. The features between 75 and 125 meV arise primarily from hydrogen–aluminum–hydrogen bending modes, with some hybridization to stretches and rotations of the tetrahedra.

5.3.7 Complementary Techniques: High-Pressure Phonon Dispersion Relation of Magnesium Oxide (MgO) and Site-Specific Phonon Density of States of Iron Obtained via Inelastic X-ray Scattering and Inelastic Nuclear Resonant Absorption Studies

Ghose et al. [87] have reported single-crystal inelastic X-ray scattering measurements (Fig. 5.21) of the longitudinal acoustic and optical phonon branches along

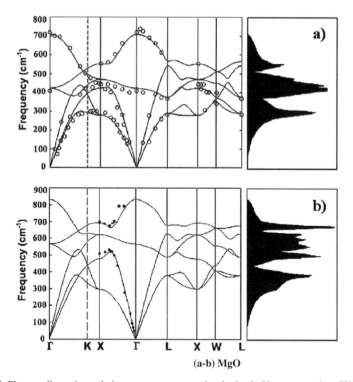

Fig. 5.21 Phonon dispersion relation measurements using inelastic X-ray scattering (IXS). (**a**, **b**) Phonon dispersion curves and phonon density of states of MgO (Ghose et al. [87]) in the NaCl structure: (**a**) at 0.1 MPa, (**b**) at 35 GPa. Experimental IXS data [87] are shown as *open circles* and *filled symbols*, while the solid lines correspond to ab initio calculations

the X direction of MgO at 35 GPa using diamond-anvil cells. As inelastic X-ray scattering studies require much smaller samples than corresponding neutron-based methods, they are better suited for such high-pressure studies. The experimentally observed phonon branches (Fig. 5.21) were found to be in remarkable agreement with ab initio lattice dynamics results. The derived thermodynamic properties such as specific heat, C_V, and the entropy, S, are in very good accord with the values obtained from a thermodynamically assessed data set involving measured data on molar volume, heat capacity at constant pressure, bulk modulus, and thermal expansion. Bosak and Krisch [89] have reported inelastic X-ray scattering measurements of the vibrational density of states of diamond (Fig. 5.22). The observed vibrational density of states of diamond (Fig. 5.22) is in good agreement with ab initio calculations [89]. The thermodynamic properties derived (Fig. 5.22) from the observed spectra are in good agreement with experimental heat capacity data in diamond.

Magnetite (Fe_3O_4) is an important mineral. It has a cubic spinel structure at room temperature. Magnetite is a mixed valent compound, and the iron atoms are located in two nonequivalent positions in the unit cell. One-third of the iron ions (Fe^{3+}) occupy the A sites, tetrahedrally coordinated, and the remaining two-thirds

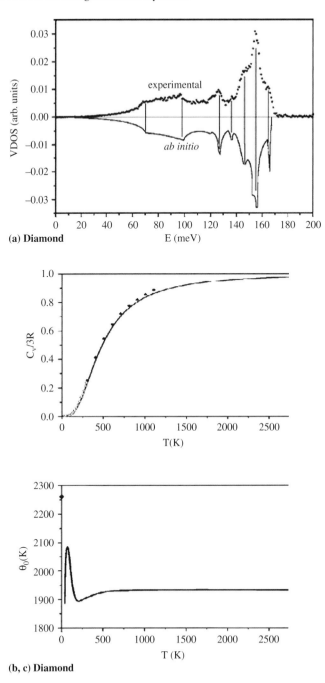

(a) Diamond

(b, c) Diamond

Fig. 5.22 (**a**) Inelastic X-ray scattering measurements of the vibrational density of states of diamond compared with ab initio calculations (Bosak and Krisch, [89]). (**b**) The specific heat and Debye temperature derived from the observed density of states are shown (*full lines*) [89]. The calculations (*lines*) compare very well with observed specific heat $C_p(T)$ data (*symbols*) [89]

of the iron ions (Fe^{3+} and Fe^{2+}) occupy the octahedrally coordinated B sites. Below $T_N = 858$ K, magnetite is ferromagnetic, with A-site moments aligned antiparallel to the B-site moments. Seto et al. [84] have reported inelastic nuclear resonant scattering of synchrotron radiation, which permits the identification of site-specific PDOS of the iron at the A and B sites (Fig. 5.23). The site-specific PDOS reveal important differences between the partial phonon densities of the iron in the distinct sites. These results are overall in good agreement with first-principles calculations.

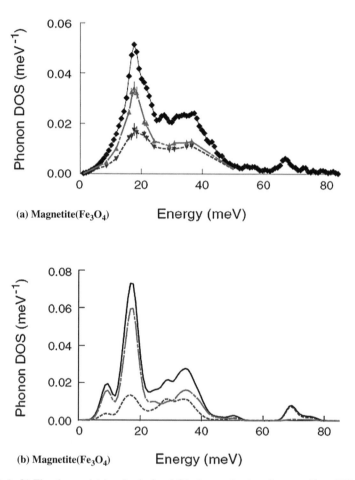

(a) **Magnetite(Fe$_3$O$_4$)**

(b) **Magnetite(Fe$_3$O$_4$)**

Fig. 5.23 (a, b) The observed (**a**) and calculated (**b**) phonon density of states of iron-57 in a magnetite (Fe$_3$O$_4$) sample obtained via nuclear resonant inelastic scattering of synchrotron radiation and first-principles calculations (Seto et al. [84]). In (**a**), the observed phonon density of states of all iron [84] is shown in *closed diamond symbols*, while the partial phonon density of states of the A and B sites obtained experimentally are shown as *downward* and *upward triangles*, respectively. In (**b**), the ab initio calculated phonon density of states [84] of all iron is shown as a *solid line*, while the calculated partial phonon density of states of the A and B sites are shown as *dashed* and *dash-dotted lines*, respectively

5.4 Conclusions and Future Directions

A combination of INS measurements and lattice dynamics calculations have been successfully used to study the phonon spectra, thermodynamic quantities, and phase transitions of several geophysically important minerals. Inelastic neutron scattering, which is not subject to tight selection rules on mode symmetries and wave vectors, can be used to determine phonon dispersion curves and PDOS, which provide fundamental information for the prediction of mineral behavior and phase transformations under pressures and temperatures like those of the Earth's interior. In this chapter we have reviewed the current understanding of the dynamical behavior of key mantle components like olivine and pyroxene, important silica polymorphs, zircon, magnesium oxide, and other complex silicates and recent studies on water and ice; hydrogen storage materials like clathrates; and high pressure–high temperature studies on quartz, ice, iron, etc.

Detailed INS measurements supported by extensive theoretical calculations have extended our knowledge of the nature of PDR and density of states of minerals and their variations in various mineral phases in the Earth's crust, mantle, and core. These provide microscopic insights on novel phenomena like the high-pressure phonon softening associated with structural phase transitions [25], phase diagrams, melting, mechanisms of fast-ion diffusion [147, 148], origin of low thermal conductivity etc. [149], etc. *State-of-the-art* quantum-mechanical calculations have now been used to successfully study the PDR and thermoelastic properties of several mantle minerals. Theoretical studies essential for the interpretation and analysis of complex neutron data still use atomistic approaches for complex minerals like garnets and pyroxenes, whose complex crystallographic structures pose significant problems for theory. The neutron experiments validate the models and the models in turn have been fruitfully used to calculate the phonon spectra and various thermodynamic properties at high pressures and high temperatures. The calculations have been very useful in the planning, execution, and analysis of the experiments and have enabled microscopic atomic-level interpretations of the observed data and thermodynamic properties. The integration of neutron scattering experiments with theoretical methods has been extremely successful in providing microscopic insights relevant to high-pressure mineral sciences, seismology, geochemistry, petrology, etc., and provides a new vehicle for exploring planetary interiors.

In addition to the studies reviewed here, studies of mineral–water interfaces using inelastic and quasielastic neutron scattering are discussed in detail in Chapter 8 of this book. Recent ab initio and INS studies have provided insights on the fundamental vibrational signatures of ferroelectric and paraelectric behavior [146] in model perovskites and suggest that vibrational spectroscopy, in general, and neutron-based methods, in particular, can play a vital role in materials design [146]. Various workers have reported studies of the intimate connections between phonon dispersion relations, vibrational spectroscopy and thermal conductivity [Chapter 12, 149, 150]. Contributions to lattice thermal conductivity from various phonon modes were uncovered using calculated group velocities and Gruneisen constants [Chapter 12, 149] in simple fluorite-structured oxides [149]. Design of new materials tuning thermal conductivity was proposed to build more efficient fuels for modern nuclear industries [149].

Several complementary techniques like inelastic X-ray scattering and inelastic nuclear resonant scattering from select isotopes like iron-57 have been used to study the high pressure–high temperature PDR, density of states, etc. As inelastic X-ray scattering studies require much smaller samples than the corresponding neutron-based methods, they are better suited for high-pressure studies. Nevertheless, neutron scattering techniques are particularly suited to study the structure and dynamics of low-Z components like hydrogen and oxygen, which are important constituents of most minerals. Neutron-based methods will therefore continue to be the techniques of choice for experimental investigations in mineral and earth sciences. Current advances and development of next-generation spallation sources offer excellent prospects for extensions to higher pressures and temperatures, which will facilitate in situ studies of the effects of these environmental variables on phonon and magnetic excitations, dispersion relation, and density of states, which in turn will lead to greater understanding of the microscopic origins of anomalous phase transitions and thermodynamic properties. Applications of INS to in situ studies offer a unique opportunity for solving fine structural details, which can permit modeling and interpretation of fundamental thermodynamic parameters. Extreme pressure conditions radically alter the properties of materials [6], and such studies have important implications for a wide range of problems, from fundamental physics to earth and planetary sciences, chemistry, and materials science. The larger flux of proposed next-generation neutron sources would imply studies using smaller crystals and samples, making it easier to study important minerals like perovskite, stishovite, SiO_2, etc., which can be synthesized in very small amounts.

References

1. D. L. Anderson, *Theory of the Earth* (Blackwell Scientific Publications, Boston, 1989)
2. A. M. Dzeiwonski and D. L. Anderson, Phys. Earth Planet. Inter. **25**, 297–356 (1981)
3. J. P. Poirier, *Introduction to the Physics of the Earth's Interior* (Cambridge University Press, Cambridge, 1991)
4. R. Jeanloz, *Mantle Convection Plate Tectonics and Global Dynamics*, edited by W. R. Peltier (Gordon and Breach, New York, 1986)
5. R. Jeanloz and A. B. Thompson, Rev. Geophys. Space Phys. **21**, 51–74 (1983)
6. H.-K. Mao and R. J. Hemley, Proc. Nat. Acad. Sci. **104**, 9114 (2007)
7. W. Marshall and S. W. Lovesay, *Theory of Thermal Neutron Scattering* (Oxford University Press, Oxford, 1971)
8. B. Dorner, *Coherent Inelastic Neutron Scattering in Lattice Dynamics* (Springer-Verlag, Berlin, 1982)
9. P. Bruesch, *Phonons: Theory and Experiments I* (Springer-Verlag, Berlin, Heidelberg, New York, 1982)
10. M. T. Dove, *Introduction to Lattice Dynamics* (Cambridge University Press, Cambridge, 1993)
11. G. Venkataraman, L. Feldkamp, and V. C. Sahni, *Dynamics of Perfect Crystals* (MIT Press, Cambridge, 1975)

12. M. P. Allen and D. J. Tildesley, *Computer Simulation of Liquids* (Oxford University Press, Oxford, 1989)
13. H. Gould, J. Tobochnik, and W. Christian, *An Introduction to Computer Simulation Methods; Applications to Physical Systems, 3rd edition* (Addison-Wesley, Boston, MA, 2006)
14. S. L. Chaplot, N. Choudhury, S. Ghose, M. N. Rao, R. Mittal, and P. Goel, Eur. J. Mineral. **14**, 291–329 (2002)
15. N. Choudhury, S. L. Chaplot, S. Ghose, M. N. Rao, and R. Mittal in *Energy Modelling in Minerals*, edited by C. M. Gramaccioli (Eötvös University Press, Budapest, 2002), European Mineralogical Union (EMU) *Notes in Mineralogy*, Vol. 4, Chapter 8, pp. 211–243
16. R. Mittal, S. L. Chaplot, and N. Choudhury, Prog. Mater Sci. **51**, 211–286 (2006)
17. S. Ghose, N. Choudhury, S. L. Chaplot, and K. R. Rao, in *Advances in Physical Geochemistry: Thermodynamic Data*, edited by S. K. Saxena (Springer-Verlag, New York, 1992), pp. 283–314
18. C.-K. Loong, in *Neutron Scattering in Earth Sciences*, edited by H.-R. Wenk (Geochemical Society/Mineralogical Society of America, USA, 2006), *Rev. Mineral. Geochem.*, Vol. 63, pp. 233–254
19. S. M. Bennington, J. Mat. Sci. **39**, 6757–6779 (2004)
20. D. A. Neumann, Mater. Today **9**, 34 (2006)
21. V. V. Struzhkin, B. Militzer, W. L. Mao, H. K. Mao, and R. J. Hemley, Chem. Rev. **107**, 4133 (2007)
22. K. R. Rao, S. L. Chaplot, N. Choudhury, S. Ghose, and D. L. Price, Science **236**, 64–65 (1987)
23. K. R. Rao, S. L. Chaplot, N. Choudhury, S. Ghose, J. M. Hastings, L. M. Corliss, and D. L. Price, Phys. Chem. Miner. **16**, 83–97 (1988)
24. S. L. Chaplot, L. Pintschovius, N. Choudhury, and R. Mittal, Phys. Rev. B **73**, 094308-1-8 (2006)
25. N. Choudhury and S. L. Chaplot, Phys. Rev. B **73**, 094304-1-11 (2006)
26. N. Choudhury, S. Ghose, C. P. Chowdhury, C. K. Loong, and S. L. Chaplot, Phys. Rev. B **58**, 756–765 (1998)
27. N. Choudhury and S. L. Chaplot, Solid State Commun. **114**, 127–132 (2000)
28. N. Choudhury, S. L. Chaplot, K. R. Rao, and S. Ghose, Indian J. Pure Appl. Phys. **35**, 690–698 (1997)
29. M. N. Rao, S. L. Chaplot, N. Choudhury, K. R. Rao, R. T. Azuah, W. T. Montfrooij, and S. M. Bennington, Phys. Rev. B **60**, 12061–12068 (1999)
30. S. L. Chaplot and N. Choudhury, Am. Mineral. **86**, 752–761 (2001)
31. S. L. Chaplot, N. Choudhury, and K. R. Rao, Am. Mineral. **83**, 937–941 (1998)
32. S. L. Chaplot and N. Choudhury, Am. Mineral. **86**, 195–196 (2001)
33. N. Choudhury, S. L. Chaplot, and K. R. Rao, Phys. Chem. Miner. **16**, 599–605 (1989)
34. P. Goel, M. N. Rao, N. Choudhury, S. L. Chaplot, S. Ghose, and M. Yethiraj, Appl. Phys. A **74**, S1149–S1151 (2002)
35. P. Goel, N. Choudhury, and S. L. Chaplot, Phys. Rev. B **70**, 174307 (2004)
36. R. Mittal, S. L. Chaplot, N. Choudhury, and C.-K. Loong, Phys. Rev. B **61**, 3983–3988 (2000)
37. R. Mittal, S. L. Chaplot, R. Parthasarathy, M. J. Bull, and M. J. Harris, Phys. Rev. B **62**, 12089 (2000)
38. R. Mittal, S. L. Chaplot, and N. Choudhury, Phys. Rev. B **64**, 94320 (2001)
39. R. Mittal, S. L. Chaplot, H. Schober, and T. A. Mary, Phys. Rev. Lett. **86**, 4692 (2001)
40. S. Ghose, J. M. Hastings, N. Choudhury, S. L. Chaplot, and K. R. Rao, Physica B **174**, 83–86 (1991)
41. D. L. Price, S. Ghose, N. Choudhury, S. L. Chaplot, and K. R. Rao, Physica B **174**, 87–90 (1991)
42. P. Bose, R. Mittal, S. L. Chaplot, and N. Choudhury, *Proc. Of the International Symposium on Neutron Scattering* (2008), Pramana-Journal of Physics (India) (In Press)

43. M. N. Rao, R. Mittal, N. Choudhury, and S. L. Chaplot, Pramana **63**, 73 (2004)
44. M. N. Rao, S. L. Chaplot, K. N. Prabhatashree, and S. Ghose, App. Phys. A **74**, S1152 (2002)
45. A. Sen, M. N. Rao, R. Mittal, and S. L. Chaplot, J. Phys. Condens. Matter **17** 6179 (2005)
46. P. Goel, R. Mittal, S. L. Chaplot, and N. Choudhury, Solid State Physics (India) **50**, 619–620 Proceedings of the DAE Solid State Physics Symposium, (Edited by V. K. Aswal, K. V. Bhushan and J. V. Yakhmi), Prime Time Education Publishers, Mumbai (2005)
47. S. L. Chaplot and S. K. Sikka, Phys. Rev. B **47**, 5710 (1993)
48. S. L. Chaplot and S. K. Sikka, Phys. Rev. Lett. **71**, 2674 (1993)
49. S. L. Chaplot, Phys. Rev. Lett. **83**, 3749 (1999)
50. R. M. Wentzcovitch, C. da Silva, J. R. Chelikowsky, and N. Bingelli, Phys. Rev. Lett. **80**, 2149 (1998)
51. L. Li, R. M. Wentzcovitch, D. J Weidner, and C. R. S. Da Silva, J. Geophys. Res. **112**, B05206 (2007)
52. B. B. Karki, R. M. Wentzcovitch, S. de Gironcoli, and S. Baroni, Phys. Rev. B **62**, 14750–14756 (2000)
53. B. B. Karki, L. Stixrude, and R. M. Wentzcovitch, Rev. Geophys. **39**, 507 (2001)
54. R. Martoňák, D. Donadio, A. R. Oganov, and M. Parrinello, Nat. Mater. **5**, 623–626 (2006)
55. R. Martoňák, A. Lio, and M. Parinello, Phys. Rev. Lett. **90**, 075503 (2003)
56. A. R. Oganov and C. W. Glass, J. Chem. Phys. **124**, 244704 (2006)
57. A. R. Oganov, J. P. Brodholt, and G. D. Price, in *Energy Modelling in Minerals*, edited by C. M. Gramaccioli (Eötvös University Press, Budapest, 2002), European Mineralogical Union (EMU), *Notes in Mineralogy*, Vol. 4, pp. 83–170
58. T. Strassle, S. Klotz, G. Hamel, M. M. Koza, and H. Schober, Phys. Rev. Lett. **99**, 175501 (2007)
59. R. Heid, D. Strauch, and K.-P. Bohnen, Phys. Rev. B **61**, 8625–8627 (2000)
60. X. Sha and R. E. Cohen, Phys. Rev. B **73**, 104303 (2006)
61. R. M. Martin, *Electronic Structure: Basic Theory and Practical Methods* (Cambridge University Press, Cambridge, 2004)
62. S. Baroni, S. de Gironcoli, A. D. Corsco, and P. Giannozzi, Rev. Mod. Phys. **73**, 515 (2001)
63. S. Baroni, P. Giannozzi, and A. Testa, Phys. Rev. Lett. **58**, 1861 (1987)
64. P. Giannozzi, S. de Gironcoli, P. Pavone, and S. Baroni, Phys. Rev. B **43**, 7231 (1991)
65. X. Gonze, D. C. Allan, and M. P. Teter, Phys. Rev. Lett. **68**, 3603 (1992)
66. X. Gonze, J. C. Charlier, D. C. Allan, and M. P. Teter, Phys. Rev. B **50**, 13035 (1994)
67. C. Lee and X. Gonze, Phys. Rev. B **51**, 8610 (1995)
68. C. Lee and X. Gonze, Phys. Rev. Lett. **72**, 1686 (1994)
69. C. Lee and X. Gonze, Phys. Rev. B **56**, 7321 (1997)
70. X. Gonze, J.-M. Beuken, R. Caracas, F. Detraux, M. Fuchs, G.-M. Rignanese, L. Sindic, M. Verstraete, G. Zerah, F. Jollet, M. Torrent, A. Roy, M. Mikami, Ph. Ghosez, J.-Y. Raty, and D. C. Allan, Comput. Mater. Sci. **25**, 478 (2002)
71. J. D. Axe and G. Shirane, Phys. Rev. B **1**, 342–348 (1970)
72. J. Bethke, G. Dolino, G. Eckold, B. Berge, I. M. Vallad, C. M. Zeyen, T. Hahn, H. Arnold, and F. Moussa, Europhys. Lett. **3**(Z), 207–212 (1987)
73. Y. Tezuka, S. Shin, and M. Ishigame, Phys. Rev. Lett. **66**, 2356 (1991)
74. S. Klotz and M. Braden, Phys. Rev. Lett. **85**, 3209 (2000)
75. S. Klotz, Zeitschrift für Kristallographie **216**, 420 (2001)
76. Th. Strässle, A. M. Saitta, S. Klotz, and M. Braden, Phys. Rev. Lett. **93**, 225901 (2004)
77. V. A. Somenkov, J. Phys. Condens. Matter **17**, S2991–S3003 (2005)
78. R. J. McQueeney, M. Yethiraj, W. Montfrooij, J. S. Gardner, P. Metcalf, and J. M. Honig, Phys. Rev. B **73**, 174409 (2006)
79. A. I. Kolesnikov, J. M Zanotti, C.-K. Loong, P. Thiyagarajan, A. P. Moravsky, R. O. Loutfy, and C. J. Burnham, Phys. Rev. Lett. **93**, 035503 (2004)
80. T. Yildirim and M. R. Hartman, Phys. Rev. Lett. **95**, 215504 (2005)

81. J. Iniguez, T. Yildirim, T. J. Udovic, M. Sulic, and C. M. Jensen, Phys. Rev. B **70**, 060101 (2004)

82. B. C. Hauback, H. W. Brinks, R. H. Heyn, R. Blom, and H. Fjellvag, J. Alloys Compd. **394**, 35 (2005)

83. M. C. Aronson, L. Stixrude, M. K. Davis, W. Gannon, and K. Ahilan, Am. Mineral. **92**, 481–490 (2007)

84. M. Seto, S. Kitao, Y. Kobayashi, R. Haruki, Y. Yoda, T. Mitsui, and T. Ishikawa, Phys. Rev. Lett. **91**, 185505 (2003)

85. R. Lübbers, H. F. Grünsteudel, A. I. Chumakov, and G. Wortmann, Science **287**, 1250 (2000)

86. H. K. Mao, J. Xu, V. V. Struzhkin, J. Shu, R. J. Hemley, W. Sturhahn, M. Y. Hu, E. E. Alp, L. Vocadlo, D. Alfe, G. D. Price, M. J. Gillan, M. Schwoerer-Bohning, D. Hausermann, P. Eng, G. Shen, H. Giefers, R. Lubbers, and G. Wortmann, Science **292**, 914–916 (2001)

87. S. Ghose, M. Krisch, A. R. Oganov, A. Beraud, A. Bosak, R. Gulve, R. Seelaboyina, H. Yang, and S. K. Saxena, Phys. Rev. Lett. **96**, 035507 (2006)

88. M. Schwoerer-Böhning, A. T. Macrander, and D. A. Arms, Phys. Rev. Lett. **80**, 5572 (1998)

89. A. Bosak and M. Krisch, Phys. Rev. B **72**, 224305 (2005)

90. G. Shirane, Rev. Mod. Phys. **46**, 437–449 (1974)

91. B. Winkler, Phys. Chem. Mineral. **23**, 310–318 (1996)

92. B. Winkler and B. Hennion, Phys. Chem. Min. **21**, 539–545(1994)

93. J. M. Carpenter and D. L. Price, Phys. Rev. Lett. **54**, 441–443 (1985)

94. A. B. Belonoshko, N. V. Skorodumova, A. Rosengren, R. Ahuja, B. Johansson, L. Burakovsky, and D. L. Preston, Phys. Rev. Lett. **94**, 195701 (2005)

95. M. Murakami, K. Hirose, K. Kawamura, N. Sata, and Y. Ohishi, Science **304**, 855–858 (2004)

96. A. R. Oganov and S. Ono, Nature **430**, 445 (2004)

97. R. M. Wentzcovitch, T. Tsuchiya, and J. Tsuchiya, Proc. Nat. Acad. Sci. **103**, 543 (2005)

98. R. Caracas and R. E. Cohen, Geophys. Res. Lett. **33**, L12S05 (2006)

99. D. Strauch and B. Dorner, J. Phys. Condens. Matter **5**, 6149 (1993)

100. J. S. Tse, D. D. Klug, J. Y. Zhao, W. Sturhahn, E. E. Alp, J. Baumert, C. Gutt, M. R. Johnson, and W. Press, Nature Mater. **4**, 917 (2005)

101. S. Ghose, N. Choudhury, S. L. Chaplot, C. P. Chowdhury, and S. K. Sharma, Phys. Chem. Miner. **20**, 469–477 (1994)

102. B. Winkler and W. Buehrer, Phys. Chem. Miner. **17**, 453–461 (1990)

103. G. Artioli, A. Pavese, and O. Moze, Am. Mineral. **81**, 19 (1996)

104. A. Pavese, G. Artioli, and O. Moze, Eur. J. Mineral. **10**, 59–70 (1998)

105. J. Zhao, P. H. Gaskell, L. Cormier, and S. M. Bennington, Physica B **241**, 906 (1998)

106. R. Mittal, S. L. Chaplot, N. Choudhury, and C. K. Loong, J. Phys. Condens. Matter **19**, 446202 (2007)

107. H. Schober, D. Strauch, and B. Dorner, Zeit. fur Physik B **92**, 273 (1993)

108. T. May, D. Strauch, H. Schober, and B. Dorner, Physica B: Condens. Matter **234–236**, 133–134(1997)

109. E. R. Cowley and A. K. Pant, Phys. Rev. B **8**, 4795–4800 (1973)

110. M. T. Dove, M. E. Hagen, M. J. Harris, B. M. Powell, U. Steigenberger, and B. Winkler, J. Phys. Condens. Matter **4**, 2761–2774(1992)

111. J. C. Nipko, C.–K. Loong, M. Loewenhaupt, M. Braden, W. Reichardt, and L. A. Boatner, Phys. Rev. B **56** 11584 (1997)

112. R. J. Hemley, in *High Pressure Research in Mineral Physics: the Akimoto Volume*, edited by M. H. Manghnani and Y. Syono (American Geophysical Union, Washington, D.C., 1987), *Mineral Physics*, Vol. 39, p. 347

113. E. Gregoryanz, R. J. Hemley, H.-K. Mao, and P. Gillet, Phys. Rev. Lett. **84**, 3117 (2000)

114. R. A. Angel, D. R. Allan, R. Miletich, and L. W. Finger, J. Appl. Crystallogr. **30**, 461 (1997)

115. L. Levien, C. T. Prewitt, and D. J. Weidner, Am. Mineral. **65**, 920 (1980)

116. J. Glinnemann, H. E. King, H. Schulz, Th. Hahn, S. J. La Placa, and F. Dacol, Z. Kristallogr. **198**, 177 (1992)
117. R. M. Hazen, L. W. Finger, R. J. Hemley, and H. K. Mao, Solid State Commun. **72**, 507 (1989)
118. Th. Demuth, Y. Jeanvoine, J. Hafner, and J. G. Angyan, J. Phys. Condens. Matter. **11**, 3833 (1999)
119. R. C. Lord and J. C. Morrow, J. Chem. Phys. **26**, 230 (1957)
120. J. C. Holm et al., Geochim Cosmochim Acta **31**, 2289 (1967)
121. K. M. Krupka, B. S. Hemingway, R. A. Robie, and D. M. Kerrick, Am. Mineral. **70**, 261–271 (1985)
122. K. M. Krupka, R. A. Robie, B. S. Hemingway, D. M. Kerrick, and J. Ito, Am. Mineral. **70**, 249–260 (1985)
123. R. J. Angel and D. A. Hugh-Jones, J. Geophys. Res. **99**(B10), 19,777–19,783 (1994)
124. L. Thieblot, C. Tequi, and P. Richet, Am. Mineral. **84**, 848–855 (1999)
125. M. Akaogi, M. Ito, and E. Ito, Geophys. Res. Lett. **20**, 105–108 (1993)
126. H. K. Mao, R. J. Hemley, Y. Fei, J. F. Shu, L. C. Chen, A. P. Jephcoat, and Y. Wu, J. Geophys. Res. B **96**, 8069–8079 (1999)
127. N. Funamori, and T. Yagi, Geophys. Res. Lett. **20**, 387–391 (1993)
128. R. M. Wentzcovitch, J. L. Martins, and G. D. Price, Phys. Rev. Lett. **70**, 3947–3950 (1993)
129. A. Zerr, and R. Boehler, Science **262**, 553–555 (1993)
130. E. Ito and T. Katsura, in *High Pressure Research in Mineral Physics, Applications to Earth and Planetary Sciences*, edited by Y. Syono and M. H. Manghani (Terra Scientific, Tokyo, 1992), pp. 35–44
131. T. Kato and M. Kumazawa, Nature **316**, 803–805 (1985)
132. K. Röttger, A. Endriss, J. Ihringer, S. Doyle, and W. F. Kuhs, Acta Crystallogr. Sect. B **50**, 644 (1994).
133. R. E. G. Pacalo and T. Gasparik, J. Geophys. Res. B **95**, 15853 (1990)
134. M. Kanazaki, Phys. Chem. Miner. **17**, 726 (1991)
135. R. A. Robie, C. B. Finch, and B. S. Hemingway, Am. Mineral. **67**, 463–469 (1982)
136. R. P. Santoro, R. E. Newnham, and S. Nomura, J. Phys. Chem. Solids **27**, 655–666 (1966)
137. M. Cococcioni, A. Dal Corso, and S. de Gironcoli, Phys. Rev. B **67**, 094106 (2003)
138. W. Lottermoser, R. Muller, and H. Fuess, J. Magn. Magn. Mater. **54–57**, 1005–1006 (1986)
139. M. Hayashi, I. Tamura, O. Shimomura, H. Sawamoto, and H. Kawamura, Phys. Chem. Miner. **14**, 341–344 (1987)
140. R. T. Downs and M. H. Wallace, Am. Mineral. **88**, 247 (2003)
141. J. Haines, J. M. Léger, F. Gorelli, and M. Hanfland, Phys. Rev. Lett. **87**, 155503 (2001)
142. K. J. Kingma, R. J. Hemley, H. K. Mao, and D. R. Veblen, Phys. Rev. Lett. **70**, 3927 (1993)
143. A. R. Oganov, R. Martoňák, A. Laio, P. Raiteri, and M. Parrinello, Nature **438**, 1142–1144 (2005)
144. S. K. Mishra, N. Choudhury, S. L. Chaplot, P. S. R. Krishna, and R. Mittal, Phys. Rev. B **76**, 024110 (2007)
145. G. Dantl, Z. Phys. **166**, 115 (1962)
146. N. Choudhury, E. J. Walter, A. I. Kolesnikov, and C. K. Loong, Phys. Rev. **B77**, 134111 (2008)
147. P. Goel, N. Choudhury, and S. L. Chaplot, Phys. Rev. B **70** 1734307 (2004)
148. P. Goel, N. Choudhury, and S. L. Chaplot, J. Phys: Condens. Matter **19**, 386239 (2007).
149. Q. Yin and Y. Savrasov, Phys. Rev. Lett. **100**, 225504 (2008)
150. A. M. Hofmeister, Science **283**, 1699 (1999)
151. A. B. Belonoshko, Am. Mineral. **86**, 194 (2001)
152. P. Goel, N. Choudhury, and S. L. Chaplot, J. Nucl. Materials, **377**, 438–443 (2008)

Chapter 6
A Microscopic View of Mass Transport in Silicate Melts by Quasielastic Neutron Scattering and Molecular Dynamics Simulations

Andreas Meyer, Florian Kargl, and Jürgen Horbach

Abstract The application of quasielastic neutron scattering and molecular dynamics simulation to the study of mass transport in silicate melts is outlined. It is shown how the knowledge of atomic dynamics and structure reveals the mechanisms of mass transport. Peculiar properties of atomic diffusion and viscous flow behaviour as a function of melt composition are discussed in terms of the formation of alkali diffusion channels in the static structure. This non-homogeneous distribution of alkali ions in a disrupted tetrahedral Si–O network is investigated in binary lithium, sodium and potassium silicate melts and in ternary sodium aluminosilicates and sodium ironsilicates representing the main compositions of natural volcanic rocks.

6.1 Introduction

Understanding the structure–transport relation in silicate melts on a microscopic level is central for geosciences and glass technology. Casting processes during the production of glass materials, as well as geological processes such as the eruption of a volcano, depend crucially on mass transport on atomic scales [1–3]. This becomes evident when one considers the temperature dependence of the shear viscosity η for various silicate melts [4–7]. Pure silica (SiO_2) is the prototype of a glass-forming system that forms a tetrahedral network structure. The rigid network structure leads to a rather slow dynamics even at relatively high temperatures. Note that at its melting temperature, $T_m \approx 2000\,K$, silica has a viscosity of about $10^6\,Pa\,s$ [8] as compared to values about $mPa\,s$ in ordinary liquids. However, mixtures of SiO_2 with other oxides may have a viscosity which is orders of magnitude lower than that of pure silica [4]. As an example, Fig. 6.1a shows the shear viscosities for several binary and ternary silicates in comparison with SiO_2. One can infer from this figure that the addition of 20 mol % Na_2O decreases the viscosity by ten orders of magnitude at $T = 1400\,K$. A further increase in the Na_2O concentration yields

A. Meyer (✉)
Institut für Materialphysik im Weltraum,
Deutsches Zentrum für Luft- und Raumfahrt (DLR), 51170 Köln, Germany
e-mail: andreas.meyer@dlr.de

L. Liang et al. (eds.), *Neutron Applications in Earth, Energy and Environmental Sciences*, Neutron Scattering Applications and Techniques,
DOI 10.1007/978-0-387-09416-8_6, © Springer Science+Business Media, LLC 2009

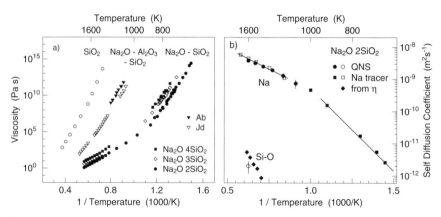

Fig. 6.1 (**a**) Arrhenius plot of viscosities for pure SiO_2 and mixtures of SiO_2 with Na_2O and Al_2O_3. "Ab" and "Jd" denote $Na_2O\ Al_2O_3\ 6SiO_2$ and $Na_2O\ Al_2O_3\ 4SiO_2$, respectively. (**b**) Sodium self-diffusion coefficients as obtained from macroscopic tracer diffusion measurements [9–11], compared with data of QENS experiments. Also shown are estimates for the Si-O diffusivity derived from viscosity data [6] and QENS. Figure taken from Ref. [12]

only a relatively minor change of the viscosity. A microscopic understanding of this drastic change in η first requires a knowledge of how the sodium ions are built into the Si–O network. In contrast to the addition of Na_2O, the addition of Al_2O_3 to an alkali silicate leads to an increase in the viscosity by several orders of magnitude. Again, this behavior can be explained through the investigation of atomic transport and structure.

In this review, we demonstrate how one can elucidate the latter issues by using a combination of quasielastic neutron scattering (QENS) experiments and molecular dynamics (MD) computer simulations. By QENS the structure and dynamics on the atomic scale can be studied, thus providing the measurement of elastic structure factors as well as time-dependent density correlation functions. This gives insight into transport mechanisms of the atoms and in some cases enables the determination of self-diffusion coefficients. The information about the latter quantities can be used to substantiate the validity of the MD simulation model. Then, if the simulation model reproduces the essential features observed in the experiment, its detailed information can give further insight into the properties of the material under consideration.

In the case of silicates, the combination of QENS experiments and MD simulations has clarified the alkali diffusion dynamics and its relation to chemical ordering of the atoms [13]. A generic feature of silicates is the appearance of characteristic length scales that go beyond that of the distance between atoms which are nearest neighbors. These length scales are often reflected by prepeaks in static structure functions. Examples of such prepeaks have been found in neutron and X-ray scattering experiments as well as in MD simulations [14]. Gaskell et al. [15] were able to measure the partial structure factors of a calcium silicate glass by means of neutron scattering and found a prepeak around a wavenumber $q = 1.3\,\text{Å}^{-1}$, while the main peak related to Si–O tetrahedra is located at $q = 1.7\,\text{Å}^{-1}$. For sodium silicate melts of different compositions, a prepeak was found around $q = 0.95\,\text{Å}^{-1}$ both in neutron scattering [13, 16] and in MD simulation [17]. Similar prepeaks have also been found for other alkali silicate melts [18, 19]. As shown by MD simulation studies, the structure factors of aluminium silicate melts $[x\,\text{Al}_2\text{O}_3\,(1-x)\text{SiO}_2]$ also display a prepeak at $q = 0.5\,\text{Å}^{-1}$ [20, 21]. All these features stem from an inhomogeneous distribution of cations on length scales of the order of 5-15Å.

In sodium silicates and other alkali silicates, the emergence of intermediate range order is of particular interest. In these systems, the tetrahedral Si–O network is partially disrupted by the sodium ions. This is accompanied by a high mobility of the alkali ions [9, 10]. Already at relatively high temperatures, the alkali ion diffusion is orders of magnitude faster than that of the Si–O network [see Fig. 6.1b]. Thus, the addition of an alkali oxide to pure silica does not only lead to an acceleration of the overall dynamics of the system, as indicated by Fig. 6.1a for the example of different sodium silicates. One also observes a decoupling of the alkali ion diffusion and the diffusive transport in the Si–O network as shown in Fig. 6.1b. It has been suggested by Angell [22], Greaves [23], Ingram [24], and others that there are "preferential pathways" through which the alkali ions can easily move, thus explaining the high mobility of the alkali ions. Here, we show that alkali atoms indeed diffuse through preferential pathways that are a property of the static structure and that are related to the prepeaks in various alkali silicates. The underlying structure is that of a percolating network of alkali-rich channels. In these channels, there are characteristic alkali ion sites that provide a hopping motion of the alkali ions from site to site. Note that the latter hopping motion has been identified in various simulations of alkali silicates [22, 25–31].

The diffusion dynamics may drastically change if further oxides are added to an alkali silicate melt. Here, we consider the addition of Al_2O_3 and Fe_2O_3. Ternary systems consisting of Na_2O, Al_2O_3, and SiO_2 are the simplest form of natural molten rock. Their properties control viscous flow in multicomponent silicate melts and therefore have a direct impact on the nature of volcanic eruptions as well as on glass production. Most natural magma contains iron oxides. We show how the addition of Al_2O_3 and Fe_2O_3 to sodium silicate melts is reflected in the channel structure and how this affects mass transport.

In the next section, we give a brief introduction to QENS and MD. The structure and dynamics of binary and ternary alkali silicates are discussed in Section 6.3. Finally, our conclusions are presented in Section 6.4.

6.2 Methods of Investigation

QENS experiments and MD simulations can be seen as complementary methods for the investigation of condensed matter systems. In the following, the two methods as well as the observables they can access are briefly introduced.

6.2.1 Molecular Dynamics Simulations

In an MD simulation, a liquid or solid is described as an assembly of atoms interacting with forces derived from a classical interatomic potential (the "force field") [32–34]. The dynamics of such a system of many interacting point particles is given by Newton's equations of motion, which are solved by a suitable numerical algorithm. A computationally efficient scheme is the so-called velocity form of the Verlet algorithm [32–34] that is very stable with respect to discretization errors, in particular due to its time reversibility property. Using this algorithm, the position $\mathbf{r}_i$ and velocity $\mathbf{v}_i$ of the ith particle are updated at time $t + \delta t$ (with δt the discretized time step):

$$\mathbf{r}_i(t + \delta t) = \mathbf{r}_i(t) + \mathbf{v}_i(t) + \mathbf{f}_i(t)\frac{(\delta t)^2}{2m_i} \tag{6.1}$$

$$\mathbf{v}_i(t + \delta t) = \mathbf{v}_i(t) + \frac{\delta t}{2m_i}\left[\mathbf{f}_i(t) + \mathbf{f}_i(t + \delta t)\right], \tag{6.2}$$

where m_i is the mass of the ith particle and $\mathbf{f}_i(t)$ is the force acting on it at time t.

For a model with purely pairwise interaction potentials $u(r_{ij})$, $r_{ij} = |\mathbf{r}_i - \mathbf{r}_j|$, the forces are given by

$$\mathbf{f}_i = -\sum_{j(\neq i)} \frac{\partial u(r_{ij})}{\partial \mathbf{r}_i}. \tag{6.3}$$

Although they do not explicitly contain interaction terms for the description of directional covalent bonding, pairwise interaction potentials have proven quite accurate in describing various static and dynamic properties of amorphous silicates. Many recent simulation studies on silicates are based on the BKS potential [35] for pure silica that consists of a Coulomb interaction term, but with effective charges q_α ($\alpha =$ Si, O) rather than the ionic charges, and a short-range Buckingham potential,

$$u_{\alpha\beta}(r) = \frac{q_\alpha q_\beta e^2}{r} + A_{\alpha\beta} \exp\left(-B_{\alpha\beta}r\right) - \frac{C_{\alpha\beta}}{r^6}, \tag{6.4}$$

where r is the distance between an ion of type α and an ion of type β ($\alpha, \beta =$ Si, O), e is the elementary charge, $q_O = -1.2$, $q_{Si} = 2.4$, and $A_{\alpha\beta}$, $B_{\alpha\beta}$, and $C_{\alpha\beta}$ are constants. The extension of the BKS model to mixtures of SiO_2 with Na_2O or Al_2O_3 is slightly more complicated. For the latter case, a modified version of the potential proposed by Kramer et al. [36] is used where the effective charges for sodium and aluminium depend on distance. This takes screening effects into account (for details see Refs. [20, 21, 26, 27]).

Although only a pair potential is used to model various silicate melts, the simulations are still technically very difficult. The long-range Coulomb interaction requires the use of time-consuming Ewald summation techniques [32–34]. Since the scale of the potential is in the eV energy range and varies rather rapidly with distance, a rather small time step, $\delta t = 1.6\,\text{fs}$, has to be used to integrate the equations of motion.

Most of the simulations that are reported below have been performed in two steps: First, they were started from initial configurations in a cubic simulation box, assuming periodic boundary conditions in all three Cartesian directions. These configurations were fully equilibrated at constant temperature T, constant volume V, and constant particle number N. Second, production runs in the microcanonical ensemble were done from which the various static and dynamic quantities of interest were calculated. By solving Newton's equations of motion with Eqs. (6.1) and (6.2), the total energy of the system is conserved and thus the microcanonical ensemble of statistical mechanics is realized. The canonical NVT ensemble can be simulated by coupling the system to an external heat bath. The typical timescales achieved were of the order of 10–20 ns for systems containing about 8000 atoms in a simulation box with a linear dimension of about $48\,\text{Å}$. In the case of sodium silicates, fully equilibrated melts around $2000\,\text{K}$ can be studied. Under these conditions, direct comparisons to recent QENS experiments are possible, which are reported in the following.

The MD simulation yields the trajectories of all the particles in the system, and thus any static or dynamic correlation function can be determined [3, 37]. Note that this assertion implies the "hypothesis of ergodicity" of statistical mechanics, which states the equivalence between time and static averages. If we consider a system of $N = \sum_{\alpha=1}^{n} N_\alpha$ particles (with N_α the number of particles of species α), the local number density fluctuations in reciprocal space for particles of type α are defined as

$$\rho_\alpha(\mathbf{q}) = \sum_{k=1}^{N_\alpha} \exp\left(i\mathbf{q} \cdot \mathbf{r}_k\right) , \tag{6.5}$$

with $\mathbf{q}$ the wavevector and $\mathbf{r}_k$ the position of particle k of type α. The partial structure factors are given by [37]

$$S_{\alpha\beta}(q) = \frac{1}{N} \langle \rho_\alpha(\mathbf{q})\rho_\beta(-\mathbf{q}) \rangle . \tag{6.6}$$

In this equation, it is assumed that the system is isotropic and homogeneous, and thus, each of the $S_{\alpha\beta}(q)$ depends only on the magnitude of wavevector $\mathbf{q}$.

The generalization of $S_{\alpha\beta}(q)$ to time-dependent density correlation functions is straightforward,

$$S_{\alpha\beta}(q, t) = \frac{1}{N} \langle \rho_\alpha(\mathbf{q}, t)\rho_\beta(-\mathbf{q}, 0) \rangle . \tag{6.7}$$

These functions reduce to the partial static structure factors Eq. (6.6) at $t = 0$. To compare the correlation functions at different q, it is convenient to introduce the normalized functions

$$F_{\alpha\alpha}(q, t) = \frac{S_{\alpha\alpha}(q, t)}{S_{\alpha\alpha}(q)}. \tag{6.8}$$

The motion of the individual (tagged) particles of species α is described by the correlation function

$$F_s^\alpha(q, t) = \frac{1}{N_\alpha} \sum_{k=1}^{N_\alpha} \langle \exp\{-i\, \mathbf{q} \cdot [\mathbf{r}_k(0) - \mathbf{r}_k(t)]\} \rangle. \tag{6.9}$$

The functions $F_{\alpha\beta}(q, t)$ and $F_s^\alpha(q, t)$ are called coherent and incoherent intermediate scattering functions [3, 37], respectively.

Another dynamic quantity that detects the self-motion of a tagged particle of species α is the mean-squared displacement

$$\langle r_\alpha^2(t) \rangle = \frac{1}{N_\alpha} \sum_{k=1}^{N_\alpha} \langle (\mathbf{r}_k(t) - \mathbf{r}_k(0))^2 \rangle. \tag{6.10}$$

In the long-time limit the displacements of the particles are expected to have a Gaussian distribution [3, 37]. In this case, they can be directly related to the incoherent intermediate scattering functions as follows:

$$F_s^\alpha(q, t) = \exp\left[-\frac{1}{6} q^2 \langle r_\alpha^2(t) \rangle\right]. \tag{6.11}$$

Note that Eq. (6.11) is valid in the limit $q \to 0$. On the other hand, the long-time limit of $\langle r_\alpha^2(t) \rangle$ yields the self-diffusion coefficient D_α via the Einstein relation [37]

$$D_\alpha = \lim_{t \to \infty} \frac{\langle r_\alpha^2(t) \rangle}{6t}, \tag{6.12}$$

and the functions $F_s^\alpha(q, t)$ are expected to decay exponentially at $t \to \infty$ and small values of q, $F_s^\alpha(q, t) = \exp(-t/\tau_\alpha(q))$ with $\tau_\alpha(q)$ q-dependent relaxation time. Thus, Eqs. (6.11) and (6.12) imply

$$D_\alpha = \lim_{q \to 0} \left[\tau_\alpha(q) q^2\right]^{-1}. \tag{6.13}$$

As explained below, self-diffusion coefficients can be determined using Eq. (6.13) in QENS. In a MD simulation, it is more convenient to determine the self-diffusion coefficient directly from the Einstein relation, Eq. (6.12).

6.2.2 Quasielastic Neutron Scattering

QENS enables one to investigate the microscopic dynamics on timescales of sub-picoseconds to nanoseconds and its relation to microscopic structure (length scales of about $0.1\text{–}60$ Å). An overview of the technique is presented in the books by Bée and Hempelmann [38, 39]. In typical QENS experiments, either the scattering law $S(q, \omega)$ (neutron time-of-flight spectroscopy and neutron backscattering) or the intermediate scattering function $S(q, t)$ (neutron spin-echo spectroscopy) are measured. The momentum transfer is denoted by q, and the energy transfer to/from the neutron is represented by ω. $S(q, \omega)$ and $S(q, t)$ are related to each other by a Fourier transform:

$$S(q, t) = \int d\omega\, S(q, \omega)\, \exp\{i\omega t\}. \qquad (6.14)$$

The measured $S^*(q, \omega)$ is a convolution of $S(q, \omega)$ with the instrumental energy resolution function $R(q, \omega)$:

$$S^*(q, \omega) = S(q, \omega) \otimes R(q, \omega). \qquad (6.15)$$

$R(q, \omega)$ is measured either with a vanadium standard or with the sample at low temperatures, where structural relaxation is frozen in. Both $S(q, \omega)$ and $S(q, t)$ simultaneously contain structural and dynamical information. Combining Eqs. (6.14) and (6.15), $S(q, t)$ can be obtained by Fourier deconvolution of $S^*(q, \omega)$, as measured in a neutron time-of-flight or backscattering experiment (see Fig. 6.2).

In alkali silicates, the measured time-of-flight spectra consist of three contributions: (1) vibrations of the atoms, which become visible in the spectra at energies above several millielectron volts; (2) structural relaxation of the alkali atoms, which leads to a broad quasielastic line; (3) structural relaxation of the Si–O network, which even at temperatures of 1600 K is too slow to cause a detectable quasielastic

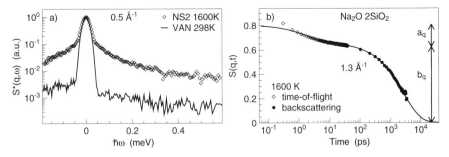

Fig. 6.2 (a) Scattering law as measured in a time-of-flight experiment. The resolution function (*solid line*) and data at high temperature showing quasielastic scattering are depicted. (b) Intermediate scattering functions, obtained after Fourier deconvolution of the measured $S^*(q, \omega)$. The line represents a fit with Eq. (6.17). Here b_q has been replaced by b_q times a second stretched exponential function. The vertical double-arrowed lines indicate the amplitudes as defined by Eq. (6.17)

line broadening and therefore appears as an elastic contribution in the spectra. Integrating $S^*(q, \omega)$ over the elastic and quasielastic contributions yields the quasielastic structure factor $S_{qel}(q)$. This quantity corresponds to a good approximation of the static structure factor $S(q)$ times the Debye-Waller factor (DWF) $DWF(q)$ [37]:

$$S_{qel}(q) = \int_{\Delta\omega} d\omega \, S(q, \omega) = S(q) \times DWF(q). \qquad (6.16)$$

The DWF is the ratio of elastic scattering (energy transfer $\omega = 0$) to inelastic scattering (energy transfer $\omega \neq 0$) due to vibrational motions.

In the intermediate scattering function $S(q, t)$, these three contributions are separated in time [Fig. 6.2b]. Vibrations cause a decay of $S(q, t)$ on a sub-picosecond timescale onto a plateau. The height of the plateau value corresponds to $DWF(q)$. The relaxation dynamics of the network modifiers and network formers leads to a decay from this plateau to zero. In the case of alkali silicates, the relaxation of the alkali ions is decoupled from the relaxation of the Si–O network such that a two-step decay from this plateau is observed. Whereas the fast alkali relaxation causes a decay in $S(q, t)$ on a 10 ps timescale [see also Fig. 6.8a], the structural relaxation of the Si–O network is on a nanosecond timescale as has been shown in a neutron backscattering experiment [16]. This two-step decay can be described by the sum of two stretched exponential functions [solid line in Fig. 6.2b]. Stretched exponential functions are generic features of relaxation processes in viscous glass-forming liquids [3]. Since the two relaxation processes are separated in time by at least two orders of magnitude, depending on temperature and momentum transfer, the fast alkali relaxation can to a good approximation be described by the sum of a single stretched exponential function and a constant:

$$S(q, t) = b_q + a_q \, \exp\{-(t/\langle\tau(q)\rangle)^\beta\}. \qquad (6.17)$$

Such a fit to the data gives q-dependent amplitudes of relaxation a_q, relaxation times $\langle\tau(q)\rangle$, and elastic contributions to scattering b_q on this fast timescale. These quantities contain information on the mechanisms of mass transport and in the small q limit on self-diffusion coefficients [Fig. 6.1b]. By comparing self-diffusion coefficients measured in tracer experiments with those derived by scaling $\langle\tau(q)\rangle$ according to Eq. (6.13), it has been shown experimentally that QENS is able to measure self-diffusion coefficients on an absolute scale [40–43]. These experiments were conducted on multicomponent metallic melts. As shown in Fig. 6.1b, the sodium diffusion coefficients obtained from QENS and tracer diffusion experiments also agree within experimental errors.

Since neutrons carry no charge, they are directly scattered by the nuclei. The scattering probability varies greatly between the different atoms and also their isotopes [44]. Moreover, the different atoms have two different scattering cross sections associated with them, namely an incoherent and a coherent cross section. Coherent scattering gives rise to interference effects and therefore probes the correlation between pairs of nuclei. Incoherent scattering on the other hand probes only the self-correlation of the nuclei of the system. Therefore, a means is provided by isotope or atom substitution (see Section 6.3.1.2) to probe different dynamical and structural properties of a system. The second intriguing property stems from the fact that the neutron bears a magnetic moment. In a scattering experiment, the neutron can therefore interact with the unpaired electron spins of magnetic atoms, probing both the distribution of these spins in real space (spin-density) as well as their dynamical behavior. If the spins are localized on the magnetic atoms, information about the spatial distribution of these atoms can be derived as shown for iron silicates in Section 6.3.2.2.

6.3 Results and Discussion

The static structure and dynamics of binary and ternary silicates are discussed in the following subsections. To avoid lengthy denominations of the samples, we introduce the following abbreviations: $Na_2O\,x\,SiO_2 = NSx$, $x\,Na_2O\,y\,Al_2O_3\,z\,SiO_2 = NxAySz$, or $NxFySz$ for Fe_2O_3 replacing Al_2O_3. If Na_2O is replaced by Li_2O, the first letter in the sample descriptors is replaced by L.

6.3.1 Binary Alkali Silicates

In this subsection we present the results of investigations of atomic dynamics and structure of binary alkali silicates by means of MD simulations and quasielastic neutron scattering.

6.3.1.1 Static Structure: Experiment vs. Simulation

As we have mentioned in the introduction, the sodium ions partially disrupt the tetrahedral Si–O network. This can be directly recognized in the partial structure factors $S_{\alpha\beta}(q)$ from simulation, defined by Eq. (6.6). Three of them, that is, those for the NaNa, NaO, and SiNa correlations, are shown in Fig. 6.3a for NS_2 at $T = 2100\,K$. They differ from the structure factors for pure SiO_2, the peak around $1.7\,Å^{-1}$ is absent, which would indicate the tetrahedral network structure. Note that a peak at $1.7\,Å^{-1}$ is also observed in the SiSi, SiO, and OO correlations in NS_2. However, it has a much smaller amplitude than in SiO_2 [26, 27]. An interesting feature in Fig. 6.3a is an additional peak around $q_1 = 0.95\,Å^{-1}$, which is not present in the static structure factor of pure SiO_2. The length scales that correspond to q_1 are of the order of 6–7 Å, and this is about the distance between next–nearest Na–Na or Si–Na neighbors [26, 27]. In NS20 [Fig. 6.3b], one can still identify peaks around q_1, although they have moved to a slightly smaller q in $S_{NaNa}(q)$ and to a slightly larger q in $S_{SiNa}(q)$ and $S_{NaO}(q)$. The emergence of the prepeak at q_1 reveals a Na–O

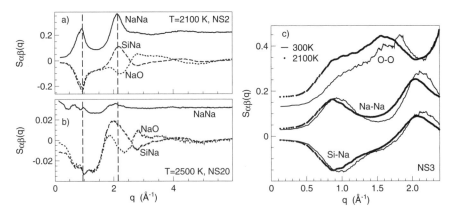

Fig. 6.3 Partial structure factors $S_{NaNa}(q)$ (*solid lines*), $S_{SiNa}(q)$ (*dotted lines*), and $S_{NaO}(q)$ (*dashed lines*) for (**a**) NS2 at $T = 2100$ K and for (**b**) NS20 at $T = 2500$ K. (**c**) Partial structure factors $S_{OO}(q)$, $S_{NaNa}(q)$, $S_{SiNa}(q)$ as obtained by the simulation in glassy and liquid NS3 at experimental densities. $S_{OO}(q) + 0.1$ for clarity. Figure taken from Ref. [13]

Fig. 6.4 Molecular dynamics snapshot of the structure of NS3 at 2100 K at the density $\rho = 2.2$ g/cm^3. The sodium atoms are represented by large dark spheres that are connected to each other by surfaces. Tetrahedrally connected small spheres represent the Si–O network with covalent bonds shown as sticks between silicon (light) and oxygen (dark) spheres. Figure taken from Ref. [13]

rich network that percolates through the Si–O network, and the characteristic length scale of this sodium-rich network is given by $\approx 2\pi/q_1 = 7$ Å. The latter sodium network is illustrated by a snapshot of a NS3 configuration at $T = 2100$ K (Fig. 6.4).

Now, we address the question of whether the prepeak at q_1 can be seen in QENS experiments on sodium silicate melts. To this end, we present the results of a study on NS3 using a combination of inelastic neutron scattering and MD simulations [13]. Note that the properties of NS3 are very similar to those of NS2, which is

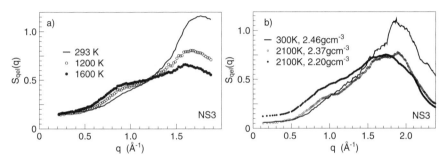

Fig. 6.5 (**a**) Quasi-elastic structure factor of NS3 from QENS for different temperatures, as indicated. (**b**) Quasi-elastic structure factor of NS3 as obtained by molecular dynamics simulation (see text) weighted with the neutron scattering lengths: Note the emerging prepeak at $\simeq 0.9\,\text{Å}^{-1}$ with decreasing density. Figures taken from Ref. [13]

evident both from simulation and from experiment [6, 13, 16, 26, 27]. Thus, the conclusions that are drawn below also hold for NS2.

The inelastic neutron scattering experiments have been performed at room temperature (i.e., $T = 293\,\text{K}$) and on liquid samples at temperatures between 1200 and 1600 K using the neutron time-of-flight spectrometer IN6 of the Institut Laue-Langevin in Grenoble [13, 16]. In these measurements the quasi-elastic structure factor $S_{\text{qel}}(q)$ was determined [see Eq. (6.16)]. This quantity can be also measured in the simulation. To this end, the time-dependent dynamic structure $S(q, t)$ has to be determined, which is defined by

$$S(q, t) = \frac{N}{\sum_\alpha N_\alpha b_\alpha^2} \sum_{\alpha, \beta} b_\alpha b_\beta S_{\alpha\beta}(q, t) \quad \alpha, \beta \in \{\text{Si, Na, O}\}, \tag{6.18}$$

with $S_{\alpha\beta}(q, t)$ given by Eq. (6.7). In Eq. (6.18), b_α denotes the experimental coherent scattering lengths of particle species α. The quasi-elastic structure factors can be estimated by the equation $S_{\text{qel}}(q) = DWF(q)S(q)$ (with $S(q) \equiv S(q, t = 0)$) where the Debye–Waller factor $DWF(q)$ can be determined by analyzing $F(q, t) = S(q, t)/S(q)$ (for details see Ref. [3, 45]).

Figure 6.5a shows the quasi-elastic structure factor of glassy and viscous NS3 as determined by QENS. As discussed before, the maximum at $1.7\,\text{Å}^{-1}$ corresponds to the disrupted tetrahedral Si–O network structure in sodium silicates. With increasing temperature, the elastic scattered intensity at this wave-vector decreases at the expense of increasing inelastic scattering. In contrast, $S_{\text{qel}}(q)$ displays an emerging prepeak around $0.95\,\text{Å}^{-1}$, which grows with increasing temperature. Although we have identified this prepeak above also in the simulation, the growth of this peak with increasing temperature seems to be in contradiction to the simulation data. For a monoatomic system, a change in the structure that is due to thermal expansion would lead—beside a slight shift of the peak positions toward a smaller q value—to a decreasing intensity of the peak height with increasing temperature. Thus, the emerging prepeak in the quasi-elastic structure factors as shown in Fig. 6.5a does not seem to be merely caused by thermal expansion, but might be related to a change in the underlying structure. However, this behavior of the experimental elastic structure factors can be clarified by the MD simulations.

The simulations for NS3 were carried out at various densities for systems of $N = 8064$ particles ($N_{Si} = 2016$, $N_{Na} = 1344$, and $N_O = 4704$). At densities of $2.2 \, g/cm^3$ and $2.37 \, g/cm^3$, systems at the temperature $T = 2100 \, K$ (i.e., in the liquid state) were equilibrated for 3.3 ns followed by microcanonical production runs of the same length. In each case, two independent runs were done in order to improve the statistics. The equilibrated systems at the density $\rho = 2.37 \, g/cm^3$ were also quenched to a glass state at $T = 300 \, K$ using a cooling rate of about $10^{12} \, K/s$. In the latter simulation, the pressure was kept constant at ambient pressure, which yielded a density of $2.46 \, g/cm^3$ at $300 \, K$. Note that the densities $\rho = 2.2 \, g/cm^3$ and $\rho = 2.46 \, g/cm^3$ are close to the experimental densities at $T = 2100 \, K$ and $T = 300 \, K$, respectively [4].

Figure 6.5b displays the neutron scattering quasi-elastic structure factor of NS3 as determined from the simulation. A similar behavior is observed as in the neutron scattering experiment. At around $1.7 \, Å^{-1}$, the quasi-elastic scattered intensity decreases with increasing temperature, and at around $0.95 \, Å^{-1}$ a shoulder is present at $T = 2100K$ which is nearly absent in the glass at $T = 300 \, K$. The elastic structure factor at a density of $2.37 \, g/cm^3$ at $T = 2100 \, K$ is also shown in Fig. 6.5b. Here the shoulder at around $0.95 \, Å^{-1}$ has a smaller amplitude than in the corresponding case at the experimental density. This gives evidence that the possible structural changes leading to a more pronounced appearance of the feature around $0.95 \, Å^{-1}$ in $S_{qel}(q)$ are not due to a change in temperature but due to the change in the density. Thus, the question arises whether the shoulder at around $0.95 \, Å^{-1}$ is related to structural features that tend to disappear with increasing density.

The puzzling behavior of the quasi-elastic structure factors can be easily clarified by the partial static structure factors $S_{\alpha\beta}(q)$. Three of these functions are shown in Fig. 6.3c (at the experimental densities for different correlations, as indicated). A well-pronounced peak is present in $S_{SiNa}(q)$ and $S_{NaNa}(q)$ at $T = 2100 \, K$ *and* at $T = 300 \, K$. Thus, the structure that leads to the peak at $q_1 = 0.95 \, Å^{-1}$ does not at all disappear with increasing density. That the feature at q_1 seems to be absent in $S_{qel}(q)$ at $300 \, K$ is due to the fact that $S_{qel}(q)$ is a linear combination of six different partial structure factors: On one hand, positive and negative contributions cancel each other [note the negative amplitude of $S_{SiNa}(q)$ at q_1], and on the other hand, due to the overlap with the peak around $q_2 = 1.7 \, Å^{-1}$, $S_{OO}(q)$ exhibits only a shoulder at q_1, which is less pronounced at $T = 300 \, K$. Because oxygen is the major component (about 60% of the particles in NS3) and the coherent scattering length of oxygen is significantly larger than that of sodium and silicon ($b_O/b_{Si} \approx 1.4$ and $b_O/b_{Na} \approx 1.6$), $S_{OO}(q)$ gives the major contribution to $S_{qel}(q)$; thus, the changes in $S_{OO}(q)$ are the main cause for the emerging prepeak in the experimental and simulated $S_{qel}(q)$ with increasing temperature and decreasing density, respectively. An experimental proof that the channel structure is also present at low temperatures (i.e., in the glass) is provided by magnetic neutron scattering in sodium ironsilicates, described in Section 3.2.2.

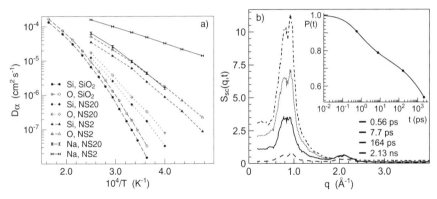

Fig. 6.6 (**a**) Arrhenius plot of self-diffusion coefficients D_α (α = Si, O, Na) for SiO$_2$, NS20, and NS2, as indicated. (**b**) "Swiss cheese" structure factor $S_{sc}(q, t)$ for the sodium-free regions at $T = 2100$ K for $t = 0.56$ ps, 7.7 ps, 164 ps, and 2.13 ns. The inset shows the probability $P(t)$ that a *cube* which is sodium free at time zero remains sodium free until time t. The *circles* on the *curve* for $P(t)$ are at the times at which $S_{sc}(q, t)$ is shown

The results presented in this section demonstrate how computer simulations can be used to elucidate experimental data. This requires a molecular dynamics model that describes the real system accurately. Then, since the MD provides more detailed information than the experiment, new insight can be gained on relevant properties of the material, in this case the intermediate range order in sodium silicates.

6.3.1.2 Channel Diffusion

We now turn our attention to the dynamics in alkali silicates. Figure 6.6a shows an Arrhenius plot of the self-diffusion coefficients D_α (α = Si, O, Na) for SiO$_2$ [46], NS20 [47], and NS2 [27]. As one can recognize from the figure, the dynamics in sodium silicates is faster than that in SiO$_2$ (see also Fig. 1). For instance, at $T = 2750$ K the silicon diffusion coefficient for NS20 is almost one order of magnitude larger than that for SiO$_2$. In the case of NS2, D_{Si} is even two orders of magnitude higher than in SiO$_2$. Moreover, a significant timescale separation between the sodium motion and that of the Si–O network is evident: even above 2000 K, D_{Na} is up to two orders of magnitude higher than D_{Si} and D_O. Whereas D_{Na} for NS2 follows an Arrhenius law (with an activation energy of 0.93 eV [26, 27]), the temperature dependence of D_{Na} for NS20 is non-Arrhenius and more similar to that of D_{Si} and D_O. This indicates that the mechanism for sodium diffusion changes from NS20 to NS2.

Now, we address the question whether the timescale separation in NS2 can be related to the structure, in particular to the prepeak seen at $q_1 = 0.95\,\text{Å}^{-1}$ in static structure factors. To this end, an idea of Jund et al. [28] is followed where the simulation box is divided into 48^3 small cubes with a linear size of about 1 Å (note that the linear dimension of the simulation box is $L = 48.653\,\text{Å}$). The probability $P(t)$ that a cube which does not contain a sodium ion at time zero is also not visited by a sodium ion until a later time t is calculated. The time dependence of $P(t)$ is shown in the inset of Fig. 6.6b. Even after 2.5 ns, more than 50% of the cubes have not yet been visited by sodium ions (note that after this time the mean squared displacement of the sodium atoms is more than 45 Å^2; thus, these atoms have already moved over a large distance). From this observation, one can infer that the sodium diffusion is restricted to a subspace within the Si–O network.

In order to analyze the structure of this subspace, a "Swiss cheese" structure factor $S_{sc}(q, t)$ is defined as follows. Each cube which has not been visited by a sodium atom until time t is assigned a point and the static structure factor of $N_{sc}(t) = P(t)(48^3 - N_{Na})$ points is computed:

$$S_{sc}(q, t) = \frac{1}{N_{sc}(t)} \sum_{k,l=1}^{N_{sc}(t)} \exp(i\mathbf{q} \cdot (\mathbf{r}_k - \mathbf{r}_l)) . \tag{6.19}$$

This quantity is shown in Fig. 6.6b for four different times: $t = 0.56$ ps, 7.7 ps, 164 ps, and 2.13 ns. It can be seen that $S_{sc}(q, t)$ has peaks at $q_1 = 0.95$ Å^{-1} and $q_2 = 2.15$ Å^{-1}, which are also the prominent features in $S_{NaNa}(q)$, the static structure factor for the Na–Na correlations [26, 27]. With increasing time, the height of the peak at q_1 in $S_{sc}(q, t)$ increases quickly. Note that this peak decreases again on timescales where the Si–O matrix starts to reconstruct itself significantly. In the previous section, the prepeak at q_1 (e.g., in the partial static structure factors) has been associated with sodium-rich regions that percolate through the Si–O network. Now, we see that this structure on intermediate length scales is reflected in the sodium dynamics. It provides a network of diffusion channels that enable the high mobility of the sodium ions.

So far it has been shown that the timescale separation in sodium silicates is intimately related to the presence of intermediate range order in the static structure. But this does not prove that the dynamics can be predicted from structural quantities such as the partial structure factors $S_{\alpha\beta}(q)$. A theoretical framework by which this

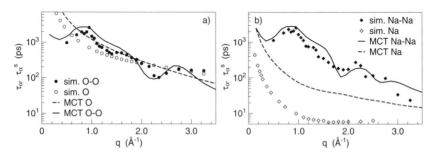

Fig. 6.7 Structural relaxation times $\tau_\alpha(q)$ for coherent density relaxation functions $F_{\alpha\alpha}(q, t)$ and $\tau_{Na}^s(q)$ for the incoherent density relaxation function $F_{Na}^s(q, t)$, a) $\tau_O(q)$ and $\tau_O^s(q)$, b) $\tau_{Na}(q)$ and $\tau_{Na}^s(q)$. As indicated, *symbols* are simulation results ($T = 2100$ K) and *lines* are MCT results ($T = 3410$ K). Figures taken from Ref. [51]

is possible (at least in principle) is the mode coupling theory (MCT) of the glass transition [3, 48–50]. Using information about the equilibrium structure as an input [i.e., essentially the $S_{\alpha\beta}(q)$], the MCT equations yield the intermediate scattering functions $F_{\alpha\beta}(q, t)$. Very recently, MCT calculations were performed for NS2 and NS20 [51], using the simulated partial structure factors as input [some of them are displayed in Fig. 6.3a and b]. This work reveals that MCT indeed reproduces the essential dynamical features of NS2 and NS20 on a qualitative level, thus confirming the structure–transport relation in sodium silicates as found by the simulation. As an example we show in Fig. 6.7 the structural relaxation times $\tau_{\alpha}^{s}(q)$ and $\tau_{\alpha}(q)$ that we have determined from incoherent and coherent intermediate scattering functions of NS2, respectively [51]. As we can infer from this figure, the values for the relaxation times from MCT are in good agreement with those from MD. The oxygen-self-relaxation time $\tau_{O}^{s}(q)$ is of the same order as the corresponding coherent one, $\tau_{O}(q)$, except for $q \to 0$. This is what one expects in a typical glass-forming liquid. On the contrary, the sodium-self-relaxation time $\tau_{Na}^{s}(q)$ is much smaller than all other structural relaxation times, both in the MCT and in the MD results. Although the MCT results for this quantity are systematically higher than the MD data, this confirms that MCT indeed predicts the fast sodium dynamics in NS2 from the equilibrium structure alone.

As previously discussed, the quasi-elastic structure factors as determined by QENS and MD simulations agree well. Intriguing properties have also been found by MD simulations in the dynamics. It was found that the alkali–alkali coherent correlation functions decay on the slower timescales of the network relaxation [17]. The separation between self- and collective motions of the alkali ions is a direct consequence of the channel structure. Based on the MD simulation, a separation of at least two orders of magnitude in time is expected for the self- and collective motions [see also Fig. 6.7b].

The quasielastic signal in neutron scattering experiments displays the sum of coherent and incoherent contributions. Since silicon and oxygen only scatter coherently, the incoherent contribution is dominated by the scattering from the alkali atoms. In the case of sodium and lithium, the coherent and incoherent cross sections are fairly equal; however, the absolute value is about a factor of two larger than that for sodium [44]. Figure 6.8a displays the $S(q, t)$ for sodium and lithium silicates on a picosecond timescale, where the fast alkali ion relaxation leads to a decay onto a plateau. The slow contributions are subtracted by a constant according to Eq. (6.17). The amplitude and the timescale of this relaxation is shown in Fig. 6.8b. As in the case of purely incoherent scattering, the relaxation time and the relaxation amplitude exhibit no coherent oscillations. For sodium atoms the relaxation amplitude and the relaxation time should exhibit a pronounced oscillation in phase with the Na–Na partial structure factor. It should also be noted that the $\langle \tau(q) \rangle$ as derived from the QENS experiment [Fig. 6.8b] and the $\tau_{Na}^{s}(q)$ of the MD simulation [Fig. 6.7b] qualitatively follow the same q dependence. This conclusion also holds for lithium and potassium silicate melts [12], where the absolute values and the ratios of incoherent to coherent contributions are different from Na. It is known from conductivity measurements [52] that potassium and lithium have transport coefficients similar

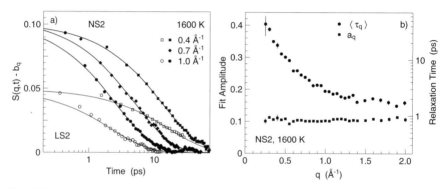

Fig. 6.8 (**a**) $S(q, t)$ of $Li_2O\ 2SiO_2$ and $Na_2O\ 2SiO_2$ from which the elastic contribution to scattering on the short picosecond timescale b_q is subtracted. The amplitudes of relaxation reflect the ratio of the incoherent scattering cross sections of Li (σ_{Li}^i) and of Na (σ_{Na}^i), respectively. (**b**) Relaxation amplitude a_q (*squares*) and relaxation time $\langle \tau(q) \rangle$ (*brackets*) obtained from fit to NS2 at 1600 K. Figures were taken from Ref. [12] and merged

to sodium at typical melt temperatures. According to their incoherent scattering contribution when compared with sodium, in lithium silicate melts the relaxation amplitude is reduced by a factor of 2 [Fig. 6.8a], and in potassium silicate melts, a relaxation cannot be observed on the picosecond timescale (negligible incoherent scattering cross section of potassium). The absence of such oscillations proves that only the alkali self-correlation decays on the picosecond timescale in agreement with the MD simulations and the MCT calculations.

6.3.2 Ternary Alkali Silicates

In the following subsections, the impact on the structure and dynamics of alkali silicates of the addition of a second network-forming component is discussed. We discuss both the addition of Al_2O_3 and Fe_2O_3 to sodium silicates of different Na_2O concentration. For the aluminosilicates we discuss both structure and dynamics and compare them to the results on macroscopic properties. For the Fe_2O_3-containing system, we only discuss the structure.

6.3.2.1 Sodium Aluminosilicates

It is well known that the addition of Al_2O_3 to a binary alkali silicate melt tremendously alters macroscopic properties. The viscosity of a ternary sodium aluminosilicate system increases by several orders of magnitude on the addition of Al_2O_3 at constant temperature [7]. The resulting viscosity lies in between the viscosity measured for a binary alkali silicate and that of pure SiO_2 [4, 6] [see Fig. 6.1a]. The increase of the viscosity is strongly nonlinear with composition. Furthermore, as shown by means of macroscopic measurements at a fixed temperature, it levels off, crossing from a system rich in Na_2O to a system rich in Al_2O_3 at constant SiO_2 content [7]. It has been suggested that Al_2O_3 reconnects the network and is built into the structure mostly in the form of tetrahedrally coordinated ^4Al. To maintain charge neutrality Na^+ ions have to charge-balance ^4Al. Therefore, it can be expected that the addition of Al_2O_3 impacts both on the channel structure and on the motion of the sodium ions. Hence, a clear footprint is expected in the structure factor as well as in the dynamics on a picosecond timescale.

QENS experiments on a number of systems containing different amounts of SiO_2, and various ratios of Na_2O to Al_2O_3, have shown that the prepeak in the quasi-elastic structure factor gradually vanishes on approaching a ratio of $Na_2O:Al_2O_3$ equal to unity [for an example see Fig. 6.9a] [53]. In addition, we have shown that the fast relaxational dynamics of the sodium ions slows down, exceeding the experimentally accessible time range for the $Na_2O:Al_2O_3 = 1$ compositions (see Fig. 6.9b for $Na_2O:Al_2O_3 = 3$). However, the increases in viscosity and sodium relaxation times are decoupled in terms of their overall magnitude at a given temperature. Therefore, the increase in Al_2O_3 impacts the channel structure, disrupting it due to reconnection of the SiO_2 network. Nonetheless, the inhomogeneous distribution of sodium must prevail since the sodium diffusion still remains fast and separated from the network motion.

6.3.2.2 Sodium Ironsilicates

A further step in the study of ternary silicates is obviously to determine how general the findings obtained for the sodium aluminosilicates are. To this end, atomic substitution can be chosen, replacing Al_2O_3 by Fe_2O_3. Addition of Fe_2O_3 to a binary silicate melt also leads to an increase in viscosity, which is, however, less pronounced [5] than that observed for the aluminosilicates.

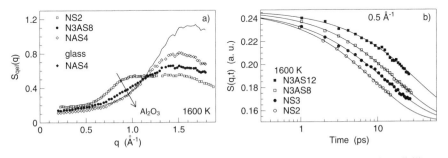

Fig. 6.9 (**a**) Elastic structure factors of $x Na_2O\ y Al_2O_3\ z SiO_2$ based on binary sodium disilicate replacing a quarter and half of the Na_2O by Al_2O_3. The *arrow* indicates the decreasing prepeak with increasing Al_2O_3 content. (**b**) Intermediate scattering function for NS2 and NS3 in comparison to samples where a quarter of Na_2O is replaced by Al_2O_3. Data are shown with the relaxation amplitude rescaled to a common arbitrary scale. The sodium dynamics is slowing down with increasing Al_2O_3 content. Figures taken from Ref. [53]

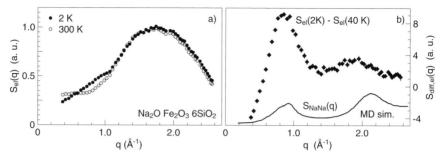

Fig. 6.10 (**a**) Quasi-elastic structure factors for $Na_2O\,Fe_2O_3\,6SiO_2$ as measured on the spectrometer IN6 (ILL). The emergence of a shoulder is observed at low temperatures. (**b**) Difference function of quasi-elastic structure factors measured at 40 K and at 2 K (symbols). The data show the magnetic contribution to the elastic structure factor and therefore reflect Fe–Fe correlations. The *solid line* represents the elastic structure factor taken from the MD-simulation on $Na_2O\,3SiO_2$ (Fig. 6.3)

For the samples considered, Mößbauer spectroscopy has shown that Fe^{3+} dominates the structure, with only minor amounts of Fe^{2+} [54]. This is in agreement with earlier studies on similar glasses [55]. Fe^{3+} is a spin $S = 5/2$ compound, and therefore the samples have magnetic properties as shown earlier [56]. Therefore, the scattering also includes contributions that arise from the interaction of the neutron spin with the magnetic moments of Fe^{3+}. At low temperatures, these spins align, and magnetic contributions to the scattering are reflected by a q-dependent increase of intensity in $S_{qel}(q)$, as shown in Fig. 6.10a. Correspondingly, the difference between the quasi-elastic structure factor measured at 40 K and that measured at 2 K only reflects the magnetic scattering contributions of the iron atoms [Fig. 6.10b]. This magnetic signal can be regarded as equivalent to the Fourier transform of the real-space density of magnetic moments that are localized on Fe^{3+}.

A first maximum is observed around 0.8 $Å^{-1}$, followed by a broad peak in the range of 1.8–2.2 $Å^{-1}$. Similar to what is reflected by the first sharp diffraction peak in the binary silicates, part of the second peak in Fig. 6.10b reflects the tetrahedral ordering of iron in these glasses [57]. A comparison to the MD simulation on $Na_2O\,3SiO_2$ shows that the first peak in Fig. 6.10b is roughly at the same position as that for the partial Na–Na structure factor in the simulation. This means that Fe^{3+} shows a correlation on the length scale that corresponds to typical sodium inter-channel distances. Taking into account the increase in viscosity upon addition of Fe_2O_3 to a binary alkali silicate, it can be argued that iron reconnects the network by bridging the percolation channels. Moreover, it can be concluded that the distribution of iron must be non-uniform on intermediate length scales, similarly to the distribution of sodium. This conclusion has been substantiated by measurements of $S_{qel}(q)$ for different iron concentrations [54]. The latter study showed that the position of this first peak is independent of the total iron content, similar to the finding for binary sodium silicates [13].

6.4 Conclusions

Alkali silicates are basic constituents of molten rock and are technologically used as glass materials. The addition of alkali oxides to pure SiO_2 causes a drastic decrease in the melt viscosity by orders of magnitude, whereas the diffusion of the alkali atoms is orders of magnitude faster than that of the Si–O network. Both features also remain as predominant characteristics in multicomponent, natural peralkaline silicate melts. Therefore, knowledge as to how the alkali atoms are built into the Si–O network structure is central to our understanding of mass transport in silicate melts. We have shown how quasielastic neutron scattering and molecular dynamics simulations are used in order to study the interplay of intermediate range order, atomic dynamics, and properties of mass transport in alkali silicate melts. The results show that the alkali ions align in alkali rich channels that disrupt the Si–O structure and form a percolation network. The alkali ions diffuse along these channels in a relatively immobile Si–O network.

The characteristic length scale of this channel structure is given by prepeaks in the quasi-elastic structure factors at about $0.9\,\text{Å}^{-1}$, corresponding to length scales of the order of $7\,\text{Å}$. The appearance of this length scale is a result of chemical ordering in sodium silicates. Therefore, the location of the prepeak does not essentially depend on the Na_2O concentration. Only at low Na_2O concentration is a slight shift of the prepeak seen, as indicated by the simulation of $Na_2O\,20\,SiO_2$. The diffusion via these channels results in a decoupling of incoherent from coherent correlations of the alkali relaxation, as shown by simulation and experiment. This finding is reproduced by calculations in the framework of mode-coupling theory (MCT), which use the partial *static* structure factors of the simulation as input. Thus, properties of mass transport, that is, the fast alkali transport and relaxation dynamics of the Si–O network, are encoded in the static melt structure.

Besides sodium and potassium oxides and SiO_2, other major components of natural rock are alumina and iron oxides. The addition of Al_2O_3 and Fe_2O_3 to an alkali silicate melt strongly affects its melt viscosity. Quasielastic neutron scattering on alkali aluminosilicate melts shows that the drastic increase in viscosity comes along with a decrease in the mobility of the alkali atoms and a decreasing prepeak in the structure factor. This indicates that the addition of Al_2O_3 disrupts the alkali channel structure. In alkali ironsilicate glasses, quasielastic neutron scattering exhibits a broad diffuse magnetic signal from scattering from the Fe^{3+} ions. At low temperature, the magnetic moments on the Fe^{3+} ions align, leading to an increase in elastic scattering, which displays a correlation on the length scale that corresponds to typical alkali inter-channel distances. It appears that in alkali aluminosilicates and alkali ironsilicates, the aluminium and Fe^{3+} atoms reconnect the Si–O network by bridging the percolation channels. However, at least for peralkaline compositions, the distribution of the alkali atoms remains inhomogeneous.

Acknowledgments It is a pleasure to thank Donald Dingwell, Kai-Uwe Hess, Walter Kob, Kurt Binder, Thomas Voigtmann, Helmut Schober, and Marek Koza for fruitful collaborations. We gratefully acknowledge support by the Deutsche Forschungsgemeinschaft (DFG) within the SPP

1055 under grant numbers ME 1958/6-1 and by Schott Glass. JH acknowledges funding by DFG under grant number HO2231/2-1. FK acknowledges the Higher Funding Council of Wales (HEFCW) for funding through the Centre for Advanced Materials and Devices (CAFMaD). Generous grants of computing time at the NIC Jülich are gratefully acknowledged.

References

1. Rev. Mineralogy **32**, *Structure, Dynamics and Properties of Silicate Melts* (1995) ed.by J. F. Stebbins, P. F. McMillan and D. B. Dingwell.
2. Y. Bottinga, D. B. Dingwell, P. Richet, and M. Toplis, edts., *7th Silicate Melt Workshop*, Chem. Geol. **213**, 1 (2004).
3. K. Binder and W. Kob, *Glassy Materials and Disordered Solids: An Introduction to their Statistical Mechanics* (World Scientific, Singapore, 2005).
4. O. V. Mazurin, M. V. Streltsina, and T. P. Shvaiko-Shvaikovskaia, *Handbook of Glass Data: Part A and Part B* (Elsevier, Amsterdam, 1983 and 1985).
5. D. B. Dingwell, D. Virgo, Geochim. Cosmochim. Acta **52**, 395 (1987).
6. R. Knoche, D. B. Dingwell, F. A. Seifert, and S. L. Webb, Phys. Chem. Minerals **116**, 1 (1994).
7. M. J. Toplis, D. B. Dingwell, and T. Lenci, Geochim. Cosmochim. Acta **61**, 2605 (1997).
8. G. Urbain, Geochim. Cosmochim. Acta **46**, 1061 (1982).
9. Y. P. Gupta and T. B. King, *Trans. Metall. Soc. AIME* **237**, 1701 (1966).
10. J. R. Johnson, R. H. Bristow, and H. H. Blau, J. Am. Ceram. Soc. **34**, 165 (1951).
11. M. Braedt and G. H. Frischat, Phys. Chem. Glasses **29**, 214 (1988).
12. F. Kargl, A. Meyer, M. M. Koza, and H. Schober, Phys. Rev. B **74**, 014304 (2006).
13. A. Meyer, J. Horbach, W. Kob, F. Kargl, and H. Schober, Phys. Rev. Lett. **93**, 027801 (2004).
14. G. N. Greaves and S. Sen, Adv. Phys. **56**, 1 (2007).
15. P. H. Gaskell, M. C. Eckersley, A. C. Barnes, and P. Chieux, Nature **350**, 675 (1991).
16. A. Meyer, H. Schober, and D. B. Dingwell, Europhys. Lett. **59**, 708 (2002).
17. J. Horbach, W. Kob, and K. Binder, Phys. Rev. Lett. **88**, 125502 (2002).
18. Y. Waseda and H. Suito, Trans. Iron Steel Inst. Japan **17**, 82 (1977).
19. O. Majerus, L. Cormier, G. Calas, and B. Beuneu, Chem. Geol. **213**, 89 (2004).
20. A. Winkler, J. Horbach, W. Kob, and K. Binder, J. Chem. Phys. **120**, 384 (2004).
21. P. Pfleiderer, J. Horbach, and K. Binder, Chem. Geol. **229**, 186 (2006).
22. C. A. Angell, P. A. Cheeseman, and S. Tamaddon, J. Phys. (Paris) **C9-43**, 381 (1982).
23. G. N. Greaves, J. Non-Cryst. Solids **71**, 203 (1985).
24. M. D. Ingram, Philos. Mag. B **60**, 729 (1989).
25. J. Oviedo and J. F. Sanz, Phys. Rev. B **58**, 9047 (1998).
26. J. Horbach and W. Kob, Phil. Mag. B **79**, 1981 (1999).
27. J. Horbach, W. Kob, and K. Binder, Chem. Geol. **174**, 87 (2001).
28. P. Jund, W. Kob, and R. Jullien, Phys. Rev. B **64**, 134303 (2001).
29. E. Sunyer, P. Jund, W. Kob, and R. Jullien, J. Non-Cryst. Solids **307–310**, 939 (2002).
30. A. N. Cormack, J. Du, and T. R. Zeitler, Phys. Chem. Chem. Phys. **4**, 3193 (2002).
31. H. Lammert, M. Kunow, and A. Heuer, Phys. Rev. Lett. **90**, 215901 (2003).
32. M. P. Allen and D. J. Tildesley, *Computer Simulation of Liquids* (Clarendon Press, Oxford, 1987).
33. D. Frenkel and B. Smit, *Understanding Molecular Simulation: From Algorithms to Applications* (Academic, San Diego, 1996).
34. K. Binder, J. Horbach, W. Kob, W. Paul, and F. Varnik, J. Phys.: Condens. Matter **16**, S429 (2004).
35. B. W. H. van Beest, G. J. Kramer, and R. A. van Santen, Phys. Rev. Lett. **64**, 1955 (1990).
36. G. J. Kramer, A. J. M. de Man, and R. A. van Santen, J. Am. Chem. Soc. **64**, 6435 (1991).

37. U. Balucani and M. Zoppi, *Dynamics of the Liquid State* (Oxford Univ. Press, Oxford, 1994).
38. M. Bée, *Quasielastic Neutron Scattering* (Hilgers, Bristol, 1988).
39. R. Hempelmann, *Quasielastic Neutron Scattering and Solid State Diffusion* (Clarendon Press, Oxford, 2000).
40. A. Meyer, Phys. Rev. B **66**, 134205 (2002).
41. V. Zöllmer, K. Rätzke, F. Faupel, and A. Meyer, Phys. Rev. Lett. **90**, 195502 (2003).
42. J. Horbach, S. K. Das, A. Griesche, M.-P. Macht, G. Frohberg, and A. Meyer, Phys. Rev. B **75**, 174304 (2007).
43. A. Griesche, M.-P. Macht, S. Suzuki, K.-H. Kraatz, and G. Frohberg, Scripta Mat. **57**, 477 (2007).
44. H. Rauch and W. Waschkowski, *Landolt-Bornstein, New Series I/*, ed. by H. Schopper, **16A** (Springer, Berlin, 2000).
45. J. Horbach and W. Kob, J. Phys.: Condens. Matter **14**, 9237 (2002).
46. J. Horbach and W. Kob, Phys. Rev. B **60**, 3169 (1999).
47. J. Horbach, W. Kob, and K. Binder, J. Phys.: Condens. Matter **15**, S903 (2003).
48. W. Götze, in *Liquids, Freezing and Glass Transition*, edited by J.-P. Hansen, D. Levesque, and J. Zinn-Justin (North-Holland, Amsterdam, 1991), p. 287.
49. W. Götze and L. Sjögren, Rep. Prog. Phys. **55**, 241 (1992).
50. W. Götze, J. Phys.: Condens. Matter **10A**, 1 (1999).
51. Th. Voigtmann and J. Horbach, Europhys. Lett. **74**, 459 (2006).
52. R. E. Tickle, Phys. Chem. Glasses **8**, 101 (1967).
53. F. Kargl and A. Meyer, Chem. Geol. **213**, 165 (2004).
54. F. Kargl, A. Meyer, M. M. Koza, and C. Strohm (in preparation).
55. J. A. Johnson, C. E. Johnson, D. Holland, A. Mekki, P. Appleyard, and M. F. Thomas, J. Non-Cryst. Solids **246**, 104 (1999).
56. Kh. A. Ziq and A. Mekki, J. Non-Cryst. Solids **293–295**, 688 (2001).
57. G. Calas and J. Petiau, Sol. St. Comm. **48**, 625 (1983).

Chapter 7
Neutron Diffraction Studies of Hydrous Minerals in Geosciences

Hermann Gies

Abstract Applications of neutron diffraction experiments for structural studies of hydrous minerals are reviewed. The analysis of hydrous componets in minerals has always played a crucial role for the understanding of chemical and physical properties of crystalline minerals. A historical perspective of the impact of neutron diffraction in the early days of structural studies is given where the analysis of crystal structures of minerals containing hydrate water was of prime interest to the community. The selected examples of the past are combined with a summary of up-to-date experiments describing current state-of-the-art experiments and problems in geosciences where crystal structure analysis from neutron diffraction data is of unique value. The examples presented here are structural studies of nominally anhydrous minerals, zeolites, layer silicates, and, finally, soluble silicates.

7.1 Structural Studies Using Neutron Diffraction in Geosciences

Although X-ray scattering experiments are the most frequently used techniques in structural characterization of earth materials, there are many obvious reasons to take advantage of neutron radiation as an analytical tool for earth science research (see Chapter1). Despite the restricted availability of and the selective access to neutron sources, structures of minerals have been studied since the very beginning of neutron diffraction experiments mainly because of the existence of sufficiently large single crystals in natural minerals. In current applications of neutron radiation, penetration of neutrons in sample containers and the materials themselves allows for diffraction studies of large samples such as for texture analysis [1, see also Chapter 10], for experiments under extreme conditions of pressure and temperature, for mineral reactions and interactions where containers are needed [2–6], and for studies of mineral melts and their properties [7, 8, see also Chapter 6]. The particularities of neutron–nucleus interactions make neutrons especially suited for the diffraction

H. Gies (✉)
Institut für Geologie, Mineralogie und Geophysik, Ruhr-Universität Bochum,
Universitätsstr. 150, D-44780 Bochum, Germany
e-mail: Hermann.Gies@ruhr-uni-bochum.de

L. Liang et al. (eds.), *Neutron Applications in Earth, Energy and Environmental Sciences*, Neutron Scattering Applications and Techniques,
DOI 10.1007/978-0-387-09416-8_7, © Springer Science+Business Media, LLC 2009

studies of minerals with magnetic properties [9, see also Chapter 4], revealing the order/disorder behavior of the magnetic moments. Neutron diffraction studies of minerals with constituents with poor (X-ray) diffraction properties, that is, light elements such as protons, lithium (e.g., [10], see also Chapters 14 and 15), and beryllium (e.g., [11]), or studies of minerals with combinations of elements with poor or too high scattering length differences such as silicon and aluminum [12], or heavy metals and oxygen, respectively [13], provide insight into, for example, the location of hydrogen, the ordering of aluminum and silicon on T-sites in silicates, and the defects of the oxygen sublattice in ion conductors. An excellent review on how earth sciences can benefit from neutron scattering experiments is provided in a recent book of the *Reviews in Mineralogy and Geochemistry* series devoted to neutron scattering in earth sciences [14]. All areas of interest for geoscientists are presented and put into context together with recent methodological developments.

Technical advances in all areas of neutron research, that is, neutron sources, wave tubes, instrumentation, detector systems, and data management and analysis, have made neutron research more and more attractive for earth science users, irrespective of the neutron source, reactor, or spallation source (see also Chapter 3). Higher beam intensities and new area detectors have made experiments possible, which were not considered a few years ago. Sample size has been reduced significantly, and at the same time, H–D exchange is no longer a strict prerequisite for experiments at high-flux diffractometers. A survey on one of these advances concerning time-resolved studies is given by Kuhs and Hansen in studying ice and gas hydrates [15]. Still, there is great competition between neutron and X-ray diffraction experiments, because with modern X-ray sources it is also possible to penetrate areas of research that formerly were the sole domain of neutron sources. With high-flux sources, magnetic [16] and inelastic [17] experiments are feasible with X-rays nowadays. Despite the inroads made by X-ray-based techniques, the neutron in the scattering experiment has unique properties that make routine studies available for condensed matter containing light elements and having elements with low or high scattering contrast in X-rays diffraction as well as studies dealing with magnetic structures of powder samples (see also Chapters 1 and 3). This analysis of how earth sciences can benefit from neutron-based structural studies will focus on the diffraction studies of hydrous minerals and present examples exploiting the power of the technique, that is, its superior sensitivity for light elements. The samples are selected from a wide range of studies available in literature ranging from early examples to state-of-the-art experiments, all demonstrating the feasibility of routine neutron diffraction experiments for the fundamental study of earth materials important to the understanding of their physical and chemical properties.

To be able to follow a systematic development of the subject in a limited space, silicate minerals were selected out of the many water-containing minerals. The field of structural studies of hydrous minerals, in particular silicate minerals, can be subdivided into studies on nominally anhydrous minerals, hydrated framework silicates, layered silicates, and, finally, soluble silicates. The review will highlight recent advances in structural studies using neutron diffraction with an emphasis on the role of hydrogen in the chemical bond system of the silicate mineral, not visible in X-ray diffraction studies. It will put the latest results into perspective with classic papers of the past, which paved the way to modern neutron diffraction analyses. The focus is also directed toward visualizing the results from structure analysis in structure plots [18] rather than discussing the technical details of the experiment or the geoscience of the study. Visualizing these results enhanced understanding of spectroscopic experiments, studies on physical and chemical properties, etc., even though the crystal structure, in general, only represents a time- and space-averaged picture of reality.

7.2 Nominally Anhydrous Minerals

In the search of water molecules or hydroxyl ions in nominally anhydrous minerals (NAMs), such as olivine, pyroxenes, or garnets, neutron diffraction experiments have been carried out, but without positive analytical success. The detection limit and the fact that diffraction probes periodic structures so far have resulted in ambiguous results. The status of NAM studies has been reviewed recently [19], also in the context of neutron diffraction experiments. Still, there is a case where hydrogen enters the garnet structure through the hydrogarnet substitution of $H_4O_4 = SiO_4$, replacing the silicate tetrahedron with the somewhat larger H_4O_4-tetrahedron. Here, the oxygen atoms are in the corners of the polyhedron with the hydrogen aligned along four of six edges. This substitution is realized in a number of rather rare minerals, demonstrating the introduction of water on periodic sites in the crystal structure into otherwise anhydrous minerals. The example of the hydrogarnet minerals has been studied in a remarkable series of successful neutron diffraction analyses.

The hydrogarnet mineral katoite, $Ca_3Al_2(O_4D_4)_3$ [20], is used as a model for incorporation of (OH) into garnets and other silicate minerals that occur in the mantle. In particular, calcium-silicate garnets are proposed as water carriers in eclogite regions of the upper mantle. For this reason G. A. Lager and his collaborators have carried out a series of ambient and non-ambient neutron diffraction studies to localize the protons in the katoite- and hibschite-type structures [21, 22]. While in katoite the $[SiO_4]$-tetrahedron is almost completely replaced by the $[H_4O_4]$-tetrahedron, hibschite, $Ca_3Al_2(SiO_4)_{2.30}(OD)_{2.8}$, still contains considerable amounts of residual $[SiO_4]$ [20, 23]. Also, oxygen disorder is observed in the garnet–hydrogarnet solid solution series. The many structure analyses that are available on this system consider hydrogen implicitly as hydroxyl. That is, no detailed information on site and occupation is available. In Fig. 7.1, two structure plots of katoite and hibschite are shown with fully determined hydrogen positions highlighted.

The unambiguous information on hydrogen positions and the hydrogen bond network immediately available from structure analyses allowed for quantitative understanding of spectroscopic experiments probing the local environment, which are complementary to diffraction analyses.

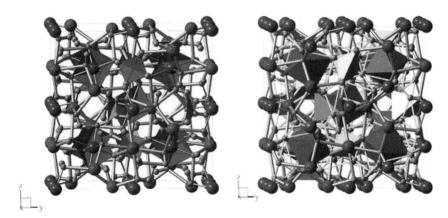

Fig. 7.1 Katoite (*left*) and hibschite (*right*), two hydrogarnet members. The big coordination poly-hedron represents the [AlO$_6$] octahedron, the small one is for the [H$_4$]-tetrahedron in katoite (*left*), and the [SiO$_4$]-tetrahedron in hibschite (*right*). The coordination polyhedra are not drawn at the border of the plot because of incompleteness. The bonding system in the ball-and-stick drawing is meant to guide the eye. In kaotite (*left*) the centre of the tetrahedron built by H-atoms of the hydroxyl groups is the T-site taken by Si in regular garnet. On the right, hibschite containing some Si shown as tetrahedra, still has considerable substitution by hydrogen. The network is shown by thin sticks

7.3 Hydrated Framework Silicates: Zeolites

Zeolites are naturally occurring framework silicates with open pores containing non-framework constituents. These are charge-balancing cations, in general alkali and earth alkali cations, and most important, hydrate water coordinating the cations. The rule of local electroneutrality leads to the distribution of cations related to the silicon/aluminum order in the silicate framework. Hydrate water is partially attached to cations and, in addition, concentrated in water clusters if pore space is very open. Structure analysis with diffraction techniques, therefore, yields a summary of the distribution of the non-framework constituents, that is, the type of cation and its occupancy factors as averaged quantities. Water is the most important component for many applications of zeolites, for example, as a drying agent or as storage material for latent heat. There are about 48 zeolite minerals with different frame-work structure types and, in addition, more than 180 framework structure types that include synthetic variants. For an up-to-date summary, a Web-based database provides excellent and exhaustive information [24].

Reversible dehydration/hydration is the property that makes zeolites useful not only as desiccants but also for the storage of latent heat. Ion exchange is another property of zeolites in general that has been exploited in commercial applications, such as water softener in washing powder formulations. In addition, the proton as cation is the source of acidity of zeolites, leading to most important catalytic prop-erties in shape-selective heterogeneous acid catalysis. All these properties rely on the crystallinity of the material, which ensures the accessibility of pore space, and on the water or hydroxyl groups in the structure responsible for the function. Both properties are difficult to analyze.

There are many structural studies on zeolite minerals and synthetic variants using neutron diffraction techniques; however, in many cases, the complicated non-framework constituents were removed from the natural samples and only dehydrated or calcined crystals were used for structural studies. For the analysis of the crystalline silicate framework, the structure information obtained from simplified experiments was sufficient. The interest in studying zeolite minerals using neutron diffraction arose from the growing need for accurate structural information not only on silicon/aluminum order/disorder but also on zeolitic water inside the voids of the silicate framework. After early studies of natrolite by Torrie et al. [25] and on laumontite by Bartl [26] in the late sixties, $\sim$15 years later Alberti et al. studied thomsonite [27] in detail; later, with a series of structural studies on the fibrous zeolites edingtonite, brewsterite, again natrolite, and laumontite [28–31], the research team of J. V. Smith used neutron diffraction with the aim to rationalize the reversible hydration/rehydration process on the atomic level. This study required precise information on the bonding of the non-framework constituents including hydrogen as key elements in hydrogen bonding. Early neutron diffraction studies required the availability of large single crystals. Therefore, only those zeolite materials for which appropriate crystals were at hand were studied, all of which were natural minerals of minor technical importance. However, the structure analyses showed convincingly the superior information on proton sites and the role of hydrogen bonding for the stability of the material. In Fig. 7.2 thomsonite, $Na_4Ca_8[Al_{20}Si_{20} O_{80}]\cdot24(H_2O)$ [27]

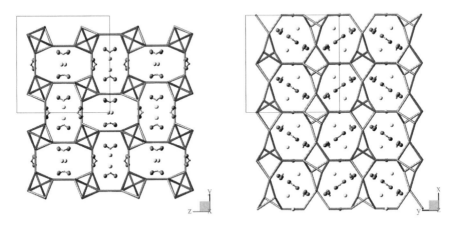

Fig. 7.2 The crystal structure of zeolite thomsonite including the position of hydrate water coordinating the charge-balancing cation. The porous silicate framework is shown in its skeletal representation. Oxygen atoms connecting Si and Al T-atoms are missing for clarity. In this way, an open view into pore space is possible, showing the arrangement of the non-framework constituents Ca (*isolated sphere*) and water molecules. Ca sits in the center of the channel and is surrounded by oxygen from water molecules completing the coordination of the cation. In thomsonite, direct electrostatic contact of the charge carrying [AlO$_4$]-tetrahedron is mediated by water

illustrates the function of water a as coordination shell for the charge-balancing sodium/calcium cations inside the channel.

An advancement in complexity concerning the organization of water inside zeolite pores presents zeolite bikitaite, a lithium-containing mineral [32]. Stahl et al. again used large bikitaite single crystals, $Li_2Al_2Si_4O_{12} \cdot 2H_2O$, and variable temperature experiments to establish the distribution and ordering of the non-framework constituents. Other than in thomsonite, the charge-balancing cation lithium is closely attached to the silicate wall of the eight-membered (8MR) ring pore. This arrangement leaves enough space for water molecules to align in a one-dimensional chain using one proton as hydrogen bridge (Fig. 7.3). The second proton occupies open space inside the channel. The periodic ordering of cation and water inside the pore also requires ordering of aluminum and silicon on T-sites of the silicate framework, establishing the periodic charge distribution. Neutron diffraction analysis then confirmed the Si/Al ordering, which was not unambiguous from previous X-ray studies.

Neutron powder diffractions analysis of fully deuterated laumontite, $Ca_4Al_8Si_{16}O_{48} \cdot 18D_2O$, by Stahl and Artioli using the Rietveld method [33] presented a new technical challenge. The rather large unit cell with $V = 1385\text{Å}^3$ at room temperature represented a difficult task for structure analysis considering the moderate resolution of the Lund diffractometer. For powder data collection, the sample was submerged in D_2O and a D_2O background had to be subtracted. Of the four water sites that were detected, two are coordinated to calcium while the other two act as bridges between the calcium-coordinated waters through hydrogen bonding (Fig. 7.4). All water sites are disordered to some degree and essentially fully occupied. They found that the local confinement to the apparently large number of possible water arrangements imposes a high degree of local water ordering.

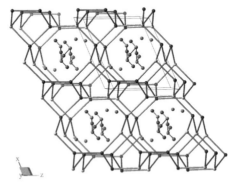

Fig. 7.3 Chain of water molecules along the 8MR channels of bikitaite. For clarity, the silicate framework is shown without oxygens bridging the silicon- (*light*) and aluminum- (*dark*) T-sites; lithium ions are shown as isolated spheres in the channels. In contrast to thomsonite, the lithium cation binds to the framework-oxygens and completes its tetrahedral coordination with hydrate water. The water molecules themselves are connected via hydrogen bridges into a one-dimensional chain. The neutron diffraction experiment nicely demonstrates its power in showing Si/Al order on T-sites and water molecules periodically ordered chains

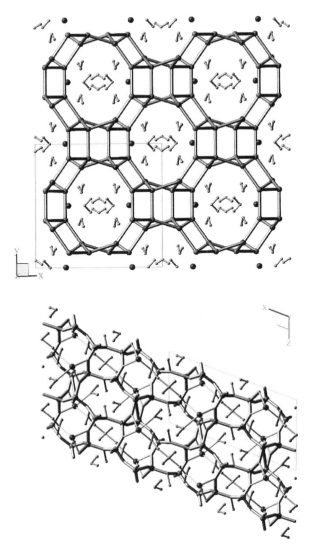

Fig. 7.4 Skeletal structure of laumontite including hydrate water molecules. On the top, a view along the channel axis parallel to z is shown. The silicate framework is presented without bridging oxygen atoms; the bonding network of non-framework constituents is not shown. At the bottom a projection along y is shown. Now, framework oxygen is included and the connectivity of non-framework constituents is highlighted. Of particular interest is the bridging link between neighboring calcium cations (*isolated in channel*) via three water molecules, which only neutron diffraction can show

Local cation and/or water vacancies will disrupt the ordering scheme and result in an average symmetry in agreement with the overall symmetry.

A recent review on structural information of zeolites from neutron diffraction is provided by Fitch and Jobic, which concentrated on synthetic materials and their structures [34]. The focus of their survey is closely related to commercial applications where Si/Al distribution and location of organic molecules inside pores and their dynamics are most important. Recently, zeolites with large cavities housing excess hydrate water attracted a lot of attention and were studied intensively, as their capability to store solar energy as latent heat stimulated much interest. When and where available, heat is used for dehydration of zeolites. This can be solar energy as well as energy generated in industrial processes, such as from, for example, cement production. Inside the storage device, zeolite water is accumulated separately from the dehydrated mineral and made available for exothermic hydration when needed. To understand the atomistic mechanisms in the dehydration/hydration processes, detailed structure analyses including all hydration intermediates are prerequisites. Interestingly, the crystal structures of the two potential candidates, synthetic zeolite Linde type A and zeolite mineral faujasite, are known in their dehydrated state; however, structure studies of fully hydrated materials are incomplete or poorly understood. This is mainly due to the fact that there is abundant void space for hydrate water. Only parts of water are directly coordinating to non-framework cations; considerable amounts are loosely bound with energies in the range of the heat of condensation of water. This leads to two effects that inhibit structure analysis: With respect to the dehydrated structure, cations move upon hydration into new sites and some water is highly dynamically disordered and almost impossible to localize. Looking at the scientific case, it is clear that only neutron diffraction analyses can answer the question of hydrogen bonding; however, considering the number of deuterons inside the cages including static and dynamic disorder, a complete structure analysis is beyond the current limits of the technique.

In addition to fully hydrated materials, the structures of partially hydrated zeolites are of particular interest. Calorimetric studies of the dehydration process show [35] that distinct energetic levels exist; that is, there are water molecules on different hydration sites with significantly different hydrations energies (Fig. 7.5).

In order to be able to understand the atomistic principles of the processes, a combination of techniques has been applied on model systems. Hunger and coworkers [36] combined DRIFT spectroscopy together with low-temperature neutron powder diffraction (5 K) of faujasites with four different loadings of water with a maximum loading of 50%, that is, 15 water molecules per structural unit. The Si/Al ratio of the NaX sample was 1.18, that is, almost 1 requiring strict alternation of aluminum and silicon on T-sites. For the refinement the aluminum T-sites were included, partially occupied by silicon, reflecting the experimental composition of the material. The following sequence of figures shows the key results of the Rietveld refinements of the LT neutron diffraction data sets. First, representations of the segments of the structure of faujasite are shown without water, showing only the distribution of sodium cations in the dehydrated state and in a hydration state reflecting ~15 water molecules. Differences can hardly be seen since all sites are present in both hydration states. The site coordinates of the various sodium cations shift only slightly. Marked differences can be observed when comparing site occupancies. Whereas at low water contents the coordination of sodium was satisfied by framework oxygen, for example, in six-membered ring sites (6MR sites) of the hexagonal prism or the sodalite cage, and near 4MR inside the supercage, higher water contents lead

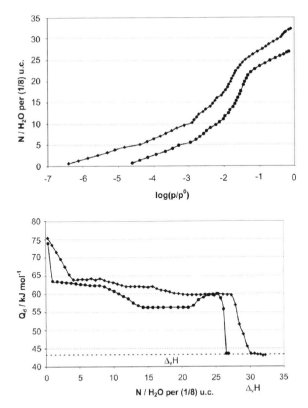

Fig. 7.5 Hydration of dehydrated zeolite faujasite. The number of molecules is normalized to 1/8 of a unit cell, describing the loading process for one supercage, sodalite cage, and two hexagonal prisms. The figure on the top shows the adsorption isotherms for a sodium faujasite zeolite, Na–FAU (Si/Al = 2.43). The lower isotherm is for NaBr-loaded material. At the bottom, the differential heat of adsorption is shown, again for Na–FAU and the lower graph for NaBr-loaded Na–FAU. Both figures clearly show the stepwise processes and, at the same time, demonstrate the ongoing exchange between different water sites. The *dashed horizontal line* represents the heat of condensation for pure water (figure adapted from reference 34)

to almost complete occupancy of the site decorating the 12MR window (Figs. 7.6 and 7.7). Here, enough space is left for hydrate water to approach sodium cations and complete its coordination.

In a recent paper, Marler et al. studied the stepwise hydration of zeolite Li-LSX, a faujasite-type material with Si/Al = 1, that is, at maximum Al content [37]. This composition state represents a maximum loading of cations with strong hydration energies, and therefore, optimum conditions for storing energy of hydration. As already reported for sodium faujasites, high-resolution isotherms for the lithium-containing sample show similar energy plateaus upon dehydration, indicating selective hydration states for lithium cations on different crystallographic sites. In a dehydrated state and at low water loadings (8 molecules per unit cell), lithium cations are close to the 6MR and 4MR near the sodalite cage, completing

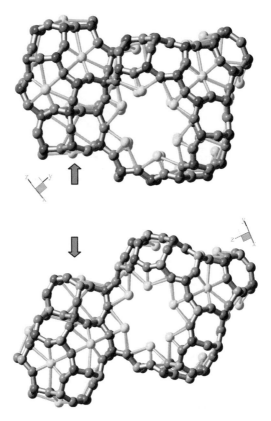

Fig. 7.6 Representative section of the crystal structure of zeolite faujasite with Si/Al ratio of 1.18, showing the distribution of charge-balancing sodium cations on non-framework sites without their hydrate water at high and low water contents. For the purpose of showing the bonding interaction, framework oxygens are included, completing the coordination sphere of the sodium cation. The specific sodium sites for the low-water (upper figure) and the high-water (lower figure) containing zeolite differ only marginally. However, the site occupancies are significantly different. Whereas at low water content the highly, by framework oxygens coordinated sodium sites are almost fully occupied, the sodium sites within the 12MR are fully occupied at high water content

the cation coordination (Fig. 7.8). As soon as more water is loaded, lithium-site SIII above the 4MR inside the supercage moves and, finally, "disappears" above 48 molecules of water per unit cell. With the many water molecules in the supercage, lithium is impossible to detect, in particular, since disorder and exchange between the different species take place. Therefore, diffraction analysis cannot unambiguously identify the cation. However, according to NMR analysis, lithium formerly occupying SIII definitely is part of the non-framework constituents inside the supercage.

Further loading of water completes the occupation of the different W3- and W4-sites until all coordination is complete. Finally, water is filling the empty space

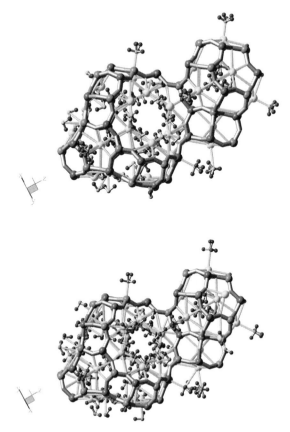

Fig. 7.7 Representative section of the crystal structure of zeolite faujasite with Si/Al ratio of 1.18 showing the distribution of charge-balancing sodium cations on non-framework sites with water hydration shell. Strong disorder of hydrogen positions is observed. Note the higher occupancy of the supercage of the higher loaded material in the lower figure

inside the supercage where water condensation in molecular clusters takes place. The final uptake of water molecules with low binding energy at around the heat of condensation of pure water can be looked at as using up residual empty space inside the faujasite supercage. Force field calculations of this process based on the experimental findings successfully simulated the hydration/dehydration process, allowing for extrapolations that are beyond the capability of neutron diffraction experiments.

It should be noted that extensive inelastic and quasielastic scattering experiments on zeolites were carried out mainly on samples of commercial interest and not on minerals, which benefited tremendously from the available structural information (e.g., [34]).

A most interesting structural behavior is observed for zeolite natrolite, $Na_2Al_2Si_3O_{10} \cdot 2H_2O$, upon pressure-induced hydration. At rather mild pressure conditions of $\sim 2\,GPa$, volume expansion of the materials is induced due to the uptake

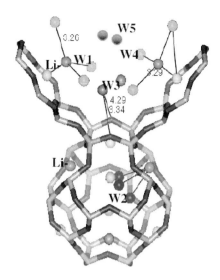

Fig. 7.8 Section of the lithium-LSX structure showing the lithium cation positions together with the various sites for hydrate water as long as it is on specific crystallographic sites

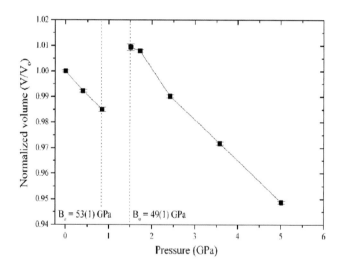

Fig. 7.9 Volume pressure diagram for the pressure-dependent super-hydration of zeolite natrolite. At a water pressure of ∼2 GPa, the volume of the unit cell suddenly increases due to the uptake of additional pore water (figure adapted from reference 38)

of additional hydrate water and cation re-ordering (Fig. 7.9) [38, 39]. The finding is reversible and was established for a number of the fibrous zeolites, mesolite, scolecite, and synthetic variants of natrolite framework structure type with gallosilicate composition.

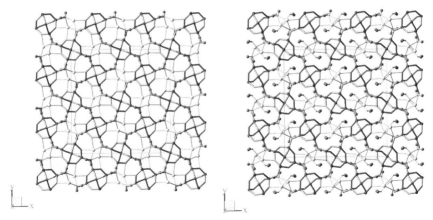

Fig. 7.10 Zeolite natrolite in its ambient-pressure (*left*) and high-pressure (*right*) state of hydration. The skeletal silicate framework is shown without the framework oxygens for clarity. The hydrate water (with *dark* hydrogens) is shown with its coordination to sodium cations (*light*) indicated by *light* connections. As can be seen from the figure on the right, water is bound to sodium cations in a continuous net at ambient pressure. At high pressure a slight rotation of the silicate strand increases the pore volume, creating additional space for more hydrate water. In the natural mineral, pressure release sets the extra water free. Synthetic gallosilicate analogs [40] retain the extra water. Only temperature increase leads back to the original hydration state (sponge property)

The explanation for the unexpected behavior was revealed from structure analyses from neutron powder diffraction data, showing the location of the water molecules including hydrogen atoms and their association with the charge-balancing sodium cations and framework oxygen atoms (Fig. 7.10). The flexibility of the silicate framework of the fibrous zeolites was already known when dehydration of samples in ambient conditions led to a sharp decrease of unit cell volume of almost 20% and loss of symmetry from F dd2 to F 112 [40]. The unusual flexibility of the silicate framework is obviously capable to accommodate even more hydrate water at higher pressures.

Comparison of results from various structure refinements allowed for a mechanistic interpretation of the volume change, the rearrangement of cations, and the accommodation of additional water within pore space. As can be seen in Fig. 7.10, the hydrogen bonding of the water protons and the completion of the coordination sphere of the extra-framework sodium cation nicely shows up from the structure refinements including the precise position of the water proton.

7.4 Layer Silicates, Micas, Clays, and Hydrous Alkali Layer Silicates

Clay minerals are layer silicates containing water molecules in interlayer space. They are another example of minerals with complicated structures suited for neutron diffraction studies, in particular if detailed analysis of water molecules is required [see also Chapter 18]. An early report by Cebula et al. [42] on the system montmorillonite-water, a 2:1 dioctahedral layer silicate with mixed occupancy of aluminum and magnesium in the octahedral layer sandwiched by two silicate layers and as charge balance hydrated sodium octahedra between the 2:1-silicate strands (cf. Fig. 7.11, structural principles are the same, however, different metal cations in oxide polyhedra), on the textural structure of the material at low water contents discussed elastic diffraction experiments. With quasielastic experiments, the nature of the intercalate water was studied. It was concluded that stacking disorder is the major feature of the structure of the solid, leading to line broadening, rather than a distribution of particle size. The simulations of the diffraction diagrams were based on an existing structural model and were confirmed. Layer silicates are notorious for their structural and chemical heterogeneity including mixed-layer structures, disordered layer structures, inter-stratification of hydrate water and isomorphous cation substitution [43]. Neutron scattering experiments in particular contributed to the structural analysis of interlayer constituents and bonding networks. The hydroxyl termination of layers plays a dominant role in this context, being responsible for hydrogen bond networks to neighboring layers, and the interlayer hydrate water of charge-compensating cations. Only neutron diffraction experiments revealed precise details of the local structure and the register of layers through the hydrogen bond network.

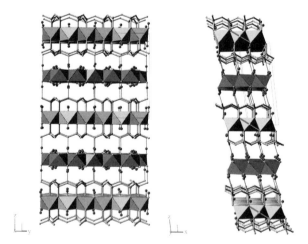

Fig. 7.11 Crystals structure of RT clinochlore showing the hydrogen-terminated hydroxyls of the octahedral layers (*dark spheres*). The room temperature model with partial occupancies for the octahedral sites for a wide variety of divalent and trivalent cations, such as Mg, Al, Fe, Cr, and Ni, are indicated with different shades of gray of polyhedra, and aluminum and silicon in a 1:3 ratio on tetrahedral sites are shown as balls and sticks. Spheres representing atoms are kept small for clarity. The strongest hydrogen bond between B-layer hydroxyls and oxygen from the tetrahedral layer are shown in *as dark sticks between layers*

High-pressure studies of minerals using neutron diffraction is an interesting combination of experimental setup since larger, homogeneous samples of the mineral can be studied while maintaining good control of the experimental parameters. However, only a small number of silicate minerals have been studied so far. A recent review on the subject by Parise [4] summarizes current state of the art. To illustrate the potential of high-pressure studies with neutrons, the high-pressure structure of clinochlore, $Mg_5AlSi_3AlO_{10}(OD)_8$ [44], a significant carrier of water into cold subduction zones, is presented here. The hydrogen bond system between the layers was then compared with the room temperature structure (Fig. 7.11).

The structure of clinochlore is built from dioctahedral 2:1 slabs of an octahedral layer with trivalent cations, in the simplest form aluminum, most often mixed with other trivalent cations such as Fe, Cr, etc., sandwiched between two tetrahedral layers with $Si/Al = 3 : 1$, introducing charge into the layer. The 2:1 slabs are separated by isolated octahedral layers of brucite type, where magnesium can also be substituted by various other divalent cations. Surprisingly, the bulk modulus of clinochlore is much higher (75 GPa) than that of brucite, a structure comparable to the isolated octahedral B-layer (47 GPa). Increasing pressure has a linear effect on the strong hydrogen bond, leading to shortening from 1.91 to 1.77 Å. The fact that brucite is not as stiff as clinochlore is attributed in equal part to the additional 2:1 layer and the hydrogen bond scheme in clinochlore, which is different from brucite. Compared with talc, a 2:1 layer silicate without interlayer hydrogen bond, the compressibility is only half as much, underlining the strength and the importance of the hydrogen bond in maintaining register between layers [44–46].

The high-pressure study of hydrogen-containing minerals to expand understanding of the physical properties of minerals is obvious; however, there are only few examples reported in literature that make use of the potential of neutron diffraction experiments for the localization of hydrogen [47–49]. With new 2D detectors as provided at Vivaldi [50] at the ILL, no deuteration of the single crystal is required anymore, even for hydrogen-rich samples. This progress also includes powder diffraction where hydrogen-containing phases might also be studied without isotope exchange using instruments at high-flux neutron sources.

As an example of more recent and more sophisticated structural studies of strongly disordered clay minerals, the crystal structure of kaolinite, $Al_2(Si_2O_5)(OD)_4$, is presented [51] (Fig. 7.12). The 1:1 dioctahedral layer silicate is decorated not only with terminal OH groups on top of the $[AlO_6]$-octahedral layer pointing toward the neighboring slab but also with OH groups that are not charge balanced by the tetrahedral silicate layer. These OH groups point straight into the 6MR pore of the silicate layer on top (Fig. 7.12, bottom). The rather weak bonding interaction between layer slabs leads to strong stacking disorder and also intercalations and, therefore, defective stacking coherence.

The smectide-type vermiculite clay shows far more irregular stacking of 2:1 silicate slabs with trivalent cations in the octahedral layer (dioctahedral clays) because of much less chemical homogeneity and strong hydration of the interlayer triva-

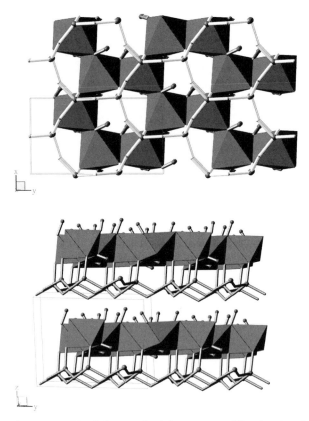

Fig. 7.12 Crystal structure of kaolinite as refined from neutron diffraction data. Layer slab with the [AlO$_6$]-octahedra in dark shading showing hydroxyl terminations between layers (top) and the excess hydroxyl between tetrahedral silicate layer and those oxygens that are not charge balanced (bottom) are shown. Neutron structure analyses clearly identify the location of the protons

lent, for example, lanthanum cations introduced through ion exchange experiments, La$_{0.65}$(Al$_{0.25}$Fe$_{0.08}$Mg$_{5.57}$) (Si$_{5.74}$Al$_{2.27}$) O$_{20}$(OH)$_4$(H$_{20}$)$_{14}$ [52]. Neutron structure analysis could not provide details about the most likely hydrogen positions because of static and dynamic disorder (Fig. 7.13).

As examples of the 2:1 type trioctahedral layered silicates, phlogopite, K$_2$Mg$_6$(Al$_2$Si$_6$O$_{20}$)(OH)$_4$, as member of the family of mica minerals, is presented where rather regular stacking between silicate layers is found through potassium cations charge balancing the excess negative charge of the 2:1 silicate layer slab [53]. For the delicate hydrogen analysis inside the [MgO$_6$]-trioctahedral layer, neutron diffraction analysis provided all the essentials for the understanding of ion exchange and charge-matching properties (Fig. 7.14). The phlogopite structure is one of the examples where static energy minimization calculations have been used to probe the proton position as obtained from neutron structure refinements [54]. As

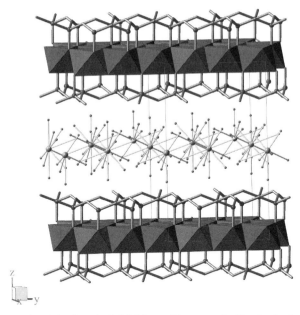

Fig. 7.13 Structure plot of the dioctahedral 2:1 layer silicate vermiculite after lanthanum exchange. The composition of the layer silicate is complex, as is obvious from the chemical formula. The hydration shell of the interlayer cation is complicated and highly disordered. Proton positions are not revealed from the neutron structure analysis

in a number of examples, experiment and simulation calculations are in good agreement, confirming the importance of the neutron experiment for scaling parameters used in numerical modeling of crystal structures.

Phlogopite is a rather rare mineral, end member of the biotite series, and often fluoride-bearing, which substitutes for OH in the octahedral layer. The fluoride substitution, which can be derived from neutron structure analysis, can reach fluorophlogopite composition and indicates that the mineral accumulates the halogenide anion. Because of its stability range, it is also an indicator for upper mantle rocks.

Hydrous alkali layer silicates without additional octahedral layer are even more flexible, soft, and instable upon dehydration. Only few structure analyses exist; detailed neutron diffraction studies are available for one hydrous sodium layer silicate. The synthetic Na-RUB-18 belongs to a family of materials [55], most of which are natural minerals and are widely used in industry, for example, as a precursor of builders for washing powders.

The negative charge of the silicate layer is partially compensated by hydrogen involved in a strong hydrogen bridge between neighboring siloxane groups (Fig. 7.15). Neutron diffractions structure analysis [56] shows the split position of hydrogens as a time average, thus confirming results from solid-state NMR experiments, which also indicate strong hydrogen bond interaction.

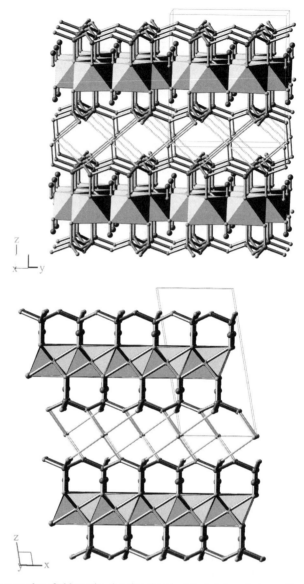

Fig. 7.14 Structure plot of phlogopite showing the trioctahedral layer between the two tetrahedral silicate layers. The OH groups from [MgO$_6$]-octahedra are shown with terminal hydrogen spheres. Register between layer slabs is maintained through bonding interaction of K interlayer cations balancing the charge of the [Si$_3$AlO$_{10}$]-silicate layer

Detailed structure analysis of the hydrate water coordinating the inter-layer sodium cations lets us assume that hydrogen atoms are strongly disordered and mobile, again confirming NMR experiments (Fig. 7.16). The temperature dependence of the displacements parameters of the hydrogen atoms shows that at lower

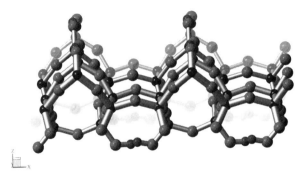

Fig. 7.15 Layer slab of the crystal structure of the sodium silicate RUB-18. Two neighboring silanol groups are connected through strong hydrogen bridges. The neutron structure refinement reveals the proton as split position atom

temperatures local exchange occurs between the silicate layer hydroxyls and the protons of axial water of the $[Na(H_2O)_6]$ coordination polyhedron, $T < 80°C$, whereas at higher temperatures all water protons, including those from apical water, are involved in 1-D proton diffusion. Low-temperature diffraction experiment, in addition, reveal phase transition at $\sim 160\,K$ of the highly symmetrical high-temperature structure to a low-temperature phase with stacking disordered arrangement of neighboring layers. Because of the high disorder of the proton positions and, in addition, the stacking disorder in the low temperature modification, no detailed structure analysis is available so far.

7.5 Outlook

The traditional neutron diffraction experiment still has many exciting facets in modern mineral science. Using hydrous minerals as examples, up-to-date experimentation allows for the study of water, for example, in minerals under extreme conditions, such as pressure and temperature, or high-resolution structure analysis of complicated and water-rich layered silicates under various pressure and temperature conditions. Combining these analyses with quasielastic or inelastic experiments extends the potential of neutron experimentation even further. With the examples given, a non-exhaustive outline of representative experiments is provided to motivate the non-neutron experimentalist to consider neutron diffraction in his/her research as a most interesting and promising possibility for the thorough investigation of problems related to the hydrous component in minerals. With the new instruments becoming available at neutron facilities worldwide in the near future, further new possibilities for mineral science research are in sight.

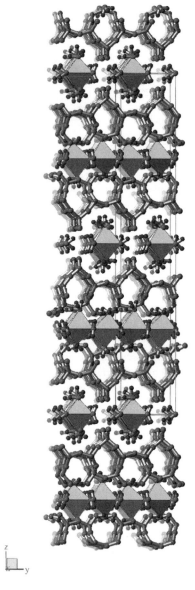

Fig. 7.16 Crystal structure of Na-RUB-18 as refined from neutron powder data at room temperature. The layer is shown as ball-and-stick representation, and the charge-balancing sodium cations with their hydrate water as chains of octahedra decorated with hydrogen atoms on split sites. Analyzing the scattering density maps reveals the strong disorder including the diffusion path of the protons

References

1. H. R. Wenk, Neutron diffraction texture analysis, Rev. Mineral. Geochem. **63**, 399 (2006)
2. A. Pavese, Neutron powder diffraction and Rietveld analysis; applications to crystal-chemical studies of minerals at non-ambient conditions, Eur. J. Mineral. **14**, 241 (2002)
3. S. A. T. Redfern, Neutron powder diffraction of minerals at high pressures and temperatures: some recent technical developments and scientific applications, Eur. J. Mineral. **14**, 251 (2002)
4. J. B. Parise, High pressure studies, Rev. Mineral. Geochem. **63**, 205 (2006)
5. B. Winkler, Applications of neutron radiography and neutron tomography, Rev. Mineral. Geochem. **63**, 459 (2006)
6. M. R. Daymond, Internal stresses in deformed crystalline aggregates, Rev. Mineral. Geochem. **63**, 427 (2006)
7. M. C. Wilding and C. J. Benmore, Structure of glasses and melts, Rev. Mineral. Geochem. **63**, 275 (2006)
8. D. R. Cole, K. W. Herwig, E. Mamontov, and J. Z. Larese, Neutron scattering and diffraction studies of fluids and fluid-solid interactions, Rev. Mineral. Geochem. **63**, 313 (2006)
9. R. J. Harrison, Neutron diffraction of magnetic materials, Rev. Mineral. Geochem. **63**, 113 (2006)
10. S. H. Park, H. Gies, B. H. Toby et al., Characterization of a new microporous lithozincsilicate with ANA topology, Chem. Mater. **14**(7), 3187 (2002)
11. D. W. Brown, M. A. M. Bourke, B. Clausen et al., A neutron diffraction and modelling study of uniaxial deformation in polycrystalline beryllium, Metall. Mater. Trans. A **34A**(7), 1439 (2003)
12. V. L. Vinograd, E. A. Juarez-Arellano, A. Lieb et al., Coupled Al/Si and O/N order/disorder in $BaYb[Si_{4-x}Al_xO_xN_{7-x}]$ sialon: neutron powder diffraction and Monte Carlo simulations, Z. Kristallogr. **222**(8), 402 (2007)
13. P. D. Battle, D. E. Cox, M. A. Green et al., Antiferromagnetism, ferromagnetism, and phase separation in the GMR system $Sr_{2-x}La_{1+x}Mn_2O_7$, Chem. Mater. **9**(4), 1042 (1997)
14. Neutron scattering *in* earth sciences, Rev. Mineral. Geochem. **63**, Jodi J. Rosso, editor, Mineralogical Society of America and Geochemical Society, Washington, D.C. (2006)
15. W. F. Kuhs and T. C. Hansen, Time resolved neutron diffraction studies with emphasis on water ices and gas hydrates, Rev. Mineral. Geochem. **63**, 171 (2006)
16. J. B. Kortright, D. D. Awschalom, J. Stohr et al., Research frontiers in magnetic materials at soft X-ray synchrotron radiation facilities, J. Magn. Magn. Mater. **207**(1–3), 7 (1999)
17. J. L. Hodeau and R. Guinebretiere, Crystallography: past and present, Appl. Phys. A—Mater. Sci. Process. **89**(4), 813 (2007)
18. T. C. Ozawa and S. J. Kang, Balls & Sticks: easy-to-use structure visualization and animation creating program, J. Appl. Cryst. **37**, 679 (2004)
19. Water in nominally anhydrous minerals, Rev. Mineral. Geochem. **62**, Jodi J. Rosso, editor, Mineralogical Society of America and Geochemical Society, Washington, D.C. (2006)
20. E. Passaglia and R. Rinaldi, Katoite, a new member of the $Ca_3Al_2(SiO_4)_3$—$Ca_3Al_2(OH)_{12}$ series and a new nomenclature for the hydrogrossular group of minerals. Bull. Mineral. **107**, 605 (1984)
21. G. A. Lager, T. Armbruster, and J. Faber, Neutron and X-ray diffraction study of hydrogarnet $Ca_3Al_2(O_4H_4)_3$, Amer. Mineral. **72**, 756 (1987)
22. G. A. Lager and R. B. VonDreele, Neutron powder diffraction study of hydrogarnet to 9 GPa, Amer. Mineral. **81**, 1097 (1996)
23. G. A. Lager, T. Armbruster, F. J. Rotella, and G. A. Rossman, OH substitution in garnets—X-ray and neutron-diffraction, infrared, and geometric-modelling studies, Am. Mineral. **74**, 840 (1989)
24. C. Baerlocher and L. B. McCusker, Database of Zeolite Structures, (2007) http://www.iza-structure.org/databases/

25. B. H. Torrie, I. D. Brown, and H. E. Petch, Neutron diffraction determination of hydrogen positions in natrolite, Canadian J. Phys. **42**, 229 (1964)
26. H. Bartl, Neutron diffraction measurements of laumontite structure, a zeolite mineral, Acta Crystallogr. **A25**, 119 (1969)
27. A. Alberti, G. Vezzalini, and V. Tazzoli, Thomsonite: a detailed refinement with cross checking by crystal energy calculations, Zeolites **1**, 91 (1981)
28. A. Kvick and J. V. Smith, A neutron diffraction study of the zeolite edingtonite, J. Chem. Phys. **79**, 2356 (1983)
29. G. Artioli, J. V. Smith, and A. Kvick, Neutron diffraction study of natrolite, $Na_2Al_2Si_3O_{10}^*2H_2O$, at 20 K, Acta Cryst. **C40**, 1658 (1984)
30. G. Artioli, J. V. Smith, and A. Kvick, Multiple hydrogen positions in the zeolite brewsterite, $Sr_{0.98}Ba_{0.05}Al_2Si_6O_{16}^*5H_2O$, Acta Cryst. **C41**, 492 (1985)
31. G. Artioli, J. V. Smith, and A. Kvivk, Single crystal neutron diffraction study of partially dehydrated laumontite at 15 K, Zeolites **9**, 377 (1989)
32. K. Stahl, A. Kvick, and S. Ghose, One-dimensional water chain in the zeolite bikitaite: Neutron diffraction study at 13 and 295 K, Zeolites **9**, 303 (1989)
33. K. Stahl and G. Artioli, A neutron powder diffraction study of fully deuterated laumontite, Eur. J. Mineral. **5**, 851 (1993)
34. A. N. Fitch and H. Jobic, Structural information from neutron diffraction, in *Molecular Sieves*, Vol. 2 (Springer-Verlag, Berlin Heidelberg, 1999), p. 31
35. B. Boddenberg, G. U. Rakhmatkariev, S. Hufnagel, and Z. Salimov, A calorimetric and statistical mechanic study of water adsorption in zeolite NaY, Phys. Chem. Chem. Phys. **4**, 4172 (2002)
36. J. Hunger, I. A. Beta, H. Böhling, C. Ling, H. Jobic, and B. Hunger, Adsorption structures of water in NaX studied by DRIFT spectroscopy and neutron powder diffraction, J. Phys. Chem. **B110**, 342 (2006)
37. A. Wozniak, B. Marler, K. Angermund, and H. Gies, Water and Cation Distribution in Fully and Partially Hydrated Li-LSX Zeolite, Chem. Mater., Web Release Date: 06-Sep-2008; (Article) DOI: 10.1021/cm703654a
38. M. Colligan, Y. Lee, T. Vogt et al., High pressure neutron diffraction study of superhydrated natrolite, J. Phys. Chem. **B109**, 18223 (2005)
39. Y. Lee, T. Vogt, J. A. Hriljac et al., Pressure induced volume expansion of zeolites in the natrolite family, J. Am. Chem. Soc. **124**(19), 5466 (2002)
40. W. H. Baur and W. Koswig, The phases of natrolite occurring during dehydration and rehydration studied by single crystal x-ray diffraction methods between room temperature and 923 K, Neues Jb. Minerl. Mh. **4**, 171 (1996)
41. Y. Lee, T. Vogt, J. A. Hriljac et al., Non-framework cation migration and irreversible pressure-induced hydration in a zeolite, Nature **420**, 485 (2002)
42. D. J. Cebula, R. K. Thomas, S. Middleton, R. H. Ottewill, and J. W. White, Neutron diffraction from clay-water Systems, Clays Clay Miner. **27**, 39 (1979)
43. V. A. Drits, Structural and chemical heterogeneity of layer silicates and clay minerals, Clay Miner. **38**, 403 (2003)
44. M. D. Welch and W. G. Marshall, High pressure behaviour of clinochlore, Am. Mineral. **86**, 1380 (2001)
45. P. F. Zanazzi, M. Montagnoli, S. Nazzareni, and P. Comodi, Structural effects of pressure on triclinic chlorite: A single-crystal study, Am. Mineral. **91**, 1871 (2006)
46. P. F. Zanazzi, M. Montagnoli, S. Nazzareni, and P. Comodi, Structural effects of pressure on monoclinic chlorite: A single-crystal study, Am. Mineral. **92**, 655–661 (2007)
47. J. R. Smyth, Hydrogen in high pressure silicate and oxide mineral structures, Rev. Mineral. Geochem. **62**, 85 (2006)
48. A. Friedrich, E. Haussühl, R. Boehler, W. Morgenroth, E. A. Juarez-Arellano, and B. Winkler, Single-crystal structure refinement of diaspore at 50 GPa, Am. Mineral. **92**, 1640 (2007)
49. A. Friedrich, G. A.. Lager, M. Kunz, B. C. Chakoumakos, J. R. Smyth, and A. J. Schultz, Temperature-dependent single-crystal neutron diffraction study of natural chondrodite and clinohumites, Am. Mineral. **86**, 981(2001)

50. Home page for VIVALDI at ILL: http://www.ill.fr/vivaldi/home/
51. E. Akiba, H. Hayakawa S. Hayashi, R. Miyawaki, S. Tomura, Y. Shibasaki, F. Izumi, H. Asano, and T. Kamiyama, Structure refinement of synthetic deuterated kaolinite by Rietveld analysis using time-of-flight neutron powder diffraction data, Clays Clay Miner. **45**, 781 (1997)
52. P. G. Slade, P. G. Self, and J. P. Quirk, The interlayer structure of La-vermiculite, Clays Clay Miner. **46**, 629 (1998)
53. J. H. Rayner, The crystal structure of phlogopite by neutron diffraction, Mineralo. Mag. **39**, 850 (1974)
54. J. E. Liang and F. C. Hawthorne, Calculation of atom positions in micas and clay minerals, Can. Mineral. **36**, 1577 (1998)
55. S. Vortmann, H. Gies, and J. Rius, Ab initio structure solution from X-ray powder data at moderate resolution: the crystal structure of a microporous layer silicate, J. Phys. Chem. **101**, 1292–1297 (1997)
56. M. Borowski, O. Kovalev, and H. Gies, Structural characterization of the hydrous layer silicate Na-RUB-18, $Na_8Si_{32}O_{64}(OH)_8^* 32H_2O$ and derivatives with XPD-, NPD- and SS NMR experiments, Microporous Mesoporous Mater. **107**, 71 (2008)

Chapter 8
Studies of Mineral–Water Surfaces

Nancy L. Ross, Elinor C. Spencer, Andrey A. Levchenko, Alexander I. Kolesnikov, David J. Wesolowski, David R. Cole, Eugene Mamontov, and Lukas Vlcek

Abstract In this chapter we discuss the application of inelastic and quasielastic neutron scattering to the elucidation of the structure, energetics, and dynamics of water confined on the surfaces of mineral oxide nanoparticles. We begin by highlighting recent advancements in this active field of research before providing a brief review of the theory underpinning inelastic neutron scattering (INS) and quasielastic neutron scattering (QENS) techniques. We then discuss examples illustrating the use of neutron scattering methods for studying hydration layers that are an integral part of the nanoparticle structure. The first investigation of this kind, namely the INS analysis of hydrated ZrO_2 nanoparticles, is described, as well as a later, complementary QENS study that allowed for the dynamics of diffusion of the water molecules within the hydration layer to be examined in detail. The diverse range of information available from INS experiments is illustrated by a recent study combining INS with calorimetric experiments that elucidated the thermodynamic properties of adsorbed water on anatase (TiO_2) nanoparticles. To emphasize the importance of molecular dynamics (MD) simulations for deconvoluting complex QENS spectra, we describe both the MD and the QENS analysis of rutile (TiO_2) and cassiterite (SnO_2) nanoparticle systems and show that, when combined, data obtained by these two complementary methods can provide a complete description of the motion of the water molecules on the nanoparticle surface. We close with a glimpse into the future for this thriving field of research.

8.1 Introduction

Water species are ubiquitous on mineral surfaces. In oxide nanopowders characterized by high surface areas, water adsorbed on the surfaces can be present in molecular and/or dissociated forms and may have a significant effect on the stability and growth of nanoparticles [1–3]. Water confined in nanoscale geometries

N.L. Ross (✉)
Department of Geosciences, Crystallography Laboratory, Virginia Tech, Blacksburg, VA 24061, USA
e-mail: nross@vt.edu

L. Liang et al. (eds.), *Neutron Applications in Earth, Energy and Environmental Sciences*, Neutron Scattering Applications and Techniques, DOI 10.1007/978-0-387-09416-8_8, © Springer Science+Business Media, LLC 2009

has also attracted much attention recently because of the large range of new phenomena observed in a confined geometry. Confinement can lead to changes in both structural and dynamical properties caused by the interaction of water with the surface. Examples include polymorphic transformations, the "slowdown" or "speedup" of dynamics of water species interacting with the nanoparticle surface [4–6], and shifts in the glass transition and crystallization temperature induced by confinement [7]. Studies of water confinement are also important to fundamental and applied science because of the widespread occurrence and variety of confinement phenomena in nature, from water and oil transport in rocks to water mobility in cells and membranes.

Inelastic neutron scattering (INS) provides an excellent method for detecting molecular vibrations and rotational and diffusive motion associated with hydrogen atoms because of the exceptionally large incoherent neutron scattering cross-section of the proton compared with the nuclei of nearly all other atoms, including deuterium. For anhydrous oxides of most metals, the incoherent signal is dominated by scattering from one or more monolayers of adsorbed surface water. INS has some advantages over Raman and infrared spectroscopies because INS is not restricted by selection rules or restricted to zone center modes or hampered by optical opacity and sample absorption. INS also has an advantage over X-ray scattering in that the latter relies on electromagnetic interaction with electrons in the sample, making the detection of hydrogen difficult, especially in the presence of heavier atoms. Although there are several neutron scattering studies of water in confined geometry in nanotubes [6], porous media [8–12], and oxides [4, 5, 9, 13, 14], there are few studies of water adsorbed on the surfaces of oxide nanoparticles.

Quasielastic neutron scattering (QENS) experiments complement INS experiments and are well suited for studying the mobility of water molecules. QENS has been extensively employed in studies of the dynamics of bulk water [15] and water confined within nanoscale pores in a variety of inorganic substrates, predominantly various silica matrices such as Vycor and Gelsil glass and MCM zeolites [16–27] and thin films on the surfaces of metal oxide nanoparticles [28, 4, 5, 14, 22, 29]. However, until recently, investigations of the dynamics of surface water on oxide nanoparticles have not included input from QENS.

In this chapter, we present a brief introduction of the INS and QENS methods applied to vibrational and diffusive motions of water molecules. Guidance is provided to the reader for more in-depth reviews of these techniques. The bulk of the chapter is devoted to reviewing the seminal studies that have identified the vibrational density of states (VDOS) of water adsorbed on oxide nanoparticles, including ZrO_2 [30, 31] and the anatase polymorph of TiO_2 [32]. We also show how QENS studies of water adsorbed on the surface of ZrO_2 nanoparticles [4, 28] and on the surfaces of isostructural rutile (TiO_2) and cassiterite (SnO_2) nanoparticles [14, 33] complement the INS studies. The latter are combined with molecular dynamics (MD) simulations to provide a comprehensive model of the diffusional dynamics of water on nanoparticle surfaces. The results from these studies are summarized and the future outlook given at the end of the chapter.

8.2 Inelastic Neutron Scattering and Quasielastic Neutron Scattering Techniques

INS refers to scattering processes that involve energy and momentum exchange between the neutron and the scatterer. Thermal neutrons have energies comparable to vibrational excitations and can be inelastically scattered from the sample being studied, exchanging part of their energy and momentum with the vibrational excitations in the solid (Chapter 3). The inelastic scattering of a monochromatic beam of neutrons is aimed at determining the scattering function, $S(Q,E)$, over wavevector, Q, and energy range, E [34, 35]. The scattering function, depending on coherent or incoherent scattering, is related to, respectively, the inter- and self-particle space-time correlation functions of the scatterer under study. Detailed analyses of the energy and momentum transfers involved in these processes provide valuable information about the vibrational properties of the solid. INS experiments on a polycrystalline sample, for example, provide information about the sample's vibrational density of states (VDOS), which can be used to calculate the thermodynamic properties of the sample. Chaplot et al. [36], Choudhury and Chaplot [37-Chapter 5], and Loong [38] provide examples of applications of INS to earth-forming minerals.

As mentioned earlier, hydrogen has an extraordinary large neutron scattering cross-section of 82.02 bn (dominated by σ_{inc} of 80.26 bn), compared with oxygen (4.23 bn) and metals such as titanium and zirconium (4.35 and 6.46 bn, respectively). Thus, the observed INS spectra of metal oxide nanoparticles with hydration layers are dominated by features relating to the motion of the hydrogen atoms. INS techniques can also give substantial insight into the structure, VDOS, and overall dynamics of water on the surface. Using appropriate models, the heat capacity of confined water can also be obtained from the VDOS.

Incoherent scattering results from the gain or loss of energy of incident neutrons interacting with the same atomic nucleus within a sample at different times (Chapter 2). Scattering from diffusional or relaxational molecular motions has energy transfers in the μeV–MeV range and characteristic time scales of nanoseconds to picoseconds. A diffuse profile of intensities appears near the elastic ($E = 0$) region and is referred to as QENS. Similar to INS experiments, QENS can best be measured using high-resolution time-of-flight, backscattering, and spin-echo spectrometers at either reactor or pulsed neutron sources.

QENS is ideally suited to studies of the dynamics of water and other hydrogen-bearing molecules because of the very large incoherent scattering cross-section of hydrogen. The time, length, and energy transfer scales sampled by QENS spectrometers are also ideally suited for direct comparison with classical MD simulations, as will be shown in Sections 8.3.1 and 8.3.3. The two approaches are linked by the Intermediate Scattering Function, ISF $\equiv I_{inc}(Q, t)$, resulting from incoherent scattering from hydrogen nuclei. The ISF can be extracted from time-space trajectories of individual atoms obtained from MD simulations. A QENS experiment yields the incoherent structure factor, $S_{inc}(Q, E)$, which is related to the ISF by

$$S_{inc}(Q, E) = \int I_{inc}(Q, t)e^{i\frac{E}{\hbar}t} dt. \tag{8.1}$$

For a regular diffusion process (either translational or rotational), the ISF is described by a simple exponential decay in the time space, and the incoherent structure factor is a Lorentzian in the energy space:

$$S_{\text{inc}}(Q, E) = \frac{1}{\pi} \frac{\Gamma(Q)}{[\Gamma(Q)]^2 + E^2}. \tag{8.2}$$

The Q-dependence of the width of the Lorentzian broadening, $\Gamma(Q)$, can often be used to determine both the spatial and the temporal characteristics of the diffusion process. Another commonly used model to represent a jump diffusion is $\Gamma(Q) \propto 1/\tau$, where τ is the characteristic time between the diffusion jumps.

8.3 Neutron Scattering Studies of Water Adsorbed on Oxide Nanoparticles

8.3.1 INS and QENS Studies of Surface Water on ZrO$_2$

The first tentative steps toward obtaining an insight into the dynamics of water on the surfaces of ZrO$_2$ nanoparticles were made by Loong and co-workers. Using INS, Loong investigated the vibrational dynamics of water adsorbed on ZrO$_2$ nanoscale powders [30, 31]. These seminal experiments were conducted with the High-Resolution Medium-Energy Chopper Spectrometer (HRMECS) at the Argonne National Laboratory.

Assuming that an average of 4.6 water molecules will cover a fully hydroxylated 100 Å^2 surface, the monoclinic ZrO$_2$ nanoparticle sample used by Loong et al. [31] in their INS study was calculated to have a submonolayer water coverage of 0.882 layers. It was noted, however, that water adsorption on zirconia surfaces does not occur in discrete steps, and consequently this value should be regarded as an approximation. To allow for comparison, a dry sample prepared by heating hydrated ZrO$_2$ at 250°C under vacuum for 2 h was also analyzed. This sample is expected to retain the chemisorbed hydroxyl groups bound to the ZrO$_2$ surface. The INS data recorded on these two samples at 15 K with incident neutron energy of 600 MeV are shown in Fig. 8.1.

Figure 8.1 displays the scattering function for the water adsorbed on the surface of the ZrO$_2$ particles. The broad librational band at ~80 MeV originates from the intermolecular vibrations of the water molecules. The bands at ~200 and ~430 MeV are due to the intramolecular H-O-H bending and the O-H stretching modes, respectively. An additional broad combination band can be observed at ~506 MeV, and this region is shown expanded in Fig. 8.2. The combination band has been fitted with a sum of several Gaussian functions and a linear background term. The band corresponding to the O-H stretching mode was centered at 459 MeV

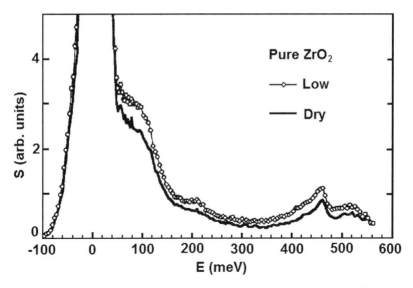

Fig. 8.1 The observed INS scattering function of water adsorbed on ZrO_2 powder [31]

with a width of ~23 MeV. The two broad components at approximately 432 and 510 MeV are assigned to the vibrational modes of water-like species formed by the association of adjacent hydroxyl groups on the ZrO_2 surface of the dry sample (Fig. 8.2a). These peaks increase in intensity and become broader for the hydrated ZrO_2 sample (Fig. 8.2b). It was postulated that the band at 432 MeV is the vibrational energy of the O-H stretch, and that the combination mode at 510 MeV results from the simultaneous breaking of hydrogen bonds between the water species and the excitation of the stretch vibration.

This INS study was the first of its kind, and it highlighted the complexity of the dynamics of the adsorbed water species on the ZrO_2 nanoparticle surface. Although this study provided information about the vibrational dynamics of the adsorbed water species, it did not allow for an evaluation of the mobility of the water on the ZrO_2 surface. This aspect of the hydration layer dynamics was investigated a decade later with QENS [28, 4].

In two separate experiments using neutron spectrometers with different resolution ranges, Mamontov was able to ascertain the nature of the movement of water molecules within the hydration layers on the surface of ZrO_2 particles that were ~15 nm in size. To ensure consistency, the same sample was employed for both QENS experiments. The average surface coverage of water on the ZrO_2 nanoparticles was determined by thermogravimetric analysis (TGA) and Fourier-transform infrared spectrometry to be ~2.3 layers; this level of hydration was later confirmed by the QENS results. The first of these experiments was conducted with a cold neutron time-of-flight Disk Chopper Spectrometer (DCS) at the National Institute of Standards and Technology (NIST) Center for Neutron Research [28]. Data were collected over a temperature range of 300–360 K. The energy resolution of the

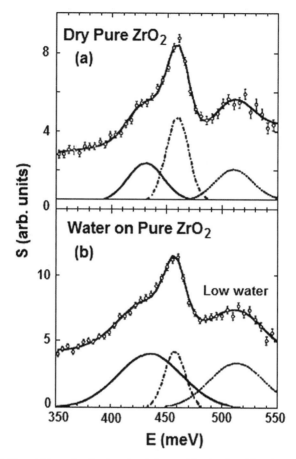

Fig. 8.2 The O-H stretching vibration band and the combination band fitted to a sum of multiple Gaussian functions and a background function for the (**a**) dry and (**b**) hydrated ZrO_2 powders [31]

instrument was $\sim$19 μeV (at half width at half maximum). At this energy resolution, one is only monitoring the behavior of the water molecules within the outer hydration layer. This is because the dynamics of the inner layer water molecules occur on a much slower timescale with respect to the outer layer (hundreds as opposed to tens of picoseconds) and can therefore be considered as immobile over the time scale of the measurement and contribute, in addition to scattering from the ZrO_2 crystallites and the surface hydroxyl groups, to the elastic scattering signal.

Figure 8.3 shows the temperature dependence of the scattering collected at $Q = 1.04\,\text{Å}^{-1}$. There are two components contributing to the QENS signal: a broad feature (Fig. 8.3a) that is ascribed to the rotational diffusion and a second, narrower feature associated with translational diffusion of the water molecules (Fig. 8.3b). As the characteristic times for these two phenomena differ by an order of magnitude, their components are well resolved. Overall, the half width at half maximum of the quasielastic signal [$\Gamma(Q)$] increases with temperature, suggesting faster translational

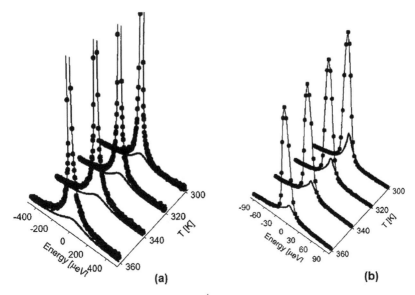

Fig. 8.3 QENS data collected at $Q = 1.04\,\text{Å}^{-1}$ (*filled symbols*) showing the overall fit obtained using the model for two-dimensional diffusion with a fixed jump length (*solid line*), (**a**) the quasielastic contribution due to the rotational diffusion component (*bold line*), and (**b**) the quasielastic contribution due to the translational diffusion component (*bold line*) [28]

diffusion at higher temperatures. This peak broadening is Q-dependent and exhibits a maximum at intermediate Q-values. The presence of this maximum is indicative of a jump-diffusion process with a fixed jump length.

Mamontov reviewed a number of possible diffusion models to explain the QENS data and concluded that a model based on a quasi-two-dimensional diffusion process with a fixed jump length gave the best agreement with the experimental data. Fitting the data with a two-dimensional diffusion model is more complex than with a three-dimensional model, as the theoretical scattering function is no longer Lorentzian. To simplify the fitting, it is legitimate to assume that the residence times for rotational (τ_R) and translational (τ_T) diffusion are not influenced by the dimensionality of the model and can be fixed at the values obtained using a three-dimensional diffusion model with a corresponding fixed jump length. By applying this model to the QENS data, Mamontov determined that the water molecules in the outer hydration layer perform two-dimensional jumps over an approximately temperature-independent distance of 4.21–4.32 Å, which is similar to the distance between water molecules in contact with the ZrO_2 surface (4.66 Å). This implies that the jump length corresponds to the distance between the potential minima of the diffusing free water molecules on the surface of the inner hydration layer.

It can be seen from Fig. 8.4 that the dynamics of the outer hydration layer are in notable contrast to the behavior of bulk water. Both the rotational and the translational components of water diffusion in the outer hydration layer exhibit Arrhenius-type behavior that, when modeled accordingly, lead to values for the activation energies for rotational diffusion of 4.48 kJ/mol (cf. 7.74 kJ/mol for bulk water at

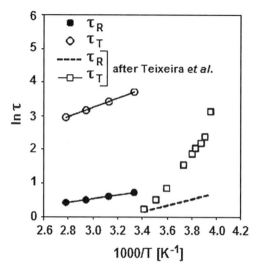

Fig. 8.4 The temperature dependence of the residence time for the translational and rotational diffusion of surface water on ZrO_2 [28]. The data for bulk water were reported by Teixeira et al. [15]

RT) and 11.38 kJ/mol for the translational motion. The rotational diffusion of the physisorbed water is a factor of 2 slower, and the residence time for the translation diffusion is ~40 times longer, than for bulk water. The dynamics of the molecules in the outer hydration layer are clearly dissimilar to those of bulk water; this deviation from bulk behavior can be attributed to the low density of molecular water within the hydration layers due to the hydroxylation of the ZrO_2 surface.

To resolve the dynamics of the inner hydration layer that is in direct contact with the ZrO_2 surface, Mamontov performed a second QENS experiment with a high-flux backscattering spectrometer (HFBS) at NIST with sub-μeV resolution [4]. At this energy resolution and at high-Q, scattering from the "fast"-moving water molecules of the outer hydration layer does not influence the quasielastic signal, as it is outside the dynamic range of the HFBS and contributes to the background. Furthermore, as rotational diffusion is expected to be influenced more by the confinement of the hydration layer than by the translational component, the separation of the time scales over which these two types of motion occur is expected to increase as the dynamics of the water slow down. For this reason, because of the narrow dynamic range of the measurement, it is not necessary to model the rotational diffusion of the water molecules comprising the inner hydration layer.

The data, collected over a temperature range of 240–300 K, are shown in Fig. 8.5. The width of the quasielastic signal is seen to increase with temperature, indicating faster diffusion at higher temperatures. The translational dynamics of the inner hydration layer are best described by a relaxation function incorporating a stretched exponential term related to a spread of relaxation times in the range of 500–1000 ps. In contrast to the dynamics of the outer hydration layer, the temperature dependence of the translational diffusion residence time of the inner water molecules strongly deviates from Arrhenius-type behavior. This implies that the inner water molecules

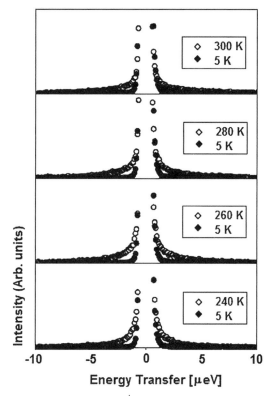

Fig. 8.5 QENS data collected at $Q = 1.11 \text{Å}^{-1}$. A linear background correction has been applied to these data, and the elastic peaks at zero energy transfer are truncated to better demonstrate the quasielastic signal. Data collected at 5 K are shown, as these represent the resolution function of the instrument [4]

are participating in a multifaceted hydrogen bond network, and consequently their translational motion is more complex than the jump-diffusion dynamics of the molecules within the outer hydration layer. Indeed, the behavior exhibited by the water molecules of the inner layer is similar to that of bulk water, and the diffusion dynamics of this layer are just one order of magnitude less than for bulk water. Mamontov proposed that the slow dynamics observed for the inner water layer are due to the presence of the outer water layer and speculated that should the outer layer be stripped from the ZrO_2 surface, then the inner layer would form fewer hydrogen bonds and would thus display the characteristic fast dynamics of the outer hydration layer prior to removal.

8.3.2 INS Studies of Surface Water on Anatase (TiO₂) Nanoparticles

Anatase is the stable polymorph of TiO_2 at the nanoscale [1, 2, 32]. Levchenko et al. investigated the energetics of hydration layers adsorbed on the surfaces of anatase nanoparticles. This important study documented how surface confinement by the nanoparticles of anatase influences water vibrations and heat capacity, as measured by INS and adiabatic calorimetry. This was the first direct comparison of INS measurements with calorimetric data on the same sample to gain insight into the dynamics of water confinement on the surface of nanoparticles.

INS experiments were conducted at 4 K on a 6.3-g sample of 7-nm anatase nanoparticles using HRMECS at the Intense Pulsed Neutron Sources at Argonne National Laboratory. The sample used in the INS experiments had a coverage of $1.5 \cdot 10^{-5}$ moles of H_2O/m^2, equivalent to 9 water molecules per nm^2, similar to the amount of surface water in the sample employed in the heat capacity measurements ($1.7 \cdot 10^{-5}$ moles of H_2O/m^2). Incident energies of 50, 150, and 600 MeV were used to measure vibrational spectra over the 0–550 MeV range, with the resolution between 2% and 3% of the incident neutron beam energy.

The INS spectrum of water adsorbed on the surface of nanophase anatase is compared with the spectrum of ice-Ih in Fig. 8.6. The typical features of the hexagonal ice-Ih spectrum are observed for the surface-confined water. The broad peak in the OH-stretching region is centered at 400 MeV (FWHM = 75 MeV), which

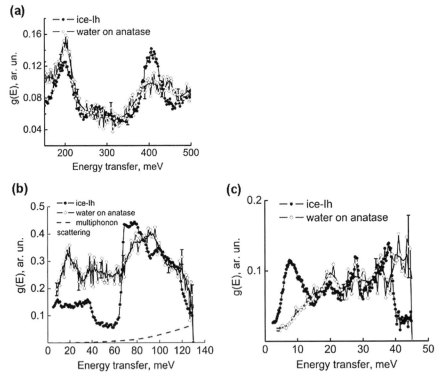

Fig. 8.6 INS spectra of 7-nm anatase nanoparticles compared with that of ice-Ih [6], obtained at $E_i = 600$ MeV (**a**), $E_i = 140$ MeV (**b**), and $E_i = 50$ MeV (**c**). The *solid thick lines* are smoothed spectra. The *dashed line* shows contributions from multiphonon scattering

is consistent with ice vibrations (Fig. 8.6a). There is a narrow band centered at 449 MeV (with FWHM = 18 MeV) superimposed on the broad peak at 400 MeV, which extends down to 350 MeV. The feature at ∼450 MeV may be related to OH-stretching modes of either an isolated hydroxyl group or water molecules without hydrogen bonds, since higher frequencies (energies) indicate shorter and stronger covalent O-H bonds. From the fit to the OH-stretching vibration band using two Gaussians, the intensity ratio for peaks at 400 and 449 MeV is approximately 1:6 and the ratio of the number of water molecules (H_2O) to the number of hydroxyls on the anatase surface is estimated to be 3:1.

The low-energy shoulder at 350–400 MeV indicates softening of the OH-stretching mode. This softening implies the existence of some longer (weaker) O-H bonds and shorter O··· O distances (stronger O···H-O hydrogen bonds) between adjacent surface OH groups. Another possible explanation is that some physically adsorbed water molecules form two hydrogen bonds with surface oxygen atoms O··· H-O-H··· O, resulting in weaker intramolecular water O-H bonds. The bending mode at ∼200 MeV, associated with molecular water, appears to be unchanged if compared with the ice-Ih spectrum (Fig. 8.6a).

The region of the spectrum assigned to librational modes for ice-Ih (50–130 MeV) and the surface water spectrum of 7-nm anatase are compared in Fig. 8.6b. The INS spectrum of ice-Ih in this range is sensitive to changes in the tetrahedral arrangement of water molecules and is typically interpreted in terms of distortions of these arrangements. The peak observed at 70 MeV for ice-Ih is not seen in the anatase spectrum, possibly because of redistribution and broadening of the peak. In addition, the spectrum of ice-Ih has no vibrational modes between 40 and 65 MeV, whereas the surface water spectrum of nanophase anatase has a rather intense band between 35 and 65 MeV. This difference might be related to the interaction of water with TiO_2, which is enhanced by intense TiO_2 optical modes occurring near these energies. The strong interactions between the water molecules or hydroxyl groups and the anatase surface contribute to the eigenvectors of H_2O or OH groups, which can be significant in the VDOS spectra of the nano-anatase.

The translational (acoustic) modes in the energy range between 15 and 40 MeV follow the ice-Ih spectrum profile (Fig. 8.6c). The absence of an ice-Ih acoustic peak at 7 MeV in the anatase spectrum is striking. The corresponding shifts of acoustic modes of surface water on anatase suggest restricted motion of water molecules near the TiO_2 surface. Strong interactions of this inner layer water via hydrogen bonds with the TiO_2 surface, which lead to acoustic vibrations at higher energies, result in acoustic modes of surface water appearing at higher frequencies than those of ice-Ih. This finding shows that translational motion of water molecules is greatly suppressed, meaning that water is strongly bound to specific sites on the TiO_2 surface.

To calculate heat capacity from VDOS, Levchenko et al. [32] normalized the INS spectra to 6 (3 degrees of freedom for each group of translational and librational modes) and approximated the heat capacity at low energies by the Debye law (Fig. 8.7). Heat capacities for bulk and confined water, calculated from the VDOS, are compared in Fig. 8.8 with the low-temperature adiabatic calorimetric data. The

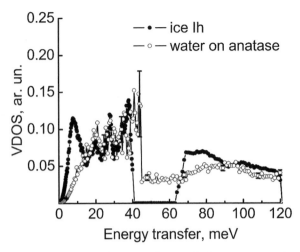

Fig. 8.7 Vibrational density of states of water confined on the 7-nm anatase surface compared with that of ice-Ih

relationships between calculated heat capacities of confined water and bulk water bear a remarkable resemblance to the trends observed for the calorimetric heat capacities (Fig. 8.8). The heat capacity of water on the anatase surface, calculated from the VDOS, is lower than that of the bulk water below 200 K, resulting in lower entropy at 298 K for confined water on the anatase surface compared with that of the bulk.

As described earlier, the librational modes of surface water on nano-anatase are sensitive to the short-range hydrogen bonding interactions that restrict rotation of the water molecules. The translational modes of ice are not observed above 45 MeV [39]; therefore, the broad feature seen between 50 and 70 MeV can be attributed

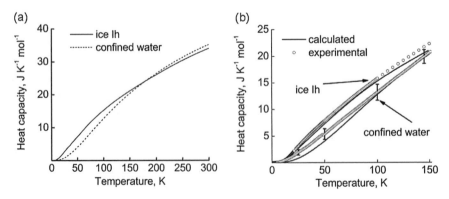

Fig. 8.8 (a) C_v for bulk ice and TiO_2 surface-confined water calculated from VDOS; (b) experimental C_p data for bulk ice and anatase-confined water, compared with calculated values from VDOS

to librations of surface-confined water. The center of gravity calculated for all librational modes of surface water (50–120 MeV) shifts to lower energies, whereas the center of modes is not changed if calculated between 70 and 120 MeV (the librational range of ice-Ih). This implies that the overall shift of librational modes to lower energies results from the presence of the vibrations that are not observed in ice-Ih. This shift reflects softening of intermolecular hydrogen bonding [39] and may be associated with distortions in water tetrahedra [40], analogous to transformations observed in amorphous ice under pressure [41, 42].

Above 200 K, the experimental confined water specific heat rises above that of bulk ice-Ih (Fig. 8.8). The calculations based on INS data show a similar crossover. Levchenko et al. [32] attributed this crossover to the presence of vibrations between 40 and 60 MeV in the hydration layers of nano-anatase that are absent in ice-Ih. It should be noted that the INS data in Levchenko et al. [32] were collected at 4 K, and there is no direct evidence for possible changes in VDOS at higher temperatures. Faraone et al. [12], Mamontov [4, 5], and Mamontov et al. [14] suggested a dynamic (fragile-to-strong liquid) transition in confined water above its glass transition (180–200 K) in a number of systems (CeO_2, rutile TiO_2, SnO_2, MCM-41) between 210 and 220 K. The observed crossover in heat capacity of water on the anatase surface and ice occurs in a similar temperature range. However, this crossover appears in VDOS models derived from the 4 K data set for the lightly hydrated anatase sample measured by INS (see Fig. 8.8a). More highly hydrated anatase samples, with hydration levels comparable to that of rutile TiO_2 [14], show a similar crossover in the experimental heat capacity relative to ice [3], but with a much larger excess molar heat capacity than that calculated from the 4 K VDOS for the lightly hydrated sample. Therefore, it is not clear at this point whether an analogous dynamic transition plays a role in the energetics of water on anatase.

The study performed by Levchenko et al. [32] has shown that INS experiments, even on a limited amount of TiO_2 sample, reveal the VDOS of adsorbed surface water. The observed vibrational modes of water molecules and surface hydroxyl groups from the INS studies confirm previous studies [1, 2], which indicate a mixed character of water adsorption (dissociative and molecular) on nanophase TiO_2. The observed shifts and redistribution of low-energy modes (translational vibrations) to higher energy and softening of librational modes indicate a very restricted dynamics of surface water and a distorted network of water tetrahedra. This combined INS and calorimetry study supports previous findings [2] that heat capacity and entropy of water molecules in confined geometry can be lower than those in the bulk as a result of stronger interactions at the interface of the nanoparticle.

8.4 QENS Study of Surface Water on Rutile (α–TiO_2) and Cassiterite (α-SnO_2) Nanoparticles

In the previous section, the influence of the hydration layers on the thermodynamics of anatase nanoparticles was discussed. This original research highlighted the fundamental importance of the hydration layers in stabilizing the particles at the nanoscale. In this section, we demonstrate how QENS data and MD simulations can be combined to provide a comprehensive model of the dynamics of water on nanoparticle surfaces, in particular on the surfaces of isostructural rutile and cassiterite nanoparticles.

Mamontov et al. [14, 33] reported a QENS and MD study of adsorbed water on rutile (α-TiO$_2$) and cassiterite (α-SnO$_2$) nanopowder surfaces. The rutile and cassiterite powders had particle sizes of 5–10 nm, and the BET surface areas were 181 and 156 m^2/g, respectively. The powders were exposed, at room temperature, to air with 70–80% humidity before being sealed in the sample containers. TGA, mass spectrometry, and water sorption studies indicated that the particle surfaces had a low level of CO$_2$ and carbon contamination and that the water coverage was approximately three structural layers.

Figure 8.9 shows the QENS energy transfer spectra, fitted with Lorentzian functions, for the hydrated rutile and cassiterite nanopowders. These spectra were obtained on the DCS and the HFBS at NIST. For this experiment the DCS had an energy resolution of 22 μeV over the energy transfer range of ±500 μeV ($\lambda = 9.0$ Å incident neutrons). The HFBS had an incident wavelength of 6.271 Å, giving an energy resolution of 0.85 μeV and dynamic range of ±11 μeV. In practical terms, this enabled the DCS to probe accurately diffusional motions that occur with characteristic time scales within the tens to hundreds of picoseconds range and enabled the HFBS to investigate motion occurring on the hundreds to thousands of picoseconds

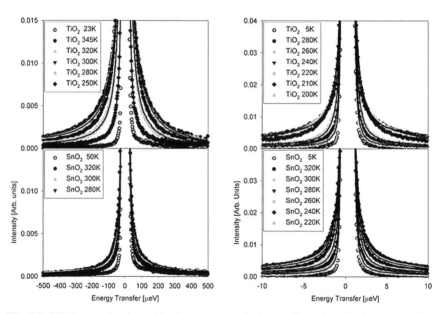

Fig. 8.9 QENS scattering intensities from water sorbed on rutile and cassiterite nanopowders collected at the NIST Center for Neutron Research using: (*left*) the Disk Chopper Spectrometer (integrated over the range of the scattering vector transfers $0.40\,\text{Å}^{-1} < Q < 1.20\text{Å}^{-1}$); and (*right*) the High-Flux Backscatter Spectrometer (integrated over $0.62\,\text{Å}^{-1} < Q < 1.60\text{Å}^{-1}$) [14]

range. Water molecules that are immobile or diffusing too slowly do not produce QENS broadening and act as purely elastic scatterers, contributing to the resolution function (represented by the lowest-temperature data for each instrument shown in Fig. 8.9). Two observations are immediately apparent: (1) QENS broadening of the DCS signal is greater for rutile and can be detected at a lower temperature than for cassiterite; and (2) in the HFBS signal for rutile, there is a noticeable discontinuity in the temperature dependence of the QENS broadening. MD simulations are invaluable for explaining the features present in QENS spectra, and it is this aspect of the data analysis to which we now turn.

The rutile and cassiterite nanoparticles that Mamontov et al. [14] used in their QENS studies predominantly express the (110) crystal face, and this surface (Fig. 8.10) has been the subject of numerous experimental and computational studies [43–46]. The atomic charges and relaxed bond lengths on the (110) surfaces of these minerals, which are in contact with a significant number of water molecules, have been determined from static ab initio discrete Fourier transform calculations that also provided potentials for the interactions between the surface atoms and the water molecules [47, 48]. Predota et al. [49, 50] and Vlcek et al. [51] carried out large-scale simulations of 40-Å-thick layers of SPC/E water [52] at liquid density $(1.0\,g/cm^3)$, sandwiched between five Ti-layer slabs [parallel to (110)] of rutile and cassiterite, respectively. As SPC/E is a non-dissociating water model, the initial configuration of the surfaces had to be constructed as shown in Fig. 8.10. The results of the simulations, extending to several nanoseconds, were compared with synchrotron X-ray crystal truncation rod studies of real liquid water in contact with annealed rutile (110) macro-crystal surfaces [53, 54] and cassiterite [51], and good agreement between the MD and the X-ray results was demonstrated at sub-angstrom resolution.

The ab initio calculations demonstrated that the "associated" or "non-hydroxylated" (110) surface is favored for rutile, on which a "terminal" water molecule chemisorbs to each under-coordinated (fivefold) metal atom that is exposed

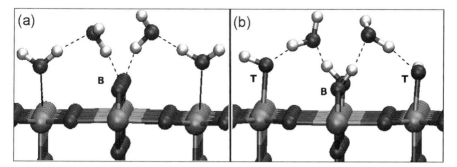

Fig. 8.10 "Non-hydroxylated" (**a**) and "hydroxylated" (**b**) surface water configurations for classical MD simulations [49, 51] of the rutile and cassiterite (110) surfaces (*dark gray*—oxygen atoms, *white*—hydrogen atoms, *light gray*—titanium or tin atoms). T and B denote "terminal" (L_1) and "bridging" oxygen atoms, as discussed in the text, which form strong hydrogen bonds (*dashed lines*) with L_2 water molecules

at the oxide surface. Cassiterite favors the "dissociated" or "hydroxylated" (110) surface, on which a terminal water molecule dissociates to form a terminal hydroxyl group bonded to the fivefold metal atom and the released proton combines with one of the "bridging" oxygen atoms protruding above the surface (bonded to two underlying sixfold metal atoms) to form another hydroxyl group. The actual (110) surfaces that are in contact with real water are expected to be a mixture of these configurations, as demonstrated by ab initio MD simulations conducted by Kumar et al. [55]. However, the model surfaces employed in this study can be regarded as a reasonable approximation to reality.

The solid curves in Fig. 8.11 show the density profiles for the oxygen atoms associated with water molecules located within a distance of 10 Å perpendicular to the (110) surfaces of both minerals, as derived from the MD simulations at 300 K. For both materials, three distinct density maxima appear within these plots, which correspond to the associated or dissociated terminal oxygen atoms within the first layer (L_1) that is closest to the surface, to a clearly defined second layer of physisorbed water molecules (L_2), and to a much less structured third layer (L_3). The curves then show a rapid approach to bulk water density beyond 7–8 Å. X-ray and MD results confirm that these layers are also laterally ordered relative to the underlying crystal structure of the mineral, as indicated in Fig. 8.10. The open symbols in Fig. 8.11 represent the results from the MD simulations when all excess

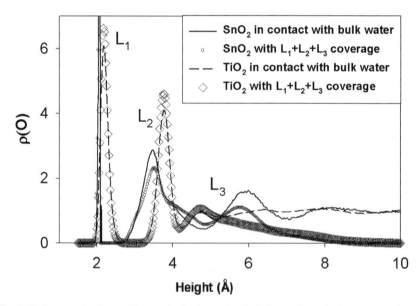

Fig. 8.11 Oxygen density profiles perpendicular to the (110) faces of non-hydroxylated rutile and hydroxylated cassiterite in contact with SPC/E water from ab initio optimized classical MD simulations at 300 K. Zero on the horizontal axis corresponds to the plane bisecting the surface titanium atoms (Fig. 8.10). The solid and dashed curves represent the density profiles for surfaces in contact with bulk SPC/E water at 1 g/cm³, and the open symbols are the profiles obtained after removing from the simulation cell all water except an amount equivalent to ~ $L_1+L_2+L_3$ coverage (see text)

water has been removed, leaving sufficient water within the 40-Å gap to account for the experimentally observed water contents of the nanopowders employed by Mamontov et al. in their QENS experiment. This corresponds to nearly complete L_1, L_2, L_3 water layers on rutile, but somewhat less than full three-layer coverage for cassiterite. Two profound observations can be made: (1) the simulations predict that water will lie in dense layers on the mineral surfaces, concurring with experimental observations of these systems; and (2) the structure of these layers is virtually identical to the water structure at the *bulk water*/oxide interface. Thus it can be assumed, with considerable confidence, that the dynamics of the bulk water/oxide interface can be probed by QENS and rationalized by MD analysis of thin layers on nanoparticle surfaces.

Figure 8.12 shows the MD-simulated trajectories over a 10-ns time period for oxygen atoms of individual water molecules that originated in layers L_1 and L_2. This figure also shows the ISF extracted from these simulations, including the total ISF and the individual contributions from layers L_1, L_2, and L_3 [33]. It is clear from Fig. 8.12 that L_1 water molecules do not undergo translational motions on the time scales accessible with the DCS and HFBS spectrometers. It is also apparent from the MD results that two distinct types of diffusional dynamics are responsible for the QENS broadening: (1) a fast component associated with localized rotation-like motions of water molecules within H-bonded "cages" in all layers, as well as a coupled rotation-translation motion of the L_3 water molecules; and (2) A slow component resulting from periodic translational jumps of L_2 water molecules out of their highly ordered sorption sites into layer L_3. The reason for the rather fast rotational-translational dynamics in L_3, which extends the range of the fast component from one to tens of picoseconds, appears to be that the L_3 water molecules lack the full hydrogen bond network that is present in bulk water. The slow translational diffusion component of the L_2 water molecules is also distributed over a wide and

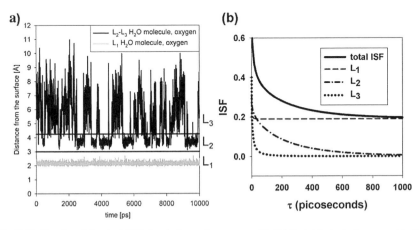

Fig. 8.12 Results of classical MD simulations of water on rutile (110): (**a**) trajectories normal to the titanium surface plane of oxygen in water molecules originally in L_1 and L_2; (**b**) total intermediate scattering function (ISF) and contributions to the ISF from individual surface layers vs time [33]

exponentially decaying range of characteristic residence times, extending from tens to many hundreds of picoseconds. In fact, it is this slow dynamic motion of the L_2 molecules that is detected as the slow DCS component and as the even slower HFBS component. The residence times (τ) for these two types of molecular motion, extracted from the DCS data, are shown as open symbols in Fig. 8.13. Fourier transformation of the MD-simulated ISF at 300 K over the $\pm 500\ \mu$eV energy range, corresponding to the dynamic range of the DCS spectrometer, gives remarkably similar residence times to those recorded experimentally, as shown by the solid points in Fig. 8.13 [14].

The simulations also help to rationalize the features that are apparent from visual inspection of the QENS spectra. The more intense QENS broadening obtained for rutile with respect to cassiterite in the DCS experiment can be explained as resulting from a nearly complete L_3 layer on the surfaces of the rutile nanoparticles. This compares with a significantly less than complete L_3 layer on the cassiterite surfaces, which results in a decrease in the contribution to the fast DCS component from L_3 for the cassiterite sample. The "gap" in the HFBS signal for rutile (Fig. 8.9) is a result of the strongly non-Arrhenius nature of the L_2 translational diffusion at the lower temperatures accessible with the HFBS, and is more pronounced for the rutile sample with respect to cassiterite because of subtle differences in the local hydrogen bonding environment of the L_2 water on this surface [14, 33]. The studies reported herein serve to emphasize that for diffusional dynamics distributed over a wide range of characteristic time scales, individual neutron spectrometers can access only portions of the dynamic ranges for each component; yet the complete range is fully

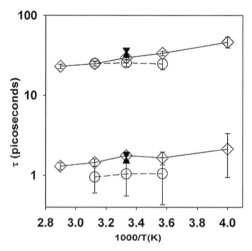

Fig. 8.13 Residence times (τ) for DCS fast (rotation-translation) and slow (translation) diffusional components for the rutile (*diamonds*) and cassiterite (*circles*) extracted from the QENS scattering (Fig. 8.9), compared with τ values extracted from the MD intermediate scattering function (Fig. 8.12), for rutile (*solid inverted triangle*) and cassiterite (*solid triangle*), Fourier-transformed over the $\pm 500\ \mu$eV energy range of the DCS spectrometer, using the same Lorentzian functions

accessible to simulation with MD. Thus, neutron scattering and MD simulations are highly complementary techniques for the interpretation of the dynamics of water on mineral surfaces.

8.5 Summary and Outlook

The principle objective of this chapter was to provide an introduction to the techniques available for investigating the energetics and dynamics of nanoparticle systems and to demonstrate, by highlighting recent advances in the field, the diverse areas of nanoparticle physics that can be accessed with these methods.

To this end, we have discussed the original research directed toward the elucidation of the structure and dynamics of adsorbed water on the surface of a number of nanoparticle systems comprising mineral oxides. The first such experiments to be presented, conducted by Loong and coworkers, were focused toward determining the energetics of water on the surface of ZrO_2 nanoparticles using INS. This research proved influential in that it verified the complex behavior of the hydration layers that are an integral part of the nanoparticle structure, paving the way to more in-depth studies of such systems.

Further insight into the fundamental physics of the nanoparticle hydration layers was later provided by Levchenko et al. [2, 32] and Boerio-Goates et al. [3], who investigated the thermodynamics of adsorbed water on anatase (TiO_2) nanoparticles by adiabatic calorimetry and INS. These studies, the first of their kind, represented the conjoining of two traditionally independent forms of analysis to proffer a comprehensive understanding of the synergetic relationship between the structure and the energetics of the nanoparticle and surface hydration layers.

Also discussed was the research of Mamontov et al. [14, 33], who have investigated extensively the dynamics of hydration layers present on the surface of oxide nanoparticles using QENS. By complementing these experimental QENS studies with MD simulations, they revealed in intricate and quantitative detail the movement of the water molecules within the various hydration layers present on the surfaces of ZrO_2, rutile (TiO_2), and cassiterite (SnO_2) nanoparticles. The unusual properties observed for confined water may be a result of the effect of sorption on surfaces that is specific to each solid phase studied, and may not be observed when the pore spaces become so large that bulk water properties dominate experimental signals. Further INS and QENS studies are needed to verify this.

Having reviewed these recent developments in nanoparticle physics, one can envisage future research as the next-generation spallation sources such as the Spallation Neutron Source (Oak Ridge, Tennessee, USA) and J-PARC (Tokai, Japan) come on-line. With the higher flux available from such sources, smaller amounts of sample and much less time will be required for neutron experiments. Attention can also be focused on INS spectra of non-hydrogenous samples and different types of adsorbates on nanoparticles. Hand in hand with these developments are advances in computer power that allow complementary calculations of vibrational spectra and dynamical simulations of adsorbed species on nanoparticles from ab initio models. Future research may also benefit from incorporating other forms of neutron spectroscopic analysis. One such technique with potential for application for studying adsorbates on nanoscale surfaces is neutron spin-echo spectroscopy (NSE, see Chapter 3). NSE promises to provide exceptionally high-resolution data enabling slow diffusion dynamics occurring on the time scale of one to hundreds of nanoseconds to be probed.

In conclusion, it can be stated with confidence that neutron scattering techniques, MD simulations, and calorimetric measurements can provide a wealth of information that may be combined to generate a complete description of the energetics, structure, and dynamics of the multiple hydration layers which are present on the surfaces of mineral nanoparticles. The research presented in this chapter demonstrates the complex behavior of these hydration layers, which are so essential to the stability of the nanoparticle that they must be regarded as an intrinsic part of the particle itself.

Acknowledgments N. L. Ross, E. C. Spencer, and A. Levchenko acknowledge support from the U.S. Department of Energy, Office of Basic Energy Sciences (DOE–BES), grant DE FG03 01ER15237. Argonne National Laboratory is supported by DOE–BES under contract #W-31-109-ENG-38 DE-AC02-06CH11357. The efforts of D. R. Cole, E. Mamontov, L. Vlcek, and D. J. Wesolowski were supported by DOE–BES (project ERKCC41; 51) and the Spallation Neutron Source at Oak Ridge National Laboratory, managed by UT-Battelle, LLC, under contract #DE-AC05-00OR22725 for DOE.

References

1. G. S. Li, L. P. Li, J. Boerio-Goates, and B. F. Woodfield, J. Am. Chem. Soc. **127**, 8659 (2005)
2. A. A. Levchenko, G. Li, J. Boerio-Goates, B. F. Woodfield, and A. Navrotsky, Chem. Mater. **18**, 6324 (2006)
3. J. Boerio-Goates, G. S. Li, L. P. Li, T. F. Walker, T. Parry, and B. F. Woodfield, Nano Lett. **6**, 750 (2006)
4. E. Mamontov, J. Chem. Phys. **123**, 024706 (2005)
5. E. Mamontov, J. Chem. Phys. **123**, 171101 (2005)
6. A. I. Kolesnikov, J.-M. Zanotti, C.-K. Loong, P. Thiyagarajan, A. P. Moravsky, R. O. Loutfy, and C. J. Burnham, Phys. Rev. Lett. **93**, 035503 (2004)
7. O. Trofymluk, A. A. Levchenko, and A. Navrotsky, J. Chem. Phys. **123**, 194509 (2005)
8. A. I. Kolesnikov, J. C. Li, and S. F. Parker, J. Mol. Liq. **96–7**, 317 (2002)
9. M. C. Bellissent-Funel, Eur. Phys. J. E **12**, 83 (2003)
10. C. Corsaro, V. Crupi, D. Majolino, S. F. Parker, V. Venuti, and U. Wanderlingh, J. Phys. Chem. A **110**, 1190 (2006)
11. J. M. Zanotti, M. C. Bellissent-Funel, S. H. Chen, and A. I. Kolesnikov, J. Phys.: Condens. Matter **18**, S2299 (2006)
12. A. Faraone, L. Liu, C. Y. Mou, C. W. Yen, and S. H. Chen, J. Chem. Phys. **121**, 10843 (2004)
13. C.-K. Loong, L. E. Iton, and M. Ozawa, Physica B **213**, 640 (1995)
14. E. Mamontov, L. Vlcek, D. J. Wesolowski, P. T. Cummings, W. Wang, L. M. Anovitz, J. Rosenqvist, C. M. Brown, and V. G. Sakai, J. Phys. Chem. C **111**, 4328 (2007)
15. J. Texeira, M.-C. Bellisent-Funel, S. H. Chen, and A. J. Dianoux, Phys. Rev. A **31**, 1913 (1985)

16. M.-C. Bellissent-Funel, K. F. Bradley, S. H. Chen, J. Lal, and J. Teixeira, Physica A **201**, 277 (1993)
17. M.-C. Bellissent-Funel, S. H. Chen, and J.-M. Zanotti, Phys. Rev. E **51**, 4558 (1995)
18. J. M. Zanotti, M.-C. Bellissent-Funel, and S. H. Chen, Phys. Rev. E **59**, 3084 (1999)
19. T. Takamuku, M. Yamagami, H. Wakita, Y. Masuda, and T. Yamaguchi, J. Phys. Chem. B **101**, 5730 (1997)
20. S. Takahara, M. Nakano, S. Kittaka, Y. Kuroda, T. Mori, H. Hamano, and T. Yamaguchi, J. Phys. Chem. B **103**, 5814 (1999)
21. S. Takahara, S. Kittaka, T. Mori, Y. Kuroda, T. Yamaguchi, and K. Shibata, J. Phys. Chem. B **106**, 5689 (2002)
22. F. Mansour, R. M. Dimeo, and H. Peemoeller, Phys. Rev. E **66**, 041307–1 (2002)
23. A. Faraone, L. Liu, C.-Y. Mou, P.-C. Shih, C. Brown, J. R. D. Copley, R. M. Dimeo, and S. H. Chen, Eur. Phys. J. E **12**, S59 (2003)
24. A. Faraone, L. Liu, C.-Y. Mou, P.-C. Shih, J. R. D. Copley, and S. H. Chen, J. Chem. Phys. **119**, 3963 (2003)
25. L. Liu, A. Faraone, C.-Y. Mou, P.-C. Shih, and S. H. Chen, J. Phys.: Condens. Matter. **16**, S5403 (2004)
26. V. Crupi, D. Majolino, P. Migliardo, and V. Venuti, Physica A **304**, 59 (2002)
27. V. Crupi, D. Majolino, P. Migliardo, and V. Venuti, J. Phys. Chem. B **106**, 10884 (2002)
28. E. Mamontov, J. Chem. Phys. **121**, 9087 (2004)
29. Y. Kuroda, S. Kittaka, S. Takahara, T. Yamaguchi, and M.-C. Bellisent-Funel, J. Phys. Chem. B **103**, 11064 (1999)
30. M. Ozawa, S. Suzuki, C.-K. Loong, and J. C. Nipko, Appl. Surf. Sci. **121/122**, 133 (1997)
31. C.-K. Loong, J. W. Richardson Jr., and M. Ozawa, J. Catalysis **157**, 636 (1995)
32. A. A. Levchenko, A. I. Kolesnikov, N. L. Ross, J. Boerio-Goates, B. F. Woodfield, G. Li, and A. Navrotsky, J. Phys. Chem. **A 111**, 12584 (2007)
33. E. Mamontov, D. J. Wesolowski, L. Vlcek, P. T. Cummings, J. Rosenqvist, W. Wang, and D. R. Cole, J. Phys. Chem. C **112**, 12334 (2008)
34. S. W. Lovesey, *Theory of Neutron Scattering from Condensed Matter* (Clarendon Press, Oxford, 1984)
35. P. C. H. Mitchell, S. F. Parker, A. J. Ramirez-Cuesta, and J. Tomkinson, *Vibrational Spectroscopy with Neutrons with Applications in Chemistry, Biology, Materials Science and Catalysis* (World Scientific Publishing Co. Pte. Ltd., Singapore, 2005)
36. S. L. Chaplot, N. Choudhury, S. Ghose, M. N. Rao, R. Mittal, and P. Goel, Eur. J. Mineral. **14**, 291 (2002)
37. N. Choudhury and S. L. Chaplot *Inelastic Neutron Scattering and Lattice Dynamics: Perspectives and Challenges in Mineral Physics in Neutron Applications in Earth, Energy, and Environmental Sciences*, edited by L. Liang, R. Rinaldi, and H. Schober (Springer–Verlag, 2009) Chapter 5
38. C.-K. Loong, Rev. Mineral. Geochem. **63**, 233 (2006)
39. A. I. Kolesnikov, V. V. Sinitsyn, E. G. Ponyatovsky, I. Natkaniec, L. S. Smirnov, and J. C. Li, J. Phys. Chem. B **101**, 6082 (1997)
40. B. Guillot and Y. Guissani, J. Chem. Phys. **119**, 11740 (2003)
41. G. P. Johari and O. Andersson, Phys. Rev. B **73**, 094292 (2006)
42. O. S. Subbotin, V. R. Belosludov, T. M. Inerbaev, R. V. Belosludov, and Y. Kawazoe, Comput. Mater. Sci. **36**, 253 (2006)
43. V. A. Gercher and D. F. Cox, Surf. Sci. **322**, 177 (1995)
44. J. Goniakowski and M. J. Gillan, Surf. Sci. **350**, 145 (1996)
45. P. J. D. Lindan, Chem. Phys. Lett. **328**, 325 (2000)
46. M. Batzill and U. Diebold, Prog. Surf. Sci. **79**, 47 (2005)
47. A. V. Bandura, J. D. Kubicki, J. Phys. Chem. B **107**, 11072 (2003)
48. A. V. Bandura, J. Sofo, and J. D. Kubicki, J. Phys. Chem. B **110**, 8386 (2006)
49. M. Predota, A. V. Bandura, P. T. Cummings, J. D. Kubicki, D. J. Wesolowski, A. A. Chialvo, and M. L. Machesky, J. Phys. Chem. B **108**, 12049 (2004)

50. M. Predota, P. T. Cummings, and D. J. Wesolowski, J. Phys. Chem. C **111**, 3071 (2007)
51. L. Vlcek, Z. Zhang, M. L. Machesky, P. Fenter, J. Rosenqvist, D. J. Wesolowski, L. M. Anovitz, M. Predota, and P. T. Cummings, Langmuir **23**, 4925 (2007)
52. H. J. C. Berendsen, J. R. Grigera, and T. P. Straatsma, J. Phys. Chem. **91**, 6269 (1987)
53. Z. Zhang, P. Fenter, L. Cheng, N. C. Sturchio, M. J. Bedzyk, M. Predota, A. Bandura, J. D. Kubicki, S. N. Lvov, P. T. Cummings, A. A. Chialvo, M. K. Ridley, P. Benezeth, L. Anovitz, D. A. Palmer, M. L. Machesky, and D. J. Wesolowski, Langmuir **20**, 4954 (2004)
54. Z. Zhang, P. Fenter, N. C. Sturchio, M. J. Bedzyk, M. L. Machesky, and D. J. Wesolowski, Surf. Sci. **601**, 1129 (2007)
55. N. Kumar, S. Neogi, P. Kent, J. D. Kubicki, D. J. Wesolowski, and J. Sofo (Personal Communication with Kumar)

Chapter 9
Neutron Diffraction and the Mechanical Behavior of Geological Materials

Stephen J. Covey-Crump and Paul F. Schofield

Abstract Establishing methods to monitor how deformation is accommodated at the grain scale within samples during mechanical tests has a key part to play in advancing our understanding of the mechanical properties of polycrystalline materials. This information is essential, both for testing the assumptions and approximations used in theoretical analyses designed to predict these properties from the properties of their constituent grains, as well as for using such analyses to interrogate the results of deformation experiments. Conventional deformation experiments (particularly those on nonporous materials) generally provide only whole sample properties, and information about how the deformation has been accommodated within the sample is usually restricted to that which can be recovered from an analysis of the experimental run products. However, the penetrating power of neutrons means that if the deformation experiment is performed on a neutron beam line, the deformation behavior within the sample can be monitored as it is being deformed from diffraction patterns collated at different stages of the deformation. Over the past 20 years, numerous studies on engineering materials have exploited this strategy. In the following contribution, parallel work on geological materials is reviewed with the aim of illustrating the potential of the approach for examining matters of mechanical property characterization that are of particular interest in the earth sciences.

9.1 Introduction

In the past decade, earth scientists have begun to use neutron diffraction techniques to seek insights into the mechanical behavior of geological materials that are not otherwise readily available using conventional rock or ice deformation experiments. For the most part, this research has used neutron facilities developed to examine the mechanical behavior of engineering materials, and has paralleled research in that field. However, minerals and rocks form a rather distinctive class of materials, both

S.J. Covey-Crump (✉)
School of Earth, Atmospheric and Environmental Sciences, University of Manchester, Manchester, M13 9PL, UK
e-mail: S.Covey-Crump@manchester.ac.uk

L. Liang et al. (eds.), *Neutron Applications in Earth, Energy and Environmental Sciences*, Neutron Scattering Applications and Techniques,
DOI 10.1007/978-0-387-09416-8_9, © Springer Science+Business Media, LLC 2009

in their mechanical properties and in the way that knowledge of those properties is used, and this has given the research different, if related, emphases.

The aim of this chapter is to illustrate the potential of neutron diffraction in experimental rock deformation research by using the examples drawn from ongoing work. The interests of the experimental rock deformation community are wide-ranging, so to keep this account manageable, it is limited to experimental defor-mation work on samples of geological materials that are on the order of a few cubic centimeters in size. This excludes related work on engineering materials [1] and the use of neutron techniques such as texture goniometry [2, 3], small-angle neutron scattering [4], and imaging [5, 6] for the recovery of mechanically relevant microstructural information. It also excludes new developments accomplished pri-marily (but not exclusively) with synchrotron X-rays in which the mechanical prop-erties of geological materials are being investigated at ultra-high confining pressures [7, 8]—the small sizes and shapes of the samples used in that work present unique methodological and data analysis challenges, which merit discussion in their own right [9–11].

9.1.1 Context Setting: The Background to Experimental Rock and Ice Deformation Research

Knowledge of the mechanical properties o f geological materials is fundamental to our understanding of the internal structure and geodynamic evolution of the Earth. Moreover, it is of practical importance in, for example, predicting the timing and effects of natural hazards such as earthquakes and volcanic eruptions, modeling the response of ice sheets to climate change, using geophysical methods to locate and exploit natural mineral resources efficiently, and in selecting construction materials and site locations for civil engineering projects. Consequently, there is a strong moti-vation to establish these mechanical properties by performing deformation experi-ments under the environmental conditions of interest.

Geological materials, like engineering materials, exhibit three basic types of deformation response: elastic, inelastic (anelastic + plastic), and brittle. The exper-imental task of characterizing these responses is essentially the same as that faced by the materials scientist who wants to use deformation experiments performed at a range of temperatures, stresses, and strain rates to characterize the mechanical prop-erties of engineering materials. However, for most geologically interesting materials the experiments must also be performed at simultaneously elevated confining pres-sures, either because these are the environmental conditions of interest or, more commonly, to suppress brittle deformation so that elastic and inelastic behavior can be examined. Such experiments are technically demanding and it has only been in the past 30 years that it has been possible to collect data of comparable quality to that obtained from room-pressure experiments on engineering materials (for summaries of the technical issues and developments see [12–15]).

The staple rock and ice deformation experiment is the axial compression test performed on a right circular cylindrical sample, typically about 10 mm in diameter and 25-mm long, at temperatures of $-100°C$ to $1200°C$ and confining pressures of 0–500 MPa. Deformation in extension, direct shear, or torsion is also possible under these conditions [16–18], as is access to higher pressures and temperatures using smaller samples or the use of bigger samples at less extreme conditions. The aim of the experiments is usually some combination of the following:

(1) to establish constitutive equations describing the elastic, inelastic, or brittle mechanical properties of the material as a function of the deformation variables, which can then be used in deformation modeling applications;
(2) to monitor the change in petrophysical properties (e.g., acoustic, electrical, fluid transport) during deformation so as to improve our capacity to interpret geophysical measurements made, for example, during mineral exploration or natural hazard monitoring surveys; and
(3) to examine the influence of deformation variables on microstructural development so as to improve our capacity to interpret the conditions and processes responsible for the microstructures found in naturally deformed materials.

The geological materials of interest are generally polycrystalline aggregates, sometimes monomineralic but more often polymineralic, which exhibit a wide range of mechanical behavior and properties. The number of volumetrically significant mineral phases is small, so much of this mechanical variability is microstructural in origin. This reflects the strong mechanical anisotropy of the important phases, which lends particular significance to the presence of any lattice-preferred orientation in the mineral phases and to the shape and spatial arrangement of the mineral phase/textural domains within the sample.

The present understanding of the influence of microstructural variables on the mechanical properties of geological materials remains rather limited. This is because assessment of the significance of such variables using the results of deformation experiments requires information about how the deformation is accommodated within the sample during the experiment. An individual grain within a polycrystal cannot deform as it would if it were a single crystal because it is constrained by how its differently oriented neighboring grains deform. Likewise, in a polymineralic aggregate, the mechanical behavior of a given phase is not necessarily the same as it would be in a monomineralic aggregate of that phase, because in the former the deformation response of each phase is constrained by how the other phases are deforming. In both cases, the nature of this constraint depends in some complex way upon the microstructure, and so to assess the effect requires information about the contribution that grains of different orientation (and/or different phase) make to the overall properties, and how this varies with the microstructural variables. Consequently, one of the focuses of current experimental rock deformation research is to establish methods of measuring in situ stresses and strains experienced by grains of different orientation and/or phase within the polycrystal during deformation.

When a sample is porous or deforms in a brittle manner, information about how the deformation is accommodated within it may be obtained by monitoring the petrophysical properties during experiments [19]. However, for nonporous, nonbrittle materials, measurements of such properties are generally insufficiently sensitive to provide useful information of this kind. Various optical techniques have been developed to determine in situ stresses [20–22], and in situ strains have been determined in both geological and engineering materials from the deformation of rectangular microgrids imprinted within samples before experiments [23, 24] or from the measurements of grain shape changes as a function of aggregate strain [25]. However, the optical techniques require materials with very specific properties, while the strain techniques require that numerous experiments be performed on suites of compositionally and microstructurally identical samples if the dependence of the in situ strains on the bulk strain is to be established. In the absence of in situ stress and/or strain information, assessment of microstructural effects can be performed only by making inferences from comparisons of the measured bulk properties with predictions from theoretical analyses. However, attempts to decouple the influences of the different microstructural variables in this way are poorly constrained unless the mechanical significance of the various assumptions used in the theoretical analyses is rigorously understood.

The theoretical problem of predicting the elastic and plastic properties of single-phase and multiphase polycrystals from the properties of their constituent grains/phases has received enormous attention (for extensive overviews see [26–28]). Rigorous bounds may be placed on the elastic properties of a polycrystal by ignoring neighboring grain constraints and by assuming that each grain experiences the same strain (thereby ensuring strain compatibility) or the same stress (thereby ensuring stress equilibrium) in response to an applied load, irrespective of its orientation and mechanical anisotropy [29]. Tighter bounds may be derived by minimizing either potential or complementary potential energy under suitable constraints [30–32]. By using the results derived by Eshelby [33] for the elastic field around an ellipsoidal inclusion embedded in an unbounded homogeneous medium, these tighter bounds have been generalized to materials in which the grains/phases have anisotropic properties and have shapes and spatial distributions that can be approximated by an ellipsoidal symmetry [34–36]. All of these bounds may be extended to the plastic deformation case by replacing the plastically deforming nonlinear material with an equivalent linear material with appropriate elastic constants [37]. The most frequently used method of deriving a specific solution for the elastic and plastic properties between these bounds employs the Eshelby analysis within a self-consistent scheme (e.g., for the elastic case [38]). This strategy forms the basis of the viscoplastic self-consistent model [39], which is one of the most widely applied methods for modeling the mechanical behavior of polycrystalline materials [40].

Schemes based on Eshelby's analysis supply aggregate properties together with the mean stress and strain for grains of similar orientations or of given phase. Obtaining information about how stresses and strains are distributed within grains requires other methods. One of the most popular analytic methods for obtaining this information for composites is the shear lag model (described further below). However, like the Eshelby-based models, the shear lag model is restricted in its ability to describe the geometry of the microstructure. This problem is overcome in finite element models by describing the microstructure as a mesh with properties that vary spatially to match the elastic and plastic properties of the composite [41, 42]. This descriptive flexibility, coupled with an enhanced flexibility in the prescription of interface and mechanical properties, has made finite element modeling increasingly popular as a method of calculating the properties of polycrystals.

These theoretical developments should not obscure the fact that the modeling problem is a complex one and that the modeling results require validation. The impact on the predicted in situ stresses and strains of the following merit particular consideration when comparing the theoretical models with experimental data:

(1) idealizations of the geometry of the microstructure (e.g., use of the ellipsoidal inclusion-matrix geometry of Eshelby-based models or the specification of a unit cell in finite element models);
(2) the specific material properties used (e.g., the yield criteria and description of the hardening behavior);
(3) the strategy used to linearize nonlinear behavior when modeling plastic behavior [43, 44]; and
(4) the description of the interfaces, which are often assumed to be perfect (i.e., so that there is continuity of the tractions and displacements across them), although analyses that allow this assumption to be relaxed are becoming available [45–47].

Since the theoretical analyses described previously provide information about the contributions that grains of given orientation or phase make to bulk properties, assessment of these four issues should be at that level. Consequently, the need for measurements of in situ stresses and strains is one that is crucial not only in allowing the theoretical treatments to be used more effectively in interrogating experimental data but also in validating and guiding the development of the theoretical analyses themselves.

9.1.2 How Neutrons Can Help

The potential for using diffraction techniques to make nondestructive measurements of elastic strains within samples has long been appreciated. If the sample of interest can be held under an applied load while the diffraction data are being collected, then by monitoring the change in lattice spacings as a function of load, the mean elastic strain in each of the diffracting phases present may be determined as a function of that load. Moreover, if the elastic properties of these phases are known a priori, then they may be used to convert the elastic strains into stresses, and so the approach can be applied to plastic as well as elastic deformation.

Diffraction data of this kind may be obtained using electrons [48], conventional X-rays [49], synchrotron X-rays [50], and neutrons [50, 51]. In the materials of interest, the penetration depths of electrons and conventional X-rays are on the order of a few microns, so their use for this purpose is restricted to the examination of surfaces. In contrast, neutrons and synchrotron X-rays have penetration depths on the order of a few millimeters to centimeters, so they can be used to monitor in situ elastic strains within the interiors of samples similar in size to those used in conventional rock and ice deformation experiments. Neutron sources have been used in this way on engineering materials since the late 1970s, with a significant surge in activity occurring when diffractometers dedicated to strain measurement came online in the 1990s. The use of synchrotron X-rays has had to wait for the introduction of third-generation synchrotron sources because penetration depths comparable to those currently obtained using neutrons require X-rays with energies of more than 150 keV. Currently, the first X-ray diffractometers dedicated to strain measurement are being commissioned.

9.2 Outline of Techniques

The requirements for neutron diffractometers that are designed to measure elastic strains, together with the practical aspects associated with the collection of neutron diffraction data for this purpose, have been described at length elsewhere [1, 50, 51]. Because a detailed consideration of how the use of geological rather than engineering materials impinges on some of these practical aspects may be found in Schofield et al. [52], the geometry of the experimental setup and data analysis procedures are only summarized here. The description is given with reference to the experimental geometry available on the ENGIN-X beam line [53–55] at the ISIS neutron facility, Rutherford Appleton Laboratory, United Kingdom, although it is easily translated to the experimental geometries available at other neutron sources.

The ISIS facility is a pulsed, time-of-flight neutron spallation source. On the ENGIN-X beam line, pulsed polychromatic neutron beams are directed at the specimen and the diffracted neutrons are recorded in two banks of detectors fixed at $\pm 90°$ to the incident beam. Because the incident neutrons are polychromatic, each detector collects a complete diffraction pattern for each pulse. Cylindrical samples (typically 10 mm in diameter and 25-mm long) are loaded in axial compression in a load frame positioned so that the loading axis is horizontal and at 45° to the incident beam. With this geometry, the detector bank to the right of the sample (as pictured in Fig. 9.1) records neutrons diffracted from lattice planes with scattering vectors parallel to the loading direction, and hence monitors lattice strains in that (axial) direction. The detector to the left of the sample records the neutrons diffracted from lattice planes with scattering vectors normal to the loading direction, and hence monitors lattice strains in this second (radial) direction. It is worth emphasizing that the lattice parameters and d-spacings obtained from a diffraction pattern collected in a given detector are the mean lattice parameters and d-spacings parallel to the scattering vector as obtained by averaging over all of the differently oriented grains in the diffraction volume that satisfy the Bragg condition for that detector. They are not the mean a, b, and c lattice parameters of some representative grain. Current practical limits on the diffracting volume of the sample are $1\,mm^3$ to about $5 \times 5 \times 20\,mm$.

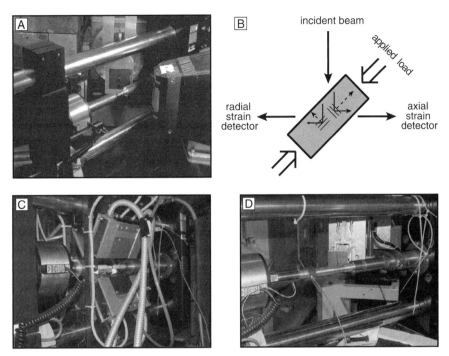

Fig. 9.1 Geometry of the experimental setup for axial compression experiments on the ENGIN-X beam line at ISIS. (**A**) Photograph of the room temperature and pressure specimen assembly on the ENGIN beam line (the forerunner to ENGIN-X) at ISIS. The sample is held within an Instron load frame at 45° to the incident neutron beam. The neutron detectors are at ±90° to the incident beam. (**B**) A schematic plan view of the experimental arrangement. In this geometry, the detector to the right of the sample detects neutrons with scattering vectors parallel to the loading direction while the detector to the left of the sample detects neutrons with scattering vectors normal to the loading direction. (**C**) Photograph of the high-temperature/room-pressure assembly in the Instron load frame showing the mirror furnaces. The assembly shown here can, in principle, attain 1000°C, although to date the authors have used it only to 500°C. (**D**) Photograph of the high-confining-pressure/room-temperature assembly in the Instron load frame showing the pressure vessel. The vessel shown here permits a maximum confining pressure of 200 MPa. Figs. 9.1A and D are reproduced from [64] with permission of the International Union of Crystallography; Fig. 9.1B is reproduced from [73] with permission of the American Geophysical Union; Fig. 9.1C is reproduced from [109] with permission from Elsevier.

A typical experimental procedure is to apply a small load to the sample and to hold it at that load while the neutron diffraction data are collected. Once a diffraction pattern of sufficient quality has been obtained (which typically takes 5–10 min on ENGIN-X), the sample is taken to a different load and held there while further diffraction data are collected. This process is repeated until diffraction patterns from a number of different loads have been collected.

Since the entire diffraction pattern is collected in each measurement direction without the need to reorient the sample or detector, the neutron data are usually analyzed using a Rietveld method to fit the whole pattern rather than individual peaks [56, 57]. Several computer codes for carrying out Rietveld refinements are available, of which one of the mostly commonly used in strain applications is the General Structure Analysis System (GSAS) [58]. Such pattern-fitting methods can refine several mineral phases simultaneously and so have the added bonus of providing accurate phase volume fractions within the actual volume element observed by the detector if the texture is weak [59]. The normal Rietveld refinement procedure makes the powder diffraction assumptions that there is a random orientation of grains, and the number of grains is sufficient for all orientations of the lattice planes to be sampled by the incident beam. Preferred orientations distort the intensities of the Bragg peaks systematically; however, if present, they can be accommodated by parameterizing the Rietveld model to approximate the preferred orientation, provided the beam line instruments being used give sufficient detector coverage of the pole figures [60]. Alternatively, they can be accommodated by relaxing the attempt to fit intensities in favor of finding a good description of peak positions [61]. The usual Rietveld refinement procedure allows the lattice parameters to vary, but only in a way that retains the original unit cell symmetry. However, the refinement procedure may be modified to accommodate elastically induced distortions in the shape of the unit cell by introducing additional parameters that describe the expected (from the single-crystal properties) variation in elastic stiffness with crystallographic direction. This strategy has been implemented within GSAS for cubic and hexagonal phases where the high symmetry means that the number of additional parameters is small [62, 63]. In addition, although the d-spacing of any particular lattice plane may be calculated using the usual formulas from the unit cell parameters obtained from the whole pattern refinements, it is frequently advantageous to fit individual peaks in the pattern. This is particularly true during plastic deformation, when the variation of the elastic strain (and hence stress) with direction and of the relative intensities of the diffraction peaks each contain information about which slip systems are operative.

The average strain parallel to the scattering vector (i.e., in the axial or radial direction) in a given lattice direction at a given load is given by

$$\varepsilon_d = -\ln(d/d_0), \qquad (9.1)$$

where d is the spacing of the set of lattice planes of interest as recovered from the diffraction pattern, and d_0 is the spacing of that set at zero load. A phase average linear strain parallel to the scattering vector is conveniently defined as

$$\varepsilon_V = -(1/3)\ln(V/V_0), \qquad (9.2)$$

where V is the unit cell volume obtained using the unit cell dimensions for that phase recovered from the diffraction pattern at given load. Because these are not the unit cell dimensions for any one particular (representative) grain, V is not a true unit cell volume. For diffraction data collected on ENGIN-X, the uncertainties of these measures of strain, which arise from the statistical uncertainties on the lattice spacings and instrument calibrations, are typically on the order of 30–100 μstrain.

The principal mechanical data recovered from one of these experiments are therefore the strains (ε_d and ε_V) for each of the peaks/phases present in the diffraction pattern as a function of the stress applied to the aggregate, σ_{agg}, as measured by a load cell. In addition, the aggregate strain, ε_{agg}, parallel to the loading direction can be monitored as a function of σ_{agg} using a capacitance extensometer, although the uncertainty on ε_{agg} is significantly greater than on the various ε_d and ε_V.

At present, experiments may be performed at room pressure in the temperature range of $-200°C$ to $1000°C$. In addition, an aluminum alloy pressure vessel equipped with an internal load cell has been successfully used to perform room-temperature deformation experiments at confining pressures of up to 150 MPa [64]. Straightforward modifications to this pressure vessel could extend its operating range to 400 MPa. However, acquiring a combined temperature and confining pressure capability would require either a pressure vessel material that, like aluminum, is relatively transparent to neutrons, but that also has good strength properties at high temperature, or a new pressure vessel design.

9.3 Investigating Elastic Properties

Lattice-preferred orientations (LPOs), the tendency toward mineralogical layering, and the presence of oriented fractures or porosity all impart a strong anisotropy on the elastic properties of geological materials, and this fact is heavily exploited in the earth sciences to recover information about the subsurface and deep interior structure of the Earth from seismological data [65–67]. Consequently, improving our understanding of the characteristic signatures of each of these sources of elastic anisotropy is of key importance, both for refining our interpretations of that structure and for developing our understanding of the processes responsible for it. Conventional methods of measuring the elastic properties of polycrystalline materials [68] are extremely sensitive to porosity, and this complicates attempts to examine the influence of the other effects. However, since neutron diffraction methods recover lattice strains, these difficulties are circumvented—at least in materials of low porosity.

The point can be demonstrated by showing that the phase average elastic strains measured by neutron diffraction lie between rigorous, theoretically derived bounds for elastically deforming two-phase nonporous composites. In this case, it is usually found that the mechanical behavior of the composite is well described by the rule of mixtures [24, 69]:

$$\sigma_{agg} = f_\alpha \sigma_\alpha + f_\beta \sigma_\beta, \tag{9.3}$$

$$\varepsilon_{agg} = f_\alpha \varepsilon_\alpha + f_\beta \varepsilon_\beta, \tag{9.4}$$

where f, σ, and ε are respectively the volume fractions, stresses, and strains of the aggregate and of the subscripted phases, α and β. At low strains, Eqs. (9.3) and (9.4) follow from force balance and strain compatibility conditions, provided that the stresses and strains of the phases are in situ stresses and strains [70, 71]. Combining Eqs. (9.3) and (9.4) and using Hooke's law ($\sigma = E\varepsilon$, where E is Young's modulus), then

$$(\varepsilon_\beta/\varepsilon_\alpha) = [f_\alpha/f_\beta][(E_{agg} - E_\alpha)/(E_\beta - E_{agg})]. \tag{9.5}$$

Hence if the Young's moduli of the two phases are known, then the observed partitioning of strains between the phases can be compared with that predicted for any theoretically prescribed E_{agg}.

The most widely used theoretical values of E_{agg} for randomly mixed two-phase composites in which the individual grains are equidimensional are those given by assuming that the average stress in each phase is the same (Reuss bound), that the average strain in each phase is the same (Voigt bound), or that potential or complementary potential energy is minimized (the two Hashin–Shtrikman bounds). Expressions giving E_{agg} in terms of E_α, E_β, and f_β for these four cases are widely quoted elsewhere [72]. The phase average elastic strains obtained using a homogeneous 46% magnesiowüstite–54% olivine composite [73] are compared with the strain partitioning predicted by these four values of E_{agg} in Fig. 9.2. The results lie between tight Hashin–Shtrikman bounds for this material, providing a strong validation of the technique.

The value of E_{agg} obtained by finding the slope of the neutron ε_β versus ε_α data and using the result in Eq. (9.5) may be compared with the theoretical treatments that

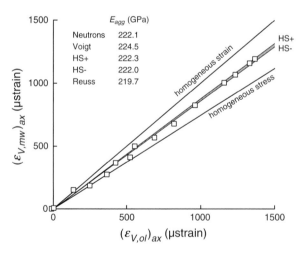

Fig. 9.2 A plot showing the phase average elastic strains parallel to the direction of loading in an elastically isotropic (randomly mixed equidimensional grains) 46% magnesiowüstite (mw)–54% olivine (ol) composite during loading in axial compression at room temperature and pressure [73]. Also shown are theoretical bounds on the behavior predicted by homogeneous stress (Reuss), homogeneous strain (Voigt), and the two Hashin–Shtrikman bounds (HS), as obtained using $E_{mw} = 260.4$ GPa and $E_{ol} = 193.9$ GPa

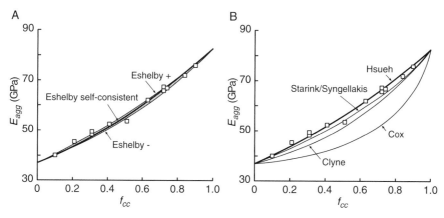

Fig. 9.3 A plot showing the volume fraction of calcite (f_{cc}) dependence of the aggregate Young's modulus (E_{agg}) as obtained using Eq. (9.5) with $E_{cc} = 82.72$ GPa and $E_{hl} = 37.04$ GPa from room temperature and pressure axial compression experiments on a range of isotropic calcite (cc)–halite (hl) composites. (**A**) Comparison of the data with the bounds predicted by the Eshelby equivalent homogeneous inclusion model and with the values of E_{agg} predicted by the self-consistent version of that model [38]. (**B**) Comparison of the data with the predictions of various developments of the shear lag model [74–77]

give E_{agg} as a function of f_β. The values of E_{agg} obtained in this way for a suite of isotropic composites of calcite and halite are compared in Fig. 9.3 with the predicted calcite volume fraction dependence of E_{agg} given by the Eshelby equivalent homogeneous inclusion model [34, 38] and by various shear lag models [74–77].

As noted above, both the Eshelby and the shear lag models allow E_{agg} to be estimated for composites in which the phases are not randomly mixed and/or the grains are not equidimensional. Perhaps more important in the present context, they also provide estimates of the phase average stresses. The Eshelby model uses an ellipsoidal inclusion-matrix representation of the microstructure and solves for the aggregate properties and the stress in the inclusion using Eshelby's solution for elastic field around an ellipsoidal inclusion embedded in a homogeneous medium. Bounds on the properties of the aggregate (which are the same as the Hashin–Shtrikman bounds [78]) are obtained by giving the inclusions the properties of the strong phase and the matrix those of the weak phase, and vice versa. A self-consistent solution of the properties may be obtained by finding a solution in which the matrix has the properties of the composite [38]. In the shear lag model, the elastically stiffer phase is treated as a cylindrical inclusion within a continuous matrix of the more compliant phase. Assuming no slip on the interfaces, then the difference in elastic properties between the phases generates shear stresses on the interfaces when the composite is loaded. These transfer load from the matrix to the inclusion. Given assumptions about the form of the radial dependence (away from axis of symmetry of the inclusion) of the shear and axial stresses, the stress distribution in inclusion and matrix may be determined. The original analysis [74] included several other approximations that had the virtue of giving a simple algebraic form to the stress

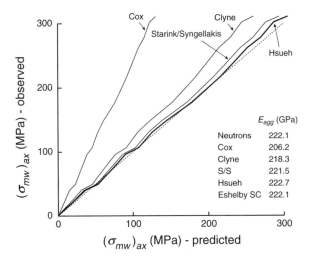

Fig. 9.4 A plot showing the phase average stresses in the magnesiowüstite parallel to the direction of loading (same sample as shown in Fig. 9.2; stresses calculated from $\varepsilon_{V,\mathrm{mw}}$ using E_{mw} and Hooke's law) compared with those predicted for this sample by various versions of the shear lag model [74–77] at the measured aggregate stresses during the experiment. The dashed line shows perfect agreement between observation and prediction. On this plot, the predicted stresses using the self-consistent version of Eshelby's equivalent homogeneous inclusion model [38] are almost indistinguishable from those predicted by Hsueh's version of the shear lag model

distribution within the inclusion; this, combined with the attractively simple physical basis for the model, has given the model an enduring popularity. Subsequent developments of the model [75–77] have succeeded in eliminating most of the original approximations while retaining the simple physical basis of the model.

In Fig. 9.4 the observed magnesiowüstite stresses in the magnesiowüstite (mw)–olivine (ol) sample shown in Fig. 9.2 (as calculated from the $\varepsilon_{\mathrm{mw}}$ using Hooke's law) are compared with the predicted stresses in that phase using various developments of the shear lag model. The results clearly validate the most recent shear lag model [77]. The stresses given by the latter model are also indistinguishable, within measurement error, from the magnesiowüstite stresses predicted by the self-consistent Eshelby model. Several measurements of the elastic properties of other mw–ol composites with different volume fractions and non-random-phase spatial distributions have been made and are currently being analyzed.

Some experiments have also been reported [79] in which a naturally deformed olivine–orthopyroxene composite was loaded in different orientations with respect to the geometry of the microstructure. The aim of the experiments was to show that the influence of the geometry of the LPO on the elastic properties could be examined in this way, and in this the experiments were successful. However, the influence of LPO geometry and intensity on thermoelastic properties is perhaps more easily investigated (in the first instance) by thermal expansion experiments on polycrystals with different LPOs (see Section 9.4)

9.4 Elastic Twinning and the Onset of Yielding in Monomineralic Carbonates

One of the most widely applied methods for estimating the magnitude and orientation of the stresses responsible for natural deformation uses the orientation and density of mechanical twinning in carbonates. Conventional methods of calibrating this piezometer rely on performing numerous deformation tests on nominally identical samples, taking each to different applied stresses, and determining the orientation and density of twins in the experimental run products [80]. However, mechanical twinning is very sensitive to the sample microstructure, so any microstructural variability within the suite of samples is carried through into the data set and induces an uncertainty about when during the loading history of any given sample the key microstructural features observed in the run products actually formed. Consequently, it is highly desirable to have a method of monitoring the progress of the twinning during the course of an experiment. Neutron diffraction offers a means of doing this.

The authors have performed several uniaxial compression experiments on Carrara marble in the temperature range of 20–500°C. Carrara marble is a monomineralic calcite rock with near random LPO and a grain size of $\sim$ 150 μm. Figure 9.5A shows the lattice strains as a function of applied stress for the sample deformed at 400°C. Examination of the phase average strains [given by Eq. (9.2)], which is all that would be measured in a conventional deformation experiment, shows

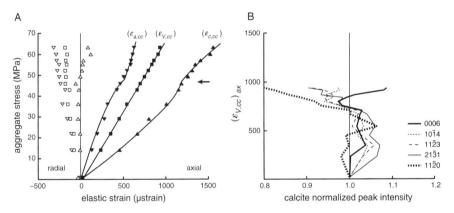

Fig. 9.5 The results of a room-pressure axial compression experiment performed on Carrara marble at 400°C. (**A**) The strains given by Eq. (9.1) in the a and c directions of the calcite hexagonal structural unit cell ($\varepsilon_{a,cc}$ and $\varepsilon_{c,cc}$ respectively) and the phase average strain ($\varepsilon_{V,cc}$) given by Eq. (9.2), as measured parallel to the loading direction (*filled symbols*) and radial direction (*open symbols*). The *solid lines* (omitted for the radial data) are shown only for visual clarity and are not fits to the data. The *arrow* marks the onset of permanent twinning. (**B**) The variation in intensity of five peaks within the diffraction patterns recorded by the detector showing axial strains during loading (as normalized by the respective intensities of these peaks obtained at the lowest load). The onset of permanent twinning is indicated by the rapid divergence of intensities for the (0006) and (11$\bar{2}$0) peaks

linear behavior up to the maximum stress measured. However, the average strains in the a and c directions of the (hexagonal) unit cell show a marked nonlinearity with a "knick point" in the stress/strain curves at the stress indicated by the arrow in the figure. This nonlinearity increases with temperature, being absent in the room temperature results and becoming very pronounced at 500°C. Measurements made during loading and unloading cycles show that the deformation up to the knick-point stress is fully elastic, whereas deformation above this stress involves some plasticity. Examination of the intensities of several of the Bragg peaks in the diffraction patterns that record strains parallel to the loading direction show a marked increase in intensity of the (0006) peak and corresponding decrease in intensity of the $(11\bar{2}0)$ peak at stresses above the knick-point stress (Fig. 9.5B). These observations are consistent with the knick-point stress being the stress required for the initiation of permanent twinning, which then results in a reorientation of the calcite c-axes toward the compression direction (with a corresponding reorientation of directions in the basal plane away from it) [23]. The curvature in the lattice strains at stresses below that required for the onset of permanent twinning almost certainly reflects the operation of elastic twinning, a phenomenon that is well documented in single crystals of calcite [81]. Results of this kind offer a much more precise method of determining the grain size and temperature dependence of the critical stress required for permanent twinning than is otherwise available (Fig. 9.6A), and hence could be used to make major improvements in the calibrations of the various twinning piezometers used in the earth sciences.

Assuming homogeneous stress within the calcite, then

$$\varepsilon_a = s_{11}\sigma_{\text{agg}}, \tag{9.6a}$$

$$\varepsilon_c = s_{33}\sigma_{\text{agg}}, \tag{9.6b}$$

$$\varepsilon_V = s_{11,\text{Reuss}}\sigma_{\text{agg}}, \tag{9.6c}$$

where the strains are in the lattice directions indicated by the subscripts, s_{ij} are the components of the elastic compliance tensor for single crystal calcite, and $s_{11,\text{Reuss}}$ is E^{-1} where E is the Young's modulus obtained by averaging the calcite s_{ij} assuming homogeneous stress. Hence

$$\varepsilon_a/\varepsilon_V = s_{11}/s_{11,\text{Reuss}}, \tag{9.7a}$$

$$\varepsilon_c/\varepsilon_V = s_{33}/s_{11,\text{Reuss}}, \tag{9.7b}$$

and departures from homogeneous stress in the calcite recorded by the neutron data can be established on a plot of ε_a or ε_c versus ε_V. Figure 9.6B shows the data plotted in Fig. 9.5A in this way, where it can be seen that the elastic twinning acts so as to facilitate homogeneous stress (parallel to the direction of loading) deformation. This is confirmed by the value of E_{cc} obtained from a linear fit to the σ_{agg} versus ε_V data

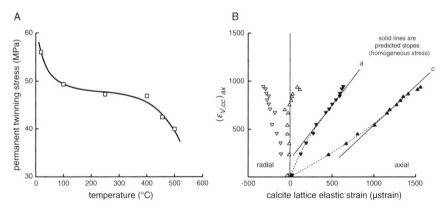

Fig. 9.6 (**A**) The temperature variation of the stress required for permanent twinning in Carrara marble during room-pressure axial compression. The reproducibility of these results remains to be established. (**B**) The lattice strains in the *a* and *c* directions of the calcite unit cell as a function of phase average strain in Carrara marble during room-pressure axial compression at 400°C. (*Filled symbols* are strains measured parallel to the loading direction; open symbols are strains measured in the radial direction.) Also shown are the slopes of these data predicted from the elastic stiffness tensor for single-crystal calcite at 400°C, assuming homogeneous stress parallel to the direction of loading [Eq. (9.7) with $s_{11} = 0.01265\,\text{GPa}^{-1}$, $s_{33} = 0.01761\,\text{GPa}^{-1}$, $s_{11,Reuss} = 0.01582\,\text{GPa}^{-1}$] [82]. From this it appears that elastic twinning at low loads acts so as to allow the aggregate to deform at approximately homogeneous stress parallel to the loading direction and that this is accommodated by markedly inhomogeneous stress in other directions

(66.27 GPa), which lies much closer to the Reuss average of the calcite single-crystal elastic constants at 400°C (63.23 GPa) than it does to the Voigt average (83.77 GPa). This is in contrast to the markedly different behavior of Solnhofen limestone loaded at 400°C (Fig. 9.7). Solnhofen limestone is a monomineralic calcite rock with a weak LPO and a grain size of ∼ 5 μm. The fine grain size inhibits twinning, and at stresses below ∼80 MPa, this results in approximately homogeneous strain (parallel to the direction of loading) deformation. At higher stresses, grains in elastically stiff orientations experience greater strains than those in elastically more compliant orientations. The origins of this strain partitioning (and hence strong stress partitioning) presumably lie in the geometry of the LPO of Solnhofen limestone and are currently under investigation.

Information about the influence of LPOs on thermoelastic properties can be gained from thermal expansion experiments on monomineralic polycrystals with different geometries and intensities of LPO [83, 84]. The authors' own thermal expansion measurements on the POLARIS beam line [85] at the ISIS facility on samples of powdered single-crystal calcite, Carrara marble, and Solnhofen limestone cores show that all three materials have a similar volumetric coefficient of thermal expansion ($\alpha_V = \partial \ln V / \partial T$) over the temperature range of 20–700°C (except for an excursion in the Carrara marble at about 300°C due to a twinning event induced by the pronounced thermal expansion anisotropy of calcite), but that

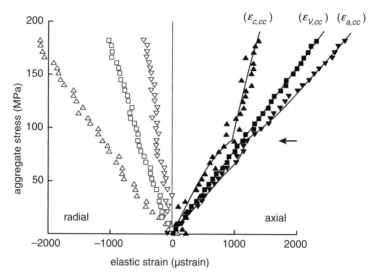

Fig. 9.7 The results of a room-pressure axial compression experiment performed on Solnhofen limestone at 400°C. The sample was loaded normal to bedding. It differs from Carrara marble only in being of much finer grain size and in having a weak lattice-preferred orientation, but the difference in behavior between the two samples (compare with Fig. 9.5A) is dramatic. Note how the elastically stiff direction (*a*) experiences more strain than the elastically more compliant direction (*c*) and that this is accommodated by the deformation in the radial direction

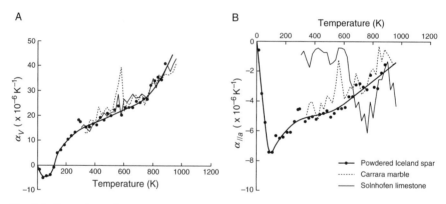

Fig. 9.8 Comparison of the room-pressure thermal expansion coefficients of powdered single-crystal calcite, Solnhofen limestone, and Carrara marble as obtained by stepwise differentiation of the lattice parameter versus temperature curves for these materials. The line ornament is common to both parts of the figure. For the Iceland spar, the curve merely shows the trends in the data for visual clarity and is not a fit to that data; for the two polycrystals, successive data points are connected by straight-line segments. (**A**) The volumetric coefficient of thermal expansion of all three materials is similar apart from an excursion for the Carrara marble at about 600 K, which is terminated when the thermally induced stresses are relieved by twinning. (**B**) The way this thermal expansion is accommodated in the different lattice directions (here shown for the *a* dimension of the hexagonal structural unit cell) in the different materials is, however, very different

the thermal expansion behavior in the a and c unit cell directions is very different in the three materials (Fig. 9.8). Further thermal expansion experiments on monomineralic samples with different LPOs is currently in progress.

9.5 Load Transfer During Plastic Yielding of Two-Phase Composites

Since the number of volumetrically significant minerals is small, it is one of the primary goals of experimental rock deformation to develop flow laws that describe the plastic deformation of polymineralic rocks in terms of the properties of their constituent minerals. Doing so requires knowledge of the contributions that each constituent mineral makes to the properties of the whole and how the contributions vary with the volume proportions of the phases and microstructural variables. The restriction of conventional deformation experiments to measurements of bulk properties means that such information is generally unavailable, so this matter is one of the more obvious ways neutron diffraction techniques can contribute to experimental rock deformation research.

In view of this, the authors have performed several room-temperature and pressure axial compression experiments on calcite–halite composites. The initial analysis of the data set has been reported elsewhere [86], and pending a more detailed account of our findings, attention here will be restricted to key parts of our analysis, which illustrate the potential of the neutron diffraction approach.

The samples were fabricated by hot isostatically pressing powders of the two phases, which had been randomly mixed in volume proportions ranging from 10% to 90% calcite. Throughout the deformation experiments, the calcite behaved elastically while the halite, after initially deforming elastically, underwent plastic yielding. The yielding was detected in the bulk strain measurements and even more precisely in a rapid increase in the rate of broadening of the halite diffraction peaks with strain. The values of E_{agg} obtained from the pre-yielding part of the experiments are shown in Fig. 9.3, where they are seen to compare well with the predictions of Eshelby's equivalent homogeneous inclusion model and Hsueh's version of the shear lag model.

Figure 9.9 shows the phase average strains for the 31 and 63% calcite samples. For samples with less than 60% calcite, halite yielding had little effect on the elastic strain (and hence stress) partitioning between the phases. However, at higher calcite contents, load transfer from the halite to the calcite was initiated when the halite yielded, with the result that the calcite experienced more elastic strain than it would have otherwise. Using the bulk strain (ε_{agg}) measurements in Eq. (9.4), the total halite strain (elastic + plastic) may be determined as a function of σ_{agg}. The halite stress at given σ_{agg} may be determined from the halite elastic strain (using E_{hl}), and so the in situ stress/total strain curve of the halite may be reconstructed. Figure 9.10A shows the stress/total strain curve calculated in this way for the 63% calcite sample, with tie-lines connecting the calcite, aggregate, and halite stresses

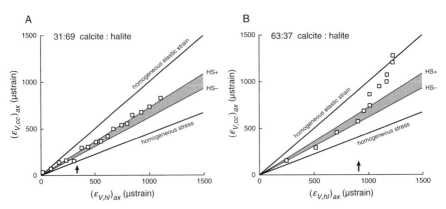

Fig. 9.9 Comparison of the phase average strains parallel to the direction of loading in two isotropic calcite–halite composites. Also shown are the theoretical bounds on the behavior predicted by homogeneous stress (Reuss), homogeneous strain (Voigt), and the two Hashin–Shtrikman bounds (HS), as obtained using $E_{cc} = 82.72$ GPa and $E_{hl} = 37.04$ GPa. The arrows show the onset of plastic yielding in the halite. At low-calcite volume fractions, this yielding has little impact on the strain partitioning between the phases; but at higher volume fractions, it leads to load transfer from the halite to the calcite. Reproduced from [86] with permission from Elsevier

and strains. Over the full range of sample compositions, the in situ behavior of the halite is variable (Fig. 9.10B): in comparison to the behavior at low-calcite contents (<25%), the halite was much weaker at intermediate compositions (30–60%) and much stronger (in the initial part of the stress/strain curve, albeit with some complexity in the shape of the curve) at high-calcite contents (>60%). This behavior may easily be rationalized by considering halite phase connectivity as a function of f_{cc}, but perhaps the more significant point in the present context is that such results demonstrate the difficulty of using the properties of monomineralic mineral phases as a basis for describing the behavior of those phases when they are in composites, thereby underscoring the need for in situ property measurements.

Examination of the calcite lattice strains provides further insight into how the two phases interact. Comparison of the a and c lattice strains with those predicted under the assumption of homogeneous stress parallel to the loading direction in calcite [using Eq. (9.7)] shows that the interactions are such as to allow the calcite to deform approximately at homogeneous stress except at high-calcite contents (>80%) when the calcite deformation switches to approximately homogeneous strain parallel to the direction of loading (Fig. 9.11). At high-calcite volume fractions, strain compatibility presumably exerts a stronger constraint on the aggregate properties than stress equilibrium, since the stress discontinuities to which it leads can be relieved by relatively small elastic strain heterogeneities close to the grain boundaries. In contrast, at lower-calcite contents, the maintenance of these elastic strain (stress) heterogeneities is unnecessary because there is sufficient highly deformable halite available to accommodate any strain incompatibilites arising from the requirements of stress equilibrium. The beauty of the neutron diffraction methods is that they allow the microstructural influences to be empirically examined.

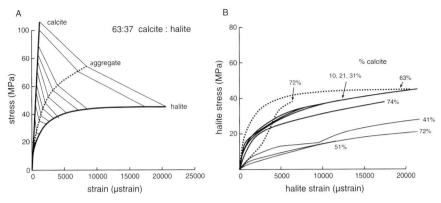

Fig. 9.10 Results from room-temperature and room-pressure axial compression experiments on isotropic calcite–halite samples. (**A**) A plot showing the stresses and strains in the two phases for the 63% calcite–37% halite sample. The calcite behaves elastically, so its stress/strain curve is given by Hooke's law; the halite strains are calculated from measurements of the aggregate strain [using Eq. (9.4)], and the halite stresses are those implied by the measured halite elastic strains. If the rule of mixtures [Eqs. (9.3) and (9.4)] is satisfied, then the tie-lines connecting the stresses and strains in the phases should be straight [25]; the small departure from linearity here probably reflects the greater uncertainty on the aggregate strain measurements (and hence the total halite strain) than on all the other variables. (**B**) Halite stress/total strain curves calculated in this way for samples of different composition. The in situ halite behavior at intermediate compositions is much weaker than at low-calcite contents, while the behavior at the halite percolation threshold (about 70% calcite) is apparently very sensitive to the connectivity of the halite

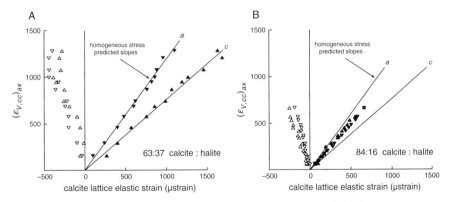

Fig. 9.11 The calcite lattice strains as a function of phase average strain during room-pressure and room-temperature axial compression of two isotropic calcite–halite samples. Also shown are the slopes of these data predicted from the elastic stiffness tensor for single-crystal calcite at 25°C, assuming homogeneous stress parallel to the direction of loading [Eq. (9.7) with $s_{11} = 0.01133\,\text{GPa}^{-1}$, $s_{33} = 0.01772\,\text{GPa}^{-1}$, $s_{11,\text{Reuss}} = 0.01404\,\text{GPa}^{-1}$] [82]. The calcite deforms at approximately homogeneous stress parallel to the direction of loading except at very high calcite contents where it deforms at approximately homogeneous strain ($\varepsilon_{a,\text{cc}} = \varepsilon_{c,\text{cc}} = \varepsilon_{V,\text{cc}}$) in this direction

9.6 Other Applications

9.6.1 Dauphiné Twinning and the α–β Transformation in Quartz

The fact that neutron diffraction methods allow the progress of any mechanically induced transformation that produces a change in a diffraction pattern to be monitored during a deformation experiment has been heavily exploited in the engineering sciences to explore controls on superelastic and shape memory properties. To date, the principal application of this strategy for geological materials has been in examining the mechanical significance of Dauphiné twinning and the α–β transformation in quartz.

Quartz is one of the most common minerals in the Earth's crust, so its rheology and the extent to which LPOs from naturally deformed quartz-bearing rocks can be used to infer deformation conditions have been matters of longstanding interest. The structural changes in quartz upon heating through the α–β transformation have been monitored in several studies (for example, [87]). The use of neutron diffraction in mechanical investigations has focused on monitoring changes in the LPO of quartz polycrystals during heating and/or loading experiments [2, 88, 89]. The observed changes have been qualitatively reproduced by finite element models that include Dauphiné twinning and the α–β transformation [90], and they have re-established the possibility of using Dauphiné twinning as a piezometer in naturally deformed rocks. Moreover, it has been found that if quartzites are repeatedly cycled through the α–β transformation, then the LPO developed after the first cycle is reproduced after subsequent cycles. Current research aimed at using this texture memory effect in the interpretation of LPOs of naturally deformed samples is focused on obtaining a better understanding of the controls on the development of in situ stresses within quartz polycrystals [89].

9.6.2 The Origin of Nonlinear Elastic Responses

Porous geological materials frequently exhibit a nonlinear elastic response, which is widely attributed to grain contact effects. However, the physical processes involved are largely unknown, primarily because it is difficult to separate the contribution to the macroscopic properties made by the grain contacts from that made by the bulk of the grain volume. This can now be done by comparing the macroscopic strains with the lattice strains obtained by neutron diffraction during loading.

Experiments on a number of high-porosity ($\sim 20\%$) quartz sandstones have shown that although the macroscopic stress/strain response may be markedly nonlinear, the lattice strains, which are much smaller, vary linearly with the applied stress [91]. One might attribute these characteristics to initial pore collapse/poroelastic effects, but analysis of the pair distribution functions (essentially the probability of two atoms being separated by given distance) obtained from the diffraction patterns shows an excess of Si–O and O–O nearest neighbor atom pairs in comparison with that predicted from purely crystalline quartz [92]. This is consistent with the presence of a small amount of an amorphous phase (since an amorphous phase will show only peaks in the pair distribution function at short distances), which is presumably located at the grain contacts. Current research is aimed at imaging this amorphous phase (using non-neutron techniques) [93]. Since even small amounts of an amorphous phase can have a large impact on elastic properties [94], the ability to explore this matter using neutron diffraction methods is highly significant.

9.6.3 Residual and Thermally Generated Stresses

Geological materials are widely quarried for use as building stones and as the raw materials for a range of art works. As such, their durability depends on the residual stresses contained within them and, if the stone is placed in an exterior environment, how these stresses are influenced by weathering and climatic variations. Neutron diffraction methods offer a means of measuring these residual stresses and of exploring how they are modified by environmental conditions. Building on neutron diffraction stress-measuring techniques developed on geological materials over a number of years [95–98], Scheffzük and colleagues have used this approach to compare the residual stresses in marble-building facade panels in varying states of disrepair and to show how such methods can be used to separate the impact of preexisting residual stresses, weathering, and thermal cycling on the deterioration of facade panels [99].

Microcracks formed in polycrystals during cooling or heating from the stresses generated by the thermal expansion anisotropy of the constituent minerals and/or the differences in thermal expansion among those minerals have a major impact on the fluid transport (and hence heat flow) properties of crustal materials. By simultaneously monitoring the buildup of lattice strains and acoustic emission activity during a heating experiment, Meredith et al. were able to determine the stresses required to initiate widespread thermal cracking in a low-porosity quartzite [100]. Such experiments demonstrate the potential of neutron diffraction methods not only for investigating the elastic and plastic properties but also for examining the microstructural controls on the initiation of brittle failure.

9.7 Conclusions and Future Outlook

By providing information at the grain scale about how deformation is accommodated within samples during mechanical tests, neutron diffraction methods have great potential for furthering our understanding of the mechanical behavior of geological materials, and in particular for furthering our understanding of how microstructural variables influence that behavior. In common with their role in research on the mechanical properties of engineering materials, neutron diffraction methods permit experimental data to be much more closely interrogated by theoretical analyses while also providing the type of experimental data that allow the significance of the various assumptions and approximations used in the theoretical analyses to be more rigorously evaluated.

Considerable scope exists for further development of most of the methods and techniques described here and extension of these techniques to other geological materials. In addition, there are other obvious applications of neutron diffraction in this field yet to be explored, including (1) the evaluation of single-crystal thermoelastic property measurements from measurements on polycrystals and (2) strain scanning. The former would be helpful in the commonly encountered situation where it is difficult to acquire single crystals of the phase of interest that are of the requisite size, shape, or perfection for conventional measurements. Some of the experimental and data analysis procedures have already been outlined [101–103]. Strain scanning is used widely in the engineering sciences to examine stress heterogeneities within engineering parts, but there have only been reconnaissance studies using geological samples [98]. A use of particular interest for earth scientists is in examining the strains around crack tips during loading and unloading, which in the material sciences has been attempted for cracks in type 316 stainless steel [104].

Extension of this kind of work to a greater range of geological materials is to some extent contingent upon developing an apparatus for neutron work, which will allow samples to be deformed at simultaneously high confining pressure and temperature. We believe that an apparatus capable of deforming cylindrical samples of conventional size at confining pressures of 300–400 MPa and temperatures of up to 400°C is a realistic prospect using the type of vessel used for room-temperature/high confining pressure experiments referred to previously [64]. However, design of an apparatus capable of higher temperatures at these pressures will require a radically different approach.

Traditionally, use of neutron powder diffraction methods for crystal structure analysis has been restricted to non-hydrogen-bearing minerals because hydrogen is an incoherent scatterer of neutrons. For powdered samples this issue can be circumvented by deuterating the sample because deuterium scatters neutrons coherently, but this approach is not possible for intact samples of natural rocks. The application of neutron methods as described here, however, is aimed primarily at the extraction of lattice parameters, which can be achieved to a high precision as long as a sufficient number of diffraction peaks are evident above the background. Consequently, minerals with a number density of hydrogen up to at least 30%—including mechanically important minerals such as amphiboles, micas, and gypsum—could be characterized by the methods described here, even though a full analysis of their atomic structures would be challenging. Alternatively, synchrotron X-rays could be used to examine these materials.

The arrival of synchrotron X-ray beam lines dedicated to strain measurement opens up an additional range of possibilities. The very high flux of synchrotron X-rays allows data to be acquired much faster than when using neutrons, although neutron data acquisition times are sufficiently fast in comparison with the time required for many other parts of the experiments (e.g., thermal equilibration, sample loading) for this to be a practical advantage only in some situations. One such situation is using X-ray tomography [105–107] to monitor pore/crack shape change during loading. Synchrotron X-rays have very low divergence in comparison with neutrons, and with the high flux, this makes it much easier to produce a small beam. This is potentially valuable in strain-scanning applications and opens up the possibility of examining the behavior of individual or small groups of grains [108]. However, if conventional powder diffraction methods are used to analyze the X-ray diffraction data, small grain sizes are required for sufficient grains to be present in the illuminated volume. Consequently, for the foreseeable future, it seems that there will be a complementary role for synchrotron X-ray and neutron diffraction methods, with the choice between the two being determined by material composition and microstructure and by the type of information required.

Acknowledgments   This contribution has benefited greatly from our interactions with instrument scientists at the ISIS facility, in particular with Kevin Knight, Mark Daymond, Ed Oliver, and Ron Smith. Some of the initial inspiration for our work came from the neutron diffraction experiments of Phil Meredith. We also acknowledge the contributions of our colleagues Iona Stretton and Rob Holloway, one or both of whom have played a major part in most of our experiments.

References

1. M. R. Daymond, in *Neutron Scattering in Earth Sciences*, edited by H.-R. Wenk (Geochemical Society/Mineralogical Society of America, USA, 2006), *Reviews in Mineralogy & Geochemistry*, Vol. 63, pp. 427–458
2. H.-R. Wenk, in *Neutron Scattering in Earth Sciences*, edited by H.-R. Wenk (Geochemical Society/Mineralogical Society of America, U.S.A., 2006), *Reviews in Mineralogy & Geochemistry*, Vol. 63, pp. 399–426
3. J. Pleuger, N. Froitzheim, J. F. Derks, W. Kurz, J. Albus, J. M. Walter, and E. Jansen, in *Neutron Applications in Earth, Energy, and Environmental Sciences*, edited by L. Liang, R. Rinaldi, and H. Schober (Springer Verlag), Chapter 10 of this volume
4. A. P. Radlinski, in *Neutron Scattering in Earth Sciences*, edited by H.-R. Wenk (Geochemical Society/Mineralogical Society of America, U.S.A., 2006), *Reviews in Mineralogy & Geochemistry*, Vol. 63, pp. 363–397
5. B. Winkler, in *Neutron Scattering in Earth Sciences*, edited by H.-R. Wenk (Geochemical Society/Mineralogical Society of America, USA, 2006), *Reviews in Mineralogy & Geochemistry*, Vol. 63, pp. 459–471
6. E. H. Lehmann, in *Neutron Applications in Earth, Energy, and Environmental Sciences*, edited by L. Liang, R. Rinaldi, and H. Schober (Springer Verlag), Chapter 11 of this volume
7. W. B. Durham, D. J. Weidner, S.-I. Karato, and Y. Wang, in *Plastic Deformation of Minerals and Rocks*, edited by S.-I. Karato and H.-R. Wenk (Geochemical Society/Mineralogical Society of America, U.S.A., 2002), *Reviews in Mineralogy & Geochemistry*, Vol. 51, pp. 21–49
8. D. P. Dobson, J. Mecklenburgh, D. Alfe, I. G. Wood, and M. R. Daymond, High Press. Res. **25**, 107–118 (2005)
9. S.-I. Karato and D. C. Rubie, J. Geophys. Res. **102**, 20111–20122 (1997)
10. W. B. Durham and D. C. Rubie, in *Properties of Earth and Planetary Materials at High Pressure and Temperature*, edited by M. H. Manghnani and T. Yagi (American Geophysical Union, Washington, D.C., 1998), pp. 63–70
11. P. Cordier and D. C. Rubie, Mater. Sci. Eng. A **309–310**, 38–43 (2001)

12. T. E. Tullis and J. Tullis, in *Mineral and Rock Deformation: Laboratory Studies, The Paterson Volume*, edited by B. E. Hobbs and H. C. Heard (American Geophysical Union, Washington, D.C., 1986), pp. 297–324

13. M. S. Paterson, in *The Brittle-Ductile Transition in Rocks, The Heard Volume*, edited by A. G. Duba, W. B. Durham, J. W. Handin, and H. F. Wang (American Geophysical Union, Washington, D.C., 1990), pp. 187–194

14. H. C. Heard, W. B. Durham, and C. O. Boro, in *The Brittle-Ductile Transition in Rocks, The Heard Volume*, edited by A. G. Duba, W. B. Durham, J. W. Handin, and H. F. Wang (American Geophysical Union, Washington, D.C., 1990), pp. 225–228

15. P. R. Sammonds, S. A. F. Murrell, M. A. Rist, and D. Butler, Cold Reg. Sci. Technol. **19**, 177–188 (1991)

16. E. H. Rutter, J. Struct. Geol. **20**, 243–254 (1998)

17. S. M. Schmid, R. Panozzo, and S. Bauer, J. Struct. Geol. **9**, 747–778 (1987)

18. M. S. Paterson and D. L. Olgaard, J. Struct. Geol. **22**, 1341–1358 (2000)

19. M. S. Paterson and T.-F. Wong, *Experimental Rock Deformation—The Brittle Field*, 2nd ed. (Springer, Berlin, 2005)

20. P. J. Withers, E. M. Chorley, and T. W. Clyne, Mater. Sci. Eng. A **135**, 173–178 (1991)

21. L. S. Schadler and C. Galiotis, Int. Mater. Rev. **40**, 116–134 (1995)

22. Q. Ma and D. R. Clarke, J. Am. Ceram. Soc. **76**, 1433–1440 (1993)

23. C. J. Spiers, Bull. Minéral. **102**, 282–289 (1979)

24. K.-M. Cho and J. Gurland, Metall. Trans. A **19A**, 2027–2040 (1988)

25. J. P. Bloomfield and S. J. Covey-Crump, J. Struct. Geol. **15**, 1007–1019 (1993)

26. U. F. Kocks, C. N. Tomé, and H.-R. Wenk, *Texture and Anisotropy: Preferred Orientations in Polycrystals and Their Effect on Materials Properties* (Cambridge University Press, Cambridge, 1998)

27. S. Nemat-Nasser and M. Hori, *Micromechanics: Overall Properties of Heterogeneous Materials*, 2nd ed. (Elsevier Science, Amsterdam, 1999)

28. G. W. Milton, *The Theory of Composites* (Cambridge University Press, Cambridge, 2002)

29. R. Hill, Proc. Phys. Soc. A **65**, 349–354 (1952)

30. Z. Hashin and S. Shtrikman, J. Mech. Phys. Solids **10**, 335–342 (1962)

31. Z. Hashin and S. Shtrikman, J. Mech. Phys. Solids **10**, 343–352 (1962)

32. Z. Hashin and S. Shtrikman, J. Mech. Phys. Solids **11**, 127–140 (1963)

33. J. D. Eshelby, Proc. Roy. Soc. Lond. A **241**, 376–396 (1957)

34. T. W. Clyne and P. J. Withers, *An Introduction to Metal Matrix Composites* (Cambridge University Press, Cambridge, 1993)

35. P. Ponte Castañeda and J. R. Willis, J. Mech. Phys. Solids **43**, 1919–1951 (1995)

36. V. M. Levin and J. M. Alvarez-Tostado, Int. J. Fract. **119(4)**, L79–L82 (2003)

37. P. Ponte Castañeda and P. Suquet, Adv. Appl. Mech. **34**, 171–302 (1998)

38. O. B. Pedersen and P. J. Withers, Phil. Mag. A **65**, 1217–1233 (1992)

39. R. A. Lebensohn, C. N. Tomé, and P. Ponte Castañeda, Phil. Mag. **87**, 4287–4322 (2007)

40. H.-R. Wenk, Modelling Simul. Mater. Sci. Eng. **7**, 699–722 (1999)

41. E. B. Marin and P. R. Dawson, Comput. Methods Appl. Mech. Eng. **165**, 23–41 (1998)

42. T.-S. Han and P. R. Dawson, Modelling Simul. Mater. Sci. Eng. **13**, 203–223 (2005)

43. C. N. Tomé, Modelling Simul. Mater. Sci. Eng. **7**, 723–738 (1999)

44. R. A. Lebensohn, Y. Liu, and P. Ponte Castañeda, Acta Mat. **52**, 5347–5361 (2004)

45. Z. Hashin, J. Mech. Phys. Solids **39**, 745–762 (1991)

46. Z. Hashin, J. Mech. Phys. Solids **40**, 767–781 (1992)

47. J. Tong, C.-W. Nan, J. Fu, and X. Guan, Acta Mech. **146**, 127–134 (2001)

48. A. J. Wilkinson, Mater. Sci. Technol. **13**, 79–84 (1997)

49. I. C. Noyan and J. B. Cohen, *Residual Stress: Measurement by Diffraction and Interpretation* (Springer-Verlag, Berlin, 1987)

50. M. E. Fitzpatrick and A. Lodini, *Analysis of Residual Stress by Diffraction Using Neutron and Synchrotron Diffraction* (Taylor and Francis, London, 2003)

51. M. T. Hutchings, P. J. Withers, T. M. Holden, and T. Lorentzen, *Introduction to the Characterization of Residual Stress by Neutron Diffraction* (Taylor and Francis, Boca Raton, Florida, 2005)
52. P. F. Schofield, S. J. Covey-Crump, I. C. Stretton, M. R. Daymond, K. S. Knight, and R. F. Holloway, Mineral. Mag. **67**, 967–987 (2003)
53. M. W. Johnson and M. R. Daymond, J. Appl. Cryst. **35**, 49–57 (2002)
54. M. R. Daymond and L. Edwards, Neutron News **15**, 24–28 (2004)
55. J. R. Santisteban, M. R. Daymond, J. A. James, and L. Edwards, J. Appl. Cryst. **39**, 812–825 (2006)
56. R. A. Young, *The Rietveld Method*, edited by R. A. Young (Oxford University Press, Oxford, 1993)
57. R. B. Von Dreele, in *Neutron Scattering in Earth Sciences*, edited by H.-R. Wenk (Geochemical Society/Mineralogical Society of America, USA, 2006), *Reviews in Mineralogy & Geochemistry*, Vol. 63, pp. 81–98
58. A. C. Larson and R. B. Von Dreele, Los Alamos National Laboratory Report LAUR 86–748, 1994 (http://www.ccp14.ac.uk/solution/gsas/ gsas_with_expgui_install.html)
59. P. F. Schofield, K. S. Knight, S. J. Covey-Crump, G. Cressey, and I. C. Stretton, Mineral. Mag. **66**, 189–200 (2002)
60. H.-R. Wenk, in *Texture and Anisotropy: Preferred Orientations in Polycrystals and Their Effect on Materials Properties*, edited by U. F. Kocks, C. N. Tomé, and H.-R. Wenk (Cambridge University Press, Cambridge, 1998), pp. 126–177
61. A. Le Bail, H. Duroy, and J. L. Fourquet, Mater. Res. Bull. **23**, 447–452 (1988)
62. M. R. Daymond, M. A. M. Bourke, R. B. Von Dreele, B. Clausen, and T. Lorentzen, J. Appl. Phys. **82**, 1554–1562 (1997)
63. M. R. Daymond, M. A. M. Bourke, and R. B. Von Dreele, J. Appl. Phys. **85**, 739–747 (1999)
64. S. J. Covey-Crump, R. F. Holloway, P. F. Schofield, and M. R. Daymond, J. Appl. Cryst. **39**, 222–229 (2006)
65. D. Mainprice, in *Mineral Physics*, edited by G. D. Price (Elsevier, Amsterdam, 2007), *Treatise on Geophysics*, Vol. 2, pp. 437–491
66. S. Crampin, E. M. Chesnokov, and R. G. Hipkin, in *Proceedings of the First International Workshop on Seismic Anisotropy*, edited by S. Crampin, E. M. Chesnokov, and R. G. Hipkin, Geophys J. R. Astr. Soc. **76**, 1–16 (1984)
67. A. Schubnel, P. M. Benson, B. D. Thompson, J. F. Hazzard, and R. P. Young, Pure Appl. Geophys. **163**, 947–973 (2006)
68. D. J. Weidner, in *Geophysics Laboratory Measurements*, edited by C. G. Sammis and T. L. Henyey (Academic Press, Orlando, Florida, 1987), *Methods of Experimental Physics*, Vol. 24A, pp. 1–30
69. Z. Fan, P. Tsakiropoulos, and A. P. Miodownik, J. Mater. Sci. **29**, 141–150 (1994)
70. R. Hill, J. Mech. Phys. Solids **15**, 79–95 (1967)
71. J. Gurland and K. Cho, Acta Stereol. **6**, 135–140 (1987)
72. J. P. Watt, G. F. Davies, and R. J. O'Connell, Rev. Geophys. Space Phys. **14**, 541–563 (1976)
73. S. J. Covey-Crump, P. F. Schofield, and I. C. Stretton, Geophys. Res. Lett. **28**, 4647–4650 (2001)
74. H. L. Cox, Br. J. Appl. Phys. **3**, 72–79 (1952)
75. T. W. Clyne, Mater. Sci. Eng. A **122**, 183–192 (1989)
76. M. J. Starink and S. Syngellakis, Mater. Sci. Eng. A **270**, 270–277 (1999)
77. C.-H. Hsueh, Compos. Sci. Technol. **60**, 2671–2680 (2000)
78. G. J. Weng, Int. J. Eng. Sci. **28**, 1111–1120 (1990)
79. S. J. Covey-Crump, P. F. Schofield, I. C. Stretton, K. S. Knight, and W. Ben Ismaïl, J. Geophys. Res. **108(B2)**, article 2092 (2003)
80. K. J. Rowe and E. H. Rutter, J. Struct. Geol. **12**, 1–17 (1990)
81. H. Kaga and J. J. Gilman, J. Appl. Phys. **40**, 3196–3207 (1969)
82. D. P. Dandekar, J. Appl. Phys. **39**, 3694–3699 (1968)

83. T. I. Ivankina, A. N. Nikitin, A. S. Telepnev, K. Ullemeyer, G. A. Efimova, S. M. Kireenkova, G. A. Sobolev, V. A. Sukhoparov, and K. Walther, Izvestiya, Phys. Solid Earth **37**, 46–58 (2001)

84. A. N. Nikitin, T. I. Ivankina, A. S. Telepnev, G. A. Sobolev, Ch. Scheffzük, A. Frischbutter, and K. Walther, Izvestiya, Phys. Solid Earth **40**, 83–90 (2004)

85. R. I. Smith, S. Hull, and A. R. Armstrong, Mater. Sci. Forum **166–169**, 251–256 (1994)

86. S. J. Covey-Crump, P. F. Schofield, and M. R. Daymond, Physica B **385–386**, 946–948 (2006)

87. M. G. Tucker, D. A. Keen, and M. T. Dove, Mineral. Mag. **65**, 489–507 (2001)

88. H.-R. Wenk, E. Rybacki, G. Dresen, I. Lonardelli, N. Barton, H. Franz, and G. Gonzalez, Phys. Chem. Mineral. **33**, 667–676 (2006)

89. H.-R. Wenk, M. Bortolotti, N. Barton, E. Oliver, and D. Brown, Phys. Chem. Mineral. **34**, 599–607 (2007)

90. N. R. Barton and H.-R. Wenk, Modelling Simul. Mater. Sci. Eng. **15**, 369–384 (2007)

91. T. W. Darling, J. A. TenCate, D. W. Brown, B. Clausen, and S. C. Vogel, Geophys. Res. Lett. **31**, L16604 (2004)

92. K. L. Page, T. Proffen, S. E. McLain, T. W. Darling, and J. A. TenCate, Geophys. Res. Lett. **31**, L24606 (2004)

93. T. W. Darling, J. A. TenCate, S. C. Vogel, T. Proffen, K. Page, C. Herrera, A. M. Covington, and E. Emmons, *American Institute of Physics Conference Proceedings* **838** (2006), pp. 19–26

94. A. S. Wendt, I. Bayuk, S. J. Covey-Crump, R. Wirth, and G. E. Lloyd, J. Geophys. Res. **108(B8)**, article 2365 (2003)

95. C. Scheffzük, A. Frischbutter, and K. Walther, Schriftenr. f. Geowiss. **6**, 39–48 (1998)

96. L. Pintschovius, M. Prem, and A. Frischbutter, J. Struct. Geol. **22**, 1581–1585 (2000)

97. A. Frischbutter, D. Neov, Ch. Scheffzük, M. Vrána, and K. Walther, J. Struct. Geol. **22**, 1587–1600 (2000)

98. Ch. Scheffzük, K. Walther, A. Frischbutter, F. Eichorn, and M. R. Daymond, Solid State Phenom. **105**, 61–66 (2005)

99. C. Scheffzük, S. Siegesmund, and A. Koch, Environ. Geol. **46**, 468–476 (2004)

100. P. G. Meredith, K. S. Knight, S. A. Boon, and I. G. Wood, Geophys. Res. Lett. **28**, 2105–2105 (2001)

101. C. J. Howard and E. H. Kisi, J. Appl. Cryst. **32**, 624–633 (1999)

102. S. Matthies, H. G. Priesmeyer, and M. R. Daymond, J. Appl. Cryst. **34**, 585–601 (2001)

103. Ch. Bittorf, S. Matthies, H. G. Priesmeyer, and R. Wagner, Intermetallics **7**, 251–258 (1999)

104. Y. Sun, H. Choo, P. K. Liaw, Y. Lu, B. Yang, D. W. Brown, and M. A. M. Bourke, Scripta Mater. **53**, 971–975 (2005)

105. V. Cnudde, B. Masschaele, M. Dierick, J. Vlassenbroeck, L. Van Hoorebeke, and P. Jacobs, Appl. Geochem. **21**, 826–832 (2006)

106. L. Babout, Automatyka **10**, 117–124 (2006)

107. L. Louis, T.-F. Wong, P. Baud, and S. Tembe, J. Struct. Geol. **28**, 762–775 (2006)

108. R. V. Martins, L. Margulies, S. Schmidt, H. F. Poulsen, and T. Leffers, Mater. Sci. Eng. A **387–389**, 84–88 (2004)

109. P. F. Schofield, S. J. Covey-Crump, M. R. Daymond, I. C. Stretton, K. S. Knight, and R. F. Holloway, Physica B **385–386**, 938–941 (2006)

Chapter 10
The Contribution of Neutron Texture Goniometry to the Study of Complex Tectonics in the Alps

Jan Pleuger, Nikolaus Froitzheim, Jan F. Derks, Walter Kurz, Jan Albus, Jens M. Walter, and Ekkehard Jansen

Abstract Textures (lattice-preferred orientations) of rocks can be analyzed by various methods, each of which has its specific qualities. Unlike any other technique, neutron texture goniometry affords true volume texture measurements of relatively large (up to several cm^3) isometric samples. Furthermore, the textures of different minerals in polyphasic rocks can be analyzed simultaneously. This is particularly attractive for applications in the geosciences where textures of ductilely deformed rocks are studied for various purposes, but mostly to gain information about the geological mechanisms of texture formation and rock deformation. Presenting two case studies from different tectonic settings within the Alps and taking quartz textures as examples, we discuss (1) some aspects of texture representation by normal and inverse pole figures, that is, the two most common diagram types, and (2) the potentials and limitations of interpreting rock textures in terms of the kinematical path, strain geometry, and physical conditions during deformation.

10.1 Introduction

The deformation of Earth's crust is concentrated at plate boundaries. It occurs by brittle faulting at shallow levels and by the activity of ductile shear zones at greater depth, where elevated temperatures promote solid-state flow by processes like intracrystalline deformation and dynamic recrystallization. Since these processes occur at depth, direct observation is not possible. In mountain belts like the Alps, however, which result from the collision of continents after the subduction of oceanic lithosphere, rocks from deep levels of the crust, including rocks from shear zones (mylonites), are exhumed to the surface and can be studied. The mineral compositions, microstructures, and lattice-preferred orientations (hereafter textures) of

J. Pleuger (✉)
Geologisches Institut der ETH Zürich, CH-8092 Zürich, Switzerland
e-mail: jan.pleuger@erdw.ethz.ch

L. Liang et al. (eds.), *Neutron Applications in Earth, Energy and Environmental Sciences*, Neutron Scattering Applications and Techniques,
DOI 10.1007/978-0-387-09416-8_10, © Springer Science+Business Media, LLC 2009

these rocks yield information about the physical conditions, kinematics, and dynamics of the deformation processes that took place at depth, allowing researchers to reconstruct the tectonic deformation and to calibrate and test models of tectonic processes at plate boundaries. The texture, that is, the orientation distribution of crystals in the deformed rocks, can be studied by a variety of methods, neutron diffraction being one of the most powerful.

The present contribution is concerned with three topics: (1) textures of rocks, (2) Alpine tectonics, and (3) the method of neutron diffraction. Since these topics are equally complex, it is beyond our scope to discuss them in great detail. Instead, we will highlight only some important qualities of neutron texture goniometry from an applicant's point of view and, after a short survey on Alpine geology (Section 10.3), present two case studies of how textures of quartzitic mylonites may be interpreted in the framework of orogenic processes in the Alps.

10.1.1 Why Look at Quartzite Textures?

Investigations of textures of polycrystalline aggregates are an established method of structural geology because rock textures form in response to various geological processes, most of which involve tectonic deformation [e.g., 1, 2]. Textures of various rock-forming minerals are more or less routinely considered in structural geological studies but, since the texture development also depends on the specific properties of each mineral, accounting for the particularities of texture development for different minerals would go beyond the scope of this text. Instead, we confine ourselves to showing quartzite textures from two tectonic settings, which may serve to exemplify how texture development correlates with large-scale tectonic processes.

On the grain scale, intracrystalline deformation and dynamic recrystallization are the most efficient among various mechanisms that can accommodate deformation, at least under the greenschist to amphibolite-facies conditions experienced by the quartzitic rocks we are concerned with in this study. In the simplest case of intracrystalline deformation (dislocation glide), the movement of defects in a crystal lattice is described by Burgers vectors defining slip systems in terms of the slip plane and the slip direction. Each slip system becomes activated only when a critical resolved shear stress (CRSS) is exceeded. The arrangement of crystals with respect to the kinematical framework depends on the active slip systems. The CRSS for each slip system in turn depends on various parameters like temperature, strain rate, and the activity of fluids. Similarly, it has been shown that the mechanisms of dynamic recrystallization, the most important of which are subgrain rotation recrystallization and grain boundary migration recrystallization, depend on temperature and strain rate [3, 4]. On the scale of a rock sample, the sum of these (and other) microscale processes will in the best case be homogeneous deformation and display a certain geometry of strain, such as constrictional, plane, or oblate strain, and a certain kinematical path, such as pure or simple shear. Based on these considerations, the physical conditions under which deformation took place, the geometry of strain, and the kinematical path are reported by mineral textures. However, in general, the relative contributions of all texture-forming processes cannot unambiguously be qualified from looking at the texture alone. Therefore, complementary information about the kinematical path and the physical conditions under which deformation occurred are usually necessary. For regional studies, such information may be supplied by structural geological field work, microstructural observations (e.g., by polarization microscopy), petrological thermo-barometry, and geochemistry.

10.1.2 What Can Be Studied in the Alps (and Elsewhere)?

Because of the overwhelming number of studies from all disciplines of geosciences, the Alps are often referred to as the best-studied orogen of the world. The wealth of data from structural geology, petrology, and geochemistry provides an excellent framework for detailed studies of rock textures and microstructures, which in turn contribute toward a more complete understanding of geological processes in Alpine orogeny. Looking at the number of publications, texture investigations of rocks from the Alps and elsewhere have become an essential part of the literature during the past two decades. They may roughly be grouped as follows:

1. Texture investigations in order to quantify anisotropic rock properties, for example, to allow for correct interpretations of seismic data (e.g., [5–9])
2. Basic investigations of the mechanisms of texture development aiming at a better understanding of the influence of physical parameters and/or how deformation is accommodated on the microscale by different minerals, for example, quartz (e.g., [4,10–14]), calcite (e.g., [15–22]), dolomite [23], plagioclase (e.g., [24–29]), garnet (e.g., [30–33]), phyllosilicates (e.g., [34–39]), amphibole (e.g., [40–43]), olivine (e.g., [44–46]), and omphacite (e.g., [47–52])
3. Numerous regional studies benefiting from the results of (2)

The two case studies described in Sections 10.4 and 10.5 shall exemplify for which purposes, according to our experience, textural studies can be applied most successfully, namely, for estimating the geometry of strain (Section 10.4), the rotational component of the kinematical path (Section 10.4), and the temperature during deformation (Section 10.5).

10.1.3 Why Use Neutron Texture Goniometry?

At the base of the upsurge in textural studies was the advent of new analytical techniques. Classical measurements of crystallographic axes with a universal stage (U-stage) mounted on an optical polarizing microscope, as compiled, for example, by B. Sander [53], are very time-consuming and generally incomplete in that for most minerals only a restricted number of crystallographic directions can be optically discerned. Probably, the greatest advantage of modern methods, such as X-ray texture goniometry, electron backscattering diffractometry (EBSD), and neutron diffraction goniometry, is that such shortcomings can be overcome. While we refer the reader to K. Ullemeyer et al. [54] for a broader description of the specific qualities of each of these three methods, and to Chapter 3 of this volume by H. Schober [55] for an introduction to neutron scattering instrumentation, we will in the following shortly summarize the most important aspects of neutron texture goniometry as compared with X-ray and EBSD techniques.

EBSD provides diffraction patterns of single grains. Like U-stage optical measurements, EBSD is therefore preferentially applied when the orientation of single grains or misorientation of neighboring grains is to be addressed. A disadvantage of these methods is that pole figures compiled from EBSD or U-stage data will generally reflect only certain crystal orientations by numbers of grains but not volume proportions, unless the grain size is independent of the crystal orientation or some correction for orientation-dependent grain size is applied. Furthermore, a large number of individual measurements is required.

Unlike the EBSD and U-stage methods, neutron texture goniometry is not position-sensitive. Instead, it is exclusively used for volume texture measurements because of the high penetration capability of neutrons, which can be regarded as the greatest advantage of the technique. Volume or statistical methods, as opposed to single-grain methods, provide integrated diffraction patterns of many grains whose intensities are directly related to grain volumes. Compared with X-ray methods, which are also used for statistical, but for surface rather than volume texture measurements, the penetration capability of neutrons is much higher [56, 57]. Therefore, large samples can be analyzed (1–$8\,cm^3$ in the case of the samples described below), and statistically meaningful results can be obtained also for coarse-grained samples [58, 59], samples with an inhomogeneous distribution of at least one out of several mineral phases, samples with different microstructural domains contributing to the overall texture, and mineral phases that have only a weak texture.

On the other hand, the experimental setup does not allow the lattice orientation of single grains in a deformed rock to be determined, which can be done by methods like EBSD. Combining both methods will therefore allow the most complete analysis of textures and microstructures.

10.1.4 Experimental Setup and Data Processing

The samples have been analyzed at the texture diffractometer SV7-b (see detailed description by E. Jansen et al. [60]), which was operated at the FRJ2 reactor of Forschungszentrum Jülich. The experiments were carried out with a monochromatic thermal neutron beam. Complete diffraction patterns were automatically scanned by a position-sensitive scintillation detector for about 500 orientations per sample. After correction for geometrical and instrumental aberrations, the diffraction peak profiles were analyzed by a flexible decomposition of selected peak clusters in order to determine orientation-dependent integrated intensities for pole-figure representations [61].

From the selected experimental pole figures, the orientation distribution function (ODF) was calculated using the WIMV method [62, 63]. The ODF represents the texture of a sample as a function of three variables, which are usually Euler angles. Rotation of an internal crystal coordinate system (e.g., the crystallographic axes) by these Euler angles around the axes of an external sample coordinate system (e.g., the principal axes of strain, see Section 10.2) will bring both coordinate systems into parallelism. There are two main reasons to calculate an ODF from the experimentally observed pole figures of different crystallographic directions. (1) In geosciences, pole figures are certainly the most common among various representations of the ODF; but other representations (for an overview see, e.g., [64]), such as inverse pole figures (Section 10.4) or σ sections [65, 66], may be favored for special purposes. (2) The pole figure (normal and inverse) of any desired crystallographic direction hkl can be calculated by integration over hkl-dependent lines of the orientation space. This is especially important for those lattice planes that cannot directly be measured owing to weak reflections or peak overlap. This is the case for the {001} pole figures of quartz, which we recalculated from the ODF by using the PCAL program (contained in the BEARTEX program package [67]).

The quality of the ODF solution can be assessed by comparing observed pole figures with the respective pole figures recalculated from the ODF. A numerical

Table 10.1 RP0, RP1, and F2 values for the textures of the samples described in Sections 10.4 and 10.5

Sample no.	Deformation phase	Average RP0 [%]	Average RP1 [%]	F2
MR48	D_1	11.9140	8.0971	8.0706
CK2	D_1	6.3297	4.7959	2.7196
B42	D_1	7.8919	5.0957	6.6235
GP40	D_1	6.7618	5.4616	4.5147
GP38	D_1	8.4443	6.3534	2.5915
B43	D_1	7.2480	4.9225	4.3161
MR146	D_1	8.5595	6.5489	5.9051
MR178	D_1	15.5724	11.3740	13.1330
MR199	D_1	9.2631	7.2822	4.6856
MR167	D_2	12.1862	7.7058	11.6049
CB32	D_2	6.9562	5.4056	2.9795
CB33	D_2	12.0431	10.2223	1.3033
MR204	D_2	15.8218	8.9983	3.1884
MR186	D_3	17.4391	11.1221	17.7002
CB14	D_3	8.8495	7.2344	1.4515
GP41	D_3	12.1088	9.0020	3.0458
JA17	D_3	6.5298	5.0091	4.4689
MR195	D_3	7.2305	5.4503	2.9780
DHM3	D_3	8.0661	6.1522	3.6167
JD20		16.1598	5.2585	8.3997
JD47		21.7386	7.5890	21.4630
JD76A		15.9481	7.6281	15.0155
JD14		21.4443	15.6036	34.5064
JD107		15.9742	6.8038	13.5124
JD103		7.5056	6.0787	4.4190

description of the discrepancy between a pair of pole figures is given by the RP0 and RP1 values [68]. These values can be thought of as the mean deviation between observed and recalculated pole figure relative to the observed pole figure for ODF values less than and greater than 1 (random), respectively. For the textures shown in our first case study (Section 10.4), RP0 and RP1 mostly range between 5% and 10% (see Table 10.1), which may be regarded as satisfactory [68]. Samples shown in our second case study mostly have values greater than 10% (Section 10.5). These are attributed to exceptionally strong textures with very sharp peaks as quantified by large texture indices F2 [69].

10.2 Representation and Interpretation of Textures

The textures shown in the following case studies mostly serve to estimate the geometry of deformation and the degree of rotation of the principal axes of strain during deformation. In homogeneous deformation, any imaginary sphere is transformed into a strain ellipsoid with the principal axes X, Y, and Z, where X is the direction of maximum elongation, Z the direction of maximum shortening, and Y the normal to X and Z (Fig. 10.1). The X and Z directions can be recognized on a deformed rock as the stretching lineation and the normal to the foliation, respectively. X, Y, and Z define the coordinate system for the pole-figure representation of textures with X and Z lying to the right and top of the pole figures and Y in the center. Inverse pole

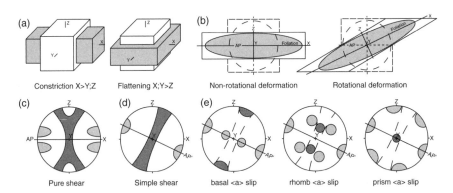

Fig. 10.1 (a) Schematic illustration of deformation of an imaginary cube leading to constrictional strain (*left*) and flattening strain (*right*). (b) Schematic illustration of two-dimensional non-rotational (pure shear, *left*) and dextral rotational (simple shear, *right*) deformation. Note that progressive simple shear will lead to a rotation of X toward parallelism with the attractor plane (AP). (c) Typical {c} crossed girdle (*dark gray*) and {a} (*light gray*) distributions of quartz textures formed by pure shear. The texture has an orthorhombic symmetry. (d) Typical {c} single girdle (*dark gray*) and {a} (*light gray*) distributions of quartz textures with monoclinic symmetry formed by dextral simple shear. The texture has a monoclinic symmetry with the {a} maximum and Y defining an apparent attractor plane (AP', see text for explanation). (e) Correspondence between active slip system of quartz and positions of {c} (*dark gray*) and {a} contributing to the {c} single-girdle texture shown in (d) (see text for explanation)

figures differ from normal pole figures in that crystallographic directions serve as a reference frame into which orientations referring to the external sample coordinate system are plotted.

The geometry of strain resulting from homogeneous deformation can be described by the values of X, Y, and Z. If these are normalized to the initial radius of the imaginary sphere and deformation was volume-constant, values <1 indicate shortening and values >1 indicate stretching in the respective directions. For homogeneous deformation without volume change, three general cases of strain geometry may be distinguished: (1) constriction with X > 1 and Y&Z < 1 (schematically illustrated in Fig. 10.1a); (2) plane strain with X > 1, Y = 1, and Z = 1/X < 1; and (3) flattening with X&Y > 1 and Z < 1.

The kinematical path may or may not involve rotation. In non-rotational deformation (Fig. 10.1b), the orientations of X, Y, and Z remain constant. For plane strain, such deformation is termed "pure shear" as opposed to simple shear. In the case of simple shear, any imaginary line will rotate according to the shear sense, except for lines parallel to the shear zone boundary whose orientation remains stable. The streamlines (particle paths) of the flow field are all parallel. Therefore, in the case of simple shear, the orientation of the streamlines can be considered as the shear direction and the plane containing the shear direction and Y can be considered as the shear plane. In non-simple-shear homogeneous deformation, a line orientation exists, which is non-rotational with respect to the instantaneous stretching and shortening directions (the fabric attractor of C. W. Passchier [70]). The streamlines converge on the fabric attractor following hyperbolic paths. Thus, there is strictly speaking no shear direction and no shear plane. Therefore, we will refer to the plane containing the fabric attractor and Y more generally as the attractor plane. In the case of non-rotational deformation, X is parallel with the fabric attractor and the foliation is parallel with the attractor plane. In the case of rotational deformation, X and the foliation plane progressively rotate toward the fabric attractor and attractor plane, respectively, without theoretically ever reaching it.

As a consequence, textures developed out of random distribution by non-rotational deformation will ideally yield pole figures with orthorhombic symmetry, while at least for many minerals, including quartz, those developed by rotational deformation will ideally yield pole figures with monoclinic symmetry. Apart from the kinematical path and the strain geometry, the geometry of a texture is determined by the crystal symmetry of a mineral and the active slip systems. In general, several different slip systems have to act simultaneously in order to maintain compatibility at the grain boundaries, and grains will never attain stable orientations during deformation (e.g., [66]). Depending on their orientation with respect to the overall stress field, a population of some grains in a rock volume will accommodate deformation by activity of a certain slip system while a different slip system will act in grains with different orientation. Also, several slip systems may act simultaneously within a grain. Although the number of favorable grain orientations and appropriate slip systems is usually limited, the texture will show a superposition of the favorable grain orientations.

In the following, we will present some particular aspects of quartz texture formation as a prerequisite for the following case studies (Sections 10.4 and 10.5). Note that while only crystallographic planes can be measured by goniometric methods, information about the orientation of crystallographic axes is required in order to discuss pole figures in terms of possible slip directions. For quartz, the most important slip directions <a> and <c> are equivalent to the plane normals {a} and {c}, which are shown in the pole figures. Figure 10.1c and d shows two typical arrangements of quartz c and a axes resulting from pure shear and simple shear, respectively. The simple shear texture example is characterized by a {c} (= {001}) girdle distribution and a prominent {a} (= {110}) maximum at the pole to the {c} girdle (Fig. 10.1d). Such textures are interpreted as the result of combined basal <a> (i.e., {c} is the slip plane and <a> the Burgers vector), rhomb <a>, and prism <a> slip, which are considered to be the most common slip systems under greenschist and lower amphibolite-facies conditions (e.g., [12]). Figure 10.1e shows which one of these slip systems will most probably be active depending on the orientation of crystals. In general, the absolute {a} maximum of a quartz texture will not coincide with the bulk fabric attractor of the rock because different grain orientations will lead to independent {a} maxima or an absolute {a} maximum that results from the superposition of the <a> directions of different grain populations. Thereby, an apparent attractor plane may be defined by an absolute {a} maximum (i.e., the apparent fabric attractor) and Y (Fig. 10.1d and e). This attractor plane is only an apparent one because the angle between an absolute {a} maximum and X may become larger or smaller than the angle between the fabric attractor of the rock and X depending on whether different slip systems act antithetically or synthetically, respectively [13]. In such cases, only the sense of shear (dextral in the simple-shear example of Fig. 10.1d) but not its magnitude can be derived from the texture, and the same is true for rotational deformation with a pure-shear component. For example, the {c} crossed girdle texture shown in Fig. 10.1c can be interpreted in terms of two straight single girdles crossing each other in Y. Both partials of the {c} crossed girdle correspond to an {a} maximum at about 25° to X, but there is no overall rotation.

As stated in the introduction, slip systems are defined by a slip plane and a slip direction. The ease of each slip system depends generally (i.e., not only for quartz) on the orientation of slip direction and slip plane with respect to the applied stress field. For each slip system, the easiest orientation is if the slip direction is at a small angle with X and close to the fabric attractor of the rock, thus perpendicular to Y, and if the slip plane contains the Y direction, that is, the bulk rotation axis of the rock. For simple-shear deformation of quartz, this has been shown by R. D. Law et al. [71]. For example, if basal <a>, rhomb <a>, and prism <a> slip act simultaneously in quartz and all of these slip systems are equally favored by the physical conditions during deformation, basal <a> slip will be the easiest slip system for grains with [c] perpendicular to Y, prism <a> slip for [c] parallel to Y, and rhomb <a> slip for intermediate orientations (Fig. 10.1e). If several slip systems act simultaneously, inverse pole figure plots of Y are therefore especially well suited to visualize how the orientation of grains deformed by a certain slip system contributes to the texture. Examples of inverse pole figures will be shown in Section 10.4.

10.3 An Overview of Alpine Geology

The Alps (Fig. 10.2) are an assembly of numerous rock units that display structural, sedimentological, petrological, and geochemical evidence for a complex orogenic evolution. This evolution is not documented by complete evidence, though, and attempts at modeling certain aspects of the Alpine orogeny by consistent reconciliation of the existing data often lead to controversial results. In the following, we will first give a short overview of the Alpine geology in general. For more comprehensive descriptions, which largely accord with what we describe in the following, we refer the reader to [72–74]. Note, however, that different scenarios for the Alpine orogeny have been proposed (for the Western and Central Alps, e.g., [75–77])

Alpine tectonics was largely determined by the southward subduction of several continental and Mesozoic oceanic ("Tethyan") domains, except for the Southern Alpine zone (Fig. 10.2), which always remained in an upper-plate position. Subduction processes started in the Middle Jurassic and led to the more or less progressive accretion of progressively external rock units forming a nappe stack. In a generalized view, the structural level of a rock unit therefore depends on its paleogeographical origin, i.e. the more to the northwest the paleogeographical origin of a nappe was, the deeper this unit is situated in the nappe stack (e.g., [78]). The oldest ocean whose traces can be found in the Alps was the Meliata–Hallstatt ocean, which, after spreading in the Middle Triassic, was subducted during the Jurassic [79]. When the Piemont–Ligurian ocean opened further to the northwest in the Bathonian to

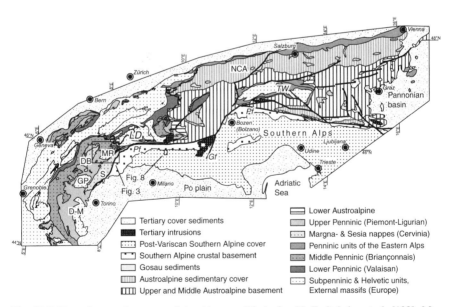

Fig. 10.2 Tectonic overview map of the Alps, modified after N. Froitzheim et al. [138]. Map areas of Figs. 10.3 and 10.9 are marked by *rectangles*. DB = Dent Blanche nappe; D − M = Dora–Maira nappe; Gf = Giudicarie fault; GP = Gran Paradiso nappe; LD = Lepontine dome; MR = Monte Rosa nappe; NCA = Northern Calcareous Alps; Pf = Periadriatic fault; S = Sesia zone; TW = Tauern window

Tithonian (see compilation of radiometric protolith ages from oceanic metabasic rocks by A. Liati et al. [80]), a northeastern promontory of the Adriatic microplate was situated between the two oceanic basins. Rock units derived from this promontory make up the present-day Austroalpine zone of the Alps (Fig. 10.2). After the Meliata–Hallstatt ocean was completely closed, opening of a third oceanic basin (Valaisan ocean) began in the Early Cretaceous [81]. The Piemont–Ligurian and Valaisan oceans formed the two branches of the Alpine Tethys south and north of the Iberia–Briançonnais microcontinent, respectively. Toward the east, rock units derived from the Briançonnais peninsula cannot be traced into the Tauern window (Fig. 10.2), indicating that the Piemont–Ligurian and Valaisan oceans merged toward the east into one single ocean. The most external units incorporated into the Alpine orogen are of European origin, that is, from northwest of the Valaisan ocean. Rock units originating from the Piemont–Ligurian, Briançonnais, and Valaisan domains as well as European basement nappes build up the Penninic zone of the Alps. The Helvetic zone comprises Permo–Mesozoic cover nappes of the former European continental shelf. The structurally lowest position is occupied by the external European basement massifs and their sedimentary cover.

10.3.1 The Penninic Zone

The Penninic zone of the Alps is commonly subdivided according to the paleogeographical origin of the rock units into Upper Penninic (Piemont–Ligurian oceanic domain), Middle Penninic (Briançonnais), Lower Penninic (Valaisan oceanic domain), and Subpenninic (European basement and its autochthonous cover). Sedimentological evidence from the most distal parts of the Austroalpine thrust wedge indicates that subduction of the Penninic domain (or, more specifically, the Piemont–Ligurian ocean) probably started already in the late Aptian/Albian [82]. Most of the Penninic high-pressure units experienced peak-metamorphic conditions only in the Eocene between around 45 and 35 Ma.

In the Swiss Central Alps, Subpenninic and Lower Penninic nappes are exposed over a large area in the core of the Lepontine dome (see [83, 84]). In the western flank of the Lepontine dome, the geometry of the Penninic nappe stack is complex because of intense large-scale folding after the primary nappe emplacement. The Dora–Maira, Gran Paradiso, and Monte Rosa nappes are often regarded as Middle Penninic (e.g., [85–87]). Other authors, by contrast, advocated a Subpenninic origin of these units [88–90]. The strongest argument for the latter opinion is that the Monte Rosa nappe is directly overlain by eclogites with magmatic protolith ages of 93.4 ± 1.7 Ma [91], which fit the Barrêmian to Cenomanian spreading ages of the Valaisan ocean rather than the Bathonian to Tithonian ones of the Piemont–Ligurian ocean (compare Section 10.3). In the cross-section through the Monte Rosa area, the Zermatt–Saas zone and the St. Bernard nappe system undoubtedly represent the Upper and Middle Penninic, respectively. In Section 10.4, we will discuss the textures of mylonitized quartzites that, together with results from geological fieldwork, geochronological data, and petrological data, have been used to develop a kinematical model for that region [92].

10.3.2 The Southern Alpine Zone

The Southern Alps are separated from the other zones of the Alps by the Periadriatic fault. Alpine regional metamorphism is largely absent or only anchizonal [93], except for a narrow greenschist-facies mylonite belt in the immediate vicinity of the Insubric fault [94]. During the Alpine orogeny, a south-verging thrust belt developed with the individual thrust sheets bounded by distinct décollement horizons [95–98]. The age of South Alpine thrusting is Late Cretaceous to Miocene [97]. Thanks to the low grade of Alpine metamorphism and the strong localization of Alpine deformation, the thrust sheets widely preserve pre-Alpine structural and lithological rock characteristics. For the Southern Alps west of the Giudicarie line, discussed in Section 10.5, this allows identification of the pre-Alpine history as follows (see [99, 100] for more comprehensive reviews): The crystalline basement of the Southern Alps exposes various crustal levels testifying to Variscan and pre-Variscan deformation typically under greenschist- to high amphibolite-facies conditions [101, 102]. A Permian event of combined magmatic activity and crustal extension is well constrained by radiometric data and is also documented in the sedimentary record of evolving basin structures like the Collio basin [103–107]. Further extensional detachment faulting was active from the Late Triassic to the Middle Jurassic [108] during rifting stages predating the opening of the Piemont–Ligurian domain north of the Adriatic micro-continent (i.e., the future Southern Alps and Austroalpine nappes).

10.4 Quartz Textures as Kinematic Indicators for the Evolution of the Penninic Nappe Stack in the Monte Rosa Area

10.4.1 Geology of the Monte Rosa Area

In our first case study, we will present an analysis of the kinematical history of the Monte Rosa nappe and neighboring units in the boundary region between the Western and the Central Alps (see also [92]). The kinematical analysis consists in a stepwise retrodeformation of structures in a cross-section through this area in which quartzite textures have been used as kinematical indicators. As a precondition for the stepwise retrodeformation, we assume that the Monte Rosa nappe is not Middle Penninic, as often stated (e.g., [86, 109, 110]), but Subpenninic according to recent findings [90, 91] that the Valaisan suture is above and to the south of the Monte Rosa nappe. Because of complex folding, however, the Lower Penninic Antrona unit is largely situated below the Monte Rosa nappe where it separates the latter from the next lower Subpenninic unit (Camughera–Moncucco unit; Figs. 10.3 and 10.4).

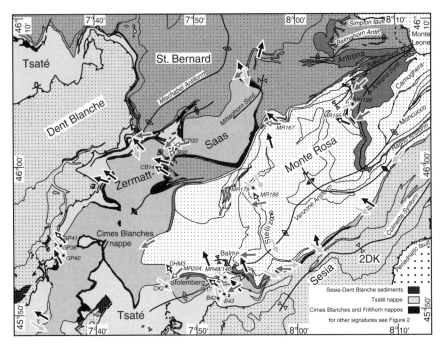

Fig. 10.3 Tectonic map of the Monte Rosa area, modified after A. Steck et al. [110]. Fold axial traces are marked according to the folds' relative ages by *black* (D$_1$), *dark gray* (D$_2$), *light gray* (D$_3$), and *white* (D$_4$) *triangles. Arrows* indicate shear senses as the hanging wall displacement directions of D$_1$ (*black*), D$_2$ (*dark gray*), and D$_3$ (*light gray*). *Arrows* with dots are derived from the textures shown in Figs. 10.4 and 10.8

Apart from sporadic Lower and Middle Penninic slivers, the next higher units above the Monte Rosa nappe are the St. Bernard nappe system (Middle Penninic) and the Zermatt–Saas zone (Upper Penninic). The Camughera–Moncucco unit, Antrona unit, and Monte Rosa nappe experienced similar metamorphic peak conditions at around 1.2–1.6 GPa/500–700°C [111–115]. For the Monte Rosa nappe, the eclogite-facies metamorphism was dated at 42.6 ± 0.6 Ma [116]. In the Zermatt–Saas zone, the metamorphic peak took place slightly earlier (44.1 ± 0.7 Ma [117]), but reached considerably higher pressures of 2.5–3.0 GPa at 550–600°C [118, 119]. Above the Zermatt–Saas zone there is a marked jump in peak pressures and temperatures toward the Combin zone (1.3–1.8 GPa/380–550°C [120]). While the age of metamorphism in the Combin zone is unknown, eclogite-facies metamorphism in the overlying Sesia–Dent Blanche nappe system was dated at 60–70 Ma [121, 122].

Several authors (e.g., [123–125]) explained the metamorphic gap between the Combin and the Zermatt–Saas zones and the exhumation of the underlying Zermatt–Saas unit by large-scale extensional faulting (Combin fault, Fig. 10.4b). One focus of our study will therefore be the Cimes Blanches nappe at the base of the Combin zone, which consists of Permo–Mesozoic metasediments [109, 126], among them strongly mylonitized quartzites. A second focus is the Stellihorn shear zone, which

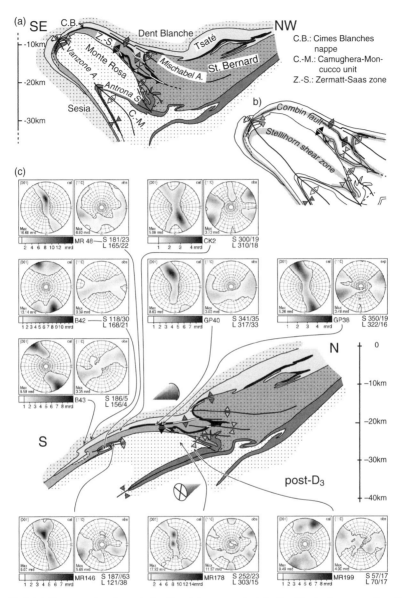

Fig. 10.4 (**a**) Cross-section through the Penninic nappe stack in the Monte Rosa area (modified after A. Escher et al. [109]) after retrodeformation of post-D$_4$ structures. See Figs. 10.2 and 10.3 for legend. Fold axial traces are marked according to the folds' relative ages by *black* (D$_1$), *dark gray* (D$_2$), *light gray* (D$_3$), and *white* (D$_4$) *triangles*. (**b**) Positions of the Combin fault and the Stellihorn shear zone in the cross-section. (**c**) Reconstructed post-D$_3$ cross-section shown together with {c} and {a} pole figures of D$_3$ quartz textures in upper-hemisphere stereographic projection (obs: observed; cal: calculated from the ODF; mrd: multiples of random distribution). Sample orientations are given by the foliation (S) and stretching lineation (L) below the {a} pole figures. Cones illustrate the overall D$_3$ shear sense. Note that the shear directions of D$_1$, D$_2$, and D$_3$ are not within the plane of the cross-section. See also Fig. 10.3 for sample localities

crosses the Monte Rosa nappe from north to south and divides it into a lower part dominated by orthogneiss and an upper part dominated by paragneiss and mica-schists (Fig. 10.4b). Additional samples for neutron texture goniometry were mostly taken from other parts of the Monte Rosa nappe, the Zermatt–Saas zone, and the Sesia–Dent Blanche nappe system. All samples, except for those from the Cimes Blanches nappe, are mylonitized foliation-parallel quartz veins from gneisses or metabasic rocks.

The macroscale structures within the study area are regionally quite consistent and can be ascribed to four successive deformation phases (see also [127, 128] for comparison): (1) northwest-verging shearing with only minor and small-scale fold-ing (D$_1$); (2) southwest-verging shearing with formation of large-scale southeast-verging folds (D$_2$); (3) southeast-verging shearing with formation of large-scale folds (D$_3$); and (4) formation of the Vanzone fold (D$_4$). The fact that all these defor-mation phases are documented in the structural record is due to heterogeneous strain distribution. Age relations between the deformation phases have been established in the field by overprinting relations of the respective structures, mostly folds refolding older stretching lineations or folds. The main foliation is in most places a composite one, and foliations that can clearly be ascribed to one of the deformation phases are restricted to fold hinges. Apart from the textures, kinematics indicators such as shear bands and asymmetric porphyroclasts have been used to determine shear senses in the field and in thin section. While overprinting relations of structures are needed

Fig. 10.5 D$_1$ stretching lineation, illustrated by a whitish mylonitized quartzitic vein within green-schist of the Combin zone, at the sampling site of JA17 (compare Fig. 10.8b; Italian Gauβ–Boaga coordinates 5079750/1412450). View is toward the northwest

to establish the relative ages of these deformation phases, we will in the following show only textures that in each case are in accord with a stretching lineation on the sample that can be ascribed to one of the deformation phases (Fig. 10.5). We do not show textures with "triclinic" geometry that probably result from a superposition of deformation phases because too little, if anything, is known about how to interpret such textures.

D_1, D_2, D_3, and D_4 structures postdate eclogite-facies metamorphism in all units. They developed under greenschist- to amphibolite-facies retrograde conditions. The boundary between greenschist- (above) and amphibolite-facies metamorphism (below) roughly corresponds with the Stellihorn shear zone. Pre-D_1 structures, that is, pre-Alpine and eclogite-facies structures, are too restricted to recognize regional deformational patterns.

10.4.2 D_3 Textures

The first step of the retrodeformation arrives at a post-D_3 situation (Fig. 10.4c) by unfolding of the Vanzone antiform (Fig. 10.4a). All quartzite textures formed during D_3 (Fig. 10.4c) have nearly monoclinic symmetries, and shear senses can be deduced both from the {a} and {c} pole figures. Top-to-the-southeast shear senses are indicated by the strongest point maxima or girdle distributions about 20–30° away from the Z direction ({c}) and the X direction ({a}) in an anticlockwise sense. Only for MR199 the pole figures indicate dextral rotation (top-to-the-northeast shear sense), because the stretching lineation of this sample dips to the northeast instead of southeast. Apart from the consistent shear sense, the D_3 textures are quite variable. {c} pole figures from the Combin fault and the Stellihorn shear zone show crossed girdles (CK2) or kinked single girdles with a more or less strong tendency to form crossed girdles. As compared with straight single girdles that are usually assumed to indicate simple shear (see Section 10.2), such {c} patterns testify a certain non-rotational component of shear [12]. This is also supported by the presence of second {a} maxima, which, compared with the respective strongest {a} maximum, also lie at the margins of the pole figures but in the opposite direction from X. Such {c} distributions can most readily be explained by antithetically acting slip systems on the grain scale.

In pole figures of quartzite textures from the uppermost part of the Tsaté nappe close to the base of the Sesia zone, {c} forms point maxima at the periphery. In B42, {c} forms a point maximum around Z containing the absolute {c} maximum and a less strongly developed sub-maximum around 20° away from Z in an anticlockwise and clockwise sense, respectively. These sub-maxima are perpendicular to {a} girdles, the stronger of which corresponds to the stronger {c} sub-maximum.

In an XZ thin-section photograph of B42 under crossed polarizers and with gypsum plate (Fig. 10.6a), quartz grain shapes indicate opposite shear senses in domains with sinistral, southeast-verging and dextral, northwest-verging shear senses. Yellow

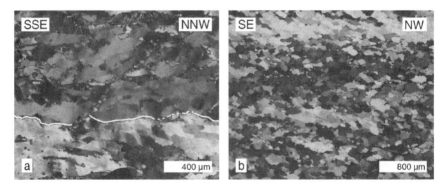

Fig. 10.6 XZ thin sections of mylonitized quartz veins from the upper part of the Tsaté nappe close to the boundary to the Sesia zone under crossed polarizers and gypsum plate. See Fig. 10.3 for sample locations. (**a**) Detail of B42 showing domains with opposite dextral and sinistral shear directions by different shape-preferred grain orientations and blue addition (*darker gray*) and yellow subtraction (*lighter gray*) colors above and below the white line, respectively. (**b**) Detail of B43 showing foliation-parallel domains comprising either mostly dextrally sheared grains (*darker gray*) or mostly sinistrally sheared grains (*light gray*)

subtraction colors (light gray in a) prevail in the domains with sinistral shear sense because <c>, that is, the positive optical axis of quartz, is preferentially nearly perpendicular to the positive optical axis of the gypsum plate trending from top right to bottom left. In dextrally sheared domains, {c} trends from top right to bottom left, resulting in blue addition colors (dark gray in a). These domains also show different shape-preferred orientations that indicate the same shear sense as the addition and subtraction colors.

In addition to normal pole figures, inverse pole figures of some commonly plotted directions of the external sample coordinate system are shown in Fig. 10.7 in order to assess the usefulness of such representations. Inverse pole figures may have advantages over normal pole figures concerning the illustration of the active slip systems during deformation of a rock sample. However, since inverse pole figures are only integrals over a certain direction in the orientation space of the ODF, they do not fully describe the texture of a sample. Therefore, care should be taken when choosing the direction that is to be plotted. For example, the Z and X inverse pole figures of sample B42 (Fig. 10.7a) mainly visualize the angle of about 20° between Z and the absolute {c} maximum and X and the absolute {a} maximum but are in no respect more demonstrative than the {c}, {m}, and {a} normal pole figures (Fig. 10.7d). Inverse pole figures of the normal to the apparent attractor plane and of the apparent fabric attractor of the sample (labeled A and B, respectively, in Fig. 10.7d) are potentially well suited to illustrate active slip systems (e.g., [12]) if the slip plane normal and the slip direction align with A and B, respectively. The A and B inverse pole figures of B42 show that A aligns preferentially with [c] and that B is almost evenly distributed within the [c] plane. Taken together, this suggests basal <a> and basal <m> slip. However, since the latter is not one of the common slip systems of quartz, it is much more probable that the equal distribution of B in

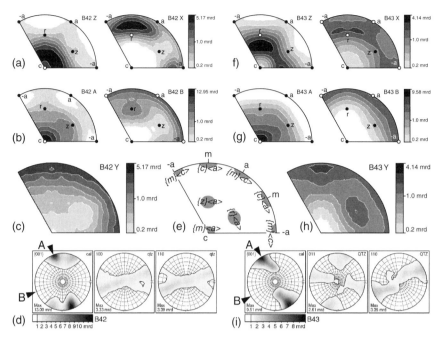

Fig. 10.7 (**a**) Inverse pole figures of Z and X of sample B42. The angles between [c] and the Z maximum and between <a> and the X maximum correspond to the angles between A and Z and B and X, respectively (d). (**b**) Inverse pole figures of A and B of sample B42 indicating dominant basal <a> slip. (**c**) Inverse pole figure of Y of sample B42. The almost even Y distribution perpendicular to [c] is the result of basal <a> slip and an almost even distribution of the rotation axis of the basal <a> slip system within the XY plane. Prism <c> slip can be ruled out since the {c} plane normals are at too high an angle with X (compare d) and therefore not in favorable orientations to accommodate prism <c> slip. (**d**) Normal {c}, {m}, and {a} pole figures of B42 in upper-hemisphere stereographic projection. A indicates the normal to the apparent attractor plane, B is parallel to the apparent fabric attractor. (**e**) Schematic inverse pole figure showing orientation of crystallographic directions defining the reference frame and how active slip systems and Y maxima will be related (see also [139]) if the rotation axes of active slip systems are aligned with Y. {r} = {101}, {z} = {011}. (**f**) Inverse pole figures of Z and X of sample B43. The angles between [c] and the Z maximum and between <a> and the X maximum correspond to the angles between A and Z and B and X, respectively (i). (**g**) Inverse pole figures of A and B of sample B43 indicating dominant basal <a> slip. (**h**) Inverse pole figure of Y of sample B43. The position of the Y maxima indicates combined basal <a> and rhomb <a> slip. (**i**) Normal {c}, {r + z}, and {a} pole figures of B42 in upper-hemisphere stereographic projection. Note that the {r + z} maximum close to Z represents rhomb planes accommodating rhomb <a> slip. A indicates the normal to the apparent attractor plane; B is parallel to the apparent fabric attractor

the [c] plane is due to exclusive activity of basal <a> slip with <a> almost evenly distributed within the attractor plane.

A shortcoming of A and B inverse pole figures is that they are only suited to visualize the active slip system during texture formation if the slip planes and slip directions of the crystallites are systematically arranged within a single attractor plane and fabric attractor. Typical examples of this are quartz textures with {c} point maxima or {c} single girdles where prism, rhomb, and basal planes are arranged in the attractor plane [12]. In the case of B42, this condition is not fulfilled because the two superimposed great circles over which {a} is distributed in the normal pole figure (Fig. 10.7d) are most readily interpreted as tracing different attractor planes of the antithetic domains of shear that can be identified in thin section. A relatively strong density of A between [c] and {z} and a B maximum around {r} in the A and B inverse pole figures do not indicate activity of a slip system other than basal <a>, but reflect the angle between the fabric attractor planes of the two antithetic shear domains, which is about 40° (note that the angle between the {r} and {m} plane normals is 38.2°).

The most appropriate way to assess active slip systems is by inverse pole figures of Y. Since Y is perpendicular to both A and B, it has the same orientation for antithetic domains of shear like in B42. More generally, inverse pole figures of Y are useful in all cases where the rotation axes of individual crystallites (which depend on the active slip system) have a preference to align with the bulk rotation axis of the rock (Y). Figure 10.7e shows the most likely positions of Y maxima depending on the active slip system. The degree of alignment of Y with the shown directions corresponds to the degree of alignment of the slip system rotation axes within Y (compare Section 10.2). In the inverse pole figure of B42 (Fig. 10.7c), Y is almost equally distributed on a great circle around {c}. The inverse pole figure thus suggests a combination of basal <a> and prism <c> slip. Looking at the normal pole figure, however, prism <c> slip can be ruled out because if {100} were the active slip plane, the {100} poles should be arranged somewhere on a great circle around Y, which they are not (Fig. 10.7d). Instead, the position of the {c} maximum at the periphery of the normal pole figure suggests that basal <a> slip was the dominant slip system. The distribution of Y in the inverse pole figure is therefore best explained in that the slip system rotation axes are distributed over the attractor planes of the antithetic slip domains with a rather weak tendency to align within Y. Taking these observations together, it is most likely that the sample consists of a conjugate set of ribbons in each of which the kinematical path only had a weak, if any, degree of non-rotational deformation. Since the bulk strain on the sample scale is a combination of the strain in the conjugate ribbons, it has a strong non-rotational component. By total volume, however, domains with sinistral top-to-the-southeast shear sense, corresponding to the absolute {c} maximum in the normal pole figure, dominate over those with dextral northwest-verging shear sense corresponding to the {c} sub-maximum. The distribution of Y in the inverse pole figure suggests overall flattening strain.

Although the normal pole figures of B43 (Fig. 10.7i) look quite different from those of B42, the texture of B43 can be explained in a very similar way. The absolute maximum of the {c} pole figure is at the periphery at a distance of about 20° to Z in an anticlockwise sense. From there, appendices extend toward an intermediate direction between Y and Z. {a} is aligned on a girdle roughly perpendicular to the absolute {c} maximum. The inverse pole figure of Y (Fig. 10.7h) suggests the combined activity of basal <a> and rhomb <a> slip while the Z and X (Fig. 10.7f) and A and B (Fig. 10.7g) inverse pole figures hardly display evidence for rhomb <a> slip. Rhomb <a> slip is in line with the position of the absolute maximum of the {101/011} poles at the periphery and close to Z in the normal pole figure. The fact that this maximum is rotated in clockwise sense from Z by a small angle, that is, opposite to the {c} maximum, indicates that basal <a> slip and rhomb <a> slip operated antithetically. This interpretation is confirmed by thin-section observations showing both sinistral (top-to-the-southeast) and dextral (top-to-the-northwest) shear senses partitioned into foliation-parallel domains, although not as distinct as in B42. While there are no obvious indications of flattening strain like in the case of B42, a strong non-rotational component for the bulk strain is evidenced also for B43 by the antithetic attitude of basal <a> and rhomb <a> slip. For one sample from the base of the Monte Rosa nappe close to the Antrona zone (MR199), the same interpretation in terms of strain geometry and rotational degree of the kinematical path may apply as for B42 because the {c} and {a} pole figures of these two samples are very similar.

10.4.3 D₂ Textures

The next step of the retrodeformation is a restoration of the post-D$_2$ nappe stack geometry. It is shown together with D$_2$ textures in Fig. 10.8a. The texture of CB33 from the Combin fault is characterized by {c} distributed on a small circle including an angle of approximately 30° with Z. {a} spreads over a broad girdle along the XY plane, which is made up of two sub-girdles lying on small circles at a high angle around Z. These two {a} small girdles represent the highest densities of intersections of (c) planes (i.e., the planes containing {a}) with {c} having a small circle distribution. The fact that the {c} and {a} pole figures are nearly rotationally symmetric with respect to the Z direction allows for two conclusions regarding the kinematical path on the one hand and the finite strain geometry on the other hand. The {c} and {a} pole figures are mirror-symmetric with respect to the XY plane; in other words, they show no obliquity with respect to X and Z, and therefore testify to a non-rotational kinematical path. Since X and Y appear to be equivalent, strain must have been flattening if volume-constant deformation is assumed. The second texture from the Combin fault, CB32, is characterized by an incomplete {c} crossed girdle with the right half almost missing. Although the reason for this distribution is unclear, non-rotational deformation must also be inferred for CB32 since there is no obliquity of the {c} and {a} pole figures with respect to X and Z. D$_2$ textures from the Stellihorn shear zone are markedly different in that {c} is distributed over incomplete single girdles whose inclination with respect to X and Z indicates a strongly rotational kinematical path with top-to-the-west (MR167) and top-to-the-southwest sense of shear. Absolute {c} maxima close to Y suggest somewhat elevated temperatures since they indicate prism <c> slip, which preferentially acts under higher temperatures than basal <a> and rhomb <a> slip (e.g., [4, 129]).

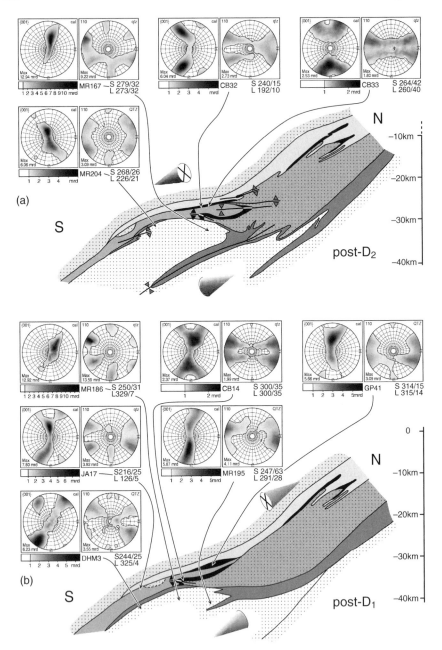

Fig. 10.8 (**a**) Reconstruction of the post-D$_2$ geometry of the Penninic nappe stack in the Monte Rosa area and {c} and {a} pole figures of D$_2$ textures in upper-hemisphere stereographic projection (obs: observed; cal: calculated from the ODF; mrd: multiples of random distribution). (**b**) Reconstruction of the post-D$_1$ geometry of the Penninic nappe stack and {c} and {a} pole figures of D$_1$ textures in upper-hemisphere stereographic projection. See also Fig. 10.3 for sample localities and Figs. 10.2 and 10.3 for legend. Fold axial traces are marked according to the folds' relative ages by *black* (D$_1$) and *dark gray* (D$_2$) *triangles*

10.4.4 D_1 Textures

With the last step of retrodeformation, a reconstruction of the post-D_1 nappe stack geometry is achieved (Fig. 10.8b). Textures from the Combin fault (CB14, GP41, JA17) are characterized by {c} crossed girdles, although sometimes incomplete ones, with little or no indication of a rotational kinematical component. In the Monte Rosa nappe (MR186, DHM3, MR195), northwest-verging shearing of D_1 is well documented by inclined {c} single girdles and {a} maxima perpendicular to {c}.

10.4.5 Retrodeformation of D_1, D_2, and D_3 Structures in a Cross-Section Through the Monte Rosa Area

The analysis of D_1, D_2, and D_3 quartz textures shows variable strain geometries and kinematical paths for the different structural levels of the nappe stack and the different deformation phases. During D_3, deformation with a considerable rotational component took place in the Stellihorn shear zone and at the Combin fault. D_3 extension in the Combin fault therefore certainly contributed significantly to the exhumation of the underlying units. However, considering that all the respective textures also indicate evidence of a non-rotational kinematical component and that D_3 deformation was not pervasive, as testified by the preservation of textures related to older deformation phases, the amount of D_3 relative movement between the Combin and the Zermatt retro Saas zones cannot have been large enough to explain the pressure gap between these two units. Assuming a pressure gap of 1.2 GPa, an average rock density of 3.0 g/cm^3, lithostatic pressure, and a 30° dip of the normal fault, the displacement amount would have to be approximately 80 km. A more realistic, albeit also crude, estimate of D_3 displacement amount at the Combin fault is to sum up the lengths of the lower limbs of D_3 folds (Fig. 10.4a), whose formation was mostly accommodated by top-to-the-southeast shearing in these limbs while the rock portions between them experienced only little deformation [130]. This suggests about 10–20 km of displacement and can be reconciled with the fact that already D_1 and D_2 took place under similar P–T conditions above and below the Combin fault. Thus, most of the exhumation of the Zermatt–Saas zone relative to the Combin zone happened during or before D_1.

The variability of D_2 textures is probably due to very heterogeneous overall strain as most obviously testified by the formation of a number of large-scale folds. The most prominent of them is the Antrona synform (Figs. 10.3 and 10.4a) between the Monte Rosa and the Camughera–Moncucco nappes, which had begun to develop during D_1 but was mostly shaped by D_2 folding. In our reconstruction, we assume that orogen-parallel stretching of D_2 was contemporaneous with orogen-perpendicular convergence in deeper levels and nappe formation of more external Subpenninic units below the Monte Rosa nappe. The area around the Antrona synform may be considered as a transition zone between the two levels, where folding was coupled with orogen-parallel (i.e., top-to-the-southwest D_2) stretching and resulted in emplacement of the Monte Rosa nappe above the Antrona and

Camughera–Moncucco units, that is, orogen-perpendicular convergence. The more pronounced rotational kinematical component indicated by D_2 textures from the Stellihorn shear zone as compared with those from the Combin fault may reflect an increase of tectonic relative movements toward lower levels of the nappe stack.

In higher levels, the present-day stacking order of the Monte Rosa nappe and overlying units was largely accomplished by the end of D_1. We interpret the textures shown in Fig. 10.8b to have formed during late stages of D_1. Like most of the D_1 macroscale structures, they are related to metamorphic conditions that increase moderately from greenschist-facies (in the Combin zone) to lower amphibolite-facies conditions (below the Stellihorn shear zone). Therefore, late stages of D_1 postdated the exhumation of eclogite-facies rocks and the activity of large-scale faults that juxtaposed units which had experienced different metamorphic peak conditions. Since there is no evidence for episodes of orogen-perpendicular extension predating D_1 structures, we assume that the exhumation of eclogite-facies rocks to greenschist- or amphibolite-facies conditions happened in the framework of orogen-perpendicular convergence, that is, during earlier stages of D_1. The most probable exhumation mechanism of the high-pressure units below the Combin fault may either be described as extrusion of the high-pressure units below the Combin fault or as downward extraction of overlying units, depending on the viewpoint [92].

10.5 Quartz Textures from a Permian-Age Extensional Detachment Fault Underlying the Collio Basin (Southern Alps)

10.5.1 Regional Geology

This is an example where the correlation between quartz texture type and temperature during deformation is particularly clear. A mylonite zone was studied, which forms the uppermost part of the pre-Permian, Variscan basement in the Orobic anticline (Figs. 10.9 and 10.10). The basement, consisting predominantly of amphibolite-facies paragneisses and micaschists, is increasingly overprinted toward the top by mylonitic shearing. The mylonitic foliation is parallel to the basement-cover contact, which was deformed into an anticline (the Orobic anticline) during south-directed Alpine (Cretaceous- or Tertiary-age) thrusting. The stretching lineation is oriented northwest–southeast, and the shear sense is consistently top-to-the-southeast as determined from shear bands, sigma-type porphyroclasts, mica fish, and oblique grain shapes in dynamically recrystallized quartz layers. The mylonites formed contemporaneously with the intrusion of several bodies of the Val Biandino quartz diorite into the basement gneisses. This is evidenced by two observations: (1) xenoliths of mylonitized basement gneisses are locally found within the quartz diorite, showing that part of the mylonitization predated part of the intrusion, and (2) portions of the quartz diorite close to the top of the basement were affected by the mylonitization as well [131]. The quartz diorite was mylonitized with the same

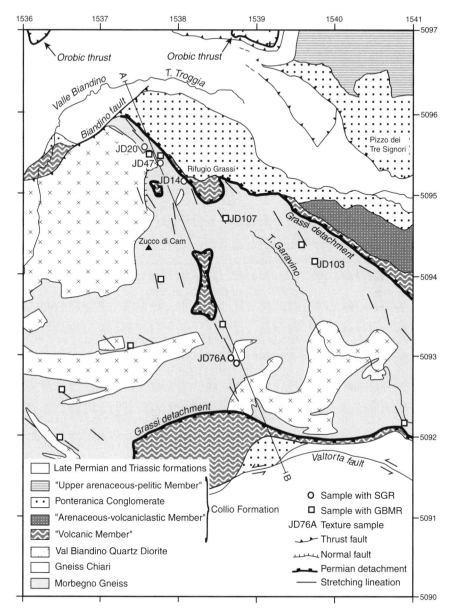

Fig. 10.9 Geological map of the study area in the Orobic anticline region (modified after [131], with kind permission of Geological Society, London). Kilometric coordinates refer to the Italian Gauβ–Boaga grid system

kinematics as the country rocks. The Val Biandino quartz diorite is dated at around 286±20 Ma (Early Permian [132]). Consequently, this is also the age of the shearing event that produced the mylonites. Along a brittle detachment fault represented by

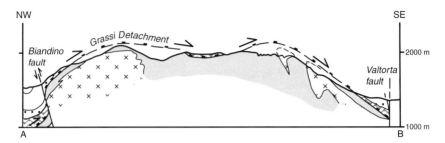

Fig. 10.10 Cross-section through the Orobic anticline in the map area of Fig. 10.9. The upper part of the basement below the Grassi detachment is a mylonitic shear zone capped by a cataclastic zone at the contact with the Collio formation. Figure reprinted from N. Froitzheim et al. [131] with kind permission of the Geological Society, London

a cataclasite layer (also of Permian age), the mylonite zone is overlain by volcanic and sedimentary rocks of the Collio basin. The volcanic rocks at the base of the succession are dated at 287–280 Ma [133]; that is, they have the same age as the quartz diorite intrusion in the footwall of the detachment.

The basement rocks in the Orobic anticline are interpreted as part of a Cordilleran-type metamorphic core complex that formed during fast extension of the crust in the Early Permian [131]. The mylonites and cataclasites represent ductile and brittle portions, respectively, of an extensional shear zone/detachment fault that unroofed the basement in this core complex (Fig. 10.11). Intrusion of granitoid rocks during unroofing and contemporaneous with calc-alkaline volcanism in the hanging wall, as observed here, is a typical feature of Cordilleran-type metamorphic core complexes, for example, in the Colorado River extensional corridor in western North America [134], suggesting a genetic link between magmatism and core-complex formation [135].

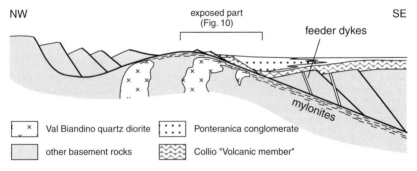

Fig. 10.11 Reconstructed cross-section of the Grassi detachment and the metamorphic core complex in its footwall for the Late Permian. The Ponteranica conglomerate is interpreted as a syntectonic delta fan. The volcanic rocks of the Collio formation are displaced toward the east relative to the Val Biandino intrusive bodies that probably were the magma chambers feeding the volcanic activity. Figure reprinted from N. Froitzheim et al. [131] with kind permission of the Geological Society, London

In thin section, the mylonites reveal two different types: A higher-temperature type with abundant reddish-brown biotite (of Permian age) and a lower-temperature type where the pre-Permian biotite, which was part of the original Variscan paragenesis [136], is completely replaced by chlorite. In both types, evidence can be found that the southeast-vergent shearing occurred when biotite and chlorite, respectively, were stable. The biotite-bearing mylonites are found in the vicinity of the quartz diorite bodies, and the chlorite-bearing mylonites farther away (>c. 1 km) from the intrusions. However, this is only a qualitative statement because the distance of the individual sample locations from the quartz diorite bodies cannot be determined exactly, owing to the irregular shape of the intrusions.

10.5.2 The Quartz Textures

Textures of almost pure quartz layers were determined from biotite-bearing mylonites (eight samples) and chlorite-bearing mylonite (one sample). The textures from the biotite-bearing mylonites are very similar, clear, and regular (Fig. 10.12a–e). They all show single {c} (= {001}) maxima parallel to the foliation and perpendicular to the stretching lineation, that is, parallel to the strain Y axis. These maxima are strong (up to 37 multiples of the random distribution) and partly round, partly slightly elongated in the plane perpendicular to the stretching lineation. From the {c} pole figure, it would be impossible to extract the shear sense. This information is, however, given by the {m} (= {010}) and {a} pole figures. Both {m} and {a} pole figures show three maxima distributed on a great circle parallel to the XZ plane, that is, on the periphery of the pole figures. These maxima are not symmetric with respect to the foliation trace but rotated anticlockwise in the case of {m} and clockwise in the case of {a}. The strongest of the {a} maxima is rotated about 10–20° clockwise from the stretching direction. The textures are clear evidence for predominant prism <a> slip. The strongest {a} maximum represents the dominant slip direction. The obliquity indicates dextral (top-to-the-southeast) shear sense, consistent with microstructural shear criteria.

In contrast, in the sample from chlorite-bearing mylonite (JD103, Fig. 10.12f), the {c} pole figure exhibits an oblique single girdle distribution, suggesting southeast-verging shear. Two maxima are on this girdle, one near the periphery, and a stronger one about 40° away from the center of the pole figure. Three {a} maxima are distributed on a great circle oblique to the foliation; the strongest of these is close to the periphery and rotated clockwise from the stretching lineation by around 45°. This angle is much larger than that for the other samples. The interpretation of this texture is difficult. One possibility is that it formed by superposition of the mylonitic shearing on some older texture of the basement gneiss. On the other hand, the weaker {c} maximum close to the periphery suggests the activity of basal <a> slip. The strong {c} maximum in the center of the pole figure, as observed in the biotite-bearing mylonites, is not found here, showing that prism <a> slip did not contribute significantly to the deformation.

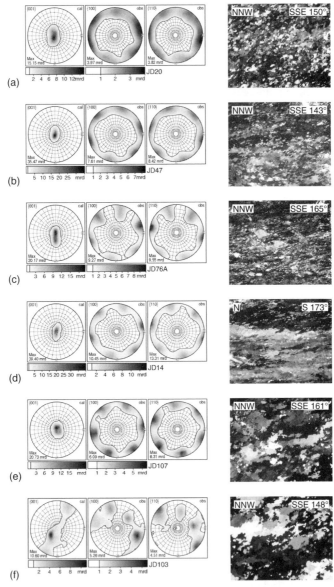

Fig. 10.12 {c}, {m}, and {a} pole figures (obs: observed; cal: calculated from the ODF; mrd: multiples of random distribution) of mylonitized quartz veins from the Morbegno gneiss (see Fig. 10.9 for sample locations) in upper-hemisphere stereographic projection shown together with photographs of XZ thin sections of the samples under crossed polarizers (width of all photographs is 3.44 mm). In all thin-section photographs, dextral (top-to-the-south-southeast) shear sense is indicated by grain-shape-preferred orientations. (**a–d**) Samples with quartz mostly deformed by sub-grain rotation recrystallization, as indicated by numerous subgrains with variable degree of misorientation among them. (**e–f**) Samples with quartz mostly deformed by grain boundary migration recrystallization as indicated by comparatively large grains with interlobate boundaries. Reproduced with kind permission of the Geological Society, London

These differences in the texture of the biotite- and chlorite-bearing mylonites are very likely to reflect different temperatures during deformation. Prism <a> slip becomes predominant in quartz mylonites under temperatures above ca. 500°C [4]. At lower temperatures, basal <a> and rhomb <a> slip are usually the active slip systems. Temperatures above 500°C during the mylonitization are in line with the observed stability of biotite. Mylonitic shearing was broadly syn-intrusive. Therefore, heat from the rising melt body is likely to have caused the elevated temperatures during mylonitization. M. Stipp et al. [4] found a correlation between the texture and the mechanism of deformation recrystallization. These authors argued that the exclusive activity of prism <a> slip is afforded by grain boundary migration recrystallization (GBMR) whereby grains with multiple slip systems, that is, a higher internal strain energy, are consumed by grains with only prism <a> slip activity. It is therefore noteworthy that some of our high-temperature textures are from samples showing evidence of subgrain rotation recrystallization (SGR; Fig. 10.12a–d), which is commonly subordinate to GBMR (evidence of which is present in all other of our texture samples; Fig. 10.12e and f) under elevated temperatures. Most, though not all, samples showing evidence of dominant SGR were taken from mylonites very close to the Val Biandino quartz diorite body (Fig. 10.9), and, therefore, it is most unlikely that SGR occurred under lower temperatures than GBMR in our samples. In experimental studies on quartz deformation, Hirth and Tullis [3] found that the temperature under which GBMR takes over from SGR is reduced by about 100°C in the presence of water or if the strain rate is reduced by one order of magnitude. Referring to these observations, two explanations for the occurrence of both SGR and GBMR under similar temperatures seem possible: (1) SGR may have been favored in the vicinity of the Val Biandino quartz diorite body as a result of the presence of a fluid phase provided by the pluton; and (2) SGR may testify to locally increased strain rates.

10.6 Conclusions and Outlook

As we tried to demonstrate in this Chapter by two examples of quartz texture studies from the Alps, analyzing the texture of deformed rocks yields information about past motions in shear zones, including shape of the finite strain ellipsoid (constriction vs flattening), strain path (pure shear vs simple shear), shear direction, shear sense, and temperature during deformation. For many rock types, such information can be obtained quantitatively from the analysis of strain markers, kinematics indicators and by petrological methods. However, the applicability of these methods is limited to certain rock types; for example, strain-marker analyses and petrological thermo-barometry often cannot be satisfactorily carried out on monomineralic or nearly monomineralic rocks. On the other hand, these are the rock types for which texture-forming processes are relatively well understood, though the information that can be obtained from texture studies is in most cases only qualitative. Thus, texture analysis and petrological methods (thermo-barometry, radiometric dating) provide complementary data that can be integrated into structural geological studies and result in more complete pictures of past tectonic processes.

One future goal of texture analysis is certainly to better understand the interaction of different minerals during texture formation in polyphase rocks (see also [137]). Another major challenge is to figure out how textures develop out of pre-existing textures (or, more generally, non-random crystallite orientation distributions), which is significant for rocks that experienced several phases of deformation.

With respect to the first issue, neutron texture goniometry will probably prove to be a valuable tool because it is especially well suited for simultaneously measuring the textures of different minerals in polyphase rocks. Regarding the second issue, the high penetration capability of neutrons may be useful because it allows large samples to be studied. This results in statistically meaningful measurements even for coarse-grained materials and for samples consisting of several shear domains, as in the examples discussed above (samples B42 and B43), and as is to be expected for samples that experienced more than one phase of deformation. Whereas neutron texture analysis was up to now mostly used on selected samples or sample suites in order to better understand deformation processes on the small scale, the techniques are now developed enough to become a standard tool for tectonic studies.

The potential of neutron diffraction for tectonic studies is at the moment used by only a few research groups. This is probably mainly due to the relatively limited number of neutron texture goniometers. Furthermore, considerable measuring time is required when large suites of samples have to be analyzed, as is usually the case in tectonic studies. New high-intensity neutron sources, application of time-of-flight techniques, and goniometer setups that are designed for texture analysis of low-symmetry minerals are therefore highly desirable, as well as increased efforts in the transfer of knowledge on neutron diffraction to the earth science community.

References

1. W. Skrotzki, Mechanisms of texture development in rocks. In: Textures of geological materials, ed. by H. J. Bunge, S. Siegesmund, W. Skrotzki, and K. Weber (DGM Informationsgesellschaft, Oberursel, pp. 167–186, 1994)
2. C. W. Passchier and R. A. J. Trouw, Microtectonics (Springer, Berlin, Heidelberg, New York, 1998)
3. G. Hirth and J. Tullis, Dislocation creep regimes in quartz aggregates. J. Struct. Geol. **14**, 145–159 (1992)
4. M. Stipp, H. Stünitz, R. Heilbronner, and S. M. Schmid, The eastern Tonale fault zone: a "natural laboratory" for crystal plastic deformation of quartz over a temperature range from 250 to 700°C. J. Struct. Geol. **24**, 1861–1884 (2002)
5. S. Siegesmund, A. Vollbrecht, and G. Nover, Anisotropy of compressional wave velocities, complex electrical resistivity and magnetic susceptibility of mylonites from the deeper crust and their relation to the rock fabric. Earth Planet. Sci. Lett. **105**, 247–259 (1991)
6. S. Siegesmund, The significance of rock fabrics for the geological interpretation of geophysical anisotropies. Geotekt. Forsch. **85**, 1–123 (1996)
7. W. Ben Ismaïl and D. Mainprice, An olivine fabric database: an overview of upper mantle fabrics and seismic anisotropy. Tectonophysics **296**, 145–157 (1998)

8. D. Mainprice, A. Tommasi, H. Couvy, P. Cordier, and D. J. Frost, Pressure sensitivity of olivine slip systems and seismic anisotropy of Earth's upper mantle. Nature **433**, 731–733 (2005)

9. K. Ullemeyer, S. Siegesmund, P. N. J. Rasolofosaon, and J. H. Behrmann, Experimental and texture-derived P-wave anisotropy of principal rocks from the TRANSALP traverse: an aid for the interpretation of seismic field data. Tectonophysics **414**, 97–116 (2006)

10. G. S. Lister and M. S. Paterson, The simulation of fabric development during plastic deformation and its application to quartzite: fabric transitions. J. Struct. Geol. **2**, 99–115 (1979)

11. G. S. Lister and B. E. Hobbs, The simulation of fabric development during plastic deformation and its application to quartzite: the influence of deformation history. J. Struct. Geol. **3**, 355–370 (1980)

12. S. M. Schmid and M. Casey, Complete fabric analysis of some commonly observed quartz c-axis patterns. In: Mineral and rock deformation: Laboratory Studies—the Paterson volume, ed. by B. E. Hobbs and H. C. Heard, Geophysical Monograph Series, vol. 36, pp. 263–286 (1986)

13. N. S. Mancktelow, Quartz textures from the Simplon Fault Zone, southwest Switzerland and north Italy. Tectonophysics **135**, 133–153 (1987)

14. B. Stöckhert, M. R. Brix, R. Kleinschrodt, A. J. Hurford, and R. Wirth, Thermochronometry and microstructures of quartz—a comparison with experimental flow laws and predictions on the temperature of the brittle-plastic transition. J. Struct. Geol. **21**, 351–369 (1999)

15. H. R. Wenk, T. Takeshita, E. Bechler, B. G. Erskine, and S. Matthies, Pure shear and simple shear calcite textures. Comparison of experimental, theoretical and natural data. J. Struct. Geol. **9**, 731–745 (1987)

16. S. M. Schmid, R. Panozzo, and S. Bauer, Simple shear experiments on calcite rocks: rheology and microfabric. J. Struct. Geol. **9**, 747–778 (1987)

17. L. Ratschbacher, H. R. Wenk, and M. Sintubin, Calcite textures: examples from nappes with strain-path partitioning. J. Struct. Geol. **13**, 369–384 (1991)

18. E. H. Rutter, Experimental study of the influence of stress, temperature, and strain on the dynamic recrystallization of Carrara marble. J. Geophys. Res. **100**, 24651–24663 (1995)

19. M. Herwegh and K. Kunze, The influence of nano-scale second-phase particles on the deformation of fine-grained calcite mylonites. J. Struct. Geol. **24**, 1463–1478 (2002)

20. G. Molli, P. Conti, G. Giorgetti, M. Meccheri, and N. Oesterling, Microfabric study on the deformational history of the Alpi Apuane marbles (Carrara marbles), Italy. J. Struct. Geol. **22**, 1809–1825 (2000)

21. A. Ebert, M. Herwegh, and A. Pfiffner, Cooling induced strain localization in carbonate mylonites within a large-scale shear zone (Glarus thrust, Switzerland). J. Struct. Geol. **29**, 1164–1184 (2007)

22. N. Oesterling, R. Heilbronner, H. Stünitz, A. Barnhoorn, and G. Molli, Strain dependent variation of microstructure and texture in naturally deformed Carrara marble. J. Struct. Geol. **29**, 681–695 (2007)

23. B. Leiss and D. J. Barber, Mechanisms of dynamic recrystallization in naturally deformed dolomite inferred from EBSP analyses. Tectonophysics **303**, 51–69 (1999)

24. S. Ji, D. Mainprice, and F. Boudier, Sense of shear in high-temperature movement zones from the fabric asymmetry of plagioclase feldspars. J. Struct. Geol. **10**, 73–81 (1988)

25. S. Ji and D. Mainprice, Recrystallization and fabric development in plagioclase. J. Geol. **98**, 65–79 (1990)

26. H. J. Dornbusch, K. Weber, and W. Skrotzki, Development of microstructure and texture in high-temperature mylonites from the Ivrea zone. In: Textures of Geological Materials, ed. by H. J. Bunge, S. Siegesmund, W. Skrotzki, and K. Weber, (DGM Informationsgesellschaft, Oberursel, pp. 187–201, 1994)

27. D. J. Prior and J. Wheeler, Feldspar fabrics in a greenschist facies albite-rich mylonite from electron backscatter diffraction. Tectonophysics **303**, 29–49 (1999)

28. R. Kruse, H. Stünitz, and K. Kunze, Dynamic recrystallization processes in plagioclase porphyroclasts. J. Struct. Geol. **23**, 1781–1802 (2001)

29. R. Kruse, H. Stünitz, and K. Kunze, Erratum to Dynamic recrystallization processes in plagioclase porphyroclasts. J. Struct. Geol. **23**, 1781–1802 (2001); J. Struct. Geol. **24**, 587–589 (2002)

30. V. Voegelé, J. I. Ando, P. Cordier, and R. C. Liebermann, Plastic deformation of silicate garnets I. High-pressure experiments. Phys. Earth Planet. Int. **108**, 305–318 (1998)

31. R. Kleinschrodt and J. P. Duyster, HT-deformation of garnet: an EBSD study on granulites from Sri Lanka, India and the Ivrea zone. J. Struct. Geol. **24**, 1829–1844 (2002)

32. D. Mainprice, J. Bascou, P. Cordier, and A. Tommasi, Crystal preferred orientations of garnet: comparison between numerical simulations and electron back-scattered diffraction (EBSD) measurements in naturally deformed eclogites. J. Struct. Geol. **26**, 2089–2102 (2004)

33. C. D. Storey and D. J. Prior, Plastic deformation and recrystallization of garnet: a mechanism to facilitate diffusion creep. J. Petrol. **46**, 2593–2613 (2005)

34. R. J. Knipe, The interaction of deformation and metamorphism in slates. Tectonophysics **78**, 249–272 (1981)

35. G. Oertel, The relationship of strain and preferred orientation of phyllosilicate grains in rocks—a review. Tectonophysics **100**, 413–447 (1983)

36. D. K. O'Brien, H. R. Wenk, L. Ratschbacher, and Z. You, Preferred orientation of phyllosilicates in phyllonites and ultramylonites. J. Struct. Geol. **9**, 719–730 (1987)

37. K. Kanagawa, Change in dominant mechanisms for phyllosilicate preferred orientation during cleavage development in the Kitakami slates of NE Japan. J. Struct. Geol. **13**, 927–943 (1991)

38. V. M. Mares and A. K. Kronenberg, Experimental deformation of muscovite. J. Struct. Geol. **15**, 1061–1075 (1993)

39. M. Herwegh and A. Jenni, Granular flow in polymineralic rocks bearing sheet silicates: new evidence from natural examples. Tectonophysics **332**, 309–320 (2001)

40. D. Gapais and J. P. Brun, A comparison of mineral grain fabrics and finite strain in amphibolites from eastern Finland. Can. J. Earth Sci. **18**, 995–1003 (1981)

41. W. Skrotzki, Defect structure and deformation mechanisms in naturally deformed hornblende. Phys. Stat. Sol. (a) **131**, 605–624 (1992)

42. A. Berger and H. Stünitz, Deformation mechanisms and reaction of hornblende: examples from the Bergell tonalite (Central Alps). Tectonophysics **257**, 149–174 (2001)

43. M. Díaz Aspiroz, G. E. Lloyd, and C. Fernández, Development of lattice preferred orientation in clinoamphiboles deformed under low-pressure metamorphic conditions. A SEM/EBSD study of metabasites from the Aracena metamorphic belt (SW Spain). J. Struct. Geol. **29**, 629–645 (2007)

44. A. Nicolas, F. Boudier, and A. M. Boullier, Mechanisms of flow in naturally and experimentally deformed peridotites. Am. J. Sci. **273**, 853–876 (1973)

45. N. M. Ribe and Y. Yu, A theory for plastic deformation and textural evolution of olivine polycrystals. Geophys. Res. **96**, 8325–8335 (1991)

46. S. Zhang, S. Karato, J. Fitz Gerald, U. H. Faul, and Y. Zhou, Simple shear deformation of olivine aggregates. Tectonophysics **316**, 133–152 (2000)

47. P. Philippot and H. L. M. van Roermund, Deformation processes in eclogitic rocks: evidence for the rheological delamination of the oceanic crust in deeper levels of subduction zones. J. Struct. Geol. **14**, 1059–1077 (1992)

48. G. Godard and H. L. M. van Roermund, Deformation-induced clinopyroxene fabrics from eclogites. J. Struct. Geol. **17**, 1425–1443 (1995)

49. B. Ábalos, Omphacite fabric variation in the Cabo Ortegal eclogite (NW Spain): relationships with strain symmetry during high-pressure deformation. J. Struct. Geol. **19**, 621–637 (1997)

50. A. Mauler, G. Godard, and K. Kunze, Crystallographic fabrics of omphacite, rutile and quartz in Vendée eclogites (Armorican Massif, France). Consequences for deformation mechanisms and regimes. Tectonophysics **342**, 81–112 (2001)

51. F. E. Brenker, D. J. Prior, and W. F. Müller, Cation ordering in omphacite and effect on deformation mechanism and lattice preferred orientation (LPO). J. Struct. Geol. **24**, 1991–2005 (2002)

52. W. Kurz, E. Jansen, R. Hundenborn, J. Pleuger, W. Schäfer, and W. Unzog, Microstructures and crystallographic preferred orientations of omphacite in Alpine eclogites: implications for the exhumation of (ultra-) high-pressure units. J. Geodyn. **37**, 1–55 (2004)

53. B. Sander, Gefügekunde der Gesteine mit besonderer Berücksichtigung der Tektonite (Springer, Wien, 1930)

54. K. Ullemeyer, G. Braun, M. Dahms, J. H. Kruhl, N. Ø. Olesen, and S. Siegesmund, Texture analysis of a muscovite-bearing quartzite: a comparison of some currently used techniques. Struct. Geol. **22**, 1541–1557 (2000)

55. H. Schober, Neutron Scattering Instrumentation. In Neutron Applications in Earth, Energy and Environmental Sciences, ed. by L. Liang, R. Rinaldi, and H. Schober (Chapter 3 of this volume)

56. H. G. Brokmeier, Application of neutron diffraction to measure preferred orientations of geological materials. In: Textures of geological materials, ed. by H. J. Bunge, S. Siegesmund, W. Skrotzki, and K. Weber (DGM Informationsgesellschaft, Oberursel, pp. 327–344, 1994)

57. R. Pynn, Neutron fundamentals. In: Neutron Applications in Earth, Energy and Environmental Sciences, ed. by L. Liang, R. Rinaldi, and H. Schober (Chapter 2 of this volume)

58. H. R. Wenk, H. Kern, W. Schaefer, and G. Will, Comparison of neutron and X-ray diffraction in texture analysis of deformed carbonate rocks. J. Struct. Geol. **6**, 687–692 (1984)

59. W. Schäfer, Neutron diffraction applied to geological texture and stress analysis. Eur. J. Mineral. **14**, 263–289 (2002)

60. E. Jansen, W. Schäfer, and A. Kirfel, The Jülich neutron diffractometer and data processing in rock texture investigations. J. Struct. Geol. **22**, 1559–1564 (2000)

61. E. Jansen, W. Schäfer, and G. Will, Profile fitting and the two-stage method in neutron powder diffractometry for structure and texture analysis. J. Appl. Cryst. **21**, 228–239 (1988)

62. S. Matthies, On the reproducibility of the orientation distribution function of texture samples from pole figures (ghost phenomena). Phys. Stat. Sol. (b) **92**, K135–K138 (1979)

63. S. Matthies and G. W. Vinel, An example demonstrating a new reproduction method for the ODF of texturized samples from pole figures. Phys. Stat. Sol. **112**, 115–120 (1982)

64. H. J. Bunge, Representation and interpretation of orientation distribution functions. In: Advances and applications of quantitative texture analysis, ed. by H. J. Bunge and C. Esling (DGM Informationsgesellschaft, Oberursel, pp. 19–48, 1991)

65. K. Helming, S. Matthies, and G. W. Vinel, ODF representation by means of σ-section. In: Proceedings of the 8th International Conference on Textures of Materials 1987, ed. by J. S. Kallend and G. Gottstein (The Metallurgical Society, Warrendale, pp. 55–60, 1988)

66. H. R. Wenk and J. M. Christie, Comments on the interpretation of deformation textures in rocks. J. Struct. Geol. **13**, 1091–1110 (1991)

67. H. R. Wenk, S. Matthies, J. Donovan, and D. Chateigner, BEARTEX: a Windows-based program system for quantitative texture analysis. J. Appl. Cryst. **31**, 262–269 (1998)

68. S. Matthies, H. R. Wenk, and G. W. Vinel, Some basic concepts of texture analysis and comparison of three methods to calculate distributions from pole figures. J. Appl. Cryst. **21**, 285–304 (1988)

69. H. J. Bunge, Texture analysis in materials science: mathematical models (Butterworths, London, 1982)

70. C. W. Passchier, The fabric attractor. J. Struct. Geol. **19**, 113–127 (1997)

71. R. D. Law, S. M. Schmid, and J. Wheeler, Simple shear deformation and quartz crystallographic fabrics: a possible natural example from the Torridon area of NW Scotland. J. Struct. Geol. **12**, 29–45 (1990)

72. G. M. Stampfli and G. D. Borel, The TRANSMED transects in space and time: constraints on the paleotectonic evolution of the Mediterranean domain. In: The TRANSMED Atlas, ed. by W. Cavazza, F. M. Roure, W. Spakman, G. M. Stampfli, and P. A. Ziegler (Springer, Berlin Heidelberg, pp. 53–80, 2004)

73. S. M. Schmid, B. Fügenschuh, E. Kissling, and R. Schuster, Tectonic map and overall architecture of the Alpine orogen. Eclogae Geol. Helv. **97**, 93–117 (2004)

74. N. Froitzheim, D. Plašienka, and R. Schuster, Alpine tectonics of the Alps and Western Carpathians. In: The Geology of Central Europe, ed. by T. McCann (Geological Society, London, (2008), pp. 1141–1232)

75. R. Polino, G. V. Dal Piaz, and G. Gosso, Tectonic erosion at the Adria margin and accretionary processes for the Cretaceous orogeny in the Alps. In: Deep structure of the Alps, ed. by F. Roure, P. Heitzmann, and R. Polino, Mém. Soc. Géol. Fr., vol. 156, pp. 345–367 (1990)

76. M. I. Spalla, J. M. Lardeaux, G. V. Dal Piaz, G. Gosso, and B. Messiga, Tectonic significance of Alpine eclogites. J. Geodyn. **21**, 257–285 (1996)

77. G. V. Dal Piaz, The Austroalpine-Piedmont nappe stack and the puzzle of Alpine Tethys. Mem. Sci. Geol. **51**, 155–176 (1999)

78. W. Frisch, Tectonic progradation and plate tectonic evolution of the Alps. Tectonophysics **60**, 121–139 (1979)

79. H. Kozur, The evolution of the Meliata–Hallstatt ocean and its significance for the early evolution of the Eastern Alps and Western Carpathians. Palaeogeogr. Palaeoclimatol. Palaeoecol. **87**, 109–135 (1991)

80. A. Liati, D. Gebauer, and C. M. Fanning, The youngest basic oceanic magmatism in the Alps (Late Cretaceous; Chiavenna unit, Central Alps): geochronological constraints and geodynamic significance. Contrib. Mineral. Petrol. **146**, 144–158 (2003)

81. G. M. Stampfli, Le Briançonnais, terrain exotique dans les Alpes? Eclogae Geol. Helv. **86**, 1–45 (1993)

82. M. Wagreich, A 400-km-long piggyback basin (Upper Aptian–Lower Cenomanian) in the Eastern Alps. Terra Nova **13**, 401–406 (2001)

83. A. G. Milnes, The structure of the Pennine zone (Central Alps): a new working hypothesis. Geol. Soc. Am. Bull. **85**, 1727–1732 (1974)

84. M. Maxelon and N. S. Mancktelow, Three-dimensional geometry and tectonostratigraphy of the Pennine zone, Central Alps, Switzerland and Northern Italy. Earth Sci. Rev. **71**, 171–227 (2005)

85. A. Michard, C. Chopin, and C. Henry, Compression versus extension in the exhumation of the Dora-Maira coesite-bearing unit, Western Alps, Italy. Tectonophysics **221**, 173–193 (1993)

86. A. Escher, J. C. Hunziker, M. Marthaler, H. Masson, M. Sartori, and A. Steck, Geologic framework and structural evolution of the western Swiss-Italian Alps. In: Deep structure of the Swiss Alps, ed. by O. A. Pfiffner, P. Lehner, P. Heitzmann, S. Mueller, and A. Steck (Birkhäuser, Basel, pp. 205–221, 1997)

87. S. Bucher, C. Ulardic, R. Bousquet, S. Ceriani, B. Fügenschuh, Y. Gouffon, and S. M. Schmid, Tectonic evolution of the Briançonnais units along a transect (ECORS-CROP) through the Italian-French Western Alps. Eclogae Geol. Helv. **97**, 321–345 (2004)

88. D. Gebauer, Alpine geochronology of the Central and Western Alps: new constraints for a complex geodynamic evolution. Schweiz. Mineral. Petrogr. Mitt. **79**, 191–208 (1999)

89. N. Froitzheim, Origin of the Monte Rosa nappe in the Pennine Alps—A new working hypothesis. Geol. Soc. Am. Bull. **113**, 604–614 (2001)

90. J. Pleuger, N. Froitzheim, and E. Jansen, Folded continental and oceanic nappes on the southern side of Monte Rosa (Western Alps, Italy): Anatomy of a double collision suture. Tectonics **24**, TC4013 (2005)

91. A. Liati and N. Froitzheim, Assessing the Valais ocean, Western Alps: U–Pb SHRIMP zircon geochronology of eclogite in the Balma unit, on top of the Monte Rosa nappe. Eur. J. Mineral. **18**, 299–308 (2006)

92. J. Pleuger, S. Roller, J. M. Walter, E. Jansen, and N. Froitzheim, Structural evolution of the contact between two Penninic nappes (Zermatt–Saas zone and Combin zone, Western Alps) and implications for exhumation mechanism and palaeogeography. Int. J. Earth Sci. **96**, 229–252 (2007)

93. R. Crespi, G. Liborio, and A. Mottana, On a widespread occurrence of stilpnomelane to the South of the Insubric line, Central Alps, Italy. Neues Jahrb. Mineral. Monatsh. **6**, 265–271 (1982)

94. S. M. Schmid, A. Zingg, and M. Handy, The kinematics of movements along the Insubric Line and the emplacement of the Ivrea Zone. Tectonophysics **135**, 47–66 (1987)

95. H. P. Laubscher, Large-scale, thin-skinned thrusting in the Southern Alps—kinematic models. Geol. Soc. Am. Bull. **96**, 710–718 (1985)

96. C. Doglioni, Tectonics of the Dolomites (Southern Alps, Northern Italy). J. Struct. Geol. **9**, 181–193 (1987)

97. G. Schönborn, Alpine tectonics and kinematic models of the Central Southern Alps. Mem. Sci. Geol. Padova **44**, 229–393 (1992)

98. M. E. Schumacher, G. Schönborn, D. Bernoulli, and H. P. Laubscher, Rifting and collision in the Southern Alps. In: Deep structure of the Swiss Alps, ed. by O. A. Pfiffner, P. Lehner, P. Heitzmann, S. Mueller, and A. Steck (Birkhäuser, Basel, pp. 186–204, 1997)

99. A. Boriani, L. Burlini, and R. Sacchi, The Cossato–Mergozzo–Brissago Line and the Pogallo Line (Southern Alps, Northern Italy) and their relationships with the late-Hercynian magmatic and metamorphic events. Tectonophysics **182**, 91–102 (1990)

100. S. M. Schmid, Ivrea Zone and adjacent Southern Alpine basement. In: Pre-Mesozoic Geology in the Alps, ed. by J. F. von Raumer and F. Neubauer (Springer, Berlin, Heidelberg, New York, pp. 567–583, 1993)

101. G. B. Vai, A. Boriani, G. Rivalenti, and F. P. Sassi, Catena ercinica e Paleozoico nelle Alpi Meridionali. In: Cento anni di geologia italiana, Vol. Giubilare, Soc. Geol. It., pp. 133–156 (1984)

102. A. Colombo and A. Tunesi, Pre-Alpine metamorphism of the Southern Alps west of the Giudicarie Line. Schweiz. Mineral. Petrogr. Mitt. **79**, 63–77 (1999)

103. R. Assereto and P. Casati, Revisione della stratigrafia permo–triassica della Val Camonica meridionale (Lombardia). Riv. Ital. Paleontol. Stratigr. **71**, 999–1097 (1965)

104. H. Wopfner, Permian deposits of the Southern Alps as product of initial alpidic taphrogenesis. Geol. Rundsch. **73**, 259–277 (1984)

105. G. G. Ori, S. Dalla, and G. Cassinis, Depositional history of the Permian continental sequence in the Val-Trompia–Passo Croce Domini area (Brescian Alps, Italy). Mem. Soc. Geol. It. **34**, 141–154 (1986)

106. G. Cassinis, F. Massari, C. Neri, and C. Venturini, The continental Permian in the Southern Alps (Italy). Z. Geol. Wiss. **16**, 1117–1126 (1988)

107. D. Sciunnach, The lower Permian in the Orobic Anticline (Southern Alps, Lombardy): a review based on new stratigraphic and petrographic data. Riv. Ital. Paleontol. Stratigr. **107**, 47–68 (2001)

108. G. Bertotti, Early Mesozoic extension and Alpine shortening in the western Southern Alps: the geology of the area between Lugano and Menaggio (Lombardy, Northern Italy). Mem. Soc. Geol. Padova **43**, 17–123 (1991)

109. A. Escher, H. Masson, and A. Steck, Nappe geometry in the Western Swiss Alps. J. Struct. Geol. **15**, 501–509 (1993)

110. A. Steck, B. Bigioggero, G. V. Dal Piaz, A. Escher, G. Martinotti, and H. Masson, Carte tectonique des Alpes de Suisse occidentale et des régions avoisinantes, 1:100,000. Carte spéc. n. 123 (4 maps), Serv. Hydrol. Géol. Nat., Bern (1999)

111. C. Chopin and P. Monié, A unique magnesio-chloritoide-bearing, high-pressure assemblage from the Monte Rosa, Western Alps; petrologic and [40]Ar-[39]Ar radiometric study. Contrib. Mineral. Petrol. **87**, 388–398 (1984)

112. A. Colombi and H. R. Pfeifer, Ferrogabbroic and basaltic meta-eclogites from the Antrona mafic-ultramafic complex and the Centovalli-Locarno region (Italy and southern Switzerland)—first results. Schweiz. Mineral. Petrogr. Mitt. **66**, 99–110 (1986)

113. A. Borghi, R. Compagnoni, and R. Sandrone, Composite P-T paths in the Internal Penninic Massifs of the Western Alps: Petrological constraints to their thermo-mechanical evolution. Eclogae Geol. Helv. **89**, 345–367 (1996)

114. M. Engi, N. C. Scherrer, and T. Burri, Metamorphic evolution of pelitic rocks of the Monte Rosa nappe: Constraints from petrology and single grain monazite age data. Schweiz. Mineral. Petrogr. Mitt. **81**, 305–328 (2001)

115. L. M. Keller, M. Hess, B. Fügenschuh, and S. M. Schmid, Structural and metamorphic evolution SW of the Simplon line. Eclogae Geol. Helv. **98**, 19–49 (2005)

116. T. J. Lapen, C. M. Johnson, L. P. Baumgartner, G. V. Dal Piaz, S. Skora, B. L. Beard, Coupling of oceanic and continental crust during Eocene eclogite-facies metamorphism: evidence from the Monte Rosa nappe, western Alps. Contrib. Mineral. Petrol. **153**, 139–157 (2006)

117. D. Rubatto, D. Gebauer, and M. Fanning, Jurassic formation and Eocene subduction of the Zermatt-Saas-Fee ophiolites: implications for the geodynamic evolution of the Central and Western Alps. Contrib. Mineral. Petrol. **132**, 269–287 (1998)

118. T. Reinecke, Very-high-pressure metamorphism and uplift of coesite-bearing metasediments from the Zermatt-Saas zone, Western Alps. Eur. J. Mineral. **3**, 7–17 (1991)

119. K. Bucher, Y. Fazis, C. de Capitani, and R. Grapes, Blueschists, eclogites, and decompression assemblages of the Zermatt-Saas ophiolite: high-pressure metamorphism of subducted Tethys lithosphere. Am. Mineral. **90**, 821–835 (2005)

120. R. Bousquet, M. Engi, G. Gosso, R. Oberhänsli, A. Berger, M. I. Spalla, M. Zucali, and B. Goffé, Explanatory notes to the map: Metamorphic structure of the Alps Transition from the Western to the Central Alps. Mitt. Österr. Miner. Ges. **149**, 145–156 (2004)

121. W. Ramsbotham, S. Inger, B. Cliff, D. Rex, and A. Barnicoat, Time constraints on the metamorphic and structural evolution of the southern Sesia Zone, Western Italian Alps. Mineral. Mag. **58A**, 758–759 (1994)

122. S. Inger, W. Ramsbotham, R. A. Cliff, and D. C. Rex, Metamorphic evolution of the Sesia-Lanzo zone, Western Alps: time constraints from multi-system geochronology. Contrib. Mineral. Petrol. **126**, 152–168 (1996)

123. M. Ballèvre and O. Merle, The Combin fault: Compressional reactivation of a Late Cretaceous-Early Tertiary detachment fault in the Western Alps. Schweiz. Mineral. Petrogr. Mitt. **73**, 205–227 (1993)

124. S. M. Reddy, J. Wheeler, and R. A. Cliff, The geometry and timing of orogenic extension: an example from the Western Italian Alps. Metamorphic Geol. **17**, 573–589 (1999)

125. M. Beltrando, G. Lister, J. Hermann, M. Forster, and R. Compagnoni, Deformation mode switches in the Penninic units of the Urtier Valley (Western Alps): evidence for a dynamic orogen. J. Struct. Geol. **30**, 194–219 (2008)

126. J. C. Vannay and R. Allemann, La zone piémontaise dans le haut Valtournanche (Val d'Aoste, Italie). Eclogae Geol. Helv. **83**, 21–39 (1990)

127. M. Sartori, Structure de la zone du Combin entre les Diablons et Zermatt (Valais). Eclogae Geol. Helv. **80**, 789–814 (1987)

128. A. Steck, Une carte des zones de cisaillement ductile des Alpes Centrales. Eclogae Geol. Helv. **83**, 603–627 (1990)

129. C. J. L. Wilson, Preferred orientation in quartz ribbon mylonites. Geol. Soc. Am. Bull. **86**, 968–974 (1975)

130. R. Müller, Die Struktur der Mischabelfalte (Penninische Alpen). Eclogae Geol. Helv. **76**, 391–416 (1983)

131. N. Froitzheim, J. F. Derks, J. M. Walter, and D. Sciunnach, Evolution of an early Permian extensional detachment fault from synintrusive, mylonitic flow to brittle faulting (Grassi Detachment Fault, Orobic Anticline, Southern Alps, Italy). In: Tectonic aspects of the Alpine–Dinaride–Carpathian system, ed. by S. Siegesmund, B. Fügenschuh, and N. Froitzheim, Geol. Soc. London Spec. Publ., vol. 298, pp. 69–82 (2008)

132. M. Thöni, A. Mottana, M. C. Delitala, L. de Capitani, and G. Liborio, The Val Biandino composite pluton: a late Hercynian intrusion into the South-Alpine metamorphic basement of the Alps (Italy). Neues Jahrb. Mineral. Abh. **12**, 545–554 (1992)

133. G. Cadel, Geology and uranium mineralization of the Collio basin (Central Southern Alps). Uranium **2**, 215–240 (1986)

134. G. A. Davis, Rapid upward transport of mid-crustal mylonitic gneisses in the footwall of a Miocene detachment fault, Whipple mountains, southeastern California. Geol. Rundsch. **77**, 191–209 (1988)

135. G. S. Lister and S. L. Baldwin, Plutonism and the origin of metamorphic core complexes. Geology **21**, 607–610 (1993)

136. G. B. Siletto, M. I. Spalla, A. Tunesi, J. M. Lardeaux, and A. Colombo, Pre-Alpine structural and metamorphic histories in the Orobic Southern Alps, Italy. In: Pre-Mesozoic geology in the Alps, ed. by J. F. von Raumer and F. Neubauer (Springer, Berlin, Heidelberg, New York, pp. 585–598, 1993)

137. S. J. Covey–Crump and P. F. Schofield, Neutron diffraction and the mechanical behavior of geological materials. In: Neutron applications in Earth, Energy and Environmental Sciences, ed. by L. Liang, R. Rinaldi, and H. Schober (Chapter 9 of this volume)

138. N. Froitzheim, S. M. Schmid, and M. Frey, Mesozoic paleogeography and the timing of eclogite-facies metamorphism in the Alps: a working hypothesis. Eclogae Geol. Helv. **89**, 81–110 (1996)

139. B. Neumann, Texture development of recrystallised quartz polycrystals unravelled by orientation and misorientation characteristics. J. Struct. Geol. **22**, 1695–1711 (2000)

Chapter 11
Neutron Imaging Methods and Applications

Eberhard H. Lehmann

Abstract Neutron imaging (NI) techniques based on direct transmission through macroscopic samples have enjoyed tremendous progress in recent years due to advances in digital imaging systems and new dedicated installations. Because of its high performance in terms of image frequency, dynamic range, and methodological variability, NI is now an option for research in many different fields.

This chapter describes the main cases where the method can be applied with a high success rate, explaining the different features that are available today for material studies with state-of-the-art tools.

Applications of NI are reported for studies in the earth sciences, mostly dealing with rock permeability, moisture transport, and porosity determination; in energy applications related to fuel cells, batteries, and nuclear fuel; and in the environmental sciences for studies related to root growth and wood research. Most of the results are based on investigations performed at the Paul Scherrer Institut beam lines for NI; further investigations carried out elsewhere are mentioned for comparison.

An outlook for future research topics, linked to further methodological improvements and scientific demands, is also given.

11.1 Principles in Modern Neutron Imaging

Compared to other neutron research devices (e.g., for neutron spectroscopy), a neutron imaging (NI) setup seems to be quite trivial: a directed "white" neutron beam is sent through an object and the two-dimensional (2D) detector behind it registers the transmitted part of the neutron distribution. Therefore, it is surprising that there are only a few NI facilities worldwide well-equipped for professional user support and optimum performance.

As shown schematically in Fig. 11.1, the main components of an NI setup are the neutron source, the collimator, and the detection system.

E.H. Lehmann (✉)
Spallation Neutron Source Division, Paul Scherrer Institut, CH-5232 Villigen, Switzerland
e-mail: eberhard.lehmann@psi.ch

L. Liang et al. (eds.), *Neutron Applications in Earth, Energy and Environmental Sciences*, Neutron Scattering Applications and Techniques,
DOI 10.1007/978-0-387-09416-8_11, © Springer Science+Business Media, LLC 2009

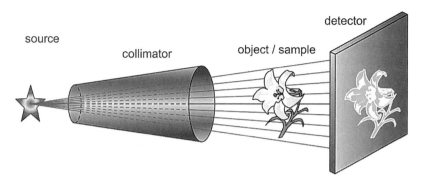

Fig. 11.1 Imaging principle: the source of (thermal or cold) neutrons illuminates the sample with a well-defined beam, which should be shaped by the collimator as a quasi-parallel neutron distribution. The detector behind the sample converts the transmitted neutrons to a "shadow image," basically a two-dimensional neutron distribution defined mostly by the inherent detector properties (resolution, energy response, sensitivity)

Additional devices may be used to manipulate samples (remotely controlled) or provide suitable conditions (experimental infrastructure). In practical applications, thermal neutrons are common, cold neutrons are rarely used, and fast neutrons are exotic. This is a result of the availability of suitable beam ports, the application range, and the detection probability. The contrast in NI mostly increases with decreasing neutron energy, but transmission and sample thickness will also be reduced accordingly.

Unlike charged particles, which can be focused or guided by electromagnetic fields, neutron performance can only be improved via a selection process. Even mirrored guides, focusing lenses, or fiber neutron optics in their present state have to be considered as "perturbing components" for optimal NI systems. The following describes the ideal case for a neutron beam with respect to imaging applications.

- A beam size as extended as possible to enable the inspection of large objects in the same image; a typical diameter is on the order of 40 cm, mostly circular.
- A homogenous illumination of the beam provided over the entire region, with the same spectral neutron distribution at each point; this is achieved with the right collimator design.
- A "clean" beam, devoid of background from gamma radiation or neutrons from other spectral ranges, can be achieved partly with filters.
- A narrow neutron spectrum to enable the best possible quantification of the sample content in a noninvasive way.
- A reasonable high intensity of the beam to enable an exposure time on the order of seconds to minutes as overly strong beams might increase the risk of activation.

In specific setups, the collimator design has to take all of these requirements into account. Frequently, compromises are necessary because some conditions compete with others. As an example, the layout of the ICON (Imaging with COld Neutrons)

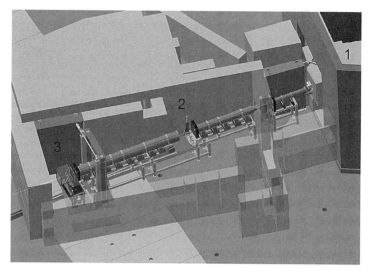

Fig. 11.2 The ICON (Imaging with COld Neutrons) facility at the SINQ spallation neutron source (Paul Scherrer Institut, Switzerland). The cold neutron beam is delivered from the source (1) after collimation inside the main shielding. Two experimental positions can be used—one for microtomography (2), another one for large-scale samples with weight up to 500 kg (3)

beam line [1] at the SINQ spallation neutron source [2] [Paul Scherrer Institut (PSI), Switzerland] is shown in Fig. 11.2, and some of its properties are summarized in Table 11.1.

Nowadays, almost all NI procedures are based on digital imaging detectors [3]. Their main advantages over the previously used film methods are much higher sensitivity, higher dynamic range, high linearity, and the direct option for quantification. Under standard conditions, a neutron image takes a few seconds compared to minutes or hours for film. This short exposure is quite important for neutrons due to the risk of activation by neutron capture in the objects. However, image quality

Table 11.1 Properties of the ICON[a] Neutron Imaging Facility (Paul Scherrer Institut, Switzerland) as an example for a State-of-the-art installation

Parameter	Value
Number of positions in the beam	3
Highest neutron intensity	$5E07\,cm^{-2}\,s^{-1}$
Largest field of view	40 by 40 cm
L/D ratio for standard inspections	About 1000
Mean neutron energy	4 meV
Number of possible apertures	6 (from 0.1 to 8 cm)
Volume of the experimental area	$100\,m^3$
Options for energy selection	Turbine type, Be filter
Option for phase contrast imaging	Pinhole, grating interferometry
Load for sample manipulator	500 kg

[a]Imaging with COld Neutrons.

Fig. 11.3 A ginseng root was studied with neutron tomography. The high contrast of the organic material in comparison to the soil enables segmentation with high quality. Some water is visible on the top of the sample due to the same attenuation behavior for thermal neutrons as for roots [data from a collaboration between the Korea Atomic Energy Research Institute's High-flux Advanced Neutron Application Reactor, HANARO, at Daejeon, South Korea, and Paul Scherrer Institut (Switzerland)]

and performance cannot be compared easily for these two methods (film and digital ones). More detailed studies are needed for this purpose.

Digital NI is a prerequisite for neutron tomography, where the whole sample volume can be studied in terms of the three-dimensional (3D) distribution of the attenuation coefficients in a voxel matrix. Usually, a few hundred projections are taken when the object is rotated around its vertical (or horizontal) axis over the range from 0° to 180° (with a parallel beam). The data quality can be improved when the full 360° range is used in the data acquisition process. A typical tomography result, obtained with neutrons, is shown in Fig. 11.3, where the advantages of neutrons (high sensitivity for hydrogenous materials) are exploited. In this measurement, the soil structure is quite transparent while the roots deliver a high contrast. The example in Fig. 11.3 also indicates the size scale of objects within normal reach with NI methods. Which material layer can be transmitted in the best possible conditions naturally depends on the composition. In Table 11.2, a summary of attenuation coefficients for thermal neutrons is given for some materials together with the maximum thickness for a transmission for at least 2%.

The spatial resolution in NI is limited by the beam geometry (divergence, distortion), scattering artifacts, and detector properties. Because it is nearly impossible to enable a magnification process in NI due to the limitations in the source intensity, the best possible case for imaging applications is a parallel beam without any perturbation.

At present, the best detection option in NI with respect to the spatial resolution is obtained by using imaging plate (IP) systems or very thin neutron-sensitive scintillators observed with special camera detectors [4]. A pixel size of about 13 μm

Table 11.2 Attenuation coefficients for thermal neutrons of selected materials (in condensed state) with relevance to geosciences [sample thicknesses for a residual beam transmission of 2% (theoretical approximation) are also shown.]

Material (natural composition)	Attenuation coefficient for thermal neutrons (cm^{-1})	Largest thickness of penetration (signal 2%) (cm)
Boron	111.60	0.04
Hydrogen	3.44	1.14
Iron	1.19	3.29
Copper	1.07	3.66
Uranium	0.82	4.77
Carbon	0.56	6.99
Lead	0.38	10.29
Oxygen	0.17	23.01
Silicon	0.11	35.56
Aluminum	0.10	39.12
Calcium	0.08	48.90

can be achieved. Therefore, NI is considered more as a "macroscopic" than a "microscopic" investigation method, especially when compared to X-ray studies with synchrotron radiation [5].

Nevertheless, NI has many new and interesting applications in science and technology, some of which are described in the following sections, where modern NI methods are presented in some detail. A forecast of future improvements and some further interesting applications are also included, for completeness.

11.2 Experimental Methods and Imaging Systems

As mentioned previously, transmission images are obtained in "standard" applications, where all layers of the object in the beam direction contribute to the final 2D distribution of the measured neutrons. Using digital imaging systems with a large number of equal pixels in both scanning directions, the image can also be considered as a matrix of discrete neutron intensity values in x and y. In first order, the intensity value for a pixel (i,j) can be described in the following manner:

$$I_{i,j} = \int \int \phi_{i,j}^0(E) \cdot e^{-\Sigma(z) \cdot z} \cdot \varepsilon(E) \mathrm{d}E \mathrm{d}z. \qquad (11.1)$$

Here it is assumed that the initial neutron flux ϕ^0 at position (i,j) with a given spectral distribution E is decreased according to Lambert's law when the neutrons are attenuated in the beam direction z perpendicular to the detector plane. Because the neutron detector also has energy sensitivity, its efficiency, $\varepsilon(E)$, has to be taken into account accordingly. The attenuation coefficient is a material property and is also energy-dependent. The relation (1) considers only a homogenous material with some structure in the beam direction z.

In the case of a composite material, all components contribute to the "efficient" attenuation coefficient (also called macroscopic cross-section). When the microscopic cross-sections, σ, of several materials are taken into account together with the atomic densities N, the relation is:

$$\Sigma_{\text{eff}}(E) = \sum_i \sigma_i(E) \cdot N_i. \tag{11.2}$$

In practical imaging applications, both neutron absorption and neutron scattering contribute to the final result in a similar way. Therefore, the tabulated *total* cross-sections have to be considered for analysis. This means that an arriving neutron which undertakes an interaction (either scattering or absorption) is considered lost for the resulting image. However, scattered neutrons have a certain probability of returning to the image—maybe not exactly along the initial direction, but a little aside and with an energy change. This effect limits the precision in quantification in some cases, especially when strongly scattering materials are under investigation. As described in Section 11.3.1, a correction algorithm has to be developed to overcome this problem, which is especially misleading in neutron tomography where a strict exponential attenuation according to Eq. (11.1) is assumed in all the reconstruction algorithms.

Modern NI detectors are based on capture either in lithium-6 or in gadolinium. Both boron-10 and helium-3 are quite uncommon in NI for several reasons for the moment. This holds also for systems containing fissile materials as converters. There are two major principles available for digital NI today, providing different properties in respect to frame rate, spatial resolution, dynamic range, and field-of-view (FOV): scintillation and excitation of semi-stable electron states in crystallites. The excitation of electrons in semiconductors has been investigated too [6], but no such device is available for routine work for the moment.

The scintillation process (e.g., in ZnS) cannot be initiated by neutrons directly. A converter is needed—mostly a strong neutron absorber as mentioned above. An optimization process is needed between neutron capture rate and excitation efficiency, resulting in the end with the best possible light output [7]. Scintillators are used in two different systems today: camera-based ones or flat panel devices with amorphous silicon matrices of photodiodes. More technical details can be found in the literature [8, 9]. These stationary systems are also in use for tomography applications, where the charge-coupled device (CCD) camera is mostly preferred.

Imaging plate detectors are used today as a replacement for films (in connection with gadolinium or dysprosium converters). Here the gadolinium converter is embedded inside the plate as a component, whereas the excitation is within special BaBrF crystallites doped with europium [10]. Due to their high linearity, wide dynamic range, and high signal-to-noise (S/N) ratio, IP detectors are also useful for quantification. However, the digital imaging process has to be carried out off-line in a readout scanner. Therefore, IPs are not useful when experiments with fixed positions (e.g., tomography, referenced imaging) have to be performed.

11.2.1 Neutron Tomography

Similar to X-ray systems (in medical or nondestructive testing devices) or installations at synchrotron radiation facilities [5], neutron tomography is able to analyze the 3D material distribution of macroscopic objects. Using a quasi-parallel, well-collimated beam, a 1:1 transmission image is produced as the basis of all such investigations, visualizing the direct attenuation by the object. Although the first attempts in neutron tomography were carried out about 10 years ago, it did not become a routine method before powerful enough computer systems became available at a reasonable price. Nowadays, setups for neutron tomography are available at different installations worldwide (see Table 11.3), although individual solutions differ in their performance.

The basic principle is the reconstruction of the information in the third direction by solving the line integral along the neutron path through the sample, according to the following:

$$I(x, y) = I_0(x, y) \cdot e^{\int \Sigma(x,y) \cdot dz},$$

and derived

$$\int \Sigma(x, y) \cdot dz = \ln \frac{I_0(x, y)}{I(x, y)}. \tag{11.3}$$

In the neutron tomography praxis, a set of projections is taken while the sample is rotated around its vertical (or horizontal) axis in sequential steps over a range of at least $180°$. This is possible because the neutron beam is assumed to be quasi-parallel; otherwise, the full $360°$ angular range must be scanned and the divergence has to be taken into account. The reconstruction algorithm is simpler if a parallel beam approximation is applied. The demands for computation time are much higher when a fan or a cone beam is to be considered.

The solution of the problem in Eq. (11.3) is obtained by a transformation of the image data of all projections into a Fourier space, where one projection layer (x,y) corresponds to one line in the frequency domain.

Using the Radon transformation of the problem for one projection in the direction $t = x \times \cos\theta + y \times \sin\theta$:

$$P_\theta(t) = \int \Sigma(x, y) \cdot ds = \int_{-\infty}^{\infty} \int_{-\infty}^{\infty} \delta(x \cdot \cos\theta + y \cdot \sin\theta - t)\Sigma(x, y)dxdy, \tag{11.4}$$

and filling up the full Fourier space by means of special interpolations procedures and a filtered back-projection from the Fourier space, it is possible to obtain the 3D matrix of the attenuation coefficients in the observed volume. These data correspond to the material characterization through the "eyes" of neutrons, because the Σ values are given for all materials involved [see Eq. (11.2)].

Table 11.3 Neutron tomography installations worldwide, showing selected performance parameters of European installations compared with prominent ones outside Europe

Installation	Country[a]	L/D-values	Flux (cm^{-2} s^{-1})	Cd-ratio	Gamma (Sv/h)	Neutron/ gamma	Beam Diameter (cm)	Beam Area (cm^2)
KFKI-channel 1	H	170	1.00E + 08	8	8.30	1.20E + 07	15	177
KFKI-channel 2	H	100	6.00E + 05	56	0.0053	1.13E + 08	7	38
AI Vienna channel 1	A	50	3.00E + 05	3	0.5	6.00E + 05	35	962
AI Vienna channel 2	A	125	1.00E + 05	16	0.01	1.00E + 07	7	38
PSI NEUTRA	CH	550	3.00E + 06	100	0.0015	2.00E + 09	40	1257
PSI ICON	CH	608	3.41E + 06	13.5				500
CEA ORPHEE	F	150	5.00E + 08	100	0.01			38
CEA-OSIRIS	F	156	6.00E + 07	4				650
JSI Ljubljana	SL	80	4.00E + 05	8.30	5.00E + 02	3.20E + 07	10	100
HMI CONRAD, Position 1	D	100	1.00E + 09				3 × 12 (guide)	36
HMI CONRAD, Position 2	D	500	1.00E + 07				10	100
ILL Lohengrin beam line	F	100	1.00E + 09				20	400
FRG-1 GKSS	D	375	5.00E + 06	100			18	254
TU Munich FRM-2	D	402	9.40E + 07	15	0.644	1.46E + 08	40 cm × 40 cm	1600
TU Munich FRM-2	D	795	2.48E + 07	15	0.17	1.46E + 08	40 cm × 40 cm	1600
JRR-3M	J	175	1.50E + 08	130	2.16	6.94E + 07	26	778
NIST	USA	300	1.70E + 07				26	531
Hanaro	K	250	1.00E + 07				20	314

Note: Data are in some cases averaged, because several options for beam modification exist.

[a]Country abbreviations: H = Hungary, A = Austria, CH = Switzerland, F = France, SL = Slovenia, D = Germany, J = Japan, K = Korea.

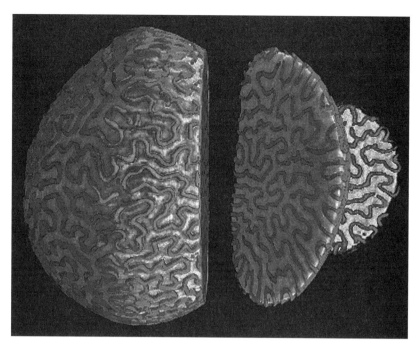

Fig. 11.4 Geology sample: stony coral with a diameter of about 12 cm. Due to the high attenuation behavior of residual organic material, a clear segmentation of the different zones is possible. Slices at arbitrary positions inside the object can be extracted, and the structure of the separated zones can be shown and measured dimensionally

After the reconstruction is done, further analysis is necessary to segment, slice, and visualize the obtained structural information. An example is given in Fig. 11.4 for a geological sample where small hydrogenous inclusions were identified and labeled.

Depending on detector performance, beam size, and sample properties in respect to the beam attenuation, objects up to about 30 cm can be investigated with neutron tomography. Because discrete pixel values from the 2D detector define the voxel dimensions in the 3D volume, the spatial resolution in the tomography results is in the best case simply the FOV divided by the number of pixels. A recently [4] developed setup with a pixel size of 13 μm, where a 2048 by 2048 pixilated sensor looks onto a 27 mm FOV, is currently the best possible in neutron tomography. Factors contributing to blurring, thus decreasing spatial resolution, include the lack of sharpness of the scintillator, scattering in the sample, and remaining divergence in the beam.

11.2.2 Real-Time Studies

Because of the much higher sensitivity of advanced NI systems today, it is now possible to inspect and investigate time-dependent phenomena in several applications.

However, the neutron beam intensity is still limited in most cases due to the intrinsic intensity of neutron sources. Therefore, temporal resolution is chosen as a trade-off with respect to the image quality, either by limiting the spatial resolution or the S/N ratio.

The presently available options for time-dependent studies are either intensified CCD camera systems optically coupled to the scintillator by a lens or amorphous silicon flat panels, where the scintillator is in close contact with the sensitive area. In the first case, the readout performance of the chip is more of a limiting factor than the real exposure time, which can range from milliseconds to seconds depending on the neutron beam intensity. The readout time, depending on the camera performance and chip size, is on the order of seconds also, when a high image quality and high dynamic range is required.

There is another option with intensified CCD detectors—their ability for gating small intervals, also in repetitive mode. Either dedicated, very narrow time frames can be selected and investigated or a repetitive process can be studied (e.g., a running combustion engine) in high quality, when corresponding narrow frames are stacked.

Recently, gated CCDs were used at a pulsed neutron source for energy-selective investigations of structural materials, where the textural behavior was directly visualized [11].

The second option for getting time-dependent information from the NI process is flat panel devices [9]. The systems are organized in such a manner that a continuous readout process is running during exposure with a given frame rate (between 1 and 30 frames/s, depending of the individual device). The usable frame rate depends on the neutron intensity and the detection efficiency and how these parameters fit into the dynamic range of the sensor.

For a more "relaxed" timing in studies of processes, a conventional setup can be used when the images to be taken have about 20-s intervals. When the study is over days or weeks, one has to guarantee that the imaging system has exactly the same setting in all investigations [e.g., plant growing (see Section 11.7.1)].

11.2.3 Phase Contrast Studies

It is well known that neutrons can be considered as "waves," and a wavelength can be derived according to the de-Brogli relation (with the neutron mass m and its velocity v):

$$\lambda = \frac{h}{m \cdot v}. \tag{11.5}$$

On the other hand, a refraction index can also be defined for neutrons, as in light optics, but this number, n, is a complex one and the deviations from 1 are very small:

$$n(\lambda) = 1 - \delta(\lambda) - i\beta(\lambda). \tag{11.6}$$

The first contribution in Eq. (11.5) is by phase effects in a material sample, whereas the second (imaginary) one is due to absorption in the material. Phase effects can be used in imaging in two ways.

For edge enhancement of weakly absorbing samples, where the wave overlap of the propagated waves behind the object in the Fresnel range is exploited [12]. A specially optimized setup is required with respect to the distances between sample and detector. Because phase effects are best exploited with coherent neutron waves, spatial coherence is enabled by small neutron apertures (e.g., 1 mm), and long exposure times are therefore required. Because the effects at the edges are narrow, a high-resolution detector is required too. With a coarse detector, the edge enhancement remains invisible. An example for the edge enhancement by phase effects is given in Fig. 11.5.

For the study of magnetic structures, where the phase shift of neutrons interacting with magnetic materials is exploited. It was shown in dedicated work [13, 14] that the phase contribution, δ, can be determined as a material property directly with imaging methods when a setup is built in such a way that a maximum number of spatially coherent neutrons are selected with a narrow absorbing grading for useful neutron intensities. After passing the object, the neutrons undertaking a phase shift are sorted in a phase grading and get analyzed within a 2D detector after passing a second analyzer grading. After demonstrating the applicability of this method to macroscopic objects with different phase properties, the main focus is now towards the study of magnetic structures because the phase shift of neutrons can also be caused by interactions with the magnetic moment of the neutron.

In both methods, cold neutrons are preferable because the phase contribution is wavelength (λ)-dependent, according to

$$\delta(\lambda) = \frac{\lambda^2}{2\pi} \cdot a_{\mathrm{coh}} \cdot \rho. \tag{11.7}$$

The coherent scattering length, a_{coh}, and the density, ρ, have to be taken into account too. The number for δ for iron in respect to cold neutrons is on the order of only 5 by 10^{-6}.

For the moment there have not been many applications in earth sciences, energy, and environmental studies, but there is great potential for the future, provided installations are optimized for routine user programs.

11.2.4 Energy-Selective Studies

Neutron sources for basic or applied science are either reactors or spallation sources, where a broad spectrum of thermal and cold neutrons is provided. The Maxwellian type energy distribution (caused mainly by the neutron moderator and its temperature) can be narrowed by different principles.

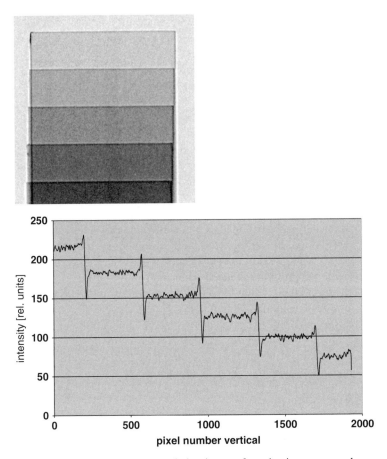

Fig. 11.5 Edge enhancement in the transmission image of an aluminum step wedge measured at the Paul Scherrer Institut cold neutron beam line (ICON) with the microtomography detection system [4]. The sample was at about 30 mm from the detector plane. Although the sample is only 20-mm wide and the steps are 0.5-mm each, the edge enhancement is very significant compared to the direct attenuation. The profile in the right figure was taken vertically in the middle of the object and demonstrates the strength of this effect

- *Single crystals* with high reflectivity, where a sharp peak at one energy can be found according to Bragg's law from lattice planes of the crystal. Because the direction of the scattered monoenergetic neutrons is also changed, there is no such signal in the forward direction. To retain the beam direction, a second monochromator crystal is installed, although the intensity of the resulting beam is further reduced. This setup is therefore called "double crystal monochromator." The wavelength can be tuned by simultaneous turning of both crystals. Such installations were tested successfully [15] for interesting imaging applications.
- *Turbines with tilted absorbing blades* can be used to narrow the energy band when only neutrons with the right velocity and direction compared to the rotation

speed of the turbine can go through the device. Compared to the crystal option, the selected energy band is broader and the intensity is higher accordingly. Nevertheless, it was shown that texture properties can be studied with neutrons of different narrow spectrum from this selector option.

- *Time-of-flight* (TOF) can be used when a pulsed beam is observed with a time-selective imaging detector [11]. Pulsed spallation sources (LANSCE, ISIS, SNS, J-PARC[1]) have the best properties with respect to intensity, but a chopper system might also be used in principle to get the required time structure: microsecond pulses would be separated into 20-ms bunches to avoid pulse overlaps. The new spallation sources (SNS, J-PARC) might enable more dedicated studies using the TOF option in the future.

There are some advantages in NI when quasimonochromatic neutrons can be used. All problems with beam hardening, spectral shift, and quantification can be handled much more easily and precisely. But the major potential will be the direct study of textures near Bragg edges and the related contrast variation when several materials are involved.

11.3 From the Image to Quantitative Evaluations

Although modern imaging systems can provide impressive and interesting "pictures" of the inner structures of investigated objects (e.g., Fig. 11.6 and [16]), the bulk properties of materials (density, thickness, composition) are often of more interest. Such data can be derived from the transmission data provided by the digital imaging detector if the attenuation in the object can be described quantitatively.

For many cases, the simple exponential law of attenuation can be applied, either directly or in the inverted mode, according to the following:

$$\Sigma \cdot d = \ln\left(\frac{I_0}{I}\right). \tag{11.8}$$

Here the intensities in front (I_0) ("open beam image") and behind (I) the object are compared for the whole area across the beam. Both intensities have to be considered with respect to their distribution in the (x,y) directions; integration is made in the z direction (see Fig. 11.7). For a homogenous material with the same attenuation coefficient Σ, the variation of the material thickness can be determined. The attenuation coefficient Σ is related to the material density, ρ, in the following manner:

$$\Sigma = \sigma \cdot \frac{\rho \cdot L}{M}. \tag{11.9}$$

---

[1] LANSCE—Los Alamos National Laboratory, United States, ISIS—Rutherford Appleton Laboratory, United Kingdom; Spallation Neutron Source (SNS)—Oak Ridge National Laboratory, United States; Japan Proton Accelerator Research Complex (J-PARC)—Japan Atomic Energy Research Institute, Tokai, Japan.

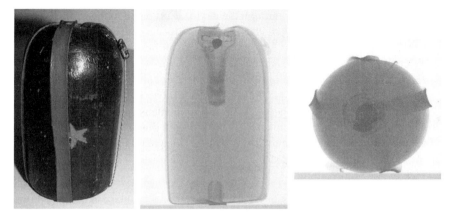

Fig. 11.6 Example of a unique neutron imaging investigation where other methods would fail: police request for inspection of a lead container of obscure origin, 55 mm in diameter, circled by a leather band (*left*). The investigation, which revealed droplets of mercury inside the container (on *top* of the hole in the *middle—center*), demonstrates the transparency of lead and the comparable high contrast for organic material (as cross-type band shown on the *right*) and for the neutron-absorbing mercury

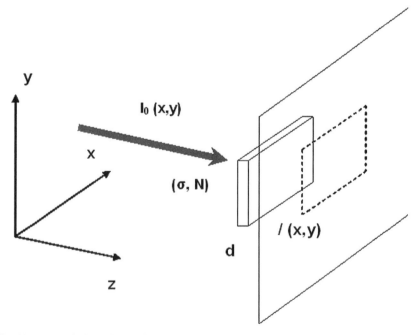

Fig. 11.7 Transmission principle in quantitative neutron imaging. The initial beam intensity, I_0, gets attenuated by the sample, with a thickness d, and the composition of the material (described by the cross-section σ and the nuclear density N) can be determined by the comparison with the transmitted beam intensity, I

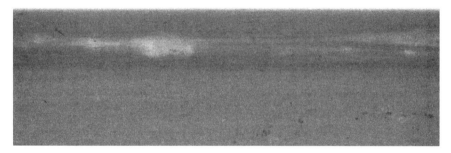

Fig. 11.8 Distribution of boron in aluminum. Such materials are frequently used for shielding purposes; however, their performance depends much on the manufacturing quality—which can easily be checked with neutron imaging methods. Even small grains of boron can be identified precisely

While Avogadro's number, L, and the molecular mass, M, of the material are known, the total microscopic cross-section averaged over the applied neutron spectrum can be determined from tabulated data libraries.

Material composition changes (inhomogeneity) can be analyzed with this method when a plate with the same thickness is investigated. An example for a boron-doped aluminum plate is shown in Fig. 11.8.

The restriction to a planar geometry can be overcome by tomography studies (see Section 11.2.1), where the sample is described by the 3D distribution of the attenuation coefficients $\Sigma(x, y, z)$ in real space. The accuracy of the determination of such data is still the subject of particular investigations [17].

11.3.1 Scattering Artifacts in Neutron Imaging

Most of the materials under investigation in NI interact more via scattering than by neutron capture reactions. Exceptions are the strong absorbers like ^{10}B, ^{6}Li, Cd, or Gd, where the scattering contribution plays only a minor role.

If a neutron gets scattered in the sample both the direction and the neutron energy can change as compared to the initial beam conditions. Because the application range in NI is more on the macroscopic scale (from millimeters to decimeters) than on the microscopic scale (as in X-ray tomography with synchrotron radiation), bulk samples are best investigated. Scattering interactions can involve both single and multiple scattering in samples. In the end, the transmitted neutron field can be described by two separate contributions, that of noncollided neutrons and that from the scattered neutrons entering the detector. As shown in Fig. 11.9 for a water layer, the scattered neutrons occur as a "sky shine" (or halo) around the object.

A direct recalculation according to Eq. (11.7) would fail by up to 100%, depending on experimental conditions and sample composition. It is necessary, therefore, to derive correction methods to overcome the scattering artifacts in neutron transmission studies. This is in particular the case for tomography studies, where the basic

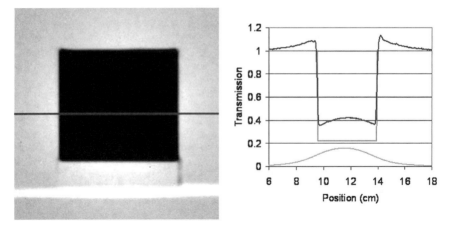

Fig. 11.9 Image of a 5-mm thick water layer in close contact with the detector plane: the scattered neutrons around the sample are clearly visible. The measured profile can be divided into the attenuated signal (*straight lines*) and the contribution of the diffuse scattered neutrons

assumption of Eq. (11.7) is used. The transmission data have to be corrected in advance of the reconstruction work with respect to the misleading scattering effects in the object, the energy shift in the transmitted spectrum, and the scattering effects from the environment. One solution is described in the next section.

11.3.2 The Quantitative Neutron Imaging Algorithm

The basic approach for the correction of scattering artifacts of a known sample composition is simulation with Monte Carlo methods. Based on such dedicated investigations, "point scattered functions" (PScFs) can be derived for all neutrons in the beam. Their superposition is then used to estimate the scattering contribution to the signal, apart from the behavior described by Eq. (11.7). If the composition is not exactly known, an iterative algorithm can be used with success.

The parameters in the neutron transmission calculations with Monte Carlo N-Particle eXtended (MCNPX) [18] are the material composition, the sample–detector distance, the neutron spectrum, the detector response function, and the beam divergence. Millions of individual neutron histories have to be analyzed to obtain the statistically precise PScFs for the given conditions. Because it is much too expensive to perform all calculations explicitly with MCNPX, some fitting functions with well-defined parameters have been derived. A typical PScF can be described in such a manner by the following:

$$\text{PScF}_{\text{approx}} = S_A \cdot \frac{d_A}{4\pi \cdot (d_A^2 + r^2)^{3/2}}, \qquad (11.10)$$

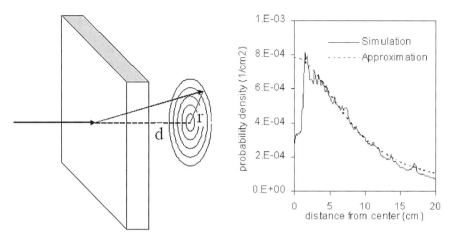

Fig. 11.10 The point scattered function is distributed around the point where the beam arrives (*left*). It can be calculated with Monte Carlo methods as shown for water in the graph on the right. An approximate description helps to simplify the correction procedure. The deviations in the simulations in the forward direction are due to data errors in the cross-section libraries

where the scattering function S_A is a material property, r describes the radial distance from the forward direction of the beam, and d_A describes the distance between the object and the detector plane (see Fig. 11.10). How well the simulated and approximated PScFs fit together is shown in Fig. 11.10 for one example. This approximation delivers better agreement to experiments than the Gaussian type previously used [19].

To implement this method in complex correction formalism for radiography and tomography applications, a tool was built using ITT Visual Information Solutions' (formerly Research Systems Inc.'s) Interactive Data Language [20]. Quantitative neutron imaging (QNI) [21] is now available for investigations with thermal neutrons. The following data are needed for its application: neutron spectrum, detector properties (composition, thickness), sample composition and thickness (as first approximation), and sample–detector distance. Based on this information, all digital NI data can be analyzed and corrected.

In addition to correcting for scattering artifacts caused by the sample, QNI also addresses background radiation contributions to scattering. How important this effect might be depends very much on the individual installation. A special measurement procedure is prescribed to determine the background radiation in investigations [17].

Further improvements to QNI are in progress, in particular, improvements for cold neutron interactions with matter, where the database in the MCNPX libraries is still weak and effects from the crystalline structure (Bragg edges) of materials of interest have to be considered in more detail. These improvements are of particular importance as cold neutron imaging, in particular with selected neutron energies, has high relevance in the future for texture analysis, where pulsed neutron sources (in TOF mode) might play an important role [11].

11.4 Applications in Earth Science

Winkler [22] gave an excellent overview of NI techniques and their applications in the geosciences, describing the status in about 2004. Therefore, this section will deal with more recent progress based on dedicated studies and further methodological improvements.

Noninvasive and nondestructive investigations are important for the analysis of processes like water migration in porous media such as soil and stones and to obtain the structural information on valuable pieces such as fossils (e.g., saurian bones and ammonites).

X-ray studies with tabletop devices or at synchrotron radiation beam lines can be performed with a spatial resolution of up to about 1 μm. Accordingly, the FOV and the sample size have to be about 1,000 times larger—on the order of either millimeters or centimeters. The magnification capability is often used to reach such high resolution.

Currently, the final limitation for spatial resolution in NI is set by the detector, because the secondary particles from the conversion reaction have a migration length on the order of 10 μm or much more. A magnification option is nearly impossible due to the limitations in the source strength, if the source spot has to be reduced to a few micrometers. Therefore, the microtomography setup at PSI [4] might represent the current limit for spatial resolution: an FOV of 27 mm with a pixel size of 13 μm. A first result with volcanic material is given in Fig. 11.11.

Regarding the attenuation behavior of neutrons and X-rays, respectively, most of the major constituents of soil and stones (Si, Ca, O, Fe, etc.) have similar properties (i.e., the attenuation coefficients have about the same ratio). This is completely different for hydrogen, which has a very high scattering probability for slow neutrons. Therefore, hydrogenous inclusions and moisture in the porous matrix deliver very high contrast in neutron investigations. This property can be exploited for the accurate study of moisture migration under different conditions. Some examples are described below.

For direct comparisons of neutron and X-ray investigations, the two kinds of radiation must be applied to the same sample alternatively. Using an X-ray tube at about the same place where the neutron beam arrives and observing the transmitted beam component with a detection system sensitive enough for both kinds of radiation, the attenuation can be analyzed and compared pixel-wise. This type of setup was implemented recently at the NEUTRA (NEUtron Transmission RAdiography) beam line at PSI [23] (the XTRA-option) and the ANTARES (Advanced Neutron Tomography and Radiography Experimental System) facility (Froschungsneutronenquelle Heinz Maier-Leibnitz, Germany) [24]. The results of investigations at these facilities can be used to enhance the contrast and to distinguish between several material components as described in the following sections.

Fig. 11.11 This volcanic material (xenolith) from southeastern Spain (diameter of the sample 12 mm) was studied with the microtomography setup at the ICON (Imaging with COld Neutrons) beam line, Paul Scherrer Institut [4]. Due to the high contrast of absorbing elements like samarium and gadolinium in monazite clusters, these inclusions, with dimensions on the order of 0.1 mm, were segmented and visualized precisely (sample courtesy of F. Ferri, Padua, Italy)

11.4.1 Structural Investigations

A very illustrative example for the difference between neutron and X-ray attenuation for stony samples is given in Fig. 11.12 [25, 26], where the hydrogenous material (probably iron hydroxide) has much less contrast for X-rays than for neutrons.

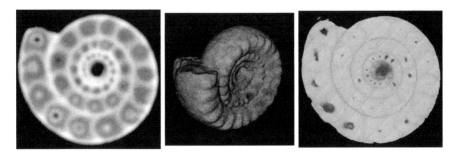

Fig. 11.12 This ammonite (diameter about 30 mm) was investigated with neutron (*left* and *center*) and X-ray tomography (*right*) [25]. The chamber walls are clearly separated in the X-ray slice, whereas the neutron tomography slice (*left*) delivers a more blurred but higher contrast image. This behavior is attributed to iron hydroxide, which has obviously a smooth distribution in the gaps between the chambers

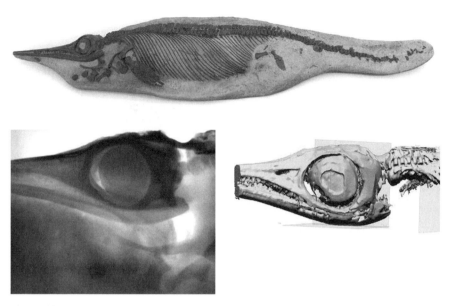

Fig. 11.13 The remains of an ichthyosaurus (sample, courtesy U. Oberli, St. Gallen, Switzerland; length about 1.4 m) were partly studied with neutron tomography methods (transmission image lower left), and a volume data set of the head was extracted. This was the basis for the production of a replica (*right*) by laser sintering

Although these regions look "blurred," this is not caused by the detection process but rather the actual material and its distribution. The neutron tomography view (center) in Fig. 11.12 demonstrates the sharp outer edges in the material distribution obtained for the whole object.

In paleontology, scientists are interested in the structure and functionality of specific finds, which may be more than 100 million years old, and of large dimensions [27]. Although no organic material is present in the fossils, typically their structures deliver quite a high contrast compared with the embedding material when subjected to NI techniques. Therefore, it is possible to differentiate fossil structures from the base material and perform reconstruction work based on virtual 3D distributions.

Such investigations have been carried out successfully with saurian fossils. For example, it was possible to distinguish the head of an ichthyosaurus from the stone slab where it was embedded and to produce a magnified replica by means of the derived 3D surface model [16]. The original rock slab containing the fossil, one transmission projection, and the 3D view of the object, constructed by a laser sintering procedure based on the neutron tomography data, are shown in Fig. 11.13.

11.4.2 Analysis of Moisture Migration Processes

Many investigations have been performed in the past few years using the high contrast characteristics of water, compared to matrix materials, in neutron transmission studies [28–34]. In addition to visualization of migration properties, the precise quantification of soil moisture has been investigated.

- Ursino et al. [28] used predetermined sand structures made from frozen cubes and basic moisture content to study the migration of additional water during artificial moistening. Here, the much lower attenuation behavior of heavy water ($\Sigma \sim 0.3 \, cm^{-1}$) as compared to light water ($\Sigma \sim 3.5 \, cm^{-1}$) was used to deliver basic moisture with low contrast. The injection of only 1 ml of light water becomes clearly visible, as shown in Fig. 11.14. The downward migration was followed for some hours following the injection during further wetting with heavy water.
- Kaestner et al. [29] compared the water distribution in well-defined quasi-soil structures with a model based on the Lattice-Boltzmann approach. Because the authors were interested in larger structures in their 3D distribution (object size 15 cm), heavy water was used as the agent. The viscosity conditions can be transposed between the two kinds of water (heavy and light).
- Carminati et al. were interested in the water migration process between wet and dry aggregates with dimensions of only a few millimeters. The authors combined the study with neutrons (determination of the local water content) by some synchrotron radiation tomography investigations, where the meniscus of the water between two aggregates was observed successfully [30].
- Koliji et al. investigated the behavior of soil structures under compression inside special cells with a determined pressure. This tomography study was performed with dry and wet soil particles. These data are compared to model considerations too.

- Water migration through three kinds of sandstone was investigated with NI [31], where the process of capillary soaking from two sources, a basin below and another one above the samples, was initiated. Because the columns of stone were 4 by 4 cm and 20-cm high, the amount of absorbed water was fairly high.

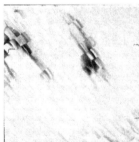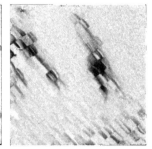

Fig. 11.14 Visualization of the distribution and migration of injected light water (two times 1 ml) in a predetermined regular sand structure with high base moisture content (heavy water) and permanent rain-type flooding with more heavy water from above. This arrangement has a thickness of 20 mm in the beam direction [28]. The results were compared to simulations of the transport of rainwater in soil

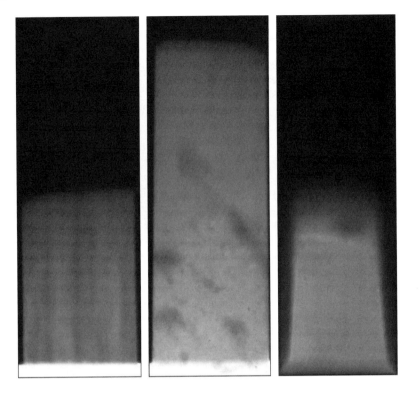

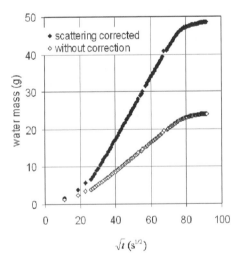

Fig. 11.15 Water uptake in three types of sandstone (column with 4 by 4 cm base line) was investigated during capillary soaking from a water reservoir below. The heterogeneity of the samples and the speed of wetting differ enormously. Quantification of the absorbed water amount was only successful after correction of the scattering artifacts using the quantitative neutron imaging correction tool [17, 34]

In such cases, quantification without a scattering correction would fail. Using the QNI correction tool (see Section 11.3.2), a good agreement with the integral water amount (measured by weighing) was found and the 3D water distribution was derived with good precision (see Fig. 11.15). In the quantification process, the time dependence of the amount of water follows, at a first approximation, a square root law ($m \sim t^{1/2}$), but the slope is very different for the investigated stone materials. The described methods can be applied to many other similar materials and processes such as impregnation (silanes [32]), diffusion, sealing strength, and evaporation.

- Lunati et al. [33] investigated the role of cracks in hard rocks for water permeability by a dedicated tomography study. A cylindrical sample was drilled out of a rock from the Grimsel region (Switzerland) and covered with an aluminum tube sealing. After application of water with pressure of several bars, the whole setup was investigated. The data were compared to the original dry state and the water migration was followed very precisely in this way. This kind of study is important for modeling the process of diffusion when disposal for radioactive water has to be evaluated.
- Solymar et al. [34] performed similar studies with greensand samples from different locations, but their main interest was the pore geometry and the mineralogical composition, which has great importance for oil exploration. They used both fast and thermal neutrons for their studies, where the thermal neutrons were found quite useful for high-resolution tomography, with satisfying resolution for this application. Attempts to perform NI studies for oil exploration procedures have also been reported by other authors [35].

Compared to other methods such as magnetic resonance imaging (MRI), X-ray inspection, or resistivity measurement, NI methods were found superior due to the high contrast of water compared to the matrix of the porous materials, the relatively high spatial resolution, and the relevant object size amenable to the inspection. Therefore, many other investigations are currently in progress.

11.5 Applications in Energy Research

The search for alternatives to the combustion of fossil fuels to prevent dramatic climate change and exhaustion of major fossil fuel reserves is currently a major topic for research. Although nuclear options were considered very promising half a century ago, their drawbacks and their risks have been criticized, and, in any event, the amount of such fuel is also limited. Neither nuclear fission nor nuclear fusion is presently considered a long-term solution to the energy supply problem. As a consequence, alternative concepts related to renewable or unlimited resources (wind, sun, biomass, tides, and earth heat) have been investigated intensively. However, a "golden" solution to the ever-growing energy demands of our times has yet to be found.

Neutron imaging can contribute to energy research when its noninvasive inspection ability is favorably applied. This section is focused on three examples: the in situ study of operating electrochemical fuel cells, the inspection of batteries, and the determination of the hydrogen content in nuclear fuel cladding under normal and accidental conditions. There are many other aspects of energy-related studies; however, a comprehensive overview would be outside the scope of this review.

11.5.1 Fuel Cell Studies

The polymer electrolyte membrane fuel cell (PEM-FC) operates according to the principle illustrated in Fig. 11.16: the streams of hydrogen and oxygen are separated by a membrane, which enables electrochemical reactions in such a manner that electrons are delivered to one electrode and a voltage is produced. The reaction product is water, which is necessary to keep the electrical process running with high performance (conductivity). On the other hand, too much water will limit performance in the opposite way. Therefore, water management is an important issue for the optimization of PEM-FCs. It is an optimization problem involving membranes; catalyzers; flow fields; and operating conditions such as temperature, moisture, and gas supply to come up with the best performance under the different environmental circumstances.

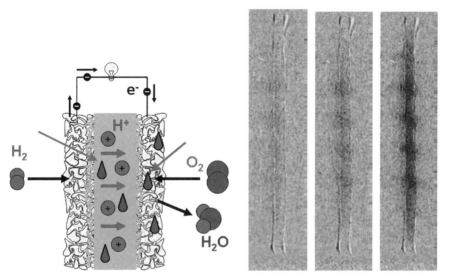

Fig. 11.16 A simplified diagram of the polymer electrolyte membrane fuel cell is shown on the left: gaseous hydrogen and oxygen come together in an electrochemical process. When charge separation takes place, electrical power can be extracted. The reaction product, water, can easily be visualized in high resolution and quantified (images *right*, membrane thickness 0.2 mm only) (courtesy P. Boillat, Paul Scherrer Institut)

Neutron imaging methods were found important for the noninvasive investigation of fuel cells because the outer structure (metal, graphite) is quite transparent to neutrons, whereas water can be well visualized and measured with satisfying spatial and temporal resolution [36]. In Fig. 11.16 (right side), images of the region around the membrane are shown under different operating conditions. The water amount can be directly determined from the image data [37, 38].

11.5.2 Battery Studies

Modern electronic devices require mobile electric support with high reliability and long-lasting performance. Batteries are among the most used components in this respect, with a long tradition but also a high potential for improvements. In the past few years, lithium-based batteries have been developed and commercialized. Their performance might be improved by further understanding of the internal processes under different operating conditions.

Neutron imaging can contribute to noninvasive examination of new battery concepts and prototypes. This was done in a study of lithium ion batteries during charging and discharging processes [38]. The change in the electrolyte distribution was visualized and measured, and the level of the electrolyte was clearly attributed in relation to the charge status. Further studies for the detection and quantification of the gas evolution at the graphite electrodes were successfully carried out [39].

11.5.3 Nuclear Fuel Inspection

Fuel elements of running nuclear power plants are rod bundles, where cylindrical pellets of uranium oxide with different enrichment in the fissile uranium isotope ^{235}U are stacked within the cladding tubes, which are made out of zirconium alloys. This relatively simple arrangement is surrounded by deionized water during power production and later during the cooling and interim storage phase. While conditions in the reactor vessel are demanding in terms of temperature (several hundred degrees) and pressure (more than 4 MPa in some), they are more relaxed in the storage stage. However, the need for the cladding to cover the spent fuel for a very long time without failure is demanding.

There are good opportunities for NI methods to contribute to the analysis and characterization of fuel elements in several respects. First of all, uranium is not as strong an absorber of neutron radiation as it is of X-rays. A few tenths of a millimeter of uranium would absorb all X-rays, even highly energetic ones. Because this is far from the case when thermal neutrons are applied, they are the ideal kind of radiation for nuclear fuel inspection. NI methods can be used to analyze defects in the pellet structure and the interaction with the fuel cladding, and the fuel enrichment can be determined directly from the neutron images because ^{235}U $[\sigma_{total}(E = 25\,\text{meV}) = 612\,\text{barn}]$ and ^{238}U $[\sigma_{total}(E = 25\,\text{meV}) = 11.8\,\text{barn}]$ differ quite strongly in their

Fig. 11.17 A dummy fuel element with zirconium cladding and uranium pellets of different enrichment was inspected in transmission mode with neutrons. The gray values correspond directly to the fuel enrichment, and quantification is possible with good precision. The black areas in the image are gadolinium "poisoned" fuel pellets. The zirconium cladding material is very transparent for thermal neutrons compared to the uranium fuel

attenuation behavior for thermal neutrons. This is shown in Fig. 11.17, where pellets with enrichment between 2% and 5% were inspected and their contents quantified.

It is quite an important task to characterize the fuel cladding too, because it is responsible for the sealing of the nuclear inventory in respect to environmental risks. Because zirconium is transparent for neutrons (as also demonstrated in Fig. 11.17), material changes can be observed very sensitively. This also holds for the accumulation of hydrogen during normal or accidental operation. The attenuation behavior of hydrogen is much higher than that of zirconium due to its large scattering cross-section, so hydrogen can be detected and quantified even in small amounts. But oxide layers and defects in the cladding can also be studied with NI methods.

Although fuel inspection has been used in different neutron labs for the past 30 years or so with the help of film methods and the dysprosium conversion technique, there are some new approaches for improvements in efficiency and quantification. Because used nuclear fuel itself emits large amounts of radiation, in particular gamma rays with dose levels of Gy/hour, a direct exposure of fuel in a neutron field is impossible. Film-type detectors would be saturated before the transmitted neutrons were detected. Therefore, the only way is to use neutron activation of either Dy, In, or Au as the primary detector and to transfer this latent image of the neutron distribution onto a radiation-sensitive area. This can be done by using IPs, but a very close contact between the activated area and the IP is needed to get high spatial resolution.

A new option is now available with dysprosium-doped IPs [40], where the dysprosium activity, embedded in the IP, grows up after strong erasure of the IP after the direct exposure. Highest spatial resolution and exact quantification options have been obtained with this method. It has also been used for the inspection of other highly activated samples like target components from spallation neutron sources [41].

11.6 Applications in Materials Research

The aerospace and automotive industries rely on technological advances such as new materials, new material combinations, and new bonding techniques. Innovations in weight reduction and bonding techniques are examples of particular economic importance; however, each advance requires additional activities for quality assurance.

Welding, soldering, and gluing are among the most widely used methods for joining metallic parts. Neutron imaging can provide higher beam transmission for Cu, Al, Fe, and even Pb than X-ray methods and easily detect the distribution of solder and, in particular, the adhesive agents used for metal bonding [42]. In this way, joints can be inspected with high precision to identify gaps and missing bonding material.

With respect to welding connections, texture analysis based on energy-selective imaging options (see Section 11.2.4) is a very promising approach that complements stress analysis by means of dedicated diffractometers (e.g., the Strain Analyser for Large- and Small-Scale engineering Applications at the Institut Laue-Langevin, France, or ENGINE-X at the ISIS neutron facility in the United Kingdom).

Because hydrogen and oxygen deliver high interaction probabilities for slow neutrons, it is quite easy to inspect metallic structures for corrosion and other chemical degradation. Again, the good transmission for metals and the high contrast for the corrosion products make NI methods preferable to other methods.

11.7 Applications in Environmental Studies

11.7.1 Plant Growth Investigations

The root is the part of plant hidden in the soil and is just as important for plant growth as the visible part. Therefore, observation of root growth behavior under diverse environmental conditions is of high importance. Furthermore, roots are responsible for water uptake and transport from the ground to the other parts of the plant, so it is useful to know in detail how water is taken up and stored by the root and the whole plant under varying conditions.

It was shown in case studies [43] that thermal neutrons can be used to transmit soil structures quite well. The thickness of the soil layer should not be larger than about 30 mm. In a plate-type arrangement perpendicular to the beam, most of the studied plants (lupine, maize, bean, etc.) showed normal, unperturbed growth behavior. Even thin roots were visible due to the high hydrogen content of the organic material. In addition to root volume, the moisture content in the soil can be measured very precisely with NI methods. The spatial resolution in such measurements can be on the order of 50 μm, which can never be achieved with methods like MRI for setups like that shown in Fig. 11.18. The comparison to X-rays is given in the same figure, where almost no organic material can be seen in the X-ray image, but the soil and organic material are both quite well defined in the neutron image.

One important goal of plant studies is the determination of the role of environmental agents, poisons, and pollutants like Ni, Pb, or B in root growth. How does the root react under the influence of specific amounts of such materials in the soil and water? Such studies have been performed successfully with the aid of NI techniques and will be continued [44].

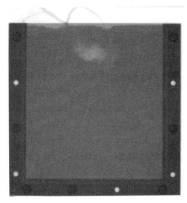

Fig. 11.18 Root growth was studied with thermal neutrons (*left*) and X-rays (*right*) (tube at 150 kV high voltages). The plant was grown in a 17-mm wide aluminum box filled with soil. Whereas the roots and the soil can be distinguished very well with neutrons, very little contrast is given for the organic material in the X-ray image

11.7.2 Wood Research

Wood, a natural material from trees and related plants, has been important as a building material for centuries. Under ideal conditions of temperature, humidity, and air quality, wood can last virtually forever due to the stability of the lignin in the cell structure. As soon as the moisture content is increased above a certain level (e.g., 60% relative humidity), wood changes dimensions by swelling, and fungi play a considerable role when the temperature is high enough too.

Neutron imaging can be used for noninvasive analysis of the moisture distribution in wooden materials in order to verify model considerations. Because wood is a natural product and not homogeneous, individual cases and specific kinds of wood have to be taken into account in the experimental design [45]. Archaeological wood and other objects that form part our cultural heritage have recently received special consideration [46].

Neutron imaging is also a very efficient tool for the investigation of wood treatment with protective agents, paints, and lacquers and methods of joining wood with adhesives and glues because all these materials deliver high imaging contrast due to their high hydrogen content. Dedicated studies on the penetration depth of different adhesive materials into the wood matrix are planned for the future.

11.8 Future Trends and Improvements

This chapter described the NI techniques in common use today and their application in the earth, energy, and environmental sciences. Most of the examples are taken from collaborations of the author's team at PSI with scientific partners at other institutions. Similar studies have been performed in different laboratories around the world; however, the number of such studies is still quite small due to the limited number of suitable NI facilities.

Therefore, it is hoped that new neutron sources will be equipped with state-of-the-art imaging beam lines and the needed experimental infrastructure. The availability of modern digital imaging systems in combination with high-performance computers for data analysis will be the driving force for enhanced use of these methods.

Other important application fields for NI can be found in the study of cultural heritage objects, inspection of engines under real-time conditions, and quality control of industrial processes such as casting, soldering, or gluing.

Experimental methods in NI continue to be improved, similar and parallel to recent developments in X-ray imaging. New topics for methodological research include use of the phase information in the transmitted neutron field for image enhancement, use of energy-selective options for material research, and use of polarized neutrons for the investigation of magnetic properties. When available at the needed performance levels, these techniques might contribute to the research areas described in this book in the future.

References

1. G. Kühne et al., Swiss Neutron News **28**, 20–29 (2005), http://sgn.web.psi.ch/sgn/snn/snn_28.pdf
2. G. S. Bauer, Nucl. Instrum. Methods Phys. Res. A **463**, 505–543 (2001)
3. E. H. Lehmann et al., Nucl. Instrum. Methods. Phys. Res. A **531**, 228–237 (2004)
4. E. H. Lehmann et al., Nucl. Instrum. Methods. Phys. Res. A **576**, 389–396 (2007)
5. M. Stampanoni, R. Abela, and G. Borchert, Nucl. Instrum. Methods. Phys. Res. A **551**, 119 (2005)
6. J. Uher et al., Nucl. Instrum. Methods. Phys. Res. A **542**, 319–323 (2005)
7. G. Frei et al., in *Proceedings of the 8th World Conference on Neutron Radiography*, Gaithersburg, page 599, 2006
8. H. Pleinert, E. Lehmann, and S. Körner, Nucl. Instrum. Methods Phys. Res. A **399**, 382–390 (1997)
9. M. Estermann et al., Paul Scherrer Institut Annual Report **III**, 188 (2003)
10. S. Tazaki et al., Nucl. Instrum. Methods Phys. Res. A **424**, 20–25 (1999)
11. W. Kockelmann et al., Nucl. Instrum. Methods Phys. Res. A **578**, 421–434 (2007)
12. B. E. Allman, P. J. McMahon, K. A. Nugent, D. Paganin, D. L. Jacobson, M. Arif, and S. A. Werner, Nature **408**, 158–159 (2000)
13. M. Strobl, W. Treimer, and A. Hilger, Nucl. Instrum. Methods Phys. Res. B **222**, 653–658 (2004)
14. F. Pfeiffer, C. Gruenzweig, O. Bunk, G. Frei, E. Lehmann, and C. David, Phys. Rev. Lett. **96**, 215505 (2006)
15. W. Treimer, M. Strobl, N. Kardjilov, A. Hilger, and I. Manke, Appl. Phys. Lett. **89**, 203504 (2006)
16. http://neutra.web.psi.ch
17. R. Hassanein, Doctor of Sciences Dissertation 16809, Swiss Federal Institute of Technology, 2006
18. *MCNPX User's Manual*, version 2.4.0, edited by L. S. Waters, Los Alamos National Laboratory Report LA-CP-02-408, 2002

19. N. Kardjilov et al., Nucl. Instrum. Methods Phys. Res. A **542**, 336–341 (2005)
20. IDL, Interactive Data Language, ITT Visual Information Solutions, Boulder, Colorado, USA (http://www.ittvis.com)
21. R. Hassanein et al., *Proceedings of the 8th World Conference on Neutron Radiography*, Gaithersburg, page 206, 2006
22. B. Winkler, in: Neutron Scattering in Earth Sciences, edited by H.-R. Wenk (Geochemical Society/Mineralogical Society of America, USA, 2006), *Reviews in Mineralogy & Geochemistry*, Vol. 63, pp. 459–471
23. E. H. Lehmann et al., Paul Scherrer Institut Annual Report, Annex (2004)
24. E. Calzada, B. Schillinger, and F. Grünauer, Nucl. Instrum. Methods Phys. Res. A, **542**, 38–44 (2005)
25. W. D. Carlson, Three-dimensional imaging of earth and planetary materials, Earth Planet. Sci. Lett. **249**, 133–147 (2006)
26. P. Vontobel et al., IEEE Trans. Nucl. Sci. **52**, 338–341 (2005)
27. D. Schwarz et al., Palaeontologia Electronica **8**, article 30A (2005)
28. N. Ursino, T. Gimmi, and H. Flühler, Adv. Water Resour. **24**, 877–885 (2001)
29. Kaestner et al., Chem. Eng. J. **130**, 79–85 (2007)
30. Carminati et al., Adv. Water Resour. **30**, 1168–1178 (2007)
31. R. Hassanein, H. O. Meyer, A. Carminati, M. Estermann, E. Lehmann, and P. Vontobel, J. Phys. D **39**, 4284–4291 (2006)
32. A. H. Gerdes, in *Transport und chemische Reaktion siliciumorganischer Verbindungen in der Betonrandzone (Aedificatio Freiburg, 2002), Building Materials Report 15*
33. F. Lunati, Doctor of Natural Sciences Dissertation 15082, Swiss Federal Institute of Technology, 2003
34. M. Solimar, Dissertation, Chalmers University, 2002
35. M. Middleton and F. De Beer, in *Proceedings of the 7th World Conference on Neutron Radiography* (Rome, 2002), p. 459
36. D. Kramer, Dissertation, Bergakademie, Freiberg, 2007
37. D. Kramer et al., Nucl. Instrum. Methods Phys. Res. A **542**, 52–60 (2005)
38. M. Lanz et al., J. Power Sources **101**, 177–181 (2001)
39. D. Goers et al., J. Power Sources **130**, 221–226 (2004)
40. M. Tamaki et al., Nucl. Instrum. Methods Phys. Res. **542**, 320–323 (2005)
41. P. Vontobel et al., J. Nucl. Mater. **356**, 162–167 (2006)
42. E. Lehmann, S. Hartmann, and M. Haller, Schweisstechnik, Issue 1, 6–9 (2008)
43. M. Menon, Doctor of Science Dissertation 16375, Swiss Federal Institute of Technology, Zurich, 2006
44. S. E. Oswald, M. Menon, E. Lehmann, and R. Schulin, Vadose Zone J. (in press)
45. D. Mannes et al., in *Proceedings of the 15th International Symposium on Nondestructive Testing of Wood* (Duluth, 2007) (in press)
46. http://www.woodculther.org

Part II
Applications: Energy

Chapter 12
Vibrational Dynamics and Guest–Host Coupling in Clathrate Hydrates

Michael M. Koza and Helmut Schober

Abstract Clathrate hydrates may turn out either a blessing or a curse for mankind. On one hand, they constitute a huge reservoir of fossil fuel. On the other hand, their decomposition may liberate large amounts of green house gas and have disastrous consequences on sea floor stability. It is thus of paramount importance to understand the formation and stability of these guest–host compounds. Neutron diffraction has successfully occupied a prominent place on the stage of these scientific investigations. Complete understanding, however, is not achieved without an explanation for the thermal properties of clathrates. In particular, the thermal conductivity has a large influence on clathrate formation and conservation. Neutron spectroscopy allows probing the microscopic dynamics of clathrate hydrates. We will show how comparative studies of vibrations in clathrate hydrates give insight into the coupling of the guest to the host lattice. This coupling together with the anharmonicity of the vibrational modes is shown to lay the foundations for the peculiar thermodynamic properties of clathrate hydrates. The results obtained reach far beyond the specific clathrate system. Similar mechanisms are expected to be at work in any guest–host complex.

12.1 Introduction

The solubility of small hydrophobic gas molecules in liquid water is low. This contrasts with the fact that the same molecules can be incorporated at rather high concentration in solid water frameworks called clathrate hydrates [1, 2]. For a long time clathrates were considered a laboratory curiosity and a comprehensive understanding of their complex structural, physical, and chemical properties a rather academic exercise. Interest from industry and government in clathrate research was triggered first in the 1930s by the clogging of natural gas pipelines as a consequence of clathrate formation. Throughout recent decades, clathrate hydrates have moved more and more into the focus of economic and technological interest. As long as the

M.M. Koza (✉)
Institut Laue-Langevin, F–38042 Grenoble Cedex, France
email: koza@ill.fr

L. Liang et al. (eds.), *Neutron Applications in Earth, Energy and Environmental Sciences*, Neutron Scattering Applications and Techniques,
DOI 10.1007/978-0-387-09416-8_12, © Springer Science+Business Media, LLC 2009

steadily growing consumption of energy by our modern societies is primarily satisfied by nonrenewable resources like fossil fuels, there will be a continuing interest in finding new energy supplies. Despite the restrictive thermodynamic formation conditions of these guest-host structures, the amount of clathrate hydrates containing hydrocarbons (primarily methane) known to exist worldwide is in the range of a few million cubic kilometers [3, 4]. Clathrates may constitute a blessing or a curse for mankind depending on whether the benefits of a potential energy resource outweigh the environmental and geological hazards related to their exploitation [5]. For example, a mole of CH_4 is about 24 times more effective in absorbing infrared light than a mole of CO_2. Improper handling of hydrate sediments in the oceans might lead to underwater gas blowout, not only affecting marine ecosystems but contributing in particular to global warming. Understanding the formation and stability of clathrates is the main contribution expected from the scientific community in this context. It is obvious that investigating all facets of structural changes is key to making progress in this area. It is thus not surprising that extensive work based on X-ray and neutron diffraction has been performed on a large variety of clathrate systems.[1] The picture must, however, remain incomplete without an explanation for the thermal properties [6] like specific heat, thermal expansion, and, in particular, thermal transport [7, 8]. Clathrate hydrates, for example, have a thermal conductivity that is only about one fifth that of ice, despite the close structural similarities. All thermal parameters influence the process of formation and conservation. They thus enter the stability of hydrate-bearing oceanic sediments and as such have an influence on global climate change and submarine slide formation [9]. Microscopically, thermal properties depend on the fluctuations within the ionic and electronic system. In a crystal in equilibrium, the basic ionic fluctuations are described by phonons with associated frequencies and lifetimes. As we hope to show in this chapter, inelastic neutron scattering is an ideal experimental tool to investigate all aspects of ionic fluctuations. Textbooks [10] often treat the crystalline world in a highly simplified way. This holds in particular for vibrational properties. At low frequencies, atomic motion is assumed to be fully accounted for by sound waves. This picture, however, breaks down as soon as a structure contains more or less loosely bound subunits that may vibrate (rattle) in a very localized fashion at very low frequencies. There is, therefore, a very fundamental reason for studying the dynamics of clathrates. They constitute a model system for loosely bound particles confined to well-defined cages a few Angstroms in size. If the cages are assumed to be completely rigid, then the dynamic problem boils down to that of a particle in a potential well. The scientific interest of such a situation is the presence of strong anharmonicities (e.g., in the case of a trough-like potential well). These anharmonicities are prone to influence the thermal expansion. An everyday manifestation of anharmonicity is thermal expansion. The elasticity of the cages introduces additional degrees of freedom that allow for the coupling of guest and host vibrations. In simple terms, the moving guest deforms the cage when bouncing into

---

[1] For a good overview, see Kuhs and Hansen [11] and references therein.

the wall, transferring kinetic into elastic energy. In clathrate hydrates this coupling is essential in the sense that it prevents the cage from collapsing and thus preserves clathrate stability [12]. The coupling may also be expected to have a strong effect on scattering of phonons and thus on thermal conductivity. Understanding the dynamics thus becomes, as mentioned before, a prerequisite for understanding stability and formation, closing the knowledge cycle from fundamental research to practical application.

12.2 Structural Properties

12.2.1 Generalities

The term clathrate was coined by H.M.J. Powell in 1948 [13].[2] Clathrates are crystalline compounds whose constituents are arranged in such a way as to form closed cavities encaging atoms or molecules referred to as guests. It should be noted that due to the closed character of the cavities clathrates behave physically quite differently from open nanoporous structures like zeolites or aerogels, where guests are chemically or physically adsorbed to the inner surfaces. The clathrate host framework is preferentially constituted by elements or molecules with strong directional bonding, however, allowing for some variation within the bond angles. The molecular or atomic guests are rather weakly interacting with the host lattice when compared with bonding forces between the host elements. However, in most cases their presence is essential for the stability of the cage structure. The most prominent monatomic clathrates are based on silicon and germanium forming different crystalline structures depending on the stoichiometric admixture of the guests, for example, earth alkali cations. The most prominent molecular clathrate structures are formed by hydrogen-bonded networks of water molecules and are therefore referred to as clathrate hydrates or gas hydrates, when containing gases as guest inclusions. As there are at least two differently sized cages in each of the clathrate structures, there exists a multitude of different ways of filling a clathrate. In addition, clathrates are formed without the requirement of strict stoichiometry of the admixed guests. Hence, the cages can be fully or partially filled, and a complete filling may even result in a multiple of guests in a single cage depending on the size of the guest inclusion and on thermodynamic conditions like temperature and pressure.

---

[2] He used the wording "There may thus arise a structural combination of two substances which remain associated not through strong attraction between them but because strong mutual binding of the molecules of one sort only makes possible the firm enclosure of the other. It is suggested that the general character of this type of combination should be indicated by the description *clathrate* compound – *clathratus*, enclosed or protected by cross bars of a grating."

12.2.2 Hydrate Structures

Clathrate hydrates were discovered in 1810 by Davy [14], and their composition was determined in 1823 by Faraday [15]. More than a century was needed to nail down the structure of hydrate clathrates. In the late 1940s and early 1950s von Stackelberg [16] and his collaborators unambiguously identified two cubic types, which are referred to as SI and SII. A hexagonal type SH was discovered later (We will concentrate in the following on the cubic structures, as they are the most relevant for geoscience.). SI is characterized by the space group Pm3n and a lattice constant of approximately 12 Å. It comprises 46 water molecules within the unit cell. SII is characterized by the space group Fd3m, with an approximate lattice constant of 17 Å and 136 water molecules in a unit cell. The polyhedral cage units forming the two hydrate structures are depicted in Fig. 12.1. A unit cell of SI is constructed by 2 pentagonal dodecahedra (5^{12}) and 6 tetrakaidecahedra, with 12 pentagonal and

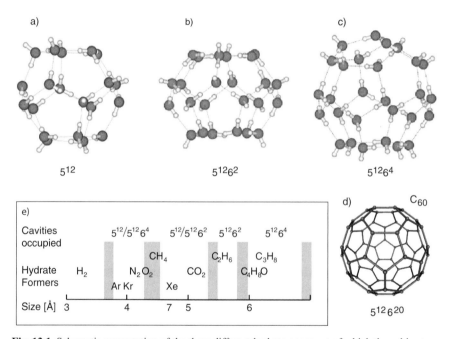

Fig. 12.1 Schematic presentation of the three different hydrate cages out of which the cubic structures SI (cages 5^{12} and $5^{12}6^2$ in a ratio 1:3) and SII (cages 5^{12} and $5^{12}6^4$ in a ratio 2:1) are formed. The cages form convex polyhedra and as such obey the Euler theorem (i.e., the number of vertices V and faces F together is exactly two more than the number of edges E). Symbolically, $V - E + F = 2$. Pentagons are an essential ingredient of the clathrate cages as a pure hexagonal network cannot close into itself. According to Euler's theorem the number of pentagons is strictly identical to 12, while the number of hexagons is in principle arbitrary. Clathrate cages are in this respect similar to fullerenes. The archetypical C_{60} is shown here in d). The clathrate cages are, therefore, equally called buckyball water clusters [18]. However, in fullerenes none of the pentagons make contact with each other. A schematic comparison of hydrate cavity size and the size of some hydrate guests is shown in e)

2 hexagonal faces ($5^{12}6^2$). In the case of SII, 16 5^{12} units and 8 hexacaidecahedra with 12 pentagonal and four hexagonal faces ($5^{12}6^4$) form a diamond like lattice. The pentagonal dodecahedron or small cage is almost spherical with an average cavity radius of about 3.95 Å. An overview of the structural properties of SI and SII is given in Table 12.1. It should be noted that the cages may be appreciably distorted (i.e., the average cage structure is not strictly dodecahedral or tetrakaidecahedral). In addition, the hydrogen atoms are disordered within the water framework like in hexagonal ice. This means that locally in a cage the symmetry is very low. It is not trivial to predict what structure a particular guest will produce. The molecules certainly have to be of a size compatible with the cage dimensions. Small molecules can fill both the small and the larger cages. Double occupancy is possible. Guests with a size on the order of 4–5.5 Å induce preferentially the formation of SI hydrate. Guests belonging to this class include Xe, H_2S, CH_4 (methane), and C_2H_6. Guests of size smaller than 4 Å or larger than 6 Å promote the formation of SII hydrate. Examples are Ar, Kr, N_2, O_2, and C_4H_8O (tetrahydrofuran). The SII structure provides more small cases, ensuring a good van der Waals contact of, for example, N_2 with the cavities. In addition, the larger cage of SII can accommodate, at least at higher pressures, two molecules in response to the high gas fugacity, thus lowering the total free energy. Figure 12.1e) indicates the size relations between the different hydrate cages and the potential guests. Hydrogen constitutes an exceptional example, as the formation of a hydrate requires a multiple occupation of the cavities resulting in an SII structure [19–21]. The large $5^{12}6^4$ cages can accommodate up to four hydrogen molecules whereby the intermolecular distance is reduced by about 30% relative to bulk solid hydrogen [21]. However, it has not been unequivocally shown yet whether the small cages can accommodate more than a single hydrogen molecule. Some discrepancy has been found not only between pure hydrates [20, 21] but also when taking into account the binary hydrate SII formed by hydrogen and tetrahydrofuran (THF, C_4H_8O) [22]. THF occupies exclusively the large cavities $5^{12}6^4$, leaving the smaller cavities free for potential hydrogen accommodation. The presence of two differently sized guests is a prerequisite for the formation of the SH structure. In general, large molecules with a diameter of the order of 7–9 Å and smaller molecular 'helpers' are required to stabilize this hydrate structure; a combination that may be found rarely in nature.

Table 12.1 Properties of the cubic hydrate structures SI and SII according to Koh [18]

Hydrate structure	I		II	
Symmetry	Pm3n		Fd3m	
Lattice constant [Å]	12		17	
Cage geometry	5^{12}	$5^{12}6^2$	5^{12}	$5^{12}6^4$
Cage symmetry	m3	$\bar{4}$2m	$\bar{3}$m	$\bar{4}$3m
Average radius [Å]	3.95	4.33	3.91	4.73
No. of cavities	2	6	16	8
No. of water molecules	46		136	
Guest	Xe, H_2S, CH_4, C_2H_6		Ar, Kr, N_2, C_4H_8O, ...	

The statistical thermodynamic model developed by van der Waals and Platteeuw [23] is the basis for predicting phase equilibria in clathrate hydrates. In this approach, one determines the chemical potentials of water μ_{empty} with respect to an idealized empty framework structure [24]. At the equilibrium of formation from water, the reduction of the chemical potential of a water molecule in the hydrate due to the cavities being occupied, $\Delta\mu_{occ.}$, must equal the difference in chemical potential between the empty lattice and water, $\Delta\mu_{water}$:

$$\Delta\mu_{occ.} = \mu_{empty} - \mu_{water}. \qquad (12.1)$$

The chemical potential is determined by the fractional cage occupancy, which is related to the fugacity of the guest and the water guest interaction. The model can be rendered more sophisticated by including the fact that larger guests slightly distort the framework structure [24].

While the cage distortion depends on the specific guest, the thermal expansion, which is the most tangible effect of anharmonicity, is only structure-dependent [25]. Both SI and SII structures show a thermal expansion coefficient α that is linearly increasing with T between 20 K and room temperature. The slope of $\alpha(T)$ for SI-type clathrates is about the same as in hexagonal ice I_h. The absolute value is about 100% higher in SI than in I_h. This indicates that despite the guest–host coupling, the anharmonicity of the framework is little influenced by the guest.

12.3 Thermal Conductivity

As already mentioned in the introduction, thermal conductivity and thermal diffusivity enter the equations of stability and formation of clathrates. Precise information on them is thus a highly desirable input for climate prediction models. The experimental problem consists in taking proper account of the porous and often inhomogeneous nature of naturally occurring clathrates. Among the inhomogeneities are untransformed or only partially converted ice and/or water grains. The thermal properties can vary appreciably when comparing samples with different degrees of porosity and homogeneity [7, 8]. For practical predictions, the thermodynamic properties pertaining to the system under consideration have to be used. From a theoretical point of view, it would be highly desirable to extract the intrinsic thermal conductivity and thermal diffusivity of an ideal nonporous clathrate, because it is that quantity that can be extracted from molecular dynamics (MD) simulations [26, 27]. Understanding the intrinsic thermal conductivity on a molecular level will be the indispensable starting point for a comprehension of the properties of inhomogeneous and highly defect prone systems.

Thermal conduction is due to heat transport via the propagation of excitations. The excitations can be of multiple origin: electronic, magnetic, or phononic. In insulators, the main transport channel is provided by propagating phonons. The thermal conductivity encounters resistance due to the scattering of the energy carriers. Sources of scattering are the surfaces of the crystallites, defects, and the excitation

Table 12.2 Van der Waals radii* (from www.webelements.com) in Å. Diameters of clathrate guest molecules in Å

	R [Å]
Ne	1.54*
Ar	1.88*
Kr	2.02*
Xe	2.16*
N_2	4.1
CH_4	4.36
H_2S	4.58
CO_2	5.12
C_2H_6	5.5
C_4H_8O	5.9 (6.3)

system itself. The latter provides intrinsic resistive processes that define the maximum thermal conductivity which cannot be surpassed even by best purification of the material. In insulators, the phonon scattering due to anharmonicities is the dominant intrinsic scattering process. The simplest expression (Debye formula) for describing thermal transport is in that case obtained via the kinetic relaxation time approximation [28–30]:

$$\kappa(T) = \frac{1}{3} \cdot C(T) \cdot v_{ph} \cdot l_{ph} = \frac{1}{3} \cdot C(T) \cdot v_{ph}^2 \cdot \tau_{ph}, \qquad (12.2)$$

with $C(T)$ the specific heat of the material and $v_{ph}(T)$, $l_{ph}(T)$, and $\tau_{ph}(T)$ the average phonon velocity, the phonon mean free path, and the mean phonon scattering (relaxation) time, respectively. The highest phonon velocities are obtained for the highly dispersive acoustic branches ($v = d\omega/dq$). The main heat transport carriers are thus the propagating acoustic phonons. If the crystal is purely harmonic, then the phonon lifetimes are infinite and the thermal conductivity as well. In the quantum mechanical picture, the phonons in a harmonic system are stationary states. Therefore, if a distribution of phonons is created which carries a thermal current that distribution will not change in time and the thermal current will remain undegraded even in the case of a vanishing temperature gradient. Although it is difficult to give general trends for the dependence of the thermal conductivity on the system parameters,[3] it is fair to say that for a given level of anharmonicity in the potential, the ideal thermal conductor is a single crystal of high density and high velocity of sound combined with a small number of atoms in the unit cell [26]. Larger unit cells lead to optical phonon modes. These lower the average propagation speed and at the same time provide a mechanism for scattering the propagating phonons.

---

[3] Changing the structure, atomic masses, or atomic interactions alters the dispersion relations. While a reduction in the mean slope of the dispersion (lower average v_{ph}) may favor a lower conductivity, the effect may in exceptional cases be compensated by a resulting reduced phase space for scattering [31].

As different scattering mechanisms lead to different temperature dependence of the thermal conductivity, the investigation of κ as a function of T contributes valuable insight into the underlying physics. The energy scale of acoustic excitations is described by the Debye temperature $\theta_D = \hbar\omega_D/k_B$. In ice I_h $\theta_D \approx 226\,\text{K}$ (i.e., $\hbar\omega_D \approx 20\,\text{meV}$). Scattering is only resistive if the momentum is at least partially reversed. At low temperatures this happens at the surface or via defects. The cross-sections for this scattering are temperature-independent and the thermal conductivity has the temperature dependence of the specific heat (i.e., $\kappa \propto T^3$). The so-called Umklapp–scattering processes that involve the backscattering of phonons via the lattice layers due to the anharmonicity of the inter-atomic potentials are frozen out.[4] In the so-called high-temperature limit (i.e., at temperatures higher than the Debye temperature of the material), Umklapp-scattering becomes the main channel for limiting the mean free path l_{ph}. Because the occupation number of a phonon with energy $\hbar\omega$ that can participate in Umklapp-scattering increases like $k_B T/\hbar\omega$ for $\hbar\omega \ll k_B T$ (see Eq. 12.3), it follows that l_{ph} is more or less $\propto 1/T$. With C and v_{ph} approximately constant for a harmonic solid, this leads according to Eq. 12.2 to $\kappa \propto 1/T$ for ($T > \theta_D$). This trend is followed in most ice phases above some $100\,\text{K}$ [28] (i.e., already below θ_D).

With respect to their canonical reference system, crystalline ice clathrates display unusual thermal conductivity characteristics. The most prominent of these is the fact that the absolute value of $\kappa(T)$ is strongly reduced. This is exemplified by the measurements performed on two grainy samples of xenon hydrate (upper panel of Fig. 12.2). The crystalline ice I_h shown for reference is obtained as the residual product of the decomposition of the respective hydrates. Despite the differences between $\kappa(T)$ of the hydrate samples due to variations in their grainy consistency, their overall thermal conductivity is reduced by one order of magnitude relative to ice I_h. The same holds for other hydrates (Figure 12.2 lower panel). The thermal conductivity close to room temperature of methane clathrate can vary from about 0.25 to 0.68 W/m/K depending on porosity [7, 8]. For comparison, the thermal conductivity of ice is about 2.23 W/m/K.

Not only the absolute value of the thermal conductivity but equally its functional dependence on temperature differ between clathrate hydrates on the one hand and crystalline ice on the other hand. A rather steadily increasing $\kappa(T)$ upon heating has been observed in some clathrate hydrates. The lower panel of Fig. 12.2 illustrates the thermal conductivities of the noble gas xenon SI and the molecular methane SI and THF SII hydrates. This property is apparently expressed in a rather distinctive way by molecular, ring-shaped guests like THF (C_4H_8O) [34, 35] or 1,3–dioxolane ($C_3H_6O_2$) and cyclobutanone (C_4H_6O) [36, 37].

At this point of the discussion it is appropriate to introduce the notion of the *glass analogy* [33], as it has found quite a large forum for discussion in the recent literature. Similar deviations of $\kappa(T)$ from the behavior expected for canonical crystals

---

[4] Only in very pure samples of large dimensions the very few remaining Umklapp processes are the limiting factor for thermal conductivity at low temperature.

Fig. 12.2 Temperature dependence of the thermal conductivity of xenon, methane, and THF hydrates. The upper panel indicates the dependence of the uncorrected, effective $\kappa(T)$ on the grainy structure of two xenon hydrate samples and their crystalline ice residuals [32]. There are marked differences between the ice and clathrate both in the absolute size and in the functional dependence of $\kappa(T)$. The thermal conductivity in ice is nearly an order of magnitude higher than in the clathrate. It decreases above 30 K. In the clathrate $\kappa(T)$ continuously increases with a sort of resonance (dip) around 100 K. The lower panel compares the corrected $\kappa(T)$ of the three hydrates [33]. The dip is very pronounced in xenon hydrate, while in the case of THF hydrate $\kappa(T)$ increases monotonously up to the decomposition temperature

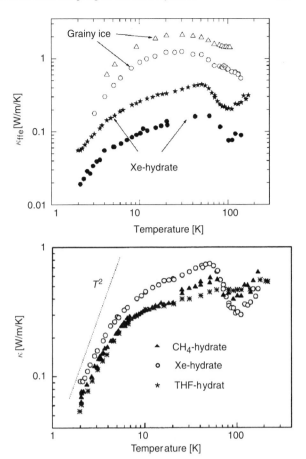

as found in clathrates are observed in disordered solids and glasses. Thermal conductivities are inferior by one to two orders of magnitude when compared to their crystalline counterparts. Like some clathrates, glasses show a persistently increasing $\kappa(T)$ at elevated temperatures. A basic understanding of this behavior is offered within the framework of Eq. (12.2) under the assumption that the *topological disorder* of glasses limits the phonon mean free path to a constant value (e.g., of the order of the next neighbor distance). In this case $\kappa(T)$ is dominated by the temperature behavior of the specific heat $C(T)$, which is supposed to approach persistently the Dulong–Petit constant with increasing T. In hydrates, the only intrinsic disorder is the proton disorder in the water framework plus the possibility of an ill-defined position of the guest in large cages.

A certain significance of the analogy with glasses is supported by the rather good approximation of $\kappa(T)$ in hydrates via models established from the characterization of the inelastic response of disordered solids like the soft-potential and resonant-scattering model [33, 34, 38, 39, 40]. An additional argument for the glassy 'picture' of hydrates is given by the phonon mean free path $l_{ph}(T)$ computed from the measured $\kappa(T)$ of Fig. 12.2 [33, 35]. $l_{ph}(T)$ is apparently reduced from the order of 10^3 Å below 10 K to only a few Å at 100 K, marking the transition from a ballistic to a diffusive thermal transport.

When it comes to the details, the glass analogy does not explain the thermal conductivity vey well. As indicated in Fig. 12.2 xenon hydrate displays a maximum in the thermal conductivity at about 50 K, which sharply decreases up to about 100 K and then recovers. This behavior is as well observed, though with a lower accuracy, in hydrates hosting the compact, tetrahedral methane molecule. Features like these are not standard signatures of glassy systems. The most obvious disordered systems to compare clathrate hydrates to are amorphous ice modifications. However, low-density amorphous ices generally show a crystal-like harmonic dynamics- and thermal conductivity [28, 41–44].

In the discussion so far, it has been assumed that the systems behave in a quasi-harmonic fashion (i.e., that the fundamental excitations are phonon-like, with frequencies that do not strongly depend on temperature). In most glasses these conditions are fulfilled even beyond the glass transition temperature. As has been demonstrated by numerous experiments, this is not always the case in clathrates, and analogies thus have to be used with due care.

The reason underlying the onset of strong anharmonicity in clathrate hydrates at rather low temperatures is the loose binding of the guest to the host framework. This allows for decoupling of certain excitations in the two subsystems. Figure 12.3 reports, for example, the results of specific heat measurements on THF hydrate in a T-range comparable to the thermal conductivity measurements [45]. It can be deduced that above 120 K THF rotates nearly freely within the large hydrate cages, whereas below 120 K it undergoes a sluggish "freezing" into hindered rotations described by librational modes over a wide temperature range. This dynamic

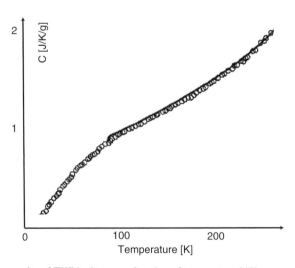

Fig. 12.3 Heat capacity of THF hydrate as a function of temperature [45]

transition from a plastic crystal to a rotational glass prevents the system from converging toward a single Dulong–Petit limit as the character of the normal modes change with T. Even though the ring-shaped guest molecule performs above 120 K a tumbling motion, it is still clathrated; that is, it is confined to the cage. This confinement leads to a *strong coupling* to the hydrate lattice with respect to the guest's translational motions. This apparent contradiction of weak guest–host interaction and strong guest–host coupling is one of the main specificities of clathrate dynamics. The guest–host coupling is demonstrated experimentally by the effect that the size of the molecules (e.g., replacing THF by HF, 1,3–dioxolane, and cyclobutanone [36]) has on the thermal conductivity. When reducing the assumptions entering the Debye formula to a constant velocity of sound while using experimental input for thermal conductivity, specific heat, and thermal expansion, a constant mean free path of about the order of the cage size is deduced for the THF hydrate [48].[5]

Summarizing, as expected for a complex solid with a large number of atoms per unit cell, the thermal conductivity of clathrates differs strongly from that of crystalline reference systems. The temperature dependence shows trends known from topologically disordered systems. The origin of this apparent glass-like response is, however, rather to be sought in the complexity of the guest dynamics, guest coupling to the vibrational degrees of freedom of the host lattice, and response to temperature changes. The study of these effects requires microscopic spectroscopy techniques that we will detail in the following sections.

12.4 Phonons

A harmonic vibration or phonon is defined via its energy $\hbar\omega_j$ and eigenvector $\mathbf{e}_j$. The eigenvector has three components $e_{j,\alpha}(k)$ for every atom k describing the amplitude of motion along three perpendicular directions α. It is in general complex, as different atoms may vibrate with an arbitrary phase with respect to each other. The density-of-states, $Z(\hbar\omega)$, enumerates these basic excitations as a function of energy. The integral over the density-of-states is equal to the number of degrees of freedom unless otherwise defined. Apart from quantum fluctuations, the population of the modes, or in classical language the amplitude of vibration, depends on temperature. As the vibrational excitations satisfy boson statistics, the occupation of a mode j at temperature T is given by

$$n(\omega_j) = \frac{1}{e^{\frac{\hbar\omega_j}{k_B T}} - 1}. \tag{12.3}$$

---

[5] For the case of 1,3–dioxolane hydrate, the observation that only rapid formation of the hydrate results in a steadily increasing $\kappa(T)$ upon heating [37] renders the interpretation even more difficult. These results could, however, not be reproduced on other systems [28].

In a purely harmonic system, neither $\hbar\omega_j$ nor $\mathbf{e}_j$ depend themselves on temperature. As demonstrated by the peculiar specific heat dependence of THF hydrate, the idea of harmonic solids has to be discarded. This introduces an explicit temperature dependence into the density-of-states $Z(\hbar\omega, T)$.[6]

So far we have not discussed translational invariance. In the case of vibrations in a crystal, symmetry requires that the phonon wavefunction in different unit cells k and l can only differ by a phasefactor of the form

$$\exp(i\mathbf{q} \cdot (\mathbf{r}_{\mathbf{k}}^o - \mathbf{r}_{\mathbf{l}}^o)), \tag{12.4}$$

with $\mathbf{q}$ lying in the first Brillouin zone of the reciprocal lattice and $\mathbf{r}_k^o$ and $\mathbf{r}_l^o$ denoting the positions of equivalent atoms in the respective cells. In a crystal we are thus dealing with a classification of the eigenvectors in terms of plane waves of wavelength $2\pi/|\mathbf{q}|$ along the direction $\hat{q}$. Such a classification scheme is not possible in an amorphous system without translational symmetry as $\mathbf{q}$ is not a good quantum number. If the crystal is harmonic, the excitations are multiple occupation of $3N$ fundamental vibrational modes described by their respective eigenvectors. There are $3r$ eigenvectors for every $\mathbf{q}$ of the first Brillouin zone, with r denoting the number of atoms in the primitive cell of the crystal. This leads to the notion of dispersion sheets $\omega_j(\mathbf{q})$ in reciprocal space. As these are periodic in reciprocal space, the complete vibrational information is contained in the first Brillouin zone. Figure 12.8 gives an example of dispersion relations for clathrate hydrates.

In a real crystal, potentials are never harmonic. In general, anharmonicities become more pronounced at higher temperatures as the atoms 'explore' regions further away from the equilibrium position. Typical manifestations of anharmonicity are thermal expansion and finite thermal conductivity. This picture is based on the assumption that the equilibrium position is well specified at any temperature. The potential has a well-defined minimum $\mathbf{r}_0$ and is described to first approximation by a parabola $V(\mathbf{r} - \mathbf{r}_0) = V_0'' \cdot |\mathbf{r} - \mathbf{r}_0|^2$ around this minimum, where V_0 may depend on the direction. Deviations from the parabolic form are felt when the atom explores regions far from the minimum (i.e., for larger amplitudes of vibration at higher temperatures). For the atoms responsible for the stability of a classical crystal, the existence of equilibrium positions is a reasonable requirement. If the atoms were not in equilibrium, then this would lead to a relaxation of the structure until equilibrium was reached. This argument applies to the clathrate framework. The guests have no real structural role apart from exerting internal pressure within the cage due to their volume and motion. Their contribution to stability can, therefore, be purely dynamic in the case where the size of the guest is smaller than the cage diameter. In that scenario, the mean position of an atom is stabilized by the motion itself, and thus, it is the more poorly defined the lower the vibrational amplitude

---

[6] It has, however, to be noted that the formalism proposed here relies on the existence of well-defined excitations, which can be enumerated as a function of energy (i.e., via a density-of-states). It is not easily applicable when the microscopic motions have a relaxation character as is the case for THF clathrate in the rotator phase.

(i.e., the lower the temperature). The situation may be described by a trough-like potential with a flat bottom and infinitely steep walls. At low temperatures the molecule is tumbling in this potential. Only at higher temperatures will it bounce periodically from wall to wall and thus acquire both a higher frequency and a mean position. We find these anomalous anharmonic effects in diatomic hydrates, for example.

In a general direction of reciprocal space, dispersion sheets avoid each other even in a purely harmonic crystal. This is called 'anti-crossing'. It was demonstrated by Neumann and Wigner [47] that the probability of crossing is zero for two phonon branches, which belong to the same irreducible representation (i.e., possess the same symmetry). An arbitrary point in the Brillouin zone has no particular symmetry, and thus all phonon branches belong to the trivial unit representation. Anti-crossing is thus an unavoidable fact for the bulk of excitations in a crystal. In the simple case of a flat optic sheet that intersects the acoustic dispersion, the acoustic modes will gain more or less optic character and vice versa. The question is, whether the avoided crossing is extended or just local in $(\mathbf{Q}, \hbar\omega)$. This is schematically shown in Fig. 12.4. If the coupling and thus mixing of eigenvectors is local in reciprocal space (i.e., restricted to the immediate neighbourhood of the avoided crossing points), the consequences for the overall distribution of modes will be minor. The experimental results accumulated on clathrates show that the coupling and thus avoided crossing properties depend crucially on the guest. Everything from a local anti crossing in the case of light and small nitrogen molecules to extended modifications of the dispersion sheets for large and heavy xenon guests is possible.

In a comprehensive formulation of the thermal conductivity, Eq. (12.2) has to be rewritten to take into consideration the details of the phonon excitations. Introducing

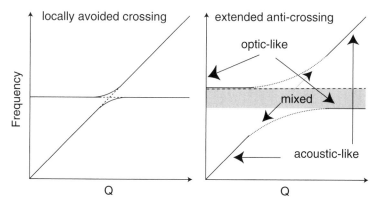

Fig. 12.4 Schematic presentation of local and extended anti-crossing. In the case of weak coupling, the branches are only marginally perturbed outside the immediate neighbourhood of the avoided crossing point. In case of strong coupling, the branches have a strongly mixed character over an extended region of the Brillouin zone. This may even lead to the opening of a gap in the excitation spectrum

the thermal conductivity tensor via the relation

$$J_\alpha = \sum_\beta \kappa_{\alpha\beta} \frac{\partial T}{\partial x_\beta}, \qquad (12.5)$$

relating the heat current J_α along the direction α to a temperature gradient $\partial T/\partial x_\beta$ along the direction β, we obtain

$$\kappa_{\alpha\beta}(T) = \frac{1}{(2\pi)^3} \sum_j \int d^3q \left[C_j(\mathbf{q}, T) \cdot v_{j\alpha}(\mathbf{q}, T) \cdot v_{j\beta}(\mathbf{q}, T) \cdot \tau_{j\beta}(T) \right], \quad (12.6)$$

where $v_{j\alpha}(\mathbf{q}, T)$ is the derivative of the dispersion sheet j along the direction α and $\tau_{j\beta}$ is the corresponding relaxation time. The sum runs over all dispersion sheets and the $\mathbf{q}$ integration over the full Brillouin zone. This covers all vibrations of the system.

$$C_j(\mathbf{q}) = \frac{(\hbar\omega)^2}{k_B T^2} n_j(\mathbf{q})(n_j(\mathbf{q}) + 1) \qquad (12.7)$$

is the contribution of the mode $(j, \mathbf{q})$ to the specific heat. The frequency - dependent specific heat derived from this expression is directly related to the distribution of characteristic frequencies, i.e., to the density of vibrational states, $Z(\hbar\omega, T)$, of the system:

$$C(\omega, T) = k_B \left(\frac{\hbar\omega}{2k_B T} \right)^2 \frac{Z(\hbar\omega, T)}{\sinh^2(\frac{\hbar\omega}{2k_B T})} . \qquad (12.8)$$

In the framework of the Debye model, the mean velocity of sound $\bar{v}_{ph}$ can be equally related to $Z(\hbar\omega, T)$.

$$\bar{v}_{ph} = \left[\frac{3}{2\pi^2} \cdot \frac{V}{N} \cdot \frac{\omega^2}{Z(\hbar\omega, T)} \right]^{1/3}, \qquad (12.9)$$

with N the number of atoms per volume V. This approximation holds for crystalline samples to a very good degree for temperatures $T \ll \theta_D$, which defines the temperature and energy range of interest. In the past, θ_D has been reported to be in the range of 200–250 K for some crystalline and amorphous ice samples [44, 48]. An equivalent energy range of interest can be deduced from the thermal conductivity of the three hydrates depicted in Fig. 12.2 to be 200 μeV–20 meV (2–200 K, 48 MHz–4.8 THz, 1.6 − 160 cm^{-1}). To access this low-energy region with the promises of sufficiently high energy resolution (e.g., 100 μeV), neutron spectrometers operating with cold and thermal neutrons (i.e., neutrons with incident energies of a few to some tens of meV) are the optimum choice (see chapter 3 on instrumentation).

To rationalize the thermal conductivity in terms of the microscopic dynamics requires, according to Eq. (12.6), the knowledge of the dispersion relations from which the velocities can be readily determined. This is, however, not sufficient. All the information concerning the phonon scattering is contained in the relaxation times $\tau_{j\beta}$. These are experimentally only accessible in single crystals via the measurement of the phonon line width. Their calculation from simulations is equally far from trivial [31]. Hints concerning τ can, however, be obtained from the temperature evolution of the phonon spectrum, which reflects the anharmonicities and thus indirectly the relaxation times.

12.5 Inelastic Neutron Scattering

We will now briefly describe how the phonon properties and in particular the density-of-states can be determined via inelastic neutron scattering (see introductory chapters). The central output of neutron scattering experiments is the double differential cross section:

$$\frac{d^2\sigma}{d\Omega d\hbar\omega} = K_P \cdot C_{PS} \cdot S(\mathbf{Q}, \hbar\omega). \tag{12.10}$$

The dynamic structure factor $S(\mathbf{Q}, \hbar\omega)$ represents the sum over all pair correlation functions of a sample's constituents in the four-dimensional reciprocal space of energy ($\hbar\omega$) and momentum ($\mathbf{Q}$). The parameters K_P and C_{PS} describe the kinematics of the neutron probe and the coupling between the probe and the sample, respectively, and $d\Omega$ characterizes the solid angle in which the signal is detected. The polycrystalline character of most of the hydrates averages the signal over the three spatial directions and, thus, reduces the information to only two dimensions, namely the energy and the modulus of momentum $S(Q, \hbar\omega)$. Therefore, in practice unique information on the physics of the polycrystalline samples is reduced from a direction in space $d\Omega$ to a single angle 2θ. Neutron time-of-flight spectrometers are designed to cover simultaneously wide 2θ ranges as it is shown in Fig. 12.5 for the time-of-flight spectrometer IN6 at Institut Laue Langevin (ILL) in Grenoble, France. The kinematics of neutrons ($E_n = (\hbar k_n)^2/2m_n$) imposes well-defined trajectories in ($Q, \hbar\omega$)-space. Figure 12.5 shows the phase space region that is accessible to neutrons of a wavelength of 9 meV at the spectrometer IN5 at ILL. The coupling C_{PS} of the probe to the sample goes via the strong interaction of the neutron with the nuclei of the material. It is described by the Fermi pseudo potential (see introductory chapters). As the neutron interaction is with the nuclei, the corresponding scattering cross sections depend on the isotope and vary in a non-systematic way as you go through the periodic table. The signal contains coherent and incoherent contributions depending on the isotope distribution and nuclear spin state of the scatterer [49]. Relevant cross sections can be viewed in

Fig. 12.5 *Upper panel:*
sketch of the cold
time–of–flight spectrometer
IN6 at the Institut Laue
Langevin (ILL) in Grenoble,
France. The detector bank of
the spectrometer comprises
337 detectors covering
scattering angles 2θ of
$10–115°$. *Lower panel:*
presentation of the $\hbar\omega–Q$
phase space accessed
simultaneously by a
time–of–flight instrument
(*gray shaded area*) with an
incident energy of 9 meV and
a 2θ coverage of $15–132°$,
like that achievable on the
IN5 instrument at ILL. The
solid line indicates
schematically the dispersion
relation of a phonon
matching the conditions of a
transverse acoustic phonon of
hexagonal ice I_h and cubic
ice I_c. The positive energy
scale corresponds to the
energy gain side of neutrons
(upscattering)

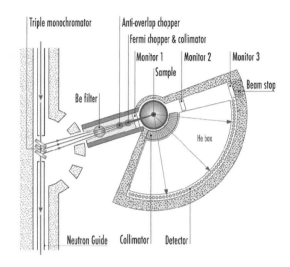

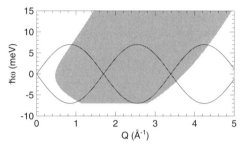

Table 12.3. As a consequence, the contributions of each of the scattering species to the measured signal $S(Q, \hbar\omega)$ is weighed with its relative concentration and cross section.

We now turn to the inelastic spectra and concentrate purely on the contribution to $S(\mathbf{Q}, \hbar\omega)$ arising from the interaction of a neutron with a single phonon.[7] It can be calculated rigorously that

$$S(\mathbf{Q}, \hbar\omega) = \frac{1}{2N < \overline{b^2} >} \sum_j \frac{|F_j(\mathbf{Q})|^2}{\omega_j} \left[(n_j + 1)\delta(\omega - \omega_j) + n_j\delta(\omega + \omega_j) \right],$$

$$(12.11)$$

---

[7] The correction of multi-phonon contributions is possible [51]. The description of the various aspects of this problem is, however, beyond the scope of the basic introduction given in this section.

Table 12.3 Coherent and incoherent scattering cross sections, atomic and molecular masses, and the resulting total scattering powers for translational modes. Of particular interest for hydrates is the strong difference in the scattering cross sections between hydrogen and deuterium. This allows a change of contrast between the scattering from the guests and the water framework. This contrast variation is a powerful standard tool in neutron scattering. For inelastic scattering the influence of the isotopic substitution on the dynamics has to be taken into consideration. The dynamics of D_2O are not identical to the one of H_2O

	$\sigma_{coh}[barn]$	σ_{inc} [barn]	m [a.m.u.]	σ_{tot}/m
H	1.76	80.27	1.0	82.03
D	5.59	2.05	2.0	3.82
C	5.55	0.00	12.0	0.46
N	11.01	0.50	14.0	0.82
O	4.23	0.00	16.0	0.26
Ar	0.46	0.23	39.9	0.02
Kr	7.68	0.01	83.8	0.09
Xe	2.96	0.00	131.3	0.02
H_2O			18.0	9.35
D_2O			20.0	0.98
CH_4			16.0	20.85
CO_2			44.0	0.32
C_4H_8O			72.0	9.48

with

$$|F_j(\mathbf{Q})|^2 = \sum_{k,l}^{N} \frac{\overline{b_k b_l}}{\sqrt{m_k m_l}} e^{-W_k(\mathbf{Q})-W_l(\mathbf{Q})} \left(\mathbf{Q}\cdot\mathbf{e}_j(k)\right)\left(\mathbf{Q}\cdot\mathbf{e}_j(l)\right) e^{i\mathbf{Q}\cdot(\mathbf{r}_k^0-\mathbf{r}_l^0)}, \quad (12.12)$$

where $\mathbf{r}_k^0$, m_k and b_k denote the average atomic position, the mass, and the scattering length of atom k, respectively and $\mathbf{e}_j(k)$ is the three-dimensional component of the 3N-dimensional eigenvector $\mathbf{e}_j$ referring to the atom k. The expression $|F_j(\mathbf{Q})|^2$ stands for the structure factor of the vibrations $\{j\}$. The *Debye-Waller factor* can be calculated as

$$W_k(Q) = \frac{1}{2} < (\mathbf{Q}\cdot\mathbf{u}_i)^2 > = \frac{\hbar}{m_i}\sum_j \frac{(\mathbf{Q}\cdot\mathbf{e}_{j(k)})^2}{\omega_j}(2n_j + 1). \quad (12.13)$$

The δ-functions in Eq. 12.11 express energy conservation. In the down-scattering case, the neutron loses energy which is taken up by the sample in the form of vibrations. Down-scattering is possible at all temperatures; however, it requires that the neutron energy exceeds the energy of the vibrations to be probed. Up-scattering involves energy transfer from the sample to the neutron and is negligibly small at temperatures $k_B T \ll \hbar\omega$.

Equations 12.11 through 12.13 imply that if all the eigenvectors and corresponding eigenfrequencies are known, then the inelastic scattering function of a harmonic system can be determined exactly. Rather complete sets of eigenvectors and eigenfrequencies can be obtained ab initio or via classical MD using modern computer codes for systems of reasonable complexity. The main art required for calculating $S(\mathbf{Q}, \hbar\omega)$ in these cases is the proper sampling of the Brillouin zone.

In a one-component system $Z(\hbar\omega)$ is directly related to the one-phonon part of the incoherent double differential cross section defined in Eq. (12.10) [50].

$$\left(\frac{d^2\sigma}{d\hbar\omega d\Omega}\right)_{\text{inc}} = \frac{N\sigma_i}{8\pi m}\frac{k_f}{k_i}Q^2 e^{-2W(\mathbf{Q})}\frac{Z(\hbar\omega)}{\omega}(n(\omega)+1), \qquad (12.14)$$

with $\sigma_i := 4\pi(\overline{b_i^2} - \overline{b_i}^2)$ denoting the incoherent scattering cross section of the nuclei. The proportionality of the incoherent one-phonon double differential cross section to the density-of-states expressed in Eq. (12.14) is rigorous if the average

$$\left\{\sum_j \sum_{k=1}^{N} |\mathbf{Q}\cdot\mathbf{e}_j(k)|^2\right\}_{\omega_j=\omega} \qquad (12.15)$$

over all modes of frequency $\omega_j = \omega$ is independent of ω. In cubic or completely isotropic systems this is the case. Equation (12.14) tells us that the incoherent one-phonon scattering takes place at all $\mathbf{Q}$ and is isotropic (i.e., independent of the direction $\hat{Q}$) in reciprocal space. For coherent scatterers, we are obliged to invoke the incoherent approximation [52, 53] to extract $Z(\hbar\omega)$ from the inelastic neutron scattering (INS) spectra. The incoherent approximation consists in assuming that

$$\int_{Q_{\min}}^{Q_{\max}} \left(\frac{d^2\sigma}{d(\hbar\omega)d\Omega}\right)_{\text{coh}} QdQ \approx \int_{Q_{\min}}^{Q_{\max}} \left(\frac{d^2\sigma}{d(\hbar\omega)d\Omega}\right)_{\text{inc}} QdQ, \qquad (12.16)$$

for a sufficiently large Q-sampling, where the right-hand side refers to a hypothetical incoherent sample. The validity of the incoherent approximation can be checked by comparing $Z(\hbar\omega)$ as obtained from different sampling regions in Q-space. Another possibility is the estimation of the error induced by the incoherent approximation using computer simulations. Good results should already be obtainable with dynamic models of a low degree of sophistication. In general, both experiment [54] and simulation [53] show that even in highly unfavorable cases errors rarely exceed 20%.

In the case of a multi-component system, we still can write the dynamic scattering function in the form [53]

$$S(\mathbf{Q}, \hbar\omega) = A(\mathbf{Q}, \hbar\omega)G(\hbar\omega), \qquad (12.17)$$

with an analogy to Eq. (12.14)

$$A(\mathbf{Q}, \hbar\omega) = \frac{\hbar\left(\overline{b}^2 e^{-2W(Q)}\right)_{\text{av}}}{2M_{\text{av}}}\frac{Q^2}{\omega}[n(\omega)+1)], \qquad (12.18)$$

with

$$M_{av} = \frac{1}{N} \sum_k m_k \quad \text{and} \quad \left(\overline{b}^2 e^{-2W(Q)}\right)_{av} = \frac{1}{N} \sum_k \overline{b}_k^2 \exp(-2W_k(Q)), \quad (12.19)$$

and the Debye-Waller factors $W_k(Q)$ averaged over $\hat{Q}$ directions as outlined above for the one-component case. $G(\hbar\omega)$ as obtained from the inelastic neutron scattering data via Eq. (12.17) is well defined. It is called the generalized vibrational density-of-state that differs from the actual vibrational density-of-states because the contributions of the different types of nuclei $\{k\}$ to the form factor of modes $\{j\}$ are weighed with their inelastic scattering powers $\{\sigma_k/M_k\}$. In cases where the aim of the experiment is a verification of a dynamical model, the proper way to proceed is to calculate from the theory the generalized density of states $G(\hbar\omega)$ or even better $S(\mathbf{Q}, \hbar\omega)$.

Without a model, one has to be aware of the errors when blindly identifying $G(\hbar\omega)$ with $Z(\hbar\omega)$. These errors can be very important as can be shown for the simple example of ice. Ice being a typical molecular crystal features several bands of excitations. At low frequencies ($\hbar\omega < 40$) we find the vibrations involving a translation of the center of gravity. Let us ignore for the moment the fact that all ice phases feature more than one molecule per primitive cell. If we assume a complete decoupling from the librations discussed below, then the system in this low-frequency region can be considered atomic (the atom is the H_2O molecule), and the weight of any one of these translational modes to the dynamic structure factor is determined by $2\sigma_i(H)/m(H_2O)$, keeping in mind that the eigenvectors e_j are normalized to unity and that the scattering is dominated by the incoherent cross section of hydrogen. The next band of excitations comprises the librations. Like the translations, the librations are external modes (i.e., they do not involve distortions of the H_2O molecule).[8] For ideal librations the center of gravity is at rest (see decoupling mentioned above). As the center of gravity is close to the oxygen atom, the eigenmodes can reasonably be considered to have nonzero components for the hydrogen atoms only. The weight of any one librational mode to the dynamic structure factor is therefore determined by $\sigma_i(H)/m(H)$. This means that although the two bands contain the same number of modes, the intensity of the librational band will be enhanced by a factor close to 9 with respect to a hypothetical translational band at the same frequency. Ignoring this fact and simply interpreting the generalized density-of-states as the true density-of-states, therefore, leads to enormous errors in this admittedly highly unfavorable case. This is confirmed by more exact calculations involving MD simulations [55]. In some cases an alternative presentation in terms of the generalized susceptibility is preferable. $\chi''(\hbar\omega)$, which stresses the temperature-corrected intensity of low-energy modes

---

[8] The external modes of the H_2O molecule are found at still higher frequencies.

$$S(Q, \hbar\omega) = \frac{1}{\pi}(n(\hbar\omega) + 1)\chi''[\hbar\omega]. \qquad (12.20)$$

For a simple harmonic oscillator the susceptibility is

$$\chi''[\hbar\omega] = \frac{\pi}{2m\omega_0} \left\{ \delta(\omega - \omega_0) - \delta(\omega + \omega_0) \right\}. \qquad (12.21)$$

$\chi''[\omega]$ for a set of uncoupled harmonic oscillators is, therefore, independent of temperature. A temperature variation in the susceptibility is thus a good indicator for anharmonicities.[9]

12.6 Clathrate Hydrogen Dynamics

12.6.1 The Dynamics of the Host Lattice

Despite the apparent complexity of the dynamics and of their experimental verification, solid water structures show some characteristics which appreciably simplify the interpretation of experimental data. First of all, eigenmodes of different character are well separated on the energy scale. Figure 12.6 shows the generalized density-of-states, $G(\hbar\omega)$, of different hydrate structures. The dominating feature in both SI and SII hydrate types are two peaks centered at about 7–10.5 meV in contrast to the single peak in hexagonal ice I_h at about 6.5 meV. This is illustrated in the upper panel of Fig. 12.6 which shows the Ar–SII and Xe–SI hydrate responses in comparison to ice I_h in the range of translational excitations. As in other ice structures, the translational excitations of both hydrate structures are well separated from the librational excitations of the water molecules by a mode gap opening up at about 40 meV [56, 57]. Librations are located at energies of $50 \geq \hbar\omega \geq 120$ meV, whereby a strong sensitivity to the density of the sample and isotope (H $\leftrightarrow$ D) exchange is to be noticed. Intramolecular vibrational modes of solid water structures display eigenfrequencies higher than 100 meV.

Depending on the encaged guests, subtle variations are observed in particular at the first pronounced peak as can be seen in Fig. 12.6. Despite the fact that the positions of the peaks do not vary greatly, the variations indicate that host–guest coupling is at work. To enhance the guest signal, all generalized densities-of-states have been measured with deuterated host material. Some of the guestmodes resolved from the strong peaks are indicated by vertical arrows in the lower panel of Fig. 12.6. These observations are in agreement with other experimental results on hydrate samples of Xe [58, 59] Kr [60], CH_4 [61] and THF [57, 62]. However, for a detailed comparison of the inelastic data reported in the literature, attention has to be payed to the sample material, because a deuteration of the sample shifts the characteristic

---

[9] This holds provided that the variation of the scattering induced by the Debye-Waller factor and higher-order phonon terms compensate, which is normally the case over quite extended temperature ranges [54].

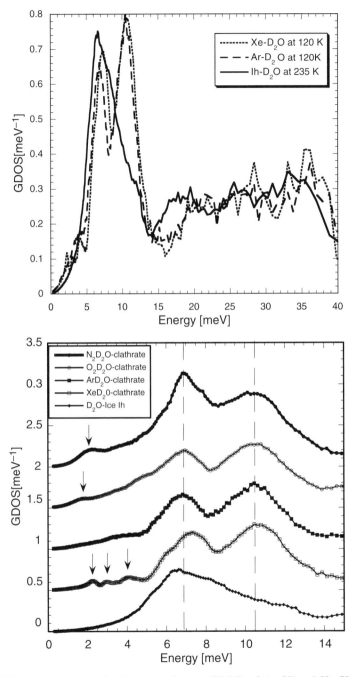

Fig. 12.6 *Upper panel*: generalized density-of-states (GDOS) of Ar–SII and Xe–SI clathrate hydrates at 120 K and of hexagonal I_h at 235 K. All data sets are normalized to 12 translational degrees of freedom corresponding to 4 molecules per unit cell of the ice I_h crystal [63]. *Lower panel*: focus on the low energy region of $G(\hbar\omega)$ of N_2, O_2, Ar and Xe clathrate hydrates. *Arrows* indicate some of the low-energy guest modes. $G(\hbar\omega)$ of ice I_h is shown for comparison [64]

frequencies of the lightweight water molecules by about 5% toward lower energies as $\omega \propto \sqrt{1/M}$. An equivalent argument has to be applied to the frequencies of the guest motions. As far as rotations and librations are concerned, the variation in the moment of inertia, θ, has to be accounted for as $\omega \propto 1/\theta$ when, for example, comparing CH_4 with CD_4.

Beyond the energy range of the two pronounced peaks, the integral number of modes is constant independent of the guests included, pointing at an unperturbed dynamics of the host lattice in this range [63, 64]. Strong support for an unaltered host dynamics above some 20 meV is given also by MD and lattice dynamics (LD) simulations on a number of other guests [60, 62, 65].

It is well established by single crystal experiments that the strong peak in ice I_h is dominated by the contribution of transverse acoustic (TA) phonon sheets at the Brillouin zone boundary and their "back folded" low-energy optic phonon sheets of tranverse polarization [58, 66, 67, 68]. The dispersion of the longitudinal acoustic (LA) phonon reaches in specific directions of the hexagonal crystal ($\Gamma \rightarrow K, \Gamma \rightarrow M$) up to energies beyond 20 meV.[10] An equivalent statement holds for other crystalline and even amorphous water structures [42, 69]. In view of the complex structure and the presence of two strong peaks in the hydrates, such a statement is less evident and single crystal experiments are missing. Nonetheless, a dispersive mode of LA character is established in powder inelastic X-ray scattering experiments exceeding the regime of the two pronounced peaks [65]. The dispersivity of the LA mode indicates an averaged longitudinal velocity of sound of 3950 ± 50 m/s, comparable with the results on different ice structures [42, 44, 69, 70, 71, 72, 73] and in agreement with other measurements on hydrates [74, 75].

12.6.2 Monoatomic Noble Gas Hydrates

The discussion of the guest dynamics is simplified in the case of monatomic guests like argon, krypton, and xenon due to the absence of rotational degrees of freedom. Figure 12.7 reports the inelastic response of the three hydrates as measured at the IN6 spectrometer at ILL.

12.6.2.1 Guest-Host Coupling

A direct comparison of Ar- and Xe-hydrate signals signifies a strong redistribution of modes within the energy range of guest vibrations. Argon displays two broad bands whose center positions can be approximately identified as 1.7 and 3.4 meV, indicated by vertical arrows in the upper panel of Fig. 12.7. Xenon, on the other hand, shows three narrow peaks at 2.2, 2.9, and 4 meV. The most significant dynamic feature, however, can be extracted from the data when comparing the density of

---

[10] We classify here the modes according to their polarization character. So a mode with in-phase motion of the atoms is considered acoustic even if it has already crossed lower lying dispersion sheets, and thus in a purely academic sense should be denoted as optic.

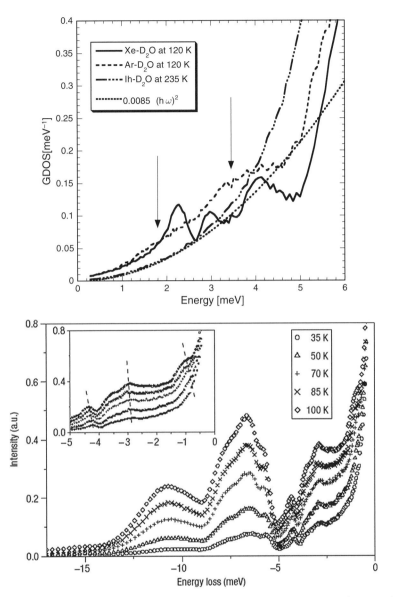

Fig. 12.7 *Upper panel*: generalized density-of-states (GDOS) of Ar and Xe hydrate in comparison to ice I_h [63]. *Dotted line* corresponds to the behavior of ice I_h as a harmonic solid within the Debye theory. *Lower panel*: inelastic intensity of krypton hydrate at different temperatures. Inset stresses the properties of the guest modes of krypton. Figure is adapted form reference [60]

states with the intensity distribution expected for ice I_h within the Debye-model of harmonic solids, which is indicated by the ω^2–line. The coupling of xenon with the water lattice is sufficiently strong to deplete the modes in the range 4.3–5.5 meV. In contrast to this mode depletion, the argon dynamics is apparently an add-on to the inelastic response of the water lattice, extending up to about 5 meV. The redistribution of xenon signal has to be compared with results on samples with a protonated water lattice, which amplify the host modes [60, 61]. Host modes at about 2.1, 2.9, and 4.0 meV are reported from the experiments performed at both the quasielastic neutron spectrometer at the Intense Pulsed Neutron Source of Argonne National Laboratory USA, and the FOCUS time-of-flight spectrometer at the Paul Scherrer Institute, Switzerland.

Results from both experimental approaches (the one enhancing the xenon signal and the other stressing the water motion) show that xenon can be considered as strongly coupling to the water framework. As the relatively heavy mass of xenon militates eigenfrequencies of low energy, we should expect the effect of the coupling to extend to the lower acoustic region. In fact, a remarkably low velocity of sound is confirmed by Brillouin light scattering and evaluated to be 76% of the velocity of sound of ice I_h [74]. Recalling relation 12.9, we may expect the density of states to ascend more steeply than in crystalline ice, which is visualized in Fig. 12.7.

As might be expected, the behavior of krypton as an intermediate size guest shows three excitations; however, they are less distinctly defined than the three modes in xenon hydrate [62]. The guest mode energies, which are stressed in the inset of the lower panel in Fig. 12.7, are located at about 1.0, 3.1, and 4.4 meV at elevated temperatures. Upon cooling, these modes soften and stabilize at lower energies. At a first glance, a softening upon cooling might appear an extraordinary feature; however, it is this very behavior we expect from a clathrate system in which the bonding forces within the host structure and between guest and host differ appreciably. As the restoring forces, f, defining the eigenfrequencies of a vibrating guest depend on the second derivative of the potential $V[u(T)]$, it follows for a symmetric but anharmonic potential $V[u(T)] = V_0 + A \cdot u^2(T) + B \cdot u^4(T) + ...$ that $f \rightarrow f([u(T)]) \sim -2A - 12B \cdot u^2(T)$ in a simple mean field approach. The quantity $u^2(T)$ is the mean square displacement of the guest, which will increase upon temperature increase and hence augment the restoring force and the eigenfrequencies by $12B \cdot u^2(T)$. From this simple one-particle picture, we may conclude that the weaker the bonding of the guest is, the more pronounced is the temperature dependence of $u^2(T)$ and, hence, the frequency shift of the eigenmodes. The physical picture behind this hardening of phonon frequencies with temperature is the fact that the guest at higher temperatures explores more and more the steep walls of the potential created by the rather stiff host framework. As reported in Fig. 12.8 the strongly coupled xenon suppresses indeed the temperature dependence. Results on protonated host material indicate a shift by 100 μeV for all modes in the temperature range 50–180 K [61]. These shifts are within the resolution of the experimental setups. As xenon fills the cages very well, the hard repulsive potential is already explored at low temperatures, and this does not change markedly with T.

12.6.2.2 Calculations

All three hydrates (Ar, Kr, and Xe) were the subject of extensive computational efforts [60, 62, 65, 76, 77] whose results matched the experimental data and

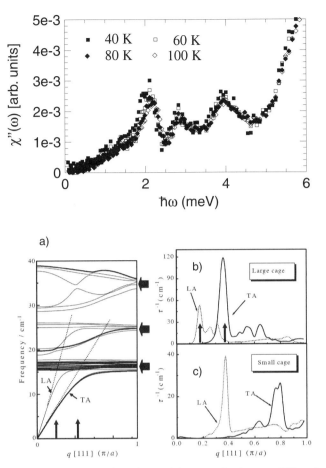

Fig. 12.8 *Upper panel*: temperature dependence of the generalized susceptibility of a deuterated xenon hydrate [56]. *Lower panel*: (**a**) phonon dispersion relation and phonon lifetimes of the transverse and longitudinal acoustic host modes in (**b**) large and (**c**) small cages as obtained from lattice dynamics calculations for H_2O xenon hydrate [60]. The *dotted lines* are guides to the eye on the resonant modes. The *vertical arrows* indicate the q-values where resonances occur in the large cages. The *horizontal arrows* indicate the vibrational bands observed in the experimental spectrum

strengthened the intuitively derived concepts. For argon hydrate it is concluded that the guest dynamics in the large cavity is centered at about 2.2 meV, whereas the non spherical small cage induces an asymmetric distribution of modes with a strong peak at the lower energy side of about 6 meV and possibly some weaker features at about 3 meV [76]. An equivalent conclusion has been drawn for the dynamics of krypton [62]. Its eigenmodes within the large cavity were calculated to center around 1.1 meV and show o split in the small cavity into two maxima at 3.2 and 4.2 meV, well matching the neutron scattering results of Fig. 12.7.

Molecular dynamics also confirmed the temperature-dependent shift to be about 0.4 meV for the mode in a large cavity and 0.7 meV for the higher energy small-cavity mode in the temperature range 30–120 K. More extensive MD and LD calculations have been performed on xenon clathrate, whose distinctiveness originates in its isostructural properties with the naturally occurring methane hydrate, however disregarding the structural and dynamic complexity of a molecular guest. An example of calculated phonon dispersion is shown in Fig. 12.8, taken from reference [58]. At the Brillouin zone boundary modes of 2.2, 3.2, and 4.4 meV are calculated matching the experimental results satisfactorily. Here, however, two modes (2.2 and 3.2 meV) are evoked by the ellipsoidal shape of the large cavity and only the mode at 4.4 meV is due to the smaller cage.

12.6.2.3 Thermal Conductivity

The calculation results have been taken as a model basis for the explanation of the reduced thermal conductivity of hydrates. Within this model it is supposed that the heat-carrying acoustic phonons of the host lattice are in resonance with so-called *localized* modes of the guests, evoking a symmetry avoided anti-crossing (see Section 4), as is illustrated in Fig. 12.8. The localized character of the guest modes is well demonstrated by the fact that they have a nearly vanishing dispersion in $\mathbf{q}$. This implies that wave packages constructed out of these modes and describing a local vibration of a guest in its cage have zero group velocity (i.e., they are not propagating). The resonance effect is supposed to limit the lifetime and hence the mean free path of the acoustic phonons and to form a dynamic barrier for the heat transport of the clathrate sample. Figure 12.8 illustrates the inverse lifetimes of the LA and TA phonons, which show indeed clear resonance peaks at their crossing points with the xenon eigenmode energies. In simple words, these peaks indicate the transformation of heat transported through the sample into a nonpropagating energy reservoir.

A shortcoming intrinsic to the calculation is the fact that the guest is treated as an isolated point defect [58]. This is certainly justified by the localized nature of the guest vibrations. However, localization is a feature of the dynamics in reciprocal space in which eigenmodes do not show a pronounced dispersive behavior. As a consequence, any optic phonon of the water framework may be termed localized. It is clear that these modes will equally reduce the thermal conductivity due to the anharmonicities of the water–water potential. It would be interesting to know the respective contributions of guest and framework phonons to thermal conductivity.

12.6.3 Diatomic Nitrogen and Oxygen Hydrates

The dynamics of diatomic molecules is apparently enriched by three additional degrees of freedom: the stretching of the molecule and the two rotational degrees perpendicular to the stretching direction. Stretching modes typically exhibit energies of a few hundred meV and are well outside the energy window of interest here. Free rotational modes can be approximately calculated from the moments of inertia to posses energies in the range of 0.2–0.5 meV for molecules of oxygen and nitrogen. Encaged, rotational modes can be locked into hindered, librational motions below some finite characteristic temperature, and the restoring forces are expected to augment their eigenfrequencies. For this reason, we expect the rotational or librational eigenfrequencies to be located in the energy range of the translational guest modes.

12.6.3.1 Guest Dynamics

Figure 12.9 shows the partial generalized density-of-states of nitrogen and oxygen in an SII hydrate. To extract the guest signal, the density-of-states of argon hydrate has been subtracted from the signal of the nitogen and oxygen samples. This approach is justified as the scattering power of the diatomic molecules is up to two orders of magnitude higher than that of argon as indicated in Table 12.3. Both guests display two wide excitation bands, centered at 1.75 and 4.75 meV in molecules in oxygen hydrate and at 2.25 and at about 7 meV in nitrogen hydrate. Indeed, this is a strong simplification of the features observed; however, it indicates that the effects of large and small cages create two energy bands of low and high energy, respectively. The strong relative shift of the bands when changing the guest is rather unexpected as it cannot be explained by the mass difference.

It is confirmed by MD simulations that the inelastic behavior of nitrogen does not differ qualitatively from that of oxygen [79]. Both diatomic molecules are highly mobile within the cages (i.e., they are not locked to any particular position on long time scales). Quantitatively, oxygen shows a larger mean square displacement than nitrogen within the large cavities, namely 2.9 $\mathring{A}^2$ compared to 2.3 $\mathring{A}^2$ for nitrogen ($T = 273$ K). Within the picture of harmonic vibrations, larger mean square displacements correspond to weaker restoring forces and lower eigenfrequencies, in agreement with the experimental data of Fig. 12.3. A recent simulation work [78] calculated the mean square displacements of nitrogen to 0.4, 2.0, and 5.5 $\mathring{A}^2$ ($T = 273$ K) for single occupation in small and large cavities and double occupation of large cavities, respectively. Figure 12.3 reports the trajectories of the center of mass of nitrogen molecules within the small and large cavities [78]. The calculated vibrational dynamics results at 80 K for the single filling case indicate energy peaks at 2.5 meV for large and 8.2 meV for small cavities in [78] and 2.3 meV for large and 7.9 meV for small cavities [64]. Both calculations are in excellent agreement with the experimental results at 120 K.

12.6.3.2 Temperature Evolution

The nitrogen modes are subject to an extreme softening upon cooling, which constitutes the strongest anharmonic effect observed in hydrates, so far. Figure 12.10 shows the temperature dependence of the generalized susceptibility of the two hydrates. The clear renormalization of intensity allows in the case of nitrogen the approximation of the first peak with a damped harmonic oscillator (DHO) model

$$\chi''[\omega] = A \frac{4\omega\Omega\Gamma}{(\omega^2 - \Omega^2)^2 - 4\omega^2\Gamma^2}, \tag{12.22}$$

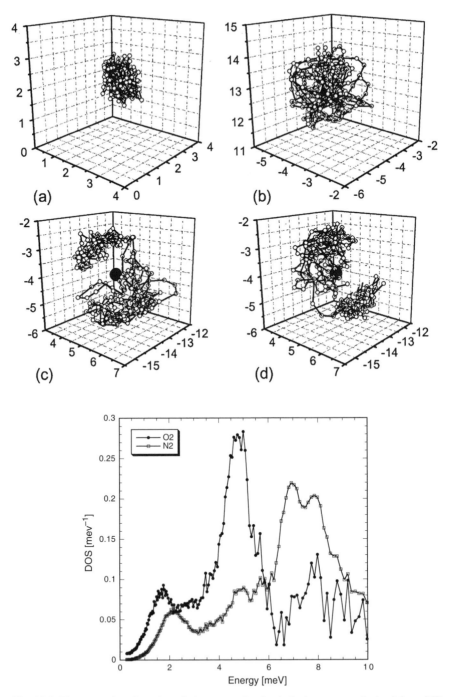

Fig. 12.9 *Upper panel*: trajectories of nitrogen molecules in hydrate cages obtained from MD simulation [78]. (**a**) single nitrogen in small cavity (**b**) single nitrogen in large cavity (**c**) and (**d**) present the trajectories of a pair of nitrogen molecules both occupying a single large cavity. *Lower panel*: Partial density of states of nitrogen and oxygen hydrates

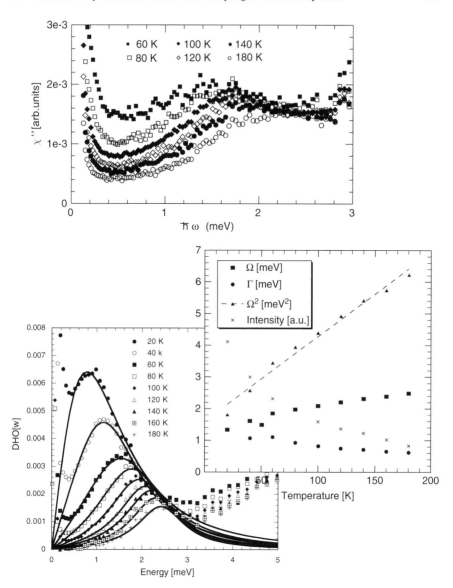

Fig. 12.10 *Lower panel*: Temperature dependence of the generalized susceptibility of oxygen hydrate. *Solid lines* correspond to fits of the strongly anharmonic low-energy peak with a damped harmonic oscillator model. The inset summarizes the fit parameters. *Upper panel*: Temperature dependence of the generalized susceptibility of oxygen hydrate

which in a simple picture describes the damping of a phonon by modes of structural relaxation. All DHO fit parameters obtained are presented in the insert. Within the model results, the peak position Ω, shifts by approximately 50% from 2.6 meV at 200 K to 1.3 meV at 20 K. A comparable behavior is observed in O_2 hydrate, with a relative peak shift matching the one in N_2 hydrate. MD results on N_2 hydrates are in excellent agreement with the experimental data identifying a shift by 1.2 meV (large cavity) and at most 0.6 meV (small cavity) between 273 and 80 K [78]. It is important to note that the mode position extrapolates in both hydrate systems to a finite energy at 0 K, indicating a well-defined vibrational or librational motion. Unfortunately, an unequivocal identification of this motion in terms of eigenmodes of the system encaged is not a straightforward task, as the supposed rotations or librations of the molecules fall into the energy range of vibrational modes. Here again, MD simulations can supply some assistance. First, there is a significant translational diffusion of the nitrogen molecules in the cages. Second, reorientational correlation times are of the order of 0.2–0.3 ps at 80 K, which corresponds to vibrational energies of 10–20 meV [78]. This, however, is only strictly applicable to single occupied large cages, since for small cages persistent reorientations of the nitrogen molecules are calculated at time scales of some picoseconds (i.e., of energies as low as 1 meV). A similar conclusion is drawn for a double occupancy of large cavities.

12.6.4 Methane Clathrate Hydrate

Being a compact molecule of tetrahedral symmetry, methane is, along with hydrogen molecule, one of the rare examples of a molecule performing nearly free quantum rotations in high symmetry environments like solid methane, argon, krypton, and xenon [80–82].

12.6.4.1 Quantum Rotations

Experiments on natural and synthetic methane hydrates confirm the presence of quantum rotations equally in these materials [61, 83–85]. This allows, contrary to the cases of diatomic molecules discussed above, to characterize the rotational dynamics of the guest and how it is influenced by the other vibrations. The methane molecule occupies two small cages of local symmetry $m\bar{3}$ and 6 size slightly larger cages of symmetry $\bar{4}2\,m$ per SI unit cell. A transition scheme in terms of the rotational quantum number J with the following energies was obtained from high-resolution experiments on natural hydrate samples on the IN5 at ILL [84].

1. $J = 0 \rightarrow 1$ at 1.1 meV (0.21 THz)
2. $J = 1 \rightarrow 2$ at 2.2 meV (0.58 THz)
3. $J = 0 \rightarrow 2$ at 3.3 meV (0.79 THz)

   The assignment of the 2.2 meV transition as $J = 1 \rightarrow 2$ is particularly unequivocal as its intensity scales with the increase in population of $J = 1$ upon an increase in temperature. Figure 12.11 illustrates the temperature dependence of methane hydrate [84]. There is a gradual transition from the quantum regime to a classical rotational diffusion as evidenced by the broadening of the peaks and the onset of

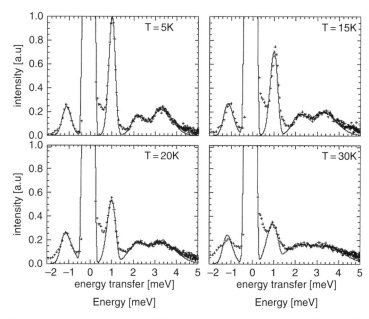

Fig. 12.11 Inelastic neutron spectra of CH_4 in the natural gas hydrate at $T = 5, 15, 20$, and 30 K [84]. The inelastic bands at energy transfers of $E = 1.1, 2.2$, and 3.3 meV are identified as the transitions between the rotational states of the almost free methane rotors. With increasing temperature, the lines broaden and the transition from quantum rotations to classical rotational diffusion is indicated by the occurrence of quasielastic scattering

quasielastic scattering. The fact that the observed rotational splitting of 1.1 meV is close to the value for free rotation of 0.66 meV is taken as an indication that the position of the guest molecule is the center of the cage. This is supported by neutron diffraction results [84, 85, 86].

All translational modes of methane are located in two bands centered at about 7 and 12 meV [61]. An assignment of these modes to one of the cages is, however, only possible for the 7 meV band, identified as a double degenerate mode with the degeneracy induced by the tetragonal symmetry of the large cavity. MD and LD calculations do not help much in assigning the second band as they show some discrepancy with the experimental data [61, 65, 83, 84, 87].

12.6.4.2 Rotational Coupling to the Framework

At temperatures below 10 K, the transition rate between the rotor states, J, decreases strongly and conversion times exceed the time scales of hours, potentially leaving the rotational system out of equilibrium with other energy reservoirs of the sample, such as the phonon heat reservoir [88, 89]. The same phenomenon is known from molecular solids like hydrogen and oxygen, where we have to make

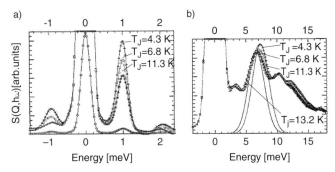

Fig. 12.12 (**a**) shows the dynamic structure factor of synthetic CH_4-methane taken at a temperature of 2.1 K, with different temperatures of the spin-system T_J. With depopulation of the excited rotational $J = 1$ state, the $J = 1 \rightarrow 0$ band at energy transfer of -1 meV (up-scattering of the neutron) weakens. Simultaneously, the intensity of the $J = 1 \rightarrow 2$ band at 2 meV (down-scattering of the neutron) is reduced. Higher rotational levels are not populated at the experimental temperatures. Thus a spin temperature can be attributed to the two lowest rotor states from the intensities of the energy gain and loss lines. The energies on the gain and loss sides are different, indicating that the excited state sees a different potential from the ground state. (**b**) shows the translational modes at $T_J = 4.3, 6.8, 11.3$, and 13.2 K. Spectra were measured using the instrument SV29 time-of-flight spectrometer at the Forschungzentrum Juelich (Germany) DIDO reactor [61]

a distinction between the para and ortho states. As a consequence, it is necessary to assign the temperatures T_J and T_{ph} to the rotation and phonon system, respectively. By a rapid cooling of a methane hydrate to 2 K, a sample can be prepared with $T_J >> T_{ph} = 2$ K because the equilibration period of the rotor system exceeds that of the phonons and of the experimental time appreciably. In a scattering experiment these temperatures can be monitored via the intensities of the spin transition lines, which are governed by thermal occupation, as is indicated in Fig. 12.12 for the $J = 0 \rightarrow 1$ transition intensity. Following the evolution of the phonon spectra in time upon equilibration of the spin system, the coupling of rotational to translational degrees of freedom can be studied. An up-shift of the low-energy vibrational mode from 6.7 to 7.2 meV upon reducing T_J is confirmed, which is indicated in Fig. 12.12 [61]. The authors ruled out that thermal contraction of the lattice is to be made responsible for this strong shift. Therefore, the coupling of the rotor states of methane with the translational degrees of freedom induces a temperature dependence of the density-of-states. The vice versa effect of the vibrations on the rotational degrees of freedom can be observed at temperatures greater than or equal to 15 K, at which a strong broadening of the rotational transition lines is reported without affecting the positions (see Fig. 12.11 [84]).

12.6.5 Hydrogen/THF Hydrate II

Molecular hydrogen and molecular deuterium are the only quantum crystals at ambient pressure. It is thus not surprising that molecular hydrogen trapped in the small cavities of SII hydrate constitutes a particular system. Although being a diatomic guest like nitrogen and oxygen, its dynamics display an intrinsic quantum behavior not only for the rotations (as is the case for methane) but equally for the translations. A series of experiments have been performed very recently at different neutron scattering spectrometers to shed light on the dynamics of hydrogen trapped in the small cages of a THF–SII hydrate [57, 90]. All measurements profitted from the opportunity of isotope substitution in the host lattice ($H_2O \rightarrow D_2O$) and the promoter (THF $\rightarrow$ deuterated terahydrofuron (TDF), enhancing the signal from the hydrogen guest. Standard dynamic SII-type features are confirmed in the host lattice as well as the location of the guest modes of the TDF promoter below the first maximum of the cage response at 7 meV [57, 62].

It is shown that hydrogen behaves like a quantum system with well-defined translational and rotational quantum energy levels. The entire energy scheme of the molecule can be constructed by superposing the excitation schemes of the translational and rotational quantum levels up to a temperature of 150 K. Hence, the system displays within the accuracy of the experiments a negligible coupling between rotational and translational degrees of freedom. The fundamental transition energy was identified experimentally as 14.4 meV for the rotor states and 9.7 meV for the vibron

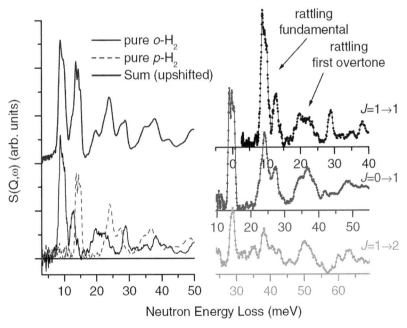

Fig. 12.13 Partial dynamic structure factor of hydrogen hydrate as obtained from a set of experiments performed on the TOSCA spectrometer at the ISIS spallation neutron source at the Rutherford Appleton Laboratory, UK [90]. Left panel: dynamic structure factors of ortho- and para-H_2 and of the sum spectrum. Right panel: excitation scheme of ortho-H_2 ($J = 1 \rightarrow 1, J = 1 \rightarrow 2$) and para-$H_2$ ($J = 0 \rightarrow 1$). $J = 1 \rightarrow 1$ corresponds to the pure vibrational scheme. $J = 0 \rightarrow 1$ and $J = 1 \rightarrow 2$ desribe the superposition of the rotational transitions with the vibrational scheme. The spectra are shifted to match the vibrational energies

states. However, this classification is not correct in detail as the degeneracy of the levels is lifted and basic modes are split into three lines. This observation marks the hydrogen guest as being an extremely sensitive probe for studying the geometry of the occupied cage. A more thorough interpretation of the dynamic features has been tried by solving numerically the Schrödinger equation of hydrogen in an anharmonic model potential of the small SII-cage [90] and by an approximation of the data by analytically solved single-particle quantum schemes [57]. Measured and calculated data from the first approach are presented in Figs. 12.13 and 12.14. A good qualitative agreement between experimental and modeled data is found in both cases although certain aspects need a more decent numerical and experimental inspection. The application of analytical solutions revealed the best compliance for a particle on a surface of a sphere with a radius of 1.05–1.10 Å, indicating an off-center position of the hydrogen molecule as well [57]. However, this possibility comes along with the presence of a low-energy mode hardly detectable in the experiments performed on the Pharos time-of-flight spectrometer at Los Alamos National Laboratory USA;

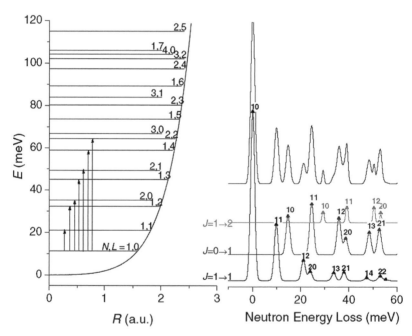

Fig. 12.14 *Left panel*: isotropic part of the potential energy for an isolated hydrogen molecule as a function of the distance, R, from the center of the dodecahedral cage. The lines indicate the calculated energy levels for the quantum numbers N, L (center-of-mass). As the potential is strongly anharmonic and rather flat in the center, the calculated energy levels labeled by the principal quantum number N are not equally spaced. *Right panel*: corresponding calculated incoherent INS spectra (*top line*). This is the sum of the three contributions shown below, where the same transition lines of the center-of-mass motion labeled by the final values of NL are shifted by the amount $\Delta_{JJ'}$ of each rotational transition from *bottom* to *top*: $\Delta_{11} = 0$, $\Delta_{01} = 14.7$ meV, and $\Delta_{12} = 29.0$ meV [90]

Disk Chopper Spectrometer at the National Institute of Standards and Technology Center for Neutron Research, USA; and TOSCA spectrometer at the ISIS spallation neutron source at the Rutherford Appleton Laboratory, UK. Within the isotropic model approach, the solution of the Schrödinger equation is not able to account for the lifted degeneracy [90]. Nonetheless, both studies show that a bulk quantum chemical calculation overestimates the splitting of the rotation levels and underestimates the vibrational energy scheme [91].

12.7 Conclusion

Clathrate hydrates are an ideal playground to study the dynamics of atoms and molecules in nanometer sized cages. Tuning the guest size allows tuning the interaction with the cage. Gradually increasing the complexity from atomic guests to diatomic molecules and finally methane allows switching on rotational degrees of freedom with different symmetry properties. In this way it is possible to develop a comprehensive picture despite the complexity of the systems.

There are certain aspects common to the dynamics of all guests, which can be understood on intuitive grounds. Such is the observation that large cavities provide weaker restoring forces on the guests than do small cavities and therefore generate eigenmodes of lower energy. This effect is observed in all hydrates in which a guest can be accommodated in both small and large cages. Monoatomic noble gas (argon, krypton, and xenon), diatomic (nitrogen and oxygen), and molecular (methane) guests prove this simple dependence.

Another common feature is the unusual anharmonic tendency of guest eigenmodes to soften upon temperature reduction. This, however, does not come as a surprise since mode softening upon cooling reflects only the effect of higher order terms in the repulsive potential confining the mobile guest in a rigid cavity. This behavior is well observed in krypton, nitrogen, and oxygen.

An appreciable mismatch of cavity to guest size in favor of the host can evoke an off-center position and diffusion in confinement of the too small guest. In such cases, a definition of eigenmodes becomes difficult since the frequency and time scale of guest vibrations and diffusion can approach each other and even coincide at certain temperatures.

The entire scenario becomes more complex for molecular guests with rotational degrees of freedom. It becomes necessary to distinguish not only whether molecules can perform free rotations or librational modes but also on which frequency scale these rotational dynamics happen. In principle, we may think of a vivid coupling between vibration-induced diffusion and librational-assisted reorientation of molecules. This coupling may of course change upon temperature changes. Molecules may overcome the energetic barrier to change from librational motion into tumbling or even free rotations.

Finally, guest and host dynamics are not fully decoupled as the cages have a finite elasticity. The extent to which the motion of the lattice influences the motion of the guests and vice versa has a wide bandwidth ranging from a light perturbation to a complete embedding of the guest motion into some eigenmodes of the host matrix. The strength of the coupling depends on the mass, size, chemical composition, and geometry of the guest. Extreme examples of guest–host coupling are hydrogen and xenon hydrates. Hydrogen embedded in the small cages of SII hydrate behaves nearly like a free quantum rotor and vibrator whose energy scheme can be well computed by using a purely static potential of rigid cavities. Xenon, on the other hand, couples strongly to the host lattice and renormalizes the vibrational states of the host in the energy range below 10 meV.

We have shown that MD and LD calculations are indispensable tools for assisting experimental investigations. Both calculation tools predict with a reasonable degree of accuracy the observed dynamic features. Beyond the frequency distribution and temperature dependence of monatomic guests, the frequency distribution, diffusion, and mode softening of nitrogen and oxygen are predicted accurately. Even in the case of methane hydrate, the vibrational dynamics is predicted with reasonable accuracy. Despite this undeniable success, several cases of pronounced discrepancy of calculations with experimental data demonstrate the necessity for the later.

The discovered dynamic features allow understanding the reduced thermal conductivity in qualitative terms. The two important ingredients—anharmonicity and flat phonon branches—are present in all cases, albeit with varying degrees of intensity. Already the framework alone is expected to show a strongly reduced thermal conductivity with respect to simpler ice structures, because the increased number of atoms per unit cells implies an increased number of flat optic phonons. These phonons are scattering vectors for the heat-propagating acoustic phonons. The filling of the cages by the guests adds additional scattering mechanisms. These are more efficient as these vibrations are more or less localized, coupled to the framework, and often anharmonic. When it comes to precisely calculating the thermal conductivity, detailed input concerning the dispersion curves and the anharmonic interactions leading to the scattering processes are necessary. This has not yet been achieved, even in the simplest cases. The quest for complete understanding of the thermal conductivity in clathrate hydrates remains, therefore, an ongoing challenge. The answer will most certainly come from improved MD calculations in combination with even more sophisticated neutron scattering experiments.

Throughout recent years, the dynamics of hydrates have often been referred to as glass-like in the literature. Clathrates share with amorphous systems the fact that the propagating acoustic phonons encounter a large number of other vibrations into which they can be scattered. Some of those are found at rather low frequencies: (1) enhanced density of states or boson peak in glasses and (2) localized guest modes in clathrate hydrates. In both cases the modes are localized and, therefore, often termed rattlers. The difference in the two cases is the fact that the frequency distribution in glasses is a result of disorder, while in clathrates it is a consequence of the complex crystal structure combined with a large range of inter atomic interactions. It is, therefore, not surprising that the microscopic dynamics of H_2, N_2, O_2, CH_4, and C_4H_8O hydrates mark those systems as highly distinct with respect to temperature changes, while in glasses there is a simple dependence of the glass dynamics on the thermal occupation characterized by the Bose–Einstein factor, hence the name of Boson peak. Therefore, it should not be stated that clathrates behave like glasses. It is far more appropriate to affirm that the factors responsible for thermal transport properties are similar in both systems. These factors have, however, different origins.

References

1. E.D. Sloan. Gas Hydrates: Relavance to World Margin Stability and Climate Change. *Geol. Soc. Lond.* 1998.
2. E.D. Sloan and C.A. Koh. *Clathrate Hydrates of Natural Gases (Chemical Industries Series).* CRC Press Boca Raton, FL 2007.
3. K. Kvenvolden. *Natural Gas Hydrate: Introduction and History of Discovery.* Kluwer Academic Publishers: Dordrecht, the Netherlands, 2000.
4. T.S. Collett and V.A. Kuuskraa. *Oil Gas J.*, 96:90. 1998.
5. M.D. Max, A.H. Johnson, and W.P. Dillon. *Economic Geology of Natural Gas Hydrates. Springer,* Berlin 2006.
6. M.D. Max. *Natural Gas Hydrate in Oceanic and Permafrost Environments (Constal Systems and Continental Margins).* Kluwer Academic Publishers: Dordrecht, The Netherlands, 2003.
7. A. Gupta, T.J. Kneafsey, G.J. Moridis, Y. Seol, M.B. Kowalsky, and E.D. Sloan Jr. *J. Phys. Chem. B*, 110(33):16384, 2006.
8. E.J. Rosenbaum, N.J. English, J.K. Johnson, D.W. Shaw, and R.P. Warzinski, *J. Phys. Chem. B*, 111(46):13194, 2007
9. C. Ruppel. *Thermal State of the Gas Hydrate Reservoir.* Kluwer Academic Publishers: Dordrecht, The Netherlands, 2000.
10. N.W. Ascroft and N.D. Mermin. *Solid State Physics.* Saunders College, Philadelphia, 1976.
11. W.F. Kuhs and T.C. Hansen. *Rev. Mineral. Geochem.*, 63:171, 2006.
12. H.Tanaka and K. Kiyohara. *J. Chem. Phys.*, 98:4098, 1993.
13. H.M. Powell. *J. Chem. Soc.*, page 61, 1948
14. H. Davy. *Philos. Trans. R. Soc. Lond.*, 101:1, 1811.
15. M. Faraday. *Philos. Trans. R. Soc. Lond.*, 113:160, 1823.
16. M. von Stackelberg and H.R. Muller. *Z. Elektrochem.*, 58:25, 1954.
17. C.A. Koh. *Chem. Soc. Rev.*, 31:157, 2002.
18. V. Chihaia, S. Adams, and W.F. Kuhs. *Chem. Phys.*, 297:271, 2004.
19. Y. Dyadin, E. Larionow, A. Manakov, F. Zhurko, E. Aladko, and T. Mikina. *Mendeleev Commun.*, 5:209, 1999.
20. W.L. Mao, H.-K. Mao, A.F. Goncharov, V.V. Struzhkin, Q. Guo, J. Hu, J. Shu, R.J. Hemley, M.Somayazulu, and Y. Zhao. *Science*, 297:2247, 2002.
21. K.A. Lokshin, Y. Zhao, D. He, W.L. Mao, H. -K. Mao, R.J. Hemley, M.V. Lobanov, and M. Greenblatt. *Phys. Rev. Lett.*, 93:125503, 2004.
22. L.J. Florusse, C.J. Peters, J. Schoonman, K.C. Hester C.A. Koh, S.F. Dec, K.N. Marsh, and E.D. *Sloan. Science*, 306:469, 2004.
23. J.H. van der Waals and J.C. Plateeuw. *Adv. Chem Phys.*, 11:1, 1959.
24. S.R. Zele, S.-Y. Lee, and G.D. Holder. *J. Phys. Chem. B*, 103:10250,1999.
25. K.C. Hester, Z.Huo, A.L. Ballard, C.A. Koh, K.T. Miller, and E.D. Sloan. *J. Phys. Chem. B*, 111:8830, 2007.
26. A.J.H. McGaughey, M.I. Hussein, E.S. Landry, M.Kaviany, and G.M. Hulbert. *Phys. Rev. B*, 74:104304, 2006.
27. J. Dong, O.F. Sankey, and C.W. Myles. *Phys. Rev. Lett.*, 86:2361, 2001.
28. O. Andersson and A. Inaba. *Phys. Chem. Chem. Phys.*, 7:1441, 2005.
29. J.M. Ziman. *Electrons and Phonons.* Claredon Press, Oxford, 1960.

30. R. Peierls. *Ann. Phys. (Leipzig)*, 3:1055,1929.
31. D.A. Broido and T.L. Reinecke. *Phys. Rev. B*, 70:081310(R), 2004.
32. A.I. Krivchikov, B.Ya. Gorodilov, O.A. Korolyuk, V.G. Manzhelii, O.O. Romantsova, H. Conrad, W. Press, J.S. Tse, and D.D. Klug. *Phys. Rev. B*, 73:64203, 2006.
33. A.I. Krivchikov, O.A. Korolyuk, and O.O Romantsova. *Low Temp. Phys.*, 33:612, 2007.
34. O. Andersson and H. Suga. *J. Phys. Chem. Solids*, 57:125, 1996
35. A.I. Krivchikov, V.G. Manzhelii, O.A. Korolyuk, B.Ya. Gorodilov, and O.O. Romantsova. *Phys. Chem. Chem. Phys.*, 7:728, 2005.
36. P. Andresson and R.G. Ross. *J. Phys. C*; Solid State phys., 16:1423, 1983.
37. N. Ahmad and W.A. Phillips. *Solid State Common.*, 63:167, 1987.
38. A.I. Krivchikov, B.Ya. Gorodilov,O.A. Korolyuk, V.G. Manzhelii, H. Conrad, and W. Press. *J. Low Temp. Phys.*, 139:693, 2005.
39. U. Buchenau, Yu.M. Galperin, V.L. Gurevich, D.A. Parshin, M.A. Ramos, and H.R. Schober. *Phys. Rev. B*, 46:2798, 1992.
40. M.A. Ramos and U. Buchenau. *Phys. Rev. B*, 55:5749, 1997.
41. H. Schober, M. Koza., A. Toelle, F. Fujara, C.A. Angell, and R. Boehmar. *Physica B*, 241:897, 1998.
42. H. Schober, M. Koza, A. Toelle, C. Masciovecchio, F. Sette, and F. Fujara. *Phys. Rev. Lett.*, 85:4100, 2000.
43. O. Andersson and H. Suga. *Phys. Rev. B*, 65:140201(R), 2002.
44. M.M. Koza. B. Geil, H. Schober, and F. Natali. *Phys. Chem. Chem. Phys.*, 7:1423, 2005.
45. M.A. White and M.T. MacLean. *J. Phys. Chem.*, 89:1380,1985.
46. M.A. White. *J. de Physique*, 48:C1-565, 1987.
47. J. Neumann and E.von Wigner. *Physik Z.*, 30:465, 1929.
48. M.M. Koza. *Phys. Rev. B*.78:064303, 2008.
49. V.F. Sears. *Neutron News*, 3:26, 1992.
50. S.W. Lovesey. *Theory of Neutron Scattering from Condensed Matter*. Oxford Science Publishers, Oxford, 1984.
51. H. Schober. *J. Phys. IV France*, 103:173, 2003.
52. M.M. Bredov, B.A. Kotov, N.M. Okuneva, V.S. Oskotskii, and A.L. Shak-Budagov. *Sov. Phys. Solid State*, 9:214, 1967.
53. S.N. Taraskin and S.R. Elliott. *Phys. Rev. B*, 55:117, 1997.
54. H. Schober, A. Toelle, B. Renker, R. Heid, and F. Gompf. *Phys. Rev. B.* 56:5937, 1997.
55. J. Dawidowski, F.J. Bermejo, and J.R. Granada. *Phys. Rev. B*, 58:706, 1998.
56. M. Prager and W. Press. *J. Chem. Phys.*, 125:214703, 2006.
57. H. Conrad, W.F. Kuhs, K. Nuenighoff, C. Pohl, M. Prager, and W. Schweika. *Physica B*, 350:e647, 2004.
58. B. Chazallon, H. Itoh, M. Koza, W.F. Kuhs, and H. Schober. *Phys. Chem. Chem. Phys.*, 4:4809. 2002.
59. H. Schober, H. Itoh, A. Klapproth, V. Chihaia, and W. F. Kuhs. *Eur. Phys. J. E*, 12:41, 2003.
60. J. Li. *J. Chem. Phys.*, 105:6733, 1996.
61. K.T. Tait, F. Trouw, Y. Zhao, C.M. Brown, and R.T. Downs. *J. Chem. Phys.*, 127:134505, 2007.
62. J.S. Tse, V.P. Shpakov, V.R. Belosludov, F. Trouw, Y.P. Handa, and W. Press. *Europhys. Lett.*, 54:354, 2001.
63. C. Gutt, J. Baumert, W. Press, J.S. Tse, and S. Janssen. *J. Chem. Phys.*, 116:3795, 2002.
64. J.S. Tse, D.D. Klug, J.Y. Zhao, W. Sturhahn, E.E. Alp, J. Baumert, C. Gutt, M.R. Johnson, and W. Press. *Nature Materials*, 4:917, 2005.
65. J. Baumert, C. Gutt, V.P. Shpakov, J.S. Tse, M. Krisch, M. Mueller, H. Requardt, D.D. Klug, S. Janssen, and W. Press. *Phys. Rev. B*, 68:174301, 2003.
66. B. Renker. *in Physics and Chemistry of Ice*. Royal Society of Canada, Ottawa, 1973.
67. F. Sette, G. Ruocco, M. Krisch, C. Masciovecchio, R. Verbeni, and U. Bergmann. *Phys. Rev. Lett.*, 77:83, 1996.

68. G. Ruocco, and F. Sette. *J. Phys.: Cond. Matter*, 11:R259, 1999.
69. M.M. Koza, H. Schober, B. Geil, M. Lorenzen, and H. Requardt. *Phys. Rev. B*, 69:024204, 2004.
70. R.E. Gagnon, H. Kiefte, and M.J. Clouter. *J. Chem. Phys.*, 92:1909, 1990.
71. G. Ruocco, F. Sette, M. Krisch, U. Bergmann, C. Masciovecchio, and R. Verbeni. *Phys. Rev. B*, 54:14892, 1996.
72. E.L. Gromnitskaya, O.V. Stal'gorova, V.V. Brazhkin, and A.G. Lyapin. *Phys. Rev. B*, 64:094205, 2001.
73. E.L. Gromnitskaya, O.V. Stal'gorova, A.G. Lyapin, V.V. Brazhkin, and O.B. Tarutin. *JETP Lett.*, 78:488, 2003.
74. H. Kiefte, M.J. Clouter, and R.E. Gagnon. *J. Phys. Chem.*, 89:3103, 1985.
75. H. Shimizu, T. Kumazaki, T. Kume, and S. Sasaki. *Phys. Rev. B*, 65:212102, 2002.
76. H. Itoh, J.S. Tse, and K. Kawamura. *J. Chem. Phys.*, 115:9414, 2001.
77. J.S. Tse, V.P. Shpakov, V.V. Murashov, and V.R. Belosludov. *J. Chem. Phys.*, 107:9271, 1997.
78. E.P. van Klaveren, J.P.J. Michels, J.A. Schouten, D.D. Klug, and J.S. Tse. *J. Chem. Phys.*, 117:6637, 2002.
79. S. Horikawa, H. Itoh, J. Tabata, K. Kawamura, and T. Hondoh. *J. Phys. Chem. B*, 101:6290, 1997.
80. W . Press and A. Kollmar. *Solid State Commun..*, 17:405, 1975.
81. W. Press. *Single Particle Rotations in Molecular Crystals*. Springer Tracts in Modern Physics Springer, Berlin 1981.
82. B. Asmussen, M. Prager, W. Press, W. Blank, and C.J. Carlile. *J. Chem. Phys.*, 97:1332, 1992.
83. J.S. Tse, C.I. Ratcliffe, B.M. Powell, V.F. Sears, and Y.P. Handa. *J. Phys. Chem. A*, 101:4491, 1997.
84. C. Gutt, B. Asmussen, W. Press, C. Merkl, H. Casalta, J. Greinert, G. Bohrmann, J.S. Tse, and A. Hueller. *Europhys. Lett.*, 48:269, 1999.
85. C. Gutt, W. Press, A. Hueller, J. S. Tse, and H. Casalta. *J. Chem. Phys.*, 114:4160, 2001.
86. C. Gutt, B. Asmussen, W. Press. M.R. Johnson, Y.P. Handa, and J.S. Tse. *J. Chem. Phys.*, 113:4713, 2000.
87. J. Baumert, C. Gutt, M. Krish, H. Requardt, M. Mueller, J. S. Tse, D. D. Klug, and W. Press. *Phys. Rev. B*, 72:054302, 2005.
88. S. Grieger, H. Friedrich, B. Asmussen, K. Guckelsberger, D. Nettling, W. Press, and R. Scherm. *Z. Phys. B: Cond. Matt.*, 87:203, 1992.
89. S. Grieger, H. Friedrich, K. Guckelsberger, R. Scherm, and W. Press. *J. Chem. Phys.*, 109:3161, 1998.
90. L. Ulivi, M. Celli, A. Giannasi, A.J. Ramirez-Cuesta, D.J. Bull, and M. Zoppi. *Phys. Rev. B*, 76:161401(R), 2007.
91. M. Xu, Y.S. Elmatad, F. Sebastianelli, J. W. Moskowitz, and Z. Bacic. *J. Phys. Chem. B*, 110:24806, 2006.

Chapter 13
Applications of Neutron Scattering in the Chemical Industry: Proton Dynamics of Highly Dispersed Materials, Characterization of Fuel Cell Catalysts, and Catalysts from Large-Scale Chemical Processes

Peter W. Albers and Stewart F. Parker

Abstract The attractiveness of neutron scattering techniques for the detailed characterization of materials of high degrees of dispersity and structural complexity as encountered in the chemical industry is discussed. Neutron scattering picks up where other analytical methods leave off because of the physico-chemical properties of finely divided products and materials whose absorption behavior toward electromagnetic radiation and electrical conductivity causes serious problems. This is demonstrated by presenting typical applications from large-scale production technology and industrial catalysis. These include the determination of the proton-related surface chemistry of advanced materials that are used as reinforcing fillers in the manufacture of tires, where interrelations between surface chemistry, rheological properties, improved safety, and significant reduction of fuel consumption are the focus of recent developments. Neutron scattering allows surface science studies of the dissociative adsorption of hydrogen on nanodispersed, supported precious metal particles of fuel cell catalysts under in situ loading at realistic gas pressures of about 1 bar. Insight into the occupation of catalytically relevant surface sites provides valuable information about the catalyst in the working state and supplies essential scientific input for tailoring better catalysts by technologists. The impact of deactivation phenomena on industrial catalysts by coke deposition, chemical transformation of carbonaceous deposits, and other processes in catalytic hydrogenation processes that result in significant shortening of the time of useful operation in large-scale plants can often be traced back in detail to surface or bulk properties of catalysts or materials of catalytic relevance. A better understanding of avoidable or unavoidable aspects of catalyst deactivation phenomena under certain in-process conditions and the development of effective means for reducing deactivation leads to

P.W. Albers (✉)
AQura GmbH, AQ-EM, Rodenbacher Chaussee 4, D-63457 Hanau, Germany
e-mail: peter.albers@aqura.de

L. Liang et al. (eds.), *Neutron Applications in Earth, Energy and Environmental Sciences*, Neutron Scattering Applications and Techniques,
DOI 10.1007/978-0-387-09416-8_13, © Springer Science+Business Media, LLC 2009

more energy-efficient and, therefore, environmentally friendly processes and helps to save valuable resources. Even small or gradual improvements in all these fields are of considerable economic impact.

13.1 Introduction

Research and development efforts are continuing in the areas of petroleum processing, the production of hydrogen, and the activation of alkanes, among others. Improvements in the corresponding processes, catalysts, and materials are of paramount economic and environmental importance. Analytical methods that are able to clarify the impact of hydrogen-containing entities in-process, to support the development of new catalytic materials and concepts, and to optimize existing industrial catalysts are highly desirable.

Neutron scattering has a long tradition in revealing hydrogen-related properties of matter. Well-known examples include studies of the dynamics of hydrogen-in-metals [1, 2] and intermetallic compounds [3, 4]. Fundamental aspects as well as applied research, such as assessing their suitability for hydrogen storage [5, 6] or hydrogen purification for a future hydrogen economy, have been studied (Chapter 14 in this book). Coupling of photovoltaic hydrogen production and its safe and efficient use in fuel cell systems for the production of energy, heat, and water in large-scale centralized and low-scale decentralized units is highly attractive.

Diffusion coefficients of hydrogen can be determined by quasielastic scattering. The site occupancy of atomic hydrogen on different interstitial sites or the formation of surface and subsurface hydrides can be measured by neutron vibrational spectroscopy [7]. Because of hydrogen embrittlement in cycles of hydrogen storage and release, the particle size of intermetallic compounds decreases.

Catalytic processes are involved in the dissociation of molecular hydrogen. Hydrogen-induced changes in materials properties can be studied with unique efficiency by means of neutron scattering. The potential of neutron scattering for studying hydrogenous entities at the surface as well as in the bulk of such materials and for investigating their dynamical changes during operation has opened many new opportunities and strategies in materials research and development. In particular, the insights into pure precious metal-hydrogen systems [6, 7] and related materials provided strong motivation to extend neutron work on pure hydrides to more complex industrial catalysts such as supported precious metal catalysts and fuel cell catalysts. Inelastic incoherent neutron scattering (IINS) has proven to be highly suitable for the investigations of heterogeneous catalysts.

Because of the growing demand for hydrogen for fuel cell operation [8] in stationary applications and, especially, environmentally friendly hydrogen-powered vehicles, the fields of research in catalytic reforming of various feedstocks (on-site or on-board), hydrogen storage, purification, and catalytic conversion for energy production were revitalized after the hydrogen fuel initiative in the United States in 2003 [9]. One target is the enhancement of volumetric and gravimetric storage capacity in new and lightweight hydrogen-storage materials. The U.S. Department of Energy's system targets for on-board vehicular hydrogen storage and hydrogen-storage materials are outlined in reference [10], and links given therein.

One active field of research that has exploited neutron scattering is the study of storing molecular hydrogen using novel materials [11] and partly carbonaceous matter, including nanotubes [12, 13]. The rotational transition of adsorbed molecular hydrogen as measured at low temperature in IINS can be used as a sensitive probe for the strength of its interaction with carbons or other porous materials, including catalysts.

Finely divided materials and products of great structural complexity are encountered in the chemical industry worldwide. These range from highly dispersed, catalytically active particles on high surface area, porous catalyst supports in heterogeneous catalysis to many finely divided industrial products themselves. The latter include—in steadily increasing volumes—advanced carbon blacks and silicas, which are used as active fillers and modifiers. In the detailed characterization of materials from applied catalysis and production and product control, neutron scattering techniques have proven to be highly complementary to well-established analytical methods of materials research and chemical characterization [14], especially when materials with a high degree of dispersity are involved. Inelastic incoherent neutron scattering complements infrared, Raman, and nuclear magnetic resonance (NMR) spectroscopies, especially when the properties of the finely divided materials, such as special grades of industrial carbons, supported catalyst particles of nanometer size, or the catalyst supports themselves, hinder or completely prevent the use of the full analytical power of well-established analytical methods [15]. Neutron scattering picks up where other analytical methods leave off.

In the following sections, these themes will be illustrated by typical applications from large-scale production technology and industrial catalysis. These include the determination of the proton-related surface chemistry of carbon blacks, which are used as reinforcing fillers in the manufacture of tires. Here the interrelations between surface chemistry, rheological properties, improved safety, and significant reduction of fuel consumption are the focus of recent developments. In addition, neutrons allow surface science studies of the dissociative adsorption of hydrogen on nanodispersed, supported precious metal particles of powder—shaped fuel cell catalysts under in situ loading at realistic gas pressures of about 1 bar. Insight into the occupation of catalytically relevant surface sites provides valuable information for tailoring better catalysts. Furthermore, the impact of deactivation phenomena (e.g., coke deposition, chemical transformation of carbonaceous deposits, and other processes that result in a significant shortening of the useful life of catalysts in large-scale plants) can often be traced back in detail to surface or bulk properties of catalysts, carbons, and other materials of catalytic relevance. A better understanding of avoidable and unavoidable aspects of catalyst deactivation phenomena under industrial conditions and the development of effective methods for reducing deactivation will lead to more energy-efficient and, therefore, environmentally friendly processes and help to save valuable resources. Even small or gradual improvements in materials performance in all of these areas are of direct, significant, and persistent economic impact.

13.2 Applications: Hydrogen on Highly Dispersed Materials and Neutrons in Catalysis

13.2.1 Inelastic Incoherent Neutron Scattering on New Grades of Reinforcing Fillers for Tires

More than 6,000,000 tons of carbon black per year are produced worldwide, mostly by the furnace process. It is used predominantly as reinforcing filler material in tires (Fig. 13.1) and mechanical rubber goods [16].

The growing demand for ultra-low-emission vehicles requires not only highly advanced management of engine operation and sophisticated exhaust gas purification systems but also reduced fuel consumption. This includes a reduction of the rolling resistance of passenger cars and, especially, of heavy trucks. Fuel consumption is the largest variable cost factor in operating a fleet on tires [17–19]. The Commission of the European Communities already recommends legal limits for rolling resistance of tires [20]. This is an essential task, since a forecast by the United Nations for the time between 2005 and 2030 indicates a doubling of the number of passenger cars and heavy trucks worldwide. This illustrates the urgent need for ultra-low rolling resistance of high-performance rubber composites containing suitable carbon black and silica materials and silanization compounds to adjust filler–filler and filler–polymer interactions [20, 21].

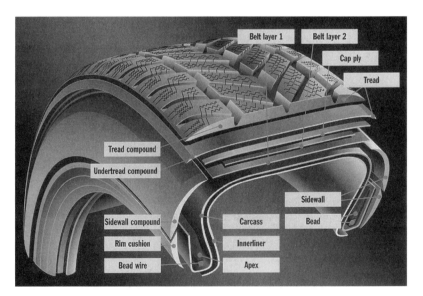

Fig. 13.1 Schematic drawing of the construction elements and internal structure of a modern tire. Various grades of carbon blacks and silicas are used as reinforcing filler material for adjusting the rheological properties of the tire tread and other regions (Reproduced with kind permission of Vredestein, B. V.)

However, in reducing rolling resistance, tread wear properties (for longevity) and good wet traction (grip) and wet skid behavior (for safety) also have to be at least maintained or, preferably, improved significantly. In modern tire technology, reduced rolling resistance can be achieved simultaneously with an adequate level of wet skid resistance and abrasion resistance. Another major need is good performance under enforced braking conditions or strong shear forces such as those that leave dark markings/skid marks on roads/runways.

The improvement of modern tires used for trucks, passenger cars, race cars, motorcycles, and aircraft requires dedicated fine tuning, not only of the polymer mixtures and chemical components of rubber but also of the carbon blacks. These are used as advanced reinforcing fillers in tire treads. Even an improvement of a few percent in reducing the rolling resistance of a tire translates into significant economic value in terms of reduced fuel consumption. And because of the steadily increasing number of tires used worldwide, even small reductions in fuel consumption can translate into significant savings on a global scale.

Finely divided carbon blacks consist of primary particles that cluster to give aggregates (Fig. 13.2). For interactions between these and the polymers and additives in rubber mixtures, the surface properties at the edges of the basic structural units of carbon black are essential. For in-rubber applications, low surface polarity, which means only small amounts of H-acidic heterofunctional surface groups such

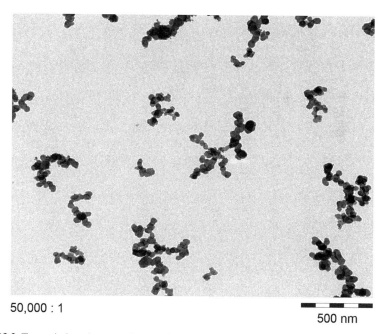

50,000 : 1

500 nm

Fig. 13.2 Transmission electron micrograph illustrating the size and shape of typical aggregates of carbon blacks. The aggregates consist of intergrown primary particles, which are chemically bonded together during formation and particle growth at high temperatures in a furnace reactor. Such blacks are used as active advanced filler materials for the reinforcement of rubber mixtures

as phenolic or carboxylic, but enhanced amounts of structural disorder and, there-fore, of chemically bound hydrogen atoms, which terminate the free valences of the polyaromatic/graphitic sheets, are required (Fig. 13.3).

Inelastic incoherent neutron scattering is uniquely suitable to study these relevant surface sites of finely divided technical carbon blacks, whereas NMR, infrared, or Raman spectroscopies are limited by electrical conductivity and strong absorption of electromagnetic radiation. In particular, the cross section for inelastic incoherent neutron scattering is large for hydrogen and small for virtually all other elements. For carbon blacks, this means that the predominant carbonaceous contribution to a carbon black is essentially transparent to neutrons and the small amounts of hydrogenous entities can be highlighted. The structural model of carbon black based on transmission electron microscopic investigations is sketched in Fig. 13.3: basic structural units of enhanced sp^2 character are arranged in a three-dimensional pri-mary particle of turbostratic disorder. These are the building blocks for the aggre-gates of carbon black.

For applications in advanced tire treads, a new generation of so-called inversion blacks was introduced. Electron microscopy and scanning tunneling microscopy show that these blacks have enhanced surface roughness of 0.8–1.3 nm [21], which translates into an enhanced concentration of terminating protons [22].

A set of typical IINS spectra for carbon blacks is depicted in Figs. 13.4 and 13.5 [23]. Figure 13.4 shows neutron spectra for a carbon black with pronounced rein-forcing properties, which is used in the tire tread of high-speed passenger cars and heavy trucks. In Fig. 13.4a, the spectrum was measured for a sample that was graphi-tized at 3,000 K under inert gas atmosphere. The spectrum in Fig. 13.4b represents the original fluffy state, as produced in the furnace reactor. The difference between the two spectra is due to a difference of about 3,500 ppm of hydrogen. Hydrogen was lost by controlled graphitization. This is accompanied by structural changes at the nanoscale. Figure 13.4a is typical of the IINS spectrum of finely divided graphitic matter, showing the low-energy graphite band at about $100 \, cm^{-1}$. The bands that are centered at about 880 and $1,170 \, cm^{-1}$ of the original material in Fig. 13.4b repre-sent the out-of-plane and in-plane deformational modes of the hydrogen-terminated small graphite platelets, the basic structural units forming the primary particles of such nanomaterials.

The IINS spectra in Fig. 13.5 are the spectra of carbon blacks with different rein-forcing performance whose total hydrogen content varies between about 3,100 and 4,400 ppm. Spectral differences are expected to be due to varying amounts of active surface sites such as conjugated double bond entities; vinylic/allylic structures; and different amounts of isolated, vicinal, or ternary carbon-hydrogen (C-H) structures. Also surface modifications of extremely finely divided carbon blacks for pigment applications can be studied by neutrons that are not accessible to other vibrational spectroscopy methods [24, 25].

In the future, IINS measurements at higher resolution and enhanced flux can be exploited to give improved depth of analysis, speed, and throughput. Computational chemistry simulations of hydrogen termination in other carbons [26] suggested that this is a useful method to further the analysis of the spectra (Fig. 13.6). The fine

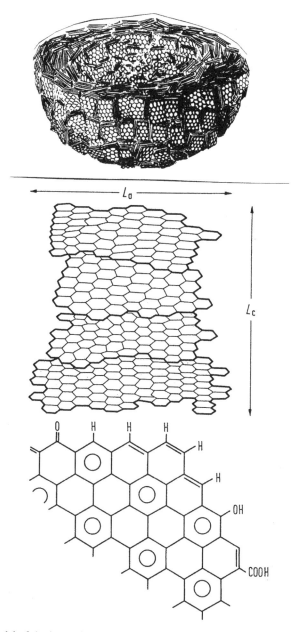

Fig. 13.3 A model of the internal structure of the primary particles of carbon blacks as derived from transmission electron microscope measurements [16, 22]. The edges of the graphite-like sheets of the basic structural units are terminated by hydrogen entities or small amounts of functional groups containing oxygen, nitrogen, etc. (Reproduced with kind permission of Taylor and Francis Group)

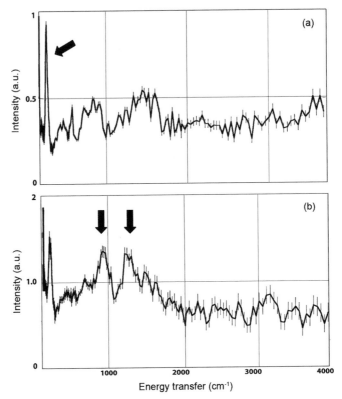

Fig. 13.4 Inelastic incoherent neutron scattering spectra of a reinforcing carbon black used in the tread of heavy trucks and high-speed passenger cars. (**a**) After graphitization; note the strong graphite band at about $100 \, \text{cm}^{-1}$; (**b**) original fluffy state, as produced in a furnace reactor [23]; C-H deformational modes centered at about 880 and $1{,}170 \, \text{cm}^{-1}$. Hydrogen content ($\mu$g/g): (**a**) 290, (**b**) 3,835. (Reproduced with kind permission of Elsevier Limited)

structure of the in-plane and out-of-plane C-H vibrations in the IINS spectra of such blacks contains quantitative information of the relevant hydrogen-containing surface sites of advanced reinforcing fillers and other carbons encountered in catalysis (Sections 13.2.2–13.2.4). The combination of ab initio methods and IINS spectroscopy will enable the extraction of these details.

Because of its selectivity to hydrogen, IINS is able to give detailed spectroscopic evidence of the hydrogen-related surface chemistry of carbon blacks and many other grades of carbons, even with differences as small as a few hundred parts per million of hydrogen. Automatically, the spectroscopic information is focused on the relevant surface sites and contains key information for future improvements in synthesis and applications of advanced fillers for tires with reduced rolling resistance and, ultimately, greater fuel economy.

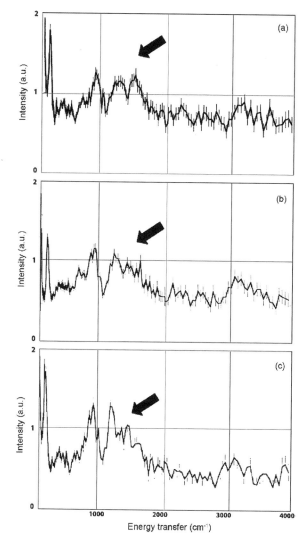

Fig. 13.5 Inelastic incoherent neutron scattering spectra of carbon blacks with different reinforcing performance and different hydrogen content showing different fine structure in the regions of
the out-of-plane (about $880\,cm^{-1}$) and in-plane (about $1,170\,cm^{-1}$) carbon–hydrogen vibrational
modes [23]. Hydrogen content ($\mu g/g$): (**a**) 4,458, (**b**) 3,103, and (**c**) 4,189. The spectral differences
are correlated with varying amounts of H-containing active surface sites and enhanced surface
roughness in the 0.8–1.3 nm scale (according to scanning tunneling microscopy [21]). (Reproduced
with kind permission of Elsevier Limited)

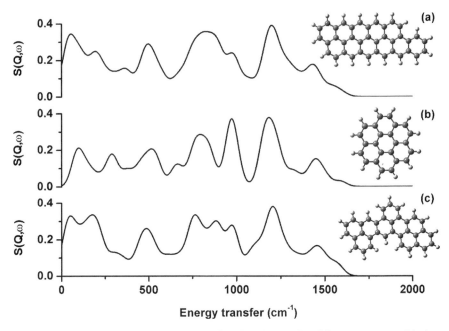

Fig. 13.6 Results from spectra simulations of hydrocarbons with different amounts of hydrogen entities: (**a**) mainly isolated hydrogen atoms, (**b**) vicinal hydrogen atoms (coronene), and (**c**) ternary neighboring hydrogen atoms [26]. (Reproduced with kind permission of the Royal Society of Chemistry)

13.2.2 Inelastic Incoherent Neutron Scattering on Fuel Cell Catalysts

Nanoparticles, in the purest sense, are readily encountered in catalysis: finely divided active ingredients are well dispersed on more or less finely divided supports to form a catalyst [27]. These are used, for example, in the synthesis of fine and specialty chemicals, in automotive exhaust gas purification systems [27, 28] and in various grades of fuel cell catalysts [8].

In a recent review [29], the authors indicated that, depending on the location of hydrogen-containing entities, IINS can be used as a surface science or bulk technique to study the proton dynamics of catalysts or materials of catalytic relevance. From a chemical engineering point of view, the attractiveness of IINS for fundamental research as well as for applied technical catalysis arises from the following.

- Catalyst samples from about 10–150 g can be analyzed in one single experiment, which can represent the catalyst equivalent of several cubic meters of reaction volume.
- It is possible to measure a catalyst under realistic gas pressures.
- Study of hydrogen-containing entities on catalyst surfaces: high sensitivity and high surface specificity allows the analysis of monolayer or submonolayer

coverage of reactants, intermediates, products, and other adsorbed species adherent to the catalyst surface such as solvent molecules, contaminants, oligomers or polymers, cokes, or poisons.

- Physico/chemical properties in the topmost atomic layers can be measured under varying catalyst conditions including hydrogenation/dehydrogenation, oxidation, activation, aging, and poisoning.
- The morphological peculiarities of many commercial catalysts, such as high porosity; complex, rugged surface; and high dispersion of supported nanoparticles, are not limiting factors.
- Existing data on well-defined reference materials from surface science work such as high-resolution electron energy loss spectroscopy (HR-EELS), X-ray photoelectron spectroscopy, and secondary ion mass spectrometry on single-crystal surfaces can significantly support the analysis of IINS spectra for the characterization of complex commercial catalysts [29].

In this section we focus on fuel cell catalysts. In the field of low-temperature fuel cells (e.g., phosphoric acid fuel cells, proton-exchange membrane fuel cells, and direct methanol fuel cells), carbon black-supported catalysts are used for the conversion of hydrogen and oxygen into water, electrical power, and heat [30, 31]. High precious metal loadings are preferable, but sufficiently low particle size and stable dispersion are also essential. For these fuel cell catalysts, extremely finely divided platinum or platinum alloy particles (e.g., 2.5-nm primary particle size), which are supported on special grades of finely divided pure carbon blacks, are used. The morphology of the particles and their composition can be studied by transmission electron microscopy (TEM) and scanning transmission electron microscopy (STEM), coupled with energy-dispersive X-ray analyses (EDXs). The degree of alloying in bimetallic or multimetallic systems can be studied by X-ray absorption fine structure spectroscopy and EDX. Nanoparticles of a few nanometers in size are at the limits of classical X-ray diffraction (XRD).

In situ information on the interaction between hydrogen and adsorption sites at the surfaces of the supported nanoparticles, under realistic gas pressures, is highly desirable. This is not available by means of NMR, HR-EELS, infrared, or Raman spectroscopies. However, quasielastic neutron scattering can probe the dissociative adsorption of H_2 on the supported platinum nanoparticles [32], and IINS is exclusively able to study the site occupation of atomic hydrogen on cubo-octahedral precious metal particles in different coordination sites. The aim is to discriminate between "spectator sites" and surface sites that are really relevant for the catalytic activity and to clarify how to generate the latter by improving catalyst preparation techniques. Interdependencies between primary particle size (2–10 nm) and relative contributions of (100) and (111) crystal orientations can be studied, for example, in the range of 2–10 nm by "surface titration with atomic hydrogen."

Typical high-resolution TEM images of a series of fuel cell catalysts are compared in Fig. 13.7 [33]: crystalline platinum particles of about 2.8 nm (a), 3.4 nm (b), 4.5 nm (c), and 7.8 nm (d) are attached to paracrystalline carbon black aggregates, which act as a support of large BET (nitrogen surface area according to Brunauer,

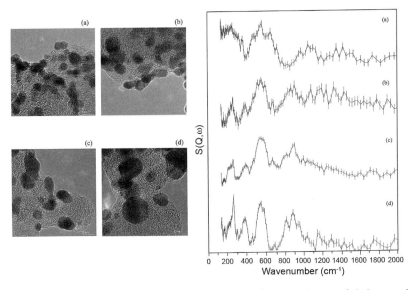

Fig. 13.7 *Left,* (**a–d**) high-resolution transmission electron microscope images of platinum-carbon black fuel cell catalysts of different average primary particle size. *Right,* (**a–d**) in situ inelastic incoherent neutron scattering spectra of the catalysts as measured under about 800 mbar of hydrogen loading. Bands at about 400–500 cm^{-1} Ptx-H vibrations, bands at about 800 cm^{-1} Pt-OH vibrations [33]. (Reproduced with kind permission of Elsevier Limited)

Emmett, Teller) surface area. The values for the average primary particle size of the precious metal were determined by statistical evaluations of TEM data. The images illustrate the quality and homogeneity of the distribution over the surface of the carbon black supports. These very finely divided particles are of cubo-octahedral shape (Fig. 13.8).

What cannot be seen in photographs such as those in Fig. 13.7 is the hydrogen that is present under operating conditions of such catalysts. It is an analytical challenge to measure the dissociative adsorption of molecular hydrogen into atomic hydrogen as well as the site occupation of the atomic hydrogen on adsorption sites at the surface of these nanoparticles. The very high dispersity of the catalysts hampers the use of surface science spectroscopies. Inelastic incoherent neutron scattering extends surface science techniques on single-crystal surfaces into the field of highly dispersed supported catalyst particles, which can be measured under about 1 bar of hydrogen operation pressure in situ.

The IINS spectra depicted in Fig. 13.7 were taken as follows: about 40 g of catalyst were sealed in flow-through sample cans and pressurized carefully up to 1 bar of hydrogen sorption equilibrium. The surface oxides were removed by cycles of hydrogen adsorption and desorption. Afterwards, the IINS spectra were taken to analyze the hydrogen-related surface properties of an active, pyrophoric fuel cell catalyst powder under in situ conditions. The bands at about 400–500 cm^{-1} are due to Pt$_x$-H vibrations, and bands at about 800 cm^{-1} represent Pt-OH vibrations

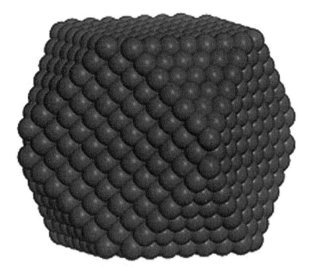

Fig. 13.8 Model of a cubo-octahedral particle with 100 and 111 faces (Reproduced with kind permission of B. Hannebauer, AQura GmbH)

(sample d) formed by diffusion-limited transport of residual traces of oxygen from the interior of platinum particles of larger size and their conversion into polar surface groups under the influence of hydrogen. The differences are due to changes in the relative amounts of (100) and (111) crystal planes and correspondingly different proportions of site occupation of hydrogen in different symmetries. In principle the changes should be pronounced, especially in the region of about 3.5-nm particle sizes [34]. According to Kinoshita [35], for such nanoparticles there is a correlation between catalytic activity and varying particle size in this nanometer regime.

Catalyst preparation may have some influence on the relative number of adsorption sites and the amount of surface disorder. The IINS technique was used to investigate the influence of varying preparation conditions on the adsorption sites of supported nanoparticles. Figure 13.9 clarifies the differences in site occupation of atomic hydrogen on ~4-nm-sized platinum particles that were produced under varying conditions. Given the similar primary particle size, the supported nanoparticles show different adsorption behavior due to structural defects [32].

Atomic hydrogen in different coordination and adsorption sites may have different catalytic activity. It is still under debate which surface sites are the active ones in catalysis and which are largely spectator sites [36, 37]. Recently, the on-top site occupation of hydrogen on platinum nanoparticles supported on carbon was measured showing the platinum-hydrogen stretch mode at about $2,080 \, cm^{-1}$ (Fig. 13.10) [32].

In the future, comparison of the vibrational frequencies obtained by HR-EELS [38] on single-crystal surfaces under ultra-high vacuum conditions and IINS data on nanodispersed catalyst particles will allow much more detailed analyses of technological catalysts under gas pressures typical for industrial applications. Therefore,

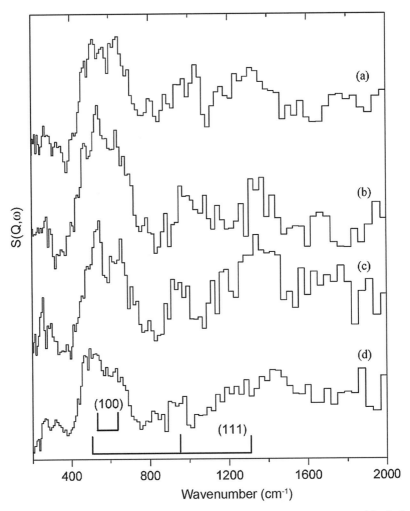

Fig. 13.9 Varying fine structure of the platinum-hydrogen vibrational band measured for fuel cell catalysts of similar platinum primary particle size as recorded under 800 mbar hydrogen pressure. Catalysts prepared under different conditions [32]. (Reproduced with kind permission of Elsevier Limited)

IINS provides a bridge across the so-called pressure gap between surface science and industrial catalysis.

Recent technological innovations and planned spectrometer optimizations will have further impact on the usefulness of IINS for catalyst research. Investigations of very low concentrations of catalytically active phases will become possible. This will allow better identification, generation, and control of the relevant active sites in developing improved catalysts, activities that will certainly not be confined to the field of fuel cell operation. To date, all IINS studies of fuel cells have been of

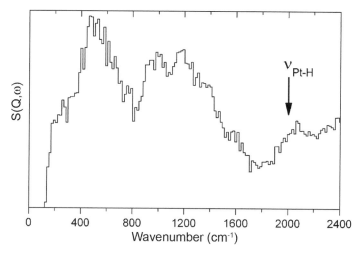

Fig. 13.10 Detection of the on-top size occupation of atomic hydrogen on platinum nanoparticles: platinum-hydrogen stretch mode at about 2,080 cm^{-1} [32]. (Reproduced with kind permission of Elsevier Limited)

the catalyst in the dry state. Under operating conditions, water is both present and essential to the fuel cell's operation. In principle, it should be possible to obtain IINS spectra from an electrochemical cell under operating conditions. Liquid water would contribute a large, but smoothly varying, background to the spectra. By measuring spectra at surface potentials where hydrogen is present and absent on the surface, the difference spectrum should give the spectrum of the adsorbed species, exactly as is done in spectroelectrochemical experiments using infrared spectroscopy. This type of experiment is highly speculative and has not yet been attempted. It would necessitate the use of a different IINS spectrometer from that used in the majority of the work described here. Under aqueous conditions, the residence time of the adsorbed hydrogen is likely to be much less than when adsorbed under a static gas pressure, but since the neutron interaction time is on the femtosecond time scale, this will not present a problem.

13.2.3 Deactivation of Palladium Catalysts from Industrial Operation

An important subject of physical/chemical analyses in industry is deactivated catalysts from chemical plants. Such studies are undertaken to understand the detrimental impact of irregular operation parameters on activity and selectivity. Avoiding deactivation saves both time and energy. Because of the broad scale of catalyst use worldwide, improvements of a few percent can have major economic impacts and save substantially on costs as well as resources.

To this end, it is essential to improve the understanding of catalyst deactivation processes by coking, poisoning, and physical degradation. The knowledge gained can be used to develop new strategies for the enhancement of the useful lifetime of catalysts and for improvement in activity and selectivity by tailoring/changing compositions, nanostructures, and operating conditions.

The blocking of active sites on industrial catalysts by cokes or other carbonaceous deposits and/or by catalyst poisons can lead to premature reduction of catalytic activity to uneconomic levels. In the worst cases, this may be irreversible. Costly premature shutdowns of production facilities can be the consequence. The mechanisms of carbon deposition and their catalytic transformations have been studied in detail to prevent catalyst deactivation and to minimize the variable costs of operating plants [39–44].

Various species of carbon can be found on and inside catalysts [45–49], including small molecular species; carbides; green oils; oligomer- or polymer-like intermediates or products; soot-like amorphous, paracrystalline, or microcrystalline species of fluffy appearance; and highly aromatic/graphitic carbons such as glossy pyrocarbons. Figure 13.11 shows some typical shapes of carbonaceous matter observed at the surface of Pd/SiO$_2$ catalysts that showed premature deactivation [50]. Cauliflower-like polymeric species were deposited in the one case (micrographs

Fig. 13.11 Various shapes of carbonaceous deposits observed at the surfaces of Pd-SiO$_2$ catalysts used in the selective catalytic hydrogenation of acetylene to ethylene in the HCl recycle gas stream of the vinyl chloride process at $T < 473$ K. *Left*: Polymer-like carbon. *Right*: Carbon filaments and amorphous carbon [50, 52]. (Reproduced with kind permission of Elsevier Limited)

on the left), whereas in the other (micrographs on the right), carbon filaments were formed at the catalyst surface by solution/precipitation mechanisms in the incorporation and catalytic transformation of deposited carbon under the influence of iron and nickel contaminations originating from dew point corrosion and deposition onto the catalyst surface.

Figure 13.12 compares the IINS spectra of low-temperature cokes, which were found after 6 months (Fig. 13.12a) and 2 years (Fig. 13.12c) of regular operation. Polymer-like matter from oligomerization reactions as shown in Fig. 13.11 (left) was expected; however, spectrum 13.12a strongly resembles the spectrum of pure chemical vapor deposited carbon [51], whereas 13.12c matches the spectrum of $[Fe(H_2O)Cl_5]^{2-}$, whose simulated spectrum is presented in Fig. 13.12d. The spectrum of a high-temperature coke, which was deposited onto the surface of a Pt/Al_2O_3

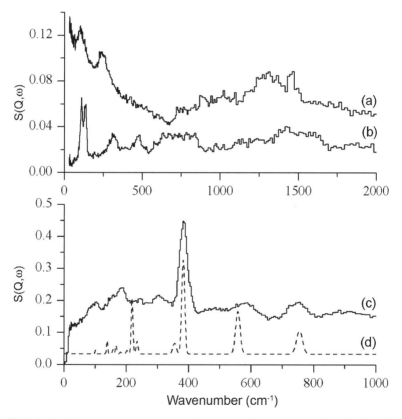

Fig. 13.12 Inelastic incoherent neutron scattering spectra of carbonaceous deposits from the surfaces of catalysts from low-temperature and high-temperature chemical processes. (a) and (c) are from the selective hydrogenation of acetylene (low temperature); (b) is from the production of HCN and H_2 from methane and ammonia (high temperature) [50, 52]; spectrum (d), which is similar to that of (c), is a simulated spectrum for $[Fe(H_2O)Cl_5]^{2-}$ and was included for comparison. (Reproduced with kind permission of Elsevier Limited)

catalyst (temperature about 1,473 K) during the synthesis of hydrogen cyanide from methane and ammonia, is completely different (13.12b). (The hydrogen cyanide is used for the synthesis of methacrylates and essential amino acids). Soot-like matter was expected, but graphitic matter was found, indicating unusually low partial pressures of hydrogen under certain operating conditions [52].

Combining such findings from different techniques is very helpful in establishing the "road maps" of possible routes to deactivation as shown in Fig. 13.13 for the deposition of low-temperature cokes. Accumulation and comparison of data from different analytical techniques allow better predictability of catalyst reactions to changes in operating conditions.

The IINS technique permits the study of surface phenomena on the contaminated particles delivered directly from the technological application, as in the following example, where a set of hydrogenation catalysts was identified and studied because of irregular behavior in a technical application over several months. The catalysts were used in the hydrogenation of C = O groups to C–OH groups. Something had blocked considerable amounts of active surface area. Results from a broad variety of characterization methods showed the following.

- No poisoning by the usual contaminants such as carbon monoxide, sulfur, metallic elements, or corrosion products.
- No loss of surface area by sintering.

What was the blocking agent? Again, the high dispersity of the catalyst material gave problems in using well-proven methods such as infrared, Raman, and NMR spectroscopies to identify the blocking agent. Other well-proven techniques such

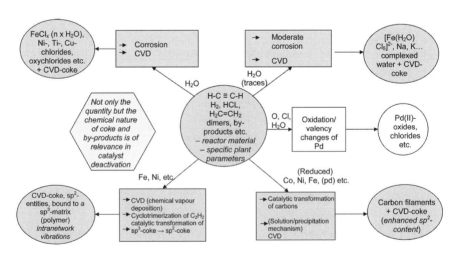

Fig. 13.13 Schematic drawing of different pathways to catalyst deactivation in the HCl recycle gas stream of the vinyl chloride process: formation of chemical vapor deposition-type coke, catalytically driven transformation of carbon to filamentous carbon, dew point corrosion, and high-speed corrosion [50]. (Reproduced with kind permission of Elsevier Limited)

as sorption analysis (BET), porosimetry, and SEM and TEM coupled with energy dispersive X-ray micro- or nanoanalyses (EDX), elution techniques coupled with organic chromatographies and mass spectrometries, and elemental analysis analyses also failed to give conclusive results. Thus, it appeared that this catalytic process was a suitable candidate for IINS.

The IINS spectrum depicted in Fig. 13.14a [53, 29] give a snapshot of the properties of about 35 g of catalyst as taken directly from the reactor and sealed under inert gas into the IINS cell. The sample includes a substantial quantity of residual solvent weakly adsorbed at the catalyst surface, accounting for the large offset in the spectra. The spectrum in Fig. 13.14b shows the results of solvent extraction on the sample whose spectrum is shown in Fig. 13.14a. Figure 13.14c shows the same sample after in situ hydrogenation up to 1.5 bar of hydrogen equilibrium pressure. The additional band at about $470 \, \text{cm}^{-1}$ is due to the reversible formation of β-palladium hydride, which could be decomposed afterward. Figure 13.14d is the IINS spectrum of a correspondingly used, but still active, catalyst, as measured after

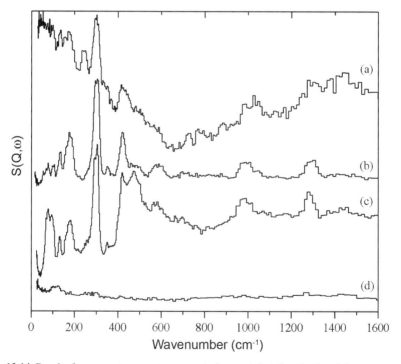

Fig. 13.14 Results from a neutron spectroscopy study on catalyst deactivation. (**a**): spectrum of a deactivated catalyst as taken directly from the hydrogenation process, sealed under argon, including solvent molecules; (**b**): same as (**a**); however, after solvent extraction for the removal of extractable components; (**c**): after hydrogen loading of the cleaned catalyst (**b**) up to 1.5 bar equilibrium partial pressure; (**d**): used but active catalyst, which was extracted under the same conditions as the deactivated catalyst (**b**) [29, 53]. (Reproduced with kind permission of Elsevier Limited)

cleaning by solvent extraction under exactly the same conditions as the deactivated sample.

The strong bands seen in Fig. 13.14b are also present in Fig 13.14a and c; thus they are not an artifact of the solvent extraction process. The species responsible for the strong bands does not interfere with hydrogen adsorption and are not readily removed by hydrogenation. In Fig. 13.14d, all the additional molecular bands observed for the deactivated catalyst are absent. Together the evidence strongly suggests that the species responsible for the spectrum in Fig. 13.14b is also responsible for the catalyst's deactivation. This was confirmed by a good correlation between the normalized IINS scattering intensities of the peaks, as measured on catalyst samples with varying degrees of deactivation.

Detailed computer evaluation revealed that the topmost atomic layer of the catalyst particles was covered with methyl groups (Fig. 13.15). From periodic discrete Fourier transform calculations [54] of CH_{0-3} on Pd(111), it is known that CH_3 is the stable species.

The situation in the topmost layer of the deactivated catalyst is visualized in Figure 13.16 for different degrees of coverage of the catalyst surface: strong shielding (Fig. 13.16a) and incomplete shielding (Fig. 13.16b). The adsorption properties of the catalyst surface for sp^2 type reactant molecules were changed by sp^3 type molecular structures. However, the absorption properties of the palladium toward hydrogen were not significantly affected.

The unexpected formation of considerable amounts of the strongly bound Pd-CH_3 species changed the adsorption properties of the catalyst's surface for the adsorption of sp^2 type reactants, presumably by a site-blocking mechanism. It

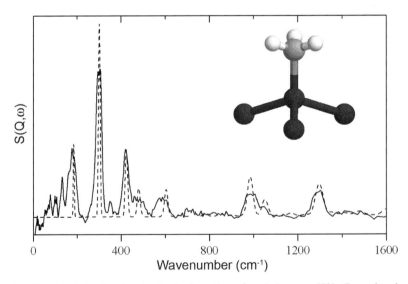

Fig. 13.15 Catalyst deactivation due to the formation of methyl groups [53]. (Reproduced with kind permission of Elsevier Limited)

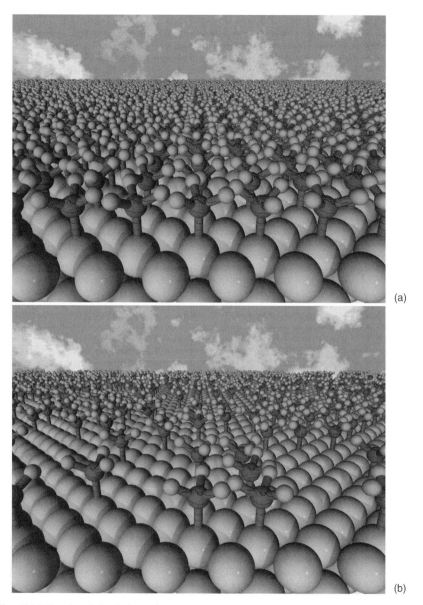

(a)

(b)

Fig. 13.16 Results of simulations of the situation in the topmost layers of a deactivated catalyst [based on the inelastic incoherent neutron scattering spectra in Figs. 13.14(b) and 13.15]: **(a)** shows strong shielding; **(b)** shows incomplete shielding. (Reproduced with kind permission of the Rutherford Appleton Laboratory)

was possible to monitor the changes at the particle surfaces over time by collecting samples from the process under controlled conditions and taking their IINS spectra.

This example from a commercial catalysis reaction shows that the surface and bulk states of hydrogen-containing entities can be determined under conditions that are close to those of the process. Information can be gained by the following.

- Surface spectroscopy of a catalyst as taken from the reaction system with adherent mixtures of solvents and products.
- Monitoring changes in the topmost atomic layers such as the formation, transformation, or degradation of adsorbed species over time at different stages of a deactivation process.

IINS, exclusively, was able to clarify the reason for deactivation of this industrial catalyst, whereas all usual methods of catalyst characterization failed due to the specific physical properties of the catalyst and the special, unexpected reason for deactivation.

13.2.4 Catalyst Supports

As pointed out in Section 13.2.1, IINS is highly suitable for evaluating the proton-related surface chemistry of finely divided carbonaceous matter. Inelastic incoherent neutron scattering allows the study of activated carbons derived from natural resources, carbon blacks, and novel grades of carbonaceous matter used as support materials for chemical catalysts for the synthesis of fine chemicals and fuel cell catalysts. It is able to contribute to the clarification of key properties of "good" catalyst supports [26] for particular processes or technical applications. Inelastic incoherent neutron scattering can reveal how, and to what extent, the surface chemistry is changed during precious metal impregnation in catalyst manufacture, reduction, or formation or during catalyst operation and elucidate the effects of hydrogen spillover and hydrogasification [55]. It supports and extends classical methods for studying the raw materials for catalyst manufacture, such as elemental analysis, porosimetry, BET, various techniques of electron microscopy, and surface analyses by, for example, X-ray photoelectron spectroscopy or secondary ion mass spectrometry.

13.3 Conclusions and Outlook

Finely divided materials are produced and used in chemical technology, fuel cell catalysis, and large-scale chemical processes. Inelastic incoherent neutron scattering is a valuable tool for research on such materials, focusing on hydrogen-related properties. It can significantly complement well-established analytical methods, which may reach their limits in analyzing highly dispersed matter or with the demands of characterization under in situ-conditions, required for some technological applications. This includes the investigation of surface sites on new grades of advanced filler particles for tires with reduced rolling resistance, investigation of the site occupation of atomic hydrogen on fuel cell catalysts, clarification of the reasons for catalyst poisoning and how to circumvent premature catalyst degradation or deactivation, and understanding the growth mechanisms in catalyst coking and the formation and catalytic transformation of carbonaceous deposits.

Identifying adequate means to prevent premature catalyst deactivation has significant benefits, which can be evaluated by techniques such as life-cycle assessment (LCA) of key figures according to the ISO 14000 and ISO 14044 environmental management standards.

This chapter has presented the advantages of neutrons for materials research. As with any technique, IINS spectroscopy has its disadvantages. Neutrons are a scarce and expensive resource, so the guiding principle is that neutrons are the technique of last resort. As has been shown throughout this article, IINS spectroscopy is complementary to other techniques and, in common with most analytical techniques, is most effective when combined with the information from other methods. The principal drawback of IINS spectroscopy is the scarcity of facilities, thus access to IINS instrumentation requires planning and patience. Most IINS measurements are made at low (<20 K) temperature, so ensuring that the data are relevant to systems that operate at room temperature or above requires care. IINS is a relatively insensitive technique, typically at least a gram of an organic material or several tens of grams of an inorganic material are needed. These are very large samples for most analytical techniques, and obtaining a sufficient quantity can be problematic, particularly for new materials that are usually only available in milligram quantities. On the other hand, the large sample size does mean that concerns about representative sampling are much less severe than for techniques that operate on the milligram or smaller scale. Measurement times are at least a few hours and may be as long as 24 h, thus kinetic measurements by IINS are not feasible (even if the measurement temperature allows kinetics to occur). The spectral resolution is modest by comparison with that obtainable by infrared or Raman spectroscopy; however, for solid-state systems, this is less of a problem because the intrinsic line widths are often larger than the instrumental resolution.

In conclusion, IINS provides a bridge across the so-called pressure gap between the surface science experiments on well-defined single-crystal surfaces at low temperatures and low pressures and the field of industrial catalysis (nanoparticles on highly disperse catalysts under much higher temperatures and gas pressures). This was illustrated by measuring, for example, the site occupation of dissociatively adsorbed atomic hydrogen at the surfaces of supported platinum particles of about 2–8 nm on fuel cell catalysts or by detecting stable methyl groups as a "surface science species" in the topmost atomic layers of a deactivated palladium catalyst, as taken directly from a chemical plant and measured under the presence of residual solvent molecules. The great, but still scarcely used, potential of neutron scattering for tackling and solving analytical questions from the chemical and petrochemical industries arises from the easy penetration of stable sample containments. This allows safe handling and spectroscopy of bulk quantities of extremely finely divided catalyst powders or pellets. It enables the study of external and internal surfaces and adsorption sites under realistic pressures of hydrogen or in the presence of contaminants, aggressive gases, or gas mixtures.

Recent technological innovations in neutron flux density at the sample, together with spectrometer optimizations, will have further impacts on the usefulness of IINS for catalyst research [29]. Investigations of very low concentrations (less than 5% and possibly less than 1%) of catalytically active phases may become possible, which would be of considerable impact. Titration of surface sites with atomic hydrogen at improved spectrometer resolution and sensitivity will provide detailed information on site occupation and particle shape. A complementary partner to the experimental work will be ab initio modeling of catalysts. This will provide unambiguous assignments of surface sites and enable quantitative information to be extracted from the data.

The possibility of probing non-hydrogen-containing entities such as CO_x, NO_x, and SO_x will open new areas of investigation, especially automotive exhaust gas catalysts. The study of catalysts for off-gas stream purification in many kinds of processes will also be feasible. Of particular relevance are reduction of emissions from power plants, petrochemical processes, and the food industry.

Phenomena responsible for catalyst formation and the development of the structures responsible for activity, selectivity, and degradation in operating catalysts may be studied in much more detail. Inelastic incoherent neutron scattering at improved performance will continuously complement and extend data from SEM, HR-TEM, STEM/EDX, TEM/EELS, HR-EELS, and XRD in catalysis research and catalyst characterization.

A better understanding of how to improve the performance of materials that are produced and used worldwide everyday in large quantities will be of significant positive economic and environmental influence. A mixture of more efficient energy use [56], the avoidance of shutdowns and carbon losses, better selectivity of processes, and improved industrial catalysts will continue to demonstrate the economic usefulness of the results from analytical investigations, including neutron spectroscopy.

References

1. T. Springer, in *Hydrogen in Metals I*, edited by G. Alefeld and J. Völkl (Springer, Berlin, 1978), Topics in Applied Physics, Vol. 28, p. 75
2. R. Wiswall, in *Hydrogen in Metals II*, edited by G. Alefeld and J. Völkl (Springer, Berlin, 1978), Topics in Applied Physics, Vol. 29, p. 201
3. A. Percheron-Guegan and J.-M. Welter, in *Hydrogen in Intermetallic Compounds I*, edited by L. Schlapbach (Springer, Berlin, 1988), Topics in Applied Physics, Vol. 63, p. 11
4. D. Richter, R. Hempelmann, and R. C. Bowman, Jr., in *Hydrogen in Intermetallic Compounds II*, edited by L. Schlapbach (Springer, Berlin, 1992), Topics in Applied Physics, Vol. 67, p. 97
5. G. Sandrock, S. Suda, and L. Schlapbach, in *Hydrogen in Intermetallic Compounds II*, edited by L. Schlapbach (Springer, Berlin, 1992), Topics in Applied Physics, Vol. 67, p. 197
6. E. Wicke, H. Brodowski, and H. Züchner, in *Hydrogen in Metals II*, edited by G. Alefeld and J. Völkl (Springer, Berlin, 1978), Topics in Applied Physics, Vol. 29, p. 73

7. D. K. Ross, in *Hydrogen in Metals III*, edited by H. Wipf (Springer, Berlin, 1997), Topics in Applied Physics, Vol. 73, p. 153
8. http://www.fuelcells.org/
9. U.S. Department of Energy, Office of Energy Efficiency and Renewable Energy, Vehicle Technologies Program, Freedom Car Web site, http://www.eere.energy.gov/vehiclesandfuels/about/partnerships/freedomcar, 2008
10. C. Read, G. Thomas, G. Ordaz, and S. Satyapal, Mater. Matters **2**, 3 (2007)
11. F. Schüth, B. Bogdanovic, and M. Felderhoff, Chem. Commun. 2249 (2004)
12. P. A. Georgiev, D. K. Ross, A. DeMonte, U. Montaretto-Marullo, R. A. H. Edwards, A. J. Ramirez-Cuesta, and D. Colognesi, J. Phys. Condens. Matter **16**, L73 (2004)
13. P. A. Georgiev, D. K. Ross, P. Albers, and A. J. Ramirez-Cuesta, Carbon **44**, 2724 (2006)
14. P. C. H. Mitchell, S. F. Parker, A. J. Ramirez-Cuesta, and J. Tomkinson, *Vibrational Spectroscopy with Neutrons* (World Scientific, Singapore, 2005), p. 508
15. P. Albers, Elements—Degussa Science Newsletter **15**, 22 (2006)
16. J.-B. Donnet, R. C. Bansal, and M.-J. Wang, Carbon Black—*Science and Technology*, 2nd edition (Marcel Dekker, New York, 1993), p. 72 (and literature cited therein)
17. B. Davis, Europ. Rubber J. **178**, 18 (1996)
18. P. Yap, Meeting of the Rubber Division, American Chemical Society, Louisville, Oct. 1996, paper no. 16
19. W. Niedermeier, P. Messer, and J. Fröhlich, Degussa Technical Report TR 814
20. *MATERIALICA Industry News* Web site, http://www.materialica.de/html/green_tire.html, 2007, alternative website: http://www.michelin.com/corporate/front/templates/affich.jsp
21. W. Niedermeyer and B. Freund, Degussa Technical Report TR 802, 2007
22. E. Koberstein, E. Lakatos, and M. Voll, Ber. Bunsenges. Phys. Chem. **75**, 1105 (1971)
23. P. Albers, G. Prescher, K. Seibold, D. K. Ross, and F. Fillaux, Carbon **34**, 903 (1996)
24. P. Albers, A. Karl, J. Mathias, D. K. Ross, and S. F. Parker, Carbon **39**, 1663 (2001)
25. P. Albers, K. Seibold, G. Prescher, B. Freund, S. F. Parker, J. Tomkinson, D. K. Ross, and F. Fillaux, Carbon **37**, 437 (1999)
26. P. W. Albers, J. Pietsch, J. Krauter, and S. F. Parker, PCCP **5**, 1941 (2003)
27. G. Ertl, H. Knözinger, and J. Weitkamp (eds.), *Preparation of Solid Catalysts*, (Wiley VCH, Weinheim, 1999)
28. E. S. J. Lox and B. H. Engler, in *Handbook of Heterogeneous Catalysis*, edited by G. Ertl, H. Knözinger and J. Weitkamp (Wiley VCH, Weinheim, 1997), Vol. 4, p. 1559
29. P. Albers and S. F. Parker, in *Advances in Catalysis*, edited by B. C. Gates and H. Knözinger (Elsevier, Netherlands, 2007), Vol. 51, pp. 99–132
30. G. J. Acres, J. C. Frost, G. A. Hards, R. J. Potter, T. R. Ralph, D. Thompsett, G. T. Burstein, and G. J. Hutchings, Catal. Today **38**, 393 (1997)
31. P. Albers, E. Auer, K. Ruth, and S. F. Parker, J. Catal. **196**, 174 (2000)
32. S. F. Parker, C. D. Frost, M. Telling, P. Albers, M. Lopez, and K. Seitz, Catal. Today **114**, 418 (2006)
33. P. W. Albers, M. Lopez, G. Sextl, G. Jeske, and S. F. Parker, J. Catal. **223**, 44 (2004)
34. R. Van Hardeveld and F. Hartog, in *Advances in Catalysis*, edited by D. D. Eley, H. Pines, and P. B. Weisz (Academic Press, New York, 1972), Vol. 22, p. 75
35. K. Kinoshita, J. Electrochem. Soc. **137**, 845 (1990)
36. G. S. McDougall and H. Yates, in *Catalysis and Surface Characterisation*, edited by T. J. Dines, C. H. Rochester, and J. Thomson (The Royal Society of Chemistry, Cambridge, 1992), Specialist Publication 114, p. 109
37. M. A. Chester, K. J. Packer, D. Lennon, and H. E: Viner, J. Chem. Soc. Faraday Trans. **91**, 2191 (1995)
38. K. Bedürftig, S. Völkening, Y. Wang, K. Wintterlin, K. Jacobi, and G. Ertl, J. Chem. Phys. **111**, 11147 (1999)
39. J. Rostrup-Nielsen and D. L. Trimm, J. Catal. **48**, 155 (1977)
40. C. H. Butt, Chem. Eng. **91**, 96 (1984)

41. J. R. Rostrup-Nielsen, in *Catalyst Deactivation by Coke Formation*, edited by R. Hughes, J. Santamaria, and A. Monzon (Elsevier 1997), Catalysis Today, Vol. 37, p. 225
42. G. C. Bond, Appl. Catal. A: General **149**, 3 (1997)
43. G. F. Froment, in *Catalyst Deactivation*, edited by C. H. Bartholomew and J. B. Butt (Elsevier, Amsterdam, 1991), *Studies in Surface Science and Catalysis*, Vol. 68, p. 53
44. D. L. Trimm, in *Handbook of Heterogeneous Catalysis*, edited by G. Ertl, H. Knözinger, and J. Weitkamp (VCH, Weinheim, 1997), Vol. 3, p. 1263
45. C. H. Bartholomew, Appl. Catal. A: Gen. **212**, 17 (2001)
46. J. Barbier, Appl. Catal. **23**, 225–243 (1986)
47. P. G. Menon, J. Mol. Catal. **59**, 207 (1990)
48. P. G. Menon, Catal. Today **11**, 161 (1991)
49. P. G. Menon, Chem. Rev. **94**, 1021 (1994)
50. P. Albers, J. Pietsch, and S. F. Parker, J. Mol. Catal. A: Chem. **173**, 275 (2001)
51. J. K. Walters, R. J. Newport, S. F. Parker, and W. S. Howells, J. Phys.: Condens. Matter **7**, 10059 (1995)
52. P. Albers, G. Prescher, K. Seibold, and S. F. Parker, Stud. Surf. Sci. Catal. **130**, 3155 (2000)
53. P. Albers, H. Angert, G. Prescher, K. Seibold, and S. F. Parker, J. Chem. Soc. Chem. Commun. (17), 1619–1620 (1999)
54. J.-F. Paul and P. Sautet, J. Phys. Chem. B **102**, 1578 (1998)
55. P. C. H. Mitchell, A. J. Ramirez-Cuesta, S. F. Parker, J. Tomkinson, and D. Thompsett, J. Phys. Chem. B, **107**, 6838 (2003)
56. M. Neelis, M. Patel, P. Bach, and K. Blok, Appl. Energy **84**, 853 (2007)

Chapter 14
Hydrogen and Hydrogen-Storage Materials

Milva Celli, Daniele Colognesi, and Marco Zoppi

Abstract Currently, neutron applications in the field of hydrogen and hydrogen-storage materials represent a large and promising research area, both from the fundamental and the applied points of view. In this chapter we review some relevant topics from this subject area, including hydrogen bulk properties (concerning both the solid and the liquid phases), hydrogen storage in nanoporous materials (mainly carbon nanotubes, zeolites, and metal organic frameworks), and hydrogen storage in solid matrices (particularly ionic hydrides). For each class of materials, the current state of neutron research on their structural and dynamic properties is presented in detail. Future perspectives in this area are also outlined.

14.1 Introduction

14.1.1 Foreword

Ten years ago (in 1998), the price of oil was around $12 per barrel. Now it runs around ten times higher and is still increasing. However, the really bad news is that, sooner or later, oil will simply run out. The age of cheap energy is on its way out, and a new age, in which mankind will have to find other energy resources than fossil fuels, is dawning.

In general, producing energy does not require a real effort; humans have done it for centuries and even millennia. However, producing energy for a world population exceeding 6 billion people is going to be a major challenge. In 2006, the overall world energy consumption was close to 1.2×10^{13} kWh. This represents 3/10,000 of the annual solar energy reaching Earth's surface. Whatever energy production method is adopted in the future, we will have to solve the problem of storage and transportation of energy from production sites to places where it is needed.

Hydrogen has been suggested as a possible mean for storing and transporting energy. However, to make this dream a reality, a great deal of scientific knowledge

M. Celli (✉)

CNR-Istituto Sistemi Complessi, Via della Madonna del Piano, 10, I-50019 Sesto Fiorentino, Italy

e-mail: milva.celli@isc.cnr.it

L. Liang et al. (eds.), *Neutron Applications in Earth, Energy and Environmental Sciences*, Neutron Scattering Applications and Techniques,

DOI 10.1007/978-0-387-09416-8_14, © Springer Science+Business Media, LLC 2009

has to be developed so that technological progress can be made. Neutrons can play an important role in this progress of basic knowledge development in allowing both study of bulk hydrogen properties and research to confirm that hydrogen compounds are suitable for hydrogen storage. As a matter of fact, hydrogen can be stored in a solid matrix (e.g. hydrides) with a number density similar to, or even greater than, that of its liquid phase. Understanding the microscopic structure of these materials and the dynamics of hydrogen in the uptake and release processes represents a fundamental step in building the background of knowledge necessary to optimize utilization and design better materials.

Among the various topics concerning neutron applications in the field of hydrogen and hydrogen-storage materials, we have chosen to concentrate on the state of the art in hydrogen bulk properties, molecular storage in nanoporous materials, and atomic storage in solid matrices. The review we present here is certainly not exhaustive, however, because we had to make choices on which material families to select. Nevertheless, we think we have provided paradigmatic examples of this class of materials, inclusive of a rich literature where the interested reader will be able to find first-hand information on selected topics of interest.

14.1.2 Hydrogen Overview

Hydrogen is the most abundant element in the universe. Its mass fraction is estimated to be 75% of the total, followed by helium (23%), oxygen (1%), and carbon (0.5%) (all other elements giving rise to the remaining 0.5% of the total estimated mass). On an atomic basis, the distribution is even more spectacular, with hydrogen contributing to 92.7% of the total number of atoms, helium to 7.2%, and all other elements adding up to the remaining 0.1% (Fig. 14.1). On the surface of the Earth, the weight distribution is different, with hydrogen in third position (5.5%) after oxygen (65.7%) and silicon (13.5), followed by aluminum (4.1%), iron (3.1%), calcium

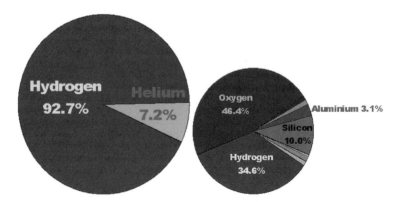

Fig. 14.1 Atomic abundance of elements in the universe (*left*) and on the Earth's surface (*right*)

(2.5%), sodium (1.7%), magnesium (1.5%), and chlorine (1.0%).[1] However, here too the atomic distribution is more favorable to hydrogen (34.6%), which holds the second place after oxygen (46.4%).

Hydrogen exists in three different isotopic species. The lightest and most abundant isotope (protium) is probably the most studied element, certainly the simplest (one proton and one electron), and forms the simplest molecule (two protons, 75 pm apart, enclosed by an electronic cloud of two electrons). Deuterium nuclei are formed by the stable association of one proton and one neutron. This is the largest possible mass ratio between two stable isotopes and gives rise to huge isotopic effects that can be easily detected by spectroscopic measurements. The nucleus of the heaviest isotope, tritium, is formed by the association of one proton and two neutrons. Tritium, with a half-life of 12.32 years, is not stable and transforms into helium-3 (β-decay). The discovery of hydrogen isotopes is relatively recent. Deuterium was spectroscopically detected in 1932, as an impurity of hydrogen [1], while tritium was artificially produced in 1934 by deuterium nuclear collisions [2].

Atomic hydrogen is chemically unstable and reacts with almost all other elements. Molecular hydrogen, by contrast, is very stable. The molecule is diatomic and can be formed by any combination of the three isotopes. In all cases, the internuclear separations do not change much, and so the electronic clouds do not either. Thus, the intermolecular interaction potential can be safely considered almost identical for all possible pair interactions (six in all), and the observed differences in the thermodynamic behavior of the various isotopic species are mainly attributed to the different quantum effects associated with the different masses.

In the following, unless a specific isotope is named, we will use the word hydrogen when referring to the lightest isotope.

This is probably the most deeply studied molecular system. Under normal conditions of temperature and pressure, molecular hydrogen is a gas whose thermodynamic behavior does not differ too much from that of an ideal gas. Molecular hydrogen was first liquefied in 1898 by Dewar, but observation of mist and liquid droplets dates back to 1877 [3]. In fact, hydrogen's critical temperature is very low, and therefore its liquid range is rather confined, in low-temperature cryogenic regions. However, solid hydrogen can be obtained, at room temperature, by applying a pressure of 5.4 GPa [4].

The relevant thermodynamic properties of fluids composed of homonuclear molecular hydrogen isotopes are reported in Table 14.1. The critical point of normal hydrogen is $T_C = 33.19$ K, while its triple point temperature is $T_{TP} = 13.96$ K; therefore, its liquid phase interval is less than 20 K. This temperature interval is so low that relevant quantum effects are detectable, although no exchange effect has ever been detected. Thus liquid hydrogen is considered the prototype of a Boltzmann liquid [5].

Molecular hydrogen exhibits some interesting quantum features that are more difficult to observe in more complex systems. In 1925, Mecke [7] spectroscopically

---

[1] Only the Earth's surface (rocks + sea) is considered.

Table 14.1 Relevant thermodynamic properties of pure isotopic molecular hydrogens in normal composition [6]

	T_C (K)[a]	p_C (bar)[a]	n_C (nm^{-3})[a]	T_{TP} (K)[a]	p_{TP} (bar)[a]	n_{TP} (nm^{-3})[b]
Hydrogen	33.19	13.15	9.00	13.96	0.072	23.06
Deuterium	38.34	16.65	10.44	18.71	0.171	26.00
Tritium[c]	40.44	18.50	10.88	20.62	0.216	27.33

[a]The labels "C" and "TP" indicate the critical point and triple point, respectively.
[b]The density n_{TP} refers to the liquid phase.
[c]Critical parameters of tritium are estimates.

detected two different species of hydrogen, and later on, Heisenberg [8, 9] gave the quantum mechanical interpretation that subsequently led to the definition of ortho-hydrogen and parahydrogen. As each proton carries a spin $\frac{1}{2}$, the total nuclear spin can assume either value $I = 0$ (parahydrogen) or $I = 1$ (orthohydrogen). This fact has interesting consequences on the electronic clouds as it affects the parity of the total eigenfunctions. Because of this, the two quantum states are energetically separated and, since the transition from one species to the other is forbidden in the absence of magnetic field gradients, experiments find, in general, a mixture of the two species. In the parahydrogen species (total nuclear spin $I = 0$), the azimuthal quantum number can assume only the $m = 0$ value. By contrast, in orthohydrogen ($I = 1$), the quantum number m may assume three different values ($m = 0, +1, -1$), and the state is degenerate with multiplicity 3. Thus, at high temperature (in practice, ambient and above), a mixture of orthohydrogen and parahydrogen, in thermodynamic equilibrium, contains three parts of orthohydrogen and one part of parahydrogen (normal hydrogen). At low temperature, parahydrogen is the energetically preferred species. However, as we mentioned previously, the nuclear spin flip transition is forbidden in the isolated molecule, and the conversion to thermodynamic equilibrium is extremely slow. This transition is catalyzed in bulk hydrogen by molecular collisions. For example, during a binary collision, two ortho molecules can make a simultaneous transition to their para states. Of course, as the number of orthohydrogen molecules decreases, the transition rate becomes smaller and smaller. Alternatively, one could use a catalyst to speed up the transition. In general, any finely dispersed paramagnetic substance can perform this task [10].

The spectroscopy of molecular hydrogen can be broadly associated with that of any homonuclear diatomic molecule. Thus, rotational transitions can be observed and, in the case of hydrogen, these are characterized by a very energy-extensive spectrum due to the small momentum of inertia of the molecule. The rotational energies of a homonuclear diatomic molecule are given by

$$E_v(J) = B_v J(J+1) - D_v J^2(J+1)^2 - H_v J^3(J+1)^3, \qquad (14.1)$$

where B_v is the rotational constant of the molecule; the parameters D_v and H_v represent the first and second centrifugal distortion correction, respectively [11]; and J is the rotational quantum number. The rotational constant, in reciprocal centimeter (cm^{-1}) units, is given by

$$B_\nu = \frac{\hbar}{8\pi^2 c\mu}\left\langle \frac{1}{R^2}\right\rangle_\nu, \tag{14.2}$$

where μ is the reduced mass, c is the speed of light, and R is the internuclear separation. The angular brackets imply an average of the internuclear separation made over the eigenfunction of the particular vibrational quantum number ν. For hydrogen, $B_0 = 7.36$ meV. Vibrational transitions can be observed too, which are ruled by the intramolecular interaction potential. This is usually approximated by a Morse-type function and therefore is definitely not harmonic. However, one can still write for the distribution of the vibrational levels

$$E(\nu) = (\nu + \frac{1}{2})F_\nu, \tag{14.3}$$

where the vibrational constant depends on the quantum number ν. For hydrogen, $F_0 = 545.7$ meV. Combination bands implying the simultaneous transition of the vibrational and rotational quantum number are, generally, easily observed.

14.1.3 Hydrogen and Neutrons

In evaluating the response function of a neutron scattering experiment, one of the basic ingredients is represented by the bound cross section of the various nuclei. From a quick search on Table 14.1 in [12], it appears immediately that the total scattering cross section of ^{1}H is almost two orders of magnitude larger than the average cross section of other elements (Fig. 14.2). Thus, one would be induced to think that, thanks to this feature, the hydrogen contribution outperforms that

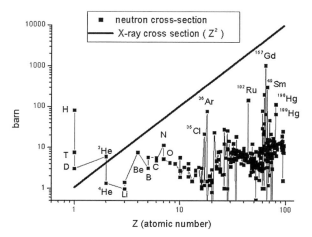

Fig. 14.2 Total scattering cross section of the elements (*squares*). The *line* represents the equivalent cross section for X-rays

of the other elements. This statement is correct, but does not represent the whole truth. As a matter of fact, neutron scattering experiments can be divided into two broad categories (Chapters 2 and 3). On one side, neutron diffraction experiments probe the microscopic structure of a sample (i.e., "where atoms are"). On the other, neutron spectroscopy experiments probe the microscopic dynamics (i.e., "what the atoms do").

In a diffraction experiment, the relevant quantity is the coherent scattering length (Chapter 2). It is this parameter that weights the interference pattern carrying information on the microscopic structure factor. Conversely, the incoherent scattering length contributes to an almost constant background in the measured cross section. However, in the case of hydrogen it is the incoherent term that mainly contributes to the total cross section, while the coherent one is only a small fraction of the total, even smaller than the average value of the various elements. Thus, measuring the structural properties of a hydrogen-containing material by a neutron diffraction experiment is a rather difficult task unless good-quality single crystals are available. The problem can be circumvented using isotopic substitution, where deuterium is substituted for hydrogen atoms. In this case, the coherent scattering length is larger than the incoherent one and quantitative diffraction experiments become possible.

The situation becomes more involved when a dynamic experiment is considered. In this case, the incoherent cross section weights the scattering contribution of the self-dynamics [13]. Therefore, the large value of the hydrogen cross section benefits the so-called inelastic incoherent neutron scattering (IINS) experiments (Chapter 2). However, dynamics experiments probing collective motions, being weighted by the coherent cross section, suffer, in general, the same problems as the diffraction ones.

We conclude this section discussing, briefly, the main differences between neutron scattering and X-ray scattering, with a particular emphasis on applications to materials for hydrogen storage. Looking at Fig. 14.2, it is possible to infer, immediately, that measuring the properties of a hydrogen-containing material for storage applications is an extremely difficult task. Due to the Z^2 dependence of the cross section, to reveal hydrogen and its host at the same time implies distinguishing between properties of different elements characterized by a cross section ratio that may range between 100 and 1,000. However, as we have already observed, the total neutron scattering cross section of hydrogen is generally greater than the average value of the heavier elements. Moreover, when structural information is needed, the problem of the small coherent cross section of hydrogen can be circumvented by isotopic substitution with deuterium.

14.2 Microscopic Structure of Condensed Hydrogen

As far as the solid phase of hydrogen is concerned, the structure has been rather well known for a long time. Crystals of solid hydrogen show, in general, the *hcp* structure, apart from a small region in the phase diagram where they assume the *fcc* form [14].

The situation is more involved as far as the structure of the liquid phase is concerned. Here, due to the lack of intense Bragg peaks, precise quantitative measurements are necessary. On the other hand, hydrogen is almost transparent to X-rays, and for several reasons, a neutron diffraction experiment is not an easy task. However, a detailed knowledge of the structural properties of liquid hydrogen was an important achievement as liquid hydrogen represents the prototype of a quantum Boltzmann (distinguishable particles) liquid. In fact, by considering the hydrogen molecule as a point particle, its de Broglie wavelength turns out, at the low temperature of the liquid phase, similar to, or even larger than, the molecular diameter [15]. However, its size does not extend so much as to enclose the first neighbor position. Therefore, quantum effects induced by the delocalization of the center of mass wave function are expected to be rather large, while no exchange effects, giving rise to quantum statistics, are expected to emerge.

Due to the aforementioned problems with the large incoherent cross section of hydrogen, the first neutron diffraction experiments were carried out on deuterium [16–18]. As expected, it was impossible to reproduce the microscopic structure factor of liquid deuterium using classical Monte Carlo computer simulations, and more involved theoretical calculations were necessary. A much better agreement was found by comparing the structural information derived from the experiments with the simulation results obtained using quantum path integral Monte Carlo simulations. In this case, using a simple pair interaction potential derived by molecular collision experiments, a satisfactory agreement was observed [19].

Nonetheless, the structure factor of liquid hydrogen was still an interesting open problem because of the different size of quantum correction deriving from the large molecular mass ratio and the challenge of the experimental difficulties. This task resulted in a lively competition between different research groups [20–23], with a final result that seems to converge toward a substantial quantitative agreement on the measured structure factors [23].

14.3 Hydrogen Microscopic Dynamics

14.3.1 Collective Excitations

The first neutron experiment aimed at measuring the collective excitations of solid parahydrogen and orthodeuterium was carried out by Nielsen [24], who used a three-axis spectrometer (Chapter 3) to determine the phonon dispersion relation in the two quantum solids and found a rather good agreement with the theoretical predictions based on the self-consistent phonon theory of Klein and Koehler [25].

Well-defined collective excitations were also found in liquid parahydrogen [26] and orthodeuterium [27]. Further experiments were carried out on parahydrogen, using both a three-axis and a time-of-flight spectrometer [20, 28], and demonstrated the presence of quantum effects in the collective liquid dynamics. As opposed to quantum solids, where a reasonable theoretical model was developed to explain the experimental results [25], a well-established theoretical framework is still lacking for Boltzmann quantum liquids. Recently, a self-consistent quantum mode-coupling theory has been suggested by Rabani and Reichman [29] and applied to the case of liquid parahydrogen. This model starts from the definition of an *exact* quantum-generalized Langevin equation that follows from the historical works of Zwanzig [30], Mori [31, 32], and Kubo [33]. Alternatively, one has to rely on simulation methods that are appearing in literature and look quite promising for evaluating the dynamic features of quantum systems [34].

14.3.2 Hydrogen Self-Dynamics

By increasing the size of momentum transfer, the response function of a neutron scattering inelastic experiment becomes less and less sensible to collective motions (coherent dynamics) and starts probing the self-motion (incoherent dynamics) of molecules. In this case, using IINS (and TOSCA, an instrument [described in Chapter 3] that can be considered the neutron equivalent of a Raman spectrometer), it is possible to take full advantage of the large incoherent scattering cross section of hydrogen. By this technique, it was possible to obtain direct experimental information on the velocity–velocity center-of-mass correlation function of liquid hydrogen [35, 36], which is equivalent, in the solid phase, to the density of phonon states [37, 38]. Further measurements were also extended to H_2–D_2 liquid mixtures [39].

If the momentum transfer is increased further, we enter the so-called Compton Neutron Scattering regime [40]. In this case, the energy and momentum transferred to the target nuclei are so large that the information on the molecular structure is lost and the neutron probes the momentum distribution of the target nucleus [41, 42].

14.4 Molecular Storage in Nanoporous Materials

Hydrogen is considered an ideal energy carrier for a future clean society, for both mobile and stationary applications. Consequently, hydrogen-storage technologies are considered fundamental challenges in developing the so-called hydrogen economy. These should be characterized by low-cost, high-safety, long-lasting materials; energetically favorable procedures; and technologically easy processes. In addition, the volumetric and gravimetric density of hydrogen in a storage material is crucial. Traditional technologies permit storage of hydrogen in high-pressure gaseous (pressure up to 800 bar) or liquid (temperature on the order of 20 K) form using pressurized or cryogenic tanks. These storage methods present practical and safety problems for on-board transport applications and do not satisfy the U.S. Department of Energy (DOE) economic and environmental requirements for hydrogen storage [43]. Hydrogen physisorption on porous materials characterized by a large specific surface area (SSA) has been proposed as an alternative method for storage. The physisorption technique satisfies an important criterion for a hydrogen-storage system: the reversibility of uptake and release. It is worth noting that if hydrogen uptake through physisorption is replaced by a real chemisorption process, implying the establishment of strong covalent carbon–hydrogen bonds, then hydrogen release becomes extremely inefficient because these bonds break up only at high temperature (close to or above 800°C), or via carbon oxidization (like in the standard hydrocarbon combustion). However, the latter case gives rise to an irreversible process unfit for hydrogen-storage applications.

Physisorption of gas molecules (adsorbate) on the surface of a solid (adsorbent) is initiated by the weak Van der Waals interaction force. In general, the potential energy felt by the adsorbate molecule is characterized by a minimum at a distance of about 1 molecular diameter from the adsorbent wall. However, due to the weak interaction force, a significant physisorption is only observed at low temperature (generally, much lower than 273 K and, typically, close to liquid nitrogen temperature). Once a monolayer of adsorbate molecules is formed, the excess gaseous molecules interact with a surface layer made by the adsorbate. The density of physisorbed hydrogen is close to that of liquid hydrogen ($\rho_{liq} = 70.8\,\text{kg m}^{-3}$) [44]. At room temperature, the hydrogen uptake in porous materials is generally small (<1 wt%) even at elevated pressures [45–48]. However, at 77 K the storage capacity is improved up to $\cong 2$ wt% in zeolites [49], $\cong 5$ wt% in carbon nanomaterials [46], and $\cong 7$ wt% in metal-organic frameworks (MOFs) [50, 51].

Investigating the microscopic properties of hydrogen adsorbed on nanoporous materials is an essential task for understanding the suitability of these systems for storing and transporting hydrogen. Just as neutron scattering is a very powerful technique for investigating the microscopic position and dynamics of hydrogen molecules, so too when the molecules are adsorbed on a porous substrate [52–60]. Other useful techniques are optical spectroscopy, to probe the roto-vibrational energies of molecular hydrogen [61–63]; X-ray scattering, to investigate possible changes in lattice parameters [64]; and thermodynamic measurements, to determine the isosteric heat of adsorption [46, 65]. Computer simulations [66, 67] and theoretical calculations [68–70] also give useful information for the interpretation of experimental results (see Fig. 14.3).

14.4.1 Zeolites

Zeolites represent the largest group of microporous compounds (pores of less than 2 nm in diameter). These are crystalline inorganic polymers based on a three-dimensional arrangement of SiO_4 and AlO_4 tetrahedra, connected through their oxygen atoms. The crystal structure defines channels and cages (i.e., micropores) of regular dimensions. Due to the presence of aluminum, zeolites exhibit a negatively charged framework that is counterbalanced by positive cations, resulting in a strong electrostatic field on the internal surface. Zeolites have been proposed for hydrogen

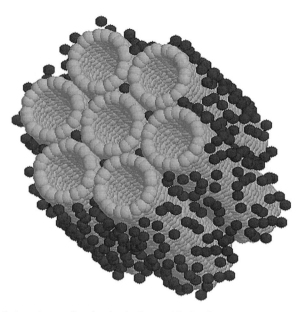

Fig. 14.3 Simulation picture of molecular hydrogen (*dark spheres*) adsorbed on the external surface of nanotubes in a bundle of seven elements

storage. In fact, by varying some parameters such as surface area, pore volume, silicon:aluminum ratio, and content and type of cations, the hydrogen adsorption capacity can be controlled and optimized [71]. Recently, a systematic investigation of hydrogen adsorption in zeolites was carried out [71, 72]. It appears that hydrogen uptake in zeolites is strongly dependent on temperature, framework topology, and cations. Moreover, micropore surface area and volume have a dominant role in hydrogen adsorption [72]. At liquid nitrogen temperature, zeolites physisorb hydrogen in proportion to the SSA of the material. In NaY zeolite (SSA $\cong 725\,\mathrm{m^2\,g^{-1}}$), a maximum of 1.8 wt% adsorbed hydrogen was found at 77 K, showing a behavior for physisorption mechanism similar to that of carbon nanomaterials [71]. The theoretical maximum storage capacity in an optimized zeolite framework is less than 3 wt%, which is consistent with the experimental results obtained at low temperature [69]. By means of recent inelastic neutron scattering measurements carried out on TOSCA [60], the interaction of H_2 molecules with a binding site in Cu-ZSM5 zeolites was investigated. Based on this experiment, hydrogen molecules appear to be strongly bound in the zeolite cavities, lying parallel to the surface. However, this interpretation is still quite controversial, as shown by Georgiev et al. [73], who suggest a strong covalent binding between H_2 and the ZSM5 zeolite framework via the formation of a metal–hydride complex.

Several theoretical studies to identify the characteristics of hydrogen sorption in zeolites have also been carried out [67,74–77].

14.4.2 Carbon Nanotubes and Nanofibers

Carbon nanoporous materials have been studied as hydrogen-storage materials for their low density, large SSA, and fast adsorption/desorption kinetics. Apart from their intrinsic academic interest, single-walled carbon nanotubes (SWCNs) possess several properties that also make them extremely important materials from an applications standpoint. In brief, a single nanotube can be described as a sheet of hexagonal carbon that is rolled to form a seamless tube and capped at each end by a fullerene hemisphere. Depending on the details of the SWCN wall structure and topology, different optical and electrical properties (insulating, semiconducting, metallic) are obtained [78].

The SWCN–H_2 system has been extensively studied in the past 10 years, with some controversies with regard to the measured storage capacity, which ranges from more than 7 wt% to less than 1 wt% [79–84]. The observed variance has been attributed to the difficulty of evaluating the storage capacity in the first measured samples due to the extremely small amount of material that was originally used and to the ill-defined purity of the samples. Now, after several years of extensive measurements by different laboratories, some aspects of the mechanism of hydrogen storage in carbon nanostructures have begun to be understood. Due to the nature of physisorption, the hydrogen-storage capacity increases with lowering temperature and/or increasing pressure. Moreover, the amount of adsorbed hydrogen at low temperature ($T = 77$ K) appears to be proportional to the SSA of the nanostructured carbon sample [85, 46]. Up to now, the curvature of the nanotubes has not seemed to affect the amount of adsorbed hydrogen too much; however, the effect on the adsorption energy has been obvious. The observed amount of adsorbed hydrogen is 2.0 wt% for SWCNs with an SSA of $1,315\,m^2g^{-1}$ at $T = 77$ K [86]. Using open nanotubes to fill the inner volume of the tubes does not appear to increase the adsorption capacity [44]. Neutron scattering has been used to measure the structural [54] and dynamical [52, 53, 55–58] properties of hydrogen physisorbed on SWCNs.

In Fig. 14.4, we show a typical diffraction spectrum for molecular deuterium adsorbed on SWCNs. The experiment was carried out on SANDALS (Small-Angle Neutron Diffractometer for Amorphous and Liquid Samples), and the difference between the two patterns contains information on the location of molecular deuterium on the surface on the nanotubes.

Carbon nanofibers can be grown by catalytic decomposition of certain hydrocarbons (e.g., ethylene) over small metal particles such as iron, cobalt, nickel, and some of their alloys. The basic microstructure consists of stacked graphite layers, which can be arranged parallel ("tubular" or "ribbon"), perpendicular ("platelet"), or with an angle of $\cong 45°$ ("herringbone") with respect to the fiber axis. The diameter of the nanostructure can vary between 5 and 200 nm, with lengths ranging from 5 to 100 μm. The spacing between graphite layers, in each case, is the same value found in conventional graphitic carbon (i.e., $\cong 3.4$ Å) [88]. The extremely high hydrogen-storage capacity reported in the literature for the herringbone nanofibers (i.e., 67 wt% after exposure at 120 atm and at 25°C) [89] has not been confirmed by

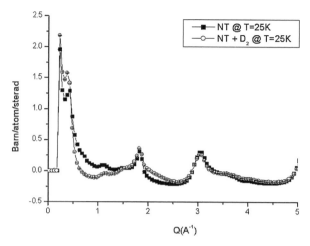

Fig. 14.4 Coherent contribution to the neutron diffraction scattering cross section of a bare single-walled carbon nanotube sample (*full squares*). The *open circles* show the same quantity when the sample is charged with deuterium [87]

the results of other laboratories, which found much lower values (from 0.08 to 3.8 wt%) [90–94].

14.4.3 Metal-Organic Frameworks

Metal-organic-framework compounds consist of metal-oxide clusters connected by organic linkers [95, 96]. Due to their properties [e.g., tunable pore size (0.5–2 nm), possibility of functional doping, low density (0.20–1.00 gcm^{-3}), and high SSA (500–4,500 m^{2}g^{-1}), this group of compounds is considered a very interesting class of nanoporous materials for hydrogen-storage applications. In 2005, Rowsell et al. reported a summary of hydrogen adsorption capacity in MOFs at 77 K and 1 bar (H$_2$ uptake: 0.2–2.48 wt%), at 77 K and 16 bar (H$_2$ uptake: 3.1–3.8 wt%), and at room temperature and pressure between 10 and 50 bar (H$_2$ uptake: 1–2 wt%) [97, 98]. The most widely studied MOF material for hydrogen storage is MOF-5 (also called IRMOF-1 or isoreticular metal-organic framework-1) [47, 48, 50, 99, 100]. This is because of the high SSA and the simple chemical constituents of this material, which consists of ZnO$_4$ clusters linked by benzene groups. Changing the organic ligand results in a series of IRMOFs with the same framework topology, but with different pore dimensions. The most recent uptake level found in MOF-5 is 4.5–5.2 wt% at about 50 bar and 77 K [47, 48, 50].

Recently, hydrogen uptake values up to 7 wt% at 77 K have been reported for MOFs with an SSA of 3,000–4,700 m^{2}g^{-1} [50, 51]. Spectroscopic investigation of the hydrogen-framework interaction has been performed using inelastic neutron scattering [99, 101]. The effect of the inorganic clusters and organic linkers with H$_2$ molecules was also investigated by single-crystal [102] and powder neutron

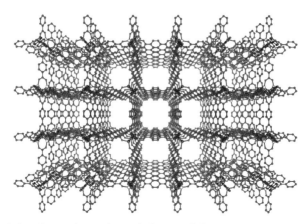

Fig. 14.5 Simulation picture of a metal-organic framework-5 type structure. The channels depicted in the figure are available for storing molecular hydrogen

diffraction [103] (Chapter 3). In MOF-5 (see Fig. 14.5), and similar MOFs, multiple adsorption site occupancy was observed. The results of some recent simulations [104, 105] and experiments [47, 48] allow us to state that the best MOFs for physisorption-based hydrogen storage should posses a combination of small pores (to increase the adsorbate–adsorbent interaction) and high SSA (responsible for a larger uptake at high pressure). Moreover, the nature of the MOF transition metal and its coordination environment can play an important role in maximizing the hydrogen-storage capacity [97, 98].

14.5 Atomic Storage in Solid Matrices

The atomic storage of hydrogen in solid matrices is commonly related to the idea of *hydride* (i.e., a negative ion of hydrogen, H^-). Actually, H^- exists in nature in some substances [e.g., alkali hydrides and $HCo(CO)_4$]; however, the term *hydride* is usually applied in a wider sense to describe compounds of hydrogen with other elements, particularly with those belonging to groups ranging from IA to IIB. Aluminum hydride (IIIB group) exists too. The variety of compounds formed by hydrogen is large, arguably greater than that of any other element [106]. Almost every element of the periodic table forms one or more stable hydrides. These may be classified into two main types by the predominant nature of their chemical bonding: interstitial metal hydrides, which may be described as having true covalent bonding, and ionic (or saline) hydrides, which have a significant ionic character. The latter class of hydrides will be further divided into two subclasses, simple ionic hydrides and complex hydrides, for our discussion.

14.5.1 Interstitial Metal Hydrides

Transition metals form binary hydrides that are often nonstoichiometric (i.e., they can accommodate a variable number of hydrogen atoms in their lattices, where they can easily migrate from one site to another). For this reason these compounds are also known as interstitial metal hydrides and are responsible, in metallurgy, for the phenomenon called hydrogen embrittlement. The most famous interstitial metal hydride is surely PdH_x (with $x = 0.62$–0.96 [107]), where palladium can absorb up to 900 times its own volume of hydrogen at room temperature. Interstitial hydrides have long been considered as possible mechanisms for safe hydrogen storage, and during the past 25 years many interstitial hydrides capable of absorbing and desorbing hydrogen at room temperature and low pressure have been developed. Such hydrides are usually based on intermetallic compounds and solid–solution alloys. However, their application is still limited to rechargeable batteries (e.g., $LaNi_5H_x$ [with $x \approx 6$], where the reversible hydrogen-storage capacity does not exceed 1.4 wt% [108]) because their hydrogen-storage capacity currently is not enough for automotive applications.

Concerning neutron scattering studies on interstitial metal hydrides, the area is extremely vast, ranging from diffraction (for refining lattice structures) to coherent inelastic scattering on deuterium-substituted samples (for measuring phonon dispersion curves), to incoherent inelastic scattering (for extracting information on the hydrogen-projected density of phonon states), to quasielastic scattering (for extracting information on the hydrogen-diffusion mechanism). A detailed review of all these subjects can be found in Chapter 5 of Fukai [109], while for IINS only, it is worthwhile to consider the review by Ross [110].

14.5.2 Simple Ionic Hydrides

Simple ionic hydrides are binary stoichiometric compounds formed by hydrogen plus an alkali or alkaline earth metal. They have received increasing interest in the past few years, mainly in connection with the hydrogen-storage problem, where, for example, magnesium dihydride (MgH_2) [111] might play a relevant role, exhibiting a hydrogen content of 7.6 wt%. In addition, it is worth mentioning the fact that lithium hydride (LiH) has always been a sort of benchmark compound for solid-state physicists [112]: it is a rock-salt crystal having only four electrons per unit cell, which makes it the simplest ionic crystal in terms of electronic structure. Moreover, there is a large isotopic effect provided by the substitution of the proton with the deuteron (LiD). Last but not least, owing to the low mass of the constituent atoms, LiH represents a good case for calculation of the zero-point motion contribution to the lattice energy and the quantum anharmonic dynamics [113]. However, different from alkali hydrides, which exhibit the same ambient-pressure rock-salt structure (i.e., *fcc, Fm-3m*) [114] along the group 1 moving from Li to Cs, alkaline earth hydrides are characterized by three distinct subsets: BeH_2, unstable and body-centered orthorhombic with an *Ibam* space group; MgH_2, well described and showing a rutile-type structure (tetragonal, $P4_2/mnm$) at low pressure [114]; and finally, calcium, strontium, and barium dihydrides, isomorphic and crystallizing with an orthorhombic lattice (at low pressure and ambient temperature), exhibiting a *Pnma* space group [114]. The last three alkaline earth hydrides were structurally studied for the first time in 1935, but only the metallic cation position was determined through standard X-ray powder diffraction: metal atoms appeared arranged in a slightly distorted hexagonal close-packed structure. As a result, further neutron scattering experiments on deuterated powder samples (CaD_2 [115], SrD_2 [116], and BaD_2 [117]) were necessary to precisely locate the hydrogen positions, and a slightly distorted $PbCl_2$-type structure was finally proposed for these three compounds.

As for the lattice dynamics of alkali hydrides, in recent papers [113, 118] inelastic neutron scattering spectra from the complete series have been reported, measured at $T = 20\,K$ in the energy transfer range $3\,meV < E < 500\,meV$. From the medium-energy region of these spectra (50–150 meV, coinciding with the fundamental optical phonon bands), accurate hydrogen-projected densities of phonon states were extracted (see Fig. 14.6). These experimental phonon distributions were then compared to the equivalent results obtained from ab initio lattice dynamics simulations based on density functional theory and pseudo-potentials. The overall agreement between neutron and ab initio data turned out to be very good, especially for the three heaviest compounds (KH, RbH, and CsH). This finding proves that, at least for this simple class of binary hydrides, a quantitative agreement between experimental and ab initio data is possible, not only for static (lattice parameters) and macroscopic quantities (bulk modulus, cohesive energy, et similia) but also for the issue of the microscopic hydrogen dynamics.

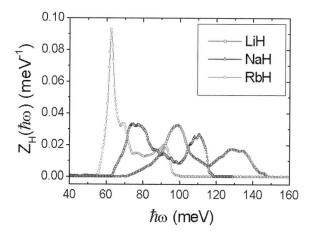

Fig. 14.6 Hydrogen-projected densities of phonon states of some selected alkali hydrides as measured by inelastic incoherent neutron scattering on TOSCA [118]

A similar situation can be envisaged for the lattice dynamics of alkaline earth hydrides (excluding the amorphous BeH_2), where three experimental studies have been reported so far, all making use of IINS to measure the hydrogen vibrational spectra in MgH_2 [119]; CaH_2 [120]; and CaH_2, SrH_2, and BaH_2 [121], respectively. In the last study listed, from the 63 to 140 meV region of the neutron spectra, precise generalized self-inelastic structure factors were derived, but not the hydrogen-projected densities of phonon states because of the existence of two nonequivalent hydrogen atoms, impossible to be singled out. The general agreement between neutron and ab initio data was satisfactory even though some discrepancies still appeared in the first optical phonon zone, especially in the case of BaH_2, where peak positions, heights, and widths were not perfectly reproduced by simulations.

14.5.3 Complex Hydrides

The term "complex hydride" is currently used to designate a solid material such as an alkali or alkaline-earth aluminum hydride (e.g., $NaAlH_4$, also known as sodium alanate), an amide (e.g., $LiNH_2$), or a borohydride (e.g., $LiBH_4$). In the present context, the adjective "complex" is not meant to imply "anionic metal complex," but rather a contrast to the "simple" ionic hydrides we have been discussing. In particular, $NaAlH_4$ has attracted significant interest since Bogdanović and Schwickardi [122] have shown that the hydrogenation reaction can be kinetically enhanced and rendered reversible after doping with transition metal salts of type MX_3, where M = Ti, Zr, Co, Ni, or Fe and X = F, Cl, or I, with TiF_3 being the most effective. The relevant reactions are as follows.

$$3\,NaAlH_4 \rightarrow Na_3AlH_6 + 2\,Al + 3\,H_2 \quad (3.7\,wt\%)$$

$$Na_3AlH_6 \rightarrow 3\,NaH + Al + 3/2\,H_2 \quad (1.8\,wt\%)$$

These may occur at moderate pressures ($p < 100$ bar) and temperatures ($T < 438$ K), so the doped $NaAlH_4$–Na_3AlH_6 system exhibits some characteristics not far from the first target set by DOE [43] for practical applications. However, a large amount of work appears to still be necessary to further improve the storage properties of alanates and, for this, a detailed understanding of the H_2 absorption/desorption process is crucial. In particular, a microscopic mechanism of the catalytic processes in titanium-doped $NaAlH_4$ has been suggested only very recently [123], and has not yet been experimentally corroborated.

Neutron scattering (using a powder diffractometer) on isotopically substituted sodium aluminum tetrahydride ($NaAlD_4$) [124] and sodium aluminum hexahydride (Na_3AlD_6) [125] has finally allowed for a precise determination of their room-temperature lattice structures, which turned out to be body-centered tetragonal with space group $I4_1/a$ and monoclinic with space group $P2_1/n$, respectively. As for the present status of inelastic neutron scattering measurements on sodium alanates (using a Be-filter spectrometer), it is worth noting that a recent study has been performed on $NaAlH_4$ (pure and 2% titanium-doped) and Na_3AlH_6 (pure) [126], with an energy resolution of 2–4.5% in the energy-transfer range 50–250 meV, showing a good agreement with ab initio lattice dynamic simulations but detecting no spectral differences between doped and undoped samples. A subsequent inelastic neutron scattering study on $NaAlH_4$ was accomplished on TOSCA by Qi Jia Fu et al. [127], making use of a larger energy-transfer range (3 meV $< E <$ 500 meV) and a narrower energy resolution (1.5–3%). This work had a particular aim mainly focusing on the doped samples and on their spectral modifications following the repetition of several H_2 charging/discharging cycles, and claimed the detection of a species of unstable aluminum hydride (AlH_3) with its oligomers. However, this result is regarded as highly controversial.

14.6 Future Perspectives and Conclusions

Hydrogen storage is, at present, an extremely active research field where it is almost impossible to distinguish between fundamental and applied physics and where the main trends are enforced by technological needs. A typical example of the present situation is the degree of international collaboration on hydrogen research represented by the International Energy Agency Hydrogen Implementing Agreement (HIA) [128] and, specifically, HIA's Task 22 [129] for international cooperation on hydrogen storage research.

In the end, we are all aware that the time of cheap oil is over and, sooner or later, the availability of fossil fuels will decrease drastically. Thus, other energy sources must be envisaged, and hydrogen, as a clean energy carrier, appears to be one of the most convincing solutions.

Acknowledgments   This work has been partially supported by Ente Cassa di Risparmio di Firenze (Project: Firenze Hydrolab) and by the European Union's Sixth Framework Program through a Marie Curie Research Training Network (MRTN-CT-2004-512443).

References

1. H. C. Urey, F. G. Brickwedee, and G. M. Murphy, Phys. Rev. **40**, 1–17 (1932)
2. M. L. E. Oliphant, P. Harteck, and E. Rutherford, Proc. R. Soc. London **144**, 692–703 (1934)
3. J. Dewar, *Collected Papers of Sir James Dewar*, edited by L. Dewar (Cambridge University Press, Cambridge, UK, 1927)
4. L. Ulivi, M. Zoppi, L. Gioè, and G. Pratesi, Phys. Rev. B **58**, 2383–2386 (1998)
5. M. Zoppi, J. Phys. Condens. Matter **15**, R1–R16 (2003)
6. H. M. Roder, G. E. Childs, R. D. McCarty, and P. E. Angerhofer, in *Survey of the Properties of the Hydrogen Isotopes Below Their Critical Temperatures* (National Bureau of Standards, Washington, D.C., 1973), NBS Technical Note N. 641
7. R. Mecke, Z. Phys. **31**, 709–712 (1925)
8. W. Heisenberg, Z. Phys. **33**, 879–893 (1925)

9. W. Heisenberg, Z. Phys. **39**, 499–518 (1926)
10. I. F. Silvera, Rev. Mod. Phys. **52**, 393–452 (1980)
11. G. Herzberg, *Spectra of Diatomic Molecules* (Van Nostrand Reinhold, New York, 1950)
12. V. F. Sears, Neutron News **3**, 26–37 (1992)
13. S. W. Lovesey, *Theory of Neutron Scattering from Condensed Matter* (Oxford University Press, Oxford, UK, 1987)
14. V. G. Manzhelii and Y. A. Freiman (Eds.) *Physics of Cryocrystals*, (AIP Press, American Institute of Physics, Woodbury, NY, 1997)
15. M. Zoppi, M. Celli, and A.K. Soper, Phys. Rev. B **58**, 11905–11910 (1998)
16. M. Zoppi, R. Magli, W. S. Howells, and A. K. Soper, Phys. Rev. A **39**, 4684–4694 (1989)
17. M. Zoppi, U. Bafile, R. Magli, and A. K. Soper, Phys. Rev. E **48**, 1000–1007 (1993)
18. M. Zoppi, A. K. Soper, R. Magli, F. Barocchi, U. Bafile, and N. W. Ashcroft, Phys. Rev. E **54**, 2773–2779 (1996)
19. M. Zoppi, U. Bafile, E. Guarini, F. Barocchi, and R. Magli, Phys. Rev. Lett. **75**, 1779–1782 (1995)
20. F. J. Bermejo, K. Kinugawa, C. Cabrillo, S. M. Bennington, B. Fåk, M. T. Fernández-Díaz, P. Verkerk, J. Dawidowski, and R. Fernández-Perea, Phys. Rev. Lett. **84**, 5359–5362 (2000)
21. M. Zoppi, M. Neumann, and M. Celli, Phys. Rev. B **65**, 092204 (2002)
22. J. Dawidowski, F. J. Bermejo, M. L. Ristig, B. Fåk, C. Cabrillo, R. Fernández-Perea, K. Kinugawa, and J. Campo, Phys. Rev. B **69**, 014207 (2004)
23. M. Celli, U. Bafile, G. J. Cuello, F. Formisano, E. Guarini, R. Magli, M. Neumann, and M. Zoppi, Phys. Rev. B **71**, 014205 (2005)
24. M. Nielsen, Phys. Rev. B **7**, 1626–1635 (1973)
25. M. L. Klein and T. R. Koehler, Phys. Lett. A **33**, 253–254 (1970)
26. K. Carneiro, M. Nielsen, and J. P. McTague, Phys. Rev. Lett. **30**, 481–485 (1973)
27. F. J. Bermejo, F. J. Mompeán, M. García-Hernández-Díaz, J. L. Martínez, D. Martin-Marero, A. Chahid, G. Senger, and M. L. Ristig, Phys. Rev. B **47**, 15097–15112 (1993)
28. F. J. Bermejo, B. Fåk, S. M. Bennington, R. Fernández-Perea, C. Cabrillo, J. Dawidowski, M. T. Fernández-Díaz, and P. Verkerk, Phys. Rev. B **60**, 15154–15162 (1999)
29. E. Rabani and D. R. Reichman, Phys. Rev. E **65**, 036111 (2002)
30. R. Zwanzig, in *Lectures in Theoretical Physics*, edited by W. E. Brittin, B. W. Downs, and J. Downs (Wiley-Interscience, New York, 1961), Vol. III, p. 106
31. H. Mori, Prog. Theor. Phys. **33**, 423–455 (1965)
32. H. Mori, Prog. Theor. Phys. **34**, 399–416 (1965)
33. R. Kubo, M. Toda, and N. Hashitsume, in *Statistical Physics II: Nonequilibrium Statistical Mechanics*, 2nd ed. (Springer, Berlin, 1995), Springer Series in Solid State Science.
34. K. Kinugawa, Chem. Phys. Lett. **292**, 454–460 (1998)
35. M. Zoppi, D. Colognesi, and M. Celli, Europhys. Lett. **53**, 34–39 (2001)
36. M. Celli, D. Colognesi, and M. Zoppi, Phys. Rev. E **66**, 021202 (2002)
37. M. Celli, D. Colognesi, and M. Zoppi, J. Low Temp. Phys. **126**, 585–590 (2002)
38. D. Colognesi, M. Celli, and M. Zoppi, J. Chem. Phys. **120**, 5657–5663 (2004)
39. D. Colognesi, M. Celli, M. Neumann, and M. Zoppi, Phys. Rev. E **70**, 061202 (2004)
40. U. Bafile, M. Celli, M. Zoppi, and J. Mayers, Phys. Rev. B **58**, 791–797 (1998)
41. J. Mayers, Phys. Rev. Lett. **71**, 1553 (1993)
42. C. Andreani, D. Colognesi, A. Filabozzi, E. Pace, and M. Zoppi, J. Phys. Condens. Matter **10**, 7091–7111 (1998)
43. U. S. Department of Energy, *Hydrogen Posture Plan; An Integrated Research, Development, and Demonstration Plan* (Department of Energy, Washington, D.C., 2004), http://www1.eere.energy.gov/hydrogenandfuelcells/pdfs/hydrogen_posture_plan.pdf
44. A. Züttel, P. Wenger, P. Sudan, T. Mauron, and S. Orimo, Mater. Sci. Eng. B **108**, 9–18 (2004)
45. S. Beyaz Kayiran and F. Lamari Darkrim, Surf. Interface Anal. **34**, 100–104 (2002)
46. B. Panella, M. Hirscher, and S. Roth, Carbon **43**, 2209–2214 (2005)
47. A. Dailly, J. J. Vajo, and C. C. Ahn, J. Chem. Phys. B **110**, 1099–1101 (2006)

48. B. Panella, M. Hirscher, H. Pütter, and U. Müller, Adv. Funct. Mat. **16**, 520–524 (2006)
49. H. W. Langmi, D. Book, A. Walton, S. R. Johnson, M. M. Al-Mamouri, J. D. Speight, P. P. Edwards, I. R. Harris, and P. A. Anderson, J. Alloys Compd. **404–406**, 637–642 (2005)
50. A. G. Wong-Foy, A. J. Matzger, and O. M. Yaghi, J. Am. Chem. Soc. **128**, 3494–3495 (2006)
51. X. Lin, J. Jia, X. Zhao, K. Mark Thomas, A. J. Blake, and G. S. Walker, Angew. Chem. Int. Ed. **45**, 7358–7364 (2006)
52. C. M. Brown, T. Yildirim, D. A. Neumann, M. J. Heben, T. Gennett, A. C. Dillon, J. L. Alleman, and J. E. Fischer, Chem. Phys. Lett. **329**, 311–316 (2000)
53. Y. Ren and D. L. Price, Appl. Phys. Lett. **79**, 3684–3686 (2001)
54. M. Muris, M. Bienfait, P. Zeppenfeld, N. Deupont-Pavlovsky, M. Johnson, O. E. Vilches, and T. Wilson, Appl. Phys. A **74**, S1293–S1295 (2002)
55. D. G. Narehood, M. K. Kostov, P. C. Eklund, M. W. Cole, and P. E. Sokol, Phys. Rev. B **65**, 233401 (2002)
56. D. G. Narehood, J. V. Pearce, P. C. Eklund, P. E. Sokol, R. E. Lechner, J. Pieper, J. R. D. Copley, and J. C. Cook, Phys. Rev. B **67**, 205409 (2003)
57. P. A. Georgiev, D. K. Ross, A. De Monte, U. Montaretto-Marullo, R. A. H. Edwards, A. J. Ramirez-Cuesta, M. A. Adams, and D. Colognesi, Carbon **43**, 895–906 (2005)
58. P. A. Georgiev, A. Giannasi, D. K. Ross, M. Zoppi, J. L. Sauvajol, and J. Stride, Chem. Phys. **328**, 318–323 (2006)
59. H. G. Schimmel, G. J. Kearley, and F. M. Mulder, Chem. Phys. Chem. **5**, 1053–1055 (2004)
60. J. Ramirez-Cuesta and P. C. H. Mitchell, Catalysis Today **120**, 368–373 (2007)
61. K. A. Williams, B. K. Pradhan, P. C. Eklund, M. K. Kostov, and M. W. Cole, Phys. Rev. Lett. **88**, 165502 (2002)
62. B. K. Pradhan, G. U. Sumanasekera, K. W. Adu, H. E. Romero, K. A. Williams, and P. C. Eklund, Physica B **323**, 115–121 (2002)
63. A. Centrone, L. Brambilla, and G. Zerbi, Phys. Rev. B **71**, 245406 (2005)
64. X. Zhao and Y. Ando, Jpn. J. Appl. Phys. **37**, 4846–4849 (1998)
65. M. Bienfait, P. Zeppenfeld, N. Dupont-Pavlovsky, M. Muris, M. Johnson, T. Wilson, M. DePies, and O. E. Vilches, Phys. Rev. B **70**, 035410 (2004)
66. D. Levesque, A. Gicquel, F. Lamari Darkrim, and S. Beyaz Kayiran, J. Phys. Condens. Matter **14**, 9285–9293 (2002)
67. M. K. Song and K. T. No, Catal. Today **120**, 374–382 (2007)
68. G. E. Froudakis, J. Phys. Condens. Matter **14**, R453–R465 (2002)
69. J. G. Vitillo, G. Ricchiardi, G. Spoto, and A. Zecchina, PCCP 7, 3948–3954 (2005)
70. D. Henwood and J. D. Carey, Phys. Rev. B **75**, 245413 (2007)
71. H. W. Langmi, A. Walton, M. M. Al-Mamouri, S. R. Johnson, D. Book, J. D. Speight, P. P. Edwards, I. Gameson, P. A. Anderson, and I. R. Harris, J. Alloys Compd. **356–357**, 710–715 (2003)
72. S. H. Jhung, J. W. Yoon, H. Kim, and J. Chang, Bull. Korean Chem. Soc. **26**, 1075–1078 (2005)
73. P. A. Georgiev, A. Albinati, B. L. Mojet, J. Ollivier, and J. Eckert, J. Am. Chem. Soc. **129**, 8086–8087 (2007)
74. F. Anderson, D. F. Coker, J. Eckert, and A. L. R. Bug, J. Chem. Phys. **111**, 7599–7613 (1999)
75. Darkrim, A. Aoufi, P. Malbrunot, and D. Levesque, J. Chem. Phys. **112**, 5991–5999 (2000)
76. E. D. Akten, R. Siriwardane, and D. S. Sholl, Energy Fuels **17**, 977–983 (2003)
77. X. Solans-Monfort, V. Branchadell, M. Sodupe, C. M. Zicovich-Wilson, E. Gribov, G. Spoto, C. Busco, and P. Ugliengo, J. Phys. Chem. B **108**, 8278–8286 (2004)
78. R. Saito, G. Dresselhaus, and M. S. Dresselhaus, *Physical Properties of Carbon Nanotubes* (Imperial College Press, London, 1998)
79. A. C. Dillon, K. M. Jones, T. A. Bekkedahl, C. H. Kiang, D. S. Bethune, and M. J. Heben, Nature **386**, 377–379 (1997)
80. M. S. Dresselhaus, K. A. Williams, and P. C. Eklund, MRS Bull. **24**, 45–50 (1999)
81. A. C. Dillon and M. J. Heben, Appl. Phys. A **72**, 133–142 (2001)
82. R. G. Ding, G. Q. Lu, Z. F. Yan, and M. A. Wilson, J. Nanosci. Nanotechnol. **1**, 7–29 (2001)

83. L. Schlapbach and A. Züttel, Nature **414**, 353–358 (2001)
84. M. Hirscher, M. Becker, M. Haluska, F. von Zeppelin, X. Chen, U. Dettlaff-Weglikowska, and S. Roth, J. Alloys Compd. **356–357**, 433–437 (2003)
85. M. Hirscher and B. Panella, J. Alloys Compd. **404–406**, 399–401 (2005)
86. A. Züttel, P. Sudan, P. Mauron, T. Kyiobaiashi, C. Emmenegger, and L. Sclapbach, Int. J. Hydrogen Energy **27**, 203–212 (2002)
87. A. Giannasi, *Characterization and Hydrogen Adsorption Properties of Single Wall Carbon Nanotubes*, PhD thesis, Department of Physics, University of Florence, Italy, (2004)
88. N. M. Rodriguez, A. Chambers, and R. T. K. Baker, Langmuir **11**, 3862–3866 (1995)
89. A. Chambers, C. Park, R. T. K. Baker, and N. M. Rodriguez, J. Phys. Chem. B **102**, 4253–4256 (1998)
90. C. C. Ahn, Y. Ye, B. V. Ratnakumar, C. Witham, R. C. Bowman Jr., and B. Fultz, Appl. Phys. Lett. **73**, 3378–3380 (1998)
91. R. Strobel, L. Jorissen, T. Schliermann, V. Trapp, W. Schütz, K. Bohmhammel, G. Wolf, and J. Garche, J. Power Sources **84**, 221–224 (1999)
92. G. G. Tibbetts, G. P. Meisner, and C. H. Olk, Carbon **39**, 2291–2301 (2001)
93. A. D. Lueking, R. T. Yang, N. M. Rodriguez, and R. T. K. Baker, Langmuir **20**, 714–721 (2004)
94. M. Marella and M. Tomaselli, Carbon **44**, 1404–1413 (2006)
95. O. M. Yaghi, H. Li, C. Davis, D. Richardson, and T. L. Groy, Acc. Chem. Res. **31**, 474–484 (1998)
96. M. Eddaoudi, H. L. Li, T. Reineke, M. Fehr, D. Kelley, T. L. Groy, and O. M. Yaghi, Top Catal. **9**, 105–111 (1999)
97. J. L. C. Rowsell, and O. M. Yaghi, Angew. Chem. Int. Ed. **44**, 4670–4679 (2005)
98. M. Hirscher and B. Panella, Scripta Materialia **56**, 809–812 (2007)
99. N. L. Rosi, J. Eckert, M. Eddadoudi, D. T. Vodak, J. Kim, M. O. Keeffe, and O. M. Yaghi, Science **300**, 1127–1129 (2003)
100. J. L. C. Rowsell, A. R. Millward, K. S. Park, and O. M. Yaghi, J. Am. Chem. Soc. **126**, 5666–5667 (2004)
101. J. L. C. Rowsell, J. Eckert, and O. M. Yaghi, J. Am. Chem. Soc. **127**, 14904–14910 (2005)
102. E. C. Spencer, J. A. K. Howard, G. J. McIntyre, J. L. C. Rowsell, and O. M. Yaghi, Chem. Commun. **3**, 278–280 (2006)
103. T. Yildirim and M. R. Hartman, Phys. Rev. Lett. **95**, 215504 (2005)
104. G. Garberoglio, A. I. Skoulidas, and J. K. Johnson, J. Phys. Chem. B **109**, 13094–13103 (2005)
105. H. Frost, T. Düren, and R. Q. Snurr, J. Phys. Chem. B **110**, 9565–9570 (2006)
106. N. N. Greenwood and A. Earnshaw, *Chemistry of the Elements* (Pergamon, Oxford, UK, 1984)
107. M. Hirabayashi and H. Asano, in *Metal Hydrides*, edited by G. Bambakidis (Plenum, New York, 1981), pp. 53–80
108. W. E. Wallace, R. F. Karlicek Jr., and H. Imamura, J. Phys. Chem. **83**, 1708–1712 (1979)
109. Y. Fukai, *The Metal-Hydrogen System* (Springer, Berlin, 2005)
110. D. K. Ross in *Hydrogen in Metals III*, edited by H. Wipf (Springer, Berlin, 1997), *Topics in Applied Physics*, Vol. 73, p. 153.
111. A. Zaluska, L. Zaluski, and J. O. Ström-Olsen, J. Alloys Compd. **288**, 217–225 (1999)
112. A. K. M. A. Islam, Phys. Status Solidi B **180**, 9–57 (1993)
113. J. Boronat, C. Cazorla, D. Colognesi, and M. Zoppi, Phys. Rev. B **69**, 174302–174310 (2004)
114. R. W. G. Wyckoff, *Crystal Structures*, Vol 1 (John Wiley & Sons, New York, London, 1963)
115. A. F. Andersen, A. J. Maeland, and D. Slotfeldt-Ellingsen, J. Solid State Chem. **20**, 93–101 (1977)
116. N. Brese, M. O'Keeffe, R. Von Dreele, J. Solid State Chem. **88**, 571–576 (1990)
117. W. Bronger, S. Chi-Chien, P. Müller, Z. Anorg. Allg. Chem. **545**, 69–74 (1987)
118. G. Auffermann, G. D. Barrera, D. Colognesi, G. Corradi, A. J. Ramirez-Cuesta, and M. Zoppi, J. Phys. Condens. Matter **16**, 5731–5743 (2004)

119. J. R. Santisteban, G. J. Cuello, J. Dawidowski, A. Fainstein, H. A. Peretti, A. Ivanov, and F. J. Bermejo, Phys. Rev. B **62**, 37–40 (2000)
120. P. Morris, D. K. Ross, S. Ivanov, D. R. Weaver, and O. Serot, J. Alloys Compd. **363**, 88–92 (2004)
121. D. Colognesi, G. Barrera, A. J. Ramirez-Cuesta, and M. Zoppi, J. Alloys Compd. **427**, 18–24 (2007)
122. B. Bogdanović and M. Schwickardi, J Alloys Comp. **253–254**, 1–9 (1997)
123. S. Chaudhuri and J. T. Muckerman, J. Phys. Chem. B **109**, 6952–6957 (2005)
124. B. C. Hauback, H. W. Brinks, C. M. Jensen, K. Murphy, and J. Maeland, J. Alloys Compd. **358**, 142–145 (2003)
125. E. Rönnebro, D. Noréus, K. Kadir, A. Reiser, and B. Bogdanović, J. Alloys Compd. **299**, 101–106 (2000)
126. J. Íñiguez, T. Yildirim, T. J. Udović, M. Sulić, and C. J. Jensen, Phys. Rev. B **70**, 060101 (2004)
127. Q. J. Fu, A. J. Ramirez-Cuesta, and S. C. Tsang, Phys. Chem. B **110**, 711–715 (2006)
128. International Energy Agency (IEA), Hydrogen Implementing Agreement (International Energy Agency, Maryland, U.S.A., 1977–), http://www.ieahia.org/
129. IEA, Task 22, Fundamental and Applied Hydrogen Storage Materials Development (IEA, Maryland, U.S.A., 2006–2009), http://www.hydrogenstorage.org/

Chapter 15
Lithium Ion Materials for Energy Applications: Structural Properties from Neutron Diffraction

Michele Catti

Abstract Cathode materials and solid electrolytes to be used in lithium batteries require a high ionic mobility of Li^+ species in their crystal structures. This in turn depends on the order–disorder state of lithium and on its bonding environment. Neutron diffraction is the choice technique to study the structural features of polycrystalline lithium materials that control their performance in ion transport processes. The basic principles of ionic mobility in solids and of the Rietveld refinement methods for neutron diffraction data are briefly reviewed. Then two important families of lithium conductors are selected from the literature and thoroughly discussed: the LLTO perovskite-type $Li_xLa_{2/3-x/3}TiO_3$ system and the $Li_{1+x}Me_2(PO_4)_3$ Nasicon phases. Accurate neutron diffraction determinations of the corresponding crystal structures have been shown to provide a considerable insight into the mechanisms of Li^+ ion transfer in such materials.

15.1 Introduction

The lithium ion rechargeable battery is one of the advanced power sources rapidly replacing conventional systems, with a leading position in miniature batteries because of its high energy density [1, 2]. Commercial versions presently available on the market are mostly based on a $CoO_2/(CoO_2)^-$ cathode and a $Li/Li^+(C)$ anode (lithium metal dispersed in graphite). The overall electrochemical reaction is

$$
\begin{aligned}
&\text{discharge} \rightarrow \\
Li(C) + CoO_2 &\leftrightarrow LiCoO_2 \\
&\leftarrow \text{charge}
\end{aligned}
$$

M. Catti (✉)
Dipartimento di Scienza dei Materiali, Università di Milano Bicocca, Via Cozzi 53, I-20125 Milano, Italy
e-mail: catti@mater.unimib.it

L. Liang et al. (eds.), *Neutron Applications in Earth, Energy and Environmental Sciences*, Neutron Scattering Applications and Techniques,
DOI 10.1007/978-0-387-09416-8_15, © Springer Science+Business Media, LLC 2009

Several other lithium/transition metal oxides have been proposed as cathode materials as an alternative to $LiCoO_2$ (with NaCl-type layer structure). The required properties include a transition metal with two oxidation states related by a convenient electrochemical potential, good electronic and lithium ion conductivity, and a crystal structure permitting easy and reversible intercalation/deintercalation processes of Li^+ ions. Examples of promising cathode materials are $LiMn_2O_4$ (spinel structure) and $LiFePO_4$ (olivine structure).

In industrially produced batteries, the electrolyte separating the cathode and anode is usually an organic liquid with inorganic salts dissolved (e.g., $LiClO_4$ in acetonitrile). Polymer electrolyte membranes containing lithium have also been tested successfully. However, it would be very attractive to manufacture a liquid-free battery resistant to thermal conditions well above room temperature. In addition to thermal stability, the major advantages over the current polymer/gel devices would be the absence of leakage and pollution, high resistance to shocks and vibrations, and a large electrochemical stability window for practical applications [3]. Thus, the research task is to develop a solid-state electrolyte with the following characteristics: (1) good lithium ion conductivity at room temperature, (2) negligible electronic conductivity, (3) absence of redox reactions in the electrochemical potential range of cathode and anode operation, and (4) very small grain-boundary resistance. The difficulty of satisfying conditions (1) and (3) at the same time is a major limitation of the systems considered so far.

From the preceding discussion, and despite the open problems emphasized, the property of high Li^+ ion conductivity (say, at least $10^{-5}\,\Omega^{-1}\,cm^{-1}$ at room temperature) appears to be a stringent requirement for both cathode materials and solid-state electrolytes to be employed in rechargeable lithium batteries. The present work is focused on the microscopic, atomistic basis underlying this property. In order to understand the fundamental mechanisms of lithium ion transport in crystalline materials, it is necessary to know the geometrical arrangement of lithium atoms in the equilibrium crystal structure. Therefore, after a short overview of the basic principles of ionic mobility in solids, neutron powder diffraction (NPD) will be presented as the choice experimental technique to accomplish this task. Then the results of NPD studies of two of the most important systems of Li^+ ion conductors will be analyzed and discussed. These are lithium lanthanum titanate $Li_xLa_{2/3-x/3}TiO_3$ (LLTO, $x \leq 0.5$), with distorted variants of the perovskite structure, and $Li_{1+x}Me_2(PO_4)_3$ (Li-Nasicon, $0 \leq x \leq 2$), with many structural modifications, among which the one with the best conductivity is related to the rhombohedral corundum structure.

15.1.1 Li⁺ Ion Conductivity

Ionic transport in crystals may simply occur by a thermally activated hopping between stable and vacant ion sites (diffusion), or it may be driven by an applied electric field. In the latter case, electric conduction based on the migration of ions occurs; it is usually some orders of magnitude smaller than the electronic conductivity typical of metallic systems. We shall briefly recall the basic principles underlying ionic transport in solids [4, 5]. The ionic conductivity σ is proportional to the number of charge carriers per unit volume, n, and to the mobility of the migrating ion, μ (e is the electron charge); when the migrating species is a cation:

$$\sigma^+ = en^+\mu^+. \tag{15.1}$$

By considering the equilibrium condition between the field-induced drift current and the opposing diffusion current due to the concentration gradient, the very important Nernst–Einstein equation relating mobility and diffusion coefficient can be derived:

$$\mu = \frac{e}{kT}D. \tag{15.2}$$

This equation provides a link between the phenomena of diffusion and of ionic conductivity. By use of the Arrhenius law, relating the diffusion coefficient D to the activation enthalpy H_a of the ion hopping, one obtains from Eqs. (15.1) and (15.2)

$$\sigma^+ = \frac{n^+e^2}{kT}D_0 \exp\left(-\frac{H_a}{kT}\right), \tag{15.3}$$

where the pre-exponential factor of the diffusion coefficient is

$$D_0 = gva^2n_v/N, \tag{15.4}$$

where g is a geometrical factor depending on the site symmetry of the ion in the crystal structure, v (vibrational frequency of the ion attempting to hop) is proportional to the probability of jumping into a neighboring vacancy, a is the distance between the occupied and the vacant ion sites, and n_v/N is the fraction of vacant sites. At low temperature, the n_v/N fraction and then D_0 are constant, whereas at high temperature, additional vacancies are formed by thermal activation processes, so that D_0 increases with T according to the Boltzmann factor $\exp\left(-\frac{\Delta H_s}{2kT}\right)$ (ΔH_s is the vacancy formation enthalpy).

From the experimental point of view, measurements of electrical conductivity are usually performed by the technique of complex impedance spectroscopy. An important feature of this experimental method is that it is able to separate the bulk from the grain boundary contribution to electrical conductivity. The bulk conductivity depends on the periodic atomic arrangement in the crystal structure, and it is the property we are interested in here. By plotting $\log(\sigma T)$ against $1/T$, one obtains a linear diagram the slope of which is equal to $-(H_a + \Delta H_s/2)/k$. Thus from these measurements, the sum of the activation enthalpy H_a for hopping of ions into vacancies, and of the formation enthalpy of point defects ΔH_s, is derived.

Here we are mainly concerned with the study of the atomistic mechanisms of ion migration in lithium conductors. For this purpose, the key experimental information should cover both the structural and the dynamical aspects. The latter ones, obtained through techniques like solid-state nuclear magnetic resonance (NMR) and inelastic neutron scattering, are not covered in this work. Here we want to deal with the structural information coming from diffraction experiments. The main aim is twofold: (1) the accurate localization of the mobile ion (lithium in the present case) in the crystal structure of the conducting material, so as to characterize in detail its crystal-chemical surroundings and bond distances with neighboring oxygen atoms, and (2) the determination of the order/disorder state of lithium atoms within the constraint of the symmetry of the average structure. This requires that the space group of the crystal structure and the site symmetries of lithium atoms be determined unambiguously; further, the occupancy factors of lithium atoms have to be refined reliably with a sufficient accuracy.

Indeed, on looking at Eqs. (15.3) and (15.4), the key quantities that control the ionic conductivity appear to be the activation enthalpy H_a, the attempt frequency ν, the jump distance a, and the vacancy concentration n_v/N. These physical parameters depend strongly on the structural features of the material, and they are not actually independent of one another. For instance, on increasing a, the favorable effect on the ion conductivity is usually more than counterbalanced by the associated decrease of frequency ν and increase of activation enthalpy H_a, so that in most cases the shorter the jump distance, the higher the ion mobility. Further, the H_a quantity depends on the energy of the chemical bonds/electrostatic interactions of lithium particles, which must be broken for the jump to occur. That energy is, in turn, closely related to the lithium-ligands' interatomic distances and angles and then to the geometrical arrangement of the crystal structure. Therefore, long lithium-ligand distances and small coordination numbers certainly favor ion mobility. Of most importance is the fraction of vacant sites n_v/N, which is linked to the degree of disorder of lithium atoms as determined by their refined occupancy factors and site multiplicities.

The reasons have been presented for which a detailed knowledge of the crystal structure arrangement, with emphasis on the lithium atom positions, is necessary to understand the lithium ion conductivity of solid materials. Of course, that knowledge is also needed to propose the atomistic pathways and theoretical mechanisms for the process of ionic migration within the structural framework.

15.2 Experimental Methods by Neutron Powder Diffraction

Most lithium-ion-conducting materials are synthesized as polycrystalline samples, so that powder diffraction methods have to be employed with them. This reduces the quality of the structure determination and refinement with respect to single-crystal samples. Thus, because of the small atomic number of lithium, the use of X-ray radiation to localize lithium atoms would be hopeless. Further, a number of investigations have proved that even the lattice geometry and symmetry may be missed by X-ray diffraction (XRD) in such materials, so that use of neutrons is required to reveal superlattice peaks due to distortions of the oxygen sublattice (see the discussion of LLTO phases in the next section).

Natural lithium is a mixture of 92.5% ^{7}Li and 7.5% ^{6}Li isotopes, yielding a coherent neutron scattering length $b = -1.90$ fm, with the advantage of a negative sign improving the contrast with respect to most other atoms with positive b [6]. However, the absolute value is modest (e.g., oxygen has $b = 5.80$ fm), and there is a significant incoherent contribution to total scattering. Further, a drawback is presented by the large absorption cross section of natural lithium (70.5 barn), which reduces the scattering efficiency significantly. For this reason, samples are often prepared, which are enriched with the ^{7}Li isotope (0.045 barn vs 940 barn of ^{6}Li). Yet this procedure may, unfortunately, be of limited usefulness taking into account the already overwhelming abundance of ^{7}Li in natural samples. Therefore, although use of neutron beams is much more convenient than use of X rays, an accurate determination of lithium atom positions by powder diffraction is still a difficult task. This is particularly true when the lithium content per formula unit (f.u.) is quite small and/or when lithium atoms are severely disordered over many crystallographic sites with small occupancies. Thus, it is not surprising that the number of structural studies of lithium-ion-conducting materials, where lithium atoms could be localized directly by difference Fourier maps and their positions were successfully Rietveld-refined, is rather small in the literature.

15.2.1 Rietveld Structure Refinement

In the powder diffraction pattern (Chapter 2), all data corresponding to different reciprocal lattice vectors are projected onto a single axis corresponding to the vector length or to its reciprocal d_{hkl}. For this reason, the superposition of Bragg peaks is a severe problem, requiring delicate numerical methods in order to deconvolute single peaks and evaluate their integrated intensities. The basic procedure is as follows. Let us recall that when diffraction takes place, the scattering vector $\mathbf{Q}$ (Chapter 2) must satisfy the von Laue condition: $\mathbf{Q} = 2\pi\mathbf{h}$, where $\mathbf{h}$ is a reciprocal lattice vector of the crystal structure. Further, $1/|\mathbf{h}| = d_{hkl}$ is the interplanar spacing appearing in Bragg's Law (Eq. 2.12). For each i-th experimental point in the pattern (corresponding to a given d_i value), the corresponding intensity I_i is considered to be the sum of contributions of Bragg peaks coming from different reciprocal lattice vectors $\mathbf{h}$: $I_i = \sum_h I_{ih}$, where the sum is extended to all Bragg peaks overlapping in the surrounding d_i. Then a sufficiently flexible function $f(X)$ (where $X = d - d_{hkl}$) is chosen to represent the profile of any Bragg peak in the pattern. The most common function employed is a linear combination of a Gaussian and a Lorentzian function (pseudo-Voigt):

$$f(X) = A_1 \exp(-\sigma X^2) + A_2/(1 + \gamma X^2). \tag{15.5}$$

This expression has a bell-shaped profile with maximum at $X = 0$. The σ and γ parameters determine the half-height width of the peak, and the A_1 and A_2 coefficients are chosen so as to satisfy the normalization condition $\int_{-\infty}^{+\infty} f(X)\mathrm{d}X = 1$.

Therefore, the h-th Bragg profile is represented by the $f(X)$ function multiplied by a scale factor, which is equivalent to the integrated intensity I_h of the peak:

$$\int_{-\infty}^{+\infty} I_h f(X)\mathrm{d}X = I_h \int_{-\infty}^{+\infty} f(X)\mathrm{d}X = I_h.$$

By adding the background radiation intensity I_i^{bkg} and introducing the peak multiplicity m_h and an overall scale factor k, the following formula is obtained for the model point intensity I_i^c:

$$I_i^c = k\Sigma_h m_h I_h f(d_i - d_{hkl}) + I_i^{bkg}. \tag{15.6}$$

On substituting I_h in Eq. (15.6) by the square of the structure factor modulus,

$$I_h = \left| \sum_{s=1}^{p} b_s \exp(2\pi i \mathbf{h} \cdot \mathbf{r}_s) \exp(-2\pi^2 |\mathbf{h}|^2 U_s) \right|^2, \tag{15.7}$$

one obtains a formula relating the model point intensity I_i^c to the structural unknowns $\mathbf{r}_s$ (position vector of the s-th atom in the unit-cell) and $U_s = \langle u_s^2 \rangle$ (displacement parameter of the same atom) for all the p atoms contained in the unit-cell. All the I_i^c-computed values are compared with the corresponding observed intensities I_i^o. Then the least-squares method is applied by minimizing the quantity $\Delta^2 = \Sigma_i (I_i^c - I_i^o)^2$ with respect to the structural parameters: x_s, y_s, and z_s atomic fractional coordinates [remembering that $\mathbf{h} \cdot \mathbf{r}_s = hx_s + ky_s + lz_s$ in Eq. (15.7)], U_s thermal factors, and unit-cell constants (Rietveld refinement, Chapter 2). Note that the profile parameters must be included in the refinement. These are the A_1/A_2 ratio and the σ, γ half-widths of the Gaussian and Lorentzian components of the pseudo-Voigt $f(X)$ function in Eq. (15.5), the k scale factor, and the parameters of a fitting function for the background intensity. In the case of time-of-flight (TOF) neutron diffraction from an energy-dispersive pulsed source, the peak shape is usually represented [7] by a convolution of the pseudo-Voigt function (sample contribution) with two back-to-back exponentials (instrumental and moderator contributions [8]). Linear dependencies of the σ and γ parameters on d_{hkl} are often assumed: $\sigma = \sigma_1 d_{hkl} + \sigma_0$, $\gamma = \gamma_1 d_{hkl} + \gamma_0$. The mixing coefficient A_1/A_2 and the full width of the pseudo-Voigt function depend on σ and γ according to the equations given in the literature [9]. Agreement factors for the profile refinement are usually defined as

$$R_p = \sum_i |I_i^c - I_i^0| \left/ \sum_i I_i^0 \right., \quad R_{wp} = \left[\sum_i w_i (I_i^c - I_i^0)^2 \left/ \sum_i w_i (I_i^0)^2 \right. \right]^{1/2},$$
$$\tag{15.8}$$

where the sums are extended to all observed data, each of which is characterized by the statistical weight w_i.

In some cases, profile and structural parameters may be strongly correlated to each other, introducing numerical instability and/or large estimated standard deviations into the least-squares procedure. More detailed discussions of the Rietveld method for structure refinement from a powder pattern can be found elsewhere (e.g., see [10]). Some of the most popular computer codes for carrying out Rietveld refinements from both X-ray and neutron diffraction measurements are FULLPROF [11], GSAS [12], and RIETAN [13].

15.3 Perovskite-Type Lithium-Lanthanum-Titanate (LLTO) Materials

Lithium perovskites have attracted a lot of attention in the search for inorganic materials displaying a high lithium ion mobility [14]. This attention is largely related to the peculiar flexibility of the ABO_3 perovskite structure type that is able to distort, in a great variety of ways, under the effect of physical parameters (temperature, pressure) and compositional changes in the A and B sites. Such distortions, coupled with charge imbalance due to chemical substitutions in the cation sites, can provide lithium ions with multiple available sites of comparable energy, so as to promote disorder and then improve the ion mobility.

The $Li_x La_{2/3-x/3} \square_{1/3-2x/3} TiO_3$ system, with perovskite-type structures and the $\square$ symbol denoting the vacant A sites, is one of those showing the highest lithium ion electrical conductivity at room temperature (about 10^{-3} Ω^{-1} cm^{-1} for $x = 0.3$) [15, 16]. Unfortunately, this excellent ionic conductor cannot be used as a solid electrolyte in the design of an ordinary lithium battery because of the high chemical reducibility of the Ti^{4+} ions when LLTO is placed in contact with lithium metal electrodes. It can, however, be employed in other electrochemical devices, and it is especially important as a model material for elucidating the reasons for its high ionic conductivity and for use in designing new conductive systems.

In Li-free $La_{2/3} \square_{1/3} TiO_3$, lanthanum occupies, on the average, two-thirds of the A sites of the ABO_3 perovskite structure, whereas one-third are vacant [17]. Lithium insertion occurs by partly replacing lanthanum atoms and partly occupying the hollows around the empty A sites; when $x = 0.5$, all the A sites are saturated by either lanthanum or lithium. Clearly, the extended disordered distribution of lithium atoms and vacancies is responsible for the high ionic mobility. However, although several structural studies of this system at a number of lithium compositions have been performed, only in a few cases could lithium be located and/or refined by neutron diffraction [18–24]. This task is particularly difficult in the case of small lithium content (say, $x < 0.25$), in which a problem of scattering sensitivity arises. In other NPD investigations [25–27], crystal structure refinements were performed, but without including the lithium positions. In all of these papers, many crystallographic variations on the theme of the basic perovskite structure were detected or hinted at, suggesting that the subject is far from being exhausted and that it is crucial to clarify the lithium distribution and then the structural basis of ionic mobility.

Many authors pointed out that conventional X-ray powder diffraction is not able to detect deviations from the standard perovskite cubic cell (with edge length $a_p \approx 3.87$ Å), whereas with electron diffraction, transmission electron microscopy (TEM), and neutron diffraction, several types of symmetry lowering, with different superlattices with respect to the basic perovskite cell, were found [14]. The lithium content (x value) and the cooling rate in the sample synthesis play a crucial role in this respect. For samples with low lithium content ($0.12 \leq x \leq 0.20$), an orthorhombic *Cmmm* or *Cm2m* phase with cell of type $2a_p \times 2a_p \times 2a_p$ was found in the case of synthesis by slow cooling from a high temperature. Within this compositional range, we shall refer to the NPD studies of the $x = 0.16$ [20] and $x = 0.18$ terms [21], based on constant-wavelength data collected at the Japan Research Reactor No. 3 (JRR-3) and the Institut Laue-Langvin (ILL) reactor sources. The *Cmmm* crystal structure of the former phase is shown in Fig. 15.1. On

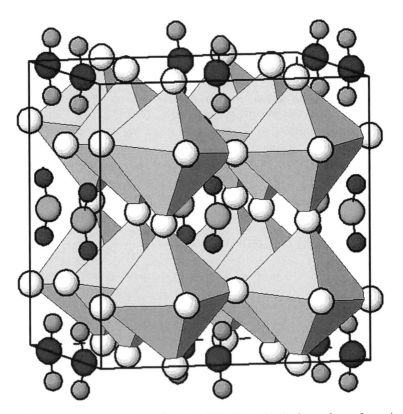

Fig. 15.1 Crystal structure of *Cmmm* $Li_{0.16}La_{0.62}TiO_3$ [20], with the $2a_p \times 2a_p \times 2a_p$ unit-cell outlined. The TiO_6 coordination octahedra are emphasized. *Dark* and *gray, large spheres* denote lanthanum atoms with major and minor occupancies, respectively; the same is true for *dark* and *gray smaller balls* (lithium atoms). Disordered lanthanum and lithium atoms that are locally incompatible are linked by *thin lines*

quenching the $x = 0.18$ sample in liquid N_2, it was claimed that the rhombohedral $R\bar{3}c$ symmetry would appear instead [21, 28], although there are indications that superstructure lines persist [29]. With the lithium-richest composition ($x = 0.5$), different results are reported again for slowly cooled and for quenched samples: in the former case, a $P4mm$ symmetry with $\sqrt{2}a_p \times \sqrt{2}a_p \times 2a_p$ unit-cell was observed by TEM [30], whereas in the second case, the $R\bar{3}c$ structure was indicated [19, 27].

Particularly interesting is the intermediate $x = 0.3$ composition to which, at first, comparatively less attention was paid by structural investigations despite its very good ionic conductivity performance. An early neutron diffraction study ($x = 0.31$, non-quenched sample) [18] proposed a $P4/mmm$ structure with a $a_p \times a_p \times 2a_p$ unit-cell, similar to that already suggested for a wider composition range on the basis of TEM results [31]. However, in this structure, lithium was assumed to lie exactly at the place of lanthanum, giving rise to Li-O bond distances of 2.7 Å, which are unrealistically long. Later a combined TEM–X-ray synchrotron radiation study of the $x = 0.35$ composition (sample quenched to room temperature) appeared [32], where the unit-cell of type $\sqrt{2}a_p \times \sqrt{2}a_p \times 2a_p$ was proposed, with a Rietveld-refined orthorhombic $Pmma$ structure (without lithium, of course, because of its low X-ray scattering factor).

Recently, two detailed neutron studies (TOF data from the ISIS pulsed spallation source, Chapter 3) of the $Li_{0.3}La_{0.567}TiO_3$ composition appeared; a slowly cooled sample [23] and one quenched into liquid nitrogen [22] were used in the investigation. Both samples gave a tetragonal $\sqrt{2}a_p \times \sqrt{2}a_p \times 2a_p$ unit-cell, but with different space group symmetries ($P4/nbm$ and $I4/mcm$, respectively) and different structure details. In particular, the lithium atom positions could be determined and refined successfully, yielding quite reasonable Li-O distances with a fourfold coordination environment. It was confirmed that XRD completely misses the superlattice peaks transforming $a_p \times a_p \times 2a_p$ into $\sqrt{2}a_p \times \sqrt{2}a_p \times 2a_p$. Another investigation, carried out on the $x = 0.36$ composition, proposed instead an orthorhombic $Cmmm$ structure similar to that found for lithium-poor terms [24].

Independently of symmetry and compositional features, two main distortions are observed, at least in unquenched LLTO samples at room temperature, with respect to the ideal perovskite structure (a_p cubic cell edge). First, a partial ordering of the La-Li-$\square$ substitution in the A cages is observed along the z axis; this causes the c cell edge to double from a_p to $2a_p$, so that alternate (001) layers of lanthanum-rich and lanthanum-poor A-type sites arise. Second, the TiO_6 octahedra tilt in anti-phase sequence around at least one of the crystallographic axes, which is z for tetragonal structures [22, 23] ($a^0a^0c^-$ scheme, according to Glazer's notation [33]) and usually y for orthorhombic structures [24, 34] ($a^0b^-c^0$ scheme). An enlargement of the unit-cell in the (001) plane from $a_p \times a_p$ to $\sqrt{2}a_p \times \sqrt{2}a_p$ (tetragonal) or to $2a_p \times 2a_p$ (orthorhombic) ensues. The partial ordering along the z axis is sensitive to temperature [31, 29]: quenched samples of the tetragonal $Li_{0.3}La_{0.567}TiO_3$ composition keep the high-temperature full disorder of the La/Li pair, which are distributed in a single A-type site ($I4/mcm$ symmetry) [22].

15.3.1 Disordered Locations of Lithium Atoms in LLTO from NPD Results

For the sake of comparison, the larger unit-cell $2a_p \times 2a_p \times 2a_p$ will be used as a common reference system for both the *Cmmm* orthorhombic phases [20, 21, 24] and the *P4/nbm* and *I4/mcm* tetragonal structures [22, 23]. The crystal structure of the *P4/nbm* term with $x = 0.3$ [23] is shown in Fig. 15.2, according to this convention. Further, the origin has been shifted to the center of the lower A cavity, so that hereafter the atomic fractional coordinates must be transformed according to $y' = y - 1/4$ (*Cmmm*) and $x' = (x - y)/2 + 1/4$, $y' = (x + y)/2 - 1/4$ (*P4/nbm* and *I4/mcm*) with respect to the published values. The two independent A sites are located at the $z = 0$ and $z = 0.5$ levels; the titanium atoms lie at $z = 0.25$ and 0.75. Six distorted, square O_4 windows surround each A site: four of them have centers lying at the same z level as A (vertical windows: 0, 0.25, 0 or 0.25, 0, 0 and equivalent locations at $z = 0.5$); and two of them have centers at 0, 0, ± 0.25 (horizontal windows). In Fig. 15.3 the NPD profile is shown [23], including experimental points determined by the high-resolution powder diffractometer (HRPD) at the ISIS pulsed source (United Kingdom), a calculated profile from Rietveld refinement, and

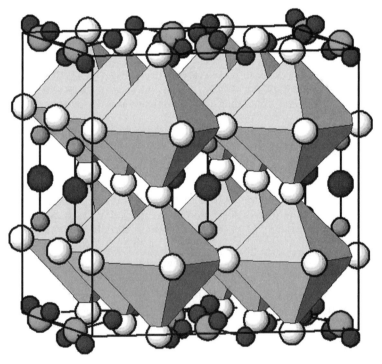

Fig. 15.2 Crystal structure of *P4/nbm* $Li_{0.3}La_{0.567}TiO_3$ [23] drawn within a $2a_p \times 2a_p \times 2a_p$ box, for comparison with the orthorhombic phase (Fig. 15.1). The real unit-cell is $\sqrt{2}a_p \times \sqrt{2}a_p \times 2a_p$

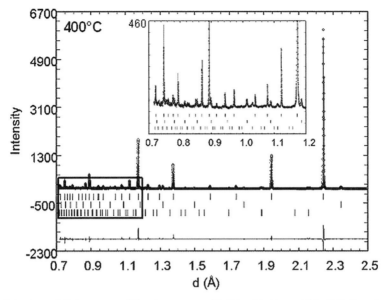

Fig. 15.3 Time-of-flight neutron diffraction pattern of *P4/nbm* $Li_{0.3}La_{0.567}TiO_3$ [23]. Experimental (*crosses*), calculated, and difference profiles are shown. The three sets of peak markers indicate *hkl* reflections with $h + k$ even and l even (*above*), $h + k$ odd and l odd (*middle*), and $h + k + l$ odd (*below*). The magnified insert shows the short-d_{hkl} region of the pattern

an observed-calculated difference profile. The agreement factors obtained for the refinement were $R_p = 0.056$ and $R_{wp} = 0.077$.

According to the few state-of-the-art neutron diffraction studies of LLTO phases, the different positions determined for lithium atoms have been reported in Table 15.1 (with respect to the previously described reference frame). All lithium sites are

Table 15.1 Unit-cell constants and fractional coordinates of lithium atoms in four phases of the LLTO $Li_xLa_{2/3-x/3}TiO_3$ system, determined from neutron diffraction data. The coordinate origin is at the center of the lower dodecahedral cage (A site) for both the *Cmmm* and the *P4/nbm* structures

	o.f./f.u.	x	y	z	Ref.
$Li_{0.16}La_{0.62}TiO_3$ *Cmmm*		$a = 7.7313$	$b = 7.7520$	$c = 7.7840$ Å	[20]
Li2 8n	0.15	0	−0.06(1)	0.602(9)	
Li1 8n	0.01	0	0.061	0.102	
$Li_{0.18}La_{0.61}TiO_3$ *Cmmm*		$a = 7.7179$	$b = 7.7397$	$c = 7.7712$ Å	[21]
Li1 8n	0.11	0	0.02(1)	0.236(8)	
Li2 4f	0.07	0.25	0	0.5	
$Li_{0.3}La_{0.567}TiO_3$ *P4/nbm*		$a = 5.4816$	$b = 5.4816$	$c = 7.7465$ Å	[23]
Li1 8m	0.30	0.142(4)	0	0.036(5)	
Li2 4h	0	0	0	0.654	
$Li_{0.36}La_{0.55}TiO_3$ *Cmmm*		$a = 7.7383$	$b = 7.7422$	$c = 7.7364$ Å	[24]
Li1 2c	0.03	0	0.25	0	
Li2 4j	0.33	0	0.067(8)	0.5	

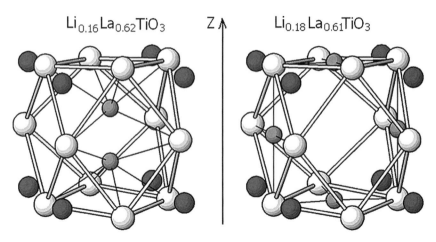

Fig. 15.4 The lithium-rich A-type cages in the *Cmmm* LLTO phases of $Li_{0.16}La_{0.62}TiO_3$ (*left* [20]) and $Li_{0.18}La_{0.61}TiO_3$ (*right* [21]), showing disordered lithium atoms (*gray balls*) inside, but not lanthanum atoms. *Dark and white, large spheres* denote titanium and oxygen atoms, respectively; Li-O bonds are indicated as *black lines*

fractionally occupied, and their occupancies per f.u. are given. In the case of the $Li_{0.30}$ term, the Li2 position (determined from Fourier difference maps) could be refined only alternatively to that of Li1. The two lithium positions of the $Li_{0.18}$ term [21] appear clearly to correspond to the centers of the horizontal (Li1) and vertical (Li2) windows bounding the A cage at $z = 0.5$, with smaller lanthanum occupancy (Fig. 15.4, right-hand side). The lithium coordination is square planar, with four fairly short and strong Li-O bonds (Table 15.2). A quite different arrangement is shown by the $Li_{0.16}$ term [20], where lithium is displaced from the cavity center along the z direction, i.e., toward the above- or below-lying (horizontal) window center (Fig. 15.4, left-hand side, and Fig. 15.1). Again a four-fold coordination appears, but with quite a distorted pyramidal rather than planar arrangement and much longer Li-O bonds (Table 15.2).

The more lithium-rich composition $Li_{0.30}$ is represented here for simplicity only by the room-temperature *P4/nbm* structure of the slowly cooled phase [23]. The primary lithium location Li1 (Figs. 15.2 and 15.5, left-hand side) occurs close to

Table 15.2 The four shortest Li-O interatomic distances (Å) from neutron diffraction for the LLTO phases of Table 15.1

$Li_{0.16}La_{0.62}TiO_3$ *Cmmm*	Li2	2.05	2.36×2	2.64	[20]
	Li1	2.05	2.36×2	2.64	
$Li_{0.18}La_{0.61}TiO_3$ *Cmmm*	Li1	1.79	1.94×2	2.10	[21]
	Li2	1.94×2	2.07×2		
$Li_{0.3}La_{0.567}TiO_3$ *P4/nbm*	Li1	2.02	2.13×2	2.43	[23]
	Li2	1.91×2	2.17×2		
$Li_{0.36}La_{0.55}TiO_3$ *Cmmm*	Li1	1.67×2	2.21×2		[24]
	Li2	2.29×2	2.62×2		

the $z = 0$ level of the A site and displaced toward the center of one of the adjacent O_4 (vertical) windows. A distorted square-pyramidal coordination ensues, with a shorter and a longer bond than the average (Table 15.2). The second lithium site reported in Table 15.1 as Li2, though detected very clearly on the Fourier difference map, could not be refined together with Li1. It is located in a position similar to that observed for the $Li_{0.16}$ term, i.e., above the A site but closer to the center of the horizontal window (Fig. 15.5, right-hand side); a fairly regular, square pyramidal coordination is observed (Table 15.2).

In summary, three disordered sites of lithium within the A cage are reported in neutron diffraction studies of LLTO phases: S1, close to the center of the O_4 (vertical) window separating adjacent A cages in the same (001) plane (Fig. 15.5, left); S2, displaced from the cavity center along the z direction, i.e., toward the above- or below-lying O_4 (horizontal) window center (Fig. 15.5, right); and S3, the window center itself (Fig. 15.4, right). Lithium was refined in S1 in the *I4/mcm* and *P4/nbm* tetragonal structures of the $x = 0.30$ phase [22, 23] and in the *Cmmm* structure of the $x = 0.16$ and $x = 0.36$ terms [20, 24]. The S2 site was also found to be populated by lithium in the former cases [22, 23], and site S3 was reported in the structures of the $x = 0.16, 0.18,$ and 0.36 terms [21, 24, 34]. On the basis of these results, some models of mobility paths of Li^+ ions were devised [22, 23], also with the help of simple bond-valence-sum evaluations [24, 35]. A primarily two-dimensional mechanism of ion hopping within the (001) plane ensues, where site S3 acts mainly as a bottleneck between the adjacent S1 sites; site S2 may provide the connection for transfer between neighboring planes. Such models were substantially confirmed by molecular dynamics calculations [36, 37].

However, although a sufficiently clear picture of the complex structural properties of LLTO phases begins presently to be outlined, this is limited to statistically

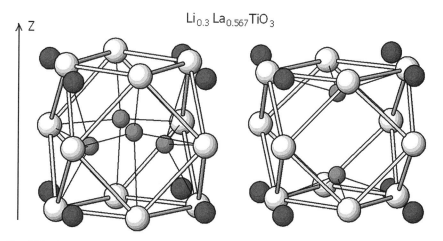

Fig. 15.5 The two independent A-type cages in the *P4/nbm* phase of LLTO-$Li_{0.3}La_{0.567}TiO_3$ [23]. The left-hand cage is rich in lithium (Li1, see Table 15.1) and poor in lanthanum, and the opposite is true for the right-hand one (Li2)

averaged configurations, such as those compatible with diffraction experiments. Solid-state nuclear magnetic resonance investigations of the Li^+ ion dynamics in LLTO were indeed performed [38], detecting the presence of ions of two kinds with slightly differing environments; but much more detailed information on a local scale would be needed to untangle the complex disorder of the La-Li-□ distribution in the A cages. This task could be accomplished by experimental studies of the short-range order (measurements of total neutron scattering) and dynamics (inelastic neutron scattering) of lithium ions. Further, theoretical simulations of the least-energy local structure of LLTO may be of great help in this respect. An example is given by a recent study of the $Li_{0.3}La_{0.567}TiO_3$ phase [39], where static quantum-mechanical calculations at the DFT-B3LYP level were performed for different locally ordered models of the LLTO structure. A primary result of the study was that the S3 sites at window centers proved to be energetically less favored than the S1 and S2 locations, which were confirmed to be possible stable sites of lithium atoms in the perovskite framework. Moreover, it is now possible to associate the two disordered lithium sites found experimentally in the average tetragonal structure of $Li_{0.3}La_{0.567}TiO_3$ to corresponding locally ordered configurations. Lithium takes the S1 or S2 site when its A-type cavity lies in an open or closed environment, i.e., it is surrounded by few or by many lanthanum-occupied adjacent hollows. From the point of view of ionic mobility, the site S2 is clearly unfavorable, as the lithium ions trapped therein have no surrounding sites available for jumping. Thus, processes of ion transfer should be based primarily on lithium hopping between the S1 sites.

15.4 Li^+ Conducting Nasicon Phases

Lithium-containing transition metal phosphates $Li_{1+x}Me_2(PO_4)_3$ ($0 \leq x \leq 2$), with the rhombohedral Nasicon-like structure, have attracted significant interest for applications in electrochemical devices. According to whether the transition metal Me can undergo a redox reaction, such phases can be used as intercalation electrode materials for reversible lithium batteries or as solid lithium ion electrolytes for a number of applications [40, 41]. In both cases, their performance relies heavily upon the mobility of Li^+ in the Nasicon structural framework. Features such as the order–disorder state of lithium, the length/strength of Li-O bonds, and the distance between adjacent lithium sites available for ion hopping strongly affect Li^+ mobility, as they directly control the attempt frequency and energy barrier for thermally activated ionic jumps. Therefore, a detailed determination of the lithium structural environment in these phases is necessary to understanding the microscopic basis of their performance as ionic materials.

The Nasicon framework (rhombohedral $R\bar{3}c$, but also symmetry subgroups such as $R\bar{3}$, $C2/c$, $C\bar{1}$, and others are possible) is built up by MeO_6 coordination octahedra and PO_4 tetrahedra sharing all vertices; it can be thought to be derived from the Al_2O_3 corundum structure by replacing all oxygen atoms by PO_4 tetrahedral groups [42]. Two main cavities are present in the structure, surrounding the M1 and M2 (sometimes named M′ and M″) sites. M1 lies at Wyckoff position 6b

$(x = 0, y = 0, z = 0)$ with site symmetry $\bar{3}$ and corresponds to the empty (one out of three) octahedral site in the (ABC)ABC...close-packed arrangement of corundum. M2 is located at 0.667, 0, and 1/4 (or at 0, 0.333, and 1/12), in position $18e$ with site symmetry 2, and the corresponding site in the corundum structure would be at the center of an O_3 triangle within a close-packed oxygen layer. Clearly, the multiplicity of M2 is three times that of M1. In Figs. 15.6 and 15.7, the rhombohedral crystal structures of corundum and Nasicon, respectively, are shown. Comparing the locations of the M1 and M2 sites in the two cases shows that the structural arrangements are closely related.

The structures of Li-Nasicon phases of type $LiMe_2(PO_4)_3$ and $Li_3Me_2(PO_4)_3$ were recently fully characterized by neutron diffraction techniques. A complex

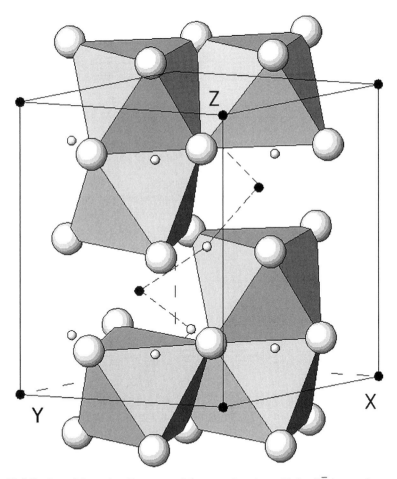

Fig. 15.6 Portion of the unit-cell content of the corundum ($\alpha - Al_2O_3$, $R\bar{3}c$) crystal structure, with the z fractional coordinate ranging from 0 to 0.5. AlO_6 coordination octahedra emphasized. The *black* and *white, small circles* indicate the M1 octahedral and M2 triangular empty sites, respectively

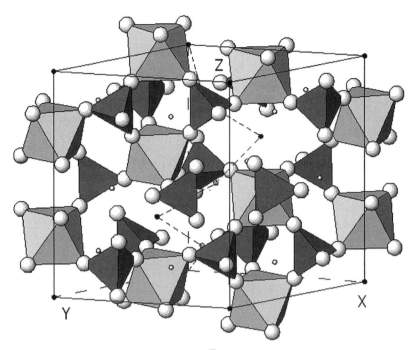

Fig. 15.7 Portion of the unit-cell content of the $R\bar{3}c$ Nasicon crystal structure, with the z fractional coordinate ranging from 0 to 0.5. MO_6 coordination octahedra and PO_4 tetrahedra are emphasized. The *black* and *white, small circles* indicate the M1 and M2 empty sites, respectively (see corundum in Fig. 15.6), connected by a *dashed line* to show a possible pathway of ion mobility

polymorphism is shown by the $LiZr_2(PO_4)_3$ term, which can be synthesized in triclinic $C\bar{1}$, rhombohedral $R\bar{3}c$, orthorhombic *Pbna*, or monoclinic $P2_1/n$ form, according to the thermal conditions [43]. The $R\bar{3}c$ phase gives the highest ionic conductivity of $\sigma = 10^{-5}\Omega^{-1}\,cm^{-1}$ close to room temperature [44, 45] and was studied by TOF neutron powder diffractometry (HRPD instrument, ISIS source) [42, 43]. Lithium atoms were mostly located in peripheral sites within the M1 cavity, disordered with a distorted tetrahedral coordination; only a minor fraction was located on the M2 site. In the isomorphous phase $LiTi_2(PO_4)_3$, instead, lithium was found to be ordered exactly on the octahedrally coordinated M1 site [46], yielding a less stable configuration with quite long Li-O bonds. On the other hand, in compounds like $Li_3In_2(PO_4)_3$, $Li_3Fe_2(PO_4)_3$, and $Li_3Ti_2(PO_4)_3$ (space group $R\bar{3}$), lithium atoms were located only within the M2 cavity (constant-wavelength, angle-dispersive NPD studies [47]). This lithium behavior contrasts with that of sodium in similar compounds, like $Na_3Fe_2(PO_4)_3$ and $Na_3Zr_2PO_4(SiO_4)_2$, where both the M1 and the M2 cavities are populated at the same time.

Li-Nasicon phases, with lithium content intermediate between one and three atoms per f.u., show a partial occupation of both the M1 and the M2 hollows by lithium, possibly improving the ionic mobility because of high disorder. Some preliminary indications in this direction came from the studies of some systems of the form $Li_{1+x}Me'_xMe_{2-x}(PO_4)_3$ with $Me = Ti$, Hf and $Me' = In$, Cr. In these systems, however, the rhombohedral Nasicon structure was shown to be stable only for $x < 1$ and to transform into an orthorhombic phase for larger lithium content [48, 49]. Also, the crystal structure of $Li_2FeTi(PO_4)_3$, orthorhombic $Pbca$ [50], for which ionic conductivity data had also been reported [51], was determined.

A full investigation of the $Li_{1+x}Fe_xTi_{2-x}(PO_4)_3$ system [52] showed that the rhombohedral $R\bar{3}c$ structure, giving the best ionic mobility, is stable only in the $x \leq 0.6$ range. Then the $Li_{1.5}Fe_{0.5}Ti_{1.5}(PO_4)_3$ phase was studied by TOF NPD at room temperature and high temperature [52]. Lithium was found to be highly disordered and distributed in similar amounts between the M1 and the M2 cavities of the Nasicon framework, unlike what had been observed in lithium Nasicon phases with one or three lithium atoms per f.u. Thus, a particularly large ion mobility should be expected. The configuration of lithium sites within the M1 hollow is strongly affected by heating, so as to displace lithium further toward the periphery of the cavity.

The lithium positions refined in representative rhombohedral Nasicon phases with $Li_{1.0}$, $Li_{1.5}$, and $Li_{3.0}$ compositions are reported in Table 15.3, including data on the occupancies and on the temperature effect. The structural arrangement observed in $Li_{1.5}Fe_{0.5}Ti_{1.5}(PO_4)_3$ (Fig. 15.8) is particularly meaningful, as it shows the tendency of lithium atoms to move off the centers of both the M1 and the M2 hollows into highly disordered sites.

The corresponding NPD pattern, obtained by data collection on the HRPD at the ISIS pulsed source, is shown in Fig. 15.9 (agreement factors: $R_p = 0.026$ and $R_{wp} = 0.035$). A quite similar configuration for Li1 appears also in $LiZr_2(PO_4)_3$, whereas Li2 goes onto the M2 site. In $L_3Ti(PO_4)_3$ (Fig. 15.10), all lithium is

Table 15.3 Unit-cell constants and fractional coordinates of lithium atoms in some Nasicon-type $Li_{1+x}MM'(PO_4)_3$ rhombohedral phases, determined from neutron diffraction data

		T (K)	o.f.	x	y	z
α-$LiZr_2(PO_4)_3$	$R\bar{3}c$	423 [42, 43]		$a = 8.8549$		$c = 22.1442$ Å
Li1	$36f$		0.15(1)	0.028(5)	0.174(5)	0.006(1)
Li2	$18e$		0.04(1)	0.90(2)	0	0.25
		673 [43]		$a = 8.8445$		$c = 22.2875$ Å
Li1	$36f$		0.15(1)	0.038(7)	0.167(7)	0.016(2)
Li2	$18e$		0.04(1)	0.88(3)	0	0.25
		873 [43]		$a = 8.8367$		$c = 22.4177$ Å
Li1	$36f$		0.13(1)	0.056(9)	0.184(9)	0.014(3)
Li2	$18e$		0.07(1)	0.93(1)	0	0.25
$L_{1.5}Fe_{0.5}Ti_{1.5}(PO_4)_3$	$R\bar{3}c$	298 [52]		$a = 8.5451$		$c = 20.8768$ Å
Li1	$36f$		0.069(5)	0.017(4)	0.200(4)	0.017(1)
Li2	$36f$		0.127(6)	0.079(2)	0.342(2)	0.0648(6)
		673 [52]		$a = 8.5459$		$c = 21.1262$ Å
Li1	$36f$		0.146(5)	0.063(3)	0.327(3)	0.004(1)
Li2	$36f$		0.104(5)	0.015(4)	0.335(4)	0.065(1)
$Li_3Ti(PO_4)_3$	$R\bar{3}$	298 [46]		$a = 8.3828$		$c = 22.873$ Å
Li1	$18f$		0.72(6)	0.030(4)	0.319(4)	0.045(1)
Li2	$18f$		0.34(6)	0.055(8)	0.373(9)	0.117(3)

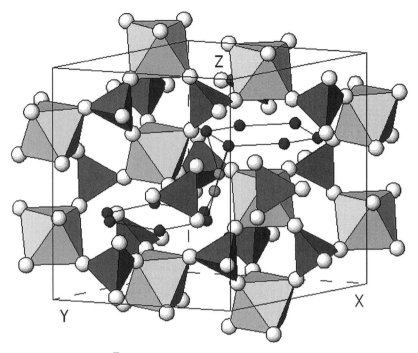

Fig. 15.8 Portion of the $R\bar{3}c$ Nasicon-type structure of $L_{1.5}Fe_{0.5}Ti_{1.5}(PO_4)_3$ at $T = 673$ K [52]. The *dark* and *pale gray spheres* indicate disordered lithium atoms with larger and smaller occupancies, surrounding the M1 and M2 sites, respectively

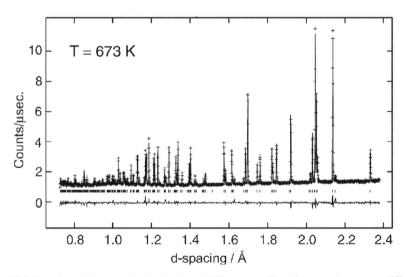

Fig. 15.9 Experimental (*crosses*), calculated, and difference profiles of the powder neutron diffraction pattern of $L_{1.5}Fe_{0.5}Ti_{1.5}(PO_4)_3$ at $T = 673$ K

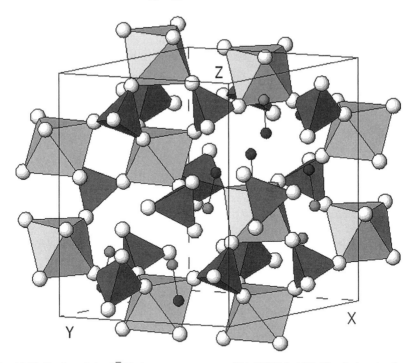

Fig. 15.10 Portion of the $R\bar{3}$ Nasicon-type structure of $Li_3Ti(PO_4)_3$ [46]. The *dark* and *pale gray spheres* indicate disordered lithium atoms with larger and smaller occupancies, respectively, surrounding the M2 site

Table 15.4 The four shortest Li-O interatomic distances (Å) from neutron diffraction for the Nasicon phases of Table 15.3

		T (K)				
α-$LiZr_2(PO_4)_3$	$R\bar{3}c$	423 [42, 43]				
Li1	36f		2.03(3)	2.30(3)	2.36(3)	2.40(4)
Li2	18e		1.892(4) × 2	1.961(9) × 2		
		673 [43]				
Li1	36f		2.10(5)	2.27(5)	2.30(4)	2.52(6)
Li2	18e		1.88(1) × 2	1.99(4) × 2		
		873 [43]				
Li1	36f		2.11(6)	2.30(6)	2.39(7)	2.44(6)
Li2	18e		1.89(2) × 2	1.97(1) × 2		
$L_{1.5}Fe_{0.5}Ti_{1.5}(PO_4)_3$	$R\bar{3}c$	298 [52]				
Li1	36f		2.04(3)	2.06(2)	2.21(3)	2.22(3)
Li2	36f		2.10(1)	2.11(2)	2.17(2)	2.38(2)
		673 [52]				
Li1	36f		1.91(2)	1.98(2)	2.20(2)	2.26(2)
Li2	36f		2.21(3)	2.33(3)	2.41(3)	2.44(3)
$Li_3Ti(PO_4)_3$	$R\bar{3}c$	298 [46]				
Li1	18f		1.95(4)	2.01(4)	2.02(4)	2.17(4)
Li2	18f		1.93(7)	2.05(8)	2.15(7)	2.53(8)

concentrated within the M2 cavity; as the symmetry is lower ($R\bar{3}$ instead of $R\bar{3}c$, there are two independent sites with different populations.

A look at Table 15.4 shows that the coordination bonding of lithium is quite different for the Li1 and Li2 sites. Although the coordination number is four in both cases, a much more regular environment with shorter Li-O bonds appears for Li2 with respect to Li1, where the bonding is looser in the case of α-LiZr$_2$(PO$_4$)$_3$.

15.5 Conclusions

A short overview of the structural features of two of the most interesting families of Li$^+$ ion conductors, LLTO and Li-Nasicon, has been presented. Such systems are important representatives of materials used as solid electrolytes and/or cathodes in lithium ion rechargeable batteries with high energy density, where a large mobility of the Li$^+$ charge carrier is urgently needed. Indeed, new materials with this property can be designed only if there is a deep understanding of the atomistic and structural basis of the phenomenon. NPD studies of the two systems considered have provided important information on the crystal-chemical environment of lithium atoms and on their order–disorder state in ionic conductors. In particular, where large cavities are present in the structural framework, lithium tends to prefer a peripheral location with distorted fourfold coordination and partial occupation of multiple disordered sites. This brings about (1) a high fraction of vacant Li sites, (2) short jump distances between adjacent locations, and (3) a small number of Li-O bonds of medium–weak strength, implying a low activation enthalpy barrier for the Li$^+$ hopping process. On the basis of the discussion on ionic mobility in solids (Section 15.1.1), it is clear that such conditions are favorable for a high lithium ion conductivity.

By extending this kind of investigation to a larger number and structural variety of Li$^+$ ion conducting systems, an exhaustive modeling of ionic transport phenomena in such materials is expected to become possible, so as to support the design of new solid electrolytes with high performance.

References

1. J. M. Tarascon and M. Armand, Nature **414**, 359 (2001)
2. M. Wakihara, Mater. Sci. Engin. **R 33**, 109 (2001)
3. V. Thangadurai and W. J. F Weppner, Ionics **12**, 81 (2006)
4. R. J. D. Tilley, *Defect Crystal Chemistry* (Blackie, London, 1987)
5. M. Catti, in *Fundamentals of Crystallography*, edited by C. Giacovazzo (Oxford University Press, Oxford, 2002), Physical Properties of Crystals, Chapter 9, pp. 599–643
6. A. Munter, NIST Center for Neutron Research, http://www.ncnr.nist.gov/ resources/n-lengths/, Neutron News **3**(3), 29–37 (1992)
7. R. B. Von Dreele, J. D. Jorgensen, and C. G. Windsor, J. Appl. Crystallogr. **15**, 581 (1982)
8. S. Ikeda and J. M. Carpenter, Nuc. Inst. and Meth. A **239**, 536 (1985)
9. P. Thompson, D. E. Cox, and J. B. Hastings, J. Appl. Crystallogr. **20**, 79 (1987)

10. L. B. McCusker, R. B. Von Dreele, D. E. Cox, D. Louer, and P. Scardi, J. Appl. Crystallogr. **32**, 36 (1999)
11. J. Rodríguez-Carvajal, FULLPROF: A Program for Rietveld Refinement and Pattern Matching Analysis, in *Abstracts of the Satellite Meeting on Powder Diffraction of the XV Congress of the IUCr*, p. 127, Toulouse, France (1990)
12. A. C. Larson and R. B. Von Dreele, *GSAS: Generalized Structure Analysis System Manual*, Los Alamos National Laboratory Report LA-UR-86-748, Los Alamos, NM (1994)
13. F. Izumi and T. Ikeda, Mater. Sci. Forum **198**, 321–324 (2000)
14. S. Stramare, V. Thangadurai, and W. Weppner, Chem. Mater. **15**, 3974 (2003)
15. Y. Inaguma, C. Liquan, M. Itoh, T. Nakamura, T. Uchida, H. Ikuta, and M. Wakihara, Solid State Commun. **86**, 689 (1993)
16. H. Kawai and J. Kuwano, J. Electrochem. Soc. **141**, L78 (1994)
17. I.-S. Kim, T. Nakamura, Y. Inaguma, and M. Itoh, J. Solid State Chem. **113**, 281 (1994)
18. A. I. Ruiz, M. L. López, M. L. Veiga, and C. Pico, J. Solid State Chem. **148**, 329 (1999)
19. J. A. Alonso, J. Sanz, J. Santamaria, C. León, A. Varez, and M. T. Fernández-Diaz, Angew. Chem. Int. Ed. **39**, 619 (2000)
20. Y. Inaguma, T. Katsumata, M. Itoh, and Y. Morii, J. Solid State Chem. **166**, 67 (2002)
21. A. Varez, Y. Inaguma, M. T. Fernández-Diaz, J. A. Alonso, and J. Sanz, Chem. Mater. **15**, 4637 (2003)
22. M. Sommariva and M. Catti, Chem. Mater. **18**, 2411 (2006)
23. M. Catti, M. Sommariva, and R. M. Ibberson, J. Mater. Chem. **17**, 1300 (2007)
24. Y. Inaguma, T. Katsumata, M. Itoh, Y. Morii, and T. Tsurui, Solid State Ionics **177**, 3037 (2006)
25. J. Sanz, J. A. Alonso, A. Varez, and M. T. Fernández-Diaz, J. Chem. Soc., Dalton Trans. **1406** (2002)
26. J. Sanz, A. Varez, J. A. Alonso, and M. T. Fernández-Diaz, J. Solid State Chem. **177**, 1157 (2004)
27. A. Varez, M. T. Fernández-Diaz, J. A. Alonso, and J. Sanz, Chem. Mater. **17**, 2404 (2005)
28. A. Rivera and J. Sanz, Phys. Rev. B **70**, 094301 (2004)
29. O. Bohnke, H. Duroy, J. L. Fourquet, S. Ronchetti, and D. Mazza, Solid State Ionics **149**, 217 (2002)
30. A. Varez, F. Garcia-Alvarado, E. Morán, and M. A. Alario-Franco, J. Solid State Chem. **118**, 78 (1995)
31. J. L. Fourquet, H. Duroy, and M. P. Crosnier-Lopez, J. Solid State Chem. **127**, 283 (1996)
32. S. Garcia-Martin, M. A. Alario-Franco, H. Ehrenberg, J. Rodriguez-Carvajal, and U. Amador, J. Am. Chem. Soc. **126**, 3587 (2004)
33. A. M. Glazer, Acta Crystallogr. B **28**, 3384 (1972)
34. M. Yashima, M. Itoh, Y. Inaguma, and Y. Morii, J. Am. Chem. Soc. **127**, 3491 (2005)
35. D. Mazza, S. Ronchetti, O. Bohnke, H. Duroy, and J. L. Fourquet, Solid State Ionics **149**, 81 (2002)
36. T. Katsumata, Y. Inaguma, M. Itoh, and K. Kawamura, Chem. Mater. **14**, 3930 (2002)
37. Y. Maruyama, H. Ogawa, M. Kamimura, and M. Kobayashi, J. Phys. Soc. Japan **75**, 064602 (2006)
38. J. Emery, O. Bohnké, J. L. Fourquet, J. Y. Buzaré, P. Florian, and D. Massiot, J. Phys.: Condens. Matter **14**, 523 (2002)
39. M. Catti, Chem. Mater. **19**, 3963 (2007)
40. L. Sebastian and J. Gopalakrishnan, J. Mater. Chem. **13**, 433 (2003)
41. A. D. Robertson, A. R. West, and A. G. Ritchie, Solid State Ionics **104**, 1 (1997)
42. M. Catti and S. Stramare, Solid State Ionics **489**, 136–137 (2000)
43. M. Catti, A. Comotti, and S. Di Blas, Chem. Mater. **15**, 1628 (2003)
44. J. Kuwano, N. Sato, M. Kato, and K. Takano, Solid State Ionics **70/71**, 332 (1994)
45. K. Nomura, S. Ikeda, K. Ito, and H. Einaga, Solid State Ionics **61**, 293 (1993)
46. A. Aatiq, M. Ménétrier, L. Croguennec, E. Suard, and C. Delmas, J. Mater. Chem. **12**, 2971 (2002)

47. C. Masquelier, C. Wurm, J. Rodriguez-Carvajal, J. Gaubicher, and L. Nazar, Chem. Mater. **12**, 525 (2000)
48. S. Hamdoune, M. Gondrand, and D. Tran Qui, Mat. Res. Bull. **21**, 237 (1986)
49. D. Tran Qui, S. Hamdoune, J. L. Soubeyroux, and E. Prince, J. Solid State Chem. **72**, 309 (1988)
50. M. Catti, J. Solid State Chem. **156**, 305 (2001)
51. M. Sugantha and U. V. Varadaraju, Solid State Ionics **95**, 201 (1997)
52. M. Catti, A. Comotti, S. Di Blas, and R. M. Ibberson, J. Mater. Chem. **14**, 835 (2004)

Part III
Applications: Environment

Chapter 16
Application of Neutron Reflectivity for Studies of Biomolecular Structures and Functions at Interfaces

Alexander Johs, Liyuan Liang, Baohua Gu, John F. Ankner, and Wei Wang

Abstract Structures and functions of cell membranes are of central importance in understanding processes such as cell signaling, chemotaxis, redox transformation, biofilm formation, and mineralization occurring at interfaces. This chapter provides an overview of the application of neutron reflectivity (NR) as a unique tool for probing biomolecular structures and mechanisms as a first step toward understanding protein–protein, protein–lipid, and protein–mineral interactions at the membrane–substrate interfaces. Emphasis is given to the review of existing literature on the assembly of biomimetic membrane systems, such as supported membranes for NR studies, and demonstration of model calculations showing the potential of NR to elucidate molecular fundamentals of microbial cell–mineral interactions and structure–functional relationships of electron transport pathways. The increased neutron flux afforded by current and upcoming neutron sources holds promise for elucidating detailed processes such as phase separation, formation of microdomains, and membrane interactions with proteins and peptides in biological systems.

16.1 Introduction

The bacterial cell membrane not only serves as a structural framework for the cell but also represents a selective permeability barrier to the environment and accommodates transport and receptor proteins. Molecular structural changes during the interaction of biomacromolecules—including proteins, lipids, and extracellular polysaccharides—are of central importance in understanding biochemical processes, such as cell signaling, chemotaxis, biofilm formation, biomineralization, and redox reactions [55, 28]. Biofilm formation, biomineralization, and redox reactions are particularly relevant to reductive biotransformation of metals (including radionuclides) to achieve effective biological remediation on contaminated lands [44, 41]. In dissimilatory metal-reducing bacteria, these processes are mediated by bacterial

A. Johs (✉)

Environmental Sciences Division, Oak Ridge National Laboratory, 1 Bethel Valley Road, MS-6038, Oak Ridge, TN 37831-6038, USA

e-mail: johsa@ornl.gov

L. Liang et al. (eds.), *Neutron Applications in Earth, Energy and Environmental Sciences*, Neutron Scattering Applications and Techniques,
DOI 10.1007/978-0-387-09416-8_16, © Springer Science+Business Media, LLC 2009

outer-membrane-associated electron transport proteins, which interact with extra-cellular electron acceptors such as iron and manganese oxides [45]. Although many of the components involved have been identified, the mechanism of the interaction at the biological–mineral interface is not well understood [96]. The lack of knowl-edge about biomolecular structures near mineral interfaces, in particular, limits our understanding of the electron transfer processes. The difficulties lie with the chal-lenge of probing protein–protein, protein–lipid, and protein–mineral interactions in living systems. However, detailed knowledge of the underlying structural organiza-tion is pivotal for understanding the molecular mechanisms and consequently for the development of enhanced bioremediation processes.

Neutron reflectivity has proved to be an effective tool to probe the structure of biological model interfaces at sub-nanometer resolution [43, 6]. Studies show that neutron reflectivity is a superior method for delineating the organization of the adsorbed protein and for the mechanisms of protein adsorption on solid surfaces [13]. Recent advances in the development of biomimetic model systems and instru-mentation, together with the availability of intense neutron sources, established the foundations for exploration of microbial–mineral interface interactions with the neutron reflectivity technique [14, 35]. Based on these developments, novel model systems can be devised to mimic the bacterial–mineral interface. We expect to see integrated methods, combining preparation of metal oxide thin films and deposition of biomimetic lipid membranes, used to create a practical framework to investi-gate the structural properties of interfacial interactions with membrane-associated proteins.

In this chapter, the applications of neutron reflectivity for studies of biological interfaces in general are reviewed. An emphasis will be placed on the existing lit-erature about the assembly of biomimetic membrane systems for investigation via neutron reflectivity. This is illustrated by examples of application of neutron reflec-tivity to reveal the interactions of proteins with lipid membranes and solid surfaces. Finally, perspectives for the application of neutron reflectivity in the environmental sciences are discussed, demonstrated by model calculations showing the potential of this technique to elucidate important molecular fundamentals of microbial–mineral interactions in the environment.

16.2 Neutron Reflectivity

Neutron reflectivity is well suited for obtaining detailed structural information from layered systems with complex compositions and buried interfaces. Unlike X-rays, neutrons generally do not cause radiation damage to biological samples. The result of a specular neutron reflectivity experiment is a laterally averaged density profile normal to the sample surface (Fig. 16.1). The thicknesses of individual layers down to fractions of a nanometer can be accurately determined (depth profiling), as well as layer composition and interfacial mixing. Using contrast variation methods (see Section 16.2.2), most often the selective substitution of deuterium (^{2}H) for hydrogen

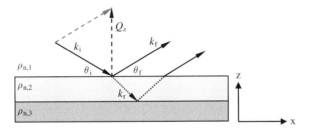

Fig. 16.1 Reflection from two interfaces; k_i and k_f are the incident and reflected wave vectors, Q_z is the momentum transfer vector perpendicular to the interface and $\rho_{n,j}$ is the neutron scattering length density of the respective layers. If the incoming wave is not totally reflected at the first interface, part of the beam k_r is refracted into the medium below the interface and, in turn, reflects and refracts from the buried interface

(^{1}H), the thickness and composition of specific interfacial layers can be detected with high accuracy. For details on the theory of neutron reflectivity and instrumentation, see Chapter 2 and Chapter 3, respectively in this volume.

For a multilayer system with n layers, the reflectivity of each interface can be described by Fresnel's relation (Eq. (2.18), Chapter 2 of this volume). Using the Parratt recursion algorithm or optical matrix methods [23], a theoretical reflectivity profile can be calculated from a stack model with n uniform layers. A starting model is typically created using a priori information about the sample under investigation, such as scattering length density and estimates for the thickness of individual layers. Additionally, a parameter accounting for interfacial roughness can be assigned to each layer and the solid substrate [56]. In an iterative procedure, the values for layer thickness, scattering length density, and volume fraction are adjusted to minimize the discrepancy between the experimental data and the calculated profile. This can be done manually or with the use of parameter optimization algorithms in combination with least squares fitting procedures like Levenberg–Marquardt or genetic algorithms. A number of analysis programs for modeling of single or multiple reflectivity data sets are available.

16.2.1 Instruments and Data Acquisition

In a neutron reflectivity experiment, the reflectivity as a function of the wave vector perpendicular to the reflecting surface can be obtained by varying either the incident angle or the wavelength. Reflectivity profiles at reactor sources are typically recorded at a fixed wavelength by scanning the incident angle θ and recording reflected intensities at the detector angle 2θ. Spallation sources operate in time-of-flight mode employing a polychromatic neutron beam. Neutrons with different energies have different wavelengths and therefore travel at different velocities:

$$\lambda = \frac{h}{mv}, \tag{16.1}$$

with m being the mass of a neutron and v its velocity. Because neutrons are produced at the same time in each neutron pulse, their energy can be determined by measuring the time of arrival at the detector (see Section 3.2 for a description of neutron sources). As a result, a broad Q-range can be covered by recording data at only a few incident angles. The resolution $\Delta Q/Q$ is derived from the angular divergence $\Delta\theta/\theta$, which depends mainly on collimation and geometry of the incident beam, and the uncertainty in the wavelength $\Delta\lambda/\lambda$ of the neutrons. Section 3.4.7 describes the characteristics of neutron reflectometers in detail including Grazing Incidence Small Angle Scattering (GISANS) and Grazing Incidence Diffraction (GID).

Reflectivity data are typically evaluated by models consisting of a stack of uniform layers. As only reflected intensity is recorded by the detector, the thickness and scattering length density of the layers in the starting model should roughly correspond to the expected result. This is possible by including a priori information about the sample, like reference layers with known thickness or scattering length density.

The reflectivity of each interface in a multilayer system with n layers can be described by the Fresnel relation:

$$F_{i,j+1} = \frac{k_j - k_{j+1}}{k_j + k_{j+1}} \qquad (j = 0, \ldots, n) \tag{16.2}$$

with k_j being the z-component of the wave vector in the layer j.

A widely used method for calculating the reflectivity of a multilayer system is the Parratt recursion algorithm [61].

$$R_{j,j+1} = \frac{F_{j,j+1} + R_{j+1}e^{2ik_{j+1}d_j}}{1 + F_{j,j+1}R_{j+1}e^{2ik_{j+1}d_j}} \tag{16.3}$$

where d_j is the thickness of layer j.

An extensive compilation of neutron reflectometers currently available worldwide, including links to data analysis software for modeling neutron reflectivity data, can be found at http://material.fysik.uu.se/Group_members/adrian/reflect.htm.

16.2.2 Contrast Variation

The goal of a neutron reflectivity experiment is to derive a coherent scattering length density profile perpendicular to the surface plane. As for any scattering method, it is not possible to record phase information for scattered neutrons. In a reflectivity experiment, multiple phase sets can correspond to a single set of amplitudes. The resulting model ambiguity can be overcome by introducing a priori information about the sample or by obtaining multiple scattering profiles from samples with varying isotopic composition. In aqueous systems this can be achieved by variation of the D_2O/H_2O ratio of the solvent, employing the large isotopic difference in the coherent scattering cross-sections of hydrogen and deuterium. The scattering length density of the solvent can be varied between $-5.60 \times 10^{-7} \text{Å}^{-2}$ and $6.36 \times 10^{-6} \text{Å}^{-2}$ by adjusting the D_2O/H_2O ratio to match the scattering of a particular component in the system. Thus, structural information can be obtained selectively either by varying the isotopic composition of components or by changing the solvent contrast. Fully and partially deuterated chemicals are commercially available, such as the most common membrane phospholipids. Sufficient quantities of deuterated proteins can be obtained by over-expression in deuterated media [91, 92, 59].

16.3 Applications of Neutron Reflectivity for Biological Interfacial Studies

16.3.1 Solid Supports

Silicon wafers are a preferred solid support for neutron reflectivity experiments because of their low absorption for neutrons, low incoherent cross-section and well-defined surface. Silicon crystals are commercially available as wafers and can be polished to have very low surface roughness (<3Å). Surface oxidation yields a reproducible hydrophilic layer of SiO_2 with a thickness of 10–20Å. It should be emphasized that properly cleaned surfaces are a stringent prerequisite for reproducible results. Residual amounts of organic materials, like traces of detergent or dust particles, can dramatically change surface properties and result in defect formation or unpredictable results. For a discussion of cleaning methods for silicon surfaces see [14]. After hydration, the surface SiO_2 layer becomes partially hydrolyzed and the isoelectric point of the silica surface is about pH 2 [60]. Therefore, most neutron reflectivity data for biological molecules at solid interfaces available today are derived from experiments at the silica–water or the silica–air interface. Other suitable materials for solid supports include quartz and sapphire.

It is possible to change surface composition and properties either by chemical modification (for an example, see Section 16.3.3.2) or by deposition of thin metal (oxide) layers, which have practical applications for studying biomolecular interactions in the environment. For example, sputtering, molecular beam epitaxy, or pulsed laser deposition are established methods of obtaining thin films on solid supports [81, 30]. A thin layer can be grown epitaxially on a crystalline solid support or deposited as a thin film, which can be polycrystalline or amorphous. The properties of the thin film, thickness, and surface roughness can be varied with deposition conditions and choice of the solid support. A requirement for high-resolution neutron reflectivity applications is to obtain highly uniform layers on large (>10 cm^2) surface areas (>10 cm^2). However, the production of such metal oxide thin films with low surface roughness and uniform thickness is not trivial.

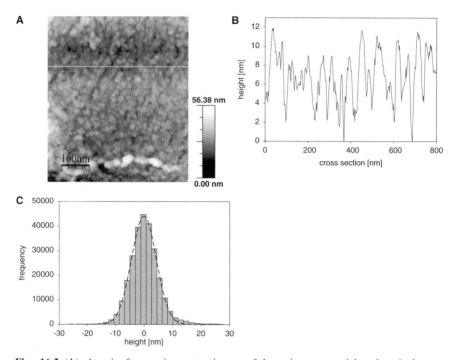

Fig. 16.2 (**A**) Atomic force microscope image of hematite nanoparticles deposited on a Formvar/carbon-coated copper EM grid by Langmuir–Blodgett (LB) deposition. The particles were treated by thermal evaporation before forming the LB film. (**B**) Cross-section profile corresponding to the horizontal line in (**A**). (**C**) Statistical roughness analysis over total image area resulting in a root mean square roughness of 5.1 nm (from Wang et al. [95])

An alternative method employs deposition of nanoparticle monolayers on a solid substrate. An example is shown in Fig. 16.2, where thin films of α-Fe_2O_3 nanoparticles have been fabricated by Langmuir–Blodgett (LB) deposition on a large surface [95].

For this particular LB film, atomic force microscopy studies showed that the LB film formed with 10-nm nanoparticles has a roughness of about 5 nm, which is significant. Well-developed methods are available to model neutron reflectivity data with rough surfaces [56]. Interfacial roughness and gradients can be experimentally quantified by recording diffuse scattering around the specular reflection. An example employing off-specular neutron reflectivity to measure buried lateral roughness is shown in Fig. 16.3. Additional information about lateral heterogeneities is essential to obtain accurate models from reflectivity data because the specular reflection contains no information about lateral organization. Generally, the roughness of the solid support should not be higher than the length scale of the layer under investigation. Although metal oxide thin films as solid supports for neutron reflectivity are relevant to studies of biomolecular interactions in subsurface environments, lateral heterogeneities and interfacial roughness can limit the accessible level of detail.

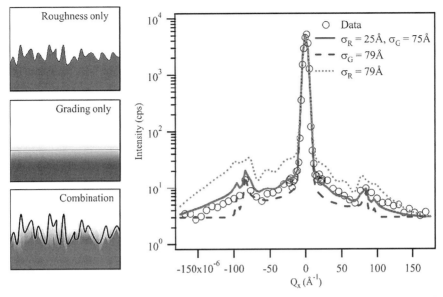

Fig. 16.3 Off-specular transverse scan and modeled data, with a combination of a smooth 75 Å wide compositional gradient and a 25-Å amplitude roughness superimposed at the leading edge of the reaction front at Q_z of 0.022 Å^{-1} with calculations for a rough interface (*dotted*), a smooth gradient interface (*dashed*), and a combination of the two (*solid*). Reprinted with permission from Lavery et al. [36]. Copyright 2008, American Institute of Physics

16.3.2 Solid-Supported Membranes for Neutron Reflectivity

Understanding the complex nature of living cell membranes can provide critical information on how cells respond to and influence their environment. One of the important applications of neutron reflectivity is the study of interfacial properties and structural arrangements in biological systems, such as cell membranes or protein adsorption at solid–liquid or solid–air interfaces. Because of the complexity of native biological membranes, it is difficult to probe these interfaces by scattering methods in vivo. Nevertheless, it is possible to address specific questions by using an appropriate simplified interface model that mimics the biological membrane as closely as possible. Model systems for living cell membranes may consist of a single lipid monolayer, a single bilayer, or multiple bilayers. The amphiphilic properties of membrane lipids, each with a hydrophobic hydrocarbons tail and a polar head group, facilitate their self-assembly into bilayer structures in aqueous environments (Fig. 16.4A–C). Many cell membrane components, such as proteins, phospholipids, sterols, and lipopolysaccharides, can be assembled or inserted into bilayer structures [53]. An important property of all biological membranes is their fluidity, by which individual components are able to diffuse freely within the membrane plane, forming functional microdomains, accommodating membrane-associated protein complexes [12].

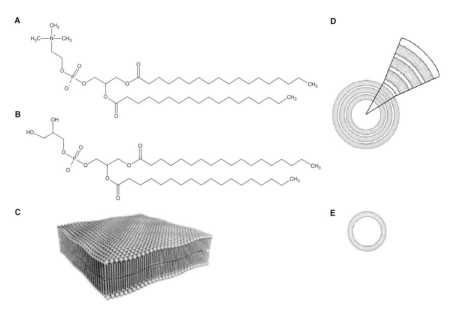

Fig. 16.4 (**A**) Example of a zwitterionic phospholipid 1,2-dipalmitoyl-*sn*-glycero-3-phosphocholine (DPPC). (**B**) An anionic phospholipid 1,2-dipalmitoyl-*sn*-glycero-3-[phospho-(1-glycerol)] (DPPG).(**C**) Illustration of a typical phospholipid bilayer. (**D**) Multilamellar vesicle. (**E**) Unilamellar vesicle

Another important aspect of the study of biomimetic membrane models is hydration and the interaction of phospholipids and lipopolysaccharides with water. Full hydration can be achieved by either 100% relative humidity or immersion in excess water. In practice, 100% humidity is difficult to achieve because it requires strict temperature and humidity control in the sample environment. This was an unresolved problem for many years and was described as the "vapor pressure paradox" [63]. Recent progress in sample cell design made it possible to overcome these technical difficulties [27, 20]. Achieving 100% humidity in a reflectivity sample environment has the advantage of better bilayer stability, particularly in the case of multilamellar systems [58, 64]. Excess water ensures the physiologically relevant condition of full hydration and offers the opportunity to modify the composition of the aqueous subphase during the experiment. This can be useful for the application of contrast variation in a neutron reflectivity experiment to overcome model ambiguity. Therefore, membrane fluidity and hydration are important parameters that need to be considered in the development of a suitable model system.

16.3.2.1 Assembly of Biomimetic Membranes on Solid Supports

The first published preparation of a planar solid-supported phospholipid bilayer membrane was performed by sequential LB deposition of two phospholipid monolayers at constant surface pressure [85]. Since then, a number of alternate preparation methods have been reported in the literature; and essential physical properties of solid supported membranes, including phase behavior, defect formation, and hydration effects, have been described [68, 73]. Supported biomimetic model membrane systems have been proved successful for the study of structures and processes at biological interfaces [72, 85].

Figure 16.5 shows some examples of phospholipid model membranes on solid supports, including single bilayers, multilamellar bilayers, polymer-cushioned bilayers, and tethered bilayers. The following sections provide a brief description of the advancement of these biomimetic models; the incorporation of proteins into the membrane systems; and the application of neutron reflectivity to analyze structure, function, and interactions among various components in the membrane and interface systems.

Solid-Supported Bilayers by Vesicle Fusion

Self-assembly of a single planar membrane bilayer by spontaneous fusion of phospholipid vesicles can be achieved on charged and hydrophilic surfaces [34, 25, 19, 93]. The steps involved in the fusion process are vesicle adsorption, rupture, and spreading into planar bilayers. Small unilamellar vesicles prepared by sonication of multilamellar vesicles or large unilamellar vesicles prepared by extrusion of

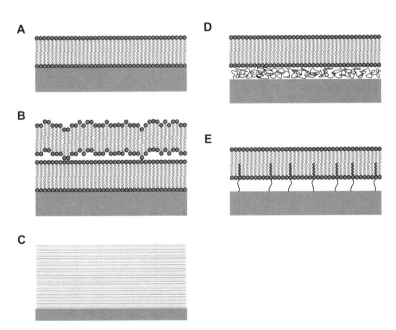

Fig. 16.5 Phospholipid model membranes on solid supports. (**A**) Single bilayer on a hydrophilic solid support. (**B**) Freely floating bilayer. (**C**) Multilamellar bilayer stack. (**D**) Polymer-cushioned bilayer. (**E**) Bilayer tethered to a gold-coated solid support by alkanethiol spacers

multilamellar vesicles through polycarbonate filters with defined pore sizes can be used for this process (Fig. 16.4D and E). Vesicle fusion is successful only for lipids above the main phase transition temperature of the lipid components used. The formation of a planar bilayer is typically complete within 20–60 min. This method has also been used to prepare uniform single phospholipid bilayers on nanoporous silica thin films [10].

A recently developed method is the deposition of phospholipid bilayers from mixed lipid-surfactant micelles in solution. Single phospholipid bilayers with high surface coverage and low numbers of defects have been obtained [90]. The method comprises two steps: in the first step, a mixed lipid-surfactant layer self-assembles on a hydrophilic solid support. In the second step, a dilution process is applied to enrich the absorbed layer in phospholipid until the surfactant is completely removed. This has been demonstrated for mixtures of dioleylphosphatidylcholine and the nonionic surfactant β-D-dodecyl maltoside. Co-adsorption with cholesterol, resulting in an increased bilayer thickness, has also been achieved [89].

Langmuir–Blodgett Deposition

The LB technique is a method of transferring monolayers of amphiphilic molecules from the liquid–air interface to a solid support (Fig. 16.6A). As mentioned earlier, the transfer of a phospholipid bilayer on solid supports by the LB method

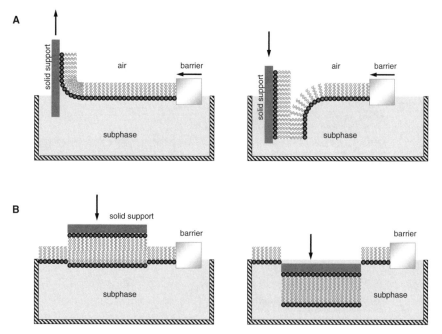

Fig. 16.6 A Illustrations of (**A**) the Langmuir–Blodgett and (**B**) the Langmuir–Schaefer method of depositing phospholipids on a solid substrate

was first demonstrated for phosphatidylcholines [85]. Most biologically relevant phospholipids can be used to deposit lamellar structures on a solid support by this technique. However, the achievable transfer efficiency depends heavily on a number of parameters, including the type of phospholipid, subphase, solid substrate surface properties, type and concentration of ions in the subphase, surface pressure, pH, and temperature. Generally, deposition of phospholipids on solid supports by LB is performed below the main phase transition temperature of the lipids at constant surface pressure.

Multilayer deposition of phospholipids by the LB method has been achieved only for a few types of phospholipids. Multilayers are most easily obtained for negatively charged phosphatidic acid using divalent cations in the subphase [42]. The LB technique (Fig. 16.6A) can be combined with other methods, like vesicle fusion or the Langmuir-Schaefer technique. Following the preparation of an LB monolayer on a solid support, small unilamellar vesicles can be fused to the hydrophobic interface of the monolayer to yield a single-supported bilayer. This method can be applied to obtain supported membranes with asymmetric lipid distribution between the membrane leaflets [26]. For a Langmuir–Schaefer deposition, the solid support with a preformed monolayer is mounted horizontally and pushed through a preformed monolayer to obtain a bilayer (Fig. 16.6B).

Freely Floating Bilayers

A disadvantage of single-supported membranes (Fig. 16.5A) is that the lower leaflet interacts with the solid support, preventing the formation of a ripple phase ($P_{\beta'}$) between the gel phase (L_{β}) and the liquid-crystalline phase (L_{α}) [54]. As a result, a decrease in the lateral diffusion rate of supported membranes by a factor of 5 has been reported [82]. However, this decrease in diffusion rate depends exclusively on the surface properties of the solid support, which is probably affecting the viscosity of the water layer between membrane and surface. It has been shown that piranha etching of glass surfaces improves diffusion rates in single phospholipid bilayers by a factor of 3, independent of the lipid used [77].

One way to avoid interaction effects with the solid support is freely floating bilayers deposited by a combination of LB and Langmuir–Schaefer techniques [5, 16, 8]. In a first step, three monolayers are transferred to a solid substrate by the LB, followed by deposition of a fourth monolayer by horizontal transfer using Langmuir–Schaefer dipping (Fig. 16.6B). It has been shown that freely floating bilayers are separated from the adsorbed bilayer by a 2–3 nm water layer, resulting in a higher membrane flexibility compared with bilayers in direct contact with the solid support (Fig. 16.5B).

Multilamellar Bilayer Stacks

Multilamellar stacks of highly oriented lipid bilayers on a solid substrate can be easily obtained by spreading the lipids from organic solution or spin-coating on a hydrophilic solid substrate [50]. Reflectivity from planar membrane arrays is equivalent to diffraction from a one-dimensional crystal as a result of Bragg reflections from the membrane planes, and is therefore useful for studying the bilayer structure perpendicular to the surface with high accuracy (Fig. 16.5C). Additional information on lateral membrane organization can be obtained from off-specular measurements. However, the intensity of off-specular scattering is generally very weak compared with the specular signal. Therefore, most in-plane scattering data on model membranes have originated from X-ray scattering experiments, which offer the advantage of high photon flux. With the advent of high-intensity neutron sources, in-plane neutron scattering may become more easily feasible in the near future. Although it is possible to reconstitute membrane proteins into multilamellar solid-supported bilayer stacks, these model systems typically offer no advantage for probing protein–lipid interactions in situ as a result of the stack structure.

Tethered Bilayers and Polymer-Cushioned Bilayers

Efforts to improve the stability of solid-supported bilayer systems led to the development of polymer-cushioned bilayers (Fig. 16.5D) and tethered bilayers (Fig. 16.5E). Long-term stability and better control over bilayer assembly is particularly important for the development of lipid bilayer–based biosensor devices, and may also be useful for the development of biological fuel cells in the future.

The advantage of tethered bilayers is an enhanced mechanical and chemical stability based on the covalent bonds between some of the lipid molecules and a tethering linker. This type of bonding also causes a spatial decoupling of membrane and substrate, allowing the incorporation of even large integral membrane proteins [32, 65]. Polymer-cushioned lipid bilayers are obtained by grafting a hydrophilic polymer film to a solid substrate and subsequent deposition of a lipid bilayer by either vesicle fusion or transfer of LB and Langmuir–Schaefer monolayers [97, 74, 94]. Applications of polymer-cushioned bilayer membranes may include the design of cell surface models to study the structure and function of membrane-associated proteins, cell adhesion, and membrane transport processes.

16.3.2.2 Reconstitution of Membrane Proteins

Purification of membrane proteins in the native state requires the use of mild detergents to screen hydrophobic parts to prevent unspecific aggregation in solution and protein denaturation [33, 37]. The functional reconstitution of these proteins is not trivial, and the process must be carefully adapted to the particular protein and detergent used. A variety of methods of studying the interaction of proteins with lipids has been described in the literature, including direct attachment of proteins to biomimetic membranes [94, 17], reconstitution into proteoliposomes by detergent removal from ternary protein–lipid–surfactant mixtures [69, 76], dilution methods [67], and two-dimensional crystallization [22, 39, 38, 80]. A common difficulty in the reconstitution process is related to the orientation of reconstituted proteins in the membrane bilayer. A practicable way to control asymmetry is the fusion of proteoliposomes to a preformed lipid monolayer [24, 66].

16.3.2.3 Structure of Bacterial Lipopolysaccharides

Lipopolysaccharides (LPS) constitute the external bilayer leaflet in the outer membrane of Gram-negative bacteria [4]. The LPS molecule consists of a polysaccharide covalently linked to a hydrophobic membrane anchor, lipid A. Because of their role as stimulators of immune response in mammalian species, the immunological implications of LPS from pathogenic bacteria are of particular interest [87]. However, little is known so far about the role of LPS with respect to molecular details of nutrient transport and mechanisms of cell adhesion to solid interfaces. Recent neutron diffraction studies (Fig. 16.7) on LPS bilayer stacks attempted to determine their permeability for water and small molecules in the hydrocarbon region [1]. The scattering length density profile derived from neutron scattering experiments shows that water permeates the entire length of the LPS bilayer, including the bilayer center. This finding has implications for the penetration of small molecules into and across the outer membrane throughout the bilayer, even though the distribution of water along the bilayer cannot be quantitatively described (Fig. 16.7).

The experiments also indicated distinct effects of monovalent and multivalent cations on structural stability and water permeability. Future integration of LPS in bacterial model membrane systems for neutron reflectivity studies has the potential to help clarify the role of LPS at the bacterial–mineral interface and to aid the study of interactions of bacterial outer membrane proteins in their native environment.

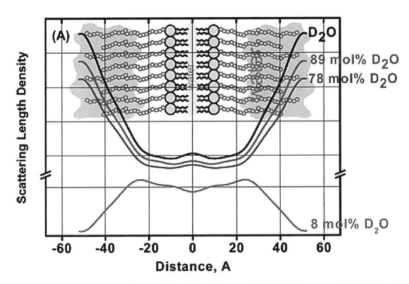

Fig. 16.7 Scattering length density profiles of liquid-crystalline LPS bilayers at 84% RH and 51°C. The 1D SLD profiles for all D_2O compositions, namely 100, 89, 78, and 8 mol% D_2O, are shown. For each profile, the center of the bilayer was placed at the origin. The bilayer is formed by two LPS monolayers with their hydrophilic groups on the outside and the hydrophobic chains of the two monolayers apposing each other (center of bilayer). For clarity, the profiles have been shifted with respect to each other. D_2O seems to penetrate the entire length of the bilayer. Reprinted with permission from Abraham et al. [1]. Copyright 2007 American Chemical Society

16.3.3 Probing Structural Properties of Proteins at Interfaces

Protein adsorption and transport at interfaces can be studied in detail by neutron reflectivity. Adsorption is often connected with major conformational changes in the structure or denaturation of adsorbed proteins [18, 47]. Investigating the mechanism of adsorption and associated structural changes is essential for understanding microbial–mineral interactions, cell adhesion, and many technological applications, such as biosensor development or food processing. It can also help in finding the strategies to prevent undesirable formation of protein layers that affect the performance of filters, membranes, reusable medical devices, or surgical implants [2]. The interaction of amphiphilic proteins is particularly difficult to predict because of the variability in the distribution of hydrophobic, ionic, and polar regions, which are frequently combined with flexible domains in the macromolecule. Interfacial effects—like surface charge, surface tension, or surface chemistry—can affect protein–surface or protein–protein interactions and their stability [83, 84]. Unlike many spectroscopic methods, neutron reflectivity can provide detailed information about the structure of protein layers at interfaces [49].

16.3.3.1 Proteins at Fluid Interfaces

The most frequently studied fluid interface relevant to proteins is the air/liquid interface.

Early neutron reflectivity studies on protein conformation and packing at an air/water interface were performed on bovine serum albumin (BSA) [11]. It was shown that the macromolecules orient themselves with the long axis parallel to the interface and the hydrophilic regions toward the aqueous phase. Additional studies on BSA confirmed this orientation along the interface and revealed a slight increase in the protein layer thickness, indicating a small conformational change at higher bulk concentrations [48]. Similar fluid interface studies on β-casein, which is an amphiphilic and highly flexible protein, revealed a two-layer model consisting of a compact layer and a more diffuse layer extending into the subphase [9, 3]. Studies of the globular protein lysozyme at the air/water interface revealed that this protein is only partially immersed into the subphase [49]. The spatial orientation at the interface was found to be dependent on the overall protein concentration in the aqueous phase (Fig. 16.8). Interfacial effects on quaternary structure were investigated on the dimeric glycoprotein glucose oxidase [18, 47]. At low ionic strengths, glucose oxidase monomers assemble into a uniform layer at the air/water interface; whereas at high ionic strengths, neutron reflectivity data indicate that the dimeric complex becomes adsorbed at the interface.

16.3.3.2 Proteins at Solid–Water Interfaces

The interaction of proteins with oxide surfaces is mainly governed by electrostatic forces, which are affected by pH and ionic strength. This was demonstrated in neutron reflectivity studies of the adsorption of the highly positively charged protein

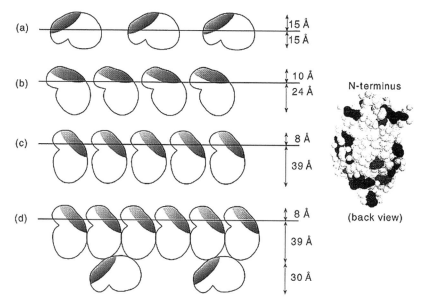

Fig. 16.8 Orientation and packing of lysozyme molecules at the air/water interface as a function of bulk concentration (**a–d**) as determined by neutron reflectivity. Lu et al. [49] - Reproduced by permission of The Royal Society of Chemistry. http://dx.doi.org/10.1039/a805731a

lysozyme, with an isoelectric point of 11, to a negatively charged silica surface. At neutral pH and low ionic strength, it was found that multiple layers of protein adsorb irreversibly to the surface while tertiary structure is maintained [83]. The addition of ions above a certain threshold concentration reduces adsorption at the solid interface, indicating that screening of charges by small ions can prevent irreversible adsorption. The effects of pH are more pronounced for proteins in solution. The adsorbed amount of protein is generally highest around the isoelectric point of the protein. In the case of lysozyme, a lower pH leads to an increased protonation of the protein, which results in a higher intermolecular repulsion that lowers the amounts of proteins being adsorbed to the surface. It was also shown by neutron reflectivity that the orientation of lysozyme molecules attached to the surface is governed by pH and protein concentration [84].

Adsorption of β-casein to hydrophilic silica is a significantly slower process [86]. The authors obtained reflectivity vs the momentum transfer curves in buffered D_2O solution before and after the injection of β-casein solution into the measurement cell. The reflectivity profiles changed considerably with time after the addition of β-casein into the sample cell. The reflectivity in the low-momentum transfer region decreased significantly, and the destructive interference moved in the direction of lower momentum transfer values with increasing adsorption time. The slow buildup of the layer most likely was caused by substantial structural rearrangement in the interfacial region. In the absence of calcium ions, the proteins self-assembled in an asymmetric bilayer structure, which was much more loosely attached to the hydrophilic surface than a β-casein monolayer at hydrophobic surfaces [86].

Neutrons also revealed the adsorption properties of human serum albumin (HSA) and fibrinogen on quartz and silica [2, 13]. These abundant components of human blood are relevant for medical devices containing glass parts, which are designed to handle blood or plasma. The adsorption characteristics of the purified protein components show that at neutral pH, fibrinogen assembles into a distinct layer that extends far into the bulk solution, whereas HSA adsorbs weakly to the surface. The results underscore the complex behavior of biological macromolecules at solid interfaces.

16.3.3.3 Proteins at Self-Assembled Monolayers

Self-assembled monolayers (SAMs) greatly expand the range of applications for the study of proteins by neutron reflectivity. SAMs are highly ordered two-dimensional layers formed on a solid surface by spontaneous organization of functional molecules from solution. Chemical modification of surfaces by alkyl trichlorosilanes or alkanethiols allows tailoring of surface properties, such as hydrophobicity, surface energy, and charge [88, 75].

This was demonstrated in a study to investigate the adsorption of β-casein to a silicon oxide surface rendered hydrophobic by deuterated octadecyltrichlorosilane (d-OTS) [15, 57]. The amphiphilic nature of casein leads to an oriented adsorption at the hydrophobic interface. It was shown that the hydrophobic portion of the β-casein forms a dense monolayer in close proximity to the d-OTS layer, whereas the hydrophilic portion extends into the aqueous subphase (Fig. 16.9).

The following example is remarkable, because it underscores the unique capability of neutron reflectivity by employing a model-independent method to retrieve phase information. Monolayers of yeast cytochrome c were covalently tethered to functionalized polar and nonpolar SAMs [31]. Well-defined thin films on the solid

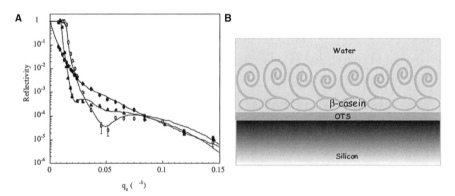

Fig. 16.9 (**A**) Neutron reflectivity profiles of β-casein adsorbed to a layer of d-OTS in (o) D_2O, (•) H_2O, and (▲) $D_2O:H_2O$ mixture, with an SLD of 4.5×10^{-6} Å^{-1} (73% D_2O) to reveal larger scale details, which are not visible because of the higher critical edge of 100% D_2O. (**B**) Schematic representation of a β-casein layer adsorbed at a hydrophobic OTS-modified silicon substrate. Reprinted with permission from Fragneto-Cusani [14]. Copyright 2001 IOP Publishing Ltd.

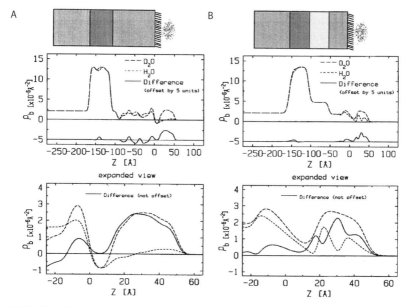

Fig. 16.10 The absolute neutron scattering length density profiles for partial hydration with D_2O and H_2O and their difference profile for both the nonpolar SAM (**A**) and the uncharged polar SAM (**B**) cases. Schematics of the composite structures are shown above their respective scattering length density profiles approximately to scale. Reprinted with permission from Kneller et al. [31]. Copyright 2001, Biophysical Society

support (Fe/Si and Fe/Au/Si, respectively) were used as a reference structure to analyze neutron reflectivity data by an interferometric phasing method. The results revealed water distribution profiles for the monolayers of oriented yeast cytochrome c with very high accuracy (Fig. 16.10).

16.4 Bacterial–Mineral Interface Models

Because of the unique properties of neutrons and the availability of numerous methods to prepare biologically relevant model interfaces, neutron reflectivity can contribute significantly to obtaining structural information about processes at the microbial–mineral interface. The supramolecular organization of electron transport proteins is considered critical for understanding dissimilatory metal reduction in bacteria. Neutron reflectivity has the potential to provide answers to currently challenging questions related to bacterial membrane organization and protein interaction mechanisms with solid oxide surfaces. Figure 16.11 shows the hypothesized localization of components in the microbial electron transport system from *Shewanella oneidensis* [96, 71]. A cytoplasmic membrane-associated cytochrome CymA [51] accepts electrons from the menaquinone pool and serves as an electron donor for a small periplasmic shuttle Stc [40] and/or a decaheme cytochrome MtrA [62].

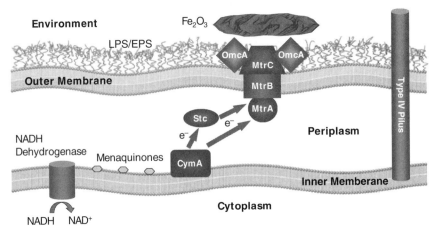

Fig. 16.11 Hypothesized mechanism of electron transfer at the outer membrane of microbial cells in *Shewanella oneidensis* MR-1

Finally, electrons are transferred to the outer membrane–associated cytochromes, OmcA and MtrC [78, 21], which were postulated to constitute the terminal Fe(III) reductase forming a functional complex located in the lipopolysaccharide leaflet of the outer membrane [52, 78, 98, 46, 71, 79].

For modeling microbe–minerals interaction, the major functional components include the mineral surface and the bacterial membrane forming a barrier and a framework for membrane-associated proteins. Iron oxides are among the most abundant minerals in soil environments. Iron atoms can assume multiple oxidation states and form a number of different crystalline mineral oxide phases and structures with varying thermodynamic properties [7]. Studies on the bioavailability of different iron oxide minerals reveal that the rate of reduction by Fe(III)-reducing bacteria is related to the accessible surface area [70]. The prevalence of micro-heterogeneities

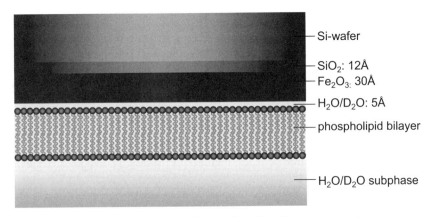

Fig. 16.12 Idealized layer model of a membrane–mineral interface in excess water

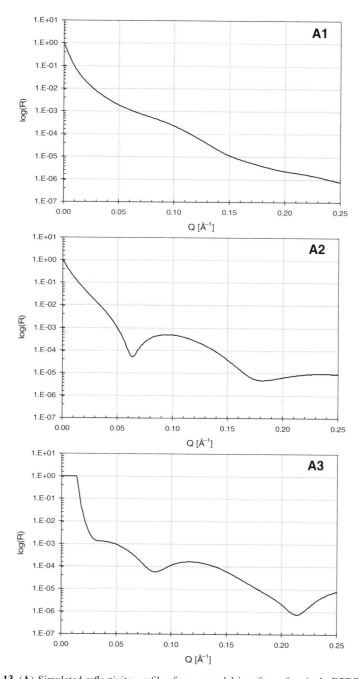

Fig. 16.13 (**A**) Simulated reflectivity profiles from a model interface of a single DPPC bilayer deposited on top of an iron oxide layer in excess water using the Parratt recursion algorithm. Assumed interfacial roughness: 5 Å. (**B**) Scattering length density profile derived from (**A**). (**A1/B1**) Hydrogenated tails, hydrogenated headgroups, 100% H_2O; (**A2/B2**) deuterated tails, deuterated headgroups, 100% H_2O; (**A3/B3**) deuterated tails, hydrogenated headgroups, 100% D_2O (See text for details)

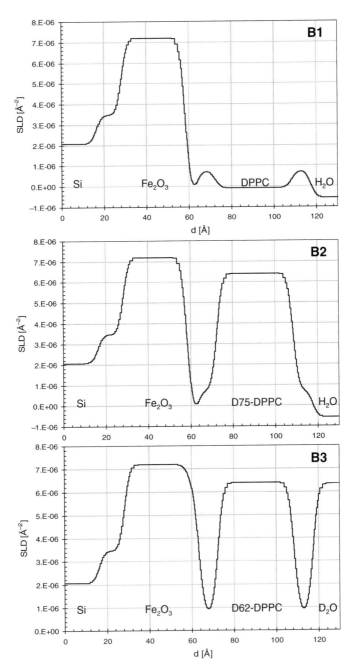

Fig. 16.13 (*Continued*)

and crystalline disorder has been found to increase the microbial reduction rate [99]. Molecular parameters determining the rate of electron transfer across the cytochrome–iron oxide interface have been evaluated using molecular dynamics simulations [29]. The authors report that the electron-transfer rate depends strongly on the configuration of heme moieties near the interface. It has been shown also that hydrogen bond acceptors in proximity to the heme confer the interaction between the protein and the mineral surface. However, there are no experimental data yet available describing the conformation of relevant cytochromes in contact with a mineral surface.

In a model system suitable for neutron reflectivity studies (Fig. 16.12), the mineral surface is ideally represented by a thin layer of the metal oxide (see Section 16.3.1).

A conceivable system consists of (1) a silicon wafer as the solid support, (2) a thin silicon oxide layer with (3) an epitaxially grown thin film of hematite, α-Fe_2O_3, (4) a thin water layer expected to form on top of the hematite, (5) a phospholipid bilayer (DPPC), and (6) the aqueous subphase. The scattering length density within a phospholipid bilayer can be further resolved into phospholipid headgroups and aliphatic tails. Figure 16.13(A1–A3) depicts simulated reflectivity profiles at three different neutron contrast conditions. The reflectivity of a sample containing hydrogenated materials exhibits only a few characteristic features, particularly at low Q. Thus crucial structural information is buried in the high-Q portion of the reflectivity profile, which is hardly accessible in practice because of incoherent background from hydrogen impairing the signal-to-noise ratio. The use of deuterated lipids and/or a deuterated subphase significantly improves the contrast situation. Figure 16.13(B1–B3) demonstrates the remarkable effect on scattering length density profiles. Thickness and position relative to the solid support of all components can be unambiguously assigned, and the 3-Å water layer between the DPPC bilayer and the mineral surface can be resolved.

Model calculations shown in Fig. 16.13 demonstrate that it is possible to use neutron reflectivity to study the interaction of biological structures and oxide mineral surfaces with high accuracy. However, the uniformity and interfacial roughness of the oxide layer need to be carefully controlled and should be taken into account. Current research efforts aim to characterize model systems that represent suitable models for the study of relevant structural aspects within the complex microbial–mineral interface.

16.5 Conclusion

Early research applying neutrons to study the structure of interfaces was limited to fairly simple systems. Nevertheless, these studies have already unveiled a wealth of new insights into membrane structure and dynamics and led to the development of novel biomimetic model membrane systems, which facilitate research on membrane biophysics and protein–lipid interactions at a molecular scale. The availability of biomimetic model systems for bacterial membranes is essential for investigating the molecular principles involved in microbial–mineral interactions. Current research efforts are directed toward the development of complex interface models suitable for the investigation by neutron reflectivity. Incorporation of charged lipids and lipopolysaccharides will greatly extend the applicability of solid-supported model membranes to the study of bacterial interfaces. The increased neutron flux provided by current and upcoming neutron sources opens up new possibilities for studying in-plane structure by off-specular scattering. This is particularly useful for the investigation of highly relevant processes such as phase separation, formation of microdomains, and membrane interaction of proteins and peptides. Detailed knowledge of membrane organization will help elucidate structure–function relationships of electron transport pathways and can facilitate the development of bio-inspired technologies such as biological fuel cells or biosensor devices.

Acknowledgments This research was supported in part by the Laboratory Directed Research and Development Program of the Oak Ridge National Laboratory (ORNL) and the Office of Biological and Environmental Research, U.S. Department of Energy (DOE). ORNL is managed by UT-Battelle, LLC, for DOE under Contract No. DE-AC05-00OR22725.

References

1. Abraham T, Schooling SR, Nieh MP, Kucerka N, Beveridge TJ, Katsaras J (Neutron diffraction study of Pseudomonas aeruginosa lipopolysaccharide bilayers. The Journal of Physical Chemistry 111:2477–2483, 2007).
2. Armstrong J, Salacinski HJ, Mu QS, Seifalian AM, Peel L, Freeman N, Holt CM, Lu JR (Interfacial adsorption of fibrinogen and its inhibition by RGD peptide: a combined physical study. Journal of Physics-Condensed Matter 16:S2483–S2491, 2004).
3. Atkinson PJ, Dickinson E, Horne DS, Richardson RM (Neutron reflectivity of adsorbed beta-casein and beta-lactoglobulin at the air/water interface. Journal of the Chemical Society-Faraday Transactions 91:2847–2854, 1995).
4. Beveridge TJ (Structures of gram-negative cell walls and their derived membrane vesicles. Journal of Bacteriology 181:4725–4733, 1999).
5. Charitat T, Bellet-Amalric E, Fragneto G, Graner F (Adsorbed and free lipid bilayers at the solid–liquid interface. European Physical Journal B 8:583–593, 1999).
6. Claesson PM, Blomberg E, Froberg JC, Nylander T, Arnebrant T (Protein interactions at solid surfaces. Advances in Colloid and Interface Science 57:161–227, 1995).
7. Cornell RM, Schwertmann U (eds.) The Iron Oxides: Structure, Properties, Reactions, Occurrences and Uses. Wiley-VCH: New York (2003).
8. Daillant J, Bellet-Amalric E, Braslau A, Charitat T, Fragneto G, Graner F, Mora S, Rieutord F, Stidder B (Structure and fluctuations of a single floating lipid bilayer. Proceedings of the National Academy of Sciences of the United States of America 102:11639–11644, 2005).
9. Dickinson E, Horne DS, Phipps JS, Richardson RM (A neutron reflectivity study of the adsorption of beta-casein at fluid interfaces. Langmuir 9:242–248, 1993).
10. Doshi DA, Dattelbaum AM, Watkins EB, Brinker CJ, Swanson BI, Shreve AP, Parikh AN, Majewski J (Neutron reflectivity study of lipid membranes assembled on ordered nanocomposite and nanoporous silica thin films. Langmuir 21:2865–2870, 2005).
11. Eaglesham A, Herrington TM, Penfold J (A neutron reflectivity study of a spread monolayer of bovine serum-albumin. Colloids and Surfaces 65:9–16, 1992).
12. Engelman DM (Membranes are more mosaic than fluid. Nature 438:578–580, 2005).

13. Forciniti D, Hamilton WA (Surface enrichment of proteins at quartz/water interfaces: a neutron reflectivity study. Journal of Colloid and Interface Science 285:458–468, 2005).
14. Fragneto-Cusani G (Neutron reflectivity at the solid/liquid interface: examples of applications in biophysics. Journal of Physics-Condensed Matter 13:4973–4989, 2001).
15. Fragneto G, Thomas RK, Rennie AR, Penfold J (Neutron reflection study of bovine beta-casein adsorbed on OTS self-assembled monolayers. Science 267:657–660, 1995).
16. Fragneto G, Charitat T, Graner F, Mecke K, Perino-Gallice L, Bellet-Amalric E (A fluid floating bilayer. Europhysics Letters 53:100–106, 2001).
17. Gamsjaeger R, Johs A, Gries A, Gruber HJ, Romanin C, Prassl R, Hinterdorfer P (Membrane binding of beta2-glycoprotein I can be described by a two-state reaction model: an atomic force microscopy and surface plasmon resonance study. The Biochemical Journal 389: 665–673, 2005).
18. Georganopoulou DG, Williams DE, Pereira CM, Silva F, Su TJ, Lu JR (Adsorption of glucose oxidase at organic-aqueous and air-aqueous interfaces. Langmuir 19:4977–4984, 2003).
19. Gutberlet T, Steitz R, Fragneto G, Klosgen B (Phospholipid bilayer formation at a bare Si surface: a time-resolved neutron reflectivity study. Journal of Physics-Condensed Matter 16:S2469–S2476, 2004).
20. Harroun TA, Fritzsche H, Watson MJ, Yager KG, Tanchak OM, Barrett CJ, Katsaras J (Variable temperature, relative humidity (0%–100%), and liquid neutron reflectometry sample cell suitable for polymeric and biomimetic materials. Review of Scientific Instruments 76: 065101, 2005).
21. Hartshorne RS, Jepson BN, Clarke TA, Field SJ, Fredrickson J, Zachara J, Shi L, Butt JN, Richardson DJ (Characterization of *Shewanella oneidensis* MtrC: a cell-surface decaheme cytochrome involved in respiratory electron transport to extracellular electron acceptors. Journal of Biological Inorganic Chemistry 12:1083–1094, 2007).
22. Hasler L, Heymann JB, Engel A, Kistler J, Walz T (2D crystallization of membrane proteins: rationales and examples. Journal of Structural Biology 121:162–171, 1998).
23. Heavens OS. Optical Properties of Thin Solid Films. London: Butterworths Scientific Publications, Ltd. (1955)
24. Hinterdorfer P, Baber G, Tamm LK (Reconstitution of membrane fusion sites. A total internal reflection fluorescence microscopy study of influenza hemagglutinin-mediated membrane fusion. The Journal of Biological Chemistry 269:20360–20368, 1994).
25. Jass J, Tjarnhage T, Puu G (From liposomes to supported, planar bilayer structures on hydrophilic and hydrophobic surfaces: an atomic force microscopy study. Biophysical Journal 79:3153–3163, 2000).
26. Kalb E, Frey S, Tamm LK (Formation of supported planar bilayers by fusion of vesicles to supported phospholipid monolayers. Biochimica et Biophysica Acta 1103:307–316, 1992).
27. Katsaras J, Watson MJ (Sample cell capable of 100% relative humidity suitable for X-ray diffraction of aligned lipid multibilayers. Review of Scientific Instruments 71: 1737–1739, 2000).
28. Kent AD, Triplett EW (Microbial communities and their interactions in soil and rhizosphere ecosystems. Annual Review of Microbiology 56:211–236, 2002).
29. Kerisit S, Rosso KM, Dupuis M, Valiev M (Molecular computational investigation of electron-transfer kinetics across cytochrome-iron oxide interfaces. Journal of Physical Chemistry C 111:11363–11375, 2007).
30. Kim YJ, Gao Y, Chambers SA (Selective growth and characterization of pure, epitaxial alpha-Fe2O3(0001) and Fe3O4(001) films by plasma-assisted molecular beam epitaxy. Surface Science 371:358–370, 1997).
31. Kneller LR, Edwards AM, Nordgren CE, Blasie JK, Berk NF, Krueger S, Majkrzak CF (Hydration state of single cytochrome c monolayers on soft interfaces via neutron interferometry. Biophysical Journal 80:2248–2261, 2001).
32. Knoll W, Frank CW, Heibel C, Naumann R, Offenhausser A, Ruhe J, Schmidt EK, Shen WW, Sinner A (Functional tethered lipid bilayers. Journal of Biotechnology 74:137–158, 2000).

33. Koebnik R, Locher KP, Van Gelder P (Structure and function of bacterial outer membrane proteins: barrels in a nutshell. Molecular Microbiology 37:239–253, 2000).

34. Koenig BW, Kruger S, Orts WJ, Majkrzak CF, Berk NF, Silverton JV, Gawrisch K (Neutron reflectivity and atomic force microscopy studies of a lipid bilayer in water adsorbed to the surface of a silicon single crystal. Langmuir 12:1343–1350, 1996).

35. Krueger S (Neutron reflection from interfaces with biological and biomimetic materials. Current Opinion in Colloid & Interface Science 6:111–117, 2001).

36. Lavery KA, Prabhu VM, Lin EK, Wu WL, Satija SK, Choi KW, Wormington M (Lateral length scales of latent image roughness as determined by off-specular neutron reflectivity. Applied Physics Letters 92:064106, 2008).

37. le Maire M, Champeil P, Moller JV (Interaction of membrane proteins and lipids with solubilizing detergents. Biochimica et Biophysica Acta 1508:86–111, 2000).

38. Levy D, Chami M, Rigaud JL (Two-dimensional crystallization of membrane proteins: the lipid layer strategy. FEBS Letters 504:187–193, 2001).

39. Levy D, Mosser G, Lambert O, Moeck GS, Bald D, Rigaud JL (Two-dimensional crystallization on lipid layer: a successful approach for membrane proteins. Journal of Structural Biology 127:44–52, 1999).

40. Leys D, Meyer TE, Tsapin AS, Nealson KH, Cusanovich MA, Van Beeumen JJ (Crystal structures at atomic resolution reveal the novel concept of "electron-harvesting" as a role for the small tetraheme cytochrome c. The Journal of Biological Chemistry 277: 35703–35711, 2002).

41. Lloyd JR, Renshaw JC (Bioremediation of radioactive waste: radionuclide-microbe interactions in laboratory and field-scale studies. Current Opinion in Biotechnology 16: 254–260, 2005).

42. Lohner K, Konovalov OV, Samoilenko II, Myagkov IV, Troitzky VI, Berzina TI (Preparation and structural analysis of Langmuir-Blodgett films of acidic and zwitterionic phospholipids. Thin Solid Films 288:262–267, 1996).

43. Losche M, Piepenstock M, Diederich A, Grunewald T, Kjaer K, Vaknin D (Influence of surface chemistry on the structural organization of monomolecular protein layers adsorbed to functionalized aqueous interfaces. Biophysical Journal 65:2160–2177, 1993).

44. Lovley DR (Cleaning up with genomics: applying molecular biology to bioremediation. Nature Reviews 1:35–44, 2003).

45. Lovley DR, Holmes DE, Nevin KP (Dissimilatory Fe(III) and Mn(IV) reduction. Advances in Microbial Physiology 49:219–286, 2004).

46. Lower BH, Shi L, Yongsunthon R, Droubay TC, McCready DE, Lower SK (Specific bonds between an iron oxide surface and outer membrane cytochromes MtrC and OmcA from *Shewanella oneidensis* MR-1. Journal of Bacteriology 189:4944–4952, 2007).

47. Lu JR, Su TJ, Thomas RK, Penfold J, Webster J (Structural conformation of lysozyme layers at the air/water interface studied by neutron reflection. Journal of the Chemical Society-Faraday Transactions 94:3279–3287, 1998).

48. Lu JR, Su TJ, Thomas RK (Structural conformation of bovine serum albumin layers at the air-water interface studied by neutron reflection. Journal of Colloid and Interface Science 213:426–437, 1999).

49. Lu JR, Su TJ, Georganopoulou D, Williams DE (Interfacial dissociation and unfolding of glucose oxidase. Journal of Physical Chemistry B 107:3954–3962, 2003).

50. Mennicke U, Salditt T (Preparation of solid-supported lipid bilayers by spin-coating. Langmuir 18:8172–8177, 2002).

51. Myers JM, Myers CR (Role of the tetraheme cytochrome CymA in anaerobic electron transport in cells of *Shewanella putrefaciens* MR-1 with normal levels of menaquinone. Journal of Bacteriology 182:67–75, 2000).

52. Myers JM, Myers CR (Role for outer membrane cytochromes OmcA and OmcB of *Shewanella putrefaciens* MR-1 in reduction of manganese dioxide. Applied and Environmental Microbiology 67:260–269, 2001).

53. Nagle JF, Tristram-Nagle S (Structure of lipid bilayers. Biochimica et Biophysica Acta 1469:159–195, 2000).
54. Naumann C, Brumm T, Bayerl TM (Phase-transition behavior of single phosphatidylcholine bilayers on a solid spherical support studied by Dsc, Nmr and Ft-Ir. Biophysical Journal 63:1314–1319, 1992).
55. Nealson KH, Saffarini D (Iron and manganese in anaerobic respiration: environmental significance, physiology, and regulation. Annual Review of Microbiology 48:311–343, 1994).
56. Nevot L, Croce P (Characterization of surfaces by grazing X-ray reflection—application to study of polishing of some silicate-glasses. Revue De Physique Appliquee 15: 761–779, 1980).
57. Nylander T, Tiberg F, Su TJ, Lu JR, Thomas RK (Beta-casein adsorption at the hydrophobized silicon oxide-aqueous solution interface and the effect of added electrolyte. Biomacromolecules 2:278–287, 2001).
58. Pabst G, Katsaras J, Raghunathan VA (Enhancement of steric repulsion with temperature in oriented lipid multilayers. Physical Review Letters 88:128101, 2002).
59. Paliy O, Bloor D, Brockwell D, Gilbert P, Barber J (Improved methods of cultivation and production of deuteriated proteins from *E. coli* strains grown on fully deuteriated minimal medium. Journal of Applied Microbiology 94:580–586, 2003).
60. Parks GA (Isoelectric points of solid oxides solid hydroxides and aqueous hydroxo complex systems. Chemical Reviews 65:177–197, 1965).
61. Parratt LG (Surface studies of solids by total reflection of x-rays. Physics Review 95:359, 1954).
62. Pitts KE, Dobbin PS, Reyes-Ramirez F, Thomson AJ, Richardson DJ, Seward HE (Characterization of the *Shewanella oneidensis* MR-1 decaheme cytochrome MtrA: expression in *Escherichia coli* confers the ability to reduce soluble Fe(III) chelates. The Journal of Biological Chemistry 278:27758–27765, 2003).
63. Podgornik R, Parsegian VA (On a possible microscopic mechanism underlying the vapor pressure paradox. Biophysical Journal 72:942–952, 1997).
64. Pozo-Navas B, Raghunathan VA, Katsaras J, Rappolt M, Lohner K, Pabst G (Discontinuous unbinding of lipid multibilayers. Physical Review Letters 91:028101, 2003).
65. Purrucker O, Fortig A, Jordan R, Tanaka M (Supported membranes with well-defined polymer tethers—incorporation of cell receptors. Chemphyschem 5:327–335, 2004).
66. Puu G, Artursson E, Gustafson I, Lundstrom M, Jass J (Distribution and stability of membrane proteins in lipid membranes on solid supports. Biosensors and Bioelectronics 15: 31–41, 2000).
67. Remigy HW, Caujolle-Bert D, Suda K, Schenk A, Chami M, Engel A (Membrane protein reconstitution and crystallization by controlled dilution. FEBS Letters 555:160–169, 2003).
68. Richter RP, Berat R, Brisson AR (Formation of solid-supported lipid bilayers: an integrated view. Langmuir 22:3497–3505, 2006).
69. Rigaud JL, Pitard B, Levy D (Reconstitution of membrane proteins into liposomes: application to energy-transducing membrane proteins. Biochimica et Biophysica Acta 1231: 223–246, 1995).
70. Roden EE, Zachara JM (Microbial reduction of crystalline iron(III) oxides: influence of oxide surface area and potential for cell growth. Environmental Science and Technology 30: 1618–1628, 1996).
71. Ross DE, Ruebush SS, Brantley SL, Hartshorne RS, Clarke TA, Richardson DJ, Tien M (Characterization of protein–protein interactions involved in iron reduction by *Shewanella oneidensis* MR-1. Applied and Environmental Microbiology 73:5797–5808, 2007).
72. Rossi C, Chopineau J (Biomimetic tethered lipid membranes designed for membrane-protein interaction studies. European Biophysics Journal 36:955–965, 2007).
73. Sackmann E (Supported membranes: scientific and practical applications. Science 271: 43–48, 1996).
74. Sackmann E, Tanaka M (Supported membranes on soft polymer cushions: fabrication, characterization and applications. Trends in Biotechnology 18:58–64, 2000).

75. Schreiber F (Self-assembled monolayers: from 'simple' model systems to biofunctionalized interfaces. Journal of Physics-Condensed Matter 16:R881–R900, 2004).

76. Seddon AM, Curnow P, Booth PJ (Membrane proteins, lipids and detergents: not just a soap opera. Biochimica et Biophysica Acta 1666:105–117, 2004).

77. Seu KJ, Pandey AP, Haque F, Proctor EA, Ribbe AE, Hovis JS (Effect of surface treatment on diffusion and domain formation in supported lipid bilayers. Biophysical Journal 92:2445–2450, 2007).

78. Shi L, Chen B, Wang Z, Elias DA, Mayer MU, Gorby YA, Ni S, Lower BH, Kennedy DW, Wunschel DS, Mottaz HM, Marshall MJ, Hill EA, Beliaev AS, Zachara JM, Fredrickson JK, Squier TC (Isolation of a high-affinity functional protein complex between OmcA and MtrC: two outer membrane decaheme c-type cytochromes of *Shewanella oneidensis* MR-1. Journal of Bacteriology 188:4705–4714, 2006).

79. Shi L, Squier TC, Zachara JM, Fredrickson JK (Respiration of metal (hydr)oxides by *Shewanella* and *Geobacter*: a key role for multihaem c-type cytochromes. Molecular Microbiology 65:12–20, 2007).

80. Signorell GA, Kaufmann TC, Kukulski W, Engel A, Remigy HW (Controlled 2D crystallization of membrane proteins using methyl-beta-cyclodextrin. Journal of Structural Biology 157:321–328, 2007).

81. Singh RK, Narayan J (Pulsed-laser evaporation technique for deposition of thin-films—physics and theoretical-model. Physical Review B 41:8843–8859, 1990).

82. Sonnleitner A, Schutz GJ, Schmidt T (Free Brownian motion of individual lipid molecules in biomembranes. Biophysical Journal 77:2638–2642, 1999).

83. Su TJ, Lu JR, Thomas RK, Cui ZF, Penfold J (The adsorption of lysozyme at the silica-water interface: a neutron reflection study. Journal of Colloid and Interface Science 203:419–429, 1998a).

84. Su TJ, Lu JR, Thomas RK, Cui ZF, Penfold J (The effect of solution pH on the structure of lysozyme layers adsorbed at the silica-water interface studied by neutron reflection. Langmuir 14:438–445, 1998b).

85. Tamm LK, Mcconnell HM (Supported phospholipid-bilayers. Biophysical Journal 47:105–113, 1985).

86. Tiberg F, Nylander T, Su TJ, Lu JR, Thomas RK (beta-Casein adsorption at the silicon oxide—aqueous solution interface. Biomacromolecules 2:844–850, 2001).

87. Ulevitch RJ, Tobias PS (Receptor-dependent mechanisms of cell stimulation by bacterial endotoxin. Annual Review of Immunology 13:437–457, 1995).

88. Ulman A (Formation and structure of self-assembled monolayers. Chemical Reviews 96:1533–1554, 1996).

89. Vacklin HP, Tiberg F, Fragneto G, Thomas RK (Composition of supported model membranes determined by neutron reflection. Langmuir 21:2827–2837, 2005a).

90. Vacklin HP, Tiberg F, Thomas RK (Formation of supported phospholipid bilayers via co-adsorption with beta-D-dodecyl maltoside. Biochimica et Biophysica Acta 1668:17–24, 2005b).

91. Vanatalu K, Paalme T, Vilu R, Burkhardt N, Junemann R, May R, Ruhl M, Wadzack J, Nierhaus KH (Large-scale preparation of fully deuterated cell components. Ribosomes from *Escherichia coli* with high biological activity. European Journal of Biochemistry/FEBS 216:315–321, 1993).

92. Venters RA, Huang CC, Farmer BT, Trolard R, Spicer LD, Fierke CA (High-level H-2/C-13/N-15 labeling of proteins for NMR-studies. Journal of Biomolecular NMR 5:339–344, 1995).

93. Wacklin HP, Thomas RK (Spontaneous formation of asymmetric lipid bilayers by adsorption of vesicles. Langmuir 23:7644–7651, 2007).

94. Wagner ML, Tamm LK (Tethered polymer-supported planar lipid bilayers for reconstitution of integral membrane proteins: Silane-polyethyleneglycol-lipid as a cushion and covalent linker. Biophysical Journal 79:1400–1414, 2000).

95. Wang W, Liang L, Johs A, Gu B (Thin films of uniform hematite nanoparticles: controls on surface hydrophobicity and self-assembly, Journal of Materials Chemistry, DOI: 10.1039/B810164G, in press, 2008).
96. Weber KA, Achenbach LA, Coates JD (Microorganisms pumping iron: anaerobic microbial iron oxidation and reduction. Nature Reviews 4:752–764, 2006).
97. Wong JY, Majewski J, Seitz M, Park CK, Israelachvili JN, Smith GS (Polymer-cushioned bilayers. I. A structural study of various preparation methods using neutron reflectometry. Biophysical Journal 77:1445–1457, 1999).
98. Xiong Y, Shi L, Chen B, Mayer MU, Lower BH, Londer Y, Bose S, Hochella MF, Fredrickson JK, Squier TC (High-affinity binding and direct electron transfer to solid metals by the *Shewanella oneidensis* MR-1 outer membrane c-type cytochrome OmcA. Journal of the American Chemical Society 128:13978–13979, 2006).
99. Zachara JM, Fredrickson JK, Li SM, Kennedy DW, Smith SC, Gassman PL (Bacterial reduction of crystalline Fe3+ oxides in single phase suspensions and subsurface materials. American Mineralogist 83:1426–1443, 1998).

Chapter 17
Pollutant Speciation in Water and Related Environmental Treatment Issues

Gabriel J. Cuello, Gabriela Román-Ross, Alejandro Fernández-Martínez, Oleg Sobolev, Laurent Charlet, and Neal T. Skipper

Abstract Neutron scattering and complementary techniques are extremely useful in the investigation of pollutant speciation in water and aqueous environments, as is shown in this chapter for both heavy metal and organic contaminants. The use of neutron diffraction, in conjunction with isotopic substitution and difference analysis, makes it possible to study the local structure developed around ions and other species in solution and in the pore spaces of minerals such as natural clays. As illustrations, the first-order difference method is applied to the hydration of mercury in aqueous solution, and the second-order difference method is used to determine the solvation of lanthanides in clay minerals. The isotopic substitution of hydrogen for deuterium is a powerful method with which to study both the structure and the dynamics of, for example, organic pollutants. In many cases the combination of neutron and X-ray diffraction is necessary, as shown for the incorporation of arsenic into the structures of minerals such as gypsum and calcite. Finally, some general conclusions and perspectives regarding the application of neutron techniques in environmental issues are drawn.

17.1 Speciation, Toxicity, Bioavailability, and Mobility

Maintaining and remediating the quality of water, air, and soil, so that Earth will be able to sustainably support human population growth, is one of the great challenges of our generation. The scarcity of water, in terms of both quantity and quality, poses a significant threat to human well-being, especially in developing countries. Environmental technology is set to play a key role in the shaping of current environmental engineering and policy. The ability to study materials at the nanoscale has stimulated the understanding of environmental systems, as well as the development and use of novel and cost-effective technologies for remediation, pollution detection, and catalysis, to name but three. There is now a huge expectation that these applications and products will lead to a cleaner and healthier environment.

G.J. Cuello (✉)
Diffraction Group, Institut Laue Langevin, F-38042 Grenoble Cedex 9, France
e-mail: cuello@ill.fr

L. Liang et al. (eds.), *Neutron Applications in Earth, Energy and Environmental Sciences*, Neutron Scattering Applications and Techniques,
DOI 10.1007/978-0-387-09416-8_17, © Springer Science+Business Media, LLC 2009

Groundwater and soils are resources that are both important and at risk, and they are of great relevance to the global environment, the economy, and the health of citizens. The protection of groundwater, in terms of quantity and quality, has become a major political priority at regional and national levels. This is increasingly reflected in developing policies, legislation, and standards on water and soil management. As an example, the European Water Framework Directive, adopted in December 2000, provides a competitive framework for the protection of all European waters. It sets as one of its objectives for groundwater the achievement of "good" chemical status by 2015. More recently, proposals for a Groundwater Daughter Directive and for a Soil Daughter Directive emphasize the need to delineate protection and remediation strategies and to determine background threshold values.

These proposals, as well as similar ones developed in the United States [1], put a strong emphasis on the development of tools and technologies for monitoring, preventing, and mitigating environmental pressures and risks and on the need for verification and testing programs. This requires a better understanding of containment behavior and pathways in the context of complex interactions between soil and groundwater systems. In Europe, the research infrastructure is fragmented and scattered with poor dissemination of knowledge and a growing gap between researchers and users. To improve the situation, the soil sciences must be made more appealing for young scientists and the cooperation of national and regional research programs must be strengthened. Recently, some initiatives [2] have been established to bring politicians and scientists interested in transport, energy, materials science, and the environment together with experts on advanced analytical techniques provided by modern synchrotron and neutron radiation sources. A recent book [3] also lists new developments in the neutron sciences to enhance the understanding of environmentally relevant concepts such as the bioavailability and toxicity of single molecules, heavy metals, and nanoparticles and to aid in the development of technologically oriented processes such as transformation and sequestration of these pollutants. Neutron scattering techniques play a key role in these studies, thanks to the fact that light atoms have good contrast in the presence of heavy atoms (see Chapter 2). This property is of fundamental importance when researchers are interested in the interaction of heavy elements with minerals, where light elements are the main components.

17.1.1 Bioavailability, Toxicity, and the HSAB Principle

Toxic heavy metals in soil and water are global problems that are directly and indirectly threatening humanity. The most toxic metals for living organisms, and their respective relative toxicities, are listed in Table 17.1 [4]. Industrial and agricultural activity has increased the release of heavy metals and made them available to fish and wildlife in aquatic and terrestrial ecosystems around the globe [5]. Heavy metal pollution is therefore often the result of human (anthropogenic) activities such as mining, metallurgy, and metal surface treatment.

Table 17.1 Toxicity sequences for metal ions in a range of organisms (after Nieboer and Richardson [4])

Organisms	Sequence
Algae *Chorella vulgaris*	Hg > Cu > Cd > Fe > Cr > Zn > Ni > Mn
Fungi	Ag > Hg > Cu > Cd > Cr > Ni > Pb > Co > Zn > Fe > Ca
Flowering plants Barley	Hg > Pb > Cu > Cd > Cr > Ni > Zn
Protozoa Paramecium	Hg > Pb > Cu, Cd > Ni, Ca > Mn > Zn
Platyhelminthe policelis, a planarian	Hg > Ag > Au > Cu > Cd > Zn > H > Ni > Co > Cr > Pb > Al > K > Mn > Mg > Ca > Sr > Na
Annelida neanthes, a polychaete	Hg > Cu > Zn > Pb > Cd
Vertebrata stickleback	Ag > Hg > Cu > Pb > Cd > Au > Al > Zn > H > Ni > Cr > Co > Mn > K > Ba > Mg > Sr > Ca > Na
Mammalia rat, mouse, rabbit	Ag, Hg, Tl, Cd > Cu, Pb, Co, Sn, Be > In, Ba > Mn, Zn, Ni, Fe, Cr > Y, La > Sr, Se > Cs, Li, Al

Pollution and contamination refer to the presence of a substance at a higher level than would normally occur, which leads to some kind of adverse effect [5]. For instance, mercury pollution comes from industrial emissions, although volcanoes and oceans also release mercury into the atmosphere. The industrial input of mercury has increased since the 19th century, first with gold mining activities (e.g., the California "gold rush") and then with the burning of wastes and fossil fuels—particularly high-sulfur coal. Other industrial sources include mercury smelting, chloralkali process plants, use of organic mercurial pesticides, and improper disposal of mercury-based batteries or fluorescent light bulbs. Mercury exposure usually occurs through consumption of mercury-contaminated carnivorous fishes as a result of the magnification of methyl mercury in the food chain. Plants and livestock may also contain mercury, but in lesser amounts, because of uptake by soil, water, and atmosphere and by the ingestion of other mercury-containing organisms [6].

Geogenic contaminants, i.e., naturally occurring pollutants, include fluorine, selenium, arsenic, lead, chromium, and manganese in soil and water environments. Significant adverse impacts of these contaminants (e.g., arsenic) on environmental and human health have been recorded in Bangladesh, West Bengal, India, Vietnam, Argentina, Chile, and China [7]. Heavy metal contamination also results from natural phenomena such as volcanic eruptions, weathering, pedogenic transformations, or climate change [8].

The toxicity of a given element to an organism in contact with "pollution" or "contamination" depends on (1) the bioavailability of the toxic element, i.e., the fraction of the total soil, air, or aqueous concentration of the element that can be taken up (e.g., absorbed, ingested, or inhaled) by the organism, and (2) its interaction with biologically active molecules such as enzymes or ion channels. Agreement on the definition of bioavailability and related terms (bioaccessibility, toxicity, mobility, geoavailability) seems unlikely in the near future [9], as it involves many chemical, physical, and biological aspects. In general terms, "bioavailability studies" deal with processes occurring at the interface between the environmental matrix and the surface of an organism. The most important factors controlling bioavailability are (1) metal concentrations in solution, (2) solute and particle metal speciation, (3) metal partitioning among solid and aqueous or solid ligands, (4) pH, (5) redox potential, (6) influence of other cations, and (7) temperature [10].

Many chemicals, both inorganic and organic, can exist in different forms, known as "species," of a given chemical. In the aqueous phase, species include hydrated free ions, inorganic and organic complexes, and species associated with heterogeneous colloidal dispersion and organometallic compounds. In the particulate phase, metals can interact with the particles in a range of processes, from weak (e.g., electrostatic) adsorption to chemical covalent binding to the mineral matrix. The dependence of bioavailability, toxicity, and geochemical mobility on speciation has been widely reported [11–13]. The toxicity sequence shown in Table 17.1 can be very broadly explained in terms of the principle of hard and soft acids and bases (HSAB). The HSAB principle states that hard acids (e.g., cations) tend to make complexes with hard bases (e.g., anions or water molecules), usually outer-sphere types of complexes (i.e., ion pairing governed by electrostatic forces); whereas soft acids tend to form complexes with soft bases, usually of an inner-sphere type with more covalent bonding. The toxic heavy metal ions are soft or near-soft acids, and the toxicity sequences listed in Table 17.1 closely follow the increasing order of the Misono softness parameter [14].

According to the HSAB, these heavy metals will therefore form more stable complexes and insoluble precipitates with soft or borderline bases (such as soluble Cl^- and S^{2-} ions) and organic functional groups containing S, P, and N donors (such as those present in enzymes) than with surface and aqueous hydroxyl groups containing oxyanions or organic carboxyl groups (which are hard bases). In the following example, we shall discuss the complexation of two very soft Lewis acids (i.e., heavy metal ions), Cd^{2+} and Hg^{2+}, with aqueous OH^-, surface OH^-, and aqueous Cl^-. We shall show how the change in speciation they experience from the mine tailings where they are released to the ocean determines the relative toxicity of the waters in which they travel.

17.1.2 Metal Speciation in River Networks

Estuaries are key transition zones between continental and oceanic environments and are characterized by intense redistribution of trace metals between particulate and dissolved phases. The Gironde estuary (France) is known to be impacted by an important historical polymetallic pollution caused by former mining and ore-treatment activities in the upstream Lot River catchment. Cadmium and mercury input to the Lot river comprises wet deposition from the atmosphere, molecular diffusion at the sediment–water interface, and surface-water runoff, but is mostly due to the discharge from the leaching of wastes from former mines and zinc-refining plants. Approximately 85% of the cadmium in the Lot River is derived from the upstream Decazeville area where, in the 19th century, coal mines and zinc refineries were active. The mine drainage is very acidic; its name, the "Riou Mort" (or Dead River), means that no upper organism can live in such an acidic (and cadmium- and mercury-rich) aqueous environment. In this stream, 80% of the cadmium is present as dissolved species (mostly free metal ion species) and 20% as particulate cadmium (Fig. 17.1). As the Riou Mort waters are discharged into the Lot River, which runs in a limestone-dominated watershed, their pH dramatically increases and the cadmium speciation changes. The cadmium, which originates not only from the Riou Mort but also from the Lot sediment and from atmospheric input, is then mostly present as adsorbed species on suspended particles, i.e., complexed by $\equiv OH^0$ surface hydroxyl groups (Fig. 17.1). This change of speciation (i.e., the complexation of Cd^{2+} by particle surface functional groups denoted $\equiv OH$) reflects the cadmium "adsorption edge" isotherm. Cadmium is present as free aqueous Cd^{2+} ion species in acidic environments and as particulate surface complex species ($\equiv OCd^+$) at neutral pHs, according to the following reaction [16, 17]:

$$Cd^{2+} + \equiv OH \leftrightarrow \equiv OCd^+ + H^+. \tag{17.1}$$

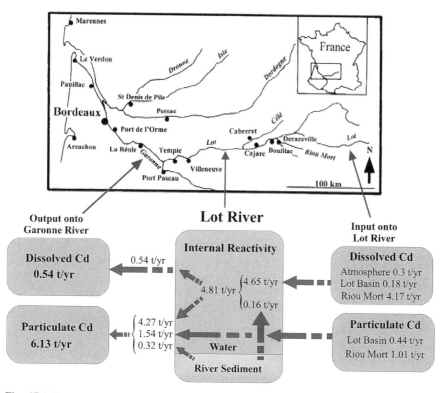

Fig. 17.1 Box model of the cadmium budget for the Lot–Garonne study basin (after Blanc et al. [15])

As a result, the Lot River water is "decontaminated," and fishing and other open-air activities are encouraged in the beautiful tourist area in which it is located.

As waters from the Lot River and then from the Garonne Rivers enter the "Gironde" estuary, the speciation of metal ions changes dramatically. In the estuary, one observes a slight pH increase from 7.0 to 8.2, but a very large increase in Cl^- total concentration (Fig. 17.2). Since chloride ions are borderline soft Lewis bases, they tend to out-compete dissolved and particulate hydroxyl ions (hard Lewis bases) in the complexation of soft acids such as Hg^{2+} or Cd^{2+}. The $\equiv OHg^+$ surface complex tends to dissociate according to the reaction

$$\equiv OHg^+ + H^+ + 2Cl^- \leftrightarrow \equiv OH + HgCl_2^0. \tag{17.2}$$

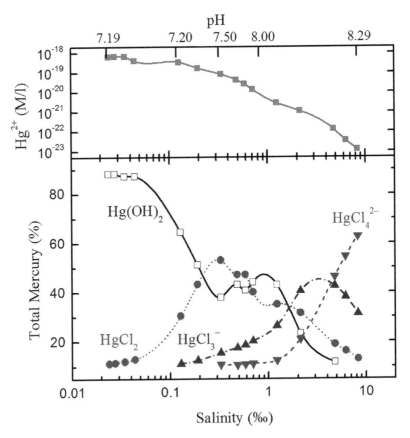

Fig. 17.2 Inorganic speciation of mercury in an estuary as a function of salinity. This example has been calculated by considering the mixture of fresh water ($[Cl^-] = 5 \times 10^{-4}$ M; Alk $= 1.3 \times 10^{-4}$ M; pH $= 7.2$) with seawater (salinity $= 36\%$, pH $= 8.20$) and maintaining the total mercury concentration at 1 nM (after Morel and Hering [18])

Whereas the dominant aqueous mercury species in rivers such as the Garonne is $Hg(OH)_2{}^0$, the dominant species in the estuary becomes $HgCl_2{}^0$, then $HgCl^{3-}$, and then finally $HgCl_4{}^{2-}$ in the ocean (Fig. 17.2). As a result, filter organisms such as mussels and oysters tend to accumulate large amounts of mercury and cadmium (which make strong complexes with their sulfur-rich components). Consumption of seafood such as oysters and mussels from the Gironde estuary is therefore forbidden.

The effect of exposure to elemental mercury or some of its compounds depends on their absorption and metabolism [19]. However, with sufficient exposure, all mercury-based toxins damage the central nervous system and other organs or systems, such as the liver or the gastrointestinal tract. Mercury and many of its chemical compounds, especially organomercury compounds, can also be readily absorbed through direct contact with bare skin or, in some cases (such as in the case of dimethyl mercury), insufficiently protected skin.

17.1.3 Radionuclides and Clays in the Environment

As shown in the Lot River, metal contaminants may exist in different chemical forms, in both the dissolved and the particulate phases of a system. In the particulate phase, metals can interact with the particles in various associations, ranging from weak adsorption to binding to the mineral matrix.

One of the main concerns related to the safe storage of nuclear waste in an underground repository is the migration of radiotoxic elements through the geosphere [20, 21]. The interaction between solute and solid surfaces, which is mainly governed by sorption/desorption processes, is a crucial issue in underground geological nuclear waste performance assessment. Therefore, a detailed knowledge of the mineral–water interface is of fundamental interest since the corresponding interfacial reactions play a significant role in the geochemical regulation of radionuclides. In this respect, neutron scattering is the best adapted technique to study the water structure near mineral surfaces. These reactions are, however, inherently complex and it is necessary to quantify several physicochemical parameters to build up a predictive model. These phenomena depend mainly on parameters such as pH, redox potential, ionic strength of the aqueous medium, concentration and speciation of the radionuclides of interest (taking into account the solid precipitates), specific surface area, and the density of functional surface groups of the substrate (e.g., clays).

Radioactive waste material confinement sites are typically engineered as multiple clay barriers surrounding metallic waste canisters. Whereas the near-field clay barrier is made of bentonite (a mixture of mostly smectite, with some calcite and quartz), the far-field barrier is made of argillite, a mixture of illite, calcite, and some iron sulfide. These barriers are classified first into layer types, differentiated by the number of tetrahedral and octahedral sheets that have combined, and then into groups differentiated by the kind of isomorphic cation substitution that has occurred [22]. These substitutions create a permanent negative charge on the basal planes that is equilibrated by the counter-ions placed in a water interlayer between the atomic

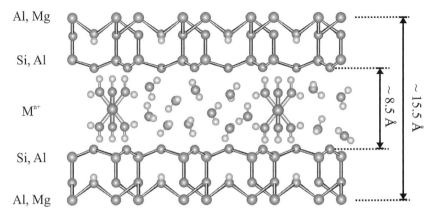

Fig. 17.3 Schematic cross-section perpendicular to the clay platelets in montmorillonite show-ing two half-layers and the interlayer region consisting of charge balancing cations, M^{n+} and water. The dimensions shown correspond approximately to a "two-layer" hydrate (after Pitteloud et al. [23])

planes (see Fig. 17.3). Smectite and illite have two tetrahedral and one octahedral sheet per crystal platelet. The specific surface of these platelets is very large: up to 80–100 meq of cations can be hosted per 100 g of clay in the clay interlayer in the case of smectites. As a result, understanding the sorption of radionuclide species on clays and calcite is a strategic issue in radioactive waste disposal. In addition, the precipitation/solubility phenomena of the radionuclides must be taken into account.

Radionuclide fate is dependent on the conditions prevailing in the clay barrier media. Radionuclides are often present as cationic species that, once sorbed on clays, have been shown to diffuse extremely slowly through and out of the clay barriers. The strength of their adsorption depends on the charge of the radionuclides (the higher the charge, the stronger the adsorption) and the structure of the sur-face complex, structure that has been intensively investigated by neutron scattering. The radionuclides are often present in the macropores as hydrolyzed low-charge species; however, they could be dehydrolyzed, i.e., present as higher-charge (e.g., +3) species, in the clay interlayer and thereafter very strongly retained. It is there-fore of strategic importance to understand the water structure and the hydrolysis of model radionuclides in the smectite interlayers, and neutrons are extremely useful for this purpose [23].

17.1.4 Macroscopic Versus Microscopic Processes

The above-mentioned barriers, in addition to clay minerals that retain mostly cationic species, are very rich in calcite. Calcite is a sink of various oxyanions; therefore, it is critical to understand the retention mechanisms of anionic radionuclides, such as selenium oxyanions. These oxyanions were long thought to behave as tracers and not to be retained at all by the barrier. Selenium and other oxyanions like arsenic are another source of contamination in industrial and natural sites: the Bangladesh-Bramapoutre delta upper aquifers [24], the Argentinean Pampa aquifers [25], eutrophic lakes [26], and contaminated industrial sites [7, 27–29] are examples of sites where water bodies are contaminated by arsenic and are at equilibrium, or slightly supersaturated, with respect to calcite.

Wet chemical methods have classically been used to study macroscopic adsorption, surface precipitation, or co-precipitation phenomena: the simple measurement of the concentration of a hazardous element in solution and the pH of this solution give information about the solubility, chemical speciation, and mobility of the contaminant. Adsorption processes are usually described through isotherms, functions that connect the amount of adsorbate on/in the adsorbent at constant temperature with its pressure for gas species or its concentration for liquid species (Fig. 17.4) [30, 31].

Macroscopic As(III) sorption isotherms obtained with calcite depict an S-shaped curve (Fig. 17.4). The figure shows a fraction of $As(OH)_3^0$ being sorbed at low sorbate concentrations, in a linear way (surface adsorption), and then reaching a plateau (depicted by dots) and corresponding to the saturation of calcite surface sites. At a higher concentration, all available sorption sites are occupied and co-precipitation of calcium carbonate–arsenite solid solution occurs [30]. At even higher concentration, precipitation of pure phase occurs (vertical part of the isotherm). These data could

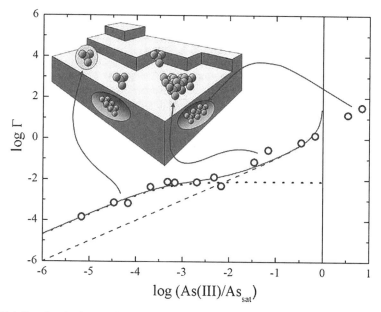

Fig. 17.4 Sorption isotherm of As(III) on calcite. The experimental data (symbols) are compared with an adsorption/precipitation model with ST = 6 sites/nm², $^c K_c = 1.5 \times 10^{-5}$, $^c K_a = 5.9 \times 10^{-9}$, as defined by Román-Ross et al. [30]. The inset shows a sketch of adsorption, co-precipitation, and surface phenomena [31], and the relationships with the isotherm

be adequately described by a model originally derived by Farley et al. [32] for metal oxides, which includes both adsorption and precipitation of an ideal solid solution as the two possible sorption mechanisms involved. The model was improved to be adapted to describe anion sorption and co-precipitation onto calcite [30]. The inflexion point of the S-shaped isotherm was assumed to correspond to surface site saturation.

In the last half-century, the advent of powerful spectroscopic methods, of neutron scattering techniques in particular, has allowed researchers to extract information from the macroscale to the bonding scale. These methods have opened new horizons in the microscopic study of the interactions between a pollutant and a mineral phase. Neutron scattering techniques are well established in the study of the structure and dynamics of condensed matter, in particular "amorphous" solid matter [33], but they are relatively new techniques in the earth and environmental sciences. In this chapter, we will focus on the use of neutron diffraction methods for studying the local order of ions in calcite, in water, or in clay interlayers.

17.2 Neutron Scattering

Thermal neutrons are a uniquely powerful probe of the microscopic-scale structure and dynamics of condensed matter systems. Their wavelengths and energies are well matched to the atomic separations and excitations, respectively, while the lack of charge makes them deeply penetrating and nondestructive. Moreover, the neutron interacts primarily with the nuclei, rather than with electrons. Light atoms such as hydrogen are therefore relatively strong scatterers of neutrons, in contrast to the situation when X-ray techniques are used; and in many cases, the different isotopes of the same element have very different neutron scattering properties. This last characteristic forms the basis of isotopic labeling techniques, whereby one structure around one particular species can be isolated from all others. For example, in a complex system such as a clay–water–cation–organic, one can use neutron diffraction in conjunction with isotope labeling to distinguish between the mineral, water, and organic molecules.

Over the past decades, neutron scattering has become an important method for the study of geological materials. To a significant extent, this development is due to the ability of neutrons to locate hydrogen, an important and common element in minerals: because of its large coherent scattering cross section, hydrogen can be resolved much more clearly with neutrons than with X-ray based techniques such as extended X-ray absorption fine structure (EXAFS) [34]. Neutron diffraction experiments yield data up to much larger scattering vectors, because the atomic form factor (scattering length) is independent of the scattering vector. This in turn provides inherently higher real-space resolution than X-ray scattering, particularly in analyzing the complex low-symmetry crystal structures of many minerals and in looking at thermal movement and atomic occupancy of lattice sites. Furthermore, incoherent neutron scattering allows us to study the dynamics of hydrogen atoms (protons) over a very wide range of time scales. In Table 17.2, the neutron properties of the water components are given. In comparison with other isotopic substitutions, the use of heavy water (D_2O) instead of light or normal water (H_2O) is relatively easy and low in cost. In addition, modern chemical synthesis gives us access to a very wide range of cheap and high-purity deuterated organic compounds—for example, alcohols and phenols. This provides a major opportunity for us to obtain new insight into the behavior of organic molecules of environmental importance.

Neutron scattering techniques therefore complement X-ray techniques and are rapidly broadening their use in environmental-related studies. For instance, neutron diffraction with isotopic substitution (NDIS) is a very useful technique for the study of the water structure around a specific cation, using contrast variation through H_2O/D_2O exchange. This provides a detailed knowledge of the atomic positions and number of both oxygen and hydrogen atoms around a given metal ion and thus allows distinguishing surface oxygen atoms (usually not protonated) from hydration

Table 17.2 Scattering lengths (in fm) and cross sections (in barns) for some relevant elements, isotopes, and compounds. The values are taken from Sears [34]. The absorption cross section is given for neutrons of 2200 m/s

System	b_{coh}	b_{inc}	σ_{coh}	σ_{inc}	σ_{sca}	σ_{abs}
H	-3.7390	–	1.7568	80.26	82.02	0.3326
^{1}H	-3.7406	25.274	1.7583	80.27	82.03	0.3326
^{2}H (or D)	6.671	4.04	5.592	2.05	7.64	0.000519
O	5.803	–	4.232	0.0008	4.232	0.00019
H_2O	0.5583	–	2.5819	53.507	56.089	0.2218
D_2O	6.3817	–	5.1387	1.367	6.505	0.000409
Li	-1.90	–	0.454	0.92	1.37	70.5
^{6}Li	2.00–0.261i	$-1.89 + 0.26i$	0.51	0.46	0.97	940.
^{7}Li	-2.22	-2.49	0.619	0.78	1.4	0.0454
N	9.36	–	11.01	0.5	11.51	1.9
^{14}N	9.37	2.0	11.03	0.5	11.53	1.91
^{15}N	6.44	-0.02	5.21	0.00005	5.21	0.000024
V	-0.3824	–	0.0184	5.08	5.1	5.08
As	6.58	-0.69	5.44	0.06	5.5	4.5
Ti	-3.438	–	1.485	2.87	4.35	6.09
Zr	7.16	–	6.44	0.02	6.46	0.185
$Ti_{68}Zr_{32}$	0.047	–	3.071	1.958	5.029	4.2
Hg	12.692	–	20.24	6.6	26.8	372.3
^{199}Hg	16.9	(±)15.5	36.	30.	66.	2150.
Nd	7.69	–	7.43	9.2	16.6	50.5
^{143}Nd	14.	(±)21.	25.	55.	80.	337.
^{144}Nd	2.8	0	1.	0	1.	3.6
Sm	0.80–1.65i	–	0.422	39.	39.	5922.
^{152}Sm	-5.0	0	3.1	0	3.1	206.
^{154}Sm	9.3	0	11.	0	11.	8.4
Yb	12.43	–	19.42	4.	23.4	34.8
^{172}Yb	9.43	0	11.2	0	11.2	0.8
^{174}Yb	19.3	0	46.8	0	46.8	69.4

water (or hydroxyl atoms). This kind of information is very valuable in the study of surface complexation of pollutant molecules on mineral surfaces [35].

There are two kinds of standard neutron sources: nuclear reactors and spallation sources. The former are steady-state sources producing neutrons in a continuous way, whereas the latter are pulsed sources that use the time-of-flight technique. These sources are equipped with a large variety of public instruments, as is the case for the Institut Laue Langevin (ILL) [36], on which scientists can apply for beam time to perform their neutron scattering experiments.

In a standard scattering experiment, the measured intensity can be related to the dynamical structure factor $S(\vec{Q}, \omega)$,[1] the magnitude containing all the information about the structure and dynamics of the system. The variables are the energy exchange $\hbar\omega$ and the momentum transfer $\hbar\vec{Q} (= \hbar\vec{k} - \hbar\vec{k}_0)$, where $\vec{k}$ and $\vec{k}_0$ are the initial and final wave vectors of the probing particles. For a set of N identical atoms of a general system, the dynamical structure factor can be related to the time-dependent correlation function $G(\vec{r}, t)$ [37]:

$$S(\vec{Q}, \omega) = \frac{1}{2\pi} \int \int \mathrm{d}\vec{r}\mathrm{d}t \, e^{i(\vec{Q}\cdot\vec{r}-\omega t)} G(\vec{r}, t). \tag{17.3}$$

This function is also known as the Van Hove correlation function, which is related to the probability of finding a particle at position $\vec{r}$ and time t, provided that another particle was at the origin at $t = 0$. The knowledge of this function containing the position of the atoms in the system as a function of time gives all the information needed to understand the physical properties of a given system. Because of their energy and finite mass, thermal neutrons can work out where the particles are and how they interact with each other in a simultaneous way. However, since this article is focused on the structure of minerals containing pollutants and the local structure of aqua ions, our discussion will be restricted to diffraction methods. Among them, small-angle neutron scattering (SANS) is very useful for studying large structures such as those of clay minerals, polymers, biological systems, or colloids. This technique is described in Chapter 2 and is applied to clay swelling in Chapter 18. Higgins and Benoit [38] apply SANS to polymers and colloidal systems. This chapter is focused on the short-range order, where wide-angle diffraction is used.

17.2.1 Diffraction Techniques

In diffraction experiments, the detectors integrate all the neutrons regardless of their energy exchanges with the sample—there is no discrimination in the final energy. In mathematical language, this is equivalent to integrating the dynamical structure factor over all energies, so,

---

[1] In Chapter 2 this function was introduced as the scattered intensity, related to the dynamical structure factor by the expression $I(\vec{Q}, \omega) = \frac{Nb^2}{\hbar} \frac{k}{k_0} S(\vec{Q}, \omega)$.

$$S(\vec{Q}) = \int_{-\infty}^{\infty} d\omega \, S(\vec{Q}, \omega) = \int d\vec{r} \, e^{i\vec{Q}\cdot\vec{r}} \, G(\vec{r}, 0). \tag{17.4}$$

In a real experiment, where the energy exchange is limited, this integral must be considered an approximation, the so-called static approximation. Under this approximation, the correlation function is evaluated at $t = 0$, which is equivalent to a snapshot of the system. In this case, the correlation function has a very simple expression; it is just $G(\vec{r}, 0) = \delta(\vec{r}) + \rho \, g(\vec{r})$, i.e., a delta function at the origin representing the autocorrelation plus the space-dependent density variations. Then the static structure factor can be easily calculated:

$$S(\vec{Q}) - 1 = \rho \int_V d\vec{r}[g(\vec{r}) - 1]e^{i\vec{Q}\cdot\vec{r}}; \tag{17.5}$$

and conversely

$$g(\vec{r}) - 1 = \frac{1}{(2\pi)^3} \int d\vec{Q}[S(\vec{Q}) - 1]e^{-i\vec{Q}\cdot\vec{r}}. \tag{17.6}$$

As usual in disordered systems, it can be assumed that the scattering is isotropic and the three-dimensional Fourier transformations can be reduced to their one-dimensional equivalents as follows:

$$S(Q) - 1 = \frac{4\pi\rho}{Q} \int_0^{\infty} r[g(r) - 1]\sin(Qr)dr; \tag{17.7}$$

and

$$g(r) - 1 = \frac{1}{2\pi^2\rho r} \int_0^{\infty} Q[S(Q) - 1]\sin(Qr)\,dQ. \tag{17.8}$$

The radial distribution function RDF $(r) = 4\pi\rho r^2 g(r)$ is the number of atoms per unit length on the surface of a sphere of radius r. With such a definition, the integral of this function directly gives the number of atoms in a given coordination shell (the coordination number).

In the case of a system constituted of n different atomic species, the total static structure factor can be generalized as follows

$$\bar{b}^2[S(Q) - 1] = \sum_{\alpha=1}^{n}\sum_{\beta=1}^{n} c_\alpha c_\beta b_\alpha b_\beta \, [S_{\alpha\beta}(Q) - 1], \tag{17.9}$$

where $S_{\alpha\beta}(Q)$ are the partial static structure factors ($n_e = n(n+1)/2$ independent functions in total) and $\bar{b}$ is the mean value of the coherent scattering length:

$$\bar{b}^2 = \sum_{\alpha=1}^{n} \sum_{\beta=1}^{n} c_\alpha c_\beta b_\alpha b_\beta. \qquad (17.10)$$

The equivalent expression can be written for the structure factor $F(Q)(= S(Q) - 1)$ in terms of the partial ones:

$$\bar{b}^2 F(Q) = \sum_{\alpha=1}^{n} \sum_{\beta=1}^{n} c_\alpha c_\beta b_\alpha b_\beta \, F_{\alpha\beta}(Q). \qquad (17.11)$$

The main problem that an experimentalist should face is the fact that all the partial structure factors are added up in the same expression; i.e., a single measured diffractogram contains all the partials that one would like to extract individually. To solve the problem, we need a set of n_e independent linear equations. The question is how to obtain this set for a given system. From a mathematical point of view, just by changing the coefficients in Eq. (17.11), we could solve our problem; but in fact we must do this carefully. If we change the composition c_α, we are changing the sample, and this is unacceptable. The only possibility is changing the scattering lengths included in the weighting factors. This can be done using the technique of isotopic substitution under the assumption that the structural properties of the system are not changed by changing isotopes. In structural studies, this is a well-founded assumption; but when we are interested in dynamical aspects, we must be careful with this technique. Other techniques like isomorphic substitution [39–43] and anomalous dispersion [44, 45] have been applied recently in order to improve the contrast in neutron diffraction for some particular cases.

The linear system (Eq. 17.11) can always be expressed as a matricial equation, $\vec{F}_S(Q) = \mathbf{A} \vec{F}_P(Q)$ where $\vec{F}_S(Q)$ and $\vec{F}_P(Q)$ are two vectors of n_e components, containing the independent measured structure factors and the partial correlations, respectively. The $n_e \times n_e$ matrix $\mathbf{A}$ must be inverted in order to obtain the partial structure factors (see for instance Salmon et al. [46, 47] where a binary system is solved). The determinant of $\mathbf{A}$ is an important parameter that should be considered when the experiment is designed: the bigger the determinant, the smaller will be the uncertainties in the final results. Larger values of this determinant are obtained by choosing isotopes with good contrast, i.e., with scattering lengths that are appreciably different [34], or by combining neutron and X-ray diffraction data.

Even when isotopes are available, it is not always possible to prepare all the samples required to obtain the complete set of correlation functions. This is particularly true for systems with more than two atomic species. In these cases, we can apply the first difference method with isotopic substitution [48]. If only one element can be substituted (γ), we can write Eq. (17.11) for each of the two samples, one with the $\gamma 1$ isotope and the other with the $\gamma 2$ isotope. The difference between these two equations is that all the correlations where the substituted atom is not included are cancelled. In this way, we can write the following expression for the first difference spectrum:

$$\frac{\bar{b}^2 \Delta F_\gamma(Q)}{c_\gamma^2 (b_{\gamma 1}^2 - b_{\gamma 2}^2)} = F_{\gamma\gamma}(Q) + \frac{\sum_{\alpha \neq \gamma}^n c_\alpha b_\alpha F_{\alpha\gamma}(Q)}{c_\gamma (b_{\gamma 1} + b_{\gamma 2})}, \qquad (17.12)$$

where it has been possible to separate the $\gamma-\gamma$ correlation and the second term is usually small.

Under the given circumstances, it is possible to perform a second substitution, the atomic species δ, for example. Then Eq. (17.12) can be rewritten, isolating the γ- and δ-correlations, for the two isotopes $\delta 1$ and $\delta 2$. The difference between these two first difference spectra contains only the $\gamma-\delta$ correlations

$$F_{\gamma\delta}(Q) = \frac{\bar{b}^2 \Delta^2 F_{\gamma\delta}(Q)}{c_\gamma c_\delta (b_{\gamma 1} - b_{\gamma 2})(b_{\delta 1} - b_{\delta 2})}, \qquad (17.13)$$

where $\Delta^2 F_{\gamma\delta}(Q)$ represents the experimental double-difference spectrum.

Using Eq. (17.11) and the Fourier transformation [8], it is possible to generalize the expressions for the correlation functions in real space:

$$\bar{b}^2 [g(r) - 1] = \sum_{\alpha=1}^n \sum_{\beta=1}^n c_\alpha c_\beta b_\alpha b_\beta [g_{\alpha\beta}(r) - 1]. \qquad (17.14)$$

17.2.2 Technical Aspects

On a seemingly more technical note, this subsection is intended to describe some practical issues. First, there is the question of the quantity and state of the sample. In contrast with X-ray techniques and because of the weak interaction of neutrons with matter, a large amount of sample (typically a few grams) is required for neutron scattering experiments. This is the main disadvantage of neutron diffraction, but it could be overcome using instruments at high-intensity sources; for instance, on instruments like D20 [49] or D4 [50] at ILL, it is possible to perform experiments even with a few milligrams of sample.

The samples can be in different states; usually in the environmental sciences, they are liquids or powders, which can be placed in thin-walled vanadium containers. Vanadium is the typical material for containers because its coherent scattering length is very small and it can be considered as a quasi-incoherent scatterer (Table 17.2) producing an almost flat diffraction spectrum. It is worth nothing that in many neutron diffraction experiments, the samples are contained in cans made of a 68:32 ratio alloy of Ti:Zr (Table 17.2). In addition to excellent resistance to chemical attack, this material has the remarkable property of being almost transparent to neutrons: the coherent neutron cross-section of the alloy is zero![2] Therefore, thick-walled Ti:Zr

---

[2] In fact, the exact ratio for a null coherent scattering length is 67.56:32.44 for the Ti:Zr alloy.

neutron sample cans can be made, which allow us to access the elevated pressures and temperatures encountered in geological settings.

Concerning the sample environment, there is almost standard equipment that allows experimenters to control humidity, temperature, or pressure in situ. Helium cryostats with vanadium cold chambers are usually used, and they allow control over a temperature range from $-250°C$ to $50°C$. For higher temperatures, cryofurnaces or furnaces can be used. With standard vanadium and niobium furnaces, samples can be heated from room temperature up to $1100°C$ and $1550°C$, respectively. For higher temperatures, special devices like tungsten furnaces [51, 52] or levitation furnaces [53] are necessary. For high pressures, the device used most often is the standard Paris-Edinburgh cell in which the sample is placed between two hard anvils made from either tungsten-carbide or sintered industrial diamond. It allows samples to be subjected to pressures of $10\,GPa$ [54].

17.3 Water and Cation Speciation

We have already noted that neutron diffraction with isotopic substitution has proved to be a pivotal technique in developing our understanding of the structure of water and aqueous solutions [55]. The widely different coherent neutron scattering lengths of hydrogen (H) and deuterium (D) given in Table 17.2 have enabled us to obtain detailed insight into the hydrogen bonded networks in liquid water and ices, which are all invisible to X-ray scattering. Likewise, the NDIS method is very well suited to probing the local coordination of many solute organic molecules, where modern synthesis methods can routinely label specific proton sites with deuterium. We also see from Table 17.2 that many important environmental elements have different isotopes with sufficient neutron scattering contrast to make NDIS feasible. Of course, neutron scattering is a relatively expensive technique, and it is almost always preferable to use it as one of a set of complementary measurements. We have already commented that X-ray scattering can often provide an experimental foil to neutron scattering. In addition, when founded on high-quality experimental data, molecular modeling can provide an efficient and readily interpretable means for us to explore parameter space [56, 57]. We illustrate these possibilities in the following sections.

17.3.1 Structure of the Free Ion Aqua Complex and Its Implications

Knowledge of the detailed nature of the ion–water coordination in the vicinity of the ion is crucial to the problem of aqueous solutions. This part of the problem is, in its most general sense, an aspect of coordination chemistry. The neutron first-order difference method described by Soper et al. [48] allows detailed ion–water conformations to be obtained directly and without recourse to modeling techniques. When the suitable isotopes are available, it avoids all the difficulties associated with total neutron or X-ray scattering data. This method can also be applied for X-ray diffraction [58], as was done for nitrogen and magnesium by Skipper et al. [59].

Salts of Hg^{2+} are much more common and stable in the environment than Hg^+ salts, and water is the main medium for Hg^{2+} transport and interaction with biological objects. Taking into account the neurotoxic effect of mercury, the study of the interaction between divalent mercury and water molecules becomes important. Indeed, the function of the nervous system is based on the electric signals caused by ions moving in and out through the ion channels in cell membranes. The ion channels are able to admit only certain ions, i.e., only those for which the distance to the oxygen atoms in the ion channel is the same as in the hydration shell of the cation tailored to the channel (e.g., Ca^{2+}) [60]. Thus, it is important to compare the parameters of hydration of mercury with other divalent ions like Ca^{2+} in order to understand how mercury can penetrate to the nerve's living cells.

The structural parameters of Hg^{2+} hydration have been studied by different experimental methods [55], and most of them give a water coordination of six molecules. However, there is significant uncertainty about Hg–O distance, with the experimental results varying in the range 2.33–2.42 Å. Furthermore, no experimental data are available on the Hg–H coordination, which is crucial for determining the orientation of water molecules around mercury and for detailed comparison with other divalent cations. With the aim of exploring the water coordination around the Hg^{2+} ion, Sobolev et al. [61] have conducted neutron diffraction experiments at the D4 diffractometer [50] at room temperature. In that work, two samples with ^{nat}Hg and ^{199}Hg (0.225 mol/l solutions) were prepared from HgO in 1 M solution of DNO_3 in D_2O.

The isotopic substitution of mercury allows obtaining the pair correlation function $\Delta G(r)$, which is a measure of the probability of finding any atom at a distance r from the mercury atom (Fig. 17.5). The first two peaks of the $\Delta G(r)$ are attributed to the Hg–O and Hg–D correlations in the first hydration shell of Hg^{2+}. Integration of $\Delta G(r)$ under these peaks gives the number of oxygen and hydrogen atoms around mercury, as shown in Fig. 17.5. The combination of n_O and n_D gives the number 6 ± 1 of water molecules in the first hydration shell of the Hg^{2+} cation. The centers of the first two peaks of the $\Delta G(r)$ correspond to the distances Hg–O and Hg–D. The Hg–O distance observed in the neutron diffraction experiment is larger by ≈ 0.1 Å than that obtained by X-ray diffraction [55]. This difference is caused by the shift of the electronic shell of the oxygen toward the mercury cation. Since X-rays are scattered by the electronic clouds of atoms and neutrons are scattered only by atomic nuclei, then the latter find the oxygen atom in a slightly different position relative to the mercury atom. This shift of the electronic density means that the Hg–O interaction is not simply electrostatic but is partially covalent.

The parameters obtained for the hydration sphere can be compared with those for Ca^{2+} [62] and Ni^{2+} [55] cations. The angle between the plane of the water molecule and the cation–water oxygen axis, which can be calculated from the observed cation–oxygen and cation–hydrogen distances, is approximately the same for Hg^{2+} and for Ca^{2+} and Ni^{2+} ($\approx 35°$). The Hg–O and Hg–H distances (see Fig. 17.5) are very close to those obtained for Ca^{2+} cations (2.40 and 3.03 Å for Ca–O and Ca–H,

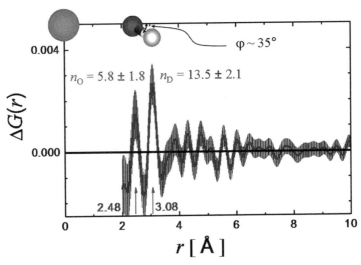

Fig. 17.5 The pair correlation function $\Delta G(r)$ centered at the Hg^{2+} ion. The water molecule position is shown, as well as the tilt angle (after Sobolev et al. [61])

respectively) [63]. This is a key result for understanding mercury toxicity: mimicking the Ca^{2+} hydration allows the Hg^{2+} ion to penetrate cell membranes through the Ca^{2+} ionic channels.

17.3.2 Solvation of Simple Organic Molecules

Just as water is susceptible to H/D isotope labeling, so too are many organic compounds. For example, NDIS has been exploited in seminal studies of methanol and other alcohols [43, 44,64–67] and of species such as alkylammonium ions in clays [68, 69]. The systematic application of this technique to the environmental sciences is now long overdue and is likely to have a major impact on our understanding of molecules such as organic contaminants.

As an example of the startling possibilities of neutron scattering in this context, consider phenol, C_6H_5OH. NDIS can be applied directly to C_6H_5OH and C_6D_5OH and other site-specific labeled analogs such as $C_6H_3D_2OH$ [70] (note that the $-OH$ is in fast exchange with the water itself and so cannot be labeled independently). We can therefore use NDIS to extract the solute–solvent and solute–solute structure for this molecule in a wide variety of environmentally relevant situations. An example of this is shown in Fig. 17.6a, where computer simulation of clay–phenol–water has generated a structural model [71]. Moreover, by placing C_6H_5OD in a deuterated solvent such as D_2O, we can exploit the very large incoherent neutron scattering cross-section of hydrogen (Table 17.2) via quasielastic neutron scattering to determine the diffusion rate and mechanism of the molecule. This approach has already

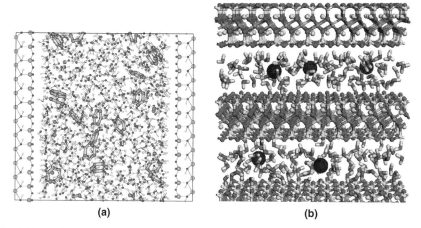

(a)   (b)

Fig. 17.6 (a) Structural model of clay–phenol–water generated by computer simulation [71]. (b) Snapshot from a molecular dynamics simulation of a hydrated montmorillonite clay. Sm^{3+} cations (big dark spheres) lie on the clay interlayer surrounded by water molecules. The layers are formed by silicon and aluminum atoms, with some isomorphic substitutions of magnesium in the aluminum octahedral layer

been applied to simpler alcohols [72] and could equally well be directed toward molecules such as the phenolic herbicides and pesticides.

17.3.3 Structure of Water at Clay Mineral Interfaces

Liquid water is ubiquitous as a solvent in the environmental sciences. If we are to understand surface complexation and speciation, it is therefore very important that we be able to probe the structure of water–mineral interfaces; here again NDIS is a unique and powerful technique [73]. For example, NDIS has provided the definitive structural data for a number of clay–water–cation systems under ambient and non-ambient conditions.

NDIS studies of water structure in Li-, Na-, K-, and Cs-smectites [74–77] have argued against a very highly structured interlayer water region in rigid association with the silicate sheets. Instead, the data support a much more liquid-like model in which hydrogen bonding is similar to that found in bulk aqueous solutions. Higher-resolution studies of single para-crystals of sodium- and alkylammonium-vermiculites [68, 69, 78] support this view, notwithstanding the higher (tetrahedral) layer charge of the clay sheets. Note that in this latter (vermiculite) case, the macroscopic samples were oriented relative to the incident and scattered neutron beams, thereby resolving the in- and out-of-(clay)-plane structures. In Fig. 17.7 [79], we can see that strongly solvated interlayer cations provide highly ordered structural motifs. NDIS of interlayer water has shown unambiguously that more ordered interlayers are found in Li-, Ni-, and Ca-vermiculites [78–80]. This, of

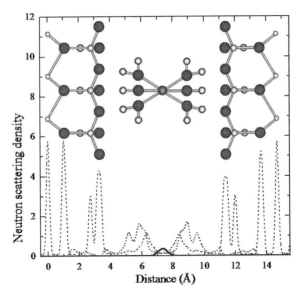

Fig. 17.7 Neutron scattering density profiles for lithium-substituted vermiculite. Oxygen plus clay layer is the *dashed line*, hydrogen is the *dotted line*, and lithium is the *solid line*. The molecular model shows two sections of clay surface and an undistorted octahedral $Li^+(H_2O)_6$ complex. In this model, all six water molecules are hydrogen-bonded directly to the clay surface; in practice we find that, on average, two of the six molecules are less strongly oriented toward the surface (taken from Skipper et al. [79])

course, has immediate implications for the mechanical and chemical strength and cation exchange timescales of clay-rich soils and formations.

Using the TiZr cells described above, Skipper et al. [81, 82] were able to conduct in situ neutron diffraction studies of clay–fluid interactions in Na, Mg-, and Ca-substituted smectites and vermiculites, up to 360°C and 1.8 kbar. In a typical sedimentary basin, such conditions would correspond to a burial depth of about 12 km [83]. These experiments confirm that interlayer water is denser than the bulk and reveal the structural changes that lead to reversible hydration/dehydration reactions under geological conditions.

17.4 Decontamination and Solid Speciation

One of the main concerns in the environmental sciences is the understanding of the physicochemical processes driving the retention or trapping of pollutant molecules by mineral phases. Processes like adsorption, surface precipitation, or co-precipitation control the bioavailability of pollutants in the environment. As shown in Fig. 17.4, wet chemical methods have classically been used for this type of study: the simple measurement of the concentration of a hazardous element and of the pH in the solution that contains it gives information about its solubility, chemical speciation, and mobility.

17.4.1 Surface Processes

Adsorption of contaminants on important soil constituents as iron or aluminum oxides [84, 85] and (oxy)hydroxides like clays [86], carbonates [87], or sulfides [88] has been widely studied over the past decades. Surface complexation modeling studies are today complemented with spectroscopic techniques that provide an atomistic description of the mineral surfaces and adsorption sites [89]. X-ray absorption spectroscopy has become a well-known, widely used technique within the environmental sciences, with dedicated synchrotron beam lines [90, 91].

The use of neutron scattering techniques in environmental studies is limited by the fact that usually "mineral particles" have very low surface-to-bulk ratios (on the order of 10^{-2}), a characteristic that makes the signals coming from correlations between the atoms on the surfaces very weak. However, some studies have been performed on the structure and dynamics of water on the surfaces of nanometric metal oxides where surface-to-bulk ratios are higher [35]. The advent of more powerful neutron sources and improvements in the instrumentation should help in the development of neutron scattering techniques for the study of water on the more heterogeneous surfaces of mineral particles.

The NDIS technique has attracted much interest in the study of ion uptake mechanisms in clay interlayers. The water structure and counter-ion coordination in the interlayer region of smectite clays such as montmorillonite are fundamental to many of the important properties of the clays and important for understanding applications such as radioactive waste disposal. But the small size of the clay crystallites and the relatively poor degree of preferential orientation achievable in aggregates render the application of conventional crystallographic methods to the study of interlayer structure difficult. The NDIS method has allowed the study of the hydration of hazardous cations in the interlayer of clay minerals [23, 77, 92, 93]. The hydration shell structure of the aqua-ions may change when they are adsorbed on the clay interlayer. Such a change was shown by Pitteloud et al. [23] in the case of Nd^{3+}- and Yb^{3+}-exchanged Wyoming montmorillonite. The lanthanide(III)-exchanged forms are of interest in the context of the migration of radionuclides in clay containment barriers and in the environment. This is possible assuming that these lanthanides and their equivalent actinides (just below in the periodic table) are isomorphous [23, 94]. In this context, other studies using EXAFS (equivalent to NDIS with X-rays) have been done [95–97]. Neodymium and ytterbium were chosen because of their positions among the early, larger, lanthanides and later, smaller lanthanides, respectively, and for the availability of isotopes with contrasting neutron scattering lengths [34]. In aqueous solution, Nd^{3+} and Yb^{3+} are found by neutron diffraction to be hydrated by nine and eight water molecules, respectively [98]. Using the NDIS technique with ^{172}Yb, ^{174}Yb, ^{144}Nd, ^{143}Nd, H_2O, and D_2O substitutions, it was possible to probe in detail the local environment of water molecules and cations in the interlayer region of smectite clays. The ability to separate the ion–hydrogen and ion–oxygen pair distribution functions enabled the researchers to distinguish between binding to water molecules and binding to other oxygen atoms. In that work, Pitteloud et al. [23] showed that a mild heat treatment produces a dehydration of the ions, suggesting that the lanthanide(III) ions may bind to the clay layers under burial conditions.

The samarium–clay interaction is of interest as this lanthanide can be considered a chemical analog of Pu^{3+}, which comes out in nuclear waste as a fission decay product of uranium. A NDIS and molecular dynamics study by Sobolev et al. [93] reports data on the sorption mechanism of Sm^{3+} on the montmorillonite clay interlayer, through isotopic exchange of ^{152}Sm with ^{154}Sm and H_2O with D_2O. The hydration of Sm^{3+} was simulated by means of molecular dynamics (Fig. 17.6b), which helped in the interpretation of the diffraction data. The results of the study show a decrease in the coordination number for hydrogen and oxygen atoms, with respect to the Sm^{3+} aqua-ion, which would indicate that the Sm^{3+} ion is binding to the clay surface via oxygen siloxane atoms and is probably partially hydrolyzed. This sorption mechanism, known as inner-sphere complexation, indicates a higher degree of stability for the complex with respect to the Sm^{3+} aqua-ion, affecting the mobility and transport of the Sm^{3+}. Other neutron techniques, such as quasielastic neutron scattering, can be used as dynamical probes for this kind of adsorption mechanism. Quasielastic neutron scattering study of water and organic mobility in clay interlayers is currently an active field of research [72, 99, 100] that undoubtedly has applications in the studies of contaminant retention by clays, as water mobility may be affected by the degree of hydration of the ions in the interlayer.

17.4.2 Pollutant Trapping in Bulk Mineral Phases

Less attention has been paid to the retention of pollutants in the bulk of mineral phases. Uptake of contaminants in solid phases can remove ions from solution, retarding their transport. When a contaminant is incorporated in the bulk, rather than simply adsorbed at the surface, it is less available and can be considered immobilized in the environment, at least until the host phase is dissolved. Some recent works underline the importance of having a sound knowledge of the mineral species derived from processing waste and the vital role that mineralogy plays in waste management [101]. Precise definition of the intrinsic nature of the pollutant (molecule type, degree of oxidation) and of the existing relationship with the mineral trap (nature of the bonding, location in the host lattice) enables sound predictive management of the waste problems associated with industrial and natural processes. A sound example of pollution is groundwater and drinking water contamination by arsenic [8]. Arsenic is a highly toxic element that has been classified in Group 1 as a carcinogenic species for human beings by the International Research Center for Cancer. The maximum permissible concentration in drinking water is 10 ppb [102]. The average crustal abundance of arsenic is 1.5 mg kg^{-1} (1.5 ppm). Typical concentrations in natural waters are much lower, with higher levels seen in river and seawaters than in rain or lake waters. Much greater levels are found locally in waters, soils, and sediments where arsenic associated with mining, smelting, or other industrial activities, or with geothermal systems, can dramatically enhance concentrations (by factors of 100–1000 or even more). Arsenic does not readily substitute into the structures of the major rock-forming minerals. It is found mainly as minerals where the arsenic occurs as the anion or dianion (As_2) or the sulfarsenide anion (AsS); these anions are bonded to metals such as iron, cobalt, or nickel (e.g., $FeAs_2$, FeAsS, CoAsS). There are also simple arsenic sulfide minerals such as realgar (AsS) and orpiment (As_2S_3).

17.4.2.1 Pollutant Trapping in Geogenic Minerals

When these primary minerals are exposed at or near the surface of the earth, reaction with the atmosphere and surface waters causes alteration; simple arsenic oxides or more complex phases containing various metals combined with arsenic, oxygen, and other anion species are formed. A large number of secondary arsenic minerals are produced in this way. Natural arsenic contamination occurs in the Bangladesh-Bramapoutre delta upper aquifers [24], the Argentinean Pampa aquifers [103], and some eutrophic lakes. Waters from these sites are at the same time at equilibrium or slightly supersaturated with respect to calcite. Calcite has been shown to be an effective scavenger of a variety of trace elements, including heavy metals, rare earth elements, actinides, and oxyanions such as selenate and phosphate; and it is capable of retaining such trace elements via adsorption (surface) reactions as well as through co-precipitation reactions (see Fig. 17.4) [104, 105].

Neutron scattering techniques have been proved to be very useful for quantifying the degree of substitution of arsenic anions in the bulk of mineral phases. The non-electrical character of the neutron–nuclei interaction makes the neutron an ideal probe for bulk properties. This is the case for the study of arsenite ($HAsO_3^{2-}$) incorporation into calcite crystals. Following several studies on the adsorption of arsenic in calcite surfaces (very relevant surfaces for environmental geochemistry in a huge variety of sites), Román-Ross et al. [30] have shown the mechanism of incorporation of arsenite anions into the bulk of calcite, through the substitution of carbonate molecules (see Fig. 17.8). A volume expansion has been found for synthetic calcite samples co-precipitated with As(III), as revealed by neutron diffraction. This study was complemented by density functional theory (DFT) simulations, proving that this substitution produces an expansion of the lattice cell compatible with the expansion observed in the experimental data [106]. In a previous study, Cheng et al. [107] demonstrated experimentally with the use of X-ray standing waves how arsenite molecules can adapt to the surface of calcite by replacing a carbonate. The incorporation of arsenite into the bulk of calcite plays a very important role in the quantification and immobilization of arsenic in natural environments.

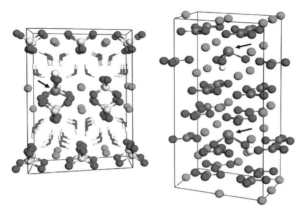

Fig. 17.8 Incorporation of an arsenate molecule $HAsO_4^{2-}$ in the gypsum unit cell (*left*) and of an arsenite molecule $HAsO_3^{2-}$ in the calcite unit cell (*right*). The arrows show arsenic atoms

17.4.2.2 Industrial-Bearing Mineral Phases

Several mineral phases are released to the environment as byproducts of industrial processes, making them susceptible to interacting with pollutant molecules and playing an important role in their transport and retention. Familiar examples are industrial activities in which mineral ores are smelted, generating arsenic-rich gypsum sludges produced by neutralization of arsenic-rich acidic solutions [108, 109]. Often, gypsum from those sludges appears associated with calcium arsenates [110]. Typically, however, little information is obtained from the incorporation of arsenic into the bulk of gypsum, which may potentially lead to its long-term immobilization in the mineral structure. A combined neutron diffraction and X-ray absorption spectroscopy study has revealed the ability of gypsum to host up to 1 mol As/kg [106, 111]. Rietveld refinement of neutron powder diffraction data combined with DFT-based calculations revealed the structure and properties of the process by which a sulfate (SO_4^{2-}) is substituted for an arsenate ($HAsO_4^{2-}$) molecule [106, 111]. The gypsum structure is formed by layers of ionic bonded sulfate molecules (SO_4^{2-}) and calcium atoms (Ca^{2+}). Water molecules are intercalated between the layers, holding them together through hydrogen bonds. The trapping process relies on the substitution of a sulfate by an arsenate molecule (see Fig. 17.8). Mulliken population analyses have demonstrated the ability of gypsum to adapt its local structure around an arsenic center to accommodate the host arsenate molecule. Neutron powder diffraction experiments have revealed an expansion of the unit cell volume proportional to the concentration of arsenic present in the gypsum samples. Two factors make neutron diffraction the better adapted technique to follow this volume expansion: its neutral electronic character makes the neutron an ideal probe for probing the bulk of the material under study, and the fact that deuterium is a good scatterer of neutrons makes it possible to determine the exact atomic positions for all the deuterium atoms of a highly hydrated structure such as gypsum.

17.5 Perspectives

Neutron radiation is increasingly being used to characterize complex environmental materials, in particular when those materials are in unusual or extreme environments (e.g., particles in reactive media, biological growth media, high pressure and temperature, and salinity). Indeed, environmental scientists need to develop their fundamental understanding of interactions between particles and liquids (or gases) at various length scales. With increasingly tighter regulation, environmental sustainability will also demand intelligent design of techniques such as confinement and immobilization, site remediation, and water treatment. The key for many of these future challenges lies in the nanoscale regime, and it is therefore imperative that environmental science and technology have the appropriate analytical tools available to explore the nanostructure of contaminant and engineered decontamination processes. This task requires the development and use of advanced analytical technologies.

Although electron microscopy often involves highly specific sample treatments prior to "post-mortem" structure visualization, as well as risk of radiation damage, more recently developed experimental methods—such as neutron scattering in conjunction with advanced microscopies (scanning tunneling microscopy/atomic force microscopy)—will revolutionize future progress in the environmental sciences. It is worth noting that neutron scattering has for many years been a bedrock of condensed matter science, but it has been applied only relatively recently to environmental and earth sciences. In line with this, there are highly promising developments in neutron reflectivity for probing surfaces and interfaces, which is clearly necessary in studying the interaction of minerals and pollutants. Likewise, small-angle neutron scattering is another important technique that can be used to determine structures and morphologies at the micrometer size level. As always in the field of structural studies, developments are necessary in contrast variation, where at present the two most extensively used techniques are NDIS and element-specific EXAFS. Major historical concerns with NDIS have been the somewhat limited suite of suitable isotopes and their high cost (though this is now being addressed, particularly in the case of specific deuteration). For EXAFS, a concern is of course the inability to see light atoms (e.g., oxygen or hydrogen) in the presence of heavier atoms. It is worth mentioning that the technique of neutron diffraction with isomorphic substitution has been successfully applied studying the structure of phosphate glasses [40–42], opening the possibility of less expensive experiments as compared with NDIS. In some cases, where the concerned elements have absorption resonances for neutron scattering, the anomalous dispersion could be very useful [45, 46]. Dynamical environmental studies are set to benefit from promising techniques such as neutron time-of-flight and neutron spin-echo, which allow one to probe diffusion processes at the picosecond or nanosecond timescales. Neutron backscattering technique could fill the gap in the time scale between time-of-flight (ps) and spin-echo (ns).

Besides the high potential of neutron scattering to visualize structures on the nanometer scale, this technique is unique in that it can observe structural changes in real time (down typically to milliseconds), under environmental-like "stopped-flow" conditions (water, flow, pressure, temperature), and in a non-destructive manner. Under the same conditions, neutron scattering is also the ideal tool to study redox change, e.g., during oxidation of organic molecules.

Several topics concerning environmental materials need to be addressed in a concerted way, allowing the results from each area to influence the development of the others. Among these topics, those we see as worth a special mention are (1) understanding the occurrence, behavior, and reactivity of natural particles; (2) development of new absorbent materials with superior reactivity and behavior; (3) fundamental studies of structure–function relationships for environmental particles; (4) interaction of particles with organisms, determining effects on health and ecosystems; (5) risk assessment and life cycle analysis; (6) time-resolved studies of dynamics in complex systems with competing interactions; (7) kinetic studies with high spatial resolution of chemical reactions that occur at the surfaces of environmental particles; (8) in situ studies by static and dynamic methods of particle structure, aggregation, and texture to account at the nanoscale for environmental and toxicological properties; and (9) microscopic understanding of properties and organization of the cell interfaces in terms of molecular permeability, diffusion, and reactivity to arrive at models for the mechanism of toxicity from nanosize particles.

Future achievements will depend strongly on the development of new tools capable of performing in situ real-time measurements at the nanoscale under realistic conditions. Effective communication between environmental scientists and industry may lead to new analytical concepts. New developments are also required in software: current software is hard-pressed to cope with vast amounts of data, which certainly result from time-resolved experiments and in some cases even from static measurements. Moreover, it would be a great advantage for on-the-fly optimization of time-resolved experiments if the software that enabled real-time interpretation was available. The interaction between staff at large facilities and new users from environmental science should be strengthened, as the latter often come from backgrounds that are not related to neutron science at all. From the instrumental point of view, increased flux at neutron sources coupled with advances in optics and counting devices could facilitate specific kinetic experiments (involving contrast variation) under conditions mimicking realistic natural processes. In this sense, the development of appropriate in situ cells should reproduce environment-like conditions. Because environmental research is by its very nature multidisciplinary, single or isolated efforts cannot address all the key challenges. It is at the nanoscale that physical changes and interactions with contaminants determine the designed decontamination process efficiency. So an integrated and multidisciplinary approach to key problems in the environmental research and industry is required.

In order to make progress, partnerships between environmental industries and agencies and neutron scientists are mandatory. The environment can be studied successfully only when research is extended from structural and compositional studies to those of environmental dynamics. This work will involve time-resolved structural studies by means of neutron small-angle scattering and diffraction. Mostly, parallel use of different techniques and the application of complicated external conditions based on concentration and hydraulic potential gradients will be necessary.

These conclusions should at least materialize in (1) instrument development at large-scale facilities for multi-technical analysis, in real time, under conditions simulating appropriate engineering strategies, dedicated to decontamination problems and waste management; (2) a network of interdisciplinary researchers involved with the study of the interplay between micro- and nanostructures made of environmental particles, cells, and upper organisms; (3) enhanced cooperation between environmental experts from the waste and water industry and academia, and scientists at large-scale facilities; and (4) a high priority on environmental research in the allocation of beam line access at research facilities.

Acknowledgments The work was partially supported by ANDRA (National Radioactive Waste Management Agency, France), European Union projects FUNMIG (Fundamental Process of Radioactive Mitigation) and RECOSY (Redox Phenomena Controlling Systems), and Natural Environment Research Council (United Kingdom). We wish to dedicate this chapter to G. Sposito. We thank M. Johnson for the helpful discussions on numerical simulations and the C-Lab at ILL for the support on computing time. We gratefully thank D. Tisserand for her technical assistance for analytical determinations and F. Bardelli from INFM-GILDA (ESRF) for his help for EXAFS experiments.

References

1. *Basic Research Needs for Geosciences: Facilitating 21st Century Energy Systems* (U.S. Department of Energy–Office of Basic Energy Sciences, 2007), pp. 287
2. *Gennesys—Grand European Initiative on Nanoscience and Nanotechnology using Neutron and Synchrotron Sources* (European Commission, 2005) (www.sinq.web.psi.ch/sinq/gennesys/gennesys.html)
3. *Neutron Scattering in Earth Science: Reviews in Mineralogy and Geochemistry*, Vol. 63, edited by H.-R. Wenk (Geochemical Society and Mineralogical Society of America, Berkeley, 2006)
4. E. Nieboer and D. H. S. Richardson, Environ. Pollut. B **1**, 3 (1980)
5. *Heavy Metals in the Environment: Origin, Interaction and Remediation*, Vol. 6, edited by H. Bradl, Interface Science and Technology series, edited by A. Hubbard (Academic Press, Orlando, Florida, 2005)
6. L. Charlet and A. Boudou, La Recherche **359**, 52 (2002)
7. L. Charlet and D. Polya, Elements **2**, 91 (2006)
8. *Energy: A Geoscience Perspective*, edited by A. M. Macfarlane, Elements, **3** (2007)
9. R. J. Reeder, M. A. A. Schoonen, and A. Lanzirotti, *Medical Mineralogy and Geochemistry: Reviews in Mineralogy and Geochemistry*, Vol. 64, edited by N. Sahai and M. A. A. Schoonen (Geochemical Society and Mineralogical Society of America, Berkeley, 2006), p. 59
10. S. N. Luoma, Sci. Total Environ. **28**, 1 (1983)
11. J. McKinney and R. Rogers, Environ. Sci. Technol. **26**, 1298 (1992)
12. J. S. Meyer, Mar. Environm. Res. **53**, 417 (2002)
13. Z. L. He, X. E. Yang, and P. J. Stoffella, J. Trace Elem. Med. Biol. **19**, 125 (2005)
14. G. Sposito, *The Surface Chemistry of Soils* (Oxford University Press, Oxford, 1984)
15. G. Blanc, Y. Lapaquellerie, N. Maillet, and P. Anschutz, Hydrobiologia **410**, 331 (1999)
16. L. Spadini, P. W. Schindler, L. Charlet, A. Manceau, and K. V. Ragnarsdottir, J. Colloid Interface Sci. **266**, 1 (2003)
17. D. A. Dzomback and F. M. M. Morel, *Surface Complexation Modeling: Hydrous Ferric Oxide* (Wiley Interscience, New York, 1990)

18. F. M. M. Morel and J. G. Hering, *Principles and Applications of Aquatic Chemistry* (John Wiley and Sons, New York, 1993)

19. L. I. Sweet and J. T. Zelikoff, J. Toxicol. Environm. Health **4**, 161 (2001)

20. F. Seby, M. Potin-Gautier, E. Giffaut, and O. F. X. Donard, Analusis **26**, 193 (1998)

21. M. Dozol and R. Hagemann, Pure Appl. Chem. **65**, 1081 (1993)

22. G. Sposito, N. T. Skipper, R. Sutton, S. H. Park, A. K. Soper, and J. A. Greathouse, Proc. Nat. Acad. Sci. **96**, 3358 (1999)

23. C. Pitteloud, D. H. Powell, M. A. González, and G. J. Cuello, Colloids Surf. A: Physicochem. Eng. Aspects **217**, 129 (2003)

24. *Arsenic Contamination of Groundwater in Bangladesh*, Technical Report WC/00/19, Vol. 1, edited by D. G. Kinniburgh and P. L. Smedley (British Geological Survey, Keyworth, UK, 2002)

25. P. L. Smedley, H. B. Nicolli, D. M. J. Macdonald, A. J. Barros, and J. O. Tullio, Appl. Geochem. **17**, 259 (2002)

26. F. Ramisch, M. Dittrich, C. Mattenberger, B. Wehrli, and A. Wüest, Geochim. Cosmochim. Acta **63**, 3349 (1999)

27. J. Plant, D. Smith, B. Smith, and L. Williams, Appl. Geochem. **16**, 1291 (2001)

28. F. Juillot, Ph. Ildefonse, G. Morin, G. Calas, A. M. de Kersabiec, and M. Benedetti, Appl. Geochem. **14**, 1031 (1999)

29. L. Charlet, S. Chakraborty, C. A. J. Appelo, G. Román-Ross, B. Nath, A. A. Ansari, M. Lanson, D. Chatterjee, and S. Basu Mallik, Appl. Geochem. **22**, 1273 (2007)

30. G. Román-Ross, G. J. Cuello, X. Turrillas, A. Fernández-Martínez, and L. Charlet, Chem. Geol. **233**, 328 (2006)

31. L. Charlet and A. Manceau, *Environmental Particles*, Vol. 2, edited by J. Buffle and H. P. Van Leeuwen, IUPAC Environmental Analytical and Physical Chemistry Series (Lewis Publishing, Chelsea, Michigan, 1993), p. 117

32. K. J. Farley, D. A. Dzombak, and F. M. M. Morel, J. Colloid Interface Sci. **106**, 227 (1985)

33. G. J. Cuello, J. Phys.: Cond. Matter **20**, 244109 (2008)

34. V. F. Sears, Neutron News **3**, 26 (1992)

35. E. Mamontov, L. Vlcek, D. J. Wesolowski, P. T. Cummings, W. Wang, L. M. Anovitz, J. Rosenqvist, C. M. Brown, and V. G. Sakai, J. Phys. Chem. C **111**, 4328 (2007)

36. G. J. Cuello, M. Jiménez-Ruiz, and M. A. González, J. Non-Cryst. Solids **353**, 724 (2007)

37. G. L. Squires, *Introduction to the Theory of Thermal Neutron Scattering* (Cambridge University Press, Cambridge, 1978)

38. J. S. Higgins and H. C. Benoit, *Polymers and Neutron Scattering*, Vol. 8, Oxford Series on Neutron Scattering in Condensed Matter (Clarendon Press, Oxford, 1994)

39. R. A. Martin, P. S. Salmon, H. E. Fischer, and G. J. Cuello, Phys. Rev. Lett. **90**, 185501 (2003)

40. R. A. Martin, P. S. Salmon, H. E. Fischer, and G. J. Cuello, J. Phys. Condens. Matter **15**, 8235 (2003)

41. R. A. Martin, P. S. Salmon, C. J. Benmore, H. E. Fischer, and G. J. Cuello, Phys. Rev. B **68**, 054203 (2003)

42. G. J. Cuello, C. Talón, C. Cabrillo, and F. J. Bermejo, Appl. Phys. A: Mat. Sci. Process. **74**, S552 (2002)

43. C. Talón, F. J. Bermejo, C. Cabrillo, G. J. Cuello, M. A. González, J. W. Richardson, Jr., A. Criado, M. A. Ramos, S. Vieira, F. L. Cumbrera, and L. M. González, Phys. Rev. Lett. **88**, 115506 (2002)

44. A. C. Wright, J. M. Cole, R. J. Newport, C. E. Fisher, S. J. Clarke, R. N. Sinclair, H. E. Fischer, and G. J. Cuello, Nucl. Instr. Meth. Phys. Res. A **571**, 622 (2007)

45. J. M. Cole, A. C. Wright, R. J. Newport, R. N. Sinclair, H. E. Fischer, and G. J. Cuello, J. Phys. Condens. Matter **19**, 056002 (2007)

46. P. S. Salmon, A. C. Barnes, R. A. Martin, and G. J. Cuello, J. Phys. Condens. Matter **19**, 415110 (2007)

47. P. S. Salmon, R. A. Martin, P. E. Mason, and G. J. Cuello, Nature **435**, 75 (2005)

48. A. K. Soper, G. W. Neilson, J. E. Enderby, and R. A. Howe, J. Phys. C: Solid St. Phys. **10**, 1793 (1977)
49. T. C. Hansen, P. F. Henry, H. E. Fischer, J. Torregrossa, and P. Convert, Meas. Sci. Technol. **19**, 034001 (2008)
50. H. E. Fischer, G. J. Cuello, P. Palleau, D. Feltin, A. C. Barnes, Y. S. Badyal, and J. M. Simonson, Appl. Phys. A **74**, S160 (2002)
51. G. J. Cuello, R. Fernández-Perea, C. Cabrillo, F. J. Bermejo, and G. Román-Ross, Phys. Rev. B **69**, 094201 (2004)
52. G. J. Cuello, R. Fernández-Perea, F. J. Bermejo, G. Román-Ross, and J. Campo, J. Non-Cryst. Solids **353**, 2987 (2007)
53. L. Hennet, I. Pozdnyakova, A. Bytchkov, V. Cristiglio, P. Palleau, H. E. Fischer, G. J. Cuello, M. Johnson, P. Melin, D. Zanghi, S. Brassamin, J. F. Brun, D. L. Price, and M. L. Saboungi, Rev. Scient. Instr. **77**, 053903 (2006)
54. S. Klotz, Th. Strässle, G. Rousse, G. Hamel, and V. Pomjakushin, Appl. Phys. Lett. **86**, 031917 (2005)
55. H. Ohtaki and T. Radnai, Chem. Rev. **93**, 1157 (1993)
56. D. Bougeard and K. S. Smirnov, Phys. Chem. Chem. Phys. **9**, 226 (2007)
57. N. T. Skipper, Min. Mag. **62**, 657 (1998)
58. J. E. Enderby, S. Cummings, G. J. Herdman, G. W. Neilson, P. S. Salmon, and N. T. Skipper, J. Phys. Chem. **91**, 5851 (1987)
59. N. T. Skipper, S. Cummings, G. W. Neilson, and J. E. Enderby, Nature **321**, 52 (1986)
60. J. J. R. Fraústo da Silva and R. J. P. Williams, *The Biological Chemistry of the Elements* (Oxford University Press, New York, 2001)
61. O. Sobolev, G. J. Cuello, G. Román-Ross, N. T. Skipper, and L. Charlet, J. Phys. Chem. A **111**, 5123 (2007)
62. N. A. Hewish, G. W. Neilson, and J. E. Enderby, Nature **297**, 138 (1982)
63. Y. S. Badyal, A. C. Barnes, G. J. Cuello, and J. M. Simonson, J. Phys. Chem. A **108**, 11819 (2004)
64. S. Dixit, J. Crain, W. C. K. Poon, J. L. Finney, and A. K. Soper, Nature **416**, 829 (2002)
65. D. T. Bowron and S. D. Díaz Moreno, J. Phys.: Condens. Matter **15**, S121 (2003)
66. G. J. Cuello, F. J. Bermejo, R. Fayos, R. Fernández-Perea, A. Criado, F. R. Trouw, C. Tam, H. R. Schober, E. Enciso, and N. G. Almarza, Phys. Rev. B **57**, 8254 (1998)
67. G. J. Cuello, R. Fayos, F. J. Bermejo, R. Fernández-Perea, C. Tam, and F. R. Trouw, Mol. Phys. **93**, 341 (1998)
68. G. D. Williams, A. K. Soper, N. T. Skipper, and M. V. Smalley, J. Phys. Chem. B **102**, 8945 (1998)
69. G. D. Williams, N. T. Skipper, M. V. Smalley, A. K. Soper, and S. M. King, Faraday Disc. **104**, 295 (1997)
70. K. S. Sidhu, J. M. Goodfellow, and J. Z. Turner, J. Chem. Phys. **110**, 7943 (1999)
71. P. A. Lock and N. T. Skipper, Eur. J. Soil Sci. 58, 958 (2007)
72. N. T. Skipper, P. A. Lock, J. O. Titiloye, J. Swenson, Z. A. Mirza, W. S. Howells, and F. Fernández-Alonso, Chem. Geol. **230**, 182 (2006)
73. H. Thompson, A. K. Soper, M. A. Ricci, F. Bruni, N. T. Skipper, J. Phys. Chem. B **111**, 5610 (2007)
74. D. J. Cebula, R. K. Thomas, S. Middleton, R. H. Ottewill, and J. W. White, Clays and Clay Miner. **27**, 39 (1979)
75. R. K. Hawkins and P. A. Egelstaff, Clays and Clay Miner. **28**, 19 (1980)
76. D. H. Powell, K. Tongkhao, S. J. Kennedy, and P. G. Slade, Clays and Clay Miner. **45**, 290 (1997)
77. D. H. Powell, H. E. Fischer, and N. T. Skipper, J. Phys. Chem. B **102**, 10899 (1998)
78. N. T. Skipper, A. K. Soper, and J. D. C. McConnell, J. Chem. Phys. **94**, 5751 (1991)
79. N. T. Skipper, M. V. Smalley, G. D. Williams, A. K. Soper, and C. H. Thompson, J. Phys. Chem. **99**, 14201 (1995)
80. N. T. Skipper, M. V. Smalley, and A. K. Soper, J. Phys. Chem. **98**, 942 (1994)

81. A. V. C. Siqueira, C. Lobban, N. T. Skipper, G. D. Williams, A. K. Soper, J. Dreyer, R. J. Humphreys, and J. A. R. Bones, J. Phys.: Cond. Matter. **11**, 9179 (1999)
82. N. T. Skipper, G. D. Williams, A. V. C. Siqueira, C. Lobban, and A. K. Soper, Clay Miner. **35**, 287 (2000)
83. F. K. North, *Petroleum Geology* (Allen and Unwin, London, 1994)
84. G. A. Waychunas, B. A. Rea, C. C. Fuller, and J. A. Davis, Geoch. Cosmoch. Acta **57**, 2251 (1993)
85. G. A. Waychunas, C. S. Kim, and J. F. Banfield, J. Nano. Res. **7**, 409 (2005)
86. French National Agency for Radioactive Waste Management (ANDRA), *Dossier 2005 Argile*
87. C. A. J. Appelo, M. J. J. Van der Weiden, C. Tournassat, and L. Charlet, Env. Sci. Tech. **36**, 3096 (2002)
88. B. C. Bostick and S. Fendorf, Geoch. Cosmo. Acta **67**, 909 (2003)
89. T. Hiemstra and W. H. Van Riemsdijk, J. Coll. Int. Sci. **179**, 488 (1996)
90. T. Reich, G. Bernhard, G. Geipel, H. Funke, C. Hennig, A. Rossberg, W. Matz, N. Schell, and H. Nitsche, Radiochim. Acta **88**, 633 (2000)
91. O. Proux, X. Biquard, E. Lahera, J.-J. Menthonnex, A. Prat, O. Ulrich, Y. Soldo, P. Trévisson, G. Kapoujyan, G. Perroux, P. Taunier, D. Grand, P. Jeantet, M. Deleglise, J.-P. Roux, and J.-L. Hazemann, Physica Scripta **T115**, 970 (2005)
92. C. Pitteloud, D. H. Powell, and H. E. Fischer, Phys. Chem. Chem. Phys. **3**, 5567 (2001)
93. O. Sobolev, L. Charlet, G. J. Cuello, A. Gehin, J. Brendle, and N. Geoffroy, J. Phys: Cond. Matter **20**, 104207 (2008)
94. J. M. Trillo, M. D. Alba, R. Alvero, M. A. Castro, A. Muñoz Páez, and J. Poyato, Inorg. Chem. **33**, 3861 (1994)
95. A. Muñoz Páez, M. D. Alba, M. A. Castro, R. Alvero, and J. M. Trillo, J. Phys. Chem. **98**, 3861 (1994)
96. A. Muñoz Páez, M. D. Alba, R. Alvero, A. I. Becerro, M. A. Castro, and J. M. Trillo, Nucl. Instr. Methods B **97**, 3861 (1995)
97. A. Muñoz Páez, M. D. Alba, R. Alvero, A. I. Becerro, M. A. Castro, and J. M. Trillo, Physica B **209**, 622 (1995)
98. C. Cossy, L. Helm, D. H. Powell, and A. E. Merbach, New J. Chem. **19**, 27 (1995)
99. N. Malikova, A. Cadène, V. Marry, E. Dubois, P. Turq, J. M. Zanotti, and S. Longeville, Chem. Phys. **317**, 226 (2005)
100. L. J. Michot, A. Delville, B. Humbert, M. Plazanet, and P. Levitz, J. Phys. Chem. C **111**, 9818 (2007)
101. P. Piantone, Comp. Rend. Geo. **336**, 1415 (2004)
102. *Guidelines for Drinking Water Quality*, Vol. 1, 2nd ed., 198 (World Health Organization, Geneva, 1993)
103. P. L. Smedley and D. G. Kinniburgh, App. Geoch. **17**, 517 (2002)
104. J. M. Zachara, C. E. Cowan, and C. T. Resch, Geoch. Cosmo. Acta **55**, 1549 (1991)
105. R. J. Reeder, Geoch. Cosmo. Acta **60**, 1543 (1996)
106. A. Fernández-Martínez, G. Román-Ross, G. J. Cuello, X. Turrillas, L. Charlet, M. R. Johnson, and F. Bardelli, Physica B **385–386**, 935 (2006)
107. L. W. Cheng, P. Fenter, N. C. Sturchio, Z. Zhong, and M. J. Bedzyk, Geoch. Cosmo. Acta **63**, 3153 (1999)
108. *Active and Semi-Passive Lime Treatment of Acid Mine Drainage* (U.S. Environmental Protection Agency, Washington, D.C., May 2006)
109. T. Gominsek, A. Lubej, and C. J. Pohar, J. Chem. Techn. Biotech. **80**, 939 (2005)
110. L. Charlet, A. A. Ansari, G. Lespagnol, and M. Musso, Sci. Tot. Environ. **277**, 133 (2001)
111. A. Fernández-Martínez, G. J. Cuello, M. R. Johnson, F. Bardelli, L. Charlet, G. Román-Ross, and X. Turrillas, J. Phys. Chem. A **112**, 5159 (2008)

Chapter 18
Clay Swelling: New Insights from Neutron-Based Techniques

Isabelle Bihannic, Alfred Delville, Bruno Demé, Marie Plazanet, Frédéric Villiéras, and Laurent J. Michot

Abstract Clayey materials are complex hierarchical and deformable porous media whose structure and organization vary at different spatial scales depending on external conditions, in particular water activity. It is therefore important, on the one hand, to follow all the structural changes that are associated with the adsorption of water molecules in the interlamellar spaces (at the scale of the particles) and, on the other hand, to describe the textural modifications induced at larger scales as a result of the swelling of individual particles. Neutron-based techniques are important to achieving this multiscale description, thanks to some special features of neutrons [e.g., specific interaction with hydrogen atoms, with in addition differential interaction with isotopes (H and D), and high penetration length of neutron beams, which allows easy preparation of versatile sample-cells container]. Finally, water dynamics in the interlayers can be investigated because of the unique interaction of neutrons with hydrogen.

After a brief discussion of the crystal chemistry of swelling clays, some examples of the application of neutron techniques to the study of clays will be given, including application of neutron diffraction to the study of the structure evolution of various expandable clays upon hydration and investigation of interlayer water dynamics by quasielastic neutron scattering (QENS) experiments, illustrating water molecule mobility as a function of hydration states. The difficulties in applying these techniques to materials with such complex crystal chemistry and anisotropic shape will be pointed out.

Clay hydration and swelling induce modification of aggregates at larger spatial scale than the nanometric one investigated by diffraction. Clay fabric and particles organization at the sub-micronic scale can be investigated by small-angle neutron scattering (SANS) experiments. Examples of SANS measurements will be given in the so-called crystalline swelling domain, for water activity below 1. In the clay–water system, as the solid/liquid ratio decreases, gels formation is observed in which individual clay layers are now separated by distances larger than a few

I. Bihannic (✉)

Laboratoire Environnement et Mineralurgie, UMR 7569 CNRS-INPL-ENSG, F-54501 Vandoeuvre Lès Nancy cedex, France

e-mail: isabelle.bihannic@ensg.inpl-nancy.fr

L. Liang et al. (eds.), *Neutron Applications in Earth, Energy and Environmental Sciences*, Neutron Scattering Applications and Techniques,
DOI 10.1007/978-0-387-09416-8_18, © Springer Science+Business Media, LLC 2009

nanometers. Experimental studies performed on clay suspensions or gels will be presented, demonstrating that clay layers are equilibrated in gels via electrostatic repulsions.

18.1 Introduction

Clay minerals are finely divided silicates that are ubiquitous in soils and sediments and are widely used in many industrial applications [1 and references within]. The following are among the numerous uses of one class of clay minerals, the bentonites:

- as plasticizing agents in ceramics;
- as binding agents in the pellitization of ores or other finely divided solids (application in foundries for moldings) or in the preparation of animal feed;
- as additives to improve the viscosity and rheological properties of drilling muds (petroleum industry);
- as soil additives in the construction of diaphragm walls or engineered barriers for waste management (civil engineering);
- as thickening agents for paints, varnishes, and cosmetics;
- as adsorbents in the food and beverage industries (e.g., for clarification of wines, beer stabilization, and, after acid activation, decolorizing vegetable and mineral oils); and
- as catalysts and catalyst supports in many organic syntheses.

Most of these uses are linked to the unique properties of clay minerals in terms of rheological behavior, water retention, and ionic exchange capacities—properties that are connected to their physicochemical characteristics.

One of the main features of clay minerals is their ability to accommodate important amounts of water in their structure, a phenomenon referred to as "clay swelling." The associated processes first occur at the nanometric scale within interlayer spaces and have strong consequences for various macroscopic behaviors such as mud rheology, swelling and shrinkage of soils, and slope stability. Furthermore, in applications where clay minerals are used as barrier materials, the transport of water and ions at large scale are somehow linked to the dynamics of water at the atomic level.

Scrutinizing changes in the structure of clays and in the dynamics of interlayer fluids as a function of water content obviously represents a mandatory prerequisite for unraveling swelling processes. It requires the use of techniques spanning several orders of magnitude in length and time to relate molecular properties to hierarchical structure, both evolving with water content. In that respect neutron scattering techniques are particularly relevant, as outlined by Dove in a special issue of the *European Journal of Mineralogy* on neutron scattering in mineral sciences [2]: "Neutron scattering methods have an astonishing versatility. . . . Since neutron scattering probes both the length and energy scales of atomic processes, it is possible to design experiments that focus on both spatial and dynamical processes at the same time, or else which focus on one or other." It can be added that the particular sensitivity of neutrons to hydrogen atoms makes them powerful probes for analyzing water features (Chapter 2).

The importance of neutron techniques for studying clay minerals has long been recognized. Neutron diffraction [3–23], SANS [24–34], and quasielastic and inelastic scattering [23,35–50] have been applied to the study of different clay structures over the past 30 years.

In this chapter we will try to illustrate the versatility and specificity of neutron-based techniques for studying water–clay interactions on various scales. We will first focus on the use of diffraction and QENS to describe the structure and dynamics of interlayer water in different clay samples. We will then show how SANS can be used to analyze the evolution with increasing water content of the multiscale structure of compacted clay samples (i.e., in conditions that are compatible with the use of clay minerals in civil engineering and in barrier materials). Finally, we will show how SANS can be used to understand the structure of clay suspensions and clay gels.

18.2 Clay Structure and Organization

Clay minerals are phyllosilicates whose unit structure is a layer built from the stacking of sheets of tetrahedrally coordinated cations (tetrahedral layer T) and octahedral (O) sheets of oxygens and cations [51]. Tetrahedra are mainly occupied by silicon, while octahedra are mainly occupied by either aluminum or magnesium. Various clay minerals may be defined depending on the nature of the aforementioned assemblage. The entity of one tetrahedral sheet and one octahedral sheet yields the so-called 1:1 layer silicate structure or TO structure. It is the one encountered for kaolinite (Fig. 18.1). In the 2:1 layer silicate structure (also called TOT structure), two silicon tetrahedral layers sandwich an octahedral layer. This is the case for smectite (swelling clays) and for mica and illite (Fig. 18.1).

Cations other than silicon, aluminum, or magnesium can fit into the voids of the tetrahedral and octahedral polyhedra. When these isomorphic substitutions lead to charge unbalance, electroneutrality is ensured by the counter-ions (cations) located

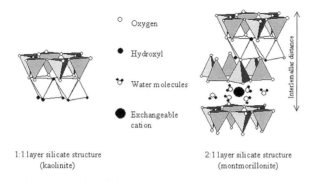

○ Oxygen

● Hydroxyl

♡ Water molecules

● Exchangeable cation

1:1 layer silicate structure
(kaolinite)

2:1 layer silicate structure
(montmorillonite)

Fig. 18.1 Schematic representation of the layer structure of clay minerals

Table 18.1 General classification of 2:1 phyllosilicates

Charge/ unit cell	Dioctahedral clays	Trioctahedral clays
0	Pyrophyllite $(Si_8)(Al_4)O_{20}(OH)_4$	Talc $(Si_8)(Mg_6)O_{20}(OH)_4$
0.4–1.2	**Montmorillonite** $(Si_8)(Al_{4-x}Mg_x)O_{20}(OH)_4CE^{z+}_{x/z}$	**Hectorite** $(Si_8)(Mg_{6-x}Li_x)O_{20}(OH)_4CE^{z+}_{x/z}$
	Beidellite $(Si_{8-x}Al_x)(Al_4)O_{20}(OH)_4CE^{z+}_{x/z}$	**Saponite** $(Si_{8-x}Al_x)(Mg_6)O_{20}(OH)_4CE^{z+}_{x/z}$
1.2–1.8	Illite $(Si_{8-x}Al_x)(Al_{4-y}Mg_y)O_{20}(OH)_4K^+_{x+y}$	Vermiculite $(Si_{8-x}Al_x)(Mg_{6-y}Cat^{3+}_y)O_{20}(OH)_4CE^{z+}_{(x+y)/z}$

Note: Mineral names in bold correspond to smectites.

in the interlamellar spaces. Most properties of swelling clays (e.g., ion exchange properties or water retention) result from the presence of such mobile exchangeable and hydratable cations. A brief classification of clay minerals is given in Table 18.1, which defines the structure and formula of clays mentioned in this chapter. The generic name "smectite" corresponds to the swelling clays reported on row 2 of Table 18.1 (i.e., when the charge of the unit cell is not too high to prevent swelling). Even though vermiculite presents swelling properties under some conditions, this mineral does not belong to the "smectite group," as swelling is limited due to the higher charge of vermiculite compared to that of the smectites.

The basic unit of clays comprises the TO layer (or TOT layer) and associated interlayer space. These units are platelet-shaped and can be considered two-dimensional (2D), with a thickness around 1 nm and a lateral extension that is extremely variable (from a few nanometers to a few microns). The stacking of these units then forms particles.

Cohesion between layers inside particles is strongly dependent on parameters such as charge density and location (tetrahedral or octahedral sheet) and the nature and valence of the exchangeable cations. Cohesion affects crystallite morphology. For example, in micas or illite, where attraction between layers is strong and counter-ions are abundant, the resulting stacking is regular. Also, adjacent 2:1 layers are keyed into positions that prevent mobility and exchange of interlayer counter-ions. Conversely, smectites, which have a lower charge compared to micas, are characterized by extremely disordered and irregular stacking. In such cases, there are no rules for defining the relative positions and orientations of adjacent layers, and the stacking thus formed is referred to as turbostratic. Considering the high polydispersity of the in-plane dimension of the layers, the notion of particle becomes difficult to picture, the layer stacking leading to a more or less continuous and disordered assembly.

The assembly of layers and particles in such a hierarchical structure generates voids, beginning with the interlamellar spaces. Within particles, stacking defaults such as irregularities in the distances sequence also give rise to lenticular pores, with thickness down to a few nanometers. At larger scales, voids existing between particles or aggregates of particles correspond to mesopores or macropores.[1] The morphology of the porous network is strongly related to the solid phase geometry and is also hierarchically organized [53].

This general picture of the clay fabric is strongly modified in the presence of water, as swelling (or water uptake) corresponds to an expansion of interlayer spaces due to penetration of water molecules into the interlamellar regions. Since the pioneering work of Norrish [54] on sodium–montmorillonite, clay scientists commonly distinguish two swelling stages: intracrystalline and osmotic swelling. Intracrystalline or crystalline swelling occurs for the lowest water contents (under-saturated systems) and corresponds to the first hydration steps of interlamellar spaces, mainly through interlayer cation hydration [55–66]. This process occurs in a stepwise manner as a function of water activity. The apparent interlamellar distance (d001) increases step by step from a value around 9.5 Å for the dry state to roughly 12.5 Å for a state referred to as water monolayer state (the increment of 3 Å corresponds to the size of the water molecule). As water activity increases, the interlamellar distance reaches 15.5 Å for the so-called bilayer state. The domains of existence of the different states, as well as the exact d001 value, strongly depend on the cation nature and on the amount and localization of charges in the TOT layer [62, 63]. On the basis of experimental work [water adsorption measurements and X-ray diffraction (XRD) studies] performed on montmorillonite and hectorite, it appears that cation hydration is the main parameter which governs water uptake and swelling. In most cases, interstratified phases of zero hydrates or one-, two-, and sometimes three-layer hydrates are encountered [61–63]. On the contrary, tetrahedral charge localization leads to more homogeneous hydration states, with marked swelling steps as seen by X-ray diffraction [58, 59], thereby revealing the importance of charge localization on water structure, in line with simulation results [67]. Close to saturation (water activities around 0.98), another regime begins and is referred to as osmotic swelling. For monovalent cations, it corresponds to a continuous expansion of clay layer spacing with increasing water content. [54, 56, 64, 65].

From a structural and textural point of view, the penetration of water molecules into the interlamellar spaces locally generates significant volume variations. For instance, the volume increase of a particle upon hydration from the dry state to the one-layer hydrate state is about 30% (ratio 9.5:12.5 Å). The expansion of the interlayer spaces occurs at the expense of the porous network as macroscopic swelling is only visible for the highest water activities. Furthermore, as hydration goes on, capillary condensation will lead to the filling of the smallest pores. The hydrated clay material is then a complex system with domains where water exists under different status: confined water in the 2D interlayer spaces and bulk water filling mesopores. Macroscopically, this will affect the transport properties of water molecules and ions.

---

[1] According to the IUAPC convention [52], pores are characterized by their diameters, where micropores (<2 nm), mesopores (26–50 nm), and macropores (>50 nm) are identified.

18.3 Crystalline Swelling: Structure and Dynamics of Water Molecules in the Interlayer Spaces

18.3.1 Structural Description of Crystalline Swelling

On the basis of the clay description provided in the previous section, swelling clearly appears as a phenomenon that is initiated at the molecular level by the solvation of interlayer cations. This is accompanied by structural changes of these semicrystalline materials (water uptake involves a few angstroms increase of the interlamellar distances). Therefore, neutron or XRD techniques where the wavelength of incident beams is of the same order as the interatomic distances are particularly relevant to following the structural changes occurring upon hydration. The applicability of XRD to studying crystalline swelling of hydrated smectites has been recognized for more than 70 years, as shown by the number of publications based on this technique (the list given here being inevitably partial) [54–65,68, 69]. The frequent occurrence of the use of XRD is linked to the ubiquitous availability of the technique in laboratories, even though there are strong limitations to its use for probing the structural details associated with light elements such as hydrogen. In contrast, neutron scattering lengths are not dependent on electronic densities, and the contribution on scattering of light elements, which have scattering lengths on the same order as heavy elements, is significant (Chapter 2). In addition, neutron scattering lengths are isotope-dependent. For instance, neutron scattering lengths for H, D, Si, and Al are -3.74, 6.67, 4.15, and 3.45 fm, respectively. Thus, structural studies of water in hydrated crystals are possible and largely benefit from the isotopic dependence on scattering length. Using H_2O/D_2O exchange yields additional information that is helpful in getting the precise location of hydrated species.

The large variability in textural properties and crystal chemistry of swelling clays was emphasized in the first paragraph of Section 18.2. Depending on the clays studied, the approaches to interpreting results from neutron diffraction to get the structural configuration of water molecules in the interlayer spaces will differ. In the following subsections, we will give some examples of the work performed on different expandable clays (vermiculite, montmorillonite, and saponite).

18.3.1.1 Vermiculite

Vermiculites are 2:1 clay minerals, with a charge deficit that mainly arises from the tetrahedral layer. The relatively high charge prevents a continuous expansion of the interlayer at high water contents. At low water contents, vermiculites present well-defined hydration steps [70] along the water adsorption isotherms. Experimentally, it is then easier to control the hydration states that are studied. Furthermore, vermiculites sometimes occur as macroscopic crystals [7], making perfect orientation of samples possible.

Skipper and coworkers undertook a complete description of the interlayer structure of vermiculite, with either monovalent counter-ions (Li^+ [9], Na^+ [7]), or divalent counter-ions (Ca^{2+} [8], Ni^{2+} [7]). The method used to derive the interlayer distribution of cations and water molecules is based on an inverse Monte Carlo simulation in which the neutron scattering density profile along the z-axis perpendicular to clay layers [$\rho(z)$] is calculated from the integrated intensities of the (00l) reflections.

Intensities are related to $\rho(z)$ by the following formulae:

$$I(q) = M(q)|F(q)|^2, \tag{18.1}$$

and

$$F(q) = \int_{-c/2}^{c/2} \rho(z)e^{iqz}\,\mathrm{d}z. \tag{18.2}$$

Intensity is expressed as a function of the modulus of the scattering vector q [$q = 4\pi \sin(\theta)/\lambda$, where 2θ is the scattering angle and λ the wavelength], c is the layer spacing, $M(q)$ is a function that takes into account textural effects (size of sample and orientation), and $F(q)$ is the structure factor. Diffraction patterns were recorded after equilibrating the samples with either H_2O or D_2O. This leads to two different density profiles, one for hydrogen atoms and the second one for the clay layer plus oxygen atoms. An example of neutron scattering density profiles obtained in this way on lithium–vermiculite is given in Chapter 17 of this volume.

The experiments carried out on Li-, Ca-, and Ni-vermiculites [7–9] reveal that in the water bilayer state, these cations are solvated by six water molecules forming octahedral hydration complexes, giving rise to more ordered interlayers than for sodium–vermiculite [7].

The swelling behavior of vermiculites is strongly modified when exchangeable cations are replaced by organic counter-ions such as propylammonium [13], with which a further expansion of interlayers is observed. For such compounds, the H/D isotope labeling can also be exploited to describe in detail the colloidal behavior of such phases. The same approach was applied using other organic compounds such as polyethylene oxide or butylammonium ions [17].

Finally, the versatility of neutrons for studying samples in conditions mimicking deep environments is illustrated in the experiments of Skipper et al. [15], where the swelling of vermiculites was followed under fluid pressures of up to 1.5 kbar and temperatures of up to 300°C. The authors took advantage of the low absorption of neutrons by some alloys with high mechanical resistance to conduct experiments in temperature and pressure conditions analogous to those encountered in sedimentary and basin conditions.

18.3.1.2 Montmorillonite

Vermiculites present typical hydration properties, but are not representative of the most common swelling clays encountered in the environment or in industrial applications. The most frequently encountered smectites are montmorillonites and beidellites. Their swelling behavior cannot be directly derived from that of vermiculite due to their smaller crystallite size (less than a micrometer) and different charge amount and location, which lead to much less well-defined hydration steps. Cebula et al. [3] undertook a complete experimental diffraction study on a series of monovalent exchanged montmorillonites by combining (1) experiments in classical diffraction conditions to follow the evolution of the (00l) reflections as a function of relative humidity, (2) mosaicity evaluation by measuring the intensity of (001) Bragg peaks for different positions of samples relative to the neutron beam to get a degree of platelets ordering, and (3) small-angle scattering measurements. The main conclusion of this work is that there are no significant structural differences depending on the nature of the exchangeable cation and that all samples are characterized by a strong disorder in the arrangement of clay platelets and a spread distribution of spacings along the c-axis. A more detailed description of the water molecules in the interlayers could not be achieved as the intensities of the (00l) harmonics were very weak. This comes from the use of H_2O to hydrate samples, as will be discussed in the next section on saponite. When D_2O is used instead of H_2O, (00l) reflections present more intense diffraction peaks [4], enabling a one-dimensional (1D) Fourier transform of intensity to get scattering density profiles. This approach conducted on a sodium-montmorillonite revealed that the interlayer water was not highly structured. For some hydration rates the interstratification between different hydrated states leads to continuous diffraction curves, hindering high resolution in the treatment.

Swelling depends on the nature and valence of counter-ions present in interlamellar spaces. With neutron scattering, the differential neutron interactions with isotopes can be taken advantage of to study the cation environment in interlayer spaces. Thus, Powell et al. and Pitteloud et al. [11, 12, 16] used both H/D and $^6Li/^{nat}Li$ or $^{58}Ni/^{nat}Ni$ isotopic substitutions to describe the two-layer hydrates in lithium- and nickel-montmorillonite. In their approach, pair radial distribution functions describing the interlayer molecules are evaluated via Fourier transform of difference functions calculated by subtracting the intensities recorded in the same conditions with two isotopes. This method is explained relative to pollutant speciation in water in Chapter 17 of this volume. For nickel, it was established that the hydration shell of this cation is made up of six water molecules arranged octahedrally but with some distortions. Additional water molecules adopt a liquid-like structure, which is also observed with lithium. The structure of the coordination shell around lithium was not derived.

A detailed knowledge of the hydration behavior of montmorillonite is crucial to understanding swelling, which plays a crucial role when this expandable mineral is used as a barrier material. In the particular context of hazardous wastes disposal, where compacted clay barriers are used to inhibit pollutants and contaminants migration [71], a precise description of smectite hydration behavior is critical to assessing the engineered barrier sealing properties that may evolve depending on external conditions (humidity and temperature gradients for instance). Besides, additional constraints arise from the fact that, in barriers, clayey materials are compacted and confined. The issue to be addressed is whether crystalline swelling in such materials still occurs even when macroscopic swelling cannot take place. In that case, neutron diffraction techniques are particularly relevant, as neutron beams can penetrate through the walls of sample holders (Chapter 2).

Neutron diffraction experiments were carried out on samples compacted at different apparent densities and undergoing swelling in isochoric conditions [22]. To reach these conditions, compacted clay pellets were placed in the special cells designed to stand the high swelling pressure developed for the most hydrated samples and allowing hydration by water molecules in vapor phase, thanks to a porous ring. Samples were equilibrated during many weeks in atmospheres where the relative humidity was controlled using various saturated salt solutions.

The experiments revealed that confinement does not prevent crystalline swelling, the evolution of the (001) reflections being similar on such samples and on samples free to swell. The influence of confinement is only observable for the most hydrated bentonite pellets ($P/P_0 = 0.98$), where peaks are bimodal, corresponding to an interstratified state with two and three water layers. Additional mosaicity measurements were also performed on the two kinds of samples studied [a montmorillonite and a bentonite (montmorillonite plus accessory minerals)]. As compared to montmorillonite, the texture of bentonite is only slightly affected by compaction, owing to the accessory phases that limit the orientation of clay aggregates during compaction.

18.3.1.3 Saponite

Studies on natural montmorillonite revealed, on some domains of relative pressure, the existence of interstratified hydration states [4,61–63] whose origin may be assigned to the heterogeneity of these minerals. Tetrahedral charge localization in natural clays such as beidellite or saponite leads to more homogeneous hydration states, with marked swelling steps as seen by X-ray diffraction [58, 59], thereby revealing the importance of charge localization on the water structure, in line with simulation results [67]. In contrast to montmorillonite, saponite is easily synthesized [21, 50, 60]. Synthetic saponite then represents a valid model system for natural clays. The use of synthetic samples with well-identified structural formulas is also a clear advantage when molecular simulations are implemented.

The hydration and swelling of this synthetic saponite has been the object of a multidisciplinary experimental and numerical study [21]. Water adsorption results are presented in Fig. 18.2, together with the evolution of the d001 interlamellar distances recorded by XRD for three saponites with different charges. By carefully looking at the monolayer state that corresponds to the first plateau on Fig. 18.2B, one can see that in the first part of this plateau, for relative pressure below 0.3, the three samples exhibit similar interlamellar distances, whereas for the most charged sample, the adsorbed amount of water can be twice the quantity adsorbed for the lowest charge. For a particular relative humidity, when looking at structure, the three samples appear to be in a similar state, but that is not the case for the quantity of water adsorbed. This emphasizes the necessity of finely characterizing each sample by comparing the data obtained from independent techniques yielding either

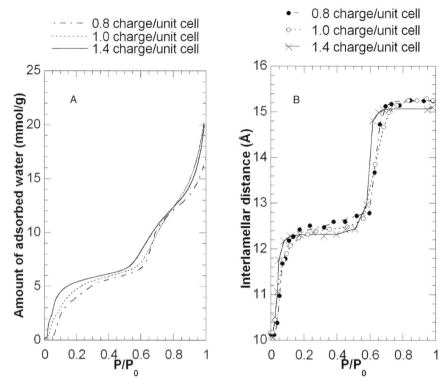

Fig. 18.2 A. Water vapor adsorption–desorption isotherms obtained on saponites with different charges. **B**. Associated evolution of the apparent d_{001} measured by X-ray diffraction (adapted from [60])

structural information or quantitative data with that for different domains of the water adsorption–desorption isotherm. For one of the samples (sodium-exchanged saponite with 1.0 charge per unit cell), neutron diffraction experiments were performed on the "small momentum transfer diffractometer D16" at Institut Laue-Langevin (ILL) [21]. The diagrams presented in Fig. 18.3 were recorded as a function of the partial water vapor pressure with either H_2O or D_2O. While for H_2O, the (001) reflection is very intense and shifts from 15.2 Å for the most hydrated state to 10 Å for the dry one, with an associated decrease in intensity due to the evolution of the structure factor, the D_2O (001) reflection vanishes (Fig. 18.3B) for a particular relative humidity ($P/P_0 = 0.38$).

By comparing the results obtained by near-infrared and Raman spectroscopy with structural results and the quantitative data obtained by water gravimetry, the following picture of this particular hydration state is obtained: interlayer cations are hydrated and interlamellar spaces are opened at 12.5 Å. Spectral results and simulations indicate the development in the interlayer of a disordered percolating hydrogen bond network [21]. To go further in the understanding of this particular

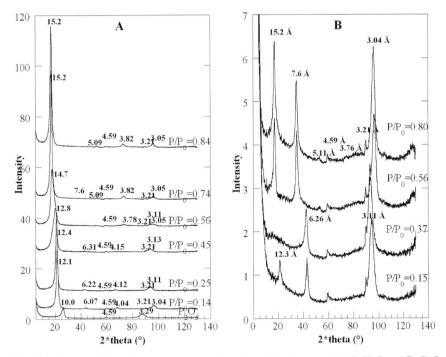

Fig. 18.3 Neutron diffraction patterns obtained in desorption for saponite. **A.** H_2O and **B.** D_2O (from [21], reprinted with permission from J. Phys. Chem. B **109**, 23745–2759 (2005). Copyright 2008 American Chemical Society)

feature, it is worth coming back to the formalism describing diffraction by lamellar structures [61, 72].

If we consider diffraction by stacks made of M identical layers distant from d_{001}, the expression of intensity for the (00l) reflections and in a direction perpendicular to the stack is as follows (z corresponds to the projection of the scattering vector q along the c-axis):

$$I00(z) \propto |F00(z)|^2 \sum_M \frac{\alpha(M)}{M} G00(z), \tag{18.3}$$

where $\alpha(M)$ is the distribution of stack thickness.

Intensity is then a product of two terms:

$$F_{00}(z) = \sum_{j=1}^{n} a_j f_j \exp(izzj). \tag{18.4}$$

$F_{00}(z)$ is the structure factor and represents the diffraction by all atoms and ions that constitute the layer plus atoms and ions of the interlayer. Each atom is characterized by its scattering length f_j and its abundance a_j at the position z_j. The structure factor corresponds to the Fourier transform of the atomic distribution of the layer:

$$G00(z) = \frac{\sin^2(M\pi z d_{001})}{\sin^2(\pi z d_{001})}. \tag{18.5}$$

$G_{00}(z)$ is the modulation function as derived by Méring in 1949 [73]. This function expresses the distribution of intensity at angles that correspond to the Bragg law. This periodic function presents peaks for z values inversely proportional to d_{001}. The peak width is inversely proportional to M, which corresponds to the size of the diffracting domain.

In the overall expression of intensity, $F_{00}(z)$ contains all the information relative to the nature of scattering elements (position, number, etc.), and $G_{00}(z)$ corresponds to the structural information (distances between planes, size of particles).

The above expression does not take into account the whole complexity of turbostratic systems, where there is no ordering between adjacent layers. More complete calculations that are not presented here can be performed for adjusting calculated and experimental patterns [61]. Still, Eq. (18.3) can be used to illustrate the influence of water content on the shape of the diffraction patterns. Results are displayed in Fig. 18.4 for both D_2O and H_2O for either 2.5 or 3.5 molecules per cation. Figure 18.4 clearly shows that the squared structure factor is very sensitive to the amount of D_2O molecules present in the interlayer. Such a sensitivity is not observed with H_2O, for which $F^2_{00}(z)$ first decreases strongly for the smallest values of z and then presents small oscillations, but without any strong dependence on the number of water molecules. The other striking feature of the evolution of $F^2_{00}(z)$ when calculated with D_2O is the presence of minima for a z value of 0.5. The minimum is partial with 3.5 water molecules and tends to zero with 2.5 water molecules. This value of z also corresponds to a maximum of the interference function when $G_{00}(z)$ is calculated with d_{001} equal to 12.55 Å (note that this function is the same for both H and D, as it is only linked to structural parameters). The resulting intensity then displays a disappearance of the 001 reflection for a particular amount of water around cations. This illustrates the high sensitivity of the shape of diffraction patterns toward the number of water molecules when samples are equilibrated with D_2O. It should also be possible to refine the amounts and positions of water molecules by calculating the diffraction patterns from the atomic positions derived from grand-canonical Monte Carlo (GCMC) simulations.

18.3.1.4 Conclusion on Diffraction

As shown in the previous sections, neutron diffraction can provide fruitful information about the atomic distributions of atoms and molecules (especially water molecules) in the interlayer spaces of clay minerals. Such information can be obtained either from a direct interpretation of the diffraction patterns or through an indirect approach using independently measured parameters (GCMC, adsorbed

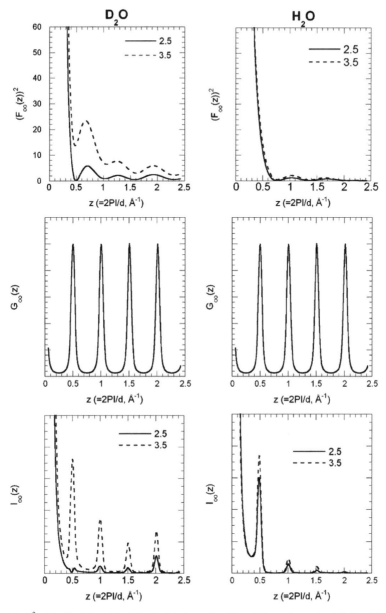

Fig. 18.4 $F^2_{00}(z)$, $G_{00}(z)$, and $I_{00}(z)$ calculated as a function of the water content (2.5 or 3.5 water molecules per cation), either with D_2O or H_2O

amounts, etc.). In any case, in carrying out either approach it is crucial to couple diffraction experiments with a precise characterization of water adsorption features and to achieve a very accurate control of humidity conditions when performing diffraction experiments.

18.3.2 Dynamics of Water—Quasielastic Neutron Scattering

As shown in the previous section, using neutron diffraction techniques in combination with other experimental methods, it is possible to obtain detailed snapshots of water and cation organization upon increase of water chemical potential. In many applications dealing with clay minerals, such a description remains incomplete because the dynamical properties of interlayer species cannot be obtained using such techniques. On the nanometric scale, and for time windows shorter than 100 ns, QENS techniques [74–76] are suitable for studying proton dynamics and therefore water mobility. The associated spatial and temporal scales also match those investigated by molecular dynamics, making cross-checking possible.

Diffusive motions like atomic translations and molecular rotations give rise to quasielastic signals, which originate from neutrons experiencing small energy transfers. Quasielastic intensity appears as a broadening of an elastic peak. Because hydrogen has a much larger incoherent cross section than any other atom, the signal of hydrogenated compounds is dominated by the self-particle dynamics of hydrogen. In QENS studies of hydrated clay minerals, the obtained spectra are therefore dominated by the motion of water molecules. Various quasielastic neutron scattering techniques are available for studying water dynamics: time of flight (TOF), backscattering (BS), and neutron spin echo (NSE) (Chapters 2 and 3). In the case of clay minerals, most studies have been performed using the TOF technique [35–42,45–50], while fewer studies have taken advantage of NSE [23, 43, 44, 46, 48] or BS spectrometers.

In a TOF spectrometer, neutron energy is deduced from the time it takes neutrons to travel a known distance (their TOF), whereas the scattered intensity is measured as a function of scattering angle (momentum transfer) and TOF (energy transfer). The instrumental resolution and accessible q-range essentially depend on the wavelength of the incident beam, which can be continuously adjusted on a TOF machine. The signals result from a convolution of a resolution (instrumental) function by the scattering function or dynamical structure factor denoted as $S_{inc}(q, \omega)$, where q is the modulus of the scattering vector and ω the energy transfer (inc refers to incoherent scattering, which is dominant for hydrogen). Characteristic rotation and translation times can then be inferred from the modeling of the evolution of the broadening and intensity of $S_{inc}(q, \omega)$ as a function of q.

To illustrate how QENS can be used for obtaining dynamical information on water motions, typical QENS spectra obtained for a synthetic saponite sample with a layer charge of 1.4 equilibrated at a relative water pressure of 0.84 for two values of the incident wavelength are given in Fig. 18.5. These spectra show the resolution function whose width depends on wavelength, various quasielastic signals modeled by Lorentzian functions, and an elastic peak present in the resolution function. This latter peak is due to the clay layer, especially the structural hydroxyl groups as well as to the water molecules whose motions are slower than the time window corresponding to the resolution function.

The quantity of probed mobile atoms can be derived from the calculation of the elastic incoherent structure factor (EISF), which is the ratio between elastic and total

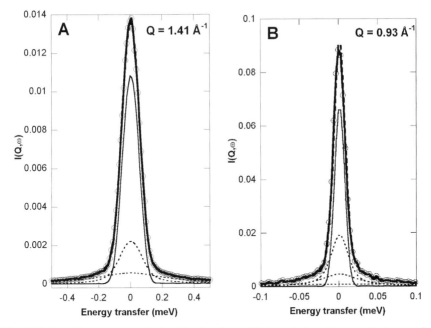

Fig. 18.5 Quasielastic spectra obtained by the time-of-flight technique for a synthetic saponite with layer charge of 1.4 per unit cell equilibrated at a relative water pressure of 0.84. **A.** $\lambda = 5\text{Å}$. **B.** $\lambda = 10\text{Å}$ (adapted from [50])

scattered intensities:

$$\text{EISF} = \frac{I_{\text{elast}}(q)}{I_{\text{elast}}(q) + I_{\text{inelast}}(q)}. \tag{18.6}$$

In the case of clay minerals, knowing the amount of water adsorbed for each point, it is easy to calculate the EISF corresponding to the structural protons of the layer. The comparison between this value and the experimental EISF then provides the proportion of adsorbed water molecules probed in the experiment. The evolution of the EISF with q also provides information about confinement geometry, which is particularly relevant for clay minerals that, as already mentioned repeatedly in this contribution, are strongly anisotropic.

Analyzing the q dependence of the width of the quasielastic Lorentzian functions, it is possible to obtain the nature and characteristic times of the various motions of interlayer water molecules and to follow the evolution of these motions with water pressure. In clay minerals, two rotational components and one translational component have generally been observed ([41] and references within, [42]). As far as rotations are concerned, the faster rotational motion (characteristic time around 2 ps) appears roughly independent of the adsorbed water amount and is assigned to water molecules bound to interlayer cations while the slower one (around 20–40 ps) is related to the planar rotation of water molecules in hydration spheres around the cation. This latter motion depends on both the hydration state and the nature of the interlayer cation.

Concerning translational diffusion, motions are significantly faster in the bilayer region as compared to the monolayer domain and are lower than that obtained for bulk water (2.3×10^{-9} m^2/s) [41]. As an example of typical values, Fig. 18.6 shows the radial translational diffusion coefficients obtained on a saponite sample together with those measured independently by molecular dynamics simulations. The values of translational diffusion coefficients reported in the literature [40–46,48–50] depend on the nature of the clay minerals and on the interlayer cation. They are also dependent on orientation, in-plane motions being faster than those perpendicular to the plane.

As noted previously, the time window accessible to TOF experiments remains limited in the high time region due to the relatively high width of the resolution function. For that reason, TOF experiments could certainly be complemented by BS experiments, which allow reaching higher resolutions (typically 1 μeV).

Another option to enlarge the time window is provided by the NSE technique, which preferentially enables the measurement of coherent signal and is the only neutron technique that gives a measurement of the intermediate function, $S(q,t)$, and covers relaxation times up to hundreds of nanoseconds (Chapter 3). Recent

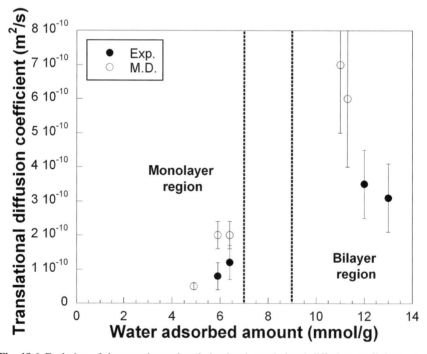

Fig. 18.6 Evolution of the experimental and simulated translational diffusion coefficients as a function of the amount of water adsorbed for a synthetic saponite clay with a layer charge of 1.4 per unit cell (adapted from [50])

publications coupling experiments and simulations [23, 46, 48] have indeed nicely shown the potential of this technique for studying water dynamics in confined interlayer spaces. However, up to very recently the diffusion coefficients obtained by TOF, NSE, and BS exhibited significant differences [42, 43, 46, 48]. In addition, the intermediate scattering functions $I(q,t)$ obtained from NSE experiments are generally better fitted using stretched exponential functions than monoexponential functions, especially to reproduce the experimental data at high q values. This behavior and discrepancy may be attributed to several phenomena.

1. At high q values, beyond $0.8\ \text{Å}^{-1}$, rotational contributions increase and lead to a decrease in the stretch exponent used to simulate the experimental data.
2. In the case of natural clay systems, structural heterogeneity in terms of chemical composition and layer charge leads to a distribution of water motions.
3. As shown very recently [23], imposing an isotropic analysis to a powder-averaged 2D diffusion model by using a stretched exponential fit results in an underestimation of diffusion coefficients by approximately 25%. By taking anisotropy into account, in the case of a synthetic fluorohectorite system, it was then possible to reconcile TOF and NSE measurements to yield similar diffusion coefficients.
4. Even taking into account all these possible reasons, microscopic simulations still reveal a non-monoexponential behavior for $I(q,t)$ [23]. This tends to reveal that a range of relaxation times is certainly present in the system and that describing the motion of interlayer water molecules with a single relaxation time may not be correct.

Despite all these limitations, and the complexity of data treatment, neutron scattering techniques still represent a unique tool for investigating interlayer water dynamics in clay minerals and their evolution with water content.

18.4 Clay Large-Scale Structure

Clayey materials are complex hierarchical and deformable systems whose structure and organization vary at different length scales depending on hydration conditions. The mesoscopic scale (i.e., a few angstroms to 1 μm) can be probed by small-angle scattering measurements (Chapter 2). In the present section, we will first recall some basic expressions of SANS and then show the application of this technique to two cases: (1) porosity changes induced by increasing water activity in confined clay materials and (2) the structure of clay suspensions and clay gels.

18.4.1 Principle of Small-Angle Scattering

Scattering phenomena result from the existence of heterogeneous domains with different scattering length densities at a submicronic length scale (for X-rays and neutrons). The incident beam is scattered at the interface between these regions with different contrast. These domains can be either particles in a solvent or pores distributed in a matrix, and the features of the scattering pattern are related to the quantity, shape, and size of the scatterers. The purpose of small-angle scattering experiments is to deduce the scatterers' characteristics from the pattern.

A comprehensive description of small-angle scattering can be found in reference text books on this subject [77–79], and only very basic expressions will be recalled here.

The expression of intensity in a system with no interaction between scatterers is

$$I(q) \propto \overline{\Delta\rho^2} F(q). \tag{18.7}$$

$\overline{\Delta\rho^2}$ describes the fluctuations of scattering length density, and in the case of a biphasic system where phases 1 and 2 are characterized by their respective scattering length density ρ_1 and ρ_2 and their respective volume fraction ϕ_1 and ϕ_2, this term can be written

$$\overline{\Delta\rho^2} = (\Delta\rho)^2\phi_1(1-\phi_1). \tag{18.8}$$

For diluted systems, ϕ_1 is negligible as compared to 1 and $\overline{\Delta\rho^2} = (\Delta\rho)^2\phi_1$. The form factor is $F(q)$, and it contains the geometrical information about the scattering phase; for instance, for bidimensional objects $F(q)$ varies as q^{-2}, whereas for linear objects, $F(q)$ scales as q^{-1}.

For interacting particles, an additional term is contained in the expression of intensity:

$$I(q) \propto \overline{\Delta\rho^2} F(q)S(q). \tag{18.9}$$

The interference function $S(q)$ describes the interaction between objects.

18.4.2 Clay Texture in the Crystalline Swelling Domain

As mentioned in Section 18.3.1.2, compacted clays are used as barrier materials in hazardous waste containment. In the context of nuclear waste disposal, compacted clay materials are placed around canisters containing radioactive wastes [80, 81]. Changes in temperature and humidity occur, leading to textural changes. Such changes can be followed by SANS on compacted clay pellets. In that case, SANS presents a definite advantage over X-ray techniques as samples can be investigated under isochoric conditions [22]. Indeed, due to the high penetrating power of neutrons, neutron beams can pass through the thick samples holder needed to keep samples in confined conditions. As such studies are rather rare, we chose to show some details of the results obtained, which also illustrate various ways of interpreting small-angle scattering data.

For a compacted dry bentonite pellet, the scattering curve recorded on line D11 at ILL displays a continuous and monotonous decrease of intensity for increasing scattering vector q (Fig. 18.7A). The background has not been subtracted and corresponds to the incoherent scattering of structural hydrogen atoms. Intensity decrease follows a power-law decay $I \propto q^{-\alpha}$, with α around 3.5 (Fig. 18.7A). In the case of rocks, such a behavior is often assigned to materials with fractal surface or to porous materials with a power-law distribution of pore sizes [82]. In the present case, the analysis of the SANS curve is based on the assumption that the system is biphasic and consists of a population of uncorrelated pores whose sizes follow a polydisperse distribution.

In that case, intensity is expressed as follows:

$$I(q) = |\Delta\rho|^2 \int\limits_0^\infty F(q, D)V^2(D)N_\mathrm{T}P(D)\mathrm{d}D,$$

and for spheres

$$F(q, D) = \left(3\frac{\sin(qD/2) - (qD/2)\cos(qD/2)}{(qD/2)^3} \right)^2, \tag{18.10}$$

where D is the dimension of the particles (for spheres D is the diameter), and $F(q, D)$ is the form factor.

The volume distribution function may be defined as

$$f(D) = V(D)N_T P(D), \tag{18.11}$$

where

  $V(D)$ is the particle volume,
  N_T is the total number of scattering particles,
  $N_T P(D)$ is the probability of occurrence of scatterers of size D (i.e., the number of pores between size D and $D + \mathrm{d}D$).

By intensity inversion, it is then possible to derive pore volume distribution functions $f(D)$ (Fig. 18.7B). Calculations were made using a software package, "IRENA," developed by Jan Ilavsky (Advanced Photon Source, Argonne National Laboratory, USA) and available on http://usaxs.xor.aps.anl.gov/staff/ilavsky/irena.html. The inversion was performed on the basis of a spherical pore shape. Even if this assumption does not really reflect the suspected pore shape, attempts made using elongated spheroids did not yield a better fit.

The distribution derived from the SANS curve was compared with the pore size distribution obtained on the same system by nitrogen adsorption using the Barret-Joyner-Halenda method [83]. The comparison, displayed on Fig. 18.7B, clearly shows the adequacy of both approaches, which rely on different principles. For

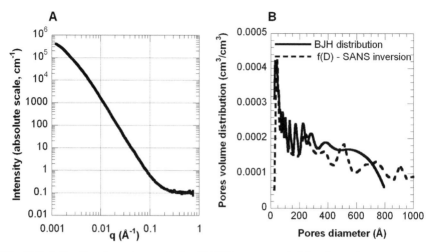

Fig. 18.7 **A**. Small-angle neutron scattering (SANS) curve recorded for a bentonite at an apparent density of 1.7 g/cm³. **B**. Comparison of the pore volume distribution derived from SANS curves and by nitrogen adsorption using the Barret-Joyner-Halenda (BJH) method

hydrated clays, the situation is more complicated as three phases are present: (1) a hydrated clayey matrix, (2) pores filled with water, and (3) macropores that are not completely filled (with water). Interpretation based on a biphasic system then does not hold anymore. However, it is still possible to get a qualitative analysis of scattering curves and to compare various experimental conditions (Fig. 18.8).

For a relative water pressure of 0.98, the scattering curves of unconstrained samples are identical whatever the density (for reasons of clarity, curves in Fig. 18.8 are not superimposed on the same graph). In contrast, those recorded in confined conditions deviate from free swelling conditions for the lowest q values. This divergence is more marked with increasing apparent density. Thus, confinement affects mesoporosity, but a thorough characterization of this effect relies on the treatment of SANS curves for hydrated samples, which is not trivial due to the coexistence of three phases.

18.4.3 Structure of Suspensions and Gels

Suspensions of smectite dispersed in water exhibit particular rheological properties: a sol–gel transition is observed even at very low solids concentration [84]. The structure of this gel has long been an object of debate, with two mechanisms proposed to explain gel formation: (1) 3D house of cards type structure with attractive interactions between the positively charged edges and the negatively charges faces of clay layers [64] and (2) oriented network of parallel layers with repulsive electrostatic interactions between double layers [54, 85]. To unravel this issue, it is necessary to establish the correlation function between clay layers inside gels

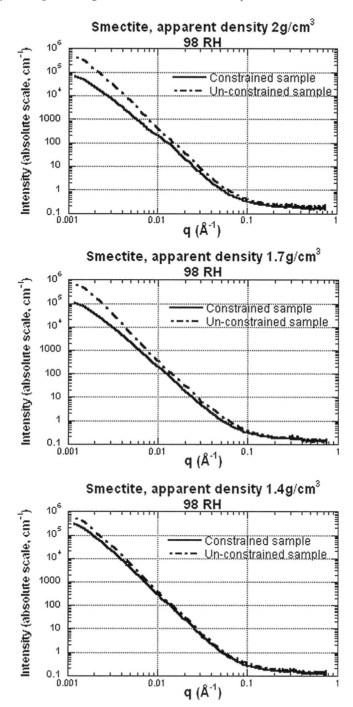

Fig. 18.8 Small-angle neutron scattering curves recorded for smectites compacted at various apparent densities, both in constrained and unconstrained swelling conditions (relative humidity of 98%)

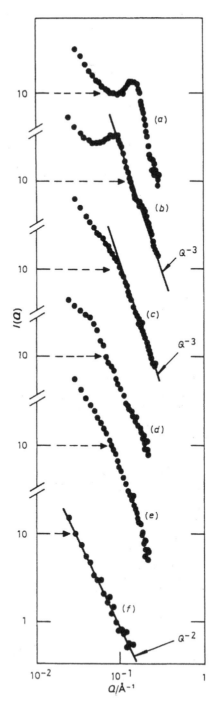

Fig. 18.9 Small-angle neutron scattering data for montmorillonite dispersed in D_2O with D_2O/montmorillonite ratios (g/g) of (a) 1.5, (b) 3.5, (c) 5.9, (d) 8.1, (e) 11.3, and (f) 22.5 (from [25]; reproduced by permission of The Royal Society of Chemistry)

and suspensions. A comprehensive SANS investigation was performed on two different smectites (montmorillonite and Laponite) as a function of solid concentration [24, 25, 27]. Examples of curves recorded on montmorillonite dispersions are shown in Fig. 18.9. For the lowest concentration, intensity follows a q^{-2} decay, which corresponds to the scattering of individual bidimensional objects without interactions. As concentration increases, a modulation appears that is represented by a well-defined diffraction peak for the highest concentrations. When plotting the evolution of this interlayer spacing as a function of water content, d spacings exhibit an inverse relationship with volume fraction, revealing a 1D swelling mechanism [25]. This demonstrates that in gels particles are equilibrated via electrostatic repulsions as suggested by Norrish [54]. This dependence of d spacing on clay concentration was confirmed with other smectites (montmorillonite [86], nontronite [87], and laponite [34]).

In cases of suspensions, it is worth mentioning that SANS patterns can be recorded under flow using various rheological devices. For example, this was implemented in the case of swelling clay gels in a Couette-type cell [27]. Other work aiming at understanding clay-based nanocomposites have also used such devices to follow the structural evolution of clay-polymer mixtures [29–32].

18.5 Conclusion

Because of their versatility and specific features, neutron-based techniques are particularly relevant for studying various aspects of clay swelling, a multiscale process extending over several orders of magnitude in space. At the atomic level, the sensitivity of neutrons to hydrogen isotopes allows a detailed refinement of both structure and amount of interlayer water, which should help in addressing the force fields used for molecular simulations of the clay–water interface. The same type of comparison between simulation and experiments can then be applied to the dynamics of interlayer water, which can be studied at various time scales and for various orientations, taking advantage of the different experimental setups (TOF, BS, NSE, triple-axis spectrometer) available in neutron facilities (Chapter 3).

From an experimental point of view, for hydration measurements for water chemical activity between 0 and 0.98, a very accurate control of the water partial pressure is needed, with also a precise analysis of the hydration properties of the sample being investigated. Indeed, it was demonstrated that the hydration behavior is strongly dependent on the clay crystal-chemistry, which should be established first. In addition, the swelling properties of each sample should be established by combining other independent techniques such as gravimetry measurements or spectroscopy experiments.

At the submicronic scale, the ability of neutrons to probe the structure and texture of large and dense samples under various experimental conditions (pressure, temperature, P_{H2O}, shear, etc.) clearly presents a key advantage for assessing the behavior of clay samples in natural or industrial situations. Due to the neutron penetration length, complex sample conditions can be adapted to carry out experiments at non-ambient temperatures and/or pressures. These conditions can be achieved either by adapting the available equipment at neutron facilities or by developing homemade devices. This versatility in sample environment represents a clear advantage of neutrons over X-rays.

Acknowledgments We thank Jean-Louis Robert (ISTO, Orléans, France) for providing us with synthetic saponite. Experimental support by Karine Devineau, Eric Ferrage, Solange Maddi, Manuel Pelletier, Angelina Razafitianamaharavo, Cedric Carteret, Emmanuel Rinnert, and Bernard Humbert during campaigns at ILL is warmly acknowledged. We thank Giovana Fragneto for technical assistance on D16. Institut Laue-Langevin is acknowledged for beam-time allocation. A part of this work was funded by ANDRA (French national agency for radioactive waste management).

References

1. Harvey C.C., Lagaly G., *Handbook of Clay Science*, edited by F. Bergaya, B. K. G. Theng, G. Lagaly (Elsevier, Oxford, 2006) pp. 501–540
2. Dove M. T., Eur. J. Mineral., **14**, 203 (2002)
3. Cebula, D. J., Thomas, R. K., Middleton, S., Ottewill, R. H., White, J. W., Clays Clay Miner., **27**, 39 (1979)
4. Hawkins, R. K., Egelstaff, P. A., Clays Clay Miner., **28**, 19 (1980)
5. Joswig W., Fuess H., Mason S.A., Clays Clay Miner., **37**, 511 (1989)
6. Skipper N. T., Soper A. K., McConnell J. D. C., Refson K., Chem. Phys. Lett., **166**, 141 (1990)
7. Skipper N. T., Soper A. K., McConnell J. D. C., J. Chem. Phys., **94**, 5751 (1991)
8. Skipper N. T., Soper A. K., Smalley M. V., J. Phys. Chem., **98**, 942 (1994)
9. Skipper N. T., Smalley M. V., Williams G. D., Soper A., Thompson C. H., J. Phys. Chem., **99**, 14201 (1995)
10. Akiba E., Hayakawa H., Hayashi S., Miyawaki R., Tomura S., Shibasaki Y., Izumi F., Asano H., Kamiyama T., Clays Clay Miner., **45**, 781 (1997)
11. Powell, D. H., Tongkhao, K., Kennedy, S. J., Slade, P. G., Clays Clay Miner., **45**, 290 (1997)
12. Powell, D. H., Tongkhao, K., Kennedy, S. J., Slade, P. G., Physica B, **243**, 387 (1998)
13. Williams G. D., Soper A. K., Skipper N. T., Smalley, M. V., J. Phys. Chem. B, **102**, 8945 (1998)
14. Sposito G, Park S.-H., Sutton R., Clays Clay Miner., **47**, 192 (1999)
15. Skipper N. T., Williams, G. D., de Siqueira A. V. C., Lobban C., Soper A. K., Clay Miner., **35**, 283 (2000)
16. Pitteloud C., Powell D. H., Fischer H. E., Phys. Chem. Chem. Phys., **3**, 5576 (2001)
17. Swenson J., Smalley M. V., Hatharasinghe, H. L. M., Fragneto G., Langmuir, **17**, 3813 (2001)
18. Beyer J, Graf von Reichenbach H., Clay Miner., **37**, 157 (2002)
19. Wasse, J. C., Stebbings, S. L., Masmanidis, S., Hayama, S., Skipper, N. T., J. Molec. Liq., **96**, 341 (2002)
20. Perdigon-Aller A. C., Aston M., Clarke S. M., J. Colloid Interface Sci., **290**, 155 (2005)
21. Rinnert E., Carteret C., Humbert B., Fragneto-Cusani G., Ramsay J. D. F., Delville A., Robert J. L., Bihannic I., Pelletier M., Michot J. L., J. Phys. Chem. B, **109**, 23745 (2005)
22. Devineau K., Bihannic I., Michot L. J., Villiéras F., Masrouri F., Cuisinier O., Fragneto G., Michau N., Appl. Clay Sci., **31**, 76 (2006)
23. Malikova N., Cadène A., Dubois E., Marry V., Durand-Vidal S., Turq P., Breu J., Longeville S., Zanotti J.-M., J. Phys. Chem. C, **111**, 17603 (2007)
24. Avery, R. G., Ramsay, J. D. F., J. Colloid Interf. Sci., **109**, 448 (1986)

25. Ramsay, J. D. F., Swanton, S. W., Bunce, J. J. Chem. Soc. Faraday Trans., **86**, 3919 (1990).
26. Allen A. J., J. Appl. Cryst., **24**, 624 (1991)
27. Ramsay, J. D. F., Lindner, P., J. Chem. Soc. Faraday Trans., **89**, 4207 (1993)
28. Hanley H. J. M., Straty G. C., Tsvetkov F., Langmuir, **10**, 3362 (1994)
29. Grillo I., Levitz P., Zemb T., Eur. Phys. J. B, **10**, 29 (1999)
30. Schmidt G., Nakatani A. I., Butler P. D., Karim A., Han C. C., Macromolecules, **35**, 7219 (2000)
31. Schmidt G., Nakatani A. I., Butler P. D., Karim A., Han C. C., Macromolecules, **33**, 4725 (2002)
32. Nettesheim F., Grillo I., Lindner P., Richtering, W., Langmuir, **20**, 3947 (2004)
33. Itakura T., Bertram W. K., Knott R. B., Appl. Clay Sci., **29**, 1 (2005)
34. Martin C., Pignon F., Magnin A., Meireles M., Lelièvre V., Lindner P., Cabane B., Langmuir, **22**, 4065 (2006)
35. Olejnik S., Stirling G. C., White J. W., Spec. Disc. Faraday Soc., **1**, 194 (1970)
36. Dianoux A. J., Volino F., Hervet H., Mol. Phys., **30**, 1181 (1975)
37. Cebula, D. J., Thomas, R. K., White, J. W., Clays Clay Min., **29**, 241 (1981)
38. Tuck J. J., Hall P., Hayes M. H. B., Ross D. K., Poinsignon C., J. Chem. Soc. Faraday Trans., **80**, 309 (1984)
39. Tuck J. J., Hall P., Hayes M. H. B., Ross D. K., Hayter J. K., J. Chem. Soc. Faraday Trans., **81**, 833 (1985)
40. Poinsignon C., Estrade-Schwarzckopf J., Conard J., Dianoux A. J., *Proc. Intl. Clay Conference*, edited by L. G. Schultz, H. Van Olphen, F. A. Mumpton (The Clay Minerals Society, Bloomington, Indiana, 1987), p. 284
41. Poinsignon C., Solid State Ionics, **97**, 399 (1997)
42. Swenson J., Bergman R., Howells W. S., J. Chem. Phys., **113**, 2873 (2000)
43. Swenson J., Bergman R., Longeville S.., Howells W. S., Physica B, **301**, 28 (1991)
44. Swenson J., Bergman R., Longeville, S., J. Chem. Phys., **115**, 11299 (2001)
45. Mamontov, E., J. Chem. Phys., **121**, 9193 (2004)
46. Malikova N., Cadene A., Marry V., Dubois E., Turq P., Zanotti J.-M., Longeville S. J., Chem. Phys., **317**, 226 (2005)
47. Chakrabarty D., Gautam S., Mitra S., Gil A., Vicente M. A., Mukhopadhyay R., Chem. Phys. Lett., **426**, 296 (2006)
48. Malikova N., Cadene A., Marry V., Dubois E., Turq P., J. Phys. Chem. B, **110**, 3206 (2006)
49. Skipper N. T., Lock P. A., Tililoye J. O., Swenson J., Mirza, Z. A., Howells, W. S., Fernandez-Alonso, F., Chem. Geol., 230, **182** (2006)
50. Michot L. J., Delville A., Humbert B., Plazanet M., Levitz P., J. Phys Chem. C, **111**, 9818 (2007)
51. Newman A. C. D., *Chemistry of Clays and Clay Minerals* (Mineralogical Society, London, 1987)
52. IUPAC Manual of symbols and terminology for physico-chemical quantities and units, Appendix 2, Definitions, Terminology, and Symbols in Colloid and Surface Chemistry. Part 1. Pure Appl Chem **31**, 578 (1972)
53. Bihannic I., Tchoubar D., Lyonnard S., Besson G., Thomas F. J., Colloid Interface Sci., **240**, 211 (2001)
54. Norrish K., Quirk J. P., Nature, **173**, 225–256 (1954)
55. Mooney R. W., Keenan A. G., Wood, L. A., J. Amer. Chem. Soc., **74**, 1371 (1952)
56. Norrish K., Raussell-Colom, J. A., Clays Clays Miner., **10**, 123 (1963)
57. Sposito G., Prost R., Chem. Rev., **82**, 553 (1982)
58. Glaeser R., Méring J., C.R. Acad. Sci. Paris, **T. 267 Série D**, 463 (1968)
59. Suquet H., Pezerat H., Clays Clay Miner., **35**, 353 (1987)
60. Michot L. J., Bihannic I., Pelletier M., Rinnert E., Robert J.-L., Am. Miner., **90**, 166 (2005)
61. Ferrage E., Lanson B., Sakharov B. A., Drits V. A., Am. Miner., **90**, 1358 (2005)
62. Bérend I., Cases J. M., François M., Uriot J. P., Michot L. J., Masion A., Thomas F., Clays Clay Miner., **43**, 324 (1995)

63. Cases J. M., Berend I., François M., Uriot J. P., Michot L. J., Thomas F., Clays Clay Miner., **45**, 8 (1997)
64. Van Olphen, H., *An Introduction to Clay Colloid Chemistry*, 2nd ed. (John Wiley & Sons, New York, 1977)
65. Güven, N., *Clay-water Interface and Its Rheological Implications*, edited by N. Güven, R. M. Pollastro (CMS Workshop Lectures, Clay Minerals Society, Boulder, CO, USA, 1992) p. 2
66. Marry V., Turq P., J. Phys. Chem. B, **107**, 1832 (2003)
67. Chang F.-R. C., Skipper N. T., Sposito G., Langmuir, **11**, 2734 (1995)
68. Nagelschmidt G., Zeit. Kristall. **93**, 481 (1936)
69. Bradley W. F., Grim R. E., Clark G. F., Zeit. Kristall. **97**, 260 (1940)
70. De la Calle C., Suquet H., Dubernat J., Pezerat, H., Clay Miner., **13**, 275 (1978)
71. Push R., *Handbook of Clay Science*, edited by F. Bergaya, B. K. G. Theng, G. Lagaly (Elsevier, Amsterdam, London, 2006) pp. 703–716
72. Drits V. A., Tchoubar C., *X-Ray Diffraction by Disordered Lamellar Structure. Theory and Applications to Microdivided Silicates and Carbons* (Springer-Verlag, Berlin, 1990)
73. Mering J., Acta Cryst., **2**, 371 (1949)
74. Bée, M.. *Quasi-Elastic Neutron Scattering* (Adam Hilger, Philadelphia, PA, 1988)
75. Bée M., Chem. Phys., **292**, 121 (2003)
76. Egelstaff P. A., *An Introduction to the Liquid State* (Clarendon, Oxford, UK, 1992)
77. Glatter O., Kratky O., *Small Angle X-ray Scattering* (Academic Press Inc., London, 1982)
78. Brumberger H., *Modern Aspects of Small-angle Scattering* (NATO ASI Series C, Vol. 451, Kluwer Academic Publishers, Dordrecht, 1995)
79. Lindner P., Zemb T., *Neutron, X-rays and Light: Scattering Methods Applied to Soft Condensed Matter* (North Holland, Amsterdam, 2002)
80. Madsen F. T., Clay Miner., **33**, 109–129 (1998)
81. Komine H., Ogata, N., Can. Geotech. J. **31**, 478 (1994)
82. Radlinski A. P., *Neutrons Scattering in Earth Sciences*, edited by J. J. Rosso (The Mineralogical Society of America, 2006)
83. Barett E. P., Joyner L. G., Halenda P. H., J. Am. Chem. Soc., **73**, 373 (1951)
84. Mourchid A., Lecolier E., Van Damme H., Levitz, P., Langmuir, **14**, 4718 (1998)
85. Callaghan I. C., Ottewill, R., Chem. Soc., **57**, 110 (1974)
86. Michot L. J., Bihannic I., Maddi S., Funari S., Baravian C., Levitz P., Davidson P., Proc. Natl. Acad. Sci., **103**, 16101 (2006)
87. Michot L. J., Bihannic I., Maddi S., Baravian C., Levitz P., Davidson P., Langmuir, **24**; 3127 (2008).

Chapter 19
Structure and Dynamics of Fluids in Microporous and Mesoporous Earth and Engineered Materials

David R. Cole, Eugene Mamontov, and Gernot Rother

Abstract The behavior of liquids in confined geometries (pores, fractures) typically differs, due to the effects of large internal surfaces and geometrical confinement, from their bulk behavior in many ways. Phase transitions (i.e., freezing and capillary condensation), sorption and wetting, and dynamical properties, including diffusion and relaxation, may be modified, with the strongest changes observed for pores ranging in size from <2 to 50 nm—the micro- and mesoporous regimes. Important factors influencing the structure and dynamics of confined liquids include the average pore size and pore size distribution, the degree of pore interconnection, and the character of the liquid–surface interaction. While confinement of liquids in hydrophobic matrices, such as carbon nanotubes, or near the surfaces of mixed character, such as many proteins, has also been an area of rapidly growing interest, the confining matrices of interest to earth and materials sciences usually contain oxide structural units and thus are hydrophilic. The pore size distribution and the degree of porosity and inter-connection vary greatly amongst porous matrices. Vycor, xerogels, aerogels, and rocks possess irregular porous structures, whereas mesoporous silicas (e.g., SBA-15, MCM-41, MCM-48), zeolites, and layered systems, for instance clays, have high degrees of internal order. The pore type and size may be tailored by means of adjusting the synthesis regimen. In clays, the interlayer distance may depend on the level of hydration. Although studied less frequently, matrices such as artificial opals and chrysotile asbestos represent other interesting examples of ordered porous structures. The properties of neutrons make them an ideal probe for comparing the properties of bulk fluids with those in confined geometries. In this chapter, we provide a brief review of research performed on liquids confined in materials of interest to the earth and material sciences (silicas, aluminas, zeolites, clays, rocks, etc.), emphasizing those neutron scattering techniques that assess both structural modification and dynamical behavior. Quantitative understanding of the complex solid–fluid interactions under different thermodynamic situations will impact both the design of better substrates for technological applications (e.g., chromatography, fluid capture, storage and release, and heterogeneous catalysis) as well

D.R. Cole (✉)
Chemical Sciences Division, Oak Ridge National Laboratory, Oak Ridge, TN 37831-6110, USA
e-mail: coledr@ornl.gov

L. Liang et al. (eds.), *Neutron Applications in Earth, Energy and Environmental Sciences*, Neutron Scattering Applications and Techniques,
DOI 10.1007/978-0-387-09416-8_19, © Springer Science+Business Media, LLC 2009

as our fundamental understanding of processes encountered in the environment (i.e., fluid and waste mitigation, carbon sequestration, etc.).

19.1 Introduction

19.1.1 Confined Fluid–Matrix Interactions

Fluids containing inorganic and organic solutes (including hydrocarbons) and gaseous species (e.g., CO_2, CH_4) can occupy the pores, grain boundaries, or fractures of numerous types of complex heterogeneous solids. This accessible porosity within the solids can span wide length scales (d as pore diameter or fracture aperture) including micro-, meso-, and macroporous regimes ($d < 20$ Å, $20 < d < 500$ Å, and $d > 500$ Å, respectively, as defined by IUPAC). Porous solid matrices include rock or soil systems that contain clays and other phyllosilicates, zeolites; coal, graphite, or other carbonaceous-rich units; and weathered or altered silicates, oxides, and carbonates. Examples of micro- and mesoporous solids, natural as well as engineered, are given in Fig. 19.1. A number of factors dictate how fluids, and with them the reactants and products of intrapore transformations, migrate into and through these nano-environments, wet, and ultimately adsorb and react with the solid sur-

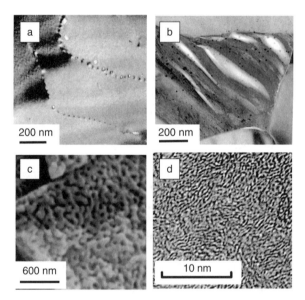

Fig. 19.1 Electron microscopy images of micro- and mesoporous earth and engineered materials: (**a**) pores along grain boundaries in weakly weathered basalt, (**b**) clay formation with large pores (*white areas*) at a grain boundary intersection in altered basalt, (**c**) controlled-pore glass, and (**d**) slit-like pores in carbon fiber monolith

faces. These include the size, shape, distribution, and interconnectedness of these confined geometries, the chemistry of the solid, the chemistry of the fluids, and their physical properties [1]. The dynamical behavior of fluids and gases contained within solids is controlled by the processes occurring at the interface between the various phases (e.g., water–water, water–solute, water–solid, solute–solid, etc.), as well as the rates of supply and removal of mobile constituents.

Compared with the effort expended to study bulk aqueous and non-aqueous fluids, an understanding of the dynamic, thermodynamic, and structural properties of fluids in confined geometries, and their influence on the properties of the porous solid, is much less evolved for earth materials. Examples of efforts relevant to the behavior of earth materials include the study of CO_2 in thin pores [2], water structure and dynamics in clays [3–6], ion adsorption into alumina mesoporous materials [7], and water within layered silicates at elevated pressure and temperature [8–11]. There is general agreement that the collective structure and properties of bulk fluids are altered by the confinement between two surfaces or in narrow pores due to the interplay of the intrinsic length scales of the fluid and the length scale due to confinement [12, 13]. Also crucial to the molecular behavior of fluids is the geometry of the pore, which can include simple planar walls such as those encountered in natural micas, clays, or graphite; slits, cylinders, and spheres; and spheres linked with cylinders as observed in zeolites [14]. Other factors that contribute to the modification of fluid properties include the randomness of the matrix and the connectivity of the pore network [15].

19.1.2 Probing Fluid Confinement Using Neutron Scattering

The richness and complexity of fluid behavior (e.g., phase transitions, molecular orientation and relaxation, diffusion, adsorption, wetting, capillary condensation, etc.) in confined geometries only underscores the need to adopt a multidisciplinary approach when trying to quantify this behavior regardless of the fluid type or nature of the porous medium. The effects of confinement have been studied using tools as diverse as computational [16], spectroscopic [17–19], and calorimetric [20] techniques. Direct structural measurements of the solid have been performed using neutron [21, 22], X-ray, and light scattering, as well an assortment of spectroscopies including nuclear magnetic resonance (NMR), Raman, Fourier transform infrared (FTIR), and dielectric relaxation. The scattering methods have the benefit of being both noninvasive and nondestructive, and are appropriate for studying the structural properties involving translational as well as orientational ordering of the confined fluids in domain sizes ranging from nanometers to microns [23].

The properties of neutrons make them an ideal probe for comparing the properties of bulk fluids with those of the fluids in confined geometries [Chapter 2]. Neutrons can be scattered either coherently or incoherently, thus providing opportunity for various kinds of analysis of both structural and dynamic properties of confined liquids. Such analysis is possible due to the fact that the wavelengths of thermal and cold neutrons are comparable with intermolecular distances in condensed phases, while the neutron energy can be tailored to probe both high- (collective and single-particle vibrational) and low-frequency (single-particle diffusive) motions in the system. Importantly, the large incoherent scattering cross section of hydrogen compared to other elements allows obtaining scattering spectra dominated by the scattering from hydrogen-containing species (see recent review article by Neumann [24]), whereas the X-ray scattering from such systems, which is virtually insensitive to hydrogen, would be dominated by the signal from the confining matrix. Last but not least, the large difference in the coherent and incoherent neutron scattering cross sections of hydrogen and deuterium allows selection of atoms to dominate the scattering signal by means of deuteration of the fragments of liquid molecules or the confining matrix.

The focus of this review is the application of neutron scattering and diffraction methods to the study of fluids and their interaction with solid matrices. By way of key examples, emphasis is placed on what neutrons can tell us about the molecular-level properties and behavior of geo-fluids and the processes attendant with their interaction with solid surfaces.

19.2 Structural Properties of Confined Fluids

The structural properties of confined liquids can be assessed using coherent scattering techniques, neutron diffraction (ND), and small-angle neutron scattering (SANS). The former allows one to measure the static structure factor, $S(Q)$, which can be then Fourier transformed to obtain the radial pair-distribution function, $g(r)$, that describes the distribution of the distances between the coherently scattering nuclei in the liquid. For hydrogen-containing species, sample deuteration (D_2O) is advantageous in order to obtain diffraction patterns from coherently scattering D nuclei, thereby avoiding the large incoherent scattering cross section from hydrogen, which can contribute large background to observed intensity patterns. While ND measurements of liquids in confinement probe structural correlations not exceeding a few molecular diameters, SANS measurements provide coverage over much broader range in the real space [22, 25, Chapter 20]. This is because SANS involves measuring neutron intensities at very low values of the scattering vector, Q (i.e., at small angles). Below we provide a few examples of how these methods have been applied to describe the behavior of water and non-polar fluids confined in silica- and clay-water systems.

19.2.1 Water in Porous Silica

There has been considerable attention focused on determining the structure of water confined in hydrophilic systems such as mesoporous silica using neutron diffraction complemented by MD simulations. Structural studies of water under confinement are typically made by one of the two techniques which emphasize either the water structure in the pores or modification of the spatial characteristics as the temperature is varied, usually in the low-temperature, supercooled regime [26]. Vycor pore glass and an assortment of mesoporous silica materials such as the MCM and SBA varieties have been used extensively to assess the behavior of fluids, especially water in pore regimes ranging from about 10–50 Å. The engineered silica materials may be viewed as reasonable proxies for silica-rich systems found in nature. Studies of water interaction with confined hydrophobic substrates have also been conducted, but will not be discussed here [27].

It is well known that water confined in mesoporous silica is significantly modified relative to bulk water. The roles of pore size, pore shape, temperature, and relative humidity (filling factor) have been explored for a variety of high-surface-area porous silicas [26–29]. Perhaps, the most noteworthy application of neutron diffraction from confined liquids is the detection of freezing–melting transitions. These transitions are known to be depressed to lower temperatures compared to the corresponding bulk liquids and to exhibit a substantial hysteresis. Further, ND results suggest the formation of a thin layer of surface water exhibiting lower density and mobility compared to bulk water. Unlike, for example, calorimetric measurements, a neutron diffraction experiment cannot only detect a freezing transition in the confined liquid but can also determine the crystal structure of the resultant solid phase. This phase may be either crystalline, but different from that of the bulk solid phase, or amorphous. Information on the average size of the crystalline particles in confinement can be obtained from broadening of the Bragg peaks.

Most of the neutron diffraction studies of freezing of the supercooled confined liquids involved water–ice transition in mesoporous silicas, silica sol-gels, or Vycor glass [30–35]. A review by Dore covers a number of structural studies of supercooled confined water [36]. It was found by several researchers that in sufficiently small pores, formation of a meta-stable cubic ice phase takes place instead of a bulk hexagonal ice phase. An example of this behavior is shown in Fig. 19.2 for water in MCM-48, which has roughly 35 Å pores. In even smaller pores, where the development of hydrogen-bonded network is strongly suppressed, formation of amorphous ice was observed. Even though the temperatures in experiments with supercooled confined water are outside the range that is typically of interest to earth and environmental sciences, such experiments are important because they elucidate the effect of confinement on hydrogen-bonded network and phase transitions.

Despite the emphasis on the behavior of water in the supercooled region, the structure of liquid water confined in Vycor glass at ambient conditions and above also has been addressed [27, 28, 37, 38]. The general approach has been to fully or partially hydrate Vycor with H_2O, D_2O, and HDO (isomolar mixture). The availability of diffraction data on three isotopic substituted samples has allowed extraction of three site–site radial correlation functions (RCF)—$g_{OwOw}(r)$, $g_{OwHw}(r)$, and $g_{HwHw}(r)$ [28]. Three types of contributions comprise the differential cross section (DCS) measurement in a diffraction experiment: two from the individual species and a third term originating from cross correlations between the porous media and the fluid [38]. The Vycor–Vycor interference term can be subtracted from

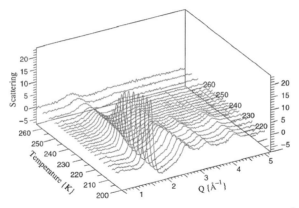

Fig. 19.2 Selected temperature difference function for D_2O in MCM silica (35 Å pore diameter) showing emergence of the cubic ice phase as a function of time and temperature, reprinted with permission from the authors and Webber and Dore [29]. Copyright 2004 Institute of Physics

the experimental data if DCS data are available for a "dry" Vycor sample. Ricci et al. [38] present a nice example of the power of combining ND results with those obtained from MD simulations. In this case they explored the comparison between simulations using the SPC/E water model with diffraction data reported by Bruni et al. [28] for hydration of Vycor at ambient conditions. The assumption was made that the individual RCFs are roughly independent of the hydration level, thereby allowing the isolation of each $g(r)$ from the cross correlation terms. Figure 19.3 presents their results for the $g_{OwOw}(r)$ comparing bulk water behavior with water in fully hydrated Vycor. In all cases, the $g(r)$ functions for confined water are considerably distorted compared to their bulk equivalents. From Fig. 19.3a we observe

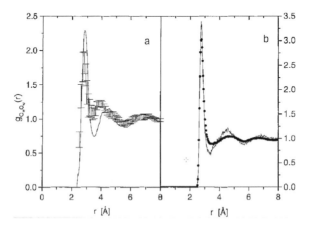

Fig. 19.3 The oxygen–oxygen RCF for **(a)** experimental data for bulk (curve) and confined water in Vycor (as *error bars*) and **(b)** MD simulations for bulk (*solid curve*) and confined SPC/E water (*dots*), reprinted with permission from the authors and Ricci et al. [38]. Copyright 2000 Institute of Physics

that the first peak is lower in amplitude than in the bulk and there is a significant increase in the peak intensity in the region near 3–4 Å. According to Ricci et al. [38], this behavior is similar to what occurs when pressure is applied to bulk water and signifies a substantial distortion of the hydrogen bond network. Despite this distortion, there is still a significant hydrogen bond peak in the $g_{OwOw}(r)$ function (not shown), although the first minimum is shifted (≈ 2.5 Å in bulk to ≈ 2.25 Å for confined water) and an additional shoulder appears at $r \approx 2.7$ Å. An estimation of the average number of hydrogen bonds (HB) per molecule indicates a reduction by approximately 50% compared to bulk water. MD simulation results of the $g_{OwOw}(r)$ shown in Fig. 19.3b exhibit the same qualitative trends observed in the ND experiments. Further, MD simulations can assess all of the properties of interest separately for the molecules wetting the pore wall and for those belonging to the inner shells. For example, MD results indicate that the distribution of the O–O–O angle between three neighboring molecules in the center of the pore is similar to bulk water, with a maximum correspondence of the angle characteristic of tetrahedral coordination. The distribution of the O–O–O angle becomes flat for molecules approaching the pore wall, confirming that the relative orientations of neighboring molecules are strongly influenced by the presence of the hydrophilic silica surface—that is, tetrahedral coordination breaks down [38].

19.2.2 Water in Clays

The structure and dynamics of water and hydrated cations in the interlayer region of clays and water behavior on clay surfaces have been the focus of numerous studies involving neutron diffraction and MD simulations [3–6,39–48]. Of particular interest has been the structure of the double layer that forms in swelling clays—for example, vermiculite and smectite, and how interlayer cations can progress from inner-sphere (non-solvated) to outer-sphere (solvated) complexes. Water molecules can be intercalated between clay layers to create an interlayer ionic solution that influences the mobility of counter-ions and the swelling behavior, both of which are related to electric double layer properties [49]. Neutron diffraction using difference methods allows the determination of the radial distribution functions that characterize the interlayer structure; in particular, D–H isotope substitution experiments have been conducted to interrogate the environment of the interlayer protons or interlayer cations [6]. Interlayer cations that have been targeted for investigation include Li, Na, K, Ca, Mg, Ni, Nd, and Yb. It is important to note that there is a fundamental difference between solvated ion interactions with tetrahedral surface sites and embedded octahedral charge sites [42].

For 2:1 swelling clays, neutron diffraction and MD simulations indicate that the coordination of the interlayer cations with water and clay surface oxygen is controlled largely by cation size and charge, in a manner similar to that observed for the ions in concentrated aqueous solution. The location of structural charge within the clay layer and the presence of hydrophobic zones on its surface modulate the binding

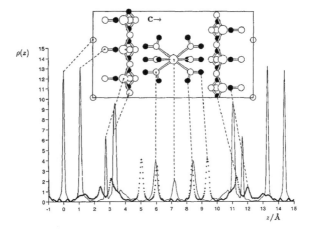

$\rho(z)$

$z/\text{Å}$

Fig. 19.4 Neutron scattering density profiles (*z*) for oxygen plus clay layer (*solid line*) and hydrogen (*crosses*) for 14.4 Å nickel-substituted vermiculite at 84% relative humidity (from [51]). For the purpose of this plot, hydrogen has been assigned the same scattering length as oxygen. The inset shows the assignment of peaks in the proposed structural model (hydrogen atoms are *solid circles*). Reprinted with permission from Skipper et al. [51]. Copyright 1990 Elsevier Pub

behavior [50]. There is a tendency for the clay mineral structure to exert more of an influence on the larger cations such as K^+, leading to inner-sphere (non-solvated) complexes. Conversely, smaller monovalent and divalent cations such as Li^+, Ca^{2+}, and Ni^{2+} tend to form outer-sphere, highly solvated complexes. An example of the ND results for a transition metal cation is shown in Fig 19.4 for a 14.4 Å nickel-substituted vermiculite reacted at 84% relative humidity [51]. This figure shows the density profiles refined from the integrated intensities extracted from corrected diffraction patterns. Each interlayer nickel is octahedrally coordinated with six water molecules, and all of the waters are oriented to form a strong hydrogen bond to the adjacent clay surface. Skipper et al. [51] also concluded that extra water is located near the clay layers with hexagonal rings of SiO_4 and AlO_4^- tetrahedra that comprise the clay surface. In general, ND studies indicate that confined water molecules form hydrogen bonds to each other and to the clay surface such that their local environment relaxes to close to the bulk water structure within two molecular layers of the clay surface [42]. As temperature and pressure are increased to sedimentary basins conditions, the hydrogen bonding of interlayer water molecules to the clay surfaces disappears as temperatures approach the critical point [45].

19.2.3 Non-polar Fluids

The behavior of non-polar fluids, particularly hydrocarbons, in confined geometries is relevant to a variety of industrial and natural processes. Recently, the mesoscopic structure of organoclays, which are widely used as flow modifiers, has been studied by a combination of X-ray scattering, SANS, and ultra-small-angle neutron scattering (USANS) by King et al. [52]. They derived a tactoid structure, that is, stacks of partially overlapping clay sheets, with an average layer number of 2 and a tactoid radius of $\approx 3\,\mu m$.

Weathering and alteration processes play an important role in many natural systems composed of inorganic and organic matter. Fractal concepts have been successfully applied to a large variety of rocks and soils. Consequently, scattering techniques have proven particularly useful for the study of these systems, since the fractal dimension is directly related to the slope of the scattering curve. The negative slope x of the scattering curve directly identifies the fractal type; it is 2–3 for a mass fractal and 3–4 for a surface fractal. For a mass fractal, the fractal dimension D_m is given by $D_m = x$, while the fractal dimension of a surface fractal, D_s, is given by the relation $D_s = 6 - x$. The diagenesis of sedimentary rocks has been studied by a combination of SANS and USANS by Radlinski et al. [21]. It has been shown that the structure of this composite material is represented by a surface fractal with a constant fractal dimension from the nanometer to micrometer length scales. Connolly et al. studied the microstructure of oil-bearing sedimentary rocks using USANS and found surface fractal behavior with a fractal dimension of $D_s = 2.7$ [53]. Fractal models have also been applied to describe the structure of water-saturated kaolinite clays [54].

A large number of studies deal with the properties of fluids and fluid mixtures imbibed in the pores of engineered nanoporous materials. Porous silica (SiO_2) is frequently chosen because it can be synthesized with well-defined pore sizes in the range of a few Angstroms up to several hundred Angstroms. SANS has widely been utilized in the study of fluid behavior in porous media and recently became the first technique capable to quantify the sorption properties of fluids in porous media in terms of the mean density and volume of the adsorbed phase [55]. In this study, Rother et al. examined the sorption properties of supercritical deuterated propane in silica aerogel with 96% porosity. SANS and neutron transmission data have been measured for fluid-saturated silica at different fluid densities and temperatures. The mean density ρ_3 and volume fraction ϕ_3 of the sorption phase were calculated from the SANS and neutron transmission data by the application of a new model, which makes use of the three-phase model by Wu [56] and a mass balance consideration of the pore fluid, which can be obtained from neutron transmission measurements or volumetric sorption measurements. It was found that the fluid is adsorbed to the porous matrix at low fluid densities, but depleted from the pore spaces at higher fluid densities (i.e., in the vicinity of the critical density and above). Figure 19.5a shows the evolution of the physical properties of the sorption phase, expressed in terms of ρ_3 and ϕ_3, as a function of the (bulk) fluid density ρ_2. The bulk critical density of deuterated propane is $\rho_c \approx 0.27\,g/cm^3$, and the critical temperature is $T_c \approx 91.0°C$. The fluid density in the adsorbed phase is up to about three times higher as ρ_2 in the low-pressure region, while it remains constant and below ρ_2 at and above the critical pressure. With the information on ρ_3 and ϕ_3, calculation of the absolute adsorption, which is the relevant quantity for the application of the equation of adsorption and molecular modeling work, is possible without the introduction of further assumptions. The calculated values for

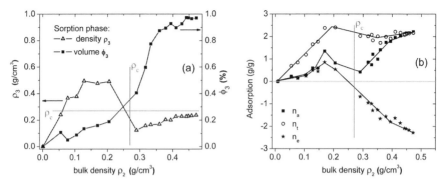

Fig. 19.5 Results of SANS on deuterated propane inside silica aerogel at a supercritical temperature of $T = 91.2°C$, from [55]; (**a**) Physical properties of the adsorbed phase, open circles: Mean density ρ_3 of the sorption phase, filled squares: volume fraction ϕ_3 of the sorption phase; (**b**) Adsorption quantities calculated from the density and volume of the sorption phase. Reprinted with permission from Rother et al. [55]. Copyright 2007 American Chemical Society

the absolute adsorption (n_a), total adsorption (n_t), and excess adsorption (n_e) are given in Fig. 19.5b. The absolute adsorption is similar to the commonly measured excess adsorption only at low fluid densities, but differentiates significantly at higher fluid densities.

Fluid depletion effects in porous media were found first by Thommes et al. [57] and named "critical depletion" for their occurrence in the close vicinity of the critical point. Similar high-pressure depletion effects have been reported since, mainly using volumetric techniques, for a number of fluids in porous materials (references given in [55]). The phenomena critical depletion and high-density depletion might be closely related effects; that is, critical depletion could in fact extend far into the region of high fluid densities. Generally, surface effects, confined volume effects, or a combination of both may cause depletion. While fundamental understanding of critical depletion is still lacking, wetting transitions leading to depletion layers were recently discussed in a related study of water sorption in hydrophilic pores [58]. The new knowledge on the nanoscopic structure of the pore fluid obtained from SANS will be used as input for molecular modeling works. The goal is to compute the density profile of the pore fluid with high spatial resolution, to determine the fluid-pore wall interaction potential, and to quantify the confinement effect on the sorption properties for different fluid–matrix combinations. The new technique for the analysis of scattering data has the potential to be employed for the study of numerous other super- and sub-critical fluids in natural and engineered porous systems.

Pore condensation is the process of liquefaction of fluids in narrow pores at pressures below the bulk condensation pressure. This effect allows soils to take up and store large amounts of water and other fluids. Pore condensation is utilized in nitrogen sorption experiments at temperatures of 77 K, which serve as a standard tool for the characterization of porous materials in terms of the BET surface area and pore width through the application of (modified) Kelvin equation. However, the details of pore condensation and correct interpretation of the sorption isotherms are still puzzling. To gain better insight into the process, Hoinkis et al. used SANS to study the pore condensation of nitrogen in Controlled Pore Glasses [59]. Liquid nitrogen has a coherent scattering length density (SLD) very similar to SiO_2, resulting in a strong decrease of the scattering from liquid-filled pores. Thus, the pore condensation process causes strong changes in the SANS signal. Sel et al. used the SLD matching effect of liquid nitrogen and silica for in situ SANS experiments of the pore filling of hierarchical SiO_2 with liquid nitrogen [60]. The pore filling of the smaller pores, which occurs at smaller relative pressure, could be separated in the SANS signals from the pore filling of the larger pores, taking place at larger relative pressures. The pore-filling process of hydrogen confined in MCM-41 materials has been studied by Floquet et al. [61]. They characterized two different sorption modes at low and higher relative hydrogen pressures. The sorption of hydrogen to nanocrystalline graphite was studied by Itoh et al. [62].

The sorption of subcritical fluids to narrow pores shows the interesting phenomenon of sorption hysteresis, that is, a shift of the desorption branch of the isotherm to lower pressures with respect to the adsorption branch. The effect occurs below a so-called pore-critical temperature, which is lower than the (bulk) critical temperature of the fluid, and is more pronounced at lower temperatures [63–65]. The origin of this at least stabilization of the condensed physisorbed phase is not fully understood yet, but ink-bottle, percolation, and kinetic effects are being discussed. Neutron scattering experiments carried out in the hysteresis region should be able to probe the structure of the pore fluid on a nanometer length scale and help to gain more insight into this interesting and longstanding problem.

SANS experiments on porous materials in contact with a fluid reservoir can also be used to test the connectivity of the pore system through differentiation of accessible and isolated pore spaces by their different scattering contrasts. Liquid nitrogen is often used in this type of experiment on porous silica, because the vanishing contrast leads to effective suppression of the scattering contributed from filled (i.e., accessible) pores. Thus, SANS measurements of the "dry" and the liquid-filled sample can reveal the amount of accessible porosity. Mixtures of H_2O and D_2O in suitable ratios can match the SLD of natural rocks and sediments of any chemical composition. Hydrophobic porous materials may not support pore filling with water, in which case mixtures of protonated and deuterated benzene can be utilized.

Enhanced oil recovery is a modern technique to increase oil production of oil wells, in which large quantities of pressurized CO_2 are pumped into the oil-bearing reservoir to replace the oil. Another recent concept is the disposal of CO_2 into a number of candidate geological compartments (e.g., subsurface brines, depleted oil or gas reservoirs, unmineable coal beds) as part of the efforts to control climate change. A detailed understanding of the underlying fluid–rock interactions is crucial for the large-scale application of these important new technologies. The sorption properties of supercritical CO_2 in highly porous silica aerogel have been studied by a combination of SANS and neutron transmission [55, 66]. Two distinct sorption regimes show in the adsorption isotherm, with positive values of the excess adsorption at low fluid densities and negative values of the excess adsorption at high fluid densities. This finding indicates that CO_2 is expelled from the pore spaces of silica aerogel at high fluid densities. The crossover from fluid adsorption to depletion takes place in the vicinity of the bulk-critical density of the fluid. A strong decrease of the CO_2 adsorption with increasing temperature has been found in the low fluid pressure region, while the strength of the high-pressure depletion effects showed little or no dependence on temperature. For CO_2 sequestration this might mean that an upper limit of fluid loading into the rock formation is given, which cannot be overcome by increasing the fluid pressure. Similar results with fluid adsorption and depletion regimes at low and high pressures have been found for the sorption of sulfur hexafluoride (SF_6) in silica aerogel. Systematic experiments with a number of fluids of different polarity imbibed in different engineered and natural porous materials, including coals and sandstones, will reveal the physical conditions under which the adsorption and depletion effects occur.

Mixing and de-mixing processes of liquid mixtures play a role in fluid enrichment and depletion processes and may be caused by local variations in temperature, pressure, and composition changes of the pore fluid and the confining matrix. The structure of partially miscible fluids of critical composition imbibed in porous materials has been extensively studied with SANS techniques [67–72]. It was found that the phase separation temperature of the binary liquid inside porous matrices is shifted toward the regime of the demixed phase, and concentration fluctuations are stabilized over a large temperature range. Macroscopic phase separation is very slow, with the size of the micro-phase-separated domains being no larger than the pore dimension in low-porosity matrices, while the picture is unclear yet for low-porosity confinements. The Single Pore model [73] was developed to describe phase-separation processes in low-porosity materials like Vycor and CPG-10, while the Random Field Ising model [74] is often applied for demixing processes in high-porosity materials (e.g., aerogels).

Recently, Schemmel et al. studied the structure of isobutyric acid (iBA) and heavy water mixtures of different off-critical compositions imbibed in Controlled Pore Glass with 75 Å nominal pore size [75, 76]. This system has a lower bulk critical-phase-separation temperature of $T_c = 44.1°C$; heavy water is preferentially adsorbed to the pore walls. SANS data were taken over a wide temperature range of (10–70°C) around the bulk T_c for two samples with different compositions of the pore liquid: iBA rich and D_2O rich. The sample containing iBA-rich pore liquid showed a strong temperature dependence of the microstructure of the pore liquid, with unique scattering patterns identifying the formation of an adsorbed water-rich layer at high temperatures, critical fluctuations over an extended temperature range, and micro-phase separation at low temperatures. The phase separation temperature of the pore liquid was shifted by 10 K into the two-phase region as compared to the bulk mixture. The SANS data of the water-rich sample showed the formation of a water-rich sorption phase with no temperature dependence of the scattering in the studied temperature range, indicating strong phase segregation over an extended temperature range.

Zeolites are microporous aluminosilicate minerals that play an important role in many natural and industrial processes, including water purification through ion exchange, catalytic hydrocarbon cracking, and separation of pollutants from natural gas. They possess pore widths of typically a few tenths of a nanometer, making X-ray and neutron diffraction suitable tools for the study of these materials and guest molecules inside their pore systems. Neutron diffractometers exist in a variety of configurations for thermal and cold neutrons optimized for resolution or flux and interrogate length scales of up to 20 Å. The ND technique has been recently used by Mentzen to study the adsorption of hydrogen and benzene in MFI-type zeolites [77]. The positions of the guest molecules in the microporous host structure were determined through Rietveld analysis to analyze the binding sites. The sorption of several hydrocarbons in silicalite-1 zeolite was studied by Floquet et al. using ND [78]. In a recent study, Floquet et al. used ND techniques to examine the sorption behavior of heptane in Silicalite-1 zeolite [79]. The results indicate that the heptane populates the straight channels of the silicalite pore network first, while the sinusoidal channels and intersections fill with heptane only above a heptane concentration of about 3.9 molecules per unit cell.

19.3 Dynamics of Confined Fluids

19.3.1 Water and Aqueous Solutions

Dynamics of confined water can be investigated by neutron spectroscopy techniques that rely mainly on the incoherent scattering of neutrons from hydrogen nuclei. Even though the actual motions of nuclei in the molecules of a liquid phase are complex, they often can be separated into fast vibrational and librational and slow diffusion (rotational and translational) motions. The more energetic (on the energy scale from several to several hundreds meV) vibrational and librational modes are typically probed using dedicated neutron spectrometers with moderate energy resolution and reasonably high incident neutron energies. On the timescale of such spectrometers, rotational and translational motions are very slow and can be neglected. This type of measurements is known as inelastic neutron spectroscopy (INS). Compared to infrared spectroscopy, INS benefits from the absence of optical selection rules and the large incoherent scattering cross section of hydrogen.

There have been numerous studies of water (or hydroxyl groups) vibration dynamics in zeolites [80–84]. More recently, INS was used to probe the dynamics of hydroxyl groups in mesoporous silica [85]. In these studies, the fundamental vibrational modes of water molecules and hydroxyl groups adsorbed at well-defined sites of the host framework are usually probed at low (helium) temperatures with neutron energy loss in order to sharpen the inelastic peaks, which suffer loss in quality due to multi-phonon scattering in ambient temperature measurements with neutron energy gain. Another recent development in the area of INS studies of confined liquids concerned measuring low-frequency (at about 100 meV and below) dynamics in supercooled water confined in Gelsil and Vycor glasses [86–88]. The emphasis in this area of research is on the properties of confined water, as opposed to studies of zeolites, where measured vibrational frequencies are typically used to characterize the adsorption sites.

The slower diffusive motions that take place on the timescale of a pico- to nanosecond, which corresponds to the energy scale from a fraction of μeV to several hundred μeV, are probed using quasielastic neutron scattering (QENS) technique [89]. QENS data are typically collected on dedicated high-energy resolution, either time-of-flight or backscattering, spectrometers [Chapter 3]. Even slower diffusion motions in liquids, on the timescale of up to tens of nanoseconds, can be probed using neutron spin-echo (NSE) technique. We cover diffusion dynamics studies of confined water in more detail because these dynamics (especially the translational component) are affected the most by a confinement: a change by two orders of magnitude in the mobility of confined water is common.

The incoherent scattering intensity from hydrogen nuclei that typically dominates QENS signal from a system containing confined water can be described as Fourier transforms of the self-correlation function $G_{\text{self}}(r, t)$, which is the probability that a particle that was at $r = 0$ at $t = 0$ is at position r at time t:

$$I_{\text{inc}}(Q, t) = \int G_{\text{self}}(r, t)e^{-iQr}dr,$$ (19.1)

$$S_{\text{inc}}(Q, E) = \int I_{\text{inc}}(Q, t)e^{i\left(\frac{E}{\hbar}\right)t}dt.$$ (19.2)

The intermediate scattering function, $I_{\text{inc}}(Q, t)$, can be measured directly in a NSE experiment and is usually the value that is computed in molecular dynamics simulations. A QENS experiment yields the incoherent structure factor, $S_{\text{inc}}(Q, E)$. For a regular diffusion process, which is characterized by the linear dependence of the particle mean-squared displacement on time, $< x^2 > \sim t$, the intermediate scattering function is described by a simple exponential decay in the time space, and the incoherent structure factor is a Lorentzian in the energy space. The Q-dependence of the width of the Lorentzian broadening, $\Gamma(Q)$, can be used to determine both the spatial and temporal characteristics of the diffusion process. For instance, a simple "jump diffusion" model that captures the essential features of diffusion in liquids is characterized by $\Gamma(Q \to 0) = \hbar DQ^2$ and $\Gamma(Q \to \infty) = \hbar/\tau_0$, where D is a diffusion coefficient and τ_0 is a time between the diffusion jumps.

Another important parameter measured in QENS experiments is the Q-dependence of the fraction of the elastic scattering intensity, often called *elastic incoherent structure factor*, EISF(Q). Scattering from the nuclei that are immobile on the timescale of the experiment ($G_{\text{self}}(r, t) = 1$) yields a flat, time-independent background in the $I_{\text{inc}}(Q, t)$ and a delta-function-like elastic peak in the $S_{\text{inc}}(Q, E)$ at all Q values. Thus, contribution from surface hydroxyl groups or immobile water molecules in direct contact with the pore surface typically yields Q-independent elastic scattering. If, on the other hand, the volume accessible to a diffusing particle is restricted by the impenetrable pore boundaries, the particle's EISF exhibits a Q-dependence that can be used to deduce the characteristic dimension and sometimes the shape of the confining pore. Regardless of the confinement geometry, EISF $(Q \rightarrow 0) = 1$ for a confined diffusion motion, reflecting the fact that for the molecule diffusing in a confined space $< x^2 >$ remains limited even at long times, unlike that for the freely diffusing particle. Concurrently, the QENS broadening resulting from a confined diffusion process does not go to zero at low Q; instead, it exhibits a practically Q-independent constant level below the Q value corresponding to the pore size.

The rotational diffusion component, also accessible in QENS measurements, is usually much less affected by confinement compared to the translational component. Confinement imposes little additional restrictions on already spatially localized rotational motions, and the characteristic rotational diffusion times for confined water usually increase by less than an order of magnitude compared to bulk water. On the other hand, the characteristic translational diffusion times and diffusion coefficients often change by two orders of magnitude or more compared to the values for bulk water. The effects of faster vibrational and librational motions on QENS signal manifest themselves in the overall reduction of scattering intensities (Debye-Waller factor). In order to simplify the analysis of QENS data, faster rotational and slower translational diffusion motions are usually assumed to be independent (decoupling approximation). Then in the time space one can write the intermediate scattering function as a product of the rotational and translational components. Consequently, the incoherent structure factor in the energy space becomes a convolution of the rotational and translational components. The landmark study of supercooled bulk water by Teixeira et al. [90] has established the framework for interpretation of QENS data from bulk water and water in various confining matrices. Both rotational and translational diffusion times in bulk water at room temperature are close to a picosecond. Upon supercooling down to 253 K, the translational jumps slow down by an order of magnitude, whereas the time between rotational jumps increases by just about a factor of two. A series of important studies of supercooled water in Vycor glass [31, 91, 92] has demonstrated the effects of confinement on water dynamics and showed some remarkable similarities between supercooled bulk and confined water. These were followed by numerous experiments on water confined in various silica matrices, such as mesoporous silicas [32,93–98] and Gelsil glass [87, 99]. As a result of these studies, two conceptually different approaches have evolved for interpretations of the QENS data on confined water. One is the "confined jump diffusion" model described above where the shape and size of the confining pore manifest themselves through the form-factor, which defines the Q-dependence of the EISF. This model assumes that water molecules perform unrestricted diffusion jumps within a pore until they encounter the pore boundaries. The second concept, commonly called "relaxing cage" model [92, 100], suggests that on the timescale of translational jumps of a water molecule, the relevant confinement size is not the size of the pore but that of the "cage" formed by the neighboring water molecules, from where the molecule can escape through the structural relaxation. In the "relaxing cage" model, the functional form of the intermediate scattering function in the time space is no longer described by the simple exponential decay. Instead, it is represented by a stretched exponential decay, $\exp[-(t/\tau)^\beta]$. A stretched exponential decay is the same functional form as one can obtain assuming a distribution of local diffusivities in the confined medium [101].

Recently, Mamontov and Cole [102] reported on the diffusion of water molecules in 2.3 molal $CaCl_2$ solution confined in 100% hydrated Vycor glass as a function of temperature from 220 to 260 K. A gradual transition was observed from the restricted diffusion regime at lower temperatures to unrestricted diffusion regime at higher temperatures. The effect of dissolved ions on the diffusion dynamics of the water molecules in the solution was amplified by confinement by at least an order of magnitude compared to bulk form, even though the dissolved ions were found to have little effect on the spatial characteristics of the restricted diffusion process of water molecules (Fig. 19.6). At 260 K, the local diffusion coefficient of water molecules in the H_2O–$CaCl_2$ confined in Vycor was only 6% of the value reported for pure water confined in Vycor.

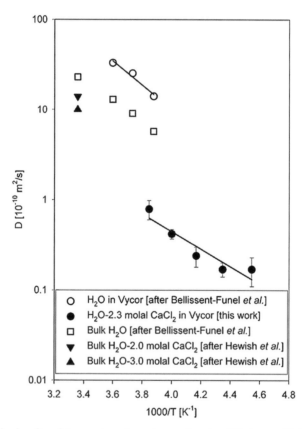

Fig. 19.6 Arrhenius fits of temperature dependence of water diffusion coefficients in Vycor-confined water [91] and Vycor-confined H_2O–$CaCl_2$ [102]. The data for bulk water [91] and H_2O–$CaCl_2$ [103] are shown for comparison

Reprinted with permission from Mamontov and Cole [102], Fig. 19.4, p. 4912, 2006 © by the Royal Society of Chemistry. URL:http://dx.doi.org/10.1039/b610674a

QENS experiments on water confined in materials other than silicas tend to focus on the influence of the matrix parameters (such as cation charge and channel size in zeolites) on the water dynamics. Various examples of matrices used to confine water include zeolites [104–107], porous alumina [108, 109], tuff [110], layered silicate AMH-3 [111], aluminosilicate glasses [112], and a silica–alumina–NaO_2 molecular sieve [113]. In these studies, ambient and elevated temperatures were of primary interest, even though it was possible to supercool water confined in small pores. In an interesting series of experiments, Fratini et al. [114, 115] and Faraone et al. [116] investigated the dynamics of water in cements (hydrated dicalcium and tricalcium silicates) as a function of aging (i.e., cement curing time) at ambient temperature. Examples of the less common systems used to study the dynamics of confined water by QENS include a layered material V_2O_5[117, 118] and chrysotile asbestos (Mg_3 $Si_2O_5(OH)_4$) fibers with macroscopically aligned one-dimensional channels [119]. Recent QENS studies of water mobility in clays [120–127] represent a rapidly developing field of significant importance to geosciences. Because oriented clay platelets are readily available, it is often possible to differentiate between in-plane and out-of-plane interlayer water dynamics by means of properly orienting the sample with respect to the scattering vector. Both ambient and supercooled temperature regimes of interlayer water diffusion in clays have been investigated. Recent studies by Malikova et al. [124, 125] provide a good example of comparison among QENS, NSE, and MD simulations to investigate the dynamics of water in montmorillonite. These studies highlight a complex nature of diffusion of water confined in clays.

The inverse relaxation times (directly related to the QENS broadening) for bihydrated montmorillonite are shown in Fig. 19.7, as a function of Q^2 obtained at room temperature using different techniques. The MD simulation and QENS experiment both yield the value for the diffusion coefficient extracted from the low-Q data of

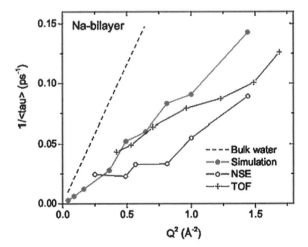

Fig. 19.7 Inverse relaxation times for hydration water in bihydrated montmorillonite [125] Reprinted with permission from Malikova et al. [125]. Copyright 2006 American Chemical Society

about a half of that for bulk water. On the other hand, the NSE experiment yields the diffusion coefficient, which is about one-half of the values obtained from the QENS and MD simulation. As discussed in the paper, the dynamics of hydration water in montmorillonite and in clays, in general, exhibits a wide distribution of relaxation times. For this particular system, only the NSE experiment probes the sufficiently long relaxation times, thus yielding more accurate diffusion coefficient. This example provides a good illustration of the well-known, yet often underappreciated, fact that underestimation of the relaxation times (that is, overestimation of water dynamics) may occur if the limitations of experiment or simulation preclude assessment of sufficiently long relaxation times. Thus, using a neutron scattering spectrometer with the energy resolution and dynamic range that provide a good match for the dynamics of interest is crucial for extracting accurate dynamic parameters, especially for systems that possess a wide distribution of the relaxation times, such as confined water.

19.3.2 Non-aqueous Fluids

Even though water has been by far the most extensively investigated medium in confinement, other liquids in confined environments have attracted significant attention. In particular, there have been numerous INS studies of molecules other than water adsorbed in zeolites, for example, alcohols [128, 129], chloroform [130, 131], methane [132], ethane and ethene [133], furan [134], ferrocene [135], and hydrogen [136]. Surveys by Jobic [137, 138] provide an overview of earlier and more recent INS studies of media confined in zeolites. A study by Rosi et al. [139] involved the investigation of hydrogen adsorbed in metal organic framework MOF-5. The INS spectra obtained in this work indicated the presence of two well-defined binding sites in this material, which is characterized by significant hydrogen uptake. As in the case of confined water, INS studies of organic molecules in confining matrices with a well-defined structure are mainly aimed at characterization of the adsorption sites. A study of propylene glycol confined in a porous glass by Melnichenko et al. [18] represents an example of research where the properties of the confined medium were of main interest.

Mobility of confined materials other than water has also been investigated in a number of QENS studies. Some examples include hydrogen [140], methane [141], toluene [142], and cyclohexane [143] in micro- and mesoporous silicas, n-hexane [144, 145], H_2/D_2 [146–148], N_2 and CO_2 [149], benzene [150], butadiene [151], propane [152] and long n-alkanes [153] in zeolites, and benzene in chrysotile asbestos fibers [154]. It should be noted that the signal from coherently scattering nuclei is dependent on collective properties (such as transport diffusivity), whereas the signal from incoherently scattering nuclei such as hydrogen, which is measured in the overwhelming majority of QENS experiments, describes self-diffusion. While both QENS and neutron spin-echo techniques can measure data from coherently and incoherently scattering samples, spin-echo is intrinsically more effective for coherent scattering measurements. This is why most spin-echo measurements involve deuterated samples and probe transport rather than self-correlation properties. Studies by Swenson et al. [123] and Jobic et al. [155] provide examples of spin-echo experiments on hydrogenated (water) and deuterated (benzene) liquids in confinement, respectively. Another study by Jobic [156] is an interesting example of investigating the dynamics of complex hydrocarbon molecules in confinement (linear and branched alkanes, for chains up to C–14, confined in ZSM-5 zeolite). Because of the relatively large size of confined molecules, the diffusion could be observed within the time window of a backscattering spectrometer only at high temperatures. Branched alkanes were found to diffuse much more slowly than linear alkanes: for example, the diffusion of $CH_3(CH_2)_6CH_3$ at 400 K was significantly faster than that of $CH(CH_3)_3$ at 570 K. A recent survey by Jobic and Theodorou [157] provides an excellent overview of QENS studies of confined media in zeolites and their synergy with molecular dynamic simulations.

The properties of organic molecules in hydrated 2:1 clays have been studied by Skipper et al. using quasielastic neutron scattering [42]. It was shown that methane interacts strongly with the clay, leading to a decrease of the methane diffusion coefficient by one order of magnitude as compared to bulk water-methane. The more hydrophilic ethylene glycol forms hydrogen bonds, and its diffusion coefficient in clay is decreased only by a factor of 3–4 as compared to the bulk.

19.4 Summary and Outlook

There can be no disputing the fact that neutron diffraction and scattering have made a clear contribution to our current understanding of the structural and dynamical characteristics of confined fluids at ambient conditions and to a somewhat lesser degree at other state conditions involving a change in temperature and pressure. Indeed, a molecular-level understanding of how fluids (e.g., water, CO_2, CH_4, higher hydrocarbons) interact with and participate in reactions with other solid earth materials is central to the development of predictive models that aim to quantify a wide array of geochemical processes. The importance of the hydrogen-bond interaction, in water and other important hydrogenous fluids, cannot be overstated. Despite the large body of work that documents the nature of hydrogen bonding and associated interactions with its local surroundings, it is premature to assume that we have a complete understanding of the mechanisms that give rise to the particular properties exhibited by water and other simple molecular fluids. This is particularly true as one goes both above and below ambient conditions. For example, there is continuing discussion on the relation between the behavior of supercooled water, the structure of the amorphous ices, the behavior of molecular fluids at interfaces, and the incorporation of simple molecules in hydrate clathrates. This is of particular interest since we have seen that nanoporous confinement of water at ambient conditions leads to structural and dynamical features that emulate the supercooled state. In the context of natural systems, interrogation of fluids and fluid–solid interactions at elevated temperatures and pressures is an area requiring much more work, particularly for complex solutions containing geochemically relevant cations, anions, and other important dissolved species such CO_2 or CH_4.

Acknowledgments Research by DRC was sponsored by the Division of Chemical Sciences, Geosciences, and Biosciences, Office of Basic Energy Sciences (OBES), U.S. Department of Energy. EM acknowledges support to SNS from the U.S. Department of Energy OBES. GR gratefully acknowledges support from the Laboratory Directed Research and Development Program. Oak Ridge National Laboratory (ORNL) is managed by UT-Battelle, LLC for the U.S. Department of Energy under Contract No. DE-AC05-00OR22725.

References

1. D. R. Cole, M. S. Gruszkiewicz, J. M. Simonson, A. A. Chialvo, Y. B. Melnichenko, G. D. Wignall, G. W. Lynn, J. S. Lin, A. Habenschuss, B. Gu, K. L. More, T. D. Burchell, A. Striolo, Y. Leng, P. T. Cummings, W. T. Cooper, M. Schilling, K. E. Gubbins, H. Frielinghaus, Influence of nanoscale porosity on fluid behavior. *Water-Rock Interaction*, Vol. 1, Editors: R. Wanty and R. Seal, Proceedings to the 11th International Symposium on Water-Rock Interaction, A. A. Balkema, Rotterdam, The Netherlands, 735-739, (2004)
2. B. Belonoshko, Geochim. Cosmochim. Acta **53**(10), 2581 (1989)
3. J. J. Tuck, P. L. Hall, M. H. B. Hayes, D. K. Ross, and J. B. Hayter, J. Chem. Soc., Faraday Trans. **81**, 833 (1985)
4. N. T. Skipper, A. K. Soper, and J. D. C. Mcconnell, J. Chem. Phys. **94**(8), 5751 (1991)
5. N. T. Skipper, M. V. Smalley, G. D. Williams, A. K. Soper, and C. H. Thompson, J. Phys. Chem. **99**(39), 14201 (1995)
6. C. Pitteloud, D. H. Powell, M. A. Gonzalez, and G. J. Cuello, Coll. Surf. A **217**(1–3), 129 (2003)
7. Y. F. Wang, C. Bryan, H. F. Xu, and H. Z. Gao, Geology **31**(5), 387 (2003)
8. J. W. Wang, A. G. Kalinichev, and R. J. Kirkpatrick, Earth Planet. Sci. Lett. **222**(2), 517 (2004)
9. J. W. Wang, A. G. Kalinichev, and R. J. Kirkpatrick, Geochim. Cosmochim. Acta **68**(16), 3351 (2004)
10. J. W. Wang, A. G. Kalinichev, R. J. Kirkpatrick, and R. T. Cygan, J. Phys. Chem. B **109**(33), 15893 (2005)
11. J. W. Wang, A. G. Kalinichev, and R. J. Kirkpatrick, Geochim. Cosmochim. Acta **70**(3), 562 (2006)
12. S. H. J. Idziak and Y. L. Li, Curr. Opin. Colloid Interface Sci. **3**(3), 293 (1998)
13. D. K. Dysthe and R. A. Wogelius, Chem. Geol. **230**(3–4), 175 (2006)
14. J. Alejandre, M. LozadaCassou, and L. Degreve, Mol. Phys. **88**(5), 1317 (1996)
15. E. Pitard, M. L. Rosinberg, and G. Tarjus, Mol. Simul. **17**(4–6), 399 (1996)
16. L. D. Gelb, K. E. Gubbins, R. Radhakrishnan, and M. Sliwinska-Bartkowiak, Rep. Prog. Phys. **62**(12), 1573 (1999)
17. J. M. Drake, J. Klafter, and P. Levitz, Science **251**(5001), 1574 (1991)
18. Y. B. Melnichenko, J. Schüller, R. Richert, B. Ewen, and C. K. Loong, J. Chem. Phys. **103**(6), 2016 (1995)
19. E. P. Gilbert, P. A. Reynolds, P. Thiyagarajan, D. G. Wozniak, and J. W. White, Phys. Chem. Chem. Phys. **1**(11), 2715 (1999)
20. T. Bellini, N. A. Clark, C. D. Muzny, L. Wu, C. W. Garland, D. W. Schaefer, B. J. Olivier, and B. J. Oliver, Phys. Rev. Lett. **69**(5), 788 (1992)
21. A. P. Radlinski, E. Z. Radlinska, M. Agamalian, G. D. Wignall, P. Lindner, and O. G. Randl, Phys. Rev. Lett. **82**(15), 3078 (1999)

22. A. P. Radlinski, Neutron Scattering In Rev. Min. Geochem. (H.-R. Wenk, ed.), **63**, 363 (2006)
23. D. R. Cole, K. W. Herwig, E. Mamontov, and J. Z. Larese, Neutron Scattering In Rev. Mineral. Geochem. (H.-R. Wenk, ed.), **63**, 313 (2006)
24. D. A. Neumann, Materials Today **9**(1–2), 34 (2006)
25. L. Feigin and D. Svergun, *Structural Analysis by Small-Angle X-Ray and Neutron Scattering* (Plenum Press, New York, 1987)
26. J. C. Dore, M. A. M. Sufi, and M. C. Bellissent-Funel, Phys. Chem. Chem. Phys. **2**(8), 1599 (2000)
27. M. C. Bellissent-Funel, J. Mol. Liq. **90**(1–3), 313 (2001)
28. F. Bruni, M. A. Ricci, and A. K. Soper, J. Chem. Phys. **109**(4), 1478 (1998)
29. B. Webber and J. Dore, J. Phys.: Condens. Matter **16**(45), S5449 (2004)
30. D. C. Steytler, J. C. Dore, and C. J. Wright, J. Phys. Chem. **87**(14), 2458 (1983)
31. M. C. Bellissent-Funel, J. Lal, and L. Bosio, J. Chem. Phys. **98**(5), 4246 (1993)
32. T. Takamuku, M. Yamagami, H. Wakita, Y. Masuda, and T. Yamaguchi, J. Phys. Chem. B **101**(29), 5730 (1997)
33. J. M. Baker, J. C. Dore, and P. Behrens, J. Phys. Chem. B **101**(32), 6226 (1997)
34. J. Dore, B. Webber, M. Hartl, P. Behrens, and T. Hansen, Physica A **314**(1–4), 501 (2002)
35. V. Venuti, V. Crupi, D. Majolino, P. Migliardo, and M. C. Bellissent-Funel, Physica B **350**, e599 (2004)
36. J. Dore, Chem. Phys. **258**(2–3), 327 (2000)
37. M. Agamalian, J. M. Drake, S. K. Sinha, and J. D. Axe, Phys. Rev. E **55**(3), 3021 (1997)
38. M. A. Ricci, F. Bruni, P. Gallo, M. Rovere, and A. K. Soper, J. Phys.: Condens. Matter **12**(8A), A345 (2000)
39. C. Poinsignon, H. Estradeszwarckopf, J. Conard, and A. J. Dianoux, Physica B **156**, 140 (1989)
40. N. T. Skipper, A. K. Soper, and M. V. Smalley, J. Phys. Chem. **98**(3), 942 (1994)
41. N. T. Skipper, G. D. Williams, A. V. C. de Siqueira, C. Lobban, and A. K. Soper, Clay Miner. **35**(1), 283 (2000)
42. N. T. Skipper, P. A. Lock, J. O. Titiloye, J. Swenson, Z. A. Mirza, W. S. Howells, and F. Fernandez-Alonso, Chem. Geol. **230**(3–4), 182 (2006)
43. G. D. Williams, N. T. Skipper, and M. V. Smalley, Physica B **234**, 375 (1997)
44. G. D. Williams, A. K. Soper, N. T. Skipper, and M. V. Smalley, J. Phys. Chem. B **102**(45), 8945 (1998)
45. A. V. de Siqueira, C. Lobban, N. T. Skipper, G. D. Williams, A. K. Soper, R. Done, J. W. Dreyer, R. J. Humphreys, and J. A. R. Bones, J. Phys.: Condens. Matter **11**(47), 9179 (1999)
46. C. Pitteloud, D. H. Powell, A. K. Soper, and C. J. Benmore, Physica B **276**, 236 (2000)
47. C. Pitteloud, D. H. Powell, and H. E. Fischer, Phys. Chem. Chem. Phys. **3**(24), 5567 (2001)
48. J. Swenson, R. Bergman, and S. Longeville, J. Non-Cryst. Solids **307**, 573 (2002)
49. G. Sposito, *The Chemistry of Soils* (Oxford Univ. Press, New York, 1989)
50. G. Sposito, N. T. Skipper, R. Sutton, S. H. Park, A. K. Soper, and J. A. Greathouse, Proc. Natl. Acad. Sci. USA. **96**(7), 3358 (1999)
51. N. T. Skipper, A. K. Soper, J. D. C. McConnell, and K. Refson, Chem. Phys. Lett. **166**(2), 141 (1990)
52. H. E. King, S. T. Milner, M. Y. Lin, J. P. Singh, and T. G. Mason, Phys. Rev. E **75**(2), 021403 (2007)
53. J. Connolly, W. Bertram, J. Barker, C. Buckley, T. Edwards, and R. Knott, J. Petrol. Sci. Eng. **53**(3–4), 171 (2006)
54. T. Itakura, W. K. Bertram, and R. B. Knott, Appl. Clay Sci. **29**(1), 1 (2005)
55. G. Rother, Y. B. Melnichenko, D. R. Cole, H. Frielinghaus, and G. D. Wignall, J. Phys. Chem. C **111**(43), 15736 (2007)
56. W. Wu, Polymer **23**(13), 1907 (1982)
57. M. Thommes, G. H. Findenegg, and M. Schoen, Langmuir **11**(6), 2137 (1995)
58. I. Brovchenko and A. Oleinikova, J. Phys. Chem. C **111**, 15716 (2007)
59. E. Hoinkis and B. Rohl-Kuhn, J. Colloid Interface Sci. **296**(1), 256 (2006)

60. O. Sel, A. Brandt, D. Wallacher, M. Thommes, and B. Smarsly, Langmuir **23**(9), 4724 (2007)
61. N. Floquet, J. P. Coulomb, and G. André, Microp. Mesopor. Mater. **72**, 143 (2004)
62. K. Itoh, Y. Miyahara, S. Orimo, H. Fujii, T. Kamiyama, and T. Fukunaga, J. Alloys Compd. **356–357**, 608 (2003)
63. A. Schreiber, S. Reinhardt, and G. H. Findenegg, Characterization of Porous Solids Vi **144**, 177 (2002)
64. K. Morishige and Y. Nakamura, Langmuir **20**(11), 4503 (2004)
65. K. Morishige, M. Tateishi, F. Hirose, and K. Aramaki, Langmuir **22**(22), 9220 (2006)
66. Y. B. Melnichenko, G. D. Wignall, D. R. Cole, and H. Frielinghaus, J. Chem. Phys. **124**(20), 204711 (2006)
67. J. V. Maher, W. I. Goldburg, D. W. Pohl, and M. Lanz, Phys. Rev. Lett. **53**(1), 60(1984)
68. A. E. Bailey, B. J. Frisken, and D. S. Cannell, Phys. Rev. E **56**(3), 3112 (1997)
69. B. J. Frisken, D. S. Cannell, M. Y. Lin, and S. K. Sinha, Phys. Rev. E **51**(6), 5866 (1995)
70. M. Y. Lin, S. K. Sinha, J. M. Drake, X. I. Wu, P. Thiyagarajan, and H. B. Stanley, Phys. Rev. Lett. **72**(14), 2207 (1994)
71. F. Formisano and J. Teixeira, J. Phys.: Condens. Matter **12**(8A), A351 (2000)
72. F. Formisano and J. Teixeira, Eur. Phys. J. E **1**(1), 1 (2000)
73. A. J. Liu, D. J. Durian, E. Herbolzheimer, and S. A. Safran, Phys. Rev. Lett. **65**, 1897 (1990)
74. P. G. de Gennes, J. Phys. Chem. **88**, 6469 (1984)
75. S. Schemmel, G. Rother, H. Eckerlebe, and G. H. Findenegg, J. Chem. Phys. **122**(24), 244718 (2005)
76. D. Woywod, S. Schemmel, G. Rother, G. H. Findenegg, and M. Schoen, J. Chem. Phys. **122**(12), 124510 (2005)
77. B. F. Mentzen, J. Phys. Chem. C **111**, 18932 (2007)
78. N. Floquet, J. P. Coulomb, G. Weber, O. Bertrand, and J. P. Bellat, J. Phys. Chem. B **107**, 685 (2003)
79. N. Floquet, J. P. Coulomb, J. P. Bellat, J. M. Simon, G. Weber, and G. Andre, J. Phys. Chem. C **111**, 18182 (2007)
80. H. Jobic, M. Czjzek, and R. A. Vansanten, J. Phys. Chem. **96**(4), 1540 (1992)
81. P. C. H. Mitchell, J. Tomkinson, J. G. Grimblot, and E. Payen, J. Chem. Soc., Faraday Trans. **89**(11), 1805 (1993)
82. H. Jobic, A. Tuel, M. Krossner, and J. Sauer, J. Phys. Chem. **100**(50), 19545 (1996)
83. I. A. Beta, H. Bohlig, and B. Hunger, Phys. Chem. Chem. Phys. **6**(8), 1975 (2004)
84. C. Corsaro, V. Crupi, F. Longo, D. Majolino, V. Venuti, and U. Wanderlingh, J. Phys.: Condens. Matter **17**(50), 7925 (2005)
85. E. Geidel, H. Lechert, J. Dobler, H. Jobic, G. Calzaferri, and F. Bauer, Micropor. Mesopor. Mater. **65**(1), 31 (2003)
86. F. Venturini, P. Gallo, M. A. Ricci, A. R. Bizzarri, and S. Cannistraro, J. Chem. Phys. **114**(22), 10010 (2001)
87. V. Crupi, D. Majolino, P. Migliardo, V. Venuti, and A. J. Dianoux, Appl. Phys. A **74**, S555 (2002)
88. V. Crupi, A. J. Dianoux, D. Majolino, P. Migliardo, and V. Venuti, Phys. Chem. Chem. Phys. **4**(12), 2768 (2002)
89. M. Bée, Chem. Phys. **292**, 121 (2003)
90. J. Teixeira, M. C. Bellissent-Funel, S. H. Chen, and A. J. Dianoux, Phys. Rev. A **31**(3), 1913 (1985)
91. M. C. Bellissent-Funel, S. Chen, and J. M. Zanotti, Phys. Rev. E **51**, 4558 (1995)
92. J. M. Zanotti, M. C. Bellissent-Funel, and S. H. Chen, Phys. Rev. E **59**(3), 3084 (1999)
93. S. Takahara, M. Nakano, S. Kittaka, Y. Kuroda, T. Mori, H. Hamano, and T. Yamaguchi, J. Phys. Chem. B **103**(28), 5814 (1999)
94. F. Mansour, R. M. Dimeo, and H. Peemoeller, Phys. Rev. E **66**(4), 041307 (2002)
95. A. Faraone, L. Liu, C. Y. Mou, P. C. Shih, J. R. D. Copley, and S. H. Chen, J. Chem. Phys. **119**(7), 3963 (2003)

96. A. Faraone, L. Liu, C. Y. Mou, P. C. Shih, C. Brown, J. R. D. Copley, R. M. Dimeo, and S. H. Chen, Eur. Phys. J. E **12**, S59 (2003)

97. L. Liu, A. Faraone, C. Mou, C. W. Yen, and S. H. Chen, J. Phys.: Condens. Matter **16**(45), S5403 (2004)

98. S. Takahara, N. Sumiyama, S. Kittaka, T. Yamaguchi, and M. C. Bellissent-Funel, J. Phys. Chem. B **109**(22), 11231 (2005)

99. V. Crupi, D. Majolino, P. Migliardo, and V. Venuti, Physica A **304**(1–2), 59 (2002)

100. S. Chen, P. Gallo, and M. C. Bellissent-Funel, Can. J. Phys. **73**, 703 (1995)

101. J. Colmenero, A. Arbe, A. Alegria, M. Monkenbusch, and D. Richter, J. Phys.: Condens. Matter **11**(10A), A363 (1999)

102. E. Mamontov and D. R. Cole, Phys. Chem. Chem. Phys. **8**(42), 4908 (2006)

103. N. A. Hewish, J. E. Enderby, and W. S. Howells, J. Phys. C: Solid State Phys. **16**(10), 1777 (1983)

104. H. Paoli, A. Methivier, H. Jobic, C. Krause, H. Pfeifer, F. Stallmach, and J. Karger, Micropor. Mesopor. Mater. **55**(2), 147 (2002)

105. V. Crupi, D. Majolino, P. Migliardo, V. Venuti, U. Wanderlingh, T. Mizota, and M. Telling, J. Phys. Chem. B **108**(14), 4314 (2004)

106. V. Crupi, D. Majolino, P. Migliardo, V. Venuti, and T. Mizota, Mol. Phys. **102**(18), 1943 (2004)

107. V. Crupi, D. Majolino, and V. Venuti, J. Phys.: Condens. Matter **16**(45), S5297 (2004)

108. S. Mitra, R. Mukhopadhyay, K. T. Pillai, and V. N. Vaidya, Solid State Commun. **105**(11), 719 (1998)

109. S. Mitra, R. Mukhopadhyay, I. Tsukushi, and S. Ikeda, J. Phys.: Condens. Matter **13**(37), 8455 (2001)

110. S. A. Maddox, P. Gomez, K. R. McCall, and J. Eckert, Geophys. Res. Lett. **29**(8), 1259 (2002)

111. S. Nair, Z. Chowdhuri, I. Peral, D. A. Neumann, L. C. Dickinson, G. Tompsett, H. K. Jeong, and M. Tsapatsis, Phys. Rev. B **71**(10), 104301 (2005)

112. S. Indris, P. Heitjans, H. Behrens, R. Zorn, and B. Frick, Phys. Rev. B **71**(6), 064205 (2005)

113. J. Swenson, H. Jansson, W. S. Howells, and S. Longeville, J. Chem. Phys. **122**(8), 084505 (2005)

114. E. Fratini, S. H. Chen, P. Baglioni, M. C. Bellissent-Funel, Phys. Rev. E **6402**(2), 020201 (2001)

115. E. Fratini, S. H. Chen, P. Baglioni, J. C. Cook, and J. R. D. Copley, Phys. Rev. E **65**(1), 010201 (2002)

116. A. Faraone, S. H. Chen, E. Fratini, P. Baglioni, L. Liu, and C. Brown, Phys. Rev. E **65**(4), 040501 (2002)

117. S. Takahara, S. Kittaka, Y. Kuroda, T. Yamaguchi, H. Fujii, and M. C. Bellissent-Funel, Langmuir **16**(26), 10559 (2000)

118. S. Kittaka, S. Takahara, T. Yamaguchi, and M. C. B. Funel, Langmuir **21**(4), 1389(2005)

119. E. Mamontov, Y. A. Kumzerov, and S. B. Vakhrushev, Phys. Rev. E **71**(6), 061502 (2005)

120. M. Gay-Duchosal, D. H. Powell, R. E. Lechner, and B. Ruffle, Physica B **276**, 234 (2000)

121. J. Swenson, R. Bergman, and W. S. Howells, J. Chem. Phys. **113**(7), 2873 (2000)

122. J. Swenson, R. Bergman, S. Longeville, and W. S. Howells, Physica B **301**(1–2), 28 (2001)

123. J. Swenson, R. Bergman, and S. Longeville, J. Chem. Phys. **115**(24), 11299 (2001)

124. N. Malikova, A. Cadene, V. Marry, E. Dubois, P. Turq, J. M. Zanotti, and S. Longeville, Chem. Phys. **317**(2–3), 226 (2005)

125. N. Malikova, A. Cadene, V. Marry, E. Dubois, and P. Turq, J. Phys. Chem. B **110**(7), 3206 (2006)

126. L. J. Michot, A. Delville, B. Humbert, M. Plazanet, and P. Levitz, J. Phys. Chem. C **111**(27), 9818 (2007)

127. N. Malikova, A. Cadene, E. Dubois, V. Marry, S. Durand-Vidal, P. Turq, J. Breu, S. Longeville, and J. M. Zanotti, J. Phys. Chem. C **111**, 17603 (2007)

128. R. Schenkel, A. Jentys, S. F. Parker, and J. A. Lercher, J. Phys. Chem. B **108**, 15013 (2004)

129. R. Schenkel, A. Jentys, S. F. Parker, and J. A. Lercher, J. Phys. Chem. B **108**, 7902 (2004)
130. C. F. Mellot, A. M. Davidson, J. Eckert, and A. K. Cheetham, J. Phys. Chem. B **102**, 2530 (1998)
131. A. Davidson, C. F. Mellot, J. Eckert, and A. K. Cheetham, J. Phys. Chem. B **104**, 432 (2000)
132. R. Stockmeyer, Zeolites **4**, 81 (1984)
133. N. J. Henson, J. Eckert, P. J. Hay, and A. Redondo, Chem. Phys. **261**, 111 (2000)
134. I. A. Beta, J. Herve, E. Geidel, H. Bohlig, and B. Hunger, Spectrochem Acta A **57**, 1393 (2001)
135. E. Kemner, A. R. Overweg, U. A. Jayasooriya, S. F. Parker, I. M. de Schepper, and G. J. Kearley, Appl. Phys. A **74**, S 1368 (2002)
136. J. Eckert, J. M. Nicol, J. Howard, and F. R. Trouw, J. Phys. Chem. **100**, 10646 (1996)
137. H. Jobic, Spectrochem. Acta A **48**, 293 (1992)
138. H. Jobic, Physica B **276**, 222 (2000)
139. N. L. Rosi, J. Eckert, M. Eddaoudi, D. T. Vodak, J. Kim, M. OKeeffe, and O. M. Yaghi, Science **300**, 1127 (2003)
140. Y. J. Glanville, J. V. Pearce, P. E. Sokol, B. Newalker, and S. Komarneni, Chem. Phys. **292**, 289 (2003)
141. N. E. Benes, H. Jobic, and H. Verweij, Micropor. Mesopor. Mater. **43**, 147 (2001)
142. C. Alba-Simionesco, G. Dosseh, E. Dumont, B. Frick, B. Geil, D. Morineau, V. Teboul, and Y. Xia, Eur. Phys. J. E **12**, 19 (2003)
143. H. Jobic, M. Bée, J. Kärger, R. S. Vartapetian, C. Balzer, and A. Julbe, J. Membr. Sci. **108**, 71 (1995)
144. A. G. Stepanov, T. O. Shegai, M. V. Luzgin, and H. Jobic, Eur. Phys. J. E **12**, 57 (2003)
145. H. Jobic, H. Paoli, A. Méthivier, G. Ehlers, J. Kärger, and C. Krause, Micropor. Mesopor. Mater. **59**, 113 (2003)
146. H. Fu, F. Trouw, and P. E. Sokol, J. Low Temp. Phys. **116**, 149 (1999)
147. N. K. Bär, H. Ernst, H. Jobic, and J. Kärger, Magn. Reson. Chem. **37**, S79 (1999)
148. H. Jobic, J. Kärger, and M. Bée, Phys. Rev. Lett. **82**, 4260 (1999)
149. G. K. Papadopoulos, H. Jobic, and D. N. Theodorou, J. Phys. Chem. B **108**, 12748 (2004)
150. H. Jobic, A. N. Fitch, and J. Combet, J. Phys. Chem. B **104**, 8491 (2000)
151. S. Gautam, S. Mitra, A. Sayeed, S. Yashonath, S. L. Chaplot, and R. Mukhopadhyay, Chem. Phys. Lett. **442**, 311 (2007)
152. S. Mitra and R. Mukhopadhyay, Curr. Sci. **84**, 653 (2003)
153. H. Jobic and D. N. Theodorou, J. Phys. Chem. B **110**, 1964 (2006)
154. E. Mamontov, Y. A. Kumzerov, and S. B. Vakhrushev, Phys. Rev. E **72**, 051502 (2005)
155. H. Jobic, M. Bée, and S. Pouget, J. Phys. Chem. B **104**, 7130 (2000)
156. H. Jobic, J. Mol. Catal. A **158**, 135 (2000)
157. H. Jobic and D. N. Theodorou, Micropor. Mesopor. Mater. **102**, 21 (2007)

Chapter 20
The Combined Ultra-Small- and Small-Angle Neutron Scattering (USANS/SANS) Technique for Earth Sciences

Roberto Triolo and Michael Agamalian

Abstract The extension of the well-known Small-Angle Neutron Scattering (SANS) technique to Ultra-Small Angles (USANS) provides a unique tool for studying hierarchical structures ranging in size from nanometers to micrometers. Hierarchical structures are common for many natural and man-made materials, which show multi-level morphology (atoms–molecules–aggregates–agglomerates), in other words, are made up of structural units encompassing the atomic, molecular, micro- and macroscopic length scales. Combining USANS and SANS data can provide complete structural information for complicated polydisperse systems, allowing the determination of their complex morphology and hence has been successfully applied to structural studies in geology, petrology, and archeology. This chapter briefly outlines the technique and provides detailed examples of the applications in the Earth Sciences.

20.1 Introduction

Basically, in all structural studies dealing with a diffraction technique the scattered radiation data are presented in units of reciprocal space and all diffractometers measure the length of the scattering vector $Q = 4\pi \sin\theta/\lambda$, where θ is the diffraction angle and λ is the wavelength of the chosen type of radiation [reference Chapter 2]. The dimension of the Q-vector or the distance in reciprocal space is [length^{-1}] and the corresponding distance in real space, or so-called diffractometric resolution, D, can be estimated as $D \sim 2\pi/Q$ [length]. Thus, structural parameters, or in particular cases even a real space image of an object under study, can finally be obtained analyzing the diffraction pattern. Characterization of materials with atomic-scale resolution definitely is one of the major accomplishments of the 20th century; however, not only the atomic but also the molecular and supra-molecular structure of condensed matter can be studied with X-ray and neutron diffraction as well as with light scattering [1]. Because the typical wavelength range for small

M. Agamalian (✉)
Neutron Scattering Science Division, Oak Ridge National Laboratory, Oak Ridge, TN 37831, USA
e-mail: magamalian@ornl.gov

L. Liang et al. (eds.), *Neutron Applications in Earth, Energy and Environmental Sciences*, Neutron Scattering Applications and Techniques,
DOI 10.1007/978-0-387-09416-8_20, © Springer Science+Business Media, LLC 2009

angle X-ray and neutron scattering instruments is $1\,\text{Å} < \lambda < 15\,\text{Å}$, the corresponding diffraction angle becomes small, $\theta < 1°$, and that is why this scientific field is named Small-Angle X-ray and Neutron Scattering (SAXS and SANS) [2]. Small-Angle Neutron Scattering has become a prominent structural tool since the mid 1970s when the majority of structural investigations in biology, polymer, and colloidal sciences were focused at the supra-atomic structural level. The conventional pin-hole geometry SANS machines [3, 4] operate effectively in the Q-range, $1 \times 10^{-3}\,\text{Å}^{-1} < Q < 1\,\text{Å}^{-1}$ and thus are capable of measuring the maximum diffraction distance $D_{max} \sim 2\pi/Q_{min} \sim 6000\,\text{Å}$. Another type of small-angle diffractometer with ultra-high angular resolution is required to extend the SANS investigations to the micro-metric length scale.

Up to now the best approach for achieving the highest resolution in reciprocal space is the Bonse–Hart Double-Crystal Diffractometers (DCDs) on multi-bounce channel-cut crystals [5]. This technique originally developed for X-rays gave rise to Ultra-Small-Angle X-ray (USAXS) measurements [6]; however, the first attempts to adapt it for neutrons made in the mid-1980s at Jülich, Germany, had limited success [7]. The Bonse–Hart technique was properly applied for USANS studies in 1996–1997 after several successful neutron dynamical diffraction experiments [8–12] carried out at the High-Flux Isotope Reactor (HFIR), Oak Ridge National Laboratory (ORNL). As a result, the Q-range of SANS was extended by two orders of magnitude to $Q_{min} \approx 2 \times 10^{-5}\,\text{Å}^{-1}$ and important scientific results were obtained on polymer blends [13], colloidal crystals [14, 15], rocks [16, 17], hydrating cement paste [18], reinforcing fillers [19], wormlike micelles [20], and colloidal gels [21]. One of the most important achievements, namely the discovery of the enormously extended surface fractal structure in sedimentary rocks, was highlighted in *Physical Review Focus* [22] and in *Science* [23]. The ORNL experience was used to upgrade the Bonse–Hart USANS instruments in Japan (Japan Atomic Energy Agency) [24, 25], Germany (GKSS Forschungszentrum, Geesthacht, Berlin Neutron Scattering Center, FRJ-2, Jülich) [26–28], and France (Institute Laue-Langevin, Grenoble) [29]. A fully optimized Bonse–Hart USANS instrument [30, 31] became operational at the National Institute of Standards and Technology (NIST) in 2000. In July 2003, the International Consortium on Ultra-Small-Angle Scattering (IConUSAS) [32] was organized at The First Workshop on Ultra-Small-Angle X-Ray & Neutron Scattering, Oak Ridge, USA. Application of USANS techniques in materials sciences has been summarized and discussed in several review papers [33–36] and lectures delivered at the 2004 Francesco Paolo Ricci School on Neutron Scattering [37] in Sardinia, Italy. Currently, several USANS instruments are under development at different institutions worldwide. A new multi-wavelength time-of-flight Bonse–Hart USANS (TOF-USANS) instrument [38, 39] is under construction at the Spallation Neutron Source (SNS), at ORNL. The Australian Nuclear Science and Technology Organization (ANSTO) plans to install a Bonse–Hart USANS instrument at the research reactor OPAL (Christine Rehm, *ANSTO, Bragg Institute*, private communication) as does the Korean Institute of Science and Technology at the HANARO reactor (Man-Ho Kim, *KIST*, private communication).

Combined USANS/SANS data obtained by a reactor-based Bonse–Hart DCD and by a conventional pin-hole SANS instrument cover the Q-range extending over four orders of magnitude, $2 \times 10^{-5}\,\text{Å}^{-1} < Q < 0.6\,\text{Å}^{-1}$. This opens up an enormously broad structural range accessible by the USANS/SANS technique, which cannot be covered either by X-ray or by light scattering. However, a synchrotron-based Bonse–Hart DCD [40, 41] with Q-resolution $\Delta Q = Q_{\min} \sim 1 \times 10^{-4}\,\text{Å}^{-1}$ has sufficient flux at the sample to enable measurements, which cover the whole range accessible with X-ray small-angle scattering, $1 \times 10^{-4}\,\text{Å}^{-1} < Q < 1.0\,\text{Å}^{-1}$, using just one instrument. The first USANS/SANS experiments on polymer blends [13], rocks [16, 17], reinforcing fillers [19], wormlike micelles [20], and colloidal gels [21] have clearly demonstrated the importance of the extended Q-range for structural characterization of objects with complex morphology. More recently, the USANS/SANS technique has been found exclusively effective in studies of hierarchical supra-molecular structures exhibiting multi-level internal organization at micro- and nanometric length scales. The extended internal surface fractality found in rocks [16, 17] has become an important characterization parameter often used nowadays in structural studies in geological, petrologic, and archeological sciences.

20.2 Theoretical Background

20.2.1 USANS Collimation Smearing Correction

The combined USANS/SANS diffraction pattern $I(Q)$ can be analyzed using the well-known modeling approach

$$I(Q) \sim N(R) \times F(QR) \times S(QD), \qquad (20.1)$$

where $N(R)$ is the size distribution function, $F(QR)$ is the form-factor of a scattering particle, and $S(QR)$ is the correlation function related to the interference between rays diffracted from individual particles in concentrated systems; here R is the particle radius and D is the correlation distance. The scattering function $I(Q)$ defined in Equation (20.1) is the theoretical intensity calculated for point geometry $I(Q) \equiv I(Q)_{\text{point/th}}$, whereas the net experimental USANS scattering curve $I(Q)_{\text{slit/exp}}$ is typically obtained in slit geometry. Hence, a slit geometry collimation correction is required for the USANS data before combining with that obtained at the conventional pin-hole geometry SANS instrument. The following mathematical expression (see reference [2] for more details) transforms the theoretical $I(Q)_{\text{point/th}} = I\{(2\pi/\lambda)[(2\theta - t)^2 + u^2]^{0.5}\}_{\text{point/th}}$ scattering function to the net experimental scattering curve $I(Q)_{\text{slit/exp}}$

$$I(Q)_{\text{slit/exp}} = \int\int W_{\text{h}}(t) \times W_{\text{v}}(u) \times I\{(2\pi/\lambda) \times [(2\theta - t)^2 + u^2]^{0.5}\}_{\text{point/th}}\, du\, dt,$$
$$(20.2)$$

where $W_h(t)$ and $W_v(u)$ are horizontal and vertical collimation functions of the USANS instrument, respectively, t and u are the horizontal and vertical angular coordinates. $W_h(t)$ is the rocking curve of the empty (no sample) USANS instrument, and $W_v(u)$ is the distribution of the primary beam intensity along the vertical axis of the detector plane, convoluted with the vertical aperture of the detector; $W_h(t)$ and $W_v(u)$ obey the normalization condition:

$$\int W_h(t)\, dt = \int W_v(u)\, du = 1. \tag{20.3}$$

The USANS data collected in slit geometry can be transferred to point geometry after fitting to the chosen modeling function defined in Equation (20.1) by using the Lake technique [42]:

$$I(Q)_{point/exp} = I(Q)_{slit/exp} \times I(Q)_{point/th} / I(Q)_{slit/th}, \tag{20.4}$$

where $I(Q)_{point/exp}$ is the desmeared USANS data. The other options of the desmearing technique are discussed in [2]. The conventional pin-hole geometry SANS data are usually given in absolute units $d\Sigma/d\Omega(Q)$ $[cm^{-1}]$; therefore, the desmeared USANS data $I(Q)_{point/exp}$ can be combined with the SANS pattern after the trivial normalization:

$$d\Sigma/d\Omega(Q)_{point/exp} = I(Q)_{point/exp} \times \left[d\Sigma/d\Omega(Q_{min}) \right] / I(Q_{max})_{point/exp}, \tag{20.5}$$

where $d\Sigma/d\Omega(Q_{min})$ is the absolute SANS profile in the range of small Qs overlapping with that for the USANS data in the range of large Qs $I(Q_{max})_{point/exp}$

20.2.2 Multiple Scattering (MS)

The neutron scattering in the ultra-small-angle range can be rather intense, and in this case the diffraction pattern is usually affected by multiple scattering. T. M. Sabine and W. K. Bertram found a theoretical solution of this problem, which allows structural information to be extracted from USANS data significantly smeared with multiple scattering [43]. Analyzing the diffraction profiles from samples that have different thicknesses but are otherwise identical, one can obtain an average size of scattering particles $<R>$ in a quasi-spherical approximation [44]. The Sabine–Bertram function can be written in the following form

$$I(Q) \sim \sum_{n=1}^{10} (b^n/n!n^m)[1 + (1/3n^m)(Q < R >)^2]^{-2}, \tag{20.6}$$

where $n = 1, 2, 3, \ldots, 10$, $m = 1.5795$, and the parameter b $(0 < b \leq 3)$ is related to the average number of scattering events in the sample. The developed theory was successfully applied to analyze the USANS profiles obtained on samples of hydrating cement paste [18] as well as in USANS experiments on phase-separated polymer blends when varying the sample thickness [13]. The theoretical and experimental results demonstrated that in the range of large Qs the power law of the scattering function is not affected by the multiple scattering. The same result has been obtained in the case of scattering from fractals [16, 17], which is very important for the application of the USANS/SANS technique in geology, petrology, and archeology.

20.2.3 Scattering from Fractals

Many processes and objects in nature happen to show the fractal behavior. The origin of the term "fractal" reflects the fact that fractals have a fractional dimension, not a whole number value [10]. For example, the basic relation of the fractal theory for mass fractals is

$$M \propto r^{-D_{\mathrm{m}}},$$

where D_{m} is the mass fractal dimension, M represents the mass, and r is the scaling factor in the units of length. The term fractal was specifically introduced for temporal and spatial phenomena that exhibit partial correlations over many scales. Examples include Brownian motion, thermal convection, earthquakes, snowflakes, coastlines, rivers, faulting, folding, and volcanic eruptions. In case of surface fractals, the specific surface area $\sigma(r)$ scales with r as

$$\sigma(r) = \sigma_0 \left(\frac{r}{\xi}\right)^{2-D_{\mathrm{s}}},$$

where D_{s} is the surface fractal dimension, ξ is the upper scale limit, and the factor σ_0 can be determined from small-angle scattering data in the range of large Qs [45, 46]:

$$\sigma_0 = \frac{\lim_{Q \to \infty}[Q^{6-D_{\mathrm{s}}} \mathrm{d}\Sigma/\mathrm{d}\Omega(Q)]}{\pi \Delta \rho^2 \rho_0 F(D_{\mathrm{s}}}),$$

where $\mathrm{d}\Sigma/\mathrm{d}\Omega(Q)$ is the scattering cross section, $\Delta\rho$ is the scattering length density contrast, ρ_0 is mass density, and

$$F(D_{\mathrm{s}}) = \frac{\Gamma(5 - D_{\mathrm{s}}) \sin[(3 - D_{\mathrm{s}})(\pi/2)]}{3 - D_{\mathrm{s}}}.$$

In the range of large Qs the asymptotic form of $\mathrm{d}\Sigma/\mathrm{d}\Omega(Q)$ can be written as [47, 48]

$$\mathrm{d}\Sigma/\mathrm{d}\Omega(Q) \to A(D_{\mathrm{s}}) \times Q^{D_{\mathrm{s}}-6} + \dots, \tag{20.7}$$

The surface fractal dimension, $2 < D_s \leq 3$, can be directly determined from Equation (20.7). The term $A(D_s) = \pi I_0 \Delta\rho^2 \sigma_x \rho_0 V F(D_s)$, where I_0 is a constant determined by the incident intensity and V is the sample volume [49, 50]. Equation (20.7) describes the scattering from real fractal objects only within a limited Q range. Assuming that the density–density correlation function decays exponentially above the upper size limit ξ of the fractal object, the following result is obtained [47]:

$$dZ/d\Omega(Q) \alpha Q^{-1} \xi^{5-D_s} [1 + (Q\xi)^2]^{(5-D_s)/2} \sin[(D_s - 1)\arctan(Q\xi)], \qquad (20.8)$$

For $Q\xi \gg 1$ scattering function (20.8) $d\Sigma/d\Omega(Q) \sim Q^{6-D_s}$ and saturates in the range of small Qs. The scattering function for mass fractals [51] is entirely analogous to Equation (20.9):

$$d\Sigma/d\Omega(Q) \propto Q^{-1} \xi^{D_m - 1} [1 + (Q\xi)^2]^{(D_m - 1)/2} \sin[(D_m - 1)\arctan(Q\xi)], \quad (20.9)$$

where D_m is the mass fractal dimension. For $Q\xi \gg 1$ (20.9) can be approximated as $d\Sigma/d\Omega(Q) \sim Q^{D_m}$. Functions (20.8) and (20.9) are the main theoretical models of scattering from fractal structures using for the data analysis of X-ray and neutron small-angle diffraction experiments.

20.3 Application in Geosciences

20.3.1 Macro-Scale Self-Similarity of Sedimentary Rocks

USANS/SANS studies on sedimentary rocks have shown that the pore–rock fabric interface is a surface fractal with the dimension $D_s = 2.82$ extended over 3 orders of magnitude in the length scale [16, 52]. Owing to the limited size range over which the fractal properties are usually observed, the question of the apparent fractal geometry of various natural objects is a contentious one. Analyzing ~100 reports on the fractality in physical systems, Avnir et al. pointed out the contradiction between the narrow range of the appropriate scaling properties for declared fractal objects (centered around 1.3 orders of magnitude) and the public image of the status of experimental fractals [53], which for rocks has previously been based on limited experimental evidence (about 1.5 decades in length scale). A notable exception is the X-ray study of Bale and Schmidt on coals (2 decades in the length scale) [49]. The internal surface fractality of sedimentary rocks studied in [16, 52] is in fact one of the most extensive fractal systems found in nature.

Sedimentary rocks are formed from a mixture of organic and inorganic origin deposited in an aqueous environment, buried and compacted over geological periods of time. Remarkably, there is no percolation threshold observed in sedimentary rocks, which indicates a microstructure more complex than one originating from just a collection of compacted grains. According to the anti-sintering hypothesis of Cohen, the rock/pore interface evolves by maximizing the internal surface area in response to the secular equilibrium between the rock matrix and the formation brine [54]. Various studies performed on rocks of different origin and lithology over length scales in the range 20 Å to 100 mm have shown that sedimentary rocks are often effective fractals [55]. Experimental tools used in these studies include molecular adsorption [56], microscopy [57, 58], and small-angle scattering (SAS). Previous small-angle neutron and X-ray scattering (SANS and SAXS) studies on shales [59–61] and sandstones [60, 62] demonstrated the surface fractal geometry of the pore–matrix interface in the scale range from 20 Å to about 2000 Å.

In the study of sedimentary rocks [16, 52] the pin-hole SANS instrument D11 at ILL (λ = 4.5, 7, and 14 Å), the USANS DCD at ORNL (λ = 2.59 Å), and the 30-m SANS machine at ORNL (λ = 4.75 Å) were used. The chosen instruments cover the range $2.5 \times 10^{-5} \leq Q \leq 0.3 \, \text{Å}^{-1}$, which offers an opportunity to study the microstructure of a natural rock in the continuous range of sizes $2 \, \text{nm} \leq R \leq 5 \, \mu\text{m}$. The SANS, USANS, and SEM data were obtained for a solid hydrocarbon source rock U116, originated from 342.7 m depth in the Urapunga 4 well (Velkerri Formation, MacArthur Basin, Northern Territory, Australia 65). Since shales, even with significant organic matter content, are perceived by neutrons as two phase [61], function (20.8) can be applied to interpret SANS data for organic-rich sedimentary rocks. In order to estimate the contribution of Multiple Scattering (MS), the preliminary SANS measurements were performed on samples of different thicknesses. The SANS data obtained with instrument D11 using long-wavelength neutrons (λ = 14 Å) show significant dependence on sample thickness (Fig. 20.1), which indicates pronounced MS effects. For strongly scattering sedimentary rocks MS may be misleading in the small-Q region due to a similarity between the saturation of the scattering curve caused by MS and by the finite size of fractal inhomogeneities [Equation (20.8)]. However, it is possible to reduce the thickness of rock samples making MS practically irrelevant. Figure 20.2 shows the absolute scattering cross section for a 0.1-mm thick sample measured at three different neutron wavelengths: 4.5, 7, and 14 Å. The three experimental curves coincide in the overlapping Q range and MS is evidently absent in this Q-range. The SANS experiments at λ = 5 Å and USANS experiments at λ = 2.59 Å were performed using samples about 1-mm thick with no significant contribution from MS.

Figure 20.3 shows the combined USANS/SANS data for rock U116, obtained at ORNL and ILL; the USANS data have been transformed to the point geometry using formulas (20.2, 20.4). The region of the power-law scattering in Fig. 20.3 extends over 3 orders of magnitude of length scale ($6 \, \text{nm} \leq 2\pi/Q \leq 6\mu\text{m}$) and over 10 orders of magnitude of the scattering cross section ($10^{-1} \leq d\Sigma/d\Omega \leq 10^9 \text{cm}^{-1}$). Such an extent of fractal microstructure in a rock is remarkable, in particular when compared with numerous other reports on the fractal properties of natural systems [53]. The slope of—3.18 obtained from a straight-line fit in the $10^{-4} < Q < 10^{-1} \text{Å}^{-1}$ region corresponds to a surface fractal with the dimension D_s = 2.82,

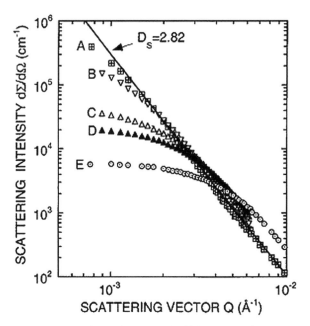

Fig. 20.1 SANS data acquired for various sample thicknesses t (instrument D11, 1–14 Å). (A) $t = 0.63$ mm; (B) $t = 1.20$ mm; (C) $t = 3.09$ mm; (D) $t = 4.23$ mm; and (E) $t = 7.4$ mm. Copied from [53]; Copyright permit # 6544

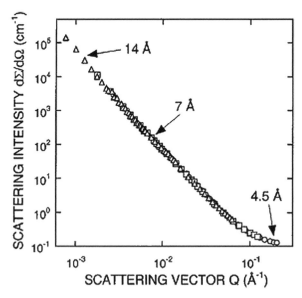

Fig. 20.2 SANS data acquired from a thin sample, $t = 0.1$ mm, using various neutron wavelengths at ILL D11-SANS instrument: $\lambda = 4.5$ Å (*circles*), $\lambda = 7$ Å (*squares*), and $\lambda = 14$ Å (*triangles*). Copied from [53]; Copyright permit # 6544

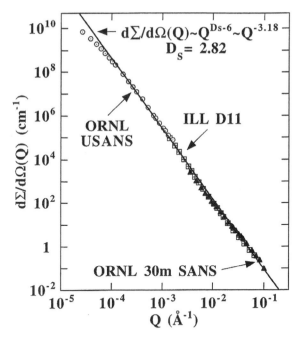

Fig. 20.3 Combined USANS/SANS data for the sedimentary rock U116 obtained by means of ORNL Bonse-Hart USANS, ILL D11-SANS, and ORNL 30 m SANS instruments. The *solid line* shows the Q-range of power law

which is consistent with the fractal dimensions found previously for Urapunga 4 source rocks [61]. The departure of the scattering curve from straight line in the ultra-small-Q region is real, although minimal. The shape of the scattering curve in this region varies for samples originating from various depths in the Urapunga 4 core indicating differences in the large-scale structure for the samples from various depths of burial.

SEM images of the surface of sample U116 cleaved in the bedding plane are shown in Fig. 20.4. The surface texture is dominated by the illite clay particles and appears rough at any length scale. For smallest magnification, however, one can see images of roughly spherical objects about 10–20 μm in diameter (Fig. 20.4A). Image C has been obtained from image A by digital enhancement (increasing contrast) of the outlines of some of these objects and serves as a guide for the eye only. The object size is close to the value of ξ estimated from small-angle scattering and, therefore, it is possible that these objects are images of individual fractal inhomogeneities. In order to independently estimate the fractal dimension and the upper cut-off characteristic size ξ in sample U116 we used the manual "feature" counting technique. For surface fractals one expects a power-law relationship between the average number of "features" per unit length, N/L, and the "feature size," R: $N(R)/L \sim R^{2-D_{sl}}$, where D_{sl} is the fractal dimension characterizing the analyzed one-dimensional region. It has been argued that for high-porosity rocks, which break

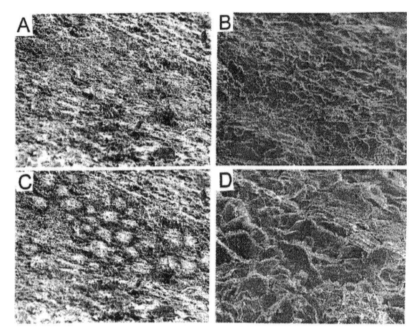

Fig. 20.4 SEM images for U116 samples cleaved in-bedding plane. (**A**) Magnification 31 000; (**B**) magnification 33 000; (**C**) digitally enhanced image (**A**); (**D**) magnification 310 000. Copied from [53]; Copyright permit # 6544

up mostly along the preexisting pore–matrix interface, the fractal dimension measured by SANS, D_s, and that obtained from the image analysis of cleaved rock surface, D_{sl}, probe the same structural features and, therefore, should be identical. The feature counting results illustrated in Fig. 20.5 are consistent with the conclusions based on neutron scattering data. D_{sl} is about 2.8–2.9 for length scales smaller than about 4 μm and, importantly, there is a marked drop-off of the number of features for length scales above 4 μm, indicating a breakdown of the fractal properties, which gives an estimated value of $\xi \sim 4$ μm. This is consistent with the estimate of several micrometers deduced from the onset of cut-off effects in USANS data.

In conclusion, the combined USANS/SANS demonstrates that the pore–matrix interface in a natural hydrocarbon source rock is a surface fractal over three decades of the length scale, from 6 nm to about 4 μm. Such an extent of the fractal properties in a natural system is remarkable. In the region from 0.7 to 7 μm, the fractal analysis based on neutron scattering data is consistent with the results of SEM image processing for the same rock sample.

20.3.2 Fractal Approach in Petrology

Petrology is a branch of geosciences that describes the origin and evolution of rocks. Petrology deals with changes that occur naturally in rocky complexes, namely,

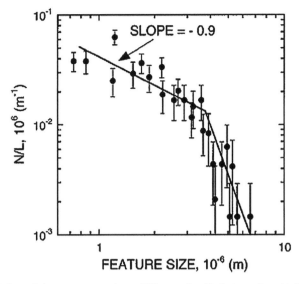

Fig. 20.5 Variation of the average number of "features" with feature size obtained from SEM images. Copied from [53]; Copyright permit # 6544

sediments undergoing physical and chemical alterations, solidifying magmatic fluids, and rocks undergoing partial or total melting. A rocky formation, as we know it, suggests that there are abundant small events, occurring close to one another in time or space, with fewer large events occurring in the same temporal or spatial region. Such relations are indicative of power law scaling, which in turn suggests that fractal geometry might prove useful for quantifying petrological patterns. Fractal geometry does not describe self-similarity, but it is the natural mathematical representation of experimental observations, which helps elucidate possible explanations. In particular, the interpretation of the textures of rocks, that is the description and analysis of crystal distributions, crystal morphology, crystal faces, crystal intergrowth, and zoning, is essential to understand the history of a rock. Although much has been accomplished in quantifying rock and mineral chemistry, comparatively much less effort has been devoted to determine textures. Only recently, after the original descriptive approach classifying the texture of rocks through optical microscopy typical of nineteenth-century researchers, some fundamental quantitative work has appeared, also thanks to neutron scattering (see Chapters 9 and 10). Nowadays, it is actually possible to define accurately the structure of rocks using electron microscopy (EBSD) and neutron scattering. We describe here studies in which the fractal theory is applied to the USANS/SANS data obtained on various samples of rocks collected from different localities.

The SANS data were obtained at the ISIS pulsed neutron source of the EPSRC Rutherford Appleton Laboratory (UK), using the "LOQ" diffractometer [62]. The experiments were performed using wavelengths $2 < \lambda < 10$, Å and the small angle data was supplemented by the data taken simultaneously in a wide-angle detector bank in the Q-range $0.01 < Q < 1.5 \, \text{Å}^{-1}$. The data were corrected for the beam transmission and the background and then transferred to the absolute units $d\sum/d\Omega$ [cm^{-1}]. The USANS results were obtained at the HFIR USANS instrument, ORNL in the Q-range $2 \times 10^{-5} < Q < 0.003 \, \text{Å}^{-1}$. The USANS and SANS data were combined using the technique described in paragraph 20.2.1. The samples of rocks of different thicknesses were measured with the USANS instrument to determine the magnitude of multiple scattering. Figure 20.6 demonstrates that the multiple scattering is significant in the range of the upper cut-off; however, it does not change the power law, which is consistent with the Sabine–Berthram theory. Therefore, the sample thickness was minimized for all the samples of rocks in this study to decrease the parasitic effect of multiple scattering as much as possible.

Figure 20.7 shows a typical combined USANS/SANS profile obtained for the rock sample of benmoreitic inclusion in hyaloclastite from Linosa Island, Italy. The upper cut-off range $3 \times 10^{-3} < Q < 8 \times 10^{-4} \, \text{Å}^{-1}$ follows by the region of power law $8 \times 10^{-4} < Q < 8 \times 10^{-2} \, \text{Å}^{-1}$, which transforms into transitional to wide-angle scattering $Q > 8 \times 10^{-2} \, \text{Å}^{-1}$. The best fit to Equation (20.8) for surface fractals was found for the fractal dimension $D_s = 2.3$ and the upper cut-off dimension $\xi \sim 3.9 \times 10^3 \, \text{Å}$. The benmoreitic inclusion in hyaloclastite rock shows a fractal behavior over two orders of magnitude in the length scale. The combined USANS/SANS data obtained for the other rock sample of rhyolitic inclusion in soda-trachytic ignimbrite from Pantelleria Island, Italy, also shows a surface fractal structure. For this rock the best fits to Equation (20.8) is obtained for the same fractal dimension $D_s = 2.3$ and a cut-off distance $\xi \sim 3.0 \times 10^3 \, \text{Å}$. Similar to the benmoreitic inclusion in hyaloclastite, the rhyolitic inclusion in soda-trachytic

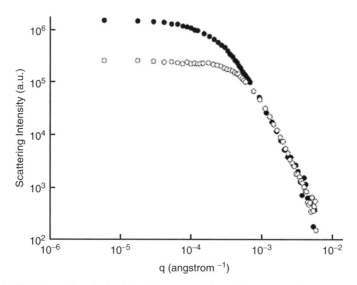

Fig. 20.6 USANS profiles obtained for the two samples of the same rock: 1-mm thick (*open circles*) and 3-mm thick (*closed circles*)

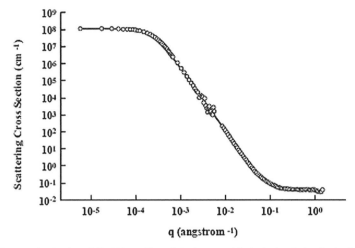

Fig. 20.7 Combined USANS/SANS profiles of the benmoreitic inclusion in hyaloclastite, Linosa Island, Italy. The solid line is the fitting curve calculated for the function (20.8)

ignimbrite rock also shows an extension of surface fractal structure over two orders of magnitude in the length scale. The surface fractal structure of soda-rhyolitic ignimbrite from Pantelleria Island, Italy, has slightly different parameters: $D_s = 2.2$ and $\xi \sim 2.7 \times 10^3$ Å, but the same extension of fractality over two orders of magnitude in the length scale. However, the USANS data measured for the sample of a granitoid inclusion in soda-trachytic lava from Pantelleria Island, Italy (Fig. 20.8), fits to the theoretical scattering curve calculated from Equation (20.9) valid for

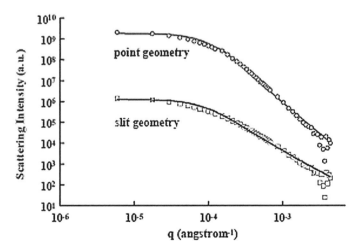

Fig. 20.8 USANS profiles (slit and point geometry) obtained for the sample of granitoid inclusion in soda-trachytic lava. The solid lines show the theoretical USANS curved calculated from Equations (20.9)

mass fractals. The mass fractal dimension $D_m = 2.8$ and the upper cut-off distance $\xi \sim 1.15\,\mu m$ are the structural parameters determined for this rock from the USANS experiment. Unfortunately, the SANS data have not been obtained for this sample and thus the length scale range of mass fractal structure has not been measured.

The USANS/SANS neutron scattering experiments on the samples of different rocks from different locations clearly demonstrate that the internal structure of rocks in the extended range of distances (from $\sim$10,000 Å to $\sim$50–70 Å) shows a fractal behavior. The self-similarity of rocks appears in the form of surface and mass fractals depending on the origin and the chemical composition. Hence, the USANS/SANS technique provides important structural information for the petrological characterization of rocks from diverse geological environments.

20.3.3 Mesoscopic Structure of Marbles

The effect of metamorphism on the formation and the structural characteristics of marble is not yet fully understood. Marble is the result of the isochemical metamorphic evolution, in different conditions, of a parent rock (protolith): normally, a sedimentary carbonate rock with a highly variable calcitic–dolomitic composition, or a pre-existing marble. This involves the destruction of the originating minerals (mostly calcite and dolomite) and their recycling through a further crystallization process. Metamorphic re-crystallization basically proceeds through the progressive building up of structural units of increasing size from starting elementary cells (Å), to intermediate aggregates (μm), to final individual crystals (mm). The process may be continuous, with abundant small events occurring close to one another in time or space, or discontinuous, with fewer large events occurring in the same temporal or spatial region. Such relations are possibly indicative of power law scaling, typical of fractal structures [63]. Fractal geometry does not describe the mechanism that produces the fractal scaling, but it nonetheless helps to elucidate the possible mechanisms or explanations [16, 17].

At the micro-scale (Å) level, the metamorphic conditions (i.e., temperature and pressure) are expected to control the size of the elementary cell, the basic building unit of the re-crystallizing carbonates. However, recent investigations on a large number of prevailingly calcitic marbles (i.e., containing less than 3% molar ratio magnesite dispersed in solid solution) showed that this is mostly dependent on the chemical composition of the evolving rocks and any eventual interacting fluids, with special reference to the amount and distribution of the elements substituting for Ca (especially Mg) in the crystal structure of calcite [64]. In other words, at this scale (and in the most common range of metamorphic conditions) the structural characteristics are mainly dependent on the ionic radius of the carbonate-forming elements more than the formation temperature. The effect of metamorphism is much more important at the macroscopic scale, i.e., as far as the size and kind of aggregation of the final crystals are concerned. This is easily observed by means of an optical microscope with transmitted and polarized light. Generally speaking, apart from the effect of other controlling factors (crystal chemistry, amount and kind of interacting fluids, time, pressure, etc.), the dimension of the newly formed carbonate grains roughly correlates with the crystallization temperature [65].

Although much has been accomplished in quantifying marbles and their crystal chemistry in particular, comparatively little effort has been spent to quantify their mesostructures. However, the microscopic and macroscopic observations, either separately or together, do not seem to be able to uniquely mark the differences between the various marbles. Surprisingly, little is known about the structural effects of metamorphism at the meso-scale level. If the mesostructure is the bridge between the micro- and the macro-structure, interpretation of the texture of a marble at this intermediate scale should be consistent with its metamorphic history [66]. In particular, the mesostructural characteristics should be related to the type and degree of metamorphism undergone by the originating protolith. This assumption has been examined on a set of white marble samples representing a wide array of metamorphic conditions [66] using USANS/SANS techniques. Marbles are strictly multi-phase systems; however, they can be considered as a two-phase system in neutron small-angle scattering experiments, since the contrast between the inorganic components (the largely prevailing carbonate grains) and the voids is dominant [16, 17]. In the past two decades, several attempts have been made to describe small-angle scattering from geological systems in terms of mass and surface fractals [60, 67–69]; however, these attempts met considerable limitations related to the absence of experimental data in the range of small Qs.

Marbles are rocks originated from the building up of primary particles undergoing a continuous series of interaction processes mainly controlled by the metamorphic degree. The study provided here as an example focused on the interpretation of neutron small-angle scattering data from marbles, with special attention to the temperature [65]. Therefore, the combined USANS/SANS data are analyzed in terms of the hierarchical structure model, which takes into account the existence of a network of fractal aggregates formed by monodisperse solid primary particles of radius r [70, 71]. On the other hand, the availability of an additional set of (meso-)structural parameters could be particularly relevant also for archaeometric information, with special reference to the characterization and provenance identification of ancient marbles, which is often a rather difficult task [65]. Provenance attributions are in fact of key importance to archaeology because technological and commercial exchange patterns may correlate with historical events. The authentication of works of art (and the identification of eventual copies or fakes) is also an important issue.

A limited selection of samples representing a wide array of metamorphic conditions was chosen; samples are from different outcrops in central and northern Italy (Table 20.1). Sample A (low grade) comes from Musso, along the western shore of Lake Como (central Italian Alps), samples B and C (medium-low and medium grades, respectively) from different parts of the Carrara marble formation (northern Apennines), sample D (medium-high grade) from the southern shore of the Island of Elba (northern Tyrrhenian Sea), and sample E (high grade) from the facing mainland (Campiglia Marittima, Tuscany). Standard ~20-mm thick slices were prepared for magnified observations in transmitted polarized light. These observations allowed

Table 20.1 The samples considered and the analytical results are discussed in the text. Numbers in parentheses for D and D_s are estimated errors on the last digit of the fit parameters

Samples	Av. grain size [mm]	r [Å]	D	D_s	$10^3 D/r$ [Å^{-1}]	Metamorphic degree (thermal ranges in °C)
(A) Musso	0.05	2.0	2.50(5)	2.45(5)	1.225	Low (100–200)
(B) Carrara I	0.10	3.1	2.35(5)	2.00(5)	0.758	Medium-low (200–300)
(C) Carrara II	0.35	4.6	2.28(5)	2.00(5)	0.495	Medium (300–400)
(D) Elba Island	1.50	8.6	2.70(5)	2.33(5)	0.314	Medium-high (400–500)
(E) Campiglia M.	0.25	24.0	2.95(5)	2.25(5)	0.013	High (500–600)

the determination of the macrostructure (i.e., fabric) of the samples in terms of sizes and types of aggregation of the crystals in their final stage and the regularity of their margins. (Fig. 20.9). The approximate averaged grain sizes are given in Table 20.1. The metamorphic degree has been evaluated as outlined in the following paragraph, while the mineralogical composition, with special reference to the crystal chemistry of the prevailing carbonates (i.e., variable calcitic–dolomitic composition), difficult to define optically, was obtained by determining the concentration of Ca and Mg with an atomic absorption spectrophotometer (flame mode). The analyses were carried out at the Geochemical Laboratory of the University of Modena, with a Mod. 603 Perkin-Elmer spectrophotometer. SANS measurements were performed at the KWSII instrument of the neutron scattering facility FRJ-2 of the Forschungszentrum Jülich (Germany), in the range $10^{-3} < Q < 0.2$ Å^{-1} ($\lambda = 7$ Å). USANS experiments were carried out at the neutron optical bench instrument S18 installed at the 58 MW high-flux reactor at the Institut Laue-Langevin (Grenoble, France) [72]. In the double crystal diffractometer configuration, triple bounce channel cut perfect silicon crystals are used both as monochromator and analyzer, covering the range $10^{-5} < Q < 10^{-2}$ Å.

In principle, the maximum temperature involved in a metamorphic event can be calculated by using different mineralogical and geochemical thermometers. These derive from the study of one or more of the following:

(a) trace minerals (silicates, etc.) acting as thermal markers according to their stability fields;

(b) the solid solution of magnesite in the calcite crystal structure (Mg substituting for Ca), of iron sulfide into sphalerite (Fe substituting for Zn), etc., which are in turn dependent on the crystallization temperature;

(c) the fluid inclusions that remain trapped in a number of crystals, and any eventual floating minerals;

(d) the isotopic fractionations concerning different mineral pairs (and nuclides): carbonate–carbonate (oxygen and carbon), carbonate–silicate, or carbonate–oxide (oxygen), etc.

In the absence of specific geothermometric investigations, the metamorphic temperature involved in the formation of a marble can be roughly estimated according to a number of petrographic and geological indications, with special reference to:

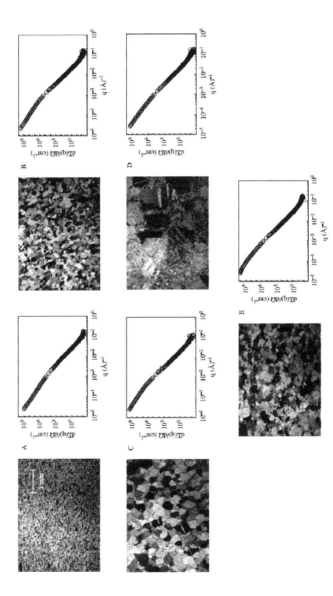

Fig. 20.9 Thin-section photographs showing comparatively, at the same magnification, the macrostructural characteristics (i.e., fabric) of the marble samples (polarized light and crossed Nicols). Alongside are the corresponding measured neutron scattering cross sections

(e) the study of the other associated rocks (silicatic, etc.) forming the metamorphic
    series of interest, and its general geological setting;
(f) the simple observation of the marble in thin section (grain size, kind of aggre-
    gation, etc.).

The thermal ranges for the various samples reported in Table 20.1 were estimated
based on the information available in the literature [73–78]. Concerning samples
B and C from Carrara, a rather restricted variability (roughly from 200 to 400°C)
is found in general for the marble formation outcropping in this area [73–78], in
spite of its great thickness and extension. The samples come from distinct positions
with values close to the extremes of the range indicated, respectively. As for sample D, the metamorphic series in the southern part of Elba is characterized by a
wide thermal range (300–600°C) passing from the eastern to the western part [75].
The values indicated (400–500°C) refer to the central section, where this sample
was collected. The metamorphic sequence in the Campiglia Marittima area shows
a similarly large thermal spread [76]. The values for sample E (500–600°C), first
deduced from the mineralogy of the silicatic rocks accompanying the marble bed, is
confirmed by field evidence, with special reference to the relatively short distance
from a genetically related magmatic body. This is in fact one of those "granitic"
plutons whose emplacement (intrusion) temperature is typically around 700–800°C.
The situation for sample A from Musso, coming from the central-southern section of
the Italian Alps, is rather complex. The general mineropetrographic and geological
features point toward a complex evolution for this section, where various metamorphic episodes characterized by very different conditions can be recognized. Precisely, one may acknowledge at least one generating phase developed at a very high
temperature, up to 600°C and more, followed by a low temperature re-elaboration
[77, 78]. According to the thin-section observations (Fig. 20.1A), such a genetic
complexity is still shown by this sample at the macro-scale level, indicating that
low-temperature thermal re-equilibration was not complete. In fact, a matrix of
very fine crystals can be recognized (Fig. 20.1A), which is the product of the final
low-temperature event (100–200°C, the net thermal range considered), whilst a few
much larger calcite grains dispersed in this matrix are the relics of the previous
high-temperature marble. This can be seen in Fig. 20.2, which is an enlarged view
of a portion of Fig. 20.1A. Having assessed the metamorphic temperatures of the
samples, we now turn to the results of the chemical analyses, which indicate that
all samples are mainly calcitic, with only a minor, although variable, magnesitic
(i.e., dolomitic) component. Such evidence is important in order to exclude any
significant mesostructural effects to be eventually related to highly variable crystal
chemistry (i.e., variable calcitic–dolomitic composition).

The thin-section photographs shown in Fig. 20.9 all taken at the same magnification for comparative evaluation show the macroscopic structure of the samples.
Their fabric is rather homogeneously granoblastic (i.e., homeoblastic) in almost all
cases. Only sample A from Musso has a preferential orientation (i.e., lineation)
of the tiny calcite grains forming the matrix in which a number of rather dispersed, larger crystals are included. Samples B and C from Carrara show a more

markedly homogeneous fabric and very regular crystal boundaries, giving an overall polygonal to mosaic configuration. This is due to metamorphic equilibrium reached over very long time at the metamorphic temperature [65]. Generally speaking, the rough temperature–grain size correlation indicated above is confirmed by these pictures, with the exception of sample E. In fact, its high thermal range (500–600°C, Table 20.1) is not consistent with the average grain size (0.25 mm), which is apparently indicative of a medium to medium-low metamorphic degree. Figure 20.9 also shows the combined curves of the USANS/SANS experiments. Symbols are experimental points and lines are fit of the data to a hierarchical structure model, which takes into account the existence of a network of fractal aggregates of size R formed by monodisperse solid primary particles of radius r [70]:

$$I(Q) \sim (1 + 0.471 Q^2 r^2)^{(D_s - 6)/2} \times [1 + D/(Q r^D)](1 + 1/Q R^2)^{(1-D)/2} sin[arctan(Q R)].$$
(20.10)

Here D_s is the dimensionality of the interfacial region of the primary particles, and its value must be between 2 and 3 ($D_s = 2$ is for smooth particles following the Porod law, $I(Q) \sim Q^{-4}$). D is the power law exponent of the fractal aggregates. Fit parameters are reported in Table 20.1. It appears clearly that the radius r of the primary structural units increases homogeneously as a function of the thermal history of the samples, thus confirming the general control of the temperature on the dimension of the structural units. However, in the case of the scattering data, the correlation is perfectly shown by all the marbles analyzed, including sample E, meaning that the temperature–size relationship is better defined at the mesoscopic level compared to the macroscopic one. With the exception of sample A, the dimensionality of the interfacial region of the primary particles (D_s) roughly increases in the same direction, passing from 2.00 (very smooth surfaces) for the low and medium-low metamorphic degrees to around 2.30 for the higher grades. This would seem to indicate that the surface of these primary particles becomes rougher with the increasing crystallization temperature. In such a context, it is interesting to notice that samples B and C from Carrara are the ones characterized by the smoothest possible surfaces of the structural units at both the meso- and macro-scale. The anomalously high D_s value for sample A could be explained by considering, in addition to the temperature, the effect of other parameters and features on the smoothness of the surface of the mesostructural units: (i) the amount (and kind) of fluids and the related control on the rheology of the system; (ii) the importance of time (at a given temperature) for the attainment of equilibrium or disequilibrium conditions, similar to what is found at the macro-scale level; and (iii) the possibility of a metamorphic evolution that is more complex than usual, as already indicated. The latter point will be better discussed in what follows. The fractal dimension (D) shows a more homogeneous increase with the temperature, again with the exception of sample A. Apart from this, it starts from an overall average value around 2.3 (typical of surface fractals) for the low and medium metamorphic degrees, and finally reaches a quite high value of about 3. This implies that the kind of aggregation of the intermediate units hardly allows fractal structures to form at the highest metamorphic degrees. The situation for sample A from Musso is rather peculiar. On one hand, the prevailing matrix formed by very fine calcite crystals (Fig. 20.9A) is indicative of low crystallization temperatures (100–200°C, Table 20.1). On the other hand, the large calcite grains found dispersed into the fine matrix (Fig. 20.10) have to be interpreted as the relics of a much coarser (parent) marble. This anomalous feature has to be related to a rather complex evolutionary history, as indicated above, with special reference to the occurrence of those "retrograde phenomena" involving one or more further dynamometamorphic (low temperature) re-elaboration(s) of a previous higher-grade marble, a process not unusual in the Alps, including this central-southern section [76, 77]. Such an evolutionary complexity is probably responsible for a sort of memory effect. In particular, the huge tectonic (i.e., mechanical–frictional) stimulation, and the related brittle deformation, involved in these (late) low-temperature dynamo–metamorphic processes may explain the anomalies found in sample A for what concerns both the surface properties (D_s) and the kind of aggregation (D) of the mesostructural units. Indeed, the size of the primary particles should also be affected, at least to some extent, meaning that the value found for r probably represents an average of the differently sized calcite crystals related to the various metamorphic steps. On the other hand, the general temperature–size correlation found at the meso-scale level is apparently consistent, so that more detailed analyses are needed to better understand this critical point. At the moment, the temperature range assigned to this anomalous sample was based on the much larger amount of fine matrix with respect to the coarse relic individuals dispersed in it, and thus hypothesizing an almost complete re-equilibration under the new (low) thermal conditions. The relatively straightforward examination

Fig. 20.10 Magnified view of Fig. 20.1A showing the effect of retrograde phenomena in sample A

of the data already points out to a correlation between the fit parameters of the scattering data and the metamorphic conditions. As discussed above, the anomalies found hamper the use of simple correlation graphs, also on account of the limited (although representative) number of samples considered. However, the binary correlation becomes much more significant when preceded by further processing of the data, i.e., by a normalization procedure. In particular, the D/r vs. r correlation graph shown in Fig. 20.11 clearly indicates a hyperbolic trend, which can be easily explained considering the effect that the different metamorphic conditions have on the general mesostructural characteristics of the marbles under study.

A hierarchical structural model was used to analyze combined USANS/SANS patterns of white marble samples in order to relate their genesis with the features observed in the scattering experiments. By properly processing the fit parameters, it was possible to correlate the characteristic parameters of the mesoscopic structure to the metamorphic evolution. In particular, the dimension r of the building units shows better correlation with the crystallization temperature than determined for the whole crystals at a macroscopic level. A similar but somewhat poorer correlation is found from aggregation (fractal dimension, D) and the surface characteristics (dimensionality of the interfacial region, D_s) of these intermediate clusters. The observed anomalies of sample A have probably to be related to additional parameters and features. Crystal chemistry, the amount and kind of fluids, the time, mechanical stresses and complexity of the metamorphic evolution, may have had an influence. More detailed investigations are therefore needed to throw light on the effect of other (meso) structural control parameters; however, the analyses still in progress on a larger number of marbles from different Mediterranean areas seem to confirm the proposed model.

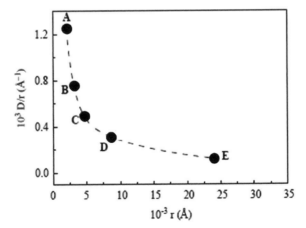

Fig. 20.11 D/r vs. r plot for the five samples under study and the corresponding metamorphic degrees

20.4 Conclusion

Most of the rocks and minerals are multi-componential complex materials with hierarchical internal structure; thus it appears to be very difficult and even impossible to study these objects using traditional small-angle diffraction techniques. In spite of this impression, previous experimental studies and theoretical simulations have shown that usually these complex objects can be considered in SANS studies as simple two-phase systems because of the dominant contrast between two easily recognizable components. However, the limited Q-range and sensitivity of conventional SANS instruments did not allow to take full advantage of this peculiarity. Broad application of the neutron diffraction microscopy (SANS) in geosciences started a decade ago when these two important parameters (Q-range and sensitivity) were significantly improved by combining the USANS and SANS techniques. The experimental studies highlighted here reveal a variety of interesting discoveries starting from the enormously extended surface fractality in sedimentary rocks and finishing with the hierarchical structure of marbles correlated with their metamorphic history. Therefore, the two-phase approximation combined with the improved experimental technique became a new powerful tool for structural investigations in geology, petrology, and archeology.

References

1. P. Debye & A. M. Bueche, *J. Appl. Phys.*, **20**, 518, 1949.
2. L. A. Feigin & D. I. Svergun, *Structural Analysis by Small-Angle X-ray and Neutron Scattering*, Plenum, New York, 1987.
3. C. J. Glinka, J. G. Barker, B. Hammouda, S. Krueger, J. Moyer & W. J. Orts. *J. Appl. Cryst.*, **31**, 430, 1998.
4. C. D. Dewhurst, *Measurement Sci. & Tech.*, **19**, 034007, 2008.
5. U. Bonse & M. Hart, *Appl. Phys. Lett.*, **7**, 238, 1965.
6. U. Bonse & M. Hart, *Zeitschrift für Physik*, **189**, 151, 1966.
7. D. Schwahn, A. Miksovsky, H. Rauch, E. Seidl & G. Zugarek, *Nucl. Instrum. Meth.*, **A239**, 229, 1985.
8. M. Agamalian, G. D.Wignall & R.Triolo, *J. Appl. Cryst.*, **30**, 345, 1997.
9. M. Agamalian, D. K. Christen, A. R. Drews, C. J. Glinka, H. Matsuoka & G. D. Wignall, *J. Appl. Cryst.*, **31**, 235, 1998.
10. M. Agamalian, C. J. Glinka, E. Iolin, L. Rusevich & G. D. Wignall, *Phys. Rev. Lett.*, **81**, 602, 1998.
11. M. Agamalian, G. D. Wignall & R. Triolo, *Neutron News*, **9**, 24, 1998.
12. M. Agamalian, E. Iolin & G. D. Wignall, *Neutron News*, **10**, 24, 1999.
13. M. Agamalian, R. G. Alamo, J. D. Londono, L. Mandelkern & G. D. Wignall, *Macromolecules*, **32**, 3093, 1999.
14. H. Matsuoka, T. Ikeda, H. Yamaoka, M. Hashimoto, T. Takahashi, M. Agamalian & G. D. Wignall, *Langmuir*, **15**, 293, 1999.
15. T. Harada, H. Matsuoka, T. Yamamoto, H. Yamaoka, J. S. Lin, M. Agamalian & G. D. Wignall, *Colloids & Surfaces A*, **190**, 2001.
16. A. P. Radlinski, E. Z. Radlinska, M. Agamalian, G. D. Wignall, P. Lindner & O. G. Randl, *Phys. Rev. Lett.*, **82**, 3078, 1999.

17. F. Triolo, A. Triolo, M. Agamalian, J. S. Lin, R. K. Heenan, G. Lucido & R. Triolo, *J. Appl. Cryst.*, **33**, 863, 2000.
18. T. M. Sabine, *Concrete in Australia*, **75**, 21, 1999.
19. D. W. Schaefer, T. Rieker, M. Agamalian, J. S. Lin, D. Fischer, S. Sukumaran, C. Chen, G. Beaucage, C. Herd & J. Ivie, *J. Appl. Cryst.*, **33**, 587, 2000.
20. Y.-Y. Won, H. T. Davis, F. Bates, M. Agamalian & G. D. Wignall, *J. Chem. Phys. B*, **104**, 7134, 2000.
21. C. D. Muzny, B. D. Butler, H. J. M. Hanley & M. Agamalian, *J. Phys.: Condens. Matter*, **11**, L295, 1999.
22. D. Mackenzie, Rocks as Fractals. *Phys. Rev. Focus.*, April 15, 1999 http://focus.aps.org/story/v3/st22
23. Fractals Spotted in Rocks. *Academic Press*, April 19, 1999 http://www.academicpress.com/insight/04191999/grapha.htm
24. T. Takahashi, M. Hashimoto & S. Nakatani, *J. Phys. & Chem. Solids*, **60**, 1591, 1999.
25. K. Aizawa & H. Tomimitsu, *Physica B*, **213 & 214**, 884, 1995.
26. D. Bellmann, P. Staron & P. Becker, *Physica B*, **276**, 124, 2000.
27. W. Treimer, M. Strobl & A. Hilger, *Physics Letters*, **A 289**, 151, 2001.
28. S. Borbèly, M. Heiderich, D. Schwahn & E. Seidl, *Physica B*, **276–278**, 138, 2000.
29. M. Hainbuchner, M. Villa, G. Kroupa, G. Bruckner, M. Baron, H. Amenitsch, E. Seidl & H. Rauch, *J. Appl. Cryst.*, **33**, 851, 2000.
30. A. R. Drews, J. G. Barker, C. J. Glinka & M. Agamalian, *Physica B*, **241–243**, 189, 1998.
31. J. G. Barker, C. J. Glinka, J. Moyer, M. H. Kim, A. R. Drews & M. Agamalian. *J. Appl. Cryst.*, **38**, 1004, 2005.
32. http://www.sns.gov/iconusas_workshop/iconusas_2003.htm/
33. M. Agamalian, *Neutron News*, **14**, 11, 2003.
34. D. W. Schaefer & M. Agamalian, *Curr. Opt. Solid State Mat. Sci.*, **8**, 39, 2004.
35. M. Agamalian, *Notiziario Neutroni e Luce di Sinchrotrone*, **10**(2), 22, 2005.
36. Y. B. Melnichenko & G. D. Wignall, *J. Appl. Phys.*, **102**, 021101, 2007.
37. R. Triolo, *Notiziario Neutroni e Luce di Sinchrotrone*, **10**(1), 39, 2005.
38. J. M. Carpenter, M. Agamalian, K. C. Littrell, P. Thiyagarajan & Ch. Rehm, *J. Appl. Cryst.*, **36**, 763, 2003.
39. M. Agamalian, J. M. Carpenter, K. C. Littrell, J. W. Richardson & A. Stoica, *Proceedings of ICANS-XVII Meeting, Santa Fe, April 24-29, 2005*, **3**, 762, 2006.
40. J. Ilavsky, A. Allin, G. Long & P. Jemian, *Rev. Sci. Instr.*, **73**, 1660, 2002.
41. http://usaxs.xor.aps.anl.gov/docs/overview/
42. J. A. Lake, *Acta Crystallogr.*, **23**, 191, 1967.
43. T. M. Sabine, W. K. Bertram, *Acta Cryst. A*, **55**, 500, 1999.
44. T. M. Sabine, W. K. Bertram & L. P. Aldridge, *Mat. Res. Soc. Symp. Proc.*, **376**, 499, 1995.
45. A. J. Hurd, D. W. Schaefer, D. M. Smith, S. B. Ross, A. Le Méhauté & S. Spooner, *Phys. Rev. B*, **39**, 9742, 1989.
46. P.-Z. Wong and A. J. Bray, *J. Appl. Cryst.*, **21**, 786, 1988.
47. D. F. R. Mildner and P. L. Hall, *J. Phys. D*, **19**, 1535, 1986.
48. J. E. Martin, *J. Appl. Cryst.*, **19**, 25, 1986.
49. H. D. Bale & P. W. Schmidt, *Phys. Rev. Lett.*, **53**, 596, 1984.
50. P.-Z. Wong and A. J. Bray, *J. Appl. Cryst.*, **21**, 786, 1988.
51. D. W. Schaefer & K. D. Keefer, *Phys. Rev. Lett.*, **53**, 1383, 1984.
52. A. P. Radlinski, E. Z. Radlinska, M. Agamalian, G. D. Wignall, P. Lindner & O. G. Randl, *J. Appl. Cryst.*, **33**, 860, 2000.
53. D. Avnir, O. Biham, D. Lidar & P. Malcai, *Science*, **279**, 39, 1998.
54. M. H. Cohen, in *Physics and Chemistry of Porous Media—II*, AIP Conf. Proc. No. 154, New York, 1987.
55. A. H. Thompson, *Annu. Rev. Earth Planet. Sci.*, **19**, 237, 1991.
56. D. Avnir & D. Farin, *J. Chem. Phys.*, **79**, 3566, 1983.

57. A. J. Katz & A. H. Thompson, *Phys. Rev. Lett.*, **54**, 1325, 1985.
58. A. H. Thompson, A. J. Katz & C. E. Krohn, *Adv. Phys.*, **36**, 625, 1987.
59. D. F. R. Mildner & P. L. Hall, *J. Phys. D*, **19**, 1535, 1986.
60. P.-Z. Wong, J. Howard & J.-S. Lin, *Phys. Rev. Lett.*, **57**, 637, 1986.
61. A. P. Radlin'ski, C. J. Boreham, G. D. Wignall & J.-S. Lin, *Phys. Rev. B*, **53**, 14152, 1996.
62. R. K. Heenan, J. Penfold & S. M. King, *J. Appl. Cryst.*, **30**, 1140, 1997.
63. B. B. Mandelbrot, *The Fractal Geometry of the Nature*, Freeman, S. Francisco, 1982.
64. A. Viani, A. F. Gualtieri, C. Gorgoni, P. Pallante & G. Cruciani, *Neues Jarbuch fur Mineralogie*, **7**, 311, 2001.
65. C. Gorgoni, L. Lazzarini, P. Pallante & B. Turi, *Interdisciplinary Studies On Ancient Stone*, Archetype, London, 2002, p. 115.
66. R. Triolo, F. Lo Celso, C. Gorgoni, P. Pallante, D. Schwahn & M. Baron, *J. Neutron Res.*, **14**, 56, 2006.
67. G. Lucido, R. Triolo & E. Caponetti, *Phys. Rev. Lett. B*, **38**, 9031, 1988.
68. G. Lucido, E. Caponetti & R. Triolo, *Miner. Petrogr. Acta*, **32** 185, 1989.
69. G. Lucido, E. Caponetti & R. Triolo, *Geol. Carpathica*, **42**, 85, 1991.
70. A. Emmerling, R. Petricevic, P. Wang, H. Scheller, A. Beck & J. Fricke, *J. Non-Cryst. Solids*, **185**, 240, 1994.
71. P. W. Schmidt, *J. Appl. Crystallogr.*, **24**, 414, 1991.
72. M. Hainbuchner, M. Villa, G. Kroupa, G. Bruckner, M. Baron, H. Amenitsch, E. Seidl & H. Rauch, *J. Appl. Crystallogr.*, **33**, 851, 2000.
73. A. Di Pisa, M. Franceschelli, L. Leoni & M. Meccheri, *Neues Jahrbuch für Mineralogie*, **151**, 197, 1985.
74. G. Cortecci, G. Leone & A. Pochini, *Miner. Petrogr. Acta.*, **37**, 51, 1994.
75. P. C. Pertusanti, G. Raggi, C. A. Ricci, S. Duranti & R. Palmeri, *Mem. Soc. Geol. It.*, **49**, 297, 1993.
76. L. Leoni & M. Tamponi, *Neues Jahrbuch für Mineralogie*, **4** 145, 1991.
77. V. Diella, M. J. Spalla & A. Tunesi, *J. Metam. Geol.*, **10**, 203, 1992.
78. G. B. Siletto, M. J. Spalla, A. Tunesi, M. Nardo & L. Soldo, *Mem. Soc. Geol. It.*, **45**, 93, 1990.

Chapter 21
Biosynthesis of Magnetite by Microbes

Sarah S. Staniland, Bruce Ward, and Andrew Harrison

A number of organisms are able to produce inorganic materials, often in a nanostructured form. Such materials usually perform functions intrinsic to the life of the organism, but also have the potential to be harvested for technological applications. In many cases the mechanism of formation of such materials is not well understood and such knowledge may hold the key to improving their functionality for applications. We use the case of magnetotactic bacteria to illustrate some of the issues involved in elucidating this form of biosynthesis, which yields nanoparticles of magnetite, Fe_3O_4, or greigite, Fe_3S_4. Each step in the formation process, from the transport of iron ions from the surrounding aqueous medium, to specific regions of the cell, the synthesis of specific proteins for catalytic reactions and the controlled growth of particles of a specific size and shape is highly regulated. A variety of physical methods have been used to follow these processes, including neutron scattering. We review such measurements and in the case of neutron scattering indicate how they might develop in future.

21.1 Introduction: Magnetic Particles

Nanoscience and technology provide a focus for much current research, as well as a subject for considerable public interest and debate. However, the field has existed by other names for millennia [1] as mankind has discovered and tried to exploit materials whose properties depend on their state of division [2]. One key area of interest has been fine magnetic particles, with applications that range from recording media and ferrofluids to medical imaging and drug delivery [3–6]. Further, such materials are found in nature in geological systems, providing insights into environmental change, as well as in organisms, and have been implicated in navigation (for example, in homing birds and bees) [7] and degenerative diseases [8]. Consequently, there is a very lively and diverse scientific activity devoted to producing, characterizing, understanding, and exploiting such materials. This chapter focuses on a particular

A. Harrison (✉)
Institut Laue Langevin (ILL), F-38042 Grenoble Cedex 9, France
e-mail: Harrison@ill.fr

L. Liang et al. (eds.), *Neutron Applications in Earth, Energy and Environmental Sciences*, Neutron Scattering Applications and Techniques,
DOI 10.1007/978-0-387-09416-8_21, © Springer Science+Business Media, LLC 2009

class of nanomagnetic materials, as well as the experimental techniques that have been used to acquire such knowledge, and these include neutron scattering. The focus that has been chosen is nanomagnets that grow inside particular organisms, which is a subject that is still young yet offers great potential as a source of new materials. Neutron scattering also has a great potential to tackle key questions in this field, and we explore this point later in the chapter.

The nanomagnetic materials that are most commonly studied are iron oxides [9]; such substances still provide the basis of most magnetic recording media, largely on account of highly developed and relatively cheap methods of producing well-defined, non-toxic particles with a controlled size and shape. Most commonly, the iron is exclusively or predominantly in its trivalent oxidation state; and in this form, the size of the atomic magnetic moment is optimum and large, with the consequence that materials can have a relatively large magnetization per gram—an advantage in most applications. However, materials that contain iron in this oxidation state are often magnetically "soft"—that is, they have a low coercivity. This is a disadvantage for recording applications because information stored as grains with a specific magnetization is then relatively volatile; however, it is a benefit in other applications—for instance, ferrofluids and particles for magnetically directed drug delivery. Coercivity is one of the key parameters that it is desirable to control, and the factors that determine its value include the oxidation state of the iron, the size and shape of the particles, and the presence of other metal atoms whose oxides have a much higher coercivity.

Much effort has been put into refining methods of controlling the size, shape, and composition of iron oxide particles, particularly those of magnetite (Fe_3O_4), maghemite (γ-Fe_2O_3), and hematite (α-Fe_2O_3). Of these, the first two have a spontaneous magnetization below their magnetic ordering temperatures, although precise values of these temperatures have been found to vary with the way the materials have been made or treated. Magnetite has an inverse spinel structure that may be formally written as $Fe^{3+}[Fe^{2+}Fe^{3+}]O_4$, indicating that one of the iron sites (A) has Fe^{3+} tetrahedrally coordinated by oxygen, whereas the other (B) has Fe^{2+} and Fe^{3+} octahedrally coordinated by oxygen. Below the Curie temperature, T_c, of about 850 K for stoichiometric samples [10, 11], incomplete cancellation of the net moment at the A sites by those at the B sites leads to ferrimagnetism. At relatively high temperatures, electron hopping between the two different iron ions makes them structurally equivalent and gives the material a moderate electrical conductivity. However, on cooling through the Verwey transition at $\sim$120 K [12], charge ordering occurs, with a sharp rise in electrical resistance and the observation of distinct sites in the lattice for Fe^{2+} and Fe^{3+} [13]. Magnetite may be converted to maghemite on careful heating in oxygen, transforming all the Fe^{2+} to Fe^{3+}, while retaining a similar cubic structure. Inconsistencies in this procedure in much of the magnetic work on maghemite lead to a range of observed values of T_c, spanning at least 200 K, although most estimates fall in the range of 860–920 K [11, 14].

Hematite is the most thermodynamically stable of these three iron oxides under most conditions and possesses a rhombohedral structure. It is an antiferromagnet with only a weak magnetization arising from a small canting of moments along a particular direction in the crystal, and then only between the temperature at which it orders on cooling from the paramagnetic state (variously reported in the range 960–1000 K [15, 16] and the Morin transition at approximately 260 K. However, hematite is thermodynamically the most stable iron oxide under most of the conditions under which such materials form, and many routes to magnetite and maghemite use hematite as a stepping stone.

The majority of methods to make such materials involve hydrothermal synthesis, which involves heating iron salts in water; the heating leads to hydrolysis and the precipitation of oxyhydroxide which, after further heating and drying, produce an oxide. The precise conditions of the reaction—temperature, presence of specific anions, length of heating, and even the nature of the reaction vessel—can lead to different compounds in a very great variety of sizes and shapes [17, 18]. Further control of size and shape has been attempted through the growth of particles within droplets produced, for example, with an aerosol or in micelles and microemulsions [19–24].

This review concerns the exploration of a different method of producing such particles. It has been increasingly recognized over recent decades that many organisms can produce inorganic materials by various processes. Here, we focus on the way iron oxide particles are produced in certain bacteria, describing what has been discovered to date about the mechanisms of the process and how a variety of physical techniques have been employed. Neutron scattering has not in fact been used extensively in this field so far, but it has the potential to illuminate the subject further and we discuss such possibilities in relation to future improvements in techniques. The implications of work in this field for a wider range of organisms and biosynthesized materials are also examined.

21.1.1 Biomineralization: Biogenic Material

Biomineralization is the controlled formation of solid inorganic compounds by biological systems. This process occurs in a variety of organisms ranging from bacteria to humans. For example, calcium is deposited as calcium phosphate in bones and as calcium carbonate (mostly calcite and/or aragonite) in shells, while a few organisms such as diatoms create shells composed of silicates [25]. A widespread function of calcium-based minerals is to provide a cytoskeleton.

Iron is also mineralized by organisms into iron storage proteins, such as ferritins. This process is carried out by a wide range of organisms. Ferritins enable the sequestration of large numbers of iron atoms as an oxyhydroxide (ferrihydrite) in the hollow core of a multimeric protein. In bacteria, two other types of iron-binding protein are found: the heme-containing bacterioferritins and the *Dps* proteins (DNA-binding proteins from starving cells) involved in protection against free radicals [26].

Ferritins are more crystalline and have a lower phosphate content than bacterio ferritins. By contrast, the occurrence of intracellular magnetite is limited to very few bacteria and is highly controlled (see next section). It provides an excellent model for the study of biologically controlled mineralization in which a number of specific processes are well regulated: uptake and transport of chemicals (often ions), formation of specific solid-phase chemicals, ordered deposition, control of particle size and crystallinity, and assembly into complex structures. Discussion of intracellular formation of magnetite nanoparticles in bacteria forms the main topic of this chapter.

However, two other areas of iron biomineralization should be mentioned. First, a number of bacteria such as *Geobacter metallireducens* can promote extracellular magnetite formation through redox processes at the cell surface. This process is less tightly controlled and is referred to as *biologically induced mineralization*. However, single-domain particles can be formed under particular conditions [27]. In vitro formation of magnetite may also be catalyzed by bacteria or fungi [28]. Second, magnetite has been found in many higher organisms, including honeybees, pigeons, pelagic fish, bats, rodents, and humans [29]; there are at least 5 million crystals per gram of tissue in human brain tissue. More recent work is investigating the possible associations between increased levels of magnetite and neurodegenerative diseases such as Alzheimer's disease and Parkinson's disease [30].

21.2 Magnetotactic Bacteria and Magnetosomes

Magnetotactic bacteria were noted first in 1963 by Bellini [31] and then more widely reported by Blakemore in 1975 [32]. It was observed that bacteria in an environmental sample consistently moved north and responded to the movement of a magnet. Magnetic bacteria can vary in shape, size, bacterial family, and preferred habitat; but what unites them all is their magnetotaxis (the ability to recognize and align with a magnetic field). This is due to the presence of single-domain magnetic nanoparticulate iron mineral inclusions within the bacterial cell. The nanomagnets that are biomineralized within lipid vesicles are termed *magnetosomes* and are usually arranged in chains approximately 20–30 magnetosomes long, lying approximately parallel to the principal axis of the cell (Fig. 21.1). There have been many extensive reviews of magnetic bacteria and magnetosomes [33, 34] and some more specific literature detailing aspects such as the bacterial phylogeny [35], biomolecular magnetosome formation process [36–39], magnetosome characterization [40], crystal habits [41, 42], magnetism [43], and geological and environmental significance [44], which we can only summarize in this short chapter. Magnetotactic bacteria live in chemically stratified aquatic and sedimental environments where the magnetic bacteria are isolated from the microaerobic region at the oxic-anoxic transition zone (OATZ) and the anoxic region. All magnetotactic bacteria found so far can navigate to and orientate themselves in such environments because they are very motile, swimming at speeds of 40–1000 μm s^{-1}, depending on the species [45]. For example, *Magnetospirillum magnetotacticum* MS-1 ((hereafter referred to as MS-1) moves at 40 μm s^{-1}, whereas the coccus ARB-1 swims at 1000 μm s^{-1}

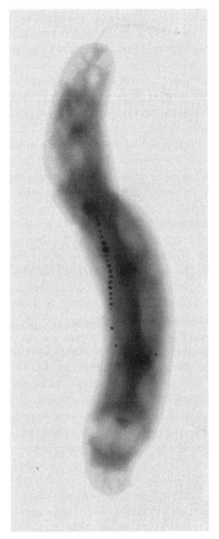

Fig. 21.1 Cell of *Magnetospirillum gryphiswaldense* MSR-1 showing a linear magnetosome chain [55]

(respectively, 2- and 500-fold faster than *Escherichia coli*). This movement is made possible by means of flagella—long filaments that protrude from the cell poles and rotate to propel the motion of bacteria—which are present either singularly or in bundles. Magnetotaxis also enables more efficient navigation, as it transforms three-dimensional "random walk" motion into a simpler one-dimensional trajectory along the local magnetic inclination.

Isolated magnetotactic bacteria are all Gram negative and have been found in a range of morphologies, such as round (coccoid), rod-like (bacillus), curved rod-like (vibrio), and spiral (spirilla) shaped, and in diverse bacterial families such as Nitrospira Phylum, δ-proteobacteria, and γ-proteobacteria, although most come from the α-proteobacterial group. Most magnetotactic bacteria biomineralize magnetite, although some form greigite (Fe_3S_4). The particles are formed with a highly defined morphology and narrow size distribution for each strain, indicating a large degree of genetic control over the process. The size of magnetosomes varies from strain to strain from 35–120 nm, and shapes range from cubo-octahedral, elongated hexagonal prisms to bullet shapes (Table 21.1). Generally, it has been found that magnetic bacteria from the δ-proteobacterial group are sulfate-reducing, with some uncharacterized many-celled magnetotactic prokaryotes producing greigite particles while other bacteria but still produce magnetite magnetosomes that have an irregular elongated shape 30–40 nm long; magnetic γ-proteobacteria form greigite elongated bullet-shaped 60–80 nm magnetosomes, and all the α-proteobacteria biomineralize magnetite. Of the α-proteobacteria, *Magnetospirilla* are the most readily culturable and contain cubo-octahedral magnetite particles 40–60 nm in size. *Magnetococci* are the most abundant in the environment and form two chains of elongated 100 nm-long hexagonal prism magnetosomes down each side of their cells. *Magnetovibrios*, isolated from a marine environment, also form elongated hexagonal prism magnetosomes approximately 50–60 nm long in a single chain. Belonging to the Nitrospira Phylum, the more exotic *Magnetobacterium bavaricum* has several chains of several hundred tooth-shaped magnetite magnetosomes. It was thought that the separation of mineral type (greigite and magnetite) with bacterial grouping (γ-proteobacteria compared with all others) was due to multiple and independent biomineralization evolutionary pathways [46]; however, an unidentified magnetic bacterium has been isolated that forms both magnetite and greigite magnetosomes within the same cell and even the same chain, with the ratio of magnetite to greigite varying depending on the available oxygen and sulfide in the environment [47].

There has been much debate over the purpose of magnetosomes and magnetotaxis for these organisms. The initial question is whether a *magnetic* mineral is deliberately synthesized; i.e., does magnetoreception serve a purpose for the cell, or is it a purely coincidental magnetic by-product of iron biomineralization? The consensus seems to be that the cells do deliberately produce magnetic minerals [33]. This is thought to be the case as the particles are single domain and arranged in chains that yield the maximum magnetic moment for the quantity of magnetite present. However, this does not rule out the possibility that the biomineralization of magnetosomes has other biological roles, such as catalysis, respiration, or detoxification.

It was noticed early on that many magnetic bacteria found in the Northern Hemisphere were exclusively north-seeking [32] and those in the Southern Hemisphere south-seeking [48]. This observation led to the theory of a magnetic navigation system whereby a consistently north-swimming magnetic bacterium in the Northern Hemisphere will swim down into the sediment according to the inclination angle of the Earth's magnetic field. The farther down the bacteria swim, the nearer they will get to their preferred environment of the OATZ and anaerobic regions [49]. However, this theory has several flaws; one being that many magnetotactic bac-

Table 21.1 Cell and magnetosome dimensions for some selected magnetic bacteria

Bacterial family	Strains	Cell length (μm)	Magnetosomes			Arrangement in cell		Ref.
			Size (nm)	Shape	Mineral	Chain motif	No particles/chain	
α-Proteobacteria	*Magnetospirillum magnetotacticum* MS-1	3–5	~40	Cubo-oct	M	Linear	~22	[53–55]
	Magnetospirillum magneticum AMB-1	3–5	~50	Cubo-oct	M	Linear	~20	[55, 56]
	Magnetospirillum gryphiswaldense MSR-1	3–6	~45	Cubo-oct	M	Linear	25–40	[55, 57]
	Magnetococcus spp. *MC-1*	1–2	70–100	Elongated	M	Linear	5–14	[41, 58]
	Magnetic Vibrio MV-1	~2	~60	Trunc. hexaoct	M	Linear	15	[54]
δ-Proteobacteria	*Desulfovibrio magneticus* RS-1	3–5	~40	Bullet-shaped	M	Irregular linear	1–18	[59]
γ-Proteobacteria	Uncultured rods	3–4	~60	Variable	G	Irregular double chains	57	[60]
Nitrospira phylum	*Magnetobacterium bavaricum*	9	110–150	Irregular	M	3–5 rope-shaped bundles	Up to 200	[61, 62]
Unknown	Uncultured rods	~3	~65	Irregular /bullet	M/G	Irregular double chains	20–40	[47]

Key: Cubo-oct = cubo-octahedral; trunc.hexaoct = truncated hexaoctahedral; M = magnetite; G = greigite

teria were found to be not polar in their magnetic orientation but axial, swimming in either direction along the magnetic field line. Also, many nonmagnetic microaerophile bacteria have an aerotactic response; therefore, they grow in a band in solution at the correct oxygen tension. The same phenomenon is seen with magnetic bacteria with the band perpendicular to the oxygen gradient, rather than horizontal to the magnetic gradient, and not in a cluster at the extremes of the tube's magnetic north as the navigation theory would suggest. Thus a revised combined magneto-aerotaxis model was proposed in which the north-seeking bacteria swim persistently north until they pass through their optimum oxygen tension, then the flagella rotation reverses, reversing the direction of motion and maintaining the cells in an optimum microaerobic region. When the magnetic field is reversed to the normal oxygen gradient, no difference is seen in the behavior of axial magnetic cells; but the polar cells find it much more difficult to form a band [50]. Thus it seems that the bacteria use the magnetic field to navigate, but are controlled by their aerotaxis.

The recent discovery of south-seeking bacteria in the northern hemisphere seemed to question this magneto-aerotactic theory. However, these south-seeking bacteria are found in an environment with a redox potential close to zero, as opposed to other magnetotactic bacteria that are found in areas with large negative redox potentials [51]; one explanation is that the sensing/magnetosome/navigation system is more complex than initially thought, and here these bacteria use a different sensing mechanism, magneto-redox taxis as opposed to magneto-aerotaxis.

Further support for the magneto-aerotactic theory may come from investigation of the role of hemerythrin proteins in magnetotaxis. Hemerythrins contain an iron center that binds oxygen. It is interesting that there are very large numbers of hemerythrins encoded in the magnetotactic bacterial genomes, leading to the hypothesis that these play a key role in the oxygen-sensing mechanism [51a]. There are still many unresolved questions with respect to magnetotaxis. Cultured strains such as *Magnetospirillum magneticum* AMB-1 and *Magnetospirillum gryphiswaldense* MSR-1 (hereafter referred to as AMB-1 and MSR-1, respectively) can live at higher oxygen tensions, but are only induced to biomineralize magnetosomes when the oxygen tension is $<2\%$ [52]. So if magnetosomes are for navigation why are they only formed under microaerobic conditions? Prior to magnetosome formation, the bacteria must exclusively use aerotaxis to navigate to their preferred microaerobic environment. Yet it is under aerobic conditions the magnetosomes should be most beneficial to navigate out of this area.

The answers most probably lie in a system whereby the bacteria start producing magnetosomes only when magnetotaxis is needed. When it becomes more favorable to synthesize magnetosomes, for the benefit of a refined navigation system in more demanding environments, magnetosome production is then induced. The production would presumably be triggered by environmental conditions, be they microaerobic or varied redox potential.

21.2.1 Biological Studies of Magnetosome Formation

The process of magnetosome formation involves a number of discrete steps; although the details remain unresolved, there is a good hypothetical framework on which to base experiments. This framework has been constructed by combining three facets of biomineralization: (1) the general requirements for biomineralization, (2) study of the genes involved, and (3) study of the proteins associated with the magnetosome. Early processes in biomineralization are the concentration of reactants in a defined space, chemical reaction(s) to form the solid phase, and controlled nucleation of the solid phase to ensure correct structure and crystal morphology. For magnetosomes, the formation of small membrane lipid vesicles defines the location and possibly the final particle size, while specific magnetosome proteins are thought to act as scaffolds for nucleation and crystal growth of magnetite. Here, we review briefly the current knowledge of the biomineralization process involved in magnetosome formation; for fuller reviews, the reader should consult the articles recommended in Section 21.2.

Knowledge of the proteins associated with magnetosome formation has come from two basic sources: genes/genomic sequences and analysis of proteins associated with isolated magnetic particles. Early work was based on nonmagnetic mutants isolated either spontaneously as white colonies, compared with the brown-black color (due to iron oxide) of wild-type colonies, or by gene disruption technology (transposon mutagenesis). In MSR-1, such work characterized a 3-kb (kilobase pairs) region that contained many genes encoding known membrane-associated magnetosome proteins (Mam proteins) and magnetic particle membrane-specific proteins (Mms proteins) [63]. By analogy with bacterial pathogens, where virulence genes are often clustered together in regions termed "pathogenicity islands," this region was termed a "magnetosome island." This work, together with parallel work on other strains of magnetic bacteria, identified a number of magnetosome proteins that appeared to be conserved within magnetic bacteria. In addition, transcriptional analysis of the 35-kb region of MSR-1 identified three units (operons): the *mms* operon (5 genes), the *mamGFDC* operon (4 genes), and the *mamAB* operon (17 genes from *mamH* to *mamU*) [64]. This analysis begged several questions, including whether all magnetosome proteins were encoded within magnetosome islands. Recent progress allows these questions to be answered more fully. First, the complete genome sequences are available for four magnetic bacteria: MSR-1, AMB-1, MS-1, and *Magnetococcus* strain MC-1 (MC-1) [65, 66]. Comparison of these sequences has confirmed the presence of a *mamAB* operon in all four strains; for the three strains MSR-1, MS-1, and AMB-1, there is high conservation, with the 17 genes *mamH–mamU* being highly similar and in the same order. Similarly, the three strains have the same basic components in *mms6* and *mamGFDC* operons, but with minor variations [38]. By contrast, MC-1 lacks *mamN*, *mamR*, and *mamU* in the *mamAB* operon, and has *mamCDF* but not *mamG* or *mms6* homologs (Table 21.2). However recently, a fifth genome, fo rmarine vibrio MV-1, has been sequenced [66a] for which the arrangement of genes for magnetosome formation differs considerably from that in MSR-1, MS-1 and AMB-1. For example, there are two or more homologs of some genes, such as *mamH*, *mamK* and *mmsF*, and the *mamAB* operon lacks *mamH* and *mamJ*.

Understanding of the role of the proteins encoded by the mms and the mamGFDC transcriptional units has been complicated by several factors. A major factor is that the mature processed forms of the proteins may differ from the encoded sequence as a result of protease processing. Other factors are (1) the start point of a protein predicted by bioinformatic analysis may not correspond to the start point of the protein as found in vivo, and (2) duplicate genes or similar genes (homologs) found elsewhere in the genome can lead to a number of proteins with similar properties. Whether such proteins function in the same way has to be determined experimentally. Finally, different nomenclature has been used for these proteins, making it more difficult to compare the same protein in different strains; for example, MamD and C of MSR-1 were termed respectively Mms7 and Mms13 in AMB-1 where they were originally discovered. With the availability of the genome sequences from the gene encoding these four magnetic bacteria, it has been possible to bring some order to this topic. The available information is summarized in Table 21.2. For example, MamF, the second-most-abundant protein (14.8%) in the magnetosome membrane, is found in MSR-1, AMB-1, and MS-1 in the *mamGFDC* cluster and these proteins are closely related. However, in MC-1 this is not the case; instead, a MamF-like protein is found elsewhere, adjacent to a gene encoding a MamA-like protein; it is 55% similar to the other MamF proteins. MamF appears to be closely related to another protein, MmsF (data not shown), with the gene found in the *mms6* cluster adjacent to the *mamGFDC* cluster. It is plausible that the MamF and MmsF proteins fulfill similar functions. Recent work indicates that the four MamGFDC proteins are involved in the regulation of magnetite crystals rather than the process of biomineralization per se and have similar functions [67].

Mms6 is a key protein in magnetosome formation. In vitro studies [68, 68a] showed that this protein binds iron and promotes the formation of magnetite particles more uniform in size (20–30 nm) and shape than in its absence (1–100 nm). It was proposed that the Mms6 protein is a scaffold protein bound to the membrane through 1–2 transmembrane helices at one end (N-terminal portion) of the protein while iron binding occurs through the hydrophilic residues at the other end (C-terminal tail). The Mms6 proteins are related to both the longer MamD proteins and the small MamG proteins. These proteins together with a magnetosome protein Mms5 from AMB-1 (Amb1027) share a repetitive LG-motif [68, 69] as shown in Fig. 21.2. The consensus sequence LGLGLGLGAWGP[FILV][LIA][LV]G is present in 10 of the 11 sequences apart from one residue in Mms6 from MSR-1. It has been pointed out that similar LG-rich repetitive sequences are found in structural scaffold proteins such as fibroin proteins, mollusk shell framework proteins, elastins, and cartilage proteins [69]. This suggests these magnetosome proteins may form similar aggregated high-molecular-weight scaffold complexes that control biomineralization.

Table 21.2 Comparison of MamA and the proteins of the *mms6* and *mamGDFC* clusters of *Magnetospirillum gryphiswaldense* MSR-1 with the corresponding proteins from *Magnetospirillum magnetotacticum* MS-1, *M. magneticum* AMB-1 and *Magnetococcus* strain MC-1. ORF numbers refer to the original designation by Schübbe et al. [63]

Protein		MSR-1			MS-1			AMB1			MC1		
		Protein no.	aa	pI	Protein no.	aa	pI	Protein no.	aa	pI	Protein no.	aa	pI
(ORF1)	HemY TPR protein	MGR4070 (MgI457)	449	8.13	Magn03007324	471	6.38	amb0959	469	5.87	No homolog found*		
(ORF2)		MGR4071 (MgI458)	347	5.87	Magn03007984	279	9.09	amb0958	348	6.83	No homolog*		
MmsF	Homology with MamF	MGR4072 (MgI459)	124	9.25	Magn03007983	124		amb0957	107	9.88	Mmc1.2274	111 or 119†	
Mms6	Iron binding	MGR_4073 (MgIa26)	136	9.13	Magn03007982	134		amb0956	157	9.90	No homolog*		
MamG	Related to MamD	MGR_4075 (MgIa15)	84	8.59	103 pI 8.74	103	8.74	amb0954	103	8.74	No homolog*		
MamF	GTPase Mms16	MGR_4076	111	9.14	111 aa pI 7.73 Magn03007980	111	7.73	amb0953	111	8.05	(Mmc1.2274)	111 or 119†	
MamD	LG motif	MGR4077	314	9.84	Magn03007979	314	9.84	amb0952 (mms7)	314	10.6	Mmc1.2237	340	9.93
MamC	Abundant	MGR_4078	125	5.07	Magn03007978	124	6.57	amb0951	145	10.36	Mmc1.2265	144	4.82
MamA	TPR protein	MGR_4099	217	5.70	Magn03009046	217	5.69	amb0971 (mms24)	221	5.53	Mmc1.2253	219	6.18

* Based on current sequence available. † Start sites different in NCBI and TIGR CMR databases

```
AMB1_mms7_mamD_amb0952        VTPITAAGTGSAMLSAKGLGLGLGLGLGAWGPFLLGAAGLAGAAALYVWA  293
MS1_mms7_mamD_Magn03007979    VTPITAAGTGSAMLSAKGLGLGLGLGLGAWGPFLLGAAGLAGAAALYVWA  293
MSR1_mms7_mamD_MGR_4077       ITPVTAAAAGSAMLTAKGVGLGLGLGLGAWGPFALGAIGLAGVVALYTWA  293
AMB1_mms6_amb0956             KAAAGAKVVGGTIWTGKGLGLGLGLGLGAWGPIILGVVG---AGAVYAYM  136
MS1_mms6_Magn03007979         KAAAGAKVVGGTIWTGKGLGLGLGLGLGAWGPIILGVVG---AGAVYAYM  113
MSR1_mms6_MGR4073             KVGAGAKAVGGTIWSGKGLALGLGMGLGAWGPLILGVVG---AGAVYAYM  115
AMB1_mamG_amb0954             ASPVGTAAIGNAMLTGKGVCLGLGLGLGAWGPVLVGIAGLAGAAYLVGKL   89
MS1_mamG_Magn03007981         ASPVGTAAIGNAMLTGKGVCLGLGLGLGAWGPVLVGIAGLAGAAYLVGKL   89
MSR1_mamG_MGR4075             APPVSAAAVGSTLLAGKGVCLGLGLGLGAWGPVLLGVAGLACAASLCDYL   70
AMB1_mms5_amb1027             ASPIAAATSSSAMLSAKGVSLGLGLGLGAWGPVILGVVGVAGAIALYGYY  194
MC1_mms7_mamD_Mmc1_2237       TATATGTAVSGTIWNGGGMSLGLGLGLGVAGPVILGAALVGTGYGSWLAY  293
                                .  .::  .    . *.: ****:***.  **.  :*
```

Fig. 21.2 Alignment of MamD, MamG, Mms6, and Mms5 proteins to show the repeat LG-motif
Asterisks denote amino acid residues conserved in all 11 sequences. Consensus motif is LGLGLGLGAWGP[FILV] [LIA] [LV]G

A central question in understanding magnetosome formation is how the surrounding lipid bilayer forms and whether magnetosomes are invaginations of the inner membrane or separate intracellular vesicles. This question has implications for possible mechanisms of iron transport and oxidation/reduction. The lipid composition of the magnetosome membrane is similar to that of the inner (cytoplasmic) membrane [69, 70], and there is strong evidence that vesicles are present before magnetite formation [71]. The conclusion from early work using both freeze etching and thin-section electron microscopy [70] was that the magnetosome membranes were not joined to the cytoplasmic membrane. By contrast, electron cryotomography [71] provided strong evidence that a good proportion (34%) of magnetosomes could be visualized as invaginations of the inner membrane. The data clearly suggested that right up to full crystal growth within the vesicle, the neck of the vesicle was not completely closed. The consequence of this view is that it would only be necessary to transport metal ions across the outer membrane into the periplasmic space between the outer and the inner membranes. A further intriguing question is whether the neck of the invagination is plugged with protein(s) and whether these proteins act as recruiting agents for other magnetosome-specific proteins. Mutants lacking "early" localization proteins would be expected to show a failure to recruit "late" magnetosome proteins correctly to the magnetosome. Photosynthetic bacteria, which are also grouped in the same bacterial family (alphaproteobacteria) as *Magnetospirillia*, are known to form membrane invaginations during the formation of photosynthetic membranes [72]. For magnetic bacteria, the observation that the protein MagA is oriented in the magnetosome membrane inversely to the orientation in the cytoplasmic membrane indicates that an invagination of the cytoplasmic membrane has occurred [73].

The magnetosome chain appears to be held together via a network of filaments that make up the actin-like protein MamK [74]. Although actin-like proteins such as MreB and Mbl are found in many bacteria, their role is in cell envelope structure and organization. The MamK subgroup of actin-like proteins is found only in magnetic bacteria. In a strain of AMB-1 where the gene encoding MamK was deleted, magnetosomes were no longer organized in chains and magnetic particles appeared to be randomly distributed throughout the cell. Specific restoration of the ability to produce MamK in mutant cells restored the wild-type organization of magnetosome chains.

Thus a likely scenario for magnetosome formation (Fig. 21.3) is that an early step is membrane invagination followed by accumulation of protein(s) that recruit magnetosome proteins specifically to the magnetosome membrane and not throughout the cytoplasmic membrane. MamA-like proteins are a candidate for this role, and proteomic studies [69, 75] provide compelling evidence that such selection occurs. Iron is specifically accumulated through cation transporters such as MamB, MamM, and MamV (not present in all strains). Evidence favors accumulation of iron directly rather than via ferritin as an intermediate storage protein, though this issue is still under debate [76]. Nucleation and scaffold proteins provide control of crystal initiation and growth, while electron transport proteins (cytochromes and oxidoreductases) ensure the correct ratio of ferric and ferrous ions.

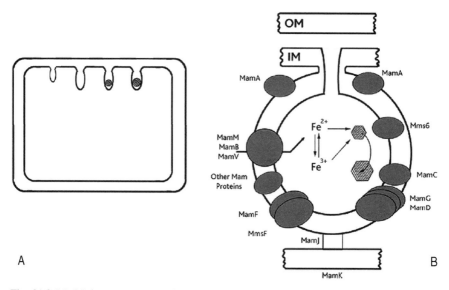

Fig. 21.3 Model for magnetosome formation (**A**) schematic of overall process—vesicles form after invagination of the cytoplasmic membrane and magnetosome proteins are recruited specifically to the magnetosome membrane, possibly by MamA and other TPR proteins. Magnetite formation occurs via iron binding to proteins such as Mms6, followed by nucleation and crystal growth, controlled by scaffold proteins. Chains of particles are organized through the actin-like protein MamK, which attaches via MamJ. MamJ is not present in all magnetic bacteria. (**B**) role of magnetosome proteins—invagination promoted possibly by MspA and MspB; MamB, MamM, and MamV transporters catalyze iron accumulation in the vesicles; the ratio of ferric to ferrous is controlled by oxidase/reductase proteins; iron binds to mms6 and other scaffold proteins that control crystal growth. The conserved integral membrane proteins MamG, MamD, MamC, MamF, and MmsF control crystal size. OM, outer membrane; IM, inner membrane

21.2.2 Initial Studies of Magnetosome Minerals

After the discovery and cultivation of magnetotactic bacteria (*M. magnetotacticum* MS-1), research followed to ascertain the nature of the bacteria [77] and the particles. The earliest research on the magnetosomes took the form of electron microscopy and Mössbauer spectroscopic studies. Early Mössbauer measurements revealed that the mineral was magnetite, but the spectrum deviates slightly from that of the pure stoichiometric material. The deviation indicates that either the biogenic magnetite has 4% vacancies in the octahedral sub-lattice, or a small percentage of a different iron-oxide phase such as γ-Fe_2O_3 is present, with some extra peaks assigned to very small superparamagnetic magnetosomes or other iron complex materials in the bacteria, such as ferritin [78]. Later, Mössbauer analysis confirmed the presence of another iron-oxide phase, and it was proposed as the precursor in the formation process [79]. Frankel et al. discuss the identity of this precursor and propose a biomineralization formation pathway. They suggest that the non-magnetite peaks could be due to ferritin or ferrihydrite and later discuss a mechanism in which ferrihydrite is the magnetite precursor [79]. This mechanism is similar to that of the biomineralization that occurs in the formation of magnetite teeth in marine mollusks known as chitons, where a ferrihydrite precursor was already known [80].

High-resolution transmission electron microscopy (HRTEM) confirmed the presence of an amorphous material on the magnetosome surface [81], although it is not seen in later studies on more mature magnetosomes, confirming that this material is most likely a precursor to magnetite [82]. Extensive HRTEM analysis of the particle morphologies was also conducted at this time, defining particle shapes and sizes for known strains and uncharacterized environmental bacteria [81–86].

Initial magnetic studies found that the biogenic magnetite nanoparticles have clearly defined and unique magnetic signals, because the particles are all single domain and arranged in chains within the cells. A characteristic Verwey transition of $T_v \approx 100$ K is also observed, as well as a distinct demagnetization of isothermal remanent magnetization and $\delta_{FC}/\delta_{ZFC} > 2$ for intact chains of magnetosomes (δ is a measure of the amount of remanence lost by warming through the Verwey transition, and FC and ZFC denote field-cooled and zero-field-cooled conditions, respectively), whereas particles not in a chain formation or oxidized have a $\delta_{FC}/\delta_{ZFC} \approx 1$ [87, 88]. It is not clear what factors have the greatest influence on magnetic effects, although it does seem unlikely that oxidation is responsible for the reduced Verwey transition, as magnetosomes are characteristically found to be of a slightly reduced form of magnetite. Further studies have compared intracellular magnetite from magnetotactic bacteria with extracellular biogenic magnetite to reveal very different magnetic profiles. The difference in profiles is due to the difference in size distribution of the particles. Whereas magnetosomes have a very narrow distribution of single-domain particles, giving a distinct Verwey transition, extracellular magnetite has a very broad distribution of mainly superparamagnetic particles that show no such transition [89]. More recently, bulk magnetic measurements of magnetosomes from the sulfur-reducing magnetic bacterium RS-1 have shown that the magnetite particles are only weakly magnetic because of their small size (30–40 nm), which puts them just inside the superparamagnetic region. Their very low $T_v = 86$ K suggests there is some degree of oxidization, and the particles show no interactive cooperation because they are not arranged closely enough together in the cell [59]. All of this means the magnetic particles are not able to exert a sufficiently strong torque to move the cell in the Earth's magnetic field, and so provide the cell with no real magnetotaxis. This suggests that the magnetic inclusions in the cells of this bacterium have a different biological purpose [59].

The ^{18}O isotopic content of magnetosomes has been analyzed in order to assess the source of oxygen within the magnetite. It was previously thought that magnetotactic bacteria used molecular O_2 from the microaerophilic environment to synthesize the magnetosomes [90]. However, this study showed that the oxygen in the magnetite was from the surrounding water, a finding that explains how magnetosomes can be formed anaerobically [91].

All of these detailed measurements of the structure and magnetic character of the magnetosomes were performed on dead organisms. The first in vivo characterization of these particles was carried out using complementary small-angle X-ray scattering (SAXS) and neutron scattering (SANS) measurements [92–94]. SANS data were taken on the bacterium MS-1 in a 30% D_2O/H_2O mixture, which had been determined to match the average scattering strength of the organic material in the cell, and provide a contrast for the magnetosomes. Further, the application of a 25 G field revealed that the larger magnetosomes were ferromagnetic, whereas smaller particles were not. When this measurement was repeated in 100% D_2O, in which case the scattering also had a strong component from the bacterium itself, an anisotropic SANS signal was observed, indicating that the bacteria had rotated in the applied field. The complementary SAXS data were dominated by the scattering from the magnetite particles and showed the same rotation in the 25 G field, whereas a lower degree of rotation was observed in the Earth's magnetic field (of the order of 0.5 G). More detailed analysis of the scattering data also revealed that on average the magnetite particles were slightly elongated in the direction of the applied field and had a size distribution compatible with that observed in TEM measurements. Further information may be provided by SANS with magnetic polarization analysis, enabling nuclear and magnetic structures to be separated; such measurements on magnetosomes that had been extracted from their parent bacteria and suspended in D_2O/H_2O showed that the particles aggregated to form chains that aligned in an applied field when the concentration was raised (Fig. 21.4) [95, 96]. Further, these studies were able to discriminate between a dense, mineral core (with a radius of approximately 16 nm) and an organic membrane (thickness of 4 nm).

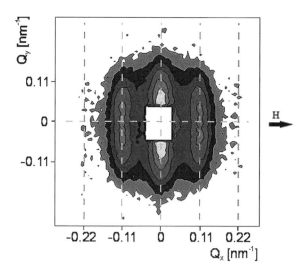

Fig. 21.4 SANS pattern of a concentrated sample of magnetosomes produced by *Magnetospirillum gryphiswaldense* taken in a horizontal field of 1 T applied perpendicular to the incident direction of the neutrons. The two-dimensional iso-intensity pattern clearly shows a periodicity in the direction of the applied field that is consistent with a chain-like arrangement of particles aligned in that direction [95]

21.2.3 Recent Magnetosome Mineral Research Developments

Magnetotactic bacteria and their magnetosomes were suddenly thrust into the lime-light in the late 1990s after it was proposed that nanoparticulate magnetite found in the Martian meteorite ALH84001 could have biogenic origins [97, 98]. The argument that this might be the case was based on the observation that the magnetite particles in ALH84001 had the same disposition as magnetosomes in terrestrial magnetotactic bacteria, strung along like a string of pearls. If this had proved to be correct, it would have provided the first direct evidence of extraterrestrial life [97]. This theory prompted extensive physical studies of the magnetosome minerals to try to identify a robust probe or signature for biogenic magnetite nanoparticles compared with abiotic magnetite. Although this theory has now been challenged [99], the renewed interest in close analysis of the size, shape, and magnetism of magnetosome minerals has shed further light on the process of formation and nature of the magnetosomes.

A part of the debate about the possible biogenic nature of the magnetite particles found in the Martian meteorite stemmed from the morphology of the individual grains, and it concluded that a minority could not be distinguished from those produced by a certain strain of terrestrial bacteria [100], with the aspect ratio (elongated form) different from that observed in geological samples. However, TEM measurements in which samples of three different strains of bacteria were tilted relative to the probe beam enabled a fuller, three-dimensional image of the magnetite crystals to be established for each source [101]. It was concluded, first, that the three strains produced crystals the shapes of which were clearly different between strains—an observation that would not have been possible from the two-dimensional projection provided by more conventional TEM measurements. Further, at least one strain produced equidimensional crystals, while it was pointed out that in fact there are examples of minerals whose crystals are elongated (and indeed, synthetic chemical routes may also produce elongated iron oxide crystallites [17]. It was suggested that a better way to discriminate between crystals that are of biogenic origin and those that are not is through consideration of the size distribution: biogenic crystals generally have a sharp cutoff at the upper size limit, whereas those that are not tend to tail off more gradually at the large-particle limit. On the basis of this criterion, the evidence that the particles in Martian meteorite ALH84001 have a biogenic origin is not very compelling.

Quite a different unconventional application of TEM measurements has been used to map out the magnetic inductance within magnetotactic bacteria [53, 54]. Off-axis holographic measurements of cultured samples of MS-1 and the marine vibroid strain MV-1 revealed that all the crystallites had single magnetic domains, and the smaller, paramagnetic crystals adopted the same alignment as the larger particles because of the magnetic interaction between them [54].

Other TEM measurements of magnetosomes from many strains of magnetotactic bacteria have shown that each strain produces particles of a very narrow asymmetric size distribution skewed toward the larger crystals and distinct morphologies [102], which is the opposite of what is more commonly observed in synthetic samples

[42, 103, 104]. Some morphologies such as bullet shapes have not yet been synthesized by conventional chemical methods, and this could be used as a pointer to a biogenic origin for magnetite crystals; however, other morphologies, such as cubo-octahedral, are common in synthetic samples [105]. Another biomarker for magnetite may be provided by the ^{18}O isotope fraction, which increases with increased temperature of formation for inorganic magnetite, but decreases with increased temperature for biogenic magnetite [104]. Unfortunately, the oxygen isotope fractions cross over between formation temperatures of 35°C and 55°C, so the method is unreliable in this zone and indeed unreliable if the formation temperature is not known at all. However, using various combinations of the analysis above, one could suggest a crystal's origin as being biogenic or non-biogenic.

Other differences between synthetic and biogenic magnetite have been shown recently through magnetic measurements: biogenic magnetite has been shown to have low-temperature remanence cycling and first-order-reversal-curves that are very different from those of synthetic magnetite [43]. A comparison of four magnetic bacterial strains with one another and with synthetic material has shown that T_v does not vary with particle size but rather with particle shape, and the sharpness of the transition depends significantly on whether the particles are arranged in a chain or not [106]. Individual particles have a much broader Verwey transition than those in chains. Further, when particles are arranged in chains, the net magnetic relaxation reflects cooperative effects with such chains, which behave as relatively large magnetic dipoles.

Since the early Mössbauer studies [79], there has been little physical analysis of the magnetosome precursor mineral; thus the assumptions made about the nature of the iron-oxide precursor being ferrihydrite have been accepted without further challenge. However, recent studies of the mechanism of magnetite film synthesis suggest that the oxyhydroxide lepidocrocite could be the magnetite precursor [107], and this suggestion is further supported by the fact that biogenic lepidocrite is also present in chitons' teeth [108]. A recent time-resolved study using Mössbauer spectroscopy suggests that the formation process begins with cellular-membrane-associated ferritin that is then transformed to ferrihydrite and then finally to magnetite in the vesicles [76]. However, a membrane-associated ferritin signal is also seen in the nonmagnetic mutated bacteria, suggesting the ferritin has another biological role, and the study still provided no direct evidence of the ferrihydrite precursor. Thus little progress can be made into elucidating the biomineralization mechanism while this precursor remains unknown. Recently, Staniland et al. employed X-ray absorption methods to analyze the magnetosomes in situ throughout the normal growth process [109]. Using a combination of X-ray absorption spectroscopy, X-ray magnetic circular dichroism (XMCD), and HRTEM with magnetic and iron uptake analysis, it was found that the process of magnetosome formation occurs within 30 min, which is much faster than previously recorded in studies of growth-restricted cells. The nonmagnetic iron-oxide precursor phase was identified in earlier samples (specifically, material produced 15 min after induction) as hematite, contributing to the longstanding debate about the precursor's identity [109].

Magnetosomes have a number of potential technological applications. The greatest interest is shown in biomedicine [22, 110], for which magnetosomes have the inherent advantage over synthetic nanomagnets in that the magnetosome lipid membrane coating offers magnetosomes unique biocompatibility. Bio-nanomedical applications range from superparamagnetic contrast agents for nuclear magnetic resonance imaging, to the treatment of tumor cells by hyperthermivity (induced by applying alternating electric fields to the particles), to the delivery of tethered drug molecules to specific sites in a patient through an external magnetic field. A recent development in tailoring coatings for biogenic magnetic particles has been through genetic engineering of the magnetosome proteins to produce so-called chimeric proteins [40]. The functionality that this may bestow on the magnetosome includes fluorescence marking—fusion tags for conjugate formation with specific biomolecules such as antibodies for immunoassays [110, 111], or with nanoparticulate metallic or semiconductor particles.

Applications could also be enhanced by modifying the particles themselves to produce objects with different or more carefully controlled shape, size, and composition. Several attempts have been made to dope magnetosomes with nonferrous metal, and recently this was achieved when magnetosomes were successfully doped with cobalt ions, increasing their coercivity [55]. The production of magnetosomes is highly regulated with respect to iron uptake, iron ion specificity, and incorporation. Therefore, only very small quantities of cobalt can be incorporated into magnetosomes ($<1.4\%$); incorporation occurs only when there are large quantities of cobalt present during growth and the iron supply is limited. XMCD studies suggest that the cobalt ions are substituted into the Fe^{2+}_{Oct} iron position and that they are mainly located near the surface of the crystal. This in turn suggests that cobalt is incorporated into the crystal only when the iron becomes very limited and the cell is flooded with excess cobalt. However, more analysis of the particles is needed in order to confirm the exact location of the cobalt.

21.3 Future Prospects for Study and Applications of Iron Oxide Magnetosomes

In the next few years, we can expect considerable progress in the understanding of the biogenic formation of magnetite. New technologies allow extremely rapid sequencing of genomes, and it is now possible to sequence bacteria that cannot be cultivated in the laboratory [112]. As many magnetic bacteria cannot be readily cultured, this capability will facilitate sequencing of additional genomes from magnetic bacteria and enable further refinement of the gene complement required for magnetosome formation. Genetic studies of mutants deficient in genes required for magnetosome formation, combined with immunolocalization and biochemical studies, will confirm the participation of specific proteins in magnetosome formation and help to define their role more exactly, as has been done for MamK, for example. The next two challenges will be to define the multiprotein complexes or combination of proteins that promote nucleation and controlled crystal growth. Bio-physical studies, including diffraction techniques, should prove useful for this task. There is currently very little structural information on the proteins associated with the magnetosome vesicles or on their interaction with membranes. X-ray crystal-lography will undoubtedly provide some high-resolution information on those pro-teins or protein fragments that can be crystallized; and where suitably large crystals can be obtained, neutron crystallography may provide detail relating to hydration, hydrogen bonding, and protonation states [113]. Together with biochemical linker and mutant studies, this should help define which proteins interact functionally.

Important information at lower resolution can also be obtained through the use of small-angle scattering (providing information on molecular shape), reflection experiments (for membrane-bound proteins or components), and fiber diffraction (for components such as the actin-like filaments of MamK [114]). In this context it should be noted that neutron methods have the potential to provide unique insights—particularly where selective deuteration can be used in conjunction with solvent contrast variation methods to model specific parts of the system while rendering others effectively invisible. Such approaches can be applied at low resolution using SANS [115], low-resolution crystallography [116], neutron reflection [117], and fiber diffraction [118, 119]. The ability to carry out novel experiments of this type is greatly enhanced by dedicated facilities for the provision of deuterated proteins.

Further insight into the evolution of *mineral phases* in the early stages of growth could also be provided by in situ studies. Recent and anticipated advances in neutron instrumentation enable the study of small samples by small-angle and diffraction measurements. For typical samples of bacteria in their growth medium, the concen-tration of the organisms lies in the range of 10^9 cells cm^{-3}, and further concentration by a factor of 200 or more may be achieved through centrifuging. This is sufficient for small-angle measurements with a time resolution of the order of a minute; in combination with contrast techniques involving the variation of D_2O/H_2O in the growth medium, together with polarization methods, the contribution to the scat-tering of the organic tissue, the particles, and the magnetization of the particles may all be separated [95, 96], leading to a picture of the evolution of the magne-tosomes from nuclei of less than a nanometer in diameter. In principle, the nuclear and magnetic structure of any crystalline precursor phase may also be determined by refinement of powder diffraction profiles, provided it corresponds to relatively simple materials whose structure is already known [94]. The current limitation here is that a useable diffraction pattern may be obtained from 1 g of a sample with a relatively simple structure in 0.1 s—with time scaling with the reciprocal of mass (typical growth media contain 0.1 $\mu g\ cm^{-3}$ biogenic magnetite/volume of medium); future powder diffractometers [120] may improve on this statistic by an order of magnitude, enabling useful data to be taken from unconcentrated media within the order of 10^4 s.

This chapter has focused on just one form of biogenic material, outlining how a combination of complementary techniques may be used to determine the mech-anism of formation and the properties of the product. Neutron methods have been underutilized to date in such work, partly because of the relatively small amounts of material that are usually involved, as well as the need to deuterate wholly or partially the biological components of the system to gain structural information. However, the current development of new instruments and sources, together with the widening availability of dedicated deuteration facilities, should greatly increase the impact of neutron methods in this growing field.

Acknowledgments The authors are grateful to Trevor Forsyth and Peter Timmins for valuable discussions during the preparation of this work.

References

1. P. Walter et al., *Nano Lett.* **6**, 2215 (2006).
2. P. Moriarty, *Reports Progr. Phys.* **64**, 297 (2001).
3. D. L. Leslie-Pelecky, R. D. Rieke, *Chem. Mater.* **8**, 1770 (1996).
4. M. M. Tartaj P, Veintemillas-Verdaguer S, Gonzalez-Carreno T, Serna CJ *J. Phys. D: Appl. Phys.* **36**, R182 (2003).
5. Q. A. Pankhurst, J. Connolly, S. K. Jones, J. Dobson, *J. Phys. D: Appl. Phys.*, **36**, R167 (2003).
6. R. Skomski, *J. Phys. Condens. Matter.* **15**, R841 (2003).
7. C. E. Diebel, R. Proksch, C. R. Green, P. Neilson, M. M. Walker, *Nature* **406**, 299 (2000).
8. J. Dobson, *FEBS Let.* **496**, 1 (2001).
9. R. M. Cornell, U. Schwertmann, *The Iron Oxides: Structure, Properties, Reactions, Occurrences and Uses* (Wiley-VCH, Weinheim, 2003).
10. D. O. Smith, *Phys. Rev.* **102**, 959 (1956).
11. A. Aharoni, E. H. Frei, M. Schieber, *J. Phys. Chem. Sol.* **23**, 545 (1962).
12. E. J. W. Verwey, *Nature* **144**, 327 (1939).
13. J. P. Wright, J. P. Attfield, P. G. Radaelli, *Phys. Rev. B* **66**, 214422 (2002).
14. O. Ozdemir, *Phys. Earth Planet. Int.* **65**, 125 (1990).
15. C. G. Shull, W. A. Strauser, E. O. Wollan, *Phys. Rev.* **83**, 333 (1951).
16. P. Gilad, M. Greenshpan, P. Hillman, H. Shechter, *Phys. Lett.* **7**, 239 (1963).
17. E. Matijevic, *Chem. Mat.* **5**, 412 (1993).
18. E. Matijevic, *Langmuir* **10**, 8 (1994).
19. L. Liz, M. A. Lopez Quintela, J. Mira, J. Rivas, *J. Mater. Sci.* **29**, 3797 (1994).
20. V. Chhabra, P. Ayyub, S. Chattopadbyay, A. N. Maitra, *Mat. Lett.* **26**, 21 (1996).
21. M. P. Pileni, *Nat. Mater.* **2**, 145 (2003).
22. A. K. Gupta, M. Gupta, *Biomaterials* **26**, 3995 (2005).
23. D. Vollath, D. V. Szabó, R. D. Taylor, J. O. Willis, *J. Mat. Res.* **12**, 2175 (1997).
24. B. Martinez, A. Roig, E. Molins, T. Gonzalez-Carreno, C. J. Serna, *J. Appl. Phys.* **83**, 3256 (1998).
25. S. Mann, *Biomineralization: Principles and Concepts in Bioinorganic Materials Chemistry* (Oxford University Press, Oxford, 2001).
26. S. C. Andrews, A. K. Robinson, F. Rodriguez-Quinones, *FEMS Microbiol. Rev.* **27**, 215 (2003).
27. H. Vali et al., *Proc. Natl. Acad. Sci. USA* **101**, 16121 (2004).
28. A. Bharde et al., *Small* **2**, 135 (2006).
29. J. L. Kirschvink, A. Kobayashi-Kirschvink, B. J. Woodford, *Proc. Natl. Acad. Sci. USA* **89**, 7683 (1992).
30. J. F. Collingwood et al., *J. Alzheimer's Dis.* **7**, 267 (2005).
31. S. Bellini, dell'Universita di Pavia (1963).

32. R. Blakemore, *Science* **190**, 377 (1975).
33. D. Schüler, Ed., *Magnetoreception and Magnetosomes in Bacteria* (Springer, Berlin, 2007).
34. D. A. Bazylinski, R. B. Frankel, *Nature Rev. Microbiol.* **2**, 217 (2004).
35. R. Amann, R. Rossello-Mora, D. Schuler, *Biomineralization* (Wiley-VCH, Weinheim, 2000), pp. 47–60.
36. T. Matsunaga, Y. Okamura, *Trends Microbiol.* **11**, 536 (2003).
37. A. Komeili, *Annu. Rev. Biochem.* **76**, 351 (2007).
38. C. Jogler, D. Schueler, in *Handbook of Biomineralization: Biological Aspects and Structure Formation*, E. Baeuerlein, Ed. (Wiley-VCH, Weinheim, 2007), pp. 145–162.
39. T. Matsunaga, T. Sakaguchi, *J. Biosci. Bioengin.* **90**, 1 (2000).
40. C. Lang, D. Schuler, *J. Phys.: Conden. Matter.* **18**, S2815 (2006).
41. B. Devouard et al., *Am. Mineral.* **83**, 1387 (1998).
42. B. Arato et al., *Am. Mineral.* **90**, 1233 (2005).
43. Y. Pan et al., *Earth Planet. Sci. Lett.* **237**, 311 (2005).
44. D. A. Bazylinski, B. M. Moskowitz, *Rev. Mineral.* **35**, 181 (1997).
45. R. B. Frankel, T. J. Williams, D. A. Bazylinksi, in *Magnetoreception and Magnetosomes in Bacteria*, D. Schüler, Ed. (Springer, Berlin, 2007), pp. 1–24.
46. E. F. DeLong, R. B. Frankel, D. A. Bazylinski, *Science* **259**, 803 (1993).
47. D. A. Bazylinski et al., *Appl. Environ. Microbiol.* **61**, 3232 (1995).
48. R. P. Blakemore, R. B. Frankel, A. J. Kalmijn, *Nature* **286**, 384 (1980).
49. R. P. Blakemore, R. B. Frankel, *Scientific Am.* **245**, 58 (1981).
50. R. B. Frankel, D. A. Bazylinski, M. S. Johnson, B. L. Taylor, *Biophys. J.* **73**, 994 (1997).
51. S. L. Simmons, D. A. Bazylinski, K. J. Edwards, *Science* **311**, 371 (2006).
51a. C. E. French, J. M. L. Bell, F. B. Ward, *FEMS Microbiol. Lett.* **279**, 131 (2008).
52. U. Heyen, D. Schuler, *Appl. Microbiol. Biotechnol.* **61**, 536 (2003).
53. R. E. Dunin-Borkowski et al., *Science* **282**, 1868 (1998).
54. R. E. Dunin-Borkowski et al., *Eur. J. Mineral.* **13**, 671 (2001).
55. S. Staniland et al., *Nat. Nanotechnol.* **3**, 158 (2008).
56. T. Matsunaga, T. Sakaguchi, F. Tadokoro, *Appl. Microbiol. Biotechnol.* **35**, 651 (1991).
57. K. H. Schleifer et al., *Syst. Appl. Microbiol.* **14**, 379 (1991).
58. F. C. Meldrum, S. Mann, B. R. Heywood, R. B. Frankel, D. A. Bazylinski, *Proc. Royal Soc. B: Biol. Sci.* **251**, 231 (1993).
59. M. Posfai et al., *Earth Planet. Sci. Lett.* **249**, 444 (2006).
60. T. Kasama, et al., *Amer. Min.* **91**, 1216 (2006).
61. M. Hanzlik, M. Winklhofer, N. Petersen, *Earth Planet. Sci. Lett.* **145**, 125 (1996).
62. S. Spring et al., *Appl. Environ. Microbiol.* **59**, 2397 (1993).
63. S. Schubbe et al., *J. Bacteriol.* **185**, 5779 (2003).
64. S. Schubbe et al., *Appl. Environ. Microbiol.* **72**, 5757 (2006).
65. T. Matsunaga et al., *DNA Res.* **12**, 157 (2005).
66. M. Richter et al., *J. Bacteriol.* **189**, 4899 (2007).
66a. D. Trubitsyn, S.S Staniland, and F. B. Ward. S., Proc. Workshop Magnetic Bacteria, Lake Balaton, Hungary (2008).
67. A. Scheffel, A. Gaerdes, K. Gruenberg, G. Wanner, D. Schueler, *J. Bacteriol.* **190**, 377 (2008).
68. A. Arakaki, J. Webb, T. Matsunaga, *J. Bio. Chem.* **278**, 8745 (2003).
68a. Y. Amemiya, A. Arakaki, S. Staniland, T. Tanaka, T. Matsunaga, *Biomaterials* **28**, 5381 (2007).
69. K. Grunberg et al., *Appl. Environ. Microbiol.* **70**, 1040 (2004).
70. Y. A. Gorby, T. J. Beveridge, R. Blakemore, *J. Bacteriol.* **170**, 834 (1988).
71. A. Komeili, H. Vali, T. J. Beveridge, D. K. Newman, *Proc. Natl. Acad. Sci. USA* **101**, 3839 (2004).
72. G. Drews, R. A. Niederman, *Photosyn. Res.* **73**, 87 (2002).
73. C. Nakamura, J. G. Burgess, K. Sode, T. Matsunaga, *J. Bio. Chem.* **270**, 28392 (1995).

74. A. Komeili, Z. Li, D. K. Newman, G. J. Jensen, *Science* **311**, 242 (2006).
75. M. Tanaka et al., *Proteomics* **6**, 5234 (2006).
76. D. Faivre, L. H. Boettger, B. F. Matzanie, D. Schueler, *Angew. Chem., Int. Ed.* **46**, 8495 (2007).
77. D. L. Balkwill, D. Maratea, R. P. Blakemore, *J. Bacteriol.* **141**, 1399 (1980).
78. R. B. Frankel, R. P. Blakemore, R. S. Wolfe, *Science* **203**, 1355 (1979).
79. R. B. Frankel, G. C. Papaefthymiou, R. P. Blakemore, W. O'Brien, *Biochim. Biophys. Acta—Mol. Cell Res.* **763**, 147 (1983).
80. K. M. Towe, H. A. Lowenstam, *J. Ultrastruc. Res.* **17**, 1 (1967).
81. S. Mann, R. B. Frankel, R. P. Blakemore, *Nature* **310**, 405 (1984).
82. S. Mann, N. H. C. Sparks, R. P. Blakemore, *Proc. Royal Soc. Lond. B: Bio. Sci.* **231**, 477 (1987).
83. T. Matsuda, J. Endo, N. Osakabe, A. Tonomura, T. Arii, *Nature* **302**, 411 (1983).
84. S. Mann, T. T. Moench, R. J. P. Williams, *Proc. Royal Soc. Lond. B: Bio. Sci.* **221**, 385 (1984).
85. S. Mann, N. H. C. Sparks, R. P. Blakemore, *Proc. Royal Soc. Lond. B: Bio. Sci.* **231**, 469 (1987).
86. B. R. Heywood, S. Mann, R. B. Frankel, *Mater. Res. Soc. Symp. Proc.* **218**, 93 (1991).
87. B. M. Moskowitz, R. B. Frankel, P. J. Flanders, R. P. Blakemore, B. B. Schwartz, *J. Magn. Magn. Mater.* **73**, 273 (1988).
88. B. M. Moskowitz, R. B. Frankel, D. A. Bazylinski, *Earth Planet. Sci. Lett.* **120**, 283 (1993).
89. B. M. Moskowitz, R. B. Frankel, D. A. Bazylinski, H. W. Jannasch, D. R. Lovley, *Geophys. Res. Lett.* **16**, 665 (1989).
90. R. P. Blakemore, K. A. Short, D. A. Bazylinski, C. Rosenblatt, R. B. Frankel, *Geomicrobiol. J.* **4**, 53 (1985).
91. K. W. Mandernack, D. A. Bazylinski, W. C. Shanks, III, T. D. Bullen, *Science* **285**, 1892 (1999).
92. S. Krueger et al., *J. Appl. Phys.* **67**, 4475 (1990).
93. S. Krueger et al., *J. Magn. Magn. Mater.* **82**, 17 (1989).
94. R. J. Harrison, *Rev. Mineral. Geochem.* **63**, 113 (2006).
95. A. Hoell, A. Wiedenmann, U. Heyen, D. Schuler, *Phys B: Condens. Matt.* **350**, e309 (2004).
96. A. Wiedenmann, M. Kammel, A. Heinemann, U. Keiderling, *J. Phys.: Condens. Matt.* **18**, s2713 (2006).
97. D. S. McKay et al., *Science* **273**, 924 (1996).
98. K. L. Thomas-Kertpa et al., *Geochim. Cosmochim. Acta.* **64**, 4049 (2000).
99. D. C. Golden et al., *Am. Mineral.* **89**, 681 (2004).
100. K. L. Thomas-Kertpa et al., *Proc. Natl. Acad. Sci. USA* **98**, 2164 (February 27, 2001).
101. P. R. Buseck et al., *Proc. Natl. Acad. Sci. USA* **98**, 13490 (2001).
102. L. Han et al., *J. Magn. Magn. Mater.* **313**, 236 (2007).
103. B. Devouard et al., *Am. Mineral.* **83**, 1387 (1998).
104. D. Faivre, P. Zuddas, *Earth Planet. Sci. Lett.* **243**, 53 (2006).
105. D. Faivre, N. Menguy, F. Guyot, O. Lopez, P. Zuddas, *Am. Mineral.* **90**, 1793 (2005).
106. R. Prozorov et al., *Phys. Rev. B: Condens. Matter Mater. Phys.* **76**, 054406/1 (2007).
107. M. Abe, T. Ishihara, Y. Kitamoto, *J. Appl. Phys.* **85**, 5705 (1999).
108. H. A. Lowenstam, *Science* **156**, 1373 (1967).
109. S. S. Staniland, B. Ward, A. Harrison, G. van der Laan, N. Telling, *Proc. Natl. Acad. Sci. USA* **104** (December 4, 2007).
110. T. Matsunaga, A. Arakaki, in *Magnetoreception and Magnetosomes in Bacteria*, D. Schüler, Ed. (Springer, Berlin, 2007).
111. T. Matsunaga, Y. Okamura, T. Tanaka, *J. Mater. Chem.* **14**, 2099 (2004).
112. P. D. Schloss, J. Handelsman, *Genome Biol.* **6**, 229 (2005).
113. M. P. Blakeley et al., *Proc. Natl. Acad. Sci. USA* **105**, 1844 (2008).
114. M. V. Petoukhov, D. I. Svergun, *Curr. Opin. Struc. Bio.* **17**, 562 (2007).
115. P. Callow, A. Sukhodub, J. E. Taylor, G. G. Kneale, *J. Mol. Biol.* **369**, 177 (2007).

116. E. Pebay-Peyroula, R. M. Garavito, J. P. Rosenbusch, M. Zulauf, P. A. Timmins, *Structure* **3**, 1051 (1995).
117. J. R. Lu, R. K. Thomas, J. Penfold, *Adv. Coll. Int. Sci.* **84**, 143 (2000).
118. D. Sapede et al., *Macromolecules* **38**, 8447 (2005).
119. K. H. Gardner, A. D. English, V. T. Forsyth, *Macromolecules* **37**, 9654 (2004).
120. A. W. Hewat, *Phys. B: Condens. Matt.* **385–386**, 979 (2006).

Index

Printed in the United States of America